"Disaster Prep 101"

the Ultimate Guide to Emergency Readiness

By Paul Purcell

READ <u>EVERYTHING</u>!

IT <u>IS</u>

<u>THAT</u> IMPORTANT!

Disaster Prep 101™

Currently distributed through InfoQuest.
InfoQuest, 6300 Powers Ferry Rd., Suite 600-294, Atlanta, GA 30339

ISBN: 0-942369-03-3

Copyright 2004, Paul Purcell

Dedications and Acknowledgements

This book is dedicated first and foremost to the reader who needs it. Of equal importance are the many people without whose help this book would never have been written, mentioned here in chronological order.

To my Mom and Dad, Audrey G. and Eugene T. Purcell, for teaching me to stand on my own two feet, and in loving memory of my Dad (1933-2003), who helped with acronyms and early review.

To Lt. Col. Ben Willis, USA Ret., my partner in crime (prevention!).

To my Sifu, Jason Lau, for the immeasurable things he taught me.

To my other half and partner in far more ways than one, Frances Carter, for her dedication not only to me, but to this and all our other endeavors.

To Col. David H. Hackworth, USA Ret., for allowing me to serve my country in my own little way.

To Chief Eddie Moody, Deputy Director Moses Ector, Dr. Robbie Friedmann, and Sgt. John Hughley of the DeKalb County Office of Homeland Security, for allowing me to serve my community and state; and to Special Agent John Lang, Georgia Bureau of Investigation, for making the introductions.

To Mr. H. Ross Perot, for his words of encouragement at the Baltimore Conference.

To Mr. Bob Sapp, Georgia Emergency Management Agency, and to Mr. Charles Stone, Georgia Bureau of Investigation, Retired, for being good sounding boards.

To Mr. Ron Carter, Frances' Dad, for his help in review and editing.

To Dr. Doug Skelton, Director of the Savannah area Public Health District 9-1, for allowing me to contribute to the 2004 G-8 Summit emergency medical reaction protocol planning.

"Disaster Prep 101"

I. INTRODUCTION

"Experience is a hard teacher because she gives the test first, the lesson afterwards." - Vernon Law

Why This Book?

Chernobyl, Bhopal, Three Mile Island, Hurricane Andrew, California earthquakes, Mississippi flooding, Texas tornadoes, western wildfires, northern blizzards, southern drought, rolling blackouts, Columbine, Mad Cow, West Nile, September 11th, Anthrax, DC snipers, SARS, asteroids, and..."Lions and tigers and bears?" It's no fun picking up a new book and becoming depressed after the first paragraph, is it?

Emergencies and disasters face us in the news every day, and once in a while they face us in our own backyards. Nature hits us from one direction and man from another, whether by intentional terrorist attack, or through negligence or accident. Global warming will ensure increased weather disasters, and the number and severity of terrorist attacks and other manmade misfortunes have increased dramatically and will continue to do so.

Much has been done over the past couple of decades to improve the safety of the general public. Early warning systems for weather have been greatly enhanced, technology has improved the reactions and capabilities of our first responders, and government has increased the funding for agencies involved in counter-terrorism, and in both disaster management and post-disaster relief and rebuilding.

However, there is still vast room for improvement in those areas, and especially in the area of personal and family emergency preparedness. There is a big difference between what is currently taught and what can actually be done. The same holds true for contributions citizens can make through organized volunteer groups. These are the reasons this book is so large; we have a lot of catching up to do.

You're probably asking, "Aren't there already a lot of books or lists that deal with emergency preparedness?" By *name*, yes. Numerous books and websites have "emergency supply" checklists and other bits and pieces, but none are as comprehensive, focused, detailed, or as training oriented as they should be. At best they'll have you sitting there all by yourself with the wrong equipment wondering what to do in the middle of an emergency that caught you completely by surprise, and then leave you clueless as to what to do afterward. There's much more to readiness than, "Listen to your radio for further instruction."

The time has come for awareness and preparedness to become a part of our daily life. Simple checklists and "how-to" pamphlets won't provide us the kind of education we need to improve our individual abilities or make the subtle lifestyle alterations necessary for adapting to the changes already thrust on our world.

This book is designed to help prepare you for the worst emergencies you may encounter. We're going to cover everything from terrorist attacks to tornadoes, and do all we can to guide you through every aspect of knowledge, training, and equipment, to make sure you've achieved the highest level of safety possible. We'll remind you here that we firmly believe in "planning without panic." Though the worst case scenario is a large-scale terrorist attack, the truth of the matter is that you're statistically more likely to experience a natural disaster. The good news is that if you're fully prepared for one you're prepared for the other.

We've also written this book to help our brave and dedicated first responders. The more you know, and the more capable you are of looking out for you and yours, the less of a burden you'll be to emergency workers, and the more you'll free them up to help others. The final section of the book will outline ways you can be part of a volunteer group and the many ways a group can further this goal.

Who is This Book For?

It's for you, and it's for everyone who realizes that the ultimate responsibility for their own safety and protection rests squarely on their own shoulders.

This book is for the individual all the way through the organized group of volunteers. It's for the young homemakers, the elderly retired couple, the single parent, the wealthy professional, the financially overextended family. It's for the members of our country's armed forces and for all our civilian first responders who have to make sure the family they leave at home while they perform their duties is well protected. It's for everyone who wants to be far safer in an increasingly dangerous world, and it's for everyone who understands that security, like charity, begins at home.

What it is and What it's Not.

We're giving you a detailed map, a guide, a directory, or a training course if you will, to aid you in your own education and preparation. We'll help you along every step of the way to determine what kinds of emergencies you may face, how prepared you already are to meet their unique challenges, and how to tailor your future preparations based on your particular set of circumstances.

What we won't do is sell you a book with a lot of fluff, reprints of generic instruction, formatting with lots of blank space, pages of unnecessary pictures and graphics, or useless information that does nothing but fill pages and make an otherwise empty book look thick and impressive on a bookstore shelf. There are too many of those out there now. We also won't give you only one set of emergency instructions and tell you "one size fits all." It doesn't work that way.

What we will do is help you find the specific kinds of information, training, and equipment that will best suit your particular needs, and we'll help you do it as economically as possible. Where other works are vague, ours focuses on specifics. The difference is in the detail, and we're going to give you the detail that makes all the difference.

Since we've written this book a little differently than most, let's look at how to get the most out of it.

How to Use This Book.

You'll notice throughout the book there will be numerous sources of background information, courses to take, books to read, suggestions to consider, checklists to complete, things to do, and data forms to fill out. In the "**Appendix**" we've given you a blank **three-month calendar** to organize all your suggested training and reading, completion of pertinent checklists, accomplishment of suggested tasks, acquisition of necessary equipment, and the attainment of all suggested goals. This book is set for action. Your action.

Since what you have is a training manual, you should use it like one. Treat this as if it were a second-hand book you got for a course in school. Read it, reread it, study it, write in it, check off the websites you've visited, dog-ear the pages, use the checklists, fill in the blanks, highlight lines and sections, fill out the forms, use the coupons, and make this book yours. (We intentionally used a "plain" format so you could highlight the parts you felt were important.)

Training also means notes and educational materials, so **here's our first suggestion**: Get a three-ring binder or a cardboard, accordion-style, divider folio. You'll be collecting a lot of information, downloading and printing pages, making notes, and filling out forms. You'll want a way to keep everything organized.

Speaking of information, many sections of the book will provide you numerous sources for additional information. (There's no such thing as _one_ source having _everything_ you need. Even ours.) Usually, these are organized into source-type categories, mainly **_Internet Sites, Periodicals, Books, Videos, and Courses_**. Let's cover how best to utilize these sources:

1. **_Internet Sites_**: If you have internet access, you should use these training resources first, as they'll all have free information. You also want to visit them **soon** as you never know when some websites will no longer be available. As you visit these websites, **download** the information they offer (saving it to disk), **print** useful pages (putting them in your 3-ring binder), and be sure to take **notes**. **You'll want to learn from these sites now rather than save them as reference**. When disaster strikes you won't have time to log on to the Internet, even if it's accessible, and you won't have time to review notes. If you don't have Internet access, team up with someone who does, or check your local library or other location that has public access and maybe a short course on how to use the Internet. The Internet is far too valuable a resource to ignore. For our friends with no computer, we've included direct contact info for companies, products, programs, etc., where available. **Note**: Though we list numerous sources of info, equipment, courses, etc., we have no financial involvement with any outside company, and these listings appear solely because we wanted you to have a variety of information. Also, their inclusion here does not necessarily constitute an endorsement of them on our part. **Note #2**: To make it easier for you to log on to all the internet sites we'll provide, we've included a links collection on our **CD** entitled "**DP1 Book Links**" which will allow you to follow along on-screen, all the links and sites listed in the book.

2. **_Periodicals_**: Magazines and other periodicals are a great source of continually updated educational material and we've done our best to find good sources that will provide you valuable information. Since we know there's no way the average family can afford to subscribe to each and every recommended publication, our suggestion is to visit your local library or bookstore / coffee shop combo to browse and take notes. If you're part of a group, the group may want to subscribe.

3. **_Books_**: We make the same recommendation about books as we do periodicals. Visit your library first. If there is a book you feel you need to own, check used bookstores. The "**Information Research**" section under "Basic Training" will have a few sources for out of print books, and the **CDs** contain **numerous manuals** and other instructional files in Adobe's "PDF" format.

4. **_Videos_**: Videos can offer some good training since many people learn better by watching than reading. Videos can be borrowed from your library (and many libraries will order titles if you ask), or rented at your local video store. Be sure to check out the "Information Research" section below.

5. ***Courses***: Experience is ALWAYS the best teacher and if you can find and afford the courses we recommend, by all means take them. If you're part of a larger volunteer group, it might be more economical to have an instructor come to you. We've also listed on-line courses we've found.

A quick word about the amount of outside information we'll give you, and about the considerable amount of educational material and suggested projects this book contains: We agree with the universal axiom that simple plans are best. In fact, we'll carry it a bit further and say that there is genius in simplicity and that complicated plans only cause confusion. We also agree with the old phrase, "K.I.S.S.," or "Keep It Simple, Stupid." However, your final reaction plans, though simple, should be comprehensive, thorough, and accurate. This kind of "simplicity" can only come from larger plans that are created through information and imagination, pared down by education and objective observation, and then customized by practice. This is why we've given you so much to think about and to learn.

In addition to all the educational information, you'll also find **checklists**, **suggestion lists**, and **data forms**.

Checklists are "definites." You should definitely acquire the item or complete the task. **Suggestion lists** are bulleted and/or numbered lists of items, concepts, or actions to be considered. **Numbered** lists have connected or interrelated items. **Bulleted** lists have related but independent concepts. **Data forms** are meant to record valuable information and will be found in the "**Appendix**" section of the book, and also on the enclosed **CDs**.

Here are what the **checklists**, **suggestion lists**, and **data forms** look like:

Checklists Look Like This:
❑ Checklists will look like this and will be lists of items to acquire, or tasks to accomplish.
❑ Some will have spaces to put due dates or completion dates, ___/___/___.
❑ In any event, checklists contain those items we strongly suggest you complete.

❑ Minor checklists might look like this.	❑ You'll find them inserted in discussion points.
❑ They'll usually be a list of items related to the topic.	❑ They're just as important as any other checklist item.

Subsections, Discussion Points, and Suggestion Lists Look Like This:

➢ Some lists will be a collection of elective considerations or discussion items.
   ● They may have sub-items of interest, ❑ or sub-items that are grouped together and recommended.
1. They may be numbered if there's a specifically related group of concepts or steps.
   ➢ Though elective, we still suggest you seriously consider them.
   ➢ Otherwise we wouldn't have included them.
   ➢ This is also how some of the sections with background information are arranged.

Data Forms Ask You To Record Valuable Information	
Insurance Agency:	Agent's Name:
Contact Phone 1:	Contact Phone 2:
Policy number:	Policy Coverage Amount: $

In many sections you'll notice we've put various words or phrases in **bold text**. We're not shouting at you, this is simply an easy way to **scan for important points**.

IMPORTANT: READ EVERYTHING VERY CAREFULLY. IN FACT, WE RECOMMEND READING THE BOOK AT LEAST TWICE. WE SUGGEST READING THROUGH EVERYTHING ONCE AND THEN GOING BACK AGAIN TO FOLLOW THE INSTRUCTIONS AND COMPLETE THE ACTIVITIES. THIS BOOK IS A "**CONDENSED MANUAL**" MEANING THAT IMPORTANT IDEAS OR ENTIRE CONCEPTS MAY BE CONTAINED IN **ONE SIMPLE SENTENCE**. DON'T MISS THE CRUCIAL DETAIL.

That being said, let's take a look at how this manual is organized and why.

Sectional Overview.

We've arranged this book's sections in a **logical progression** to take you from a beginning to a desirable goal. Please keep in mind that though you feel one particular section may apply to you far more than another, there is extremely useful information interlaced in each, so be sure to read and study everything. Emergency situations are always different, and each is fluid and dynamic with a life of its own. You need as much training and practice as possible so you can react quickly and properly as situations within a disaster unfold.

The first section following this "**Introduction**" section, is the "**Foundation**" which covers the "basic training" you'll need when facing an emergency. The "Foundation" section sets the stage by helping you get the training and information you may require, reviewing basic types of emergencies, and helping you with the most important part of all, and that is developing a "readiness" or "self-reliant" mindset.

Next we'll provide you some specific "**Reaction**" material. Here you'll learn what to do in advance, and as a primary response, to a variety of incidents and situations. These immediate reactions are intended to save your life so you can proceed with either an "Evacuation" or an "Isolation."

"**Evacuation**" covers all the subtle details of planning, procedure, and equipment to get you rapidly away from point A and safely to a pre-planned point B, within a variety of scenarios. As you'll learn in this section, evacuation means worlds more than "get out of the house."

"**Isolation**" (as from a quarantine after a biological attack) comes next. We'll help you not only survive an extended isolation, but add some level of comfort as well. Pay attention to both the "Evacuation" and "Isolation" sections as your response to an emergency may be a combination of the two.

Following these is the "**Cooperation**" section. We've called it that as it covers working with others. This is the most important section of all, in our opinion, as it goes into great detail on coordinating your efforts with others for the purposes of mutual safety, or to help your fellow citizens. "Cooperation" covers everything from organizing a simple neighborhood watch, to establishing a networked civilian volunteer group liaisoned with your local government entities to assist during or after an emergency.

Wrapping up the book is the "**Appendix**." There you'll find the data forms and worksheets you'll be asked to fill out in the course of this book.

The enclosed **CDs** contain wordprocessor and/or spreadsheet files for these data forms and worksheets, **two interactive Internet links collections** (one is "**DP1 Book Links**" which is useful for following along with the book's printed links), and numerous useful **instructional manuals**. You'll find it's table of contents in the "**Appendix**."

Ready to get started setting up your "**Foundation**?" Good. Let's look at your first checklist:

Before Turning to the "Foundation" Section:

❑ Make sure to have a **highlighter, pen, and a notepad or notebook** handy. (You'll need them.)
❑ Get out your **calendar or day planner**, or use our calendar. You'll find ours in the "**Appendix**".
❑ While in the "**Appendix**", spend a minute locating and looking through the different **worksheets** and **data forms**.
❑ Keep your **phone book** nearby because we'll be asking you to find a few **local sources**.
❑ If you have **internet access**, you might want to **log on now**, as we'll be giving you lots of sites to see.
❑ The "**DP1 Book Links**" file on our CD will make it much easier to follow along, and log on to these internet sites.
❑ Have your **3-ring binder** ready so you can store any **downloaded info, data forms, or notes**.

Notes: _____

_____

_____

_____

_____

_____

II. FOUNDATION

✓	Goals of the Foundation Section:		
	Activity	**Date Completed**	**Pg.**
	Health:		**6**
❑	Had a full **physical** and prepared a **health improvement plan**.	___/___/___	
	Basic Training:		**10**
❑	Received **First Aid <u>and</u> CPR** training from:_____	___/___/___	
❑	Completed a basic **introductory firefighting** course.	___/___/___	
❑	Gathered and studied **NBC data** materials from:_____	___/___/___	
❑	Bought a repair manual and/or received basic **auto repair** instruction.	___/___/___	
❑	Took an intro **outdoor survival / camping** course and/or went camping.	___/___/___	
❑	Opened an **Internet Service Provider & Email** account.	___/___/___	
❑	Received **elective instruction** in: _____, and _____	___/___/___	
	Frugality:		**30**
❑	Reviewed **all expenditures**, and methods of **reducing** each.	___/___/___	
❑	Researched outside financial services and discount memberships.	___/___/___	
❑	Got a **library card**.	___/___/___	
❑	Downloaded a free copy of "**Adobe Acrobat**."	___/___/___	
	Planning Ahead for Phase II:		**43**
❑	Reviewed all **insurance needs and policies** with agents.	___/___/___	
❑	Collected contact info for Gov't and private **emergency & relief services**.	___/___/___	
❑	Gathered, copied, and stored all documents necessary for our "**Info Pack**."	___/___/___	
❑	Filled out all **Personal**, **Household**, and other **Data Forms** in "**Appendix**."	___/___/___	
	Emergencies and Disasters. An Overview:		**54**
❑	Reviewed all likely **manmade and natural disasters** for our area.	___/___/___	
❑	Created a "**Threat Map**" customized to our location.	___/___/___	
	Basic Home Prep:		**63**
❑	Reviewed all **safety, security, and fire safety needs** in our home.	___/___/___	
❑	Examined **structural safety needs** in our home and made upgrade plans.	___/___/___	
❑	Created a "**safe room**" or safe area(s) in our home.	___/___/___	
	Basic Vehicle Prep:		**97**
❑	Reviewed all **equipment, maintenance, and repair** needs.	___/___/___	
	The EAS Alerts:		**105**
❑	Learned about the **Emergency Alert System** in our area.	___/___/___	
❑	Purchased a **weather alert radio**.	___/___/___	
❑	Signed up for an **alternate** Emergency Alert **service**.	___/___/___	
❑	Made **mutual EAS contact agreements** with friends and colleagues.	___/___/___	
	Communication:		**109**
❑	Reviewed, diversified, and accessorized all **electronic communications**.	___/___/___	
❑	**Budgeted** for the purchase of **additional communication equipment**.	___/___/___	
❑	Chose **codes to use** when filling out **parts** of various **data forms**.	___/___/___	
	Goal Setting:		**A-1**
❑	Entered **start dates** and **goal dates** onto our **calendar** for all the above.	___/___/___	

You don't build a house without first laying down a good foundation. Before we send you out to buy equipment, to take a long road trip to an evacuation site, or to lock yourself in the house for a three-week quarantine, we're going to help you lay the foundation that will make all the difference in the world on how you deal with emergency situations.

We're going to start with you, and provide the basics you'll need to improve your physical capabilities as well as your training, education, and knowledge base regarding emergency reactions. Knowledge is power, and the right kind of knowledge in the middle of a disaster is priceless. That's one of the reasons we've listed so many outside sources.

While we're discussing the amount of suggested reading you'll find in this section, we should remind you to not let the number of information sources intimidate you. We've provided as much redundant background data as we could since variety and multiple perspectives are always best, and with internet sites you can't count on all of them being there forever. (**Note**: If you try an Internet link that no longer works, use a search engine to locate the name of the company, course, or product. Links and sites change a lot, but most products and companies will still be around, and new sites or products will come along. More on this below, under "Information Research.")

A quick word on this "**Foundation**" section and its training suggestions: Our biggest suggestion is that you **set a goal of having everything in this book accomplished within 3 months (only 12 weekends).** That's why our calendar in the "**Appendix**" is only three months long. We felt this to be the shortest amount of time in which the average person or family could complete all necessary steps. If you don't set a specific goal, you're liable to put things off, little will get done, and the next thing you know you're in the midst of an emergency and unprepared. So, **get your calendar and mark three months from today** as the deadline for having *everything* completed (though you'll want to budget larger purchases). If you're ahead of the game and get things done sooner, you should help others get ready.

Let's start with the most important part of your foundation, **health**.

Health

All other things in this book considered, the one thing topping the list by a country mile is health. We should all do what we can to improve our general wellbeing, our immune systems, and our strength and stamina. This doesn't mean you have to go from sofa spud to superstar, but it does mean most of us could stand to improve our overall health and physical condition.

This is one of those areas where there is far more information than we can possibly include in this book. For every reader there will be a different set of health needs, and 4 or 5 good info sources on each need. Also, we disagree with the notion that any one book should be your sole source of education – even our book. So, as we'll do in other sections, we'll give you some basic suggestions and then point you in the right direction for acquiring more information on your own. You'll find lots of websites, recommended reading, sources of courses, and further education. **We want you to learn everything you can from independent sources and from a variety of perspectives.**

Goals of Health Improvement:
1. To help you **increase your energy** in order to accomplish all this book asks.
2. To help you **handle the physical rigors, emotional upsets, and general stress** of an emergency.
3. To **improve your immunity and resistance** in case of **disaster induced** injury or illness.
4. To **keep you well in general** so you have fewer mishaps to worry about.
5. To give you **an additional benefit** from this book.

Sometimes the most effective things are also the simplest. With that in mind, let's cover a few health suggestions. **Diet, rest, and exercise** are the basics of any good health program. The list below should be nothing new to anyone, but it's our reminder to you of the importance of good health in dealing with physically and emotionally stressful situations, and how **simple details make a big difference**.

Health Improvement Considerations

Interestingly enough, this will probably be the smallest freestanding section of this whole book, yet it's one of the most informative and important. Though simple and short, this portion on health illustrates just how effective a brief list of simple truths can be.

- ➤ Set appointments with your doctor **and** dentist for **full checkups**. Mark these on your calendar.
- ➤ While at the doctor's office, make sure all your **immunizations** are up to date.
- ➤ Find time to **walk or exercise at least 20 minutes a day**, at least three days a week.
- ➤ If not in school, **read more, and/or work brainteaser puzzles**. The brain needs exercise too.
- ➤ **Drink more water**, preferably filtered or distilled water (not the overpriced bottled water).
- ➤ **If you smoke, quit.**
- ➤ If you drink **alcoholic beverages** regularly, cut down or **quit**. (Red wine now and then is okay.)
- ➤ **Reduce your intake** of saturated fats, refined sugar, aspartame, dairy products, and caffeine.
- ➤ **Switch from 3 large meals a day** to 5 or 6 smaller ones, and make sure one is breakfast.
- ➤ Make **fresh fruits and vegetables** a larger part of your diet.
- ➤ Look into buying a **fruit and vegetable juicer**, and get a good book on juicing.
- ➤ Add a daily **vitamin/mineral supplement** to your diet.
- ➤ Stay away from **periodic fad diets**. Improve your health by improving your overall diet.
- ➤ If you have forced-air heating/cooling, buy the best **air filters** you can (such as HEPA).
- ➤ Consider buying a **room air filter**. (One that filters down to 1 micron and ionizes.)
- ➤ **Wash your hands** more often.
- ➤ Objectively examine your **sleep patterns** to make sure you're getting enough rest.
- ➤ Consider keeping a **diet and health improvement log** and record your goals and changes.

Another good reason to improve your health is that it improves your chances of surviving an **NBC** (Nuclear, Biological, Chemical) terrorist attack. The better your health, the stronger your condition, the more functional your immune system, the better your chances of surviving if you fall direct victim to an NBC attack. The emotional and physical rigors of dealing with such an attack notwithstanding, good health will more than likely help you survive should you be physically affected.

As promised, we're making this book a training manual that will prepare you for the worst scenario imaginable. We should **always** prep for the worst though we hope for the best.

The list below is not a hard and fast list of all the information that's available. *As with all other lists of suggestions, this is not our bibliography of sources, but guides to basic information so you can expand your education beyond the boundaries of this book.*

Health and Fitness Information

As you'll notice throughout the book, several sections will have *numerous* "additional information" sources. These are **not** bibliographies of our info sources included for your verification, but sources we have located for the express purpose of you continuing your outside education. We wish to reiterate that you don't have to find and study each and every one, though we do suggest you visit all the websites, and at least browse the listed books and magazines. We give you so many sources since it's always best to have a variety of information and perspectives, and several sources for the same material as **some won't always be available**. Also, different sources will have different things for different people. Some will have all the info necessary for your unique situation and existing level of preparedness, and others won't for you, though they'll have everything for someone else.

Internet

- ➤ For **search engines** go to http://www.google.com or http://www.about.com for health searches.
- ➤ The sites http://allhealthnet.com and http://www.webmd.com have some really good info.
- ➤ Try the government-sponsored http://www.healthfinder.gov, and http://www.healthypeople.gov.
- ➤ Another great general health site is http://www.healthcentral.com.
- ➤ Definitely see the **US National Library of Medicine**, online at: http://www.nlm.nih.gov. Direct contact is 8600 Rockville Pike, Bethesda, MD 20894, 888-346-3656, Fax: 301-402-1384. While you're on the site, check out their publications search page at http://www.ncbi.nlm.nih.gov/entrez/query.fcgi.
- ➤ You'll probably find all you'd ever need at the "Virtual Naval Hospital" on their "Patients Page" at http://www.vnh.org/Patients.html. See other categories on the main site at http://www.vnh.org.
- ➤ For general health and medical info go see: http://www.pdrhealth.com. At the bottom of their page you'll see a small link to "**Join our Email List**" to receive updates.

- ➤ Similarly, check out the online resources of the "Virtual Hospital" at http://www.vh.org.
- ➤ See the National Institute of Allergy and Infectious Diseases http://www.niaid.nih.gov.
- ➤ Health Databases and Dictionaries, online only at : http://www.little-engine.com/health.html.
- ➤ There's a **free daily email newsletter** from Dr. Andrew Weil at http://www.drweil.com.
- ➤ The AMA has a medical info site at: http://www.ama-assn.org/ama/pub/category/3158.html.
- ➤ **National Safety Council** site: http://www.nsc.org. You can also contact this organization by phone at 1-800-621-7619 or at 1121 Spring Lake Drive, Itasca, IL 60143-3201, (630) 285-1121; Fax: (630) 285-1315. Be sure to see their subpage concerning safety stats and information at http://www.nsc.org/library/facts.htm.
- ➤ Be sure to visit "Health A to Z": http://www.healthatoz.com/.
- ➤ Visit the American Dietetic Association at http://www.eatright.org.
- ➤ See the Merck Manual, accessible online at http://www.merck.com.
- ➤ The **CDC's** health information page: http://www.cdc.gov/health/default.htm.
- ➤ Good general information can be found at: http://www.medconnect.com.
- ➤ New York Online Access to Health, NOAH, online at http://www.noah-health.org.
- ➤ While you're searching government sites for health, be sure to see http://www.nutrition.gov.
- ➤ Also check out: http://www.applesforhealth.com. Good site.
- ➤ There's a lot to search through, but http://www.emedicine.com, is a highly informative site.
- ➤ Try MedScape online at: http://www.medscape.com.
- ➤ You can find good Q&A info online at: http://www.theonlinemd.com .
- ➤ AOL subscribers can find free **health "e-zines"** courtesy of AOL. Keyword is "**health**."
- ➤ Also on AOL, enter keyword "**newsletters**" to find more health and fitness related news.
- ➤ Data on **prescription medicine** and interactions: http://www.rxlist.com.
- ➤ BBC Health Information and Courses: http://www.bbc.co.uk/education/health/index.shtml .
- ➤ General health questions can be researched at: http://www.mediconline.com/.
- ➤ For a "second opinion" on symptoms see: http://www.wrongdiagnosis.com.
- ➤ MedicineNet for general medical info: http://www.medicinenet.com/script/main/hp.asp.
- ➤ "Firstgov for Consumers" health page: http://www.consumer.gov/health.htm, 1-800-688-9889.
- ➤ Information on drug or alcohol addictions can be found at http://www.health.org, or by calling 800-729-6686.
- ➤ Check out the federal **Health and Human Services** site at: http://www.hhs.gov. Their contact info is The U.S. Department of Health and Human Services, 200 Independence Avenue, SW, Washington, D.C. 20201, Telephone: 202-619-0257, Toll Free: 1-877-696-6775.
- ➤ **The President's Council on Physical Fitness and Sports**, online at http://www.fitness.gov, or contact them at PCPFS, Department W, 200 Independence Ave., SW, Room 738-H, Washington, D.C. 20201-0004, Phone: 202-690-9000, Fax: 202-690-5211. Be sure to follow the links for "**Publications**."
- ➤ Go to http://www.adtdl.army.mil/cgi-bin/atdl.dll?type=fm, and download the military training manual on **Physical Fitness Training, FM 21-20**. By the way, you'll also find this manual on our enclosed **CD**.
- ➤ See the online Marine Corps PT manual at http://www.tpub.com/content/USMC/mcrp302a.
- ➤ The CDC has its own health-related hoaxes page at http://www.cdc.gov/hoax_rumors.htm.

Periodicals
- ➤ Four good magazines covering general health are:
  - ● "**Health**," http://www.health.com, Health, PO Box 60001, Tampa, FL 33660,1-800-274-2522.
  - ● "**Longevity**," http://www.longevitymagazine.com, Longevity Magazine, 7411 114th Avenue North Suite 308, Largo , FL 33773, 727-547-5450.
  - ● "**Prevention**" http://www.prevention.com, 33 East Minor St., Emmaus, PA 18098-0099, fax: 610-967-8963, phone: 610-967-8038 .
  - ● "**Herbs for Health**," http://www.herbsforhealth.com, Ogden Publications, Inc., 1503 SW 42nd St., Topeka, KS, 66609-1265, Phone: 785-274-4300.
- ➤ Go to http://www.magazines.com to search multiple subjects and titles. (Also take a look at our "Information Research" portion of "Basic Training.")
- ➤ Start at your **library, newsstand or bookstore** to research and preview health and fitness periodicals.

Books

➢ Start with the books and magazines at your **local library**. They're plentiful and **free to read**.

➢ Before going to regular bookstores, check local **used book stores and "Chapter 11 Books."**

➢ Visit **Amazon** at http://www.amazon.com. It's a great source for books on any subject.

➢ Also visit the **US Government Printing Office** at http://www.gpoaccess.gov/, Superintendent of Documents, 732 North Capitol Street NW., Washington, DC 20401, 202-512-0000, and 888-293-6498, or P.O. Box 371954, Pittsburgh, PA 15250-7954, 1-866-512-1800.

➢ See the Medical Book search engine online at http://www.medicalengine.com.

➢ **"Dummies"** books and **"Complete Idiot's Guides"** really are good intro manuals. Go to http://www.dummies.com and http://www.idiotbooks.com, to browse.

➢ Read: **Vitamins for Dummies, Nutrition for Dummies**, and **Fitness for Dummies**, available through http://www.dummies.com, http://www.amazon.com, or your local library or bookstore.

➢ Read Dr. Andrew Weil's **Eight Weeks to Optimum Health**, and Gary Null's **7 Steps to Perfect Health** and **Get Healthy Now**. Dr. Weil's info can be found at http://www.drweil.com, and Gary Null's at http://www.garynull.com.

➢ For fruit and veggie juicing read **Fresh Vegetable and Fruit Juices** by Dr. Norman W. Walker, or **The Juiceman's Power of Juicing** by Jay Kordish.

➢ For the juicing equipment, take a look at juicers and related equipment at "Discount Juicers" online at: http://www.discountjuicers.com, or contact them directly at On the World Wide Web, 6366 Commerce Blvd., #200, Rohnert Park, CA 94928. You can also find Jack La Lanne's juicer at www.jackspowerjuicer.com, or call them at 800-260-1785.

➢ Books on **fruit and veggie juicing** are at: http://hallfood.com/drinks_beverages/8.shtml, with more found at: http://www.healingdaily.com/juicing-for-health/juicing-books.htm.

➢ **Prescription for Nutritional Healing** by James and Phyllis Balch is good nutrition material.

➢ For advanced fitness, read **Body for Life** by Bill Phillips., at: http://www.bodyforlife.com. Contact them at 555 Corporate Circle, Golden, CO 80401, Phone: 1-800-297-9776, Fax: 303-279-8057.

Videos

➢ Check your local **library and local video rental outlet** for health, nutrition, or exercise videos.

➢ Go to Google.com and enter the search string **"health and nutrition videos"** (with quotes).

➢ For the moderately fit, the "8 Minute___" videos and those on Yoga, Pilates, or Tai Chi, are good.

➢ If you're already in **rather** good shape, try the "**Power 90**" series http://www.power90.com, Billy Blanks' "**Tae Bo**" videos http://www.taebo.com, or similar cross-training instructional tapes.

Courses

➢ See if your **local college** has a **Continuing Education** program in nutrition or fitness. In continuing ed classes the general public is allowed to enroll without being a full-time student.

➢ Many **gyms or health clubs** have licensed and accredited nutrition and/or fitness advisors who'll offer good training to those getting into the fitness lifestyle for the first time or after long layoffs. **Accreditation is key.** (However, be cost conscious regarding membership fees.)

➢ Go to http://www.google.com and enter the search string *health and fitness courses online*. Type it in just like that with no quotes or punctuation. Enter the search string again with the word *nutrition* instead of *fitness* for a different grouping of courses.

➢ Check your **newspaper or leisure publication for health and fitness course** listings.

➢ Check the **Yellow Pages** for health and fitness consultants and/or practitioners and ask them about seminars, courses, and outings.

➢ For health and self-protection, look into a **good martial arts school**. All will give you good exercise, so choose one that teaches combat techniques instead of sport competition.

The above might seem like a lot of information to digest on one single subject, but as we said, it's an important topic. Your biggest asset is always going to be you. Not a piece of equipment...you. So, utilize as many of the above sources as you need in order to get the best overall information possible (and store the info in your 3-ring binder). Remember, we'll always give you a lot more information than we can contain in just this one book. Let's look at other areas of **training** you'll need in order to effectively protect you and your family from disaster and its after effects.

Be sure to record **medical and dental appointments**, and your **workout schedule**, on your **calendar**.

Basic Training

None of us are born experts at anything, and the more one knows about a particular subject, the less they've studied another. It's the nature of our lifestyles. We get so caught up in careers and everything else around which our life revolves, we wind up lacking in a few skills that can come in handy during an urgent situation.

It may seem odd that in a book about preparedness, one of the first things we do is tell you to go to other sources to find information. However, there's a good reason for it: Nothing, absolutely nothing, beats education and information, and the more sources for each you have, the better off you'll be. It would be really easy to cut and paste a few checklists and a couple of "if this happens, do this" lists, but such simplistic and shallow instruction will produce nothing but less than desirable results. Later on in the book we will give you some how-to instruction on several of these topics, but for now let's focus on education, background, outside info, and the general basics.

The following are the most important of your educational and training topics so utilize each and every source listed, bookmark sites, print pages, take notes, **practice what you learn**, and then find a **course** with a qualified instructor to take your training to the next level.

Goals of Basic Training:

1. To give you sources for improving the best disaster asset you could have; **skills in an emergency**.
2. To stress the fact that there is **more to preparedness than "duct tape and plastic sheeting."**
3. To reiterate our point that **no one source is going to have everything you need**.
4. To provide you a **variety of information** that will be useful in a number **unforeseeable situations**.
5. To open your eyes to the **multitude of factors** present in **any number of emergencies**.

Below are some categories of basic training that will be beneficial to you in reaching your preparedness goals. Here, you'll find background information sources for:

1. **First Aid****.
2. **Limited Fire Fighting****.
3. **NBC (Nuclear Biological Chemical) Data****.
4. **Auto Repair.**
5. **Outdoor Survival****.
6. **Information Research****.
7. **Miscellaneous Electives.**

Those items above marked with ****** will have additional direct information contained elsewhere in the book or on the enclosed **CD**. For the other items, most all the information you'll need can be found through the listed sources. The information provided by these sources is comprehensive, plentiful, much of it is free, and there's no reason to clog book space here, or make this book larger or more expensive.

If you're ahead of the game and adept at one or more of the following, focus on the topics you're least familiar with. When taking any of these courses or classes try to take as many family members or friends and neighbors as possible. With regards to safety, it really is "the more the merrier."

Before we get started, let us remind you how important this recommended basic training is. Remember: **IN A LARGE-SCALE EMERGENCY, OFFICIAL FIRST RESPONDERS WILL BE TEMPORARILY OVERWHELMED AND MAY NOT BE ABLE TO HELP YOU RIGHT AWAY. THE MORE SELF-SUFFICIENT YOU ARE, THE BETTER OFF YOU'LL BE. ALSO, THE LESS YOU NEED THE HELP OF OFFICIALS, THE MORE YOU'VE FREED THEM UP TO HELP THOSE LESS CAPABLE THAN YOU.** Let's proceed.

First Aid Background

First Aid heads our list of recommended training for the same reason "Health" tops our list in this **"Foundation"** section. People and lives come first. You'll find information about **First Aid kits** in the **"Evacuation"** and **"Isolation"** sections, and we've included the **military manual on first aid**, FM 4-25.11, on the enclosed **CD**.

In the meantime, let's give you some information sources to help you develop a good knowledge base. It would be an easy task to post a couple of simplistic charts on how to stop bleeding or perform CPR, but those won't teach you first aid any more than looking at pictures of road signs will teach you to drive. You need lots of detailed information, and it's always good to get as much information as possible from a variety of sources and a number of different perspectives. **Learn as much as you possibly can** because the first time you're at an automobile accident scene, or if you ever have to face the aftermath of a suicide bomber, you'll wish you had far more training than you do.

When studying the information sources listed below, we suggest you research the following:

1. **Basic first aid** to care for yourself and your family in **household mishaps and minor injuries**.
2. **Specific aid and treatment for illnesses or conditions** you or a family member may have.
3. First aid for **major wounds** and **broken bones**.
4. Look at both new, and old and unusual **substances that will stop heavy bleeding**.
5. Emergency and primary treatment of **burns, chemical burns, eye, and lung** (inhalation) injuries.
6. Primary and secondary treatment for **snake bite, insect and spider bites, bee stings, and ticks**.
7. Memorize the symptoms and treatment of **various types of shock**, and of **poisoning**.
8. Know the signs of **exposure** such as for **hypothermia, frostbite, heat stroke**, etc.
9. Learn the **threat level of various injuries and conditions** for the purposes of **triage**.
10. Find a good **First Aid and CPR course** that you attend <u>in person</u>.

Note: Books are great, but nothing beats a good hands-on course with a qualified instructor!

Internet

➤ You can download the military **First Aid manual, FM 4-25.11 (**supercedes previous FM 21-11), and numerous other manuals at: http://www.adtdl.army.mil/cgi-bin/atdl.dll?type=fm, or at: http://www.globalsecurity.org/military/library/policy/army/fm. (By the way, most of the military manuals we recommend can be found in softcopy format on the **enclosed CD**.)

➤ Free books and a host of other info can be found on the NY ER Nurse site: http://www.nyern.com.

➤ Virtual Naval Hospital Online First Aid Course: http://www.vnh.org/StandardFirstAid/toc.html.

➤ Also visit the Virtual Naval Hospital's topics list: http://www.vnh.org/Misc/SiteMap.html.

➤ First Aid information by category: http://www.healthy.net/clinic/firstaid.

➤ Public Health and Safety, and Emergency Services links: http://www.1uphealth.com/links/desc-235.html.

➤ Yahoo's emergency Services links collection: http://dir.yahoo.com/health/emergency_services.

➤ BBC's Online First Aid Course: http://www.bbc.co.uk/health/first_aid_action.

➤ Adventure Network Wilderness First Aid: http://www.adventurenetwork.com/HEALTHTemp.html.

➤ First Aid Tutorial: http://www.survival-center.com/firstaid/book.htm.

➤ Good First Aid info from an Australian EMT site: http://www.parasolemt.com.au/Manual/afa.asp.

➤ Australia's Wilderness Medicine Institute: http://www.wmi.net.au/wmi/Default.htm.

➤ Learn **CPR**: http://depts.washington.edu/learncpr.

➤ Go to http://www.emedicine.com, and do a search for "first aid." This is also a good general info site.

➤ First Aid & Safety articles (good.): http://kidshealth.org/parent/firstaid_safe.

➤ Utah Mountain Biking is a good info site: http://www.utahmountainbiking.com/firstaid/index.htm.

➤ Mayo Clinic: http://www.mayoclinic.com/findinformation/firstaidandselfcare/index.cfm.

➤ Columbia University Medical Guide: http://cpmcnet.columbia.edu/texts/guide/toc/toc14.html.

➤ IEASR's First Aid site: http://firstaid.eire.org/Firstaid_index.html.

➤ MedicineNet's First Aid info: http://www.medicinenet.com/Script/Main/Art.asp?li=MNI&ag=Y&ArticleKey=224.

➤ Treating shock: http://www.allhealthnet.com/First+Aid+-+Emergency/Emergencies/Shock.

➤ Multiple sites for treatment of shock: http://directory.google.com/Top/Health/Public_Health_and_Safety/First_Aid/Shock.

➤ Another shock collection: http://dmoz.org/Health/Public_Health_and_Safety/First_Aid/Shock.

➤ Visit Safety Campus.com for some **online courses**: http://www.safetycampus.com/.

➤ More online courses at: http://www.emszone.com/onlinecourses.

➤ The "Wound Care" site: http://www.wounds1.com.

➤ A discussion of primary assessment: http://www.techrescue.org/firstaid/firstaid-ref1.html.

➤ A discussion of Triage: http://www.techrescue.org/firstaid/firstaid-ref2.html.

- More on triage: http://www.wikipedia.org/wiki/Triage.
- Boy Scout's, First Aid Exam Online: http://ourworld.compuserve.com/homepages/havlicek/1st_Aid008.html.
- Health and Safety for Kids: http://webtech.kennesaw.edu/ewashington/health_wellness2.htm.
- Snakebite First Aid: http://www.xmission.com/~gastown/herpmed/snbite.htm.
- Free monthly email newsletter on safety: http://www.efwhomesafety.com/newsub.htm.
- "Survival Medicine": http://www.geocities.com/CapitolHill/Lobby/2276/survmed.htm.
- A few online first aid quizzes: http://www.techrescue.org/pages/Emergency_First_Aid/Quizes.
- You can find some **dental first aid** links at http://dentistry.about.com/cs/dentalemergencies.
- National Registry of EMTs: http://www.nremt.org/about/related_links.asp.
- EMS and CPR training: http://www.emsadvocate.com/, or write them at EMS Advocate, LLC, P.O. Box 1657, Pomona, CA 91769-1657.
- Extensive first aid information, instruction, tests, newsletters, and the like can be found at "MedTrng," online only at http://www.medtrng.com. An online test with further first aid instruction can be found on one of their subpages: http://www.medtrng.com/cls_dl.htm.
- The Medical Corps site offers not only a source of emergency/combat medicine courses (call them at 740-783-8009), but if offers a nice collection of links at http://www.medicalcorps.org/web_links.htm, and a good pharmaceutical info page at http://www.medicalcorps.org/pharmacy.htm.
- Massive links collection to emergency medicine and first aid info sites and publications: http://www.arkanar.minsk.by/medicine/3/Emergency_Medicine_index.htm.
- For in-depth information, read **Emergency War Surgery**, at http://www.vnh.org/EWSurg/EWSTOC.html.

Periodicals
- "**EMS**" and "**Rescue Technology**" magazines: http://www.emsmagazine.com, Summer Communications, Inc., 7626 Densmore Ave., Van Nuys, CA 91406-2042, Phone: Toll-free 800/224-4367; 818/786-4367, Fax: 818/786-9246.
- Occasional articles on first aid and emergency readiness can be found in an online-only magazine called "**Modern Survival**." It's the online version of the old "**American Survival Guide**" and can be found at: http://www.modernsurvival.net.
- "**Backwoods Home**" magazine is similar in scope, but available in hardcopy. You can find more information at: http://www.backwoodshome.com, or Backwoods Home Magazine, P.O. Box 712, Gold Beach, OR 97444, Toll free: 800-835-2418, Phone: 1-541-247-8900 Fax: 1-541-247-8600.

Books
- Check your **local library, used bookstore, and bookstore** for books on First Aid, CPR, and Paramedic instruction (as your level of expertise rises).
- Your **local camping supply store** may have a variety of wilderness first aid manuals as well.
- We suggest the **US Army Special Forces Medical Handbook ST 31-91B**, which is for sale at http://www.fetchbook.info/US_Army_Special_Forces_Medical_Handbook_ST_31_91B.html , or at military surplus stores (along with other titles) or other book sources.
- Several companies have first aid books and/or equipment (among other useful subjects).
  - **Paladin Press** at: http://www.paladinpress.com, Gunbarrel Tech Center, 7077 Winchester Circle, Boulder, Colorado 80301, Customer Service: 303.443.7250, Fax: 303.442.8741.
  - **Galls** at: http://www.galls.com/, Galls, Inc., 2680 Palumbo Drive, P.O. Box 54308, Lexington, Kentucky 40555-4308, 866-230-3385. Get the Emergency Medical Products catalog.
  - **Rescue Technology** at: http://www.resqtek.com, RESCUE TECHNOLOGY, 251 Beulah Church Road, Carrollton, GA 30117, 800-334-3368, 770-832-9694, 770-832-1676 (FAX).
  - **Nitro-Pak** at: http://www.nitro-pak.com, 13309 Rosecrans Ave. Santa Fe Springs, CA 90670-4940, 800-866-4876, 310-802-0099, Fax: 310-802-2635.
  - **Emergency Essentials** at: http://www.beprepared.com, 362 S. Commerce Loop, Suite B, Orem, UT 84054-5119, 800-999-1863.
  - **The EMS Zone** at: http://www.emszone.com, Jones and Bartlett Publishers, Inc., 40 Tall Pine Drive, Sudbury, MA 01776, 800-832-0034.
- **First Aid and Safety for Dummies**, by C. Inlander, at http://www.dummies.com, is pretty good.
- So is the book **First Aid, Responding to Emergencies** (Mosby Press) by the Red Cross. Their headquarters is at 431 18th Street, NW, Washington, DC 20006. Phone: (202) 639-3520 or log on to http://www.redcross.org.

- ➢ Find **Where There is no Doctor** by David Werner with Carol Thruman and Jane Maxwell.
- ➢ Read **Where There is no Dentist** by Murray Dickson.
- ➢ Read **The Doctor's Book of Home Remedies**, ISBN: 0-87857-873-0, by the Editors of Prevention Magazine Health Books, Rodale Press.
- ➢ EMS books and materials: http://www.hultgren.org/books/index.htm.

Videos

- ➢ "**EMS**" and "**Rescue Technology**" magazines (listed above) will have some good training videos listed in various ads and reviews.
- ➢ Check your **library and video rental store** for instructional videos.
- ➢ The **above book sources** of Paladin Press, Gall's (definitely), Rescue Technology, etc. also have good first aid videos to choose from.
- ➢ Also, ask your local Volunteer Fire Department, Hospital's EMS members, or your local technical college which video programs they use for basic and intermediate first aid.

Courses

- ➢ Your local **Fire Department or hospital may offer periodic First Aid and CPR courses,** or may know local classes and trainers that can give you not only basic, but advanced training.
- ➢ Contact the **American Red Cross** for First Aid and CPR courses in your area. Their headquarters is at 431 18th Street, NW, Washington, DC 20006. Phone: (202) 639-3520 or log on to http://www.redcross.org. The link for courses is: http://www.redcross.org/services/hss/courses.
- ➢ The **American Heart Association** offers quality CPR courses, with info at: http://www.americanheart.org, or by mail at: 7272 Greenville Ave., Dallas, TX 75231, 1-800-AHA-USA-1 (1-800-242-8721).
- ➢ **National Safety Council**: http://www.nsc.org/psg/fai/faichart.htm, 1-800-621-7619 or 630-285-1121.
- ➢ Medic ® First Aid Training: http://www.medicfirstaid.com, Medic First Aid, P.O. Box 21738, Eugene, OR 97402, PHONE: 541.344.7099 or 800.800.7099, FAX: 541.344.7429.
- ➢ "Wilderness First Aid" equipment and courses: http://www.wildmedcenter.com/wildmedcenter, P.O. Box 11, Winthrop, WA 98862, Phone: (509) 996-2502.
- ➢ First Aid and CPR instruction: http://www.rescuebreather.com/. (Click on "Instructor Directory.")
- ➢ In lieu of anything else, contact your **local Boy Scout troop** and tell them you'd like basic First Aid instruction. Most will be happy to do this "good deed." You'll find more info at: http://www.scouting.org, Boy Scouts of America, National Council · P.O. Box 152079 · Irving, Texas 75015-2079.
- ➢ EMS and CPR training: http://www.emsadvocate.com/, or write them at EMS Advocate, LLC, P.O. Box 1657, Pomona, CA 91769-1657.
- ➢ Contact **local community colleges** to see if they offer a basic First Aid or CPR course, or a basic EMT course (as your level of expertise rises). Some colleges will offer free public benefit courses, and free courses to retired persons. Call your local college and ask about both.
- ➢ **Check your newspaper,** community bulletin or publication, your place of worship bulletin, and your local leisure publication for notices of First Aid or CPR classes.
- ➢ As you and your family members learn the basics, **conduct training at home** to help everyone learn how to handle bigger emergencies. It helps to keep everyone refreshed and up to date.
- ➢ As your level of expertise grows, take **more and more advanced classes**.

Miscellaneous

- ➢ **Poison Control** Center: http://www.poison.org, 800-222-1222 (call 911 first in an emergency!).
- ➢ **American Association of Poison Control Centers:** http://www.aapcc.org, American Association of Poison Control Centers, 3201 New Mexico Avenue, Suite 330, Washington, DC, 20016, 202-362-7217 (not an emergency information center).
- ➢ Each state's poison control center: http://www.medicinenet.com/Art.asp?li=MNI&ag=Y&ArticleKey=869. Or: http://www.aapcc.org/findyour.htm.
- ➢ First Aid for Poisoning: http://www.oklahomapoison.org/general.

Pets

First aid doesn't apply to only people. Ask your vet about First Aid info on **how to care for your pets.** (Under "**Evacuation**" we'll discuss pet first aid equipment needs.)

- ➢ Contact your **Vet, Humane Society, local animal shelter, or major pet supply store** for pet first aid classes and equipment.
- ➢ Pet First Aid site: http://www.canadianliving.com/pets/pet-aid/index.asp.

- More pet first aid: http://www.preparenow.org/fa-anim.html.
- "Dog Owner's Guide" first aid site (and links): http://www.canismajor.com/dog/fstaidk.html.
- More on pet poisoning can be found online at http://www.petcaretips.net/pet_poisons.html.
- Dog and Cat First Aid Videos: http://www.firstaidforpets.com.
- More pet medical info: http://www.merckvetmanual.com.
- First Aid kit for your horse: http://petplace.netscape.com/articles/artshow.asp?artID=1251.
- American Pet Association (with vet-finder link), http://www.apapets.com, HQ - PO Box 725065 Atlanta, GA 31139-9065, 800-APA-PETS (800-272-7387), Fax: 305-294-8964.
- American Society for the Prevention of Cruelty to Animals, http://www.aspca.org, American Society for the Prevention of Cruelty to Animals (ASPCA), 424 E. 92nd St, New York, NY, 10128-6804, (212) 876-7700.
- The American Veterinary Medical Foundation: http://www.avma.org, 1931 North Meacham Road, Suite 100, Schaumburg, IL, 60173, Phone: 847-925-8070, Fax: 847-925-1329.
- Clemson University's Animals in Disasters: http://virtual.clemson.edu/groups/ep/animal.htm.
- Information on **Emergency Animal Rescue** at: "United Animal Nations" http://www.uan.org, United Animal Nations, 5892A South Land Park Drive, P.O. Box 188890, Sacramento, CA 95818, Tel: (916) 429-2457, Fax: (916) 429-2456.
- Animal Disaster Preparedness links: http://www.horsereview.com/Links_Disaster.htm.

Have you located a first aid and/or CPR course? Mark the class dates on your **planning calendar**.

Fire Fighting Training

This is some tough training to find. The vast majority of training and information readily available is polarized into either the very simplest of instructions on how to buy smoke detectors and use household fire extinguishers, or full-fledged professional firefighter training. Very little exists in between. We've intermingled what information we could throughout this book since fire is such an important topic.

Later on we're going to show you what we can about fighting a fire, or at least containing it SAFELY, along with what NOT to do. You'll find some prevention and equipment lists in the" **Basic Home Prep**" portion of this "**Foundation**" section, some reaction procedure in the "**Fire**" portion of the "**Reaction**" section, and some fire fighting procedures under "**Isolation**." Start by learning ALL the fire prevention and safety measures you can, from the sources below. Anything you can do to save lives and property is a good thing, and anything that helps reduce the burden on our first responders is icing on the cake.

While reviewing the following, we suggest you focus on:
1. Common **types and causes** of fires.
2. The **nature of fire and how it works**.
3. Basic **home fire safety**.
4. How **fire extinguishers work to put out fires**.
5. Common **mistakes made by civilians** trying to fight a fire.
6. Discussions and descriptions of **civilian firefighting equipment**.
7. **Caveats and cautions** about different types of fires.
8. Various methods of **carrying people out of harm's way** in a fire situation.

Internet
- **National Volunteer Fire Council**: http://www.nvfc.org, 1050 17th St. NW, Suite 490, Washington, DC 20036, Voice: 800-FIRE-LINE (800-347-3546), 888-ASK-NVFC (800-275-6832), or 202-887-5700, Fax: 202-887-5291. Be sure to check out their "**Fire Links**."
- **National Fire Protection Association**: http://www.nfpa.org/Home/index.asp, 1 Batterymarch Park, P.O. Box 9101, Quincy, Massachusetts 02269-9101 Phone: 1-617-770-3000. Good source of fire stats, causes, firefighting associations, etc.
- For information on fire itself, and a collection of links, studies, software, and other useful materials, visit the National Institute of Standards' "**Building and Fire Research Laboratory**" site at http://www.bfrl.nist.gov/fris, or contact them at 5285 Port Royal Road, Springfield, VA 22161, 800-553-6847, or 703-605-6000.
- Visit the **Building and Fire Research Laboratory** at http://www.fire.gov, and sign up for their newsletter.

- General (professional) firefighting info: http://www.firefighting.com. While at "Firefighting DOT Com," click on "**Learning**" and then click on "**Links**."
- More professional firefighting info and related data: http://www.firehouse.com, or **FIREHOUSE**, PO Box 820, Fort Atkinson, WI 53538-0820, 1-800-825-8577.
- Visit http://www.firetactics.com, for an extensive online library of downloadable articles and manuals.
- **National Association of Fire Equipment Distributors**: http://www.nafed.org, 104 S. Michigan Ave., Suite 300, Chicago, IL 60603, Tel (312) 263-8100, Fax (312) 263-8111. Read their links and publications.
- Good info on accidents, programs, and other stats at the **US Fire Administration**, http://www.usfa.fema.gov, 16825 South Seton Ave., Emmitsburg, MD, 21727, phone 301-447-1000, fax 301-447-1052.
- Also see **FEMA's "National Fire Incident Reporting System"** at: http://www.nfirs.fema.gov. They're a good source of fire start stats and other useful information.
- Some other online training materials regarding fire can be found on FEMA's NETC, or National Emergency Training Center, website at http://www.training.fema.gov, or by phone at 301-447-1175. The link, http://www.usfa.fema.gov/fire-service/nfa/courses/offcampus/nfa-off3.shtm, will take you to the National Fire Academy's Self Study page.
- **Fire / EMS links** collection: http://www.fire-ems.net.
- **General fire info** can be found at: http://www.ou.edu/oupd/fireprim.htm.
- **Wildfire protection tips:** http://www.firewise.org/fw99/home.html (visit the homepage too).
- Visit the **National Interagency Fire Center** at http://www.nifc.gov for more wildfire data.
- "Find Fire," a **fire fighting search** engine / portal: http://www.fire-find.com. Also see the links to training: http://www.fire-find.com/cgi-bin/ms2/firefind/search?c=TRA&q=* .
- Visit the links collection at www.firefightinglinks.com.
- See the **NIOSH** firefighting links collection: http://www.cdc.gov/niosh/firelink.html.
- Read **Fire Safety** at the National Library of Medicine: http://www.nlm.nih.gov/medlineplus/firesafety.html.

Periodicals

- "**National Fire & Rescue**" magazine at http://www.nfrmag.com, or at National Fire & Rescue, 5808 Faringdon Place, Suite 200, Raleigh, NC 27609-3930, Ph: 919-872-5040, Fax: 919-876-6531
- "**EMS**" and "**Rescue Technology**" magazines. Both found at: http://www.emsmagazine.com, Summer Communications, Inc., 7626 Densmore Ave., Van Nuys, CA 91406-2042, Phone: Toll-free 800/224-4367; 818/786-4367, Fax: 818/786-9246.
- "**Firehouse**" magazine can be ordered through http://www.firehouse.com, or directly through FIREHOUSE, PO Box 820, Fort Atkinson, WI 53538-0820, 1-800-825-8577.
- For some related info, try the "**National Fire and Rescue**" magazine, online at http://www.nfrmag.com, or contact them at National Fire & Rescue, 5808 Faringdon Place, Suite 200, Raleigh, NC 27609-3930, 919-872-5040, Fax: 919-876-6531.
- An extensive **list of periodicals** and books can be found through "**Find Firefighting's**" publications page at: http://www.fire-find.com/cgi-bin/ms2/firefind/search?c=PUB&q=*.

Books

- Though rather expensive, an excellent book on firefighting is the **Firefighter's Handbook: Essentials of Firefighting and Emergency Response**, by Delmar Publishers. It's available at http://www.amazon.com, and may be found at your local library. Direct contact for Delmar is Delmar Learning, 5 Maxwell Dr., PO Box 8007, Clifton Park, NY 12065 – 8007, 1-800-347-7707 or fax 1-800-487-8488.
- An extensive **list of periodicals** and books can be found through "**Find Firefighting's**" publications page at: http://www.fire-find.com/cgi-bin/ms2/firefind/search?c=PUB&q=*.
- Visit "**Firehouse Books**" at http://www.firehousebooks.com, or call them at 888-698-1700 for a catalog.
- Also visit "**Firebooks**" at http://www.firebooks.com, for a rather extensive collection of Fire and EMS related subjects. Contac them directly at (714) 375-4888 • (800) 727-332718281, or at Gothard St. #105 • Huntington Beach, CA • 92648-1205.

Videos

- We researched numerous video and interactive software packages. Most of them were directly oriented to professional firefighters, and were extremely expensive. However, if your volunteer group wants to know more about them, you'll find many listed in the sites mentioned above, such as from the **Firehouse Books** video selection at http://www.firehousebooks.com/videos.html .

- The **Ohio Bureau of Workers Compensation** at 30 W. Spring St. Columbus, Ohio 43215-2256, 1-800-OHIO-BWC, or http://www.ohiobwc.com/employer/programs/safety/SHVidFire.asp, seemed to have quite a few good videos on various aspects of fire prevention and fire fighting.
- Also check your library or local video store for videos on basic home fire safety and fire extinguisher usage.

Courses

- Contact both your local **municipal and volunteer fire departments**. Let them know what you'd like to learn. They can either give you the limited training you need or give you additional educational sources. Remind them that you're not looking to take on a fire by yourself, you just want to be able to safely keep a small to medium sized fire in check until the pros can arrive.
- **CERT** (**C**ommunity **E**mergency **R**esponse **T**eams) from http://www.citizencorps.gov (Directed by FEMA and coordinated through your state's Emergency Management Agency) will offer various training to include minor firefighting in some locations. Contact Citizencorps through their website, through your state's EMA, or through FEMA at: 500 C Street SW, Washington, D.C. 20472, Phone: (202) 566-1600, 202-646-2500.

Have you located a local fire safety or fire fighting course? Mark the date on your **planning calendar**.

NBC Data

NBC is the military acronym for "**Nuclear, Biological, Chemical**." NBC data is the information you need to know regarding a nuclear, biological, or chemical terrorist attack or a hazardous materials (HazMat) accident (the new comprehensive acronym is **CBRNE**, pronounced 'see burn' which stands for **C**hemical **B**iological **R**adiological **N**uclear or **E**xplosive). You'll need to know countermeasures, first aid, self-decontamination procedures, and what procedures to expect from officials. You'll also need to understand the symptoms, effects, and follow-up treatment for the type of NBC attack you're a victim of. **Note**: The 1993 attack on the World Trade Center had a container of **cyanide** hidden in the vehicle bomb. You never know what you'll be up against.

In the "**Reaction**" section, we'll go into more detail on this subject and tell you what to do should you be a direct victim of any kind of NBC attack or related incident. Also under the "**Reaction**" section we'll talk more about basic decontamination and how to put together rudimentary NBC protective measures.

When studying the information sources listed below, we suggest you focus on the following:
1. Know the **physical and chemical characteristics** of the most popular chemical weapons.
2. Know the **symptoms and effects** of both chemical and biological weapons.
3. Learn **basic decontamination** procedures to perform on yourself and others.
4. Understand the **decontamination measures that officials might use** on you.
5. Learn the **effective radius** of various types of nuclear weapons.
6. Learn the **symptoms of radiation sickness**.
7. Study the myths and reality surrounding "**dirty bombs**."

Internet

- Two REALLY good sites are: http://www.nbc-links.com/, and http://www.nbc-med.org.
- For a good collection of substance data visit: http://www.cbwinfo.com.
- A large NBC info links collection can be found on the National Association of County and City Health Officials page at http://archive.naccho.org/Documents/BT/Misc.htm.
- A comprehensive and easy-to-understand guide to bioterrorism, along with a collection of outside links, can be found online at http://www.infotoday.com/searcher/mar02/perkins.htm.
- See the **CDC "Public Health Emergency Preparedness & Response**" (bioterrorism) site: http://www.bt.cdc.gov. You may also contact the CDC at: Centers for Disease Control and Prevention, 1600 Clifton Rd., Atlanta, GA 30333, (404) 639-3311, 1-800-311-3435. Be sure to look at their extensive planning page at: http://www.bt.cdc.gov/Planning/ and for more information, you can call their public response hotline at 888-246-2675. See the CDC's list of chemicals at http://www.bt.cdc.gov/Agent/AgentlistChem.asp, and the CDC's biological weapons list at http://www.bt.cdc.gov/agent/agentlist.asp.
- Go see the **US Army Medical Research Institute of Chemical Defense** links page at: http://chemdef.apgea.army.mil/related/links.asp, or view their **online course** information site at http://ccc.apgea.army.mil. Contact them at U.S. Army Medical Research Institute of Chemical Defense,

3100 Ricketts Point Road, Aberdeen Proving Ground, MD 21010-5400, Voice 410-436-3628, Fax 410-436-1960. Be sure to visit the links page at http://ccc.apgea.army.mil/links.htm.

➢ A close cousin is the Armed Forces Radiobiology Research Institute (with downloadable publications) at http://www.afrri.usuhs.mil.

➢ Infectious disease info is on the World Health Organization's site at http://www.who.int/home-page.

➢ Johns Hopkins Center for BioDefense: http://www.hopkins-biodefense.org/index.html, or (410) 223-1667.

➢ A good source of biochemical info is at: http://www.911guide.com/reference_center.html. Be sure to visit the homepage at http://www.911guide.com as well.

➢ Visit Emergency.Com's "Emergency Response to Chemical/Biological Terrorist Incidents" page at http://www.emergency.com/cbwlesn1.htm. In fact, this company offers a lot of good info so here's their contact data: The Emergency Response & Research Institute, online at http://www.emergency.com, with direct contact at ERRI, 6348 N. Milwaukee Ave., Suite 312, Chicago, IL 60646, 773-631-ERRI Voice/Voice Mail, 773-631-4703 Fax.

➢ See the National Library of Medicine sites on chemical warfare and biological warfare at: http://www.sis.nlm.nih.gov/Tox/ChemWar.html, and http://www.sis.nlm.nih.gov/Tox/biologicalwarfare.htm.

➢ Log on to http://www.cbiac.apgea.army.mil/awareness/newsletter/intro.html to read the Army's Chemical and Biological Defense Information Analysis Center newsletters.

➢ To keep abreast with current news regarding biochemical threats and the US Government's response and involvement, take a look at the online newsletter, "Dispatch" at http://www.cbaci.org/dispatch.htm, compiled by the Chemical and Biological Arms Control Institute.

➢ A good table of chemical agents and antidotes appears on the New York Health Department's website at http://www.health.state.ny.us/nysdoh/bt/chemical_terrorism/chemical.htm.

➢ See the Virtual Naval Hospital's BioChemical info at: http://www.vnh.org/Misc/SiteMap.html, and their "Chemical Casualties Handbook" at http://www.vnh.org/FieldManChemCasu.

➢ The **CIA**'s online NBC info: http://www.odci.gov/cia/reports/cbr_handbook/cbrbook.htm.

➢ Visit the Oak Ridge National Laboratory's "Emergency Management Center" site at http://emc.ornl.gov, and click on "Publications." This will take you to a page offering downloads on a variety of chemical information.

➢ An excellent listing of industrial chemicals, their properties, cautions, and safety considerations can be found online on the International Programme on Chemical Safety's "International Chemical Safety Cards" page at http://www.inchem.org/pages/icsc.html.

➢ NBC Disaster Info: http://www4.umdnj.edu/camlbweb/disaster.html.

➢ More NBC data at: http://www.twotigersonline.com/resources.html (really good).

➢ Though primarily a publishing company site, on http://www.chem-bio.com you'll find links and resources that will be useful if you wish to extend your NBC education beyond the basics. Also visit their other page http://www.chem-bio.com/links/periodicals.html. You'll find links to various organizations, and newsletters.

➢ A good collection of **bioterrorism** links can be found at: http://www.labt.org/links.asp. You can also try contacting the Los Angeles County, Department of Public Health Services, Bioterrorism Preparedness and Response, 313 N. Figueroa Street, Los Angeles, CA 90012, Tel: (213) 240-7941, Fax: (213) 482-4856.

➢ The USAMRIID's Medical Management of Biological Casualties Handbook can be found online at http://www.usamriid.army.mil/education/bluebook.html, as well as on our enclosed **CD**.

➢ Be sure to check the US Army Medical Research Institute of Infectious Diseases (USAMRIID) online at http://www.usamriid.army.mil. Check out the links to their "Education," "Publications," and "Related Links" sections. Direct contact is Commander, USAMRIID, ATTN: MCMR-UV-ZM (CCCD), 3100 Ricketts Point Road, Aberdeen Proving Ground, MD 21010-5400, Phone: 410-436-2230, FAX: 410-436-3086.

➢ Visit a US Navy medical site, complete with disease symptoms and photos at http://www.mercy.navy.mil/bioterror/PROGRAM/CHEMICAL.HTM.

➢ Quite a number of online training courses, some free, some not, are available through Swank Health Care, online at http://www.swankhealth.com, or you can contact them for more info at Swank HealthCare, 201 South Jefferson Avenue, St. Louis, Missouri 63103-2579, phone: 800-950-4248, fax: 314-289-2187.

➢ Definitely visit **NASA's** Occupational Health links page regarding **Threats**. It's online at http://www.ohp.nasa.gov/hthreats. While you're there, be sure to follow each of the **links**.

➢ The Department of Defense Anthrax site: http://www.anthrax.osd.mil.

➢ Links to a variety of biochemical terrorism info sites can be found at http://www.ilpi.com/terrorism/bio.html.

➢ NATO's NBC Medical Manual: http://www.fas.org/nuke/guide/usa/doctrine/dod/fm8-9/toc.htm.

➢ To see what the Army's up to with NBC prep and readiness, see: http://www.apgea.army.mil.

- ➢ **Chemical weapons info** can be seen at: http://www.nbcprotect.com/new/cwagents.htm. While you're there, visit their links page at: http://www.nbcprotect.com/new/links.htm.
- ➢ The **Occupational Safety and Health Administration** at http://www.osha.gov, 800-321-OSHA (6742), will have searchable data on HazMat and related subjects.
- ➢ The **FDA's Bioterrorism** page: http://www.fda.gov/oc/opacom/hottopics/bioterrorism.html, or contact: Food and Drug Administration, 5600 Fishers Lane, Rockville, MD 20857, 1-888-INFO-FDA (1-888-463-6332).
- ➢ The **BBC's BioChemical discussion**: http://www.bbc.co.uk/science/hottopics/biochemicalweapons/index.shtml.
- ➢ The Federation of American Scientists **Terrorism and Weapons of Mass Destruction** page: http://www.fas.org/terrorism/wmd/index.html. You can also contact them via Federation of American Scientists, 1717 K St., NW, Suite 209, Washington, DC 20036, Voice: (202) 546-3300, Fax: (202) 675-1010.
- ➢ Also see the FAS "Introduction to Special Weapons": http://www.fas.org/nuke/intro/index.html.
- ➢ Visit The FAS **BioChemical Weapons** page: http://www.fas.org/bwc/index.html.
- ➢ And see the FAS **Biological Weapons** Primer: http://www.fas.org/nuke/intro/bw/index.html.
- ➢ The FAS **Chemical Weapons** Primer: http://www.fas.org/nuke/intro/cw/index.html.
- ➢ **Nuclear Weapons** FAQ: http://nuketesting.enviroweb.org/hew/Nwfaq/Nfaq0.html.
- ➢ One set of free information can be found at http://www.survive-nbc.org. Download their manual. The direct link for downloads is http://www.survive-nbc.org/download.htm.
- ➢ Basic NBC protective facts: http://www.aosafety.com/basicsafety.htm.
- ➢ Nuclear and general terror info: http://www.surviveanuclearattack.com.
- ➢ WMD links: http://nuketesting.enviroweb.org/hew/News/Bigsubpages/Bigsubpage40.html.
- ➢ Myth, fact, and opinion on "Surviving Doomsday": http://www.ki4u.com/survive/doomsday.htm.
- ➢ Numerous documents and courses on WMD are at http://www.msiac.dmso.mil/wmd/documents.asp.
- ➢ You'll find some individual chemical info files on the Army's Center for Health Promotion and Preventative Medicine at: http://chppm-www.apgea.army.mil/dts/dtchemfs.htm.
- ➢ See the University of Pittsburgh's BCW links page: http://www.pitt.edu/~super1/lecture/lec0901/021.htm.
- ➢ Find **field manuals FM 3-3** through **3-7** at http://www.adtdl.army.mil/cgi-bin/atdl.dll?type=fm, or the other sources we've listed previously. (You'll also find these manuals on our enclosed **CD**.)
- ➢ More **Army WMD** Manuals: http://www.globalsecurity.org/wmd/library/policy/army/fm/index.html.
- ➢ We suggest you read the military manual, **FM 8-285** on the **treatment of chemical agent casualties**. We've found several places you can still read the manual online, and as of this writing, the Global Security site is one of few that offers the manual in a downloadable format:
  - http://www.globalsecurity.org/wmd/library/policy/army/fm/8-285/toc.htm .
  - http://www.nbc-med.org/SiteContent/MedRef/OnlineRef/FieldManuals/fm8_285/toc.htm.
  - http://www.nbc-med.org/SiteContent/MedRef/OnlineRef/FieldManuals/fm8_285/index.htm .
  - http://www.vnh.org/FM8285/cover.html .

Periodicals

- ➢ There really aren't a lot of publicly available magazines that specialize in NBC warfare protection. However, you'll find occasional articles in "**Modern Survival**" at http://www.modernsurvival.net and in "**Backwoods Home**" magazine at http://www.backwoodshome.com, or Backwoods Home Magazine, P.O. Box 712, Gold Beach, OR 97444, Toll free: 800-835-2418, Phone: 1-541-247-8900, Fax: 1-541-247-8600.

Books

- ➢ **A caveat about purchasing "NBC-only" books:** There are a few NBC books on the market that are pointless, useless, and their contents don't really match the titles. Some of the books we've reviewed have gone into extreme, yet unnecessary detail on a very limited number of biological and chemical weapons. They dwell on areas such as history, scientific detail of different substances or organisms, detailed instructions for unobtainable or obsolete military equipment, and other slightly interesting yet superfluous data. Most of this type of detail might be somewhat interesting to first-year biology or chemistry students, but very little of it applies to the average victim, and even less data is provided on preventative measures. It's almost like seeing a book entitled "How to Defend Yourself Against Firearms," and there's nothing in the book except schematics for 3 or 4 different types of guns, and a short history on gunpowder. Some of these books might make for interesting reading on a rainy day, but we don't recommend buying any of them without objectively deciding if they have anything to offer.

- See the NBC book selections at **Paladin Press** at: http://www.paladinpress.com Gunbarrel Tech Center, 7077 Winchester Circle, Boulder, Colorado 80301, Customer Service: 303.443.7250, Fax: 303.442.8741.
- The book, **Nuclear Survival Skills** by Cresson H. Kearny, can be read online at http://www.oism.org/nwss. Of particular interest are the various appendices showing, among other things, how to build your own radiation fallout meter, and construct various forms of shelter. Guess what? It's also on our **CD**.

Videos

- Several of the internet sources listed above offer different media and will have training videos.
- NBC Disaster Videos: http://www4.umdnj.edu/camlbweb/disaster.html#Video.
- Bioterrorism Audiovisuals: http://www.tdh.state.tx.us/avlib/bioterrorism.htm. Texas Department of Health, Audiovisual Library, 1100 W. 49th Street, Austin, TX 78756-3199, Toll free: 1-888-963-7111 ext. 7260, (512) 458-7260, Fax: 512-458-7474, TDD: 512-458-7708.
- The Emergency Film Group: http://www.efilmgroup.com/films.html. Emergency Film Group, P.O. Box 1928, Edgartown, MA 02539; Phone 800-842-0999 or 508 627-8844, Fax: 508-627-8863.

Courses

- The US Army on-line NBC decontamination course can be downloaded (and it's on our **CD**) at: https://hosta.atsc.eustis.army.mil/cgi-bin/atdl.dll/accp/cm2506/cm2506_top.htm.
- The site mentioned above, http://www.survive-nbc.org, offers some course work in various cities. They'll also help you organize your own training classes and help you locate an instructor.
- For a comprehensive, online, **WMD (Weapons of Mass Destruction)** course, log on to: http://www.teexwmdcampus.com. This training program was developed in cooperation with The Texas A&M University System, Texas Engineering Extension Service (TEEX), and National Emergency Response and Rescue Training Center (NERRTC), a member of the National Domestic Preparedness Consortium.
- Another online course covering Anthrax, is offered by the National Library of Medicine. Go to: http://www.nlm.nih.gov/medlineplus/tutorials/anthrax.html.

Misc.

- Myth and Fact about NBC: http://www.marlalarue.bizland.com/survive_1.html.
- Random decontamination information: http://www.marlalarue.bizland.com/survive_2_decontamination.html.
- Several military manuals on CD: http://www.military-media.com, Military Media, Inc., 7454 Lancaster Pike, Suite 321, Hockessin, DE 19707, 1-800433-6214, or Fax: 1-610-268-0466. (Before buying anything, check our enclosed **CD**.)
- Another CD collection of military manuals can be found through http://www.hotmlmlists.com/cds/army2.html. They can also be contacted through Big-Web Development Corp, 14194 80th Lane N, Loxahatchee, FL 33470, Phone: (561) 236-7907, Fax: (561) 656-1785. (Before you buy anything, make sure the manual is not already included on our own **CD**.)

Note: You'll find additional information sources for HazMat incidents and general industrial chemical emergencies in the "**Reaction**" section under "**Chemical Attack or HazMat Incident**."

Auto Repair Training

One of the major sections in this book is "**Evacuation**," and evacuation depends heavily on transportation. Therefore you need to learn what you can to keep yours operational. It would be a shame to break down on the way to your evacuation destination because of something easily prevented or simple to repair.

There are three other good reasons to learn what you can about auto repair. One is that you may need to **help a fellow evacuee repair their vehicle**, two, it's a **barterable skill**, and three, repairs that you can make on your own will **save you money**.

Among other items pertinent to your type of vehicle, we suggest you learn:

1. **Basic maintenance:** Troubleshoot your electrical system, check fuel lines, change your oil, perform tune-ups, basic malfunction diagnostics...
2. **Minor mishaps:** Change a tire, patch a tire, patch your radiator, replace a fan belt...
3. **Staying mobile:** Jump-start a car, roll-start a car, change a battery, put on snow chains, counter snow or mud...

4. **Tricks:**   Among all the other things, make sure you learn all the impromptu repair "tricks" you can from those more experienced than yourself.   Tricks like the fact that if your car is overheating in the summer, you can use your car's **heater** as an "assistant radiator," how to patch your radiator using black **pepper**, that some fan belts can be **sewn** and sealed with duct tape to last a few more miles, which fluids can be swapped with which, etc.

Internet

➢   Google's auto repair information page: http://directory.google.com/Top/Recreation/Autos/Repair/.

➢   Jonko online auto repair questions and tutorials with message board: http://www.jonko.com.

➢   "**10W40**" online auto repair information: http://www.10w40.com.

➢   "**2 Car Pros**" FAQ page: http://www.2carpros.com/faq.htm.

➢   "**Auto Repair Guy**" online: http://www.auto-repair-guy.com.

➢   "**Do It Yourself**" dot com's auto repair and maintenance page: http://doityourself.com/auto.

➢   "**Auto Shop 101**": http://www.autoshop-online.com/auto101.html.

➢   "**Trust My Mechanic**" online newsletter: http://www.trustmymechanic.com/newsletter1.html.

➢   "**Car Stuff**" Automotive Info Links: http://www.car-stuff.com.

➢   **Autoparts compendium** at: http://www.partsamerica.com, 1-877-808-0698.

➢   **Hints and Things** site at: http://www.hintsandthings.co.uk/garage/garage.htm.

➢   **NAPA Auto Parts** offers online FAQs at http://www.napaautocare.com/education/index.jsp.   Their main site is http://www.napaonline.com, with direct contact at (877) 805-6272.

➢   **AutoZone** offers free online repair guides for most vehicles at http://www.autozone.com/servlet/UiBroker?UseCase=RG001&UserAction=beginRepairGuide.   Be sure to visit their main site at http://www.autozone.com and click on both "**My Zone**" for updates and technical publications on your vehicle, plus click on the link to "**Free Repair Guides**."   They can also be contacted at 1-800-AUTOZONE (1-800-288-6966).

➢   **PEP Boys** also has a site with a learning center.   The main site is http://www.pepboys.com, and from the main page follow the "Learning Center" links.   PEP Boys' main office can be contacted at (215) 430-9000.

Periodicals

➢   Automotive magazines are one of the most prolific in the US market.   Any location that carries magazines will more than likely have a good selection of vehicle related periodicals.   Many autoparts stores will have a limited selection as well.

➢   Go to http://www.magazines.com to search multiple titles.

➢   Car repairs and magazine links: http://www.carsandmagazines.com/links/carrepair.html.

Books

➢   The **Glove Compartment Guide to Emergency Car Repair** by Richard V. Nunn.   You can buy it online at: http://www.randmcnally.com, or contact Rand McNally Consumer Affairs, 8255 N. Central Park, Skokie, IL 60076-2970, (800) 333-0136 ext 6171.

➢   Take a look at the **Chilton Book series**, with a specific repair manual available for pretty much every make and model of motorized vehicle on the road.   Web is http://www.chiltonsonline.com, and direct contact at 1-800-347-7707.   Mailing info is: Delmar Learning, 5 Maxwell Dr., PO Box 8007, Clifton Park, NY 12065 – 8007, or fax 1-800-487-8488.   Not only will bookstores carry these, many **auto parts stores** will also.

➢   Many repair manuals and repair parts can be found through **JC Whitney**.   Web is: http://www.jcwhitney.com, and contact is: JC Whitney, 1 JC Whitney Way, LaSalle, IL 61301, (800) 469-3894 or fax at (800) 537-2700.

➢   "Gearhead Cafe" **repair manuals on CD**: http://www.gearheadcafe.com/cgi-bin/store/agora.cgi, GearHead, 3011 N Wren Ave, Meridian, ID 83642, 208-713-7125 .

➢   **Auto Repair for Dummies** by Deanna Sclar, http://www.dummies.com.

➢   **How to Make Your Car Last Almost Forever** by Jack Gillis, the Putnam Publishing Group.

➢   **The Car Owner's Survival Guide** by Robert Appel, Fawcett Columbine Publishing, ISBN: 0-449-90151-3.

Videos

➢   "**Top 10 Auto Repair Problems**" by "Bob and Ken." http://www.2carpros.com/store/index.htm.

➢   **Auto Repair Interactive Guide**: http://www.autorepaircd.com.

➢   Don't forget the usual sources: **Your local library and video rental store**.

Courses

➤ Most courses immediately available will be at your area **technical college**. However, keep an eye on **auto parts stores or auto dealerships** near you as some will offer occasional classes. Also contact your **County Extension Office** (or equivalent), and browse your **local leisure publications** for classes.

➤ Check your **yellow pages** for **"you do the work" garages**. They're basically repair garages but you do the work with tools and supervision supplied by the garage.

➤ Similarly, it probably wouldn't cost much for you or your volunteer group to **hire a reputable mechanic** to come show you how to perform basic auto maintenance, or to trouble-shoot and make moderate to detailed repairs on specific vehicles. If nothing else, a good mechanic can teach a good diagnostics course to keep you from getting ripped off by dishonest repair shops.

Misc.

➤ **AAA auto club**: http://www.aaa.com/scripts/WebObjects.dll/ZipCode. (American Automobile Association) 8111 Gatehouse Road, Falls Church, VA 22047, 1-800-AAA-HELP (800-222-4357).

➤ The **National Highway Traffic Safety Administration** will keep you abreast of consumer issues and recalls amongst other road safety issues: http://www.nhtsa.dot.gov, National Highway Traffic Safety Administration, 400 Seventh Street, S.W., Washington, DC 20590, 1-888-DASH-2-DOT (800-327-4326), or 800-424-9393.

➤ In addition to safety ratings, and recall information, you can order NHTSA's **"Traffic Safety Materials Catalog"** at http://www.nhtsa.dot.gov/people/outreach/media/catalog/Index.cfm.

Be sure to record the date of your auto-repair class on your **planning calendar**.

Outdoor Survival

While we don't foresee sending everyone out to the boonies following each and every disaster, you never know where you'll be or what will happen when disaster strikes. The more contingencies you're prepared for, the better off you'll be. What if you're at a shelter that runs out of food? Do you know which edible plants are commonly found in your area that could keep you and yours going for a while? What if you survive a plane crash out in a wilderness area? What if your car breaks down in the desert or during a blizzard, or your boat becomes disabled fifty miles from shore?

Nothing will prepare you as well or as quickly, or help instill the "readiness mindset" so necessary to preparedness as good outdoor survival skills training. You don't need to go from Teddy Bear to Tarzan, but a good skill and knowledge base coupled with a little hands-on practice through camping trips will make all the difference in the world.

When we say camping, we don't mean pulling the RV into a lot and staying there overnight, or going to a hotel with no room service. We mean honest-to-goodness, packs on the back, hike a few miles to a vacant spot in the wilderness, stay a few days, camping trips.

However, we do wish to point out that you should do what you can within whatever limitations you may be under. Not all of us are physically capable of backpacking. If *absolutely* nothing else, pitch a tent in your backyard and stay out of the house a full 24 hours. It won't be the best training by any means, but it's far better than nothing and it will open your eyes to *some* of the problems you'd face in an outdoor survival situation. If you've compiled a "Bugout" kit as outlined under "**Evacuation**," going camping will give you a chance to test your gear. The thing to remember is that experience is always the best teacher, and in an emergency, **our response will only be as good as the training we've allowed ourselves**.

Not only is this good training for those reasons, backpacking as a sport or hobby is a great way to stay in shape, and more importantly, it will teach you the logic of packing lightweight and highly useful items, which will come in handy when prepping your gear under "**Evacuation**." So if you've never considered backpacking or camping, we urge you to give it a try. Besides, if you're capable of surviving and living comfortably in a remote wilderness location with little or no creature comforts, then spending a few days in a shelter after a disaster will be a piece of cake.

For now, gather the information, go on some practice trips, and learn what you can. Of course, the best way to learn is to practice as often as you can. Go camping every chance you get. Better yet, find and take a good course with a good instructor. **Experience is always the best teacher!**

When studying the information sources listed below, we suggest you focus on the following:

1. **Water and food sources**: Drinkable water, edible plants, and basic hunting and gathering skills.
2. **Protection from the elements**: Expedient shelter, permanent shelter, and fire.
3. **Fieldcraft**: Knots, rope work, making useful tools and implements from indigenous materials.

4. **Navigation**: Using map and compass (and <u>no</u> map or compass) to find your way.

5. **Minimal gear**: Learn to make the most with the least equipment (i.e. tin can stove vs. gas grill).

6. **Location-specific info**: Learn to survive in the climate(s) you're most likely to encounter.

The sources below will have varying elements of the above.

Internet

> To download the Army **Survival Manual FM 21-76-1** (and numerous others) Go to: http://www.globalsecurity.org/military/library/policy/army/fm or to: http://www.aircav.com/survival/asurtoc.html to read it online. You'll also see a listing of other useful survival books. (**FM 21-76.1** is also on our **CD**.)

> Go to: http://www.adtdl.army.mil/cgi-bin/atdl.dll?type=fm to download the Army **Map Reading and Land Navigation** manual **FM 3-25.26**, and **FM 3-97.61** on **Mountaineering**. (See our **CD**.)

> Marine Corps Scouting and Patrolling manual online: http://www.tpub.com/content/USMC/mcwp3113.

> Visit http://www.aircav.com/survival/appb/asappbtoc.html, for an interactive source of photos and info on edible plants along with a listing of applicable books.

> "Camp Clueless" beginner's guide: http://www.totalescape.com/tripez/clueless.html. Great source of introductory materials for the camping newbie.

> Compendium of equipment providers: http://www.andinia.com/and_surven.shtml.

> Visit "Primitive Ways" at: http://www.primitiveways.com, and visit the "Primitive Ways" interesting links at: http://www.primitiveways.com/pt-interesting_links.html.

> Campsites and camping info: http://www.woodalls.com. For books and directories, contact them on the web or at Woodall Publications Corp., 2575 Vista Del Mar Drive, Ventura, CA 93001, (877) 680-6155 (Toll-Free).

> "Roper's Knot Page": http://www.realknots.com/knots/index.htm#bends.

> Go to http://www.adtdl.army.mil/cgi-bin/atdl.dll?type=fm and download the **military manual on ropes and rigging, FM 5-125**. (You'll also find this manual on our **CD**.)

> For online survival articles, visit http://www.equipped.com, and its mirror site http://www.equipped.org.

Periodicals

> "**Wilderness Way**" magazine, info at: http://www.wwmag.net, or at Wilderness Way Magazine, P.O. Box 621, Bellaire, TX 77402-0621, 713-667-0128 – Phone, 801-730-3329 - Fax.

> "**Backpacker**" magazine, info at: http://www.backpacker.com (800-666-3434).

> "**Backwoods Home**" magazine, at: http://www.backwoodshome.com, Backwoods Home Magazine, P.O. Box 712, Gold Beach, OR 97444, Toll free: 800-835-2418, Phone: 1-541-247-8900 Fax: 1-541-247-8600.

> "**Modern Survival**" at http://www.modernsurvival.net .

> "**Camping**" magazine, info at: http://www.acacamps.org/campmag, or at American Camping Association, 5000 State Road 67 North, Martinsville, IN 46151-7902, Phone: 765-342-8456, Fax: 765-342-2065.

> "**Backwoodsman**" magazine, info at: http://www.backwoodsmanmag.com, or at Backwoodsman Magazine, P.O. Box 627, Westcliffe, CO 81252, (866-820-4387).

Books

> Definitely read **<u>The Complete Walker III</u>**, by Colin Fletcher.

> Also, read **both** the Boy Scouts of America **Handbook** and **Fieldbook**.

> **<u>Outdoor Survival Skills</u>** by Larry Dean Olsen.

> **<u>Tom Brown's Field Guide to Wilderness Survival</u>** by Tom Brown and Brandt Morgan.

> **<u>The Complete Wilderness Training Book</u>** by Hugh McManners.

> **<u>Be Expert with Map and Compass</u>** by Bjorn Ajellstrom.

> **<u>Camping & Wilderness Survival</u>** by Paul Tarwell.

> Powell's Books, **<u>Guides to Edible Wild Plants</u>**: http://www.powells.com/subsection/NatureStudiesWildEdiblePlants.html.

> Another good Dummies books is **<u>Camping for Dummies</u>**, by Michael Hodgson.

> Books, videos, and assorted info: http://www.survival.com, or contact directly at **Hoods Woods**, P.O. Box 549, Garden Valley, ID 83622, (208) 462 1916, (or 888-257-2847 for orders).

> Wilderness Survival Institute, videos, books, and equipment, online at http://www.wildernesssurvival.com, and with direct contact at P.O. Box 394, Tok, AK 99780, 907-883-4243.

- ➢ For a huge listing of books on outdoor survival, go to http://www.amazon.com and in the search box enter **"outdoor survival"** just like that with quotes.
- ➢ Books on Knots: http://www.realknots.com/links/books.htm.
- ➢ To really get back to basics, check out the **Foxfire** book series by Elliot Wigginton.

Videos

- ➢ "Woodsmaster" videos: http://www.survival.com/woodsmaster.htm, available through Hoods Woods above.
- ➢ "Survival Camping": http://www.survival.com/volume-10.htm, also through Hoods Woods.
- ➢ For a huge listing of video info go to http://www.google.com and enter the search string **"survival videos"** exactly like that with quotes.
- ➢ **"Survival - Learn to Become a Survivor in the Wild"** available through Amazon.com. On the Amazon main page at http://www.amazon.com, enter the word "survival" to locate this video.

Courses

- ➢ **"Outward Bound"** has long had an amazing array of outdoor survival and confidence course offerings all over the country. Find them online at http://www.outwardbound.com, or contact them at Outward Bound USA, 100 Mystery Point Road, Garrison, NY 10524, 866-467-7651.
- ➢ Lots of information through the **Sierra Club**: http://www.sierraclub.org, 85 Second Street, 2nd Floor, San Francisco, CA 94105, Phone: 415-977-5500, Fax: 415-977-5799.
- ➢ Christopher Nyerges' **"Self-Reliance"** course: http://www.self-reliance.net, School of Self-Reliance, P.O. Box 41834, Los Angeles, CA 90041, 323-255-9502, fax: 323-402-1223 x 1939.
- ➢ The Boulder Outdoor Survival School: http://www.boss-inc.com, Boulder Outdoor Survival School, P.O. Box 1590, Boulder, CO 80306, phone: 303-444-9779, fax: 303-442-7425.
- ➢ **"Ancient Pathways"** courses: Ancient Pathways, LLC, 1931 E. Andes, Flagstaff, AZ 86004, 928-774-7522. http://www.apathways.com/Subjects/CourseAndSchedule/aboutourcourses.htm.
- ➢ You'll find the occasional course offering in **"Modern Survival"** at http://www.modernsurvival.net.
- ➢ Earth Connection http://www.earth-connection.com, P.O. Box 32, Somerville, VA 22739, 540-270-2531.
- ➢ Salmon Outdoor School, http://www.aboman.com, P.O. Box 17, Tendoy, ID 83468, 208-756-8240.
- ➢ Jack Mountain Bushcraft http://www.jackmountainbushcraft.com, P.O. Box 61, 267 Camp School Rd., Wolfeboro Falls, NH 03896, 603-569-6150.
- ➢ Wilderness Learning Center http://weteachu.com/index.htm, 435 Sanky Knoll Rd., Chateaugay, NY 12920, 518-497-3179.
- ➢ Ray Mears http://www.raymears.com, Woodlore Ltd, PO Box 3, Etchingham, East Sussex, TN19 7ZE, U.K., Telephone: 0044 (0)1580 819668, Fax: 0044 (0)1580 819668.
- ➢ Equipped to Survive: http://www.equipped.org/medschol.htm. Log on to see various courses.
- ➢ Rainbow Culture/Survival Camp http://www3.sk.sympatico.ca/neufeldd, 306-377-2125.
- ➢ Simply Survival http://www.simply-survival.com, P.O. Box 449, Stevenson, WA 98648, 509-427-4022, Fax: 509-427-4023.
- ➢ Also check out your local leisure guide and local camping supply store to see if any courses are offered. Additionally, check your local community college or extended education program.

Misc.

- ➢ For some really useful survival skills, log on to http://www.primitive.org, Society of Primitive Technology, PO Box 905, Rexburg, ID 83440, 208-359-2400.
- ➢ Similar info can be found at: http://www.hollowtop.com, Thomas J. Elpel's Hollowtop Outdoor Primitive School, LLC, PO Box 697, Pony, MT 59747-0697, 406-685-3222.
- ➢ The UK's Scouting page at http://www.scoutingresources.org.uk/knots_index.html, offers online pictures and downloadable files on various knots.
- ➢ Lots of Knots: http://www.knots.com.
- ➢ More Knots: http://www.earlham.edu/~peters/knotlink.htm.
- ➢ Most Knots: http://www.realknots.com.
- ➢ For illustrated online instruction in knots, see http://www.netknots.com/index.html.
- ➢ Animated Knot Tying online: http://www.mistral.co.uk/42brghtn/knots/42ktmenu.html.
- ➢ To download instructional pages, go to http://www.northnet.org/ropeworks/text/arch.html#ROPE.

> For some entertainment intermingled with some surprisingly useful information, check out the "**Worst Case Scenario**" books. Online info: http://www.worstcasescenarios.com/mainpage.htm.

We can't stress enough, the importance of knots. Let's go over why we suggest you learn to tie the following:

Square knot: This is the basic knot, it's used to finish off lashings, and it's good for tying bundles and gear.

Bowline: This lets you tie a non-slip loop, and therefore has numerous uses in rescue operations.

Fisherman's knot: This is the best knot for tying two ropes or bed sheets together for a no-fail connection.

Clove Hitch: This is a great knot for tying rappelling lines to posts, tow ropes to logs, and tying bags closed.

Taught-Line Hitch: This is a slip-knot that will hold in place and is great for tying cargo to your vehicle.

We have several books and pamphlets on ropework and knots on our **CD**, and camping and backpacking *equipment* sources will be listed under "**Evacuation**" to help you find good gear to include in your "Bugout Kit."

Have you set a **date** for a camping trip yet? You know where to record it.

Information Research

Information research is just as much a part of your "basic training" as anything else since one of our main goals with your preparedness is driving home the importance of continual education outside the boundaries of this book. We live in the information age and it's extremely important that you know how to find good information. Knowledge is power, and the right kind of knowledge in an emergency is priceless.

In most sections of the book we've divided info sources into internet, periodicals, books, and videos, so it only seems logical that we give you a few ways to continue researching and locating these sources on your own.

The Internet

As popular as the internet is, it's surprising that only a small percentage of the population is computerized and has internet access. If that happens to be you, and you're wanting to finally get "computerized," let's give you a few pointers to get you up and running. If you're already computer literate, or "mousebroken" as we like to call it, you still might find a new piece of info here so keep reading.

First, let's get you computerized if you're not. If you are not computer literate, you should make it a goal to become so. All other uses notwithstanding, computers are an extremely useful tool for **research** (as evidenced by the number of internet sites we've provided), and **communication**. Even if you don't own one, you should learn how to use them as you may find yourself in a situation where it's your only communication option. Numerous private companies offer introductory computer training, and your local **library** may have computers with internet access. If they do, be sure to at least learn how to "surf the web" by way of search engines, and learn how to use email.

> Our first suggestion is that you try to ignore all the marketing, and find some independent sources that will tell you what you need to know about buying a computer, or how to put the one you have to better use. Start by browsing the books at your local library and bookstore, and then migrate to your local community college or other business that provides economical classes.

> You might also look at a copy of **Buying a Computer for Dummies**, by Dan Gookin.

> The "Dummies" book company, online at http://www.dummies.com, offers a title on pretty much any software package you would ever use, and most of them are exceptionally good instructional manuals.

> Once you get a computer, you'll want a little more instruction. You'll find a few economical instructional packages for different computer software packages and general computer usage through the "**Video Professor**" at http://www.videoprofessor.com, or by calling 1-800-525-7763.

Once you're computerized, the first thing you should learn is **how to use the internet**. The Video Professor source above will have a title or two on that subject, plus we recommend the following:

> Before clicking a mouse, you might want to look at a copy of **The Internet for Dummies**, by John R. Levine, Carol Baroudi, and Margaret Levine Young. (Check Amazon.com)

> See Online Internet Tutorials and more: http://library.albany.edu/internet.

> Webteacher is also a great place to start: http://www.webteacher.org.

> How to Search the Web: http://websearch.about.com/cs/howtosearch.

> University at Albany has a great internet search tutorial site at http://library.albany.edu/internet.

> Microsoft provides one as well: http://www.microsoft.com/insider/guide/intro.asp.

- ➢ Visit the "Learn the Net" page at: http://www.learnthenet.com.
- ➢ A Journalist's Guide to the Internet: http://reporter.umd.edu/.
- ➢ Net Search: Tutorial: http://www.lib.berkeley.edu/TeachingLib/Guides/Internet/FindInfo.html.
- ➢ Net Search: 5 Steps: http://www.lib.berkeley.edu/Help/search.html.
- ➢ Search Engine Compendium: http://www.allsearchengines.com.
- ➢ Go to http://www.pandia.com, and among other useful things, click on "Search Tutorial" and "Books."
- ➢ The Journalists' Search Tools Portal, at http://powerreporting.com, not only has a great collection of subject links, but a good tutorial section as well.
- ➢ Some good research sites are compendiums, or collections of related links, that will help you do your searching faster. One example can be found at http://www.ceoexpress.com. We'll have others on our **CD**.

The heart of internet research is the "**search engine.**" These are websites that will look things up for you once you enter the "clues" they need . Here are some of the more popular search engines:

- ❑ Google: http://www.google.com
- ❑ Excite: http://www.excite.com
- ❑ Dogpile: http://www.dogpile.com
- ❑ "Ask Jeeves": http://www.ask.com
- ❑ Yahoo: http://www.yahoo.com
- ❑ Lycos: http://www.lycos.com
- ❑ AltaVista: http://www.altavista.com
- ❑ Webcrawler: http://www.webcrawler.com
- ❑ For a search engine that finds other search engines, try http://www.finderseeker.com.

The trick with using these sites lies in knowing how to give them the clues they'll need to find the information you want. As most of them operate the same way, let's look at some tips.

- ➢ Most of these pages will have a blank line prominently displayed and marked with something like "Search for:" This is called your "**search line**" and the words or clues you enter into it are called your "**search string.**" Success in searching lies in how you enter your search string.

- ➢ The problem with internet research is that there is so much information out there, you need to be very specific in order to isolate the info you need. For example if you enter the word "baseball" you're going to get about a zillion "**hits.**" Hits are webpages, or "**sites,**" that the search engine finds for you that it thinks match what you're looking for. These hits are usually displayed on a "**results**" page that pops up when you hit "enter" or "submit" after typing your search string.

- ➢ Be very specific about the type of information you're looking for. For example, if want to see a "1943 batting lineup for the New York Yankees" you'd be better off entering that instead of "baseball." Use specific words, commonly called **keywords**, that will apply to your information more than to just any loosely related sites.

- ➢ Phraseology comes next after you figure out the keywords you want to use. Crucial to phraseology is the use of **quotes (")** or **plus signs (+)**. For example, if you simply entered the words *1943 batting lineup*, just like that with no quotes, you're liable to get a ton of pages that have those words on them somewhere, but in no particular order, and with no guarantee that all the words you entered will be used. You might get some *1943* romance story that mentions a girl *batting* her eyelashes. However, if you entered the phrase in quotes like, "*1943 batting lineup*," you'll only get pages that have those words in that exact order and together. If you used plus signs such as +1943+batting+lineup, you'll get pages that have all of those words, but not necessarily in that order. Plus signs aren't as specific as quotes but work well if you have combinations of words that should appear on a site but you don't know a particular order.

- ➢ A good example of the use of plus signs can be given in one of the ways we searched for downloadable files to include on our **CD**. Take first aid for example. On a search engine we'd enter the search string: +"first aid"+pdf. This would do two things. First, it would make sure the phrase "first aid" was just that and not a random collection of the words *first* and *aid*. Secondly, the addition of *pdf* would show only sites that specifically mentioned Adobe Acrobat file names.

- ➢ Try this: Pick a search engine and enter the search string "**first aid training**" and then search again using the string **+"first aid"+training**. Compare your results.

One of the reasons we want to teach you how to do your own research is that information, and the websites where it appears, changes so quickly and so often, and we want you to always be able to find it.

Now that you're using the internet a bit, we want to point out that one of the biggest assets at your disposal are the vast number of **free "e-zines," or online magazines, newsletters, and downloadable books**. Some of these books are in an "**e-book**" format that will open themselves on your computer, and others are in an "**Adobe Acrobat**" file format. The only thing you'll need to do in order to set yourself up for success in reading these documents is to find the "viewers" that will allow you to read them. Most files are in text or html format which you can read immediately. For the others that are not, go to these sites and download their **free viewers**:

- For **Adobe**, the most popular format, go to http://www.adobe.com to download a **free viewer**.
- Numerous **"viewer" programs**, which allow you to read various file types without owning the original program, are available online at http://emd.wa.gov/site-general/viewers.htm.
- More viewers are available through Microsoft at http://www.microsoft.com/office/000/viewers.asp, and at http://office.microsoft.com/assistance/9798/viewerscvt.aspx.

While we're still on the subject of free information, the **US Government** offers quite a bit. We'll give you numerous links throughout the book and on the **CD**, but here are a few of the main "portals" to get you started:

- "FirstGov, the US Government's Official Web Portal" at: http://www.firstgov.gov, or you can call the First Gov locator service at 1-866-FIRST-GOV.
- First Gov's links at: http://www.firstgov.gov/Topics/Usgresponse/Protect_Yourself.shtml.
- Its services-oriented cousin, Government Guide, at http://www.governmentguide.com.
- The Federal Consumer Information Center maintains an online collection of government "blue pages" at http://www.pueblo.gsa.gov/specialstuff/bluepage/america5.htm, or at http://www.pueblo.gsa.gov/call/phone.htm. You can also call 1-800-FED-INFO (800-333-4636).
- Other "Blue Pages" links can be found at http://bp.fed.gov, or http://www.usbluepages.gov/gsabluepages.
- Visit The Library of Congress at http://www.loc.gov.
- The US General Accounting Office at http://www.gao.gov.
- The Consumer Links page at: http://www.consumer.gov/Links.htm.
- The civilian guide to government sites is http://govstar.com.
- An alternate source of government contact info can be found at http://www.statelocalgov.net/index.cfm.
- The US Government's Security Awareness links: http://www.dss.mil/training/salinks.htm.
- Visit the Government Printing Offices's database collection at http://www.gpoaccess.gov/databases.html.

Next, let's give you a **list of *general* disaster preparedness information sources** to get you started on web searching, research, and to tie your computer education back in to your disaster prep education. These sources don't fall under any one category, and most will cover quite a number of related topics (we call these "compendium" sites). We don't agree with everything on these sites or endorse 100% of what you'll find. However, each contains some rather useful and educational information. Among these sites are:

- One group, called Knowledge Hound, has a research website hosting a collection of online instructional web pages. Log on to http://www.knowledgehound.com and search for the subject of your choice.
- See "The Survival Ring" at http://www.survivalring.org.
- The site http://www.ringsurf.com will help you find collections of related info sites.
- Survival Forum at: http://www.survivalforum.com.
- Great links collection at: http://www.greatdreams.com/survival.htm.
- And another huge collection at: http://members.aol.com/rafleet/links1.htm.

Hint: The above links collections are so large that we recommend not trying to read them in one sitting. Save them as a source of additional information when you're learning one particular subject or preparing a specific kit. Then use them to find more information related to your current single project. Otherwise, you could surf for weeks and see a lot of interesting things, but get nothing specific accomplished. We'll also have an extended collection of these general subject, or "compendium" sites on the links collection in the **CD**. Speaking of which, don't forget that it might be easier to follow along and use these all these links by way of our "**DP1 Book Links**" collection than by trying to type them in.

Some general disaster info sites have **download pages** offering free information and publications. Among these are:

- The Survival Ring website at: http://www.survivalring.org/cd-downloads.htm. Some of these are out of date, many are rather useful, but not all have been chosen to appear on our enclosed CD. Also, while you're on this page, look for the **free email and newsletter subscription list**.
- The Survival Forum downloads site at http://www.survivalforum.com/modules.php?name=Downloads.

All info searches aside, another great function of the internet is **current news**. As news is extremely important for keeping up with potential threats, both manmade and natural, we've given you a few miscellaneous news websites below in the "Getting the Word" portion of "The EAS Alerts." All other useful websites will be listed under the section to which they apply.

Periodicals

Getting back offline and returning to the world of hardcopy and print, let's take a look at periodicals. As we said earlier, the best places to review current magazines and other periodicals are either at your **public library or your local bookstore / coffee shop**. However, they don't necessarily carry all the titles that might interest you, and can't find you the best subscription rates on the magazines you feel you need. For that you'll need these sources.

> ➢ Visit the "Mag Mall" at: http://www.magmall.com, or contact them at 6310 San Vicente Blvd, Suite 404, Los Angeles, CA 90048, 1-888-255-6247.

> ➢ Magazine Article Search Engine: http://www.magportal.com/, or contact them at Hot Neuron LLC, P.O. Box 463275 Bryn Mawr Ave. #M14, Bryn Mawr, PA 19010, (610)581-7702.

> ➢ Magazine subscription deals: http://magazinevalues.com/all.cfm, with direct contact at Magazine Values, 149 Okaloosa Dr., Winter Haven, FL 33884, Phone: 863-326-1125, Fax: 508-374-8599.

> ➢ General magazines and periodicals: http://magsonthenet.com, or contact them at Mags On The Net, Post Office Box 79, Lawrence, New York 11559, 800.665.6690.

> ➢ Even more magazines: http://magazines.com, with direct contact at Magazines.com Inc., P.O. Box 682108, Franklin, TN 337068, phones: 800-258-9558 or 1-800-929-2691, fax: 877-643-6775 or 615-778-2139.

> ➢ Ecola will let you search for magazines by subject matter: http://www.ecola.com.

> ➢ Magazines On-Line: http://www.zinos.com. This is a listing and links to online magazines.

> ➢ The site http://www.publist.com, will let you search for magazines, journals, newsletters, and more.

> ➢ Visit http://www.freetrademagazinesource.com, and http://free-ezines.com for free online magazines.

> ➢ The site http://www.newsletteraccess.com, will search available online newsletters by subject or keyword.

Books

Sometimes the hardest thing about books is finding a title, especially when it's out of print. Here are some good sources for current and hard-to-find titles. As with periodicals, always start at your local library and bookstore. In fact, many libraries and bookstores will help you locate older or out-of-print titles. After that, try local used-book stores. When you've depleted those, you'll need a few sources for help, plus you'll need ways to find books that the library and bookstore wouldn't carry anyway.

> ➢ The king of online book sources is Amazon.com at, appropriately enough, http://www.amazon.com.

> ➢ Finding out-of-print books might be a little harder. Try these sources:

> • Contact http://www.bibliofind.com for out-of-print books. It'll route you to an Amazon page as bibliofind has recently merged with Amazon to provide this online-only search service.

> • You can find new and used books at "Fetchbook": http://www.fetchbook.info .

> • More can be found at: http://www.powells.com, or by phone: 800-291-9676, or if calling from Oregon: 503-228-0540, ext. 482.

> • Contact Harvest Books for out-of-print and hard-to-find books: http://www.harvestbooks.com, 1-800-563-1222, or Mail: Harvest Book Company LLC, 260 New York Drive, Suite B, Fort Washington, PA 19034, Telephone: 1.877.512.3022, Fax: 1.215.619.0308.

> • Try BookFinder online at http://www.bookfinder.com .

> • "AddAll" has an out-of-print book search engine at: http://used.addall.com .

> • A good online tutorial on how to find out-of-print books (through online links and offline means) can be see at http://marylaine.com/bookbyte/getbooks.html.

> • Contact the Advanced Book Exchange at http://www.abebooks.com, or write to Advanced Book Exchange Inc., PMB 185, 1574 Gulf Road, Pt Roberts, WA 98281-9007.

> • You might find some out-of-print titles through the Ebay auction site at http://www.ebay.com.

> • Visit http://www.searchebooks.com to find online E-books. Some are free, others can be purchased.

> ➢ For all the publications printed by the US Government, go to the US Government Printing Office at: http://www.gpoaccess.gov, or write to the **US Government Printing Office** at: Superintendent of Documents, 732 North Capitol Street NW., Washington, DC 20401, 202-512-0000, and 888-293-6498, or PO Box 371954, Pittsburgh, PA 15250-7954, 1-866-512-1800.

> ➢ One of the best sources of disaster and emergency information can be found in various **US military training manuals**. You can find some through various dealers and through the US Government Printing Office above, but we've found a few direct links that will allow **free downloads** of numerous titles. By the way, don't forget we already have quite a few on our **CDs**, and the CDs' table of contents is in the "**Appendix.**"

> • The General Dennis J. Reimer Training and Doctrine Digital Library offers free PDF downloads from the site at http://www.adtdl.army.mil/cgi-bin/atdl.dll?type=fm.

- Global Security's site at: http://www.globalsecurity.org/military/library/policy/army/fm/.
- The Army's Electronic Publications and Forms site at http://www.army.mil/usapa/index.html, will allow you to select a category of publication and will then allow free downloads.
- Related to this is the Army's publications links site at http://www.army.mil/usapa/index.html, and the Army's Publishing Directorate site at: http://www.apd.army.mil.
- Marine Corps manuals can be found and read online at http://www.tpub.com/content/USMC.
- The "Rocky Mountain Survival Group" offers the same at: http://www.rmsg.us/mfm.htm. They also have quite a few printed copies for sale. Contact them at "RMSG"; P. O. Box 2572; Dillon, CO 80435.
- Printed manuals can be ordered from the National Technical Information Service of the US Department of Commerce. They're online at http://www.armymedicine.army.mil/publications/publications.htm, or you can contact them at National Technical Information Service, 5285 Port Royal Road, Springfield, VA 22161, 1-800-553-6847 or (703) 605-6000.

Videos

Through internet searching, you'll be able to find a ton of videos just as you're able to find a ton of book titles. However, as videos are important educational tools, we've gathered a few specialized video sources for you.

- ➤ Remember that your local library, video rental store, video sales outlet, and local bookstore can search and order various video titles for you. You can also find several on Amazon.com.
- ➤ Emergency Films Group has a wide selection of emergency-related subjects. Find them online at: http://www.efilmgroup.com, or contact them directly at: PO Box 1928, Edgartown, MA 02539, 1-800-842-0999, Fax: 508-627-8863.
- ➤ Numerous training films and other information in a multimedia format are available through the National Audiovisual Center, online at http://www.ntis.gov/products/types/audiovisual/index.asp, or contact them directly at the National Audiovisual Center, c/o the National Technical Information Service, 5285 Port Royal Road, Springfield, VA 22161, (703) 605-6000, 1-800-553-6847, Fax: (703) 605-6900, TDD: (703) 487-4639.
- ➤ To find a good number of potential sites for locating a video title or subject you're looking for, log on to your favorite **search engine** and enter the string **+"*your subject*"+video**, and of course, replace "your subject" with whatever it is you're looking for, like "first aid," etc.

Speaking of videos, many television programs are extremely useful in continuing your basic training and education. **The Discovery Channel, The Weather Channel, The History Channel**, and **The Learning Channel** each have regular and excellent features on survival stories, natural disasters, manmade disasters, and a host of others. Be sure to **note upcoming programming on your calendar**.

Miscellaneous Elective Training

Disasters and emergencies carry with them peculiar needs. These needs will be based on the type of emergency, where you are when it hits, and what kind of follow-up or recovery efforts you'll have to make in order to get your life back in order. Since we don't know what any of that will be, we need to be as prepared as possible in advance.

The list below contains a few of our recommendations for additional training or education. It's presented in no particular order.

1. **Online FEMA Training**: Several mixed topics of interest can be found on FEMA's **NETC**, or National Emergency Training Center, website at http://www.training.fema.gov, or by phone at 301-447-1175. Log on to take a look at the various topics which range from National Fire Academy subjects to Federal Emergency Management. The **"First Gov"** site offers a comprehensive collection of links to various government agencies and services, offering information regarding disaster at: http://www.firstgov.gov/Government/State_Local/Disasters.shtml.

2. **Martial Arts**: Self defense needs aside, learning a martial art is not only good to keep you in shape, as mentioned under "Health," but it can help you develop the confidence and the readiness mindset so necessary in situations you may face. Seek a combat style and not a sport style. Let's look at some information sources:
   - ➤ To learn the basics of a combat oriented martial art, look for the **"Wing Chun Today"** video series by Jason Lau, at http://www.jasonlau-wingchun.com/video.htm. Contact for the producer is: Jason Lau Training Center, 4218 West Atlanta Rd., Smyrna, GA 33080, 770-433-0821, http://www.jasonlau-wingchun.com.
   - ➤ The Asian World of Martial Arts offers videos, as well as books and equipment. They're online at http://www.awma.com, with direct contact at AWMA, Inc., 11601 Caroline Road, Philadelphia, PA 19154-2177, 1.800.345.2962, Fax: 1.800.922.2962.

3. **Rappelling and skydiving**. Yes, you're right, this is rather unusual. Besides being good confidence builders, we put this here for the millions of people working in office buildings above the 10th floor. Ladders on fire trucks can only reach 10 floors. Getting down from anything higher may be up to you. Check with your local camping supply store, leisure publications, yellow pages, etc., for courses, outings, and instruction. With these sports <u>only</u> direct instruction by qualified personnel is acceptable.

4. **Learn to swim**. While we're on the topic of athletic endeavors, if you don't already know how to swim, learn. Not only is it great exercise, you never know where you'll be or what you'll be doing when an emergency strikes. Since 2/3 of the Earth's surface is covered with water, learn to swim.

5. **Emergency Driving**. If you can find a good class at an affordable price, it wouldn't be a bad idea to enhance your driving skills. You never know what you'll run into on the road during a panicked mass evacuation. Check your yellow pages for driving schools and ask around. Also ask your local Police or Sheriff's department for sources of civilian instruction.

6. **"Industrial Arts"**: A broad category, we mention this as you may have to make some emergency or post emergency repairs yourself. We suggest learning all you can about **electrical work, carpentry, masonry, welding, plumbing, appliance repair**, etc. Many community colleges and technical schools will offer individual classes. Such training, if done now, will also help you with some of the home and vehicle maintenance and modifications we'll ask you to do, and will also provide you a **barterable skill**. Other sections in this book will list tools we recommend you acquire.

7. **"Home Ec."**: We also suggest you learn all you can about **cooking, gardening, home canning** (strongly suggested), **home hairstyling, sewing and making clothing, and general household frugality**. It's a great way to learn how to save money now, and a great set of skills to have later as we have no way of knowing what effect the next terror attack will have on our economy.

Now that we've mentioned the need for **frugality**...

Notes:

_____

_____

_____

_____

_____

_____

_____

_____

_____

_____

_____

_____

_____

_____

_____

_____

_____

_____

_____

_____

_____

_____

_____

_____

Frugality

Not everything this book recommends is going to be free. In fact, some of it may be rather expensive, and it'll be up to each and every one of us to foot our own bill. You need to consider how and where to cut corners in your normal expenditures to make financial room for your Disaster Prep activities and equipment, and for the **economic eventualities that will befall you in the wake of the next disaster or terrorist attack**. The changes you make in your family's financial lifestyle today, may protect you tomorrow as surely as any kit you put together for your physical survival. It'll also prepare you if there is any official rationing program instituted as the result of an attack or disaster.

We try to help you with frugality in the way we recommend training, information, and equipment. In each "suggested information" section we start with internet sources, most of which are free. With equipment, we'll steer you away from what you don't need, suggest the minimum you do need, help you find economical sources, and in some cases tell you how to make your own. We also recommend that you read through this entire book before spending <u>any</u> money on activities and equipment, and make sure your Disaster Prep shopping list truly matches <u>your</u> needs and budget.

As an additional tool to aid in your financial planning, you'll find a "**Family Budget**" form in the "**Appendix**." You'll also find its Excel spreadsheet counterpart on CD-2. On CD-1 you'll find a couple of other budget spreadsheets. Having a little variety in your planning tools will better help you understand the process. Let's move on.

Numerous "financial planning" magazines and websites discuss high-dollar portfolio and stock management ad nauseum. This is good for the small percentage of the population that has the assets to play with in such a manner, but for the rest of us, the best financial plan is to plug the leaks in order to hang on to what we have. So for here and now, let's look at the ways we can cut costs so we'll have more money for our Disaster Prep needs.

Some of the sources of information we've included below will also be extensive enough to help you with larger financial investments and savings. In any event, we've made this a rather lengthy section as it's such an important one. You never know what tomorrow will bring (or deny).

<u>**Goals of the Frugality section**</u>:
1. To help you become more able to **afford your Disaster Prep needs**.
2. To help you develop a **thrift-oriented attitude, and more frugal and conservative lifestyle**.
3. To help prepare you for the **inevitable economic impact** of the next disaster or terror attack.
4. To help you get your assets in line in order to help you **recover from disaster more quickly**.
5. As yet another **benefit** of investing in this book.

We've divided "Frugality" into the following sections:
1. **General Purchasing and Spending.**
2. **Consumables.**
3. **Utilities.**
4. **Insurance.**
5. **Taxes.**
6. **Medical Costs.**
7. **Legal Expenses.**
8. **Miscellaneous Considerations.**

General Purchasing and Spending

We're a world full of consumers. We're bombarded daily by relentless marketing techniques designed to have us buy products and services on impulse. "Buy now, pay later" has become the mantra for many, and "keeping up with the Joneses," a major pastime. Therefore, our main suggestions concern impulse buying and general spending.

> ➤ Since we're hitting you with about a zillion suggestions on what to do and buy regarding your Disaster Prep goals, we thought we should start this section with a lose hierarchy of **Disaster Prep Expenditure Priorities**. Everyone's needs and assets are different, but generally speaking, this planner should give you a model to help you decide where you need to put your money and where not. Draw your own on a sheet of paper and use it in conjunction with your "**Family Budget**" and your "**Disaster Prep Expense Journal / Planner**" from the "**Appendix**." Here's an example.

Disaster Prep Expenditure Priorities

	FREE	$$$
Most Important	- Organize the emergency gear and goods you already have. - Make simple safety changes at home to include fire drills, and basic safety improvements. - Stock your vehicle with items already on hand. - Organize and store all important documents. - Prepare a budget to organize your finances. - _____	- Buy the bare minimum amount of food, first aid, and supplies necessary for an evacuation or isolation. - Get a regular, prepaid, or 911 cell phone. - Update and copy all important documents. - Get a second form of ID. - Photo and ID all belongings. - _____ - _____
Important	- Download free info from the internet and research subjects at your library. - Attend free training classes in related Disaster Prep subjects. - _____ - _____ - _____	- Maintenance and structural upgrades to your home. - Instructional classes requiring a fee. - Buy or make a home electric generator. - Maintenance and safety equipment for your vehicle. - Upgrade pertinent insurance policies. - _____ - _____
Less Important	- _____ - _____ - _____ - _____ - _____ - _____	- Get a computer and internet access. - More extensive home upgrades. - Elective emergency equipment (that which <u>might</u> be necessary but might not). - Elective comfort equipment. - Elective peripheral training courses or books. - _____

➢ The above ties in to some of the reasons we've given you so many sources for finding the goods and equipment you need. In fact, there are four reasons we give so many source locations.

● We want to make it **easy for you to find things**. Though some items may be familiar to many, some folks might not know how or where to find things they'll need.

● Having different suppliers with different items and associated ads may be **educational** in and of itself.

● Sources will **change over time** so it's always better to have several.

● Most importantly, having choices allows you to **shop around for the best deal**.

➢ For your family's normal expenditures, sit down and write out a **detailed accounting** of where every penny goes. List all regular expenses as well as periodic expenses, and include EVERYTHING from mortgage or rent all the way to birthday presents for friends. This will line up all your reduction targets as well as show you where you may be truly wasting money. Keep your budget in front of you as we go through the remaining areas of thrift opportunity. (You'll find a **"Family Budget"** form in the "**Appendix**" and on the **CD**.)

➢ Though your financial planning and savings strategy may differ, the "average" plan of attack for the average family is to **organize your budgeting and savings** keeping the following in mind:

❏ Use the sections below to see where you can reduce your normal expenditures.

❏ With the money you save, increase your payments to **high interest credit cards** and other accounts to pay them off, or pay them down, as early as possible. Interest rates, finance charges, late fees, and other hidden fees associated with some of these accounts can be ten to twenty times the amount you'd earn on interest if your money was in a savings account.

❏ Though you're paying off the high-interest accounts, **don't neglect savings**. Ideally, you'd want to be able to put 10 to 20% of your income into some sort of savings, with at least 10% of <u>that</u> amount being held as **cash** in a safe location (we suggest the cash due to **accessibility** issues, which is a Disaster Prep strategy, not a financial one). Work up to this savings amount once those high-interest credit cards and loans are paid off or paid down to a reasonable amount.

- As you're using the above planner, your budget, and the hints and sources below, you'll find that your **biggest enemy** is your **impulse to spend money**. To help you keep your priorities in mind when this impulse hits you, we've put together a little acronym for the word **S.P.E.N.D.**:

  Savings   **S**ave your money wherever possible. Stifle the impulse to spend.

  Payments   **P**ay down the higher interest credit cards first, and keep your cash for other payments.

  Emergencies   **E**veryone will face emergencies. Be financially prepared to face yours.

  Needs   **N**ever put an impulse or simple desire before a true need.

  Disaster Prep   **D**elay the pleasure shopping until you've completed your disaster preparations.

  We also heard another one that goes **S**ave **P**ennies **E**very **N**ew **D**ay. Memorize whichever one works for you and helps you fight the impulse to waste money.

- We've found one educational online financial newsletter we like at http://www.investyourself.com. Though not free, its viewpoint is very much in line with this section's goals, and the newsletter contains financial advice geared for mainstream readers rather than being information only the wealthy could use.

- Budget and expenditures sometimes involve **credit cards**, which is where most families get into financial trouble. We've found sources that will help you with credit or credit repair problems:

  - The **Federal Trade Commission** regulates the activities of credit providers and has several publications for the consumer. Contact them at http://www.ftc.gov, or through Public Reference, Room 130, federal Trade Commission, Washington, DC 20580, 202-326-2222, or toll free at 877-FTC-HELP (382-4357). **Note**: **Save this number** as you may need it to report illegal price gouging during or after a disaster.

  - **National Foundation for Consumer Credit**, 8611 Second Ave., Suite 100, Silver Spring, MD 20910, 800-388-2227, or http://www.credit.org.

  - **Credit Counseling Center of America**, P.O. Box 830489, Richardson, TX 75083-0489, 800-493-2222, or http://www.ccamerica.org.

  - Definitely add the **Consumer Action** group at http://www.consumer-action.org to your bookmarks.

  - Consumer Credit Counseling Service, online at http://www.cccsinc.org, with direct contact at 100 Edgewood Avenue, Suite 1800, Atlanta, GA 30303, 1-800-251-CCCS (800-251-2227).

  - Contact the **National Consumer Council** online at http://www.thencc.org, or directly at 1025 Connecticut Ave NW, Suite 1012, Washington D.C. 20036, 1-800-990-3990.

  - Call your local **County Cooperative Extension Service** (or equivalent). You'll find their number in your phone book's blue government pages. (Call this office anyway as they usually offer an extensive list of classes, services, and benefits you'll want to be aware of.)

- You'll also want to keep track of what credit card companies are saying about you and whether or not it is accurate. Contact each of these credit bureaus for a copy of your **credit report**:

  - **Equifax**: Online at: http://www.equifax.com, with direct contact at Equifax Credit Information Services, Inc., P.O. Box 740241, Atlanta, GA 30374, 1-800-685-1111. Their fraud department can be contacted at P.O. Box 105069 (same City, State, Zip) or call 800-525-6285, 404-885-8000, Fax: 770-375-2821.

  - **Trans Union**: Online at: http://www.transunion.com, with direct contact at TransUnion, Post Office Box 2000, Chester, PA 19022, 800-888-4213. Their Fraud Victim Assistance Department is located at P.O. Box 6790, Fullerton, CA 92834, phone: 800-680-7289, fax: 714-447-6034.

  - **Experian**: Online at http://www.experian.com, or call 888-397-3742. Experian's National consumer Assistance Center (for fraud) is located at P.O. Box 2002, Allen, TX 75013.

- Keep a **daily expense journal**. Not only will it show the surprising ways in which you're nickel and diming yourself to death, it'll tie in with a **daily activity journal** we'll ask you to keep during **Orange and Red alerts**.

- If your income, expenses, and finances in general are truly a concern, visit the Community Action Partnership site, **"Managing My Money,"** at http://www.managingmymoney.com, and be sure to click on "Worksheets/Tips" (as well as their other services and resources).

- Naturally, the best way to increase your spendable cash is to **increase your income**. You can do that by adding a part-time job, home-based side business, or by changing to a better job entirely. In addition to your local help-wanted ads and employment services, we've found a few resources to help you:

  - Read the book, **What Color is Your Parachute**?, and see their online companion site at http://www.jobhuntersbible.com.

  - The Quintessential Careers site, at http://www.quintcareers.com.

  - Job-Hunt.org at http://www.job-hunt.org.

  - Monster.com at http://www.monster.com.

- One of the biggest investments you can make is the **purchase of a home**. The Fannie Mae Foundation has some comprehensive guides covering your financial lifestyle and home ownership. You can download this info from http://www.homebuyingguide.com, or call 800-611-9566. You can also write to 4000 Wisconsin Ave., NW, North Tower, Suite One, Washington, DC 20016-2804, 202-274-8000, Fax: 202-274-8100.

➤ Also get some information from the Department of Housing and Urban Development, or **HUD**. Housing and Urban Development is online at http://www.hud.gov, or directly at U.S. Department of Housing and Urban Development, 451 7th Street S.W., Washington, DC 20410, Phone: (202) 708-1112 TTY: (202) 708-1455.

➤ For homeowners, we'd like to suggest that you start a "**home improvement journal.**" You'll want to keep **receipts** from services and materials used to upgrade your home, and you might want to keep a **photo or video journal** of both results and work-in-progress. It'll help document your spending with the "home prep" changes you'll be making later, it'll help document tax-deductible expenses, and it may make your home easier to sell if you decide to move.

➤ Get a **library card** if you don't have one. Public libraries are an overlooked and undervalued resource. Most libraries will not only have **internet access,** but they'll have numerous periodicals, video tapes, audio tapes, and other useful forms of information.

➤ Speaking of **internet access**, and especially since we've included so many sites for research, we thought we'd give you information on finding **low-cost internet service providers**. To give you more options though, the first two listings for AOL and Earthlink are the "big two" internet service providers. They're a bit more expensive, but you get what you pay for. Try contacting:

● **America On Line**. AOL free startup CDs can be found pretty much anywhere, and usually offer hundreds of hours worth of free trial time. Contact them online at http://www.aol.com, or at: America On Line, 22000 AOL Way, Dulles, VA 20166. The main toll-free number to order a free startup disk is 888-265-8001, with technical support numbers being 800-827-6364, and 800-427-6218.

● **Earthlink** can be found online at http://www.earthlink.net, with direct contact at 1375 Peachtree St., Level A, Atlanta, GA 30309, 404-815-0770, or 800-EARTHLINK (327-8454).

● **"Juno" and "NetZero"** are now both offered through a company called "United." You'll find either of them online at: http://www.netzero.com, or http://www.juno.com, with direct contact at: NetZero Online Services (or Juno respectively), Inc., Attn: CD Orders, P.O. Box 10849, Terre Haute, IN 47801-0849, 1-800-333-3633, although Juno uses a different phone number of 800-TRY-JUNO (879-5866).

● **"All Free ISP"** is an online search engine for free or low-cost internet service providers, and have a state by state search and online info. They're online only at http://www.all-free-isp.com.

● A similar search services is offered online at: http://www.internet4free.net.

● If either of these sites are down, log on to http://www.google.com and do a search for "free internet service," "free email service," or variations of those phrases (with quotes).

● For a good source on computer problems and economical solutions (in layman's terms) as well as information on free items, programs, etc., sign up for the **"Kim Kommando" newsletter** by logging on to http://www.komando.com/newsletter.asp.

➤ Before making any moderate to large purchases, do some research. Research the item on the **internet**, and check out **Consumer Reports** magazine and their "**Annual Buyer's Guide.**" You can find the Buyer's Guide at libraries and bookstores, and you can read some of their articles on their website at http://www.consumerreports.org/main/home.jsp, or write to Consumer Reports Online, Department, 101 Truman Avenue, Yonkers, NY 10703, 1-800-333-0663.

➤ Also be sure to visit **Consumer Affairs**, online at http://www.consumeraffairs.com. They have a free newsletter online and you can also contact them at ConsumerAffairs.Com, The Oakton Press, Inc., 11350 Random Hills Rd., Suite 650, Fairfax, VA 22030.

➤ Take a look at **Consumer.TheLinks.com**. It's a good collection of consumer and consumer protection links. You'll find them at: http://consumer.thelinks.com .

➤ Check out "**Consumer's World**" at: http://www.consumerworld.org, and their resource page at http://www.consumerworld.org/pages/resource.htm. While you're on the main page, be sure to sign up for their **free email newsletter**.

➤ See the "**Federal Consumer's Gateway**" at: http://www.consumer.gov, or use their search page at http://www.consumer.gov/yourhome.htm. You can also call them at 1-800-688-9889.

➤ Contact the **Federal Consumer Information Center** at Dept. WWW, Pueblo, CO, 81009, 1-800-FED-INFO. Their website is: http://www.pueblo.gsa.gov. **Download their free info** and be sure to get a copy of the "**Consumer Action Handbook**" from http://www.pueblo.gsa.gov/crh/respref.htm, and you'll also find a copy of the book on our enclosed **CD**.

➤ Also look at the **US Consumer Product and Safety Commission** site at http://www.cpsc.gov, or write to U.S. Consumer Product Safety Commission, Washington, D.C. 20207-0001, 800-638-2772.

➢ A good collection of **consumer tips and links** is at http://www.imarvel.com/news/consumer.htm, Imarvel, c/o Ark Royal Software, 14332 NE 126th Ave., suite # B-203, Kirkland, WA 98034-1542, 425-825-0906. Their **personal finance** site is at http://www.imarvel.com/news/pfnov.htm.

➢ Some consumer items should be researched through "**Underwriter's Laboratory**" at: http://www.ul.com, 1285 Walt Whitman Road, Melville, NY 11747-3081 USA, phone: 1-877-ULHELPS (1-877-854-3577), or, 847-272-8800, Fax: 1-631-439-6464.

➢ An online governmental cousin to Consumer Reports is the government's product recall site, located at http://www.recall.gov, where you can look up recalls by category.

➢ For additional consumer protection, contact the **Better Business Bureau**, or BBB, at: http://www.bbb.org which has info and a locator page. You can also find your local BBB office in your phone book.

➢ Clark Howard is rapidly becoming a cross between Ralph Nader and Heloise. You'll find money-saving info online at http://clarkhoward.com, or you can call 404-892-8227.

➢ Be sure to visit the federal government's Consumer Complaint Center online at: http://www.governmentguide.com/consumer_services/complaint.adp.

➢ For protection against unscrupulous national banks, contact the "**Comptroller of the Currency**" online at http://www.occ.treas.gov, or directly at Comptroller of the Currency, US Department of the Treasury, Customer Assistance Group, 1301 McKinner St., Suite 3710, Houston, TX 77010, 800-613-6743.

➢ For protection against other banks, contact the **Federal Deposit Insurance Corporation** online at http://www.fdic.gov, or at Federal Deposit Insurance Corporation, Credit Card Center, 2345 Grand; Suite 100, Kansas City, MO 64108, 1-877-ASK-FDIC (1-877-275-3342), 800-925-4818 (TTY), Fax (816) 234-9060.

➢ For help with savings and loan banks, contact the **Office of Thrift Supervision** online at http://www.ots.treas.gov or directly at Office of Thrift Supervision, US Department of Treasury, 1700 G St. NW, Washington DC 20552, 202-906-6000, or 202-906-6237, or 800-842-6929.

➢ Additionally, you can contact your state's **Banking Commission** or similar office. Look in your government blue pages in your phone book.

➢ You can get some consumer information regarding **automobiles** at:

● **Vehix.com**, at http://www.vehix.com, or Vehix Dot Com, 1165 E. Wilmington Avenue, Suite 200, Salt Lake City, UT 84106, 801-401-6060.

● "**Carfax**" at: http://www.carfax.com, or Carfax, 10304 Eaton Place, Suite 500, Fairfax, Virginia 22030-2213, 1-888-422-7329.

● "**Kelly's Blue Book of Values**": http://www.kbb.com (copies available at most bookstores and libraries).

● **CarMax** is another good online shopping tool you can find at http://www.carmax.com.

● Be sure to look through the **Auto Trader**, online at http://www.autotrader.com, to help you shop for a used vehicle. You'll find copies at many free-publication newsstands or at auto parts stores.

➢ **General shopping info** can be found at http://www.yellowbook.com, 1-800-YB-YELLOW (800-929-3556).

➢ You can **comparison shop** online at: http://www.pricescan.com, http://www.shopathome.com, http://www.lowermybills.com, and http://www.catalogs.com, for various products and services.

➢ Since you'll need several sources in order to comparison shop, and since you'll also need several sources to find some of the things we mention in this book, try **The Catalog Handbook**, available online at http://www.busop1.com/cathand.html, or through Enterprise Magazines, Inc., 1020 N. Broadway, Suite 111, Milwaukee, Wisconsin 53202, Phone: 414-272-9977, Fax: 414-272-9973.

➢ Another frugal lifestyle info page with a good links collection, is at http://www.bettertimesinfo.org.

➢ Find **The Tightwad Gazette,** by Amy Dacyczyn. It used to be a monthly newsletter, and now there are three published volumes of the newsletter, and a newly released compendium (get the compendium). It'll really get you into thrift mode. Most bookstores and libraries will carry it.

➢ See "**The Dollar Stretcher**" at http://www.stretcher.com/index.cfm, and sign up for their email newsletters.

➢ And a close cousin of those is: http://www.jrmooneyham.com/poor.html .

➢ Similarly, be sure to visit http://www.thefrugalshopper.com .

➢ Take a look at "**Do It Yourself's**" frugality page: http://www.doityourself.com/frugal/index.htm.

➢ The "Simplitudes" site has quite a number of links and pieces of info designed to let you live a more frugal lifestyle. You'll find them online at http://www.simplitudes.com.

➢ Definitely read **Personal Finance for Dummies** by Eric Tyson, at bookstores and http://www.dummies.com.

➢ Take a look at the books put out by **Charles Givens**. Two in particular that come to mind are **Wealth Without Risk** and **More Wealth Without Risk**, which discuss savings and investment techniques applicable to us middle class folks. Try your local library.

- Relatedly, you might want to consider the financial advice of **Suze Orman** in her book **The Laws of Money, the Lessons of Life**. Her book can be found pretty much anywhere, and her other info can be found at http://www.suzeorman.com/home.asp.
- Read **The Complete Idiot's Guide to Being a Cheapskate**, by Mark Miller, at http://www.idiotbooks.com .
- Take a look at **Frugal Living for Dummies** by Deborah Taylor-Hough, at http://www.dummies.com.
- Also read the book **The Best of Living Cheap News** by Larry Roth. It can be found along with other useful info (including a links page to other frugality sites) at: http://www.livingcheap.com, Living Cheap Press, P.O. Box 8178, Kansas City, MO 64112 (no phone listed).
- Be sure to visit "Miserly Moms" at http://www.miserlymoms.com .
- Try Google's new **product and purchasing search engine**, "**Froogle**," online at: http://froogle.google.com.
- A similar site is the Thrifty Life Tips site at: http://thrifty.lifetips.com.
- Join local **discount buyer's clubs** such as Sam's Club or Costco, but stay away from some "memberships" that may give a very small percentage off high-end items you wouldn't buy anyway.
- Check out Kipplinger's **"cost of living" calculators**: http://www.kiplinger.com/tools/index.html.
- **Learn to do as much as you can for yourself.** Learn to cut your children's' hair, change your car's oil, sew and repair your own clothes, and if you followed our advice earlier about "industrial arts" training, you can probably perform some of your own appliance or household repairs. Remember too, that with such skills, you may be able to barter rather than buy some of the goods or services you need. Some good info on this "do it yourself" topic can be found through:
  - You can find a lot of patterns online at http://www.sewingpatterns.com, and you can contact them directly at 4456 Haskell Ave., Encino CA. 91436. They also have a free email newsletter.
  - "**Backwoods Home**" at: http://www.backwoodshome.com, Backwoods Home Magazine, P.O. Box 712, Gold Beach, OR 97444, Toll free: 800-835-2418, Phone: 1-541-247-8900, Fax: 1-541-247-8600.
  - "**Mother Earth News**" at: http://www.motherearthnews.com/index_js.html, 1503 SW 42d Street, Topeka, KS 66609-1265, 800-234-3368.
  - Reader's Digest Books' **Back to Basics**, ISBN: 0-89577-086-5, is an older, but highly recommended, book available at libraries, bookstores, or http://ww.amazon.com.
  - "**Countryside Magazine**" online at http://www.countrysidemag.com, or at W11564 Hwy 64, Withee, WI 54498, 715-785-7979, or 800-551-5691, Fax: 715-785-7414.
  - **Home Maintenance for Dummies**, by James & Morris Carey.
  - **Household Hints for Dummies**, by Janet Sobesky.
  - You should certainly visit http://www.doityourself.com.
  - Also visit "Hints and Things" at: http://www.hintsandthings.co.uk.
  - Relatedly, see the "Heloise Hints" site at http://www.heloise.com.
  - To REALLY get back to basics check out the **Foxfire** book series by Elliot Wigginton. (We recommend you at least go to the library or your bookstore to flip through this series.)
  - You'll find recipe and food-related info sites in the "Manual Kitchen" portion of the "**Isolation**" section.
- **Reuse and recycle.** Check local **Salvation Army** stores, **Goodwill** stores, independent **thrift** and second-hand stores, **military surplus stores**, and **garage or yard sales**. We can't begin to tell you the number of incredible deals on camping gear, tools, and other items of interest we've found for next to nothing. In addition, keep an eye on any local **"shopper"** or **"trader"** publications. Keep in mind too that with private individuals, you can probably barter as quickly as you can make a cash transaction.
- By the way, since the main goal of this book is to make you think of as many alternatives and options as possible, we'd be shirking our duties if we didn't at least mention "**dumpster diving**." You'd be surprised at the number of useful things that get thrown away.

Consumables

For this category, we'll cover items the average family would buy at a supermarket or grocery store.

- **Buy in bulk**. There's anything from **SAM'S Club** (http://www.samsclub.com, 1-888-SHOP-SAMS), **Costco** (http://www.costco.com, 800-774-2678), all the way through **local farmer's markets**, with several chains of **bulk-purchase grocery stores** in between. If there's anything you buy a lot of, or use regularly, buy it in bulk *after you've researched the price break as compared to what you normally pay*. Spend a day with a notepad, shopping various stores in your area. (Important for stocking your "**Isolation**" pantry.)

- ➤ **Coupons**. A good rule to remember is to <u>never</u> pay retail again. <u>Always</u> look for the deal, discount, or coupon. And, when using coupons, always try to **combine the coupon with a sales price** as opposed to regular price. The magazine "**Budget Living**" (at most newsstands and at http://www.magazines.com) is a good source of savings information and coupons as are these websites (be sure to try each):
    - ❑ http://www.thefrugalshopper.com. (This site has far more than just coupons.)
    - ❑ http://www.mycoupons.com.
    - ❑ http://www.upons.com.
    - ❑ http://www.valupage.com/Entry.pst.
    - ❑ http://www.hotcoupons.com.
    - ❑ http://www.myclipper.com.
    - ❑ http://www.flamingoworld.com.
    - ❑ http://www.coolsavings.com.
    - ❑ http://www.100hotcoupons.com.
    - ❑ http://www.couponsurfer.com.
    - ❑ http://www.ultimatecoupons.com/coupons/offline_coupons.htm.
    - ❑ http://www.dailybreeze.com/coupons/.
    - ❑ http://www.adobeshoppingmall.com/coupons-grocery-free.htm.

- ➤ Coupons are almost reason enough on their own to **subscribe to your local paper**. Also check the website of your favorite grocery store.

- ➤ Even when using coupons on top of a sales price, if the total is still higher than the **store brand or a suitable generic**, go for the already-lower price of the generic. Also, don't let the mere fact you have a coupon talk you into buying something you don't really need in the first place.

- ➤ If you go to http://www.google.com and enter the search string **+coupon+online** (just like that with no quotes), and you'll get more coupon sites than you could ever hope to read or use.

- ➤ One shopping service that offers small coupons and rebates to quite a large number of stores is the "Ask Mr. Rebates" site at http://www.askmrrebates.com, or their soon-to-be-revised site, http://www.mrrebates.com.

- ➤ Explore **home gardening and/or home canning**. Even if you don't grow things yourself, you can preserve some of the things you pick yourself at a "you-pick" farm or that you buy in bulk at a farmer's market. You can find more info on home canning at http://www.homecanning.com, Consumer Affairs, P.O. Box 2729, Muncie, IN 47307-0729, 1-800-240-3340. (Though you might not need to can anything now, it's a skill you might want to master, and equipment you'll want to gather, in case the day comes that's all you have. Same with gardening. Agro-terrorism is a chief concern of the Office of Homeland Security.)

- ➤ The USDA offers extensive **home garden** info at http://www.usda.gov/news/garden.htm.

- ➤ More **information on grocery savings** can be obtained from the **Federal Consumer Information Center** at Pueblo, CO 81009, 1-800-FED-INFO. Their website is http://www.pueblo.gsa.gov.

- ➤ **Identify that which you truly do not need**. Just how many <u>different</u> household cleaners do you really need anyway? Take a look at http://www.tipking.com, for some ideas. Also see the cleaning solution "recipes" at: http://www.metrokc.gov/health/asthma/greencleaning.htm, and at: http://www.ci.seattle.wa.us/util/ept/clngrn/recipes.htm.

- ➤ Health reasons aside, **quitting smoking** and/or **drinking** will save you quite a bit of money.

- ➤ Consider **consumables vs. reusables**. For example, if you already do a lot of laundry and a few extra items a week won't make any difference at all, have you considered using cloth napkins at meals instead of paper? For a really detailed study on this concept, see the previously mentioned **Tightwad Gazette**.

- ➤ If you want to learn more about sewing, you can start with The Home Sewing Association, online at http://www.sewing.org, or at PO Box 1312, Monroeville PA 15146, (412) 372-5950, (412) 372-5953 Fax.

- ➤ Two suitable works for working with clothing are **Knitting for Dummies**, by Pam Allen and **Sewing for Dummies**, by Janice Saunders Maresh.

- ➤ **Cooking at home vs. eating out**. The biggest drain on anyone's food budget comes from dining out. Cook at home more for both health and monetary reasons. If you eat out, look for coupons. (Look at http://www.entertainment.com for a coupon book for your city.) Also, at work, consider "lunch pooling." Say there are 5 of you in your office. One day a week a different person brings in a lunch they cooked at home to feed the group. You could all eat rather well and for far less than you'd spend eating out every day.

- ➤ Plastic "store and microwave" goods, like **GladWare ®,** are so inexpensive and plentiful, there's no reason you can't cook in advance and **freeze a week's worth of food for the whole family**.

- Start saving soup cans, tuna cans, larger tin cans, plastic plates from microwave dinners, plastic grocery bags, and plastic resealable soda bottles. We'll be making things out of them later.
- In the section, "The Isolation Pantry," under "**Isolation**," we'll give you sources for meal recipes that are meant for those with limited supplies and/or budgets.
- And the best rule of all? **Never go grocery shopping when you're hungry.**

Utilities

Having mentioned the **Federal Consumer Information Center** at Pueblo Colorado, if you get their catalog, you'll find a publication or two on saving on your utility bills. It's a good idea to conserve as much as you can, not only for the benefit of your pocketbook, but for the environment as well.

- When considering the **purchase of energy saving upgrades**, weigh the difference between how much money you'd spend on repairs or gadgets, and the time before you'd recoup your money. Such investments should either pay for themselves within the first year, or should add resale value to your home. Too, check with your tax prep person, or call the IRS to see which energy saving investments might be tax deductible.
- When buying any new appliances, do your homework on their **energy efficiency rating**. (Both "**Consumer Reports**," and the dealer should have all the info you need.)
- **Set your thermostat** to 78 degrees or higher in the summer, 68 degrees or lower in the winter, use **fans or space heaters** to make up the difference, and learn to **utilize or block sunlight** accordingly. If you set your thermostat to even more economical levels, not only will you save money, but you'll be better acclimated should a disaster deprive you of heating or cooling.
- Set your thermostat according to **periods of use**. For example, when everyone is gone for the day, or in bed for the night, set the thermostat to more economical levels. However, don't set your thermostat more than five degrees from where you usually keep it. Making your HVAC cool or heat the house those extra 5 degrees may cost more than you save with the lower levels.
- During spring and fall, when you don't have the doors and windows open, use only the **inexpensive space heaters or fans** for small areas rather than heating or cooling the whole house. Also, several companies make "space air conditioners" (found at most department stores) that are floor units which can be plugged into any outlet in any room, and they don't need to be window mounted.
- Though a long-term change as far as results are concerned, if you live in a particularly hot region, take a good look at your **landscaping** and see where you might plant leafy **trees or bushes** to provide **shade**.
- **Cover any roof vents** during the winter so you can retain your heat.
- Electrical outlets are notorious sources of **drafts and thermal exchange**. Use inexpensive foam insulators and plug covers to seal them.
- Get an inexpensive **insulating blanket for your hot water heater** and use pipe insulators for exposed piping coming from the hot water heaters to sinks, tubs, or appliances.
- Check all your **door and window seals** to cut down on energy wasting drafts, and consider sealing some windows with plastic in the winter. Not only will this save you money now, but if you're in an isolation situation it will help you conserve any rationed or limited heating fuel supply and will also insulate you when utilities might be cut off altogether. Also, it will help **seal your home** should you have to "**shelter in place**" due to a biochemical attack or HazMat incident, or it may help protect you if the incident comes as a surprise.
- In warmer regions, consider **tinting certain windows**. You can buy very economical tinting film at hardware stores, department stores, or through automotive supply sources. Not only will tinting some windows cut down on heat, the film itself may help protect you from flying glass during a destructive event, or make it more difficult for burglars to break that window to get in.
- Check your **water and gas lines** all the way from the meters throughout the house for leaks. Even a minor leak could increase your monthly bill by **25%**, and gas leaks are dangerous. If you call your gas company, they'll usually come check the lines from your meter to the street using a sensitive sniffer device. Gas companies want a leak less than you do.
- If you don't have a **water saving toilet**, put a brick in the tank. Also get a water-saving shower head, and make sure you have aerators and restrictors on sink faucets. These can cut your water bill by 10% on average. (**Hint**: If you use an actual brick, paint it with waterproof paint to keep it from dissolving and gumming up the valve works in your toilet's tank.)
- We found some water saving products online at http://www.watersavers.com. Contact them at The Green Culture, 23192 Verdugo Dr., Suite D, Laguna Hills, CA 92653, Phone 800-233-8438, Fax 949-360-7864.

- Don't run your **dishwasher or clotheswasher** for small batches. **Hand wash** the small batches, and in warm weather hang your clothes outside on a line rather than use the dryer.

- While running the water waiting for it to get hot, **let it run into a jug** or flower watering can. Don't waste it.

- Use a **squeeze nozzle** on your garden hose, so when you put it down for a second, water flow is restricted.

- **Reuse energy where possible**. For example, you might want to have both an outlet vent and an indoor lint trap setup for your drier. In the summer (when you're not using the clothesline) you can vent the drier to the outside. In the winter, use the indoor lint trap and take advantage of the heat the drier creates.

- As bulbs burn out, replace them with **lower wattage bulbs**, or other energy saving bulbs.

- As always, follow the old rule: "**Turn off when not in use**."

- **Call your gas, electric company, and water company** for more information on conservation and efficiency. Ask if they have publications, inspectors that can come to your house, or any other programs that will help you save money and conserve resources.

- Look into alternative energy sources both for reducing your utility bills, and for being self-sufficient during any utility outage. See "**Advanced Home Prep**" under "**Isolation**."

- For more energy saving articles and sources, log on to http://healthandenergy.com/energy_efficiency.htm, and scroll down to "Additional Sources of Information."

- For more energy saving information, contact Energy Star, online at http://www.energystar.gov, or contact this government agency directly at US EPA, Climate Protection Partnerships Division, ENERGY STAR Programs, Hotline & Distribution (MS-6202J), 1200 Pennsylvania Ave NW, Washington, DC 20460, or call 888-STAR-YES (888-782-7937).

- Contact the **Energy Efficiency and Renewable Energy Clearinghouse** at P.O. Box 3048, Merrifield, VA 22116, 800-363-3732, and ask for the publication "Energy Savers: Tips on Saving Energy and Money at Home." You can see it online at http://www.eere.energy.gov/consumerinfo/energy_savers/index.html.

- Visit the energy efficiency info page at http://www.buyenergyefficient.org/index.html, or contact the Consumer Federation of America Foundation at 1424 16th Street NW, Suite 604, Washington, DC 20036 Phone: (202) 387-6121.

- A free online booklet on various ways to reduce your utility consumption and costs is offered by the Alliance to Save Energy, online at http://www.ase.org/powersmart.

- The Home Energy Saver site at http://hes.lbl.gov, offers an online energy efficiency rating calculator.

Insurance

Insurance refers to any money you pay in order to protect against a future loss or inconvenience. This can be car insurance, homeowner's, or even extended maintenance plans for appliance purchases. While some of the sources listed under "**General Purchasing and Spending**" above will provide good insurance guidance, here are a few direct information sources and suggestions.

- Always research before you buy. Check with the **Better Business Bureau** at: http://www.bbb.org, 4200 Wilson Blvd, Suite 800, Arlington, VA 22203-1838, 1-703-276-0100.

- Also check with your state's **Insurance Bureau** at: http://www.insure.com/links/doi.html, and if this link is down, look up your state's **Insurance Commissioner** in your phone book's blue government pages.

- The "**Insurance Information Institute**": http://www.iii.org has some really good information.

- So does Consumer Insurance Information: http://www.rmis.com/sites/risconsu.htm.

- Check out **comparative insurance rates** at http://www.quoteseek.com/auto.html, http://www.insure.com, http://www.progressive.com, http://www.insurance-finder-online.com, and at: http://www.insweb.com.

- The Independent Insurance Agents of America site is at: http://www.independentagent.com.

- Check out insurance sites at: http://www.bizweb.com/categories/finance.insurance.html.

- Also be sure to check with both the **National Insurance Consumer Helpline** (1-800-942-4242), and **The Insurance Information Institute** (1-800-331-9146).

- As for automotive and/or homeowner's insurance, shop around for the best rates. Remember to **keep your deductible high**, and to use the same provider for all your insurance needs if you can, as you'll usually get a multi-policy discount.

- However, don't forget to **increase your insurance coverage** if you increase your assets. Have your policies reviewed as your home appreciates, you remodel or renovate, or as you make major purchases.

- If you're a homeowner consider getting a "**Homebuyer's Warranty**" policy. Your mortgage holder or homeowner's insurance provider can clue you in to various providers.

- Some good **information on insurance** can be found at the Federal Consumer Information Center at Dept. WWW, Pueblo, CO 81009, 1-800-FED-INFO, http://www.pueblo.gsa.gov.

- In the wake of 9/11, many insurance companies are redefining **terrorist acts as an "act of war"** and therefore not a coverable event. Pay attention to the fine print.

- The most outrageous of insurance rates are associated with **health and medical** needs. The best advice we can give you is to follow all the health suggestions we gave you and to always practice good safety measures. The sites listed above will let you investigate companies, costs, and reputation. Also, keep your deductible as high as you can afford.

- For health insurance counseling for the elderly and disabled, call your local Department of Aging or the Aged, or call the Eldercare Locator Hotline at 800-677-1116 and ask them for the **Health Insurance Counseling and Advocacy Program** (HICAP) in your area.

- Finally, ask your insurance company about all the things you can do to **reduce your premiums**, such as all the things regarding safety and fire under "**Basic Home Prep**." You'll be surprised at the things you'll learn on asking that weren't originally volunteered.

Taxes

Taxes are one of the unavoidable and yet necessary evils of life. Learn what you can on your own, as accountants, financial planners, and consultants aren't free.

- Naturally, the **IRS** is the first, and usually last, word on taxation. They do offer a lot of information and downloadable forms, etc. on their website at: http://www.irs.gov (800-829-1040). Your main post office or federal building will also have forms around tax time. Look in the "US Government" section of your phone book for the IRS office nearest you. Actually you'll find that most IRS staffers will be helpful in giving you information on how to prepare and file your taxes and on what deductions, extensions, etc. are allowable.

- A multitude of online **tax information links** can be found at: http://www.el.com/elinks/taxes, and http://www.tax.org/taxwire/taxwire.htm.

- For more tax info don't forget the **Federal Consumer Information Center** at Dept. WWW, Pueblo, CO, 81009, 1-800-FED-INFO. Their website is: http://www.pueblo.gsa.gov.

- Free electronic tax filing can be completed at http://www.turbotax.com. It even runs you through standard deductions you might otherwise have missed doing it on your own, or using just the 1040EZ forms.

- If you decide to **enlist the aid of a professional**, find one that has a proven track record, a reasonable rate, or bases their fee on a percentage of what they actually save you.

- One of the more recognized tax prep companies is **H&R Block**. Contact them at http://www.hrblock.com, to use their online service, or office locator. Also look in your phone book for an office near you. H&R Block also offers a **course** you can take on preparing your own taxes. Call them at 800-267-6880. If there is no office near you, your phone book is bound to list numerous reputable companies.

- Always be sure to **save all your receipts** and organize them by category of expenditure and then sub-organize by date. Your tax prep professional may surprise you with what's deductible.

- Also, in the "**Cooperation**" section, we'll be asking for your help in lobbying to make personal **Disaster Prep expenditures tax deductible**.

- If you're considering having a garage sale, reconsider. By **donating the items** to the Salvation Army or Goodwill, you can deduct a higher dollar amount from your taxes than you could have made at a yard sale. Also, some items might generate a better selling price on Ebay, at http://www.ebay.com, than in a yard sale.

Medical Costs

The extreme cost of medical care in this country is another of the many, many reasons we started this book with a section on your health. To help in other areas of medical costs, be sure to use the above sites to review all your insurance coverage, and these sites below if you need additional assistance.

- Call your local **Health Department or Division of Health and Human Services**. Many health related services are offered at reduced prices through your local health department. Find them in your phone book's blue government pages.

- When your physician prescribes a drug, ask if there is a **generic** alternative to the name brand, or if the doctor has sample packs that can fill the prescription or at least part of it.

- For locations offering **free immunizations** for children, free flu shots, etc. contact the CDC's "Immunization Hotline" at 800-232-2522, or log on to http://www.cdc.gov/nip.

➢ The **Center for Patient Advocacy** can help in disputes over poor care or erroneous medical bills. They're at 1350 Beverly Rd., Suite 108, McLean, VA 22101, 800-846-7444 or http://www.patientadvocacy.org.

➢ The **Center for Medicare Advocacy** offers information to those who may have been wrongly denied Medicare coverage. Contact them at The Center for Medicare Advocacy, Inc., P.O. Box 350, Willimantic, CT 06226, 860-456-7790, or at http://www.medicareadvocacy.org.

➢ The **Center for Health Care Rights**, is online at http://www.healthcarerights.org, and has a good links page at http://www.healthcarerights.org/links.html. They offer good information on programs, insurance coverage, and health related topics. You can also contact them at Center for Health Care Rights, 520 S. Lafayette Park Place, Suite 214, Los Angeles, CA 90057, (213) 383-4519; FAX: (213) 383-4598.

➢ If you have a **disease or ailment** and want to find out if there is a **special study** going on concerning possible discounted treatments, contact the following sources:

● The "Computer Retrieval of Information on Scientific Projects" at http://crisp.cit.nih.gov, or write to the Office of Reports and Analysis, Office of Extramural Research, 6700 Rockledge Dr., Room 3210, Bethesda, MD 20892, phone: 301-435-0656, fax: 301-480-2845. Ask if they list any studies involving your condition.

● Also pose the same query to the National Institute of Health's Clinical Center at http://www.cc.nih.gov, or at National Institutes of Health, Patient Recruitment, Warren Grant Magnuson Clinical Center, National Institutes of Health, Bethesda, MD 20892-2655, phone 301-496-4891 or 800-411-1222, fax: 301-480-9793. Or contact the main office of the National Institutes of Health at http://www.nih.gov, National Institutes of Health, 9000 Rockville Pike, Bethesda, MD 20892, 301-496-4000.

● For information on clinical trials, log on to http://clinicaltrials.com, or call 888-874-2511 or 800-664-5099.

● The government has a similar site at http://clinicaltrials.gov, which is a search portal run by the National Library of Medicine for the National Institutes of Health. Call the NLM for more info at 888-FIND-NLM (888-346-3656), or you can contact the NIH online at http://www.nih.gov, or at 301-496-4000, for more details.

➢ For additional consumer information about medical sites, sources, and services, log on to http://www.healthfinder.gov/healthcare, and their homepage at http://www.healthfinder.gov.

➢ The US Dept. of Health and Human Services has also initiated a program aimed at providing free or low-cost health insurance for children. You can learn more online at http://www.insurekidsnow.gov, or by calling 1-877-KIDS-NOW (877-543-7669).

➢ Some private organizations offer varying degrees of medical and pharmacy discounts. One program we've seen is **AmeriPlan**. You can find more information at http://www.deliveringonthepromise.com/mmitchell.

Legal Expenses

One thing most of these consumer protection, insurance, taxation, and medical sites have in common is that there's an interwoven thread of law. You'll find the same holds true with your normal life, and that from time to time, you're forced to deal with legal issues whether it's writing a simple will, making your insurance company pay a valid claim, or fighting off a social parasite who thinks they can get rich by suing everyone they meet. Another legal need can be found in our recommendations that you draw up wills, and mutual powers of attorney for each head of household, just in case. Fortunately, law is another expense for which there are available services to help you get what you need for less. Contact the following groups to see if they can help with your legal needs.

➢ Contact the **Legal Services Corporation**, online at http://www.lsc.gov, or directly at 3333 K Street NW, 3rd Floor; Washington, D.C., 20007, phone 202-295-1500, fax 202-336-8959. The LSC provides help with numerous civil matters to low-income individuals.

➢ The **American Bar Association** keeps a list of attorneys in each state who offer pro bono services to families in need. The site link is http://www.abanet.org/legalservices, and at: American Bar Association, 740 15th Street NW, Washington, DC 20005-1019, 202.662.1000. Also, if all else fails, the ABA can help refer you to a local attorney qualified to handle your need.

➢ The American Bar Association can also help you contact your State Bar Association which can then help you locate a local attorney specializing in your particular need.

➢ For assistance with possible civil rights violations, contact the **American Civil Liberties Union** online at http://archive.aclu.org, with their local chapter locator being at: http://archive.aclu.org/action/chapters.html. Contact the ACLU directly at American Civil Liberties Union, 125 Broad Street, 18th Floor, New York, NY 10004, 1-888-567-ACLU (2258).

➢ Self-help and self-education materials can be found on a number of law subjects and legal situations at http://www.nolo.com. You can reach **Nolo Press** at 950 Parker Street, Berkeley CA 94710-2524, Phone: 800-728-3555, or Fax: 800-645-0895, to ask about books and other services. Also check your library and bookstore for legal books put out by Nolo.

➤ **Identity theft**. Billions are lost each year due to identity theft. Be sure to always shred your trash, opt out of any sale of information by any of your account providers, and never give out personal information over the phone. Also, you might want to look into getting a Tax ID number to use instead of your Social Security Number. The IRS can help you there. Let's look at some other considerations:

● Always **shred your trash** and destroy any piece of paper that has an account number or any sort of sensitive information on it. Most identity theft cases begin with someone digging through your trash.

● Every six months, you should **review your credit report**. You can get yours through the three credit reporting agencies we listed above. Look for unknown inquiries, or new accounts you did not open. You should also take a look at the following ID theft resources:

● **Privacy Rights Clearinghouse**: Online at www.privacyrights.org, with direct contact at Privacy Rights Clearinghouse, 3100 5th Ave., Suite B, San Diego, CA 92103, Voice: 619-298-3396, Fax: 619-298-5681.

● **Identity Theft Resource Center**: Online at www.idtheftcenter.org, with direct contact at Identity Theft Resource Center, P.O. Box 26833, San Diego, CA 92196, Phone: 858-693-7935.

● **FightidentityTheft.com**: Online only at http://www.fightidentitytheft.com.

● If you think you might be a victim of identity theft, contact the **fraud departments** of the credit bureaus listed above, as well as the Federal Trade Commission at http://www.ftc.gov or 877-438-4338, and Consumer.gov's page at: http://www.consumer.gov/idtheft.

● If you suspect fraud, contact the National Fraud Information Center online at http://www.fraud.org, or by calling 1-800-876-7060.

➤ Take advantage of the **discounts or rebates offered by various credit cards, and/or club memberships**. Some credit cards offer discounts if used at certain places, and they may have a bonus points or cash back program. Make sure they have **low interest charges** and that you pay off all purchases at the end of the month without letting them accrue in your balance. It makes sense to use these cards for your normal purchases and receive the bonus points for expenditures you would have made anyway. Shop around for the best programs.

➤ Relatedly, don't forget to check the weekly sales section of your local newspaper for any "**zero total after rebate**" sales. Many stores (most notably office supply stores in our area) will use such items as a draw hoping that customers will stay and shop.

➤ For information on **Federal Programs and Services**, try: **The Citizen Information Center** at http://www.info.gov, 1-800-FED-INFO; "**First Gov**" at http://www.firstgov.gov; or the **Commerce Department's** http://www.fedworld.gov, 703-605-6000.

➤ Many of us have heard of the numerous **grant and loan programs** available today. The main source for information on *federal* grants is through the "**Catalog of Federal Domestic Assistance**" at http://www.cfda.gov. To order a copy of the catalog, call or write to the **US Government Printing Office** at: Superintendent of Documents, 732 North Capitol Street NW., Washington, DC 20401, 202-512-0000, and 888-293-6498, or PO Box 371954, Pittsburgh, PA 15250-7954, 1-866-512-1800. They're also online at http://www.gpoaccess.gov.

➤ Try **FirstGov's** benefits and grant locator: http://www.firstgov.gov/Citizen/Topics/Benefits.shtml.

➤ For grants and loans through state and private organizations, pick up a copy of **Mathew Lesko's How to Write and Get a Grant**. You can find this book through your library, local bookstore, Amazon.com, or directly from Mathew Lesko at http://www.lesko.com, Information USA, 12079 Nebel St., Rockville, MD 20852, 1-800-955-POWER. While you're at it, look at various other titles offered by Mathew Lesko.

➤ Some of you might want to find a second job, or a small home-based business that can be started on a shoestring budget. We have an older work we've published called **The Complete Guide to Homemade Income**. You'll find a coupon in the "**Appendix**."

➤ For fun and **free samples**, **free info** etc. try some of the various "freebies" sites. We've found a couple at:

● Freebies, online at http://www.freebies.com, or contact "**Freebies**" magazine at P.O. Box 21957, Santa Barbara, CA 93121, Phone 805-962-5014, Fax 805-962-6200.

● A **similar freebies site** is at: http://www.getyourfreebies.com, and there's another one at http://www.winfreestuff.com.

● Also, be sure to visit http://www.freebiehighway.com, and http://cory.myfree.com/www.

➤ For **free catalogs and magazines**: http://www.shopathome.com.

> Speaking of free, many of the websites we send you to will have **free downloadable information**. Most often, these files are in Adobe Acrobat's "PDF" file. Adobe will provide you a <u>free</u> reader that will allow you to read (but not alter) these files. Go to http://www.adobe.com to download the "**Acrobat Reader**."

> You'll find several topics on the USENET group, "Frugal-Living." Go to news:misc.consumers.frugal-living.

> As a final thought, you'll want to include your children when doing much of this work since you'll not only want to prep them for financial responsibility when they grow up, but in an emergency, you might need your children to access and utilize some of your family's financial assets if the adults are incapacitated.

Now that you've read through these suggestions, go through your own **budget** and mark those items you can **reduce**.

Financial planning provides assets that can be used before, during, and after a disaster. This is a good thing since there's always another side of the coin, or as some people like to say, the other shoe that's going to drop. We like to look at it as "phases." Your prior planning and immediate reaction to an emergency is "**Phase I**." Putting your life back together after the emergency is '**Phase II**." Let's take a look at how all this financial planning applies to your advance planning and preparation for post-disaster relief and rebuilding efforts. Let's talk about "**Phase II**."

Notes:

_____
_____
_____
_____
_____
_____
_____
_____
_____
_____
_____
_____
_____
_____
_____
_____
_____
_____
_____
_____
_____
_____
_____
_____
_____
_____
_____
_____

Planning Ahead for Phase II

Unless you plan to head permanently to the hills if your house burns down, unless you're independently wealthy or own your own contracting company, or unless you're part of one of the most well-planned civil defense groups in history, you'll have to deal with some sort of agency, office, or bureau in the process of putting your life back together if you're the victim of a destructive incident. Planning ahead to handle all the required red tape can tame a "disaster" and make it simply a "major inconvenience."

Each of the various groups you'll deal with after a disaster will have their own particular specialties and methods of doing business. Take time now to make sure you have the proper safeguards in place, that you've researched information on the types of offices you'll need to deal with in the wake of a disaster, and that you've gathered and protected the information and documents you'll need in the process.

Goals of Planning Ahead for Phase II:

1. To help you through one of the **most often overlooked aspects** of disaster preparation.
2. To examine **conventional and unconventional forms of "insurance."**
3. To examine various **agencies and services** involved in relief and rebuilding.
4. To help you with the **important data and documents** you'll need in an emergency or after a disaster.
5. To help you **recover more rapidly** after experiencing a loss in a disaster.

We've divide this section into:

1. The **safeguards** you need to have in place.
2. The **services and agencies** you may have to deal with in the aftermath of a disaster.
3. The **supporting documentation** or information (your "**Info Pack**") you'll need in the process.

Pay close attention to this section as it may well be more important than anything we've included in this entire book.

Safeguards

Most of the points listed below concern some type of "insurance" whether it's a policy, or just having your "ducks in a row." Some actual insurance policies will be a must while others will be optional. Be selective and go with what you can afford, but if you're going to err, err to the side of good coverage.

➢ **Homeowner's insurance**: If you do any one thing get good insurance. Review your policy and make sure it covers all kinds of contingencies; fire, theft, violent weather, earthquake, flood (see below), and liability. Also **find out how this company regards terrorist attack**. Many are starting to consider terrorism an "act of war" and will not cover losses.

➢ **Flood insurance**: Flood insurance is not included in most homeowner policies. The main source of flood insurance is FEMA's "National Flood Insurance Program" at www.floodalert.fema.gov and http://www.fema.gov/nfip, by phone 1-888-FLOOD-29, or 1-888-RAIN-619. You can also contact them at http://www.floodsmart.gov or at 1-888-FLOOD-05. Most insurance providers can also write a separate flood policy. Independent providers can be found at http://info.insure.com/home/flood, 1-800-556-9393.

➢ **Medical Insurance**: Under "Frugality" we discussed keeping deductibles high to keep premiums low. For the purposes of "Phase II" we want to remind you to make sure your medical coverage will cover you not only for injuries related to terrorist attacks, but for mishaps experienced while travelling. You want to make sure there are no hidden stipulations that would reduce your coverage for receiving treatment "outside your normal area" or for receiving treatment from non-specified providers. Review any **dental** policy you may have for the same stipulations.

➢ **Auto Insurance**: For many, the loss of a vehicle would be a disaster in and of itself, and you should therefore be protected. Make sure your auto or homeowner's policy has sufficient coverage for vehicle replacement in case of total loss, and for providing you a rental car while your claim is being handled. Also, many auto insurance companies will offer **additional coverage** for medical, life insurance and even **"travel club" or road safety club** coverage.

➢ **Term Life**: More affordable than "Whole Life," a Term Life policy is a straight forward purchase and covers an individual in case of death. Look through the insurance information sites listed in the "Frugality" section for sources of more info on these policies, and again, be sure to review any "terrorism" clause they've added.

> **"Micro" Life**: Numerous companies, auto clubs, banks, and various memberships, offer simple term-life insurance policies for accidental death and cost only a few dollars a month. If you have a family and have a high score on our **"Emergency Risk Checklist"** below, you might want to look into one of these policies.

> If you're not a homeowner and you rent, it's usually a good idea to get an inexpensive **"Renter's Insurance"** policy. As a renter, not only are you just as at risk of disaster as anyone else, you're also at risk from mistakes made by your neighbors in attached apartments.

> **Business Continuity**: If you're a business owner, and especially if you operate a home-based business, see which of your basic insurance policies carry a "business continuity" or business insurance rider. This may have to be a separate policy. Your insurance providers can give you sufficient information on this topic.

> **Animal Insurance**: You can get insurance on farm animals and livestock, which is more of a business continuity policy, or you can get medical insurance on your pets. If you have numerous pets, you might want to consider this type of policy anyway to offset what you already know to be considerable veterinary costs. Talk to both your vet and your insurance provider. (We've listed some providers under the "Evac Atlas" portion of the "**Evacuation**" section.)

> Find out **how** and **when** your insurance companies will pay when damage occurs, and **what kind of proof of damage or proof of loss** they'll need. Find out what kind of property is **excluded** from coverage. For example, some policies will only cover so much on specific types of items (such as jewelry or electronics) and require a rider or amendment to offer full coverage.

> **Helpful insurance hint**: Start documenting your valuables. Use our "**Household Inventory**" form from the "**Appendix**" and record the serial number, model number, manufacturer, etc. from each of your possessions. For items not already numbered and identified, **engrave your name, phone number,** or other code on them in an inconspicuous spot (your local Law Enforcement office might lend you an engraver). For non-engravables, use a small **Sharpie ® or other indelible marker** or use a label maker that makes stick-on labels and apply them in hidden spots such as inside electronics. Some companies also make a "**UV ID Kit**" that uses invisible, indelible inks that only show up under UV light. (As a side note, engraving makes property ID easier when **searching for belongings** after hurricane or tornado destruction, or after mass decontamination efforts following a BioChemical attack.) Also, take **photographs <u>and</u> video footage of <u>ALL</u> your possessions.** The small items you wouldn't inventory will add up when it comes time to shell out replacement money. Photo everything including every room, all closet interiors, your attic, storage rooms, the exterior of your home, landscaping, outbuildings, and vehicles (you can check out cameras from your local library). Next, photocopy or scan as many **receipts** (or cancelled checks and credit card statements) as you can. More on this under "**Info Pack**." In addition to insurance, these records will help you with claiming **tax deductions** for your disaster losses.

> ● We found an interesting **UV ID Kit** online at http://www.smarthome.com/7319.html. While you're there, bookmark the main **Smarthome** site.

> **Legal Insurance**: Legal insurance will pay for the use of an attorney just as medical insurance pays for a doctor. It's always nice to have one of these policies to help deal with occasional legal needs such as making wills, power of attorney (which you need), legal letters, court representation, etc. More importantly, as those who survived Hurricane Andrew will attest, you may need this insurance as a backup weapon to collect what you are rightfully due from your other insurance policies. It's unfortunate, but that's the reality of things, so BE PREPARED. Several plans from quite a few companies are available at very reasonable prices.

> **"Self Insurance"**: The greatest safeguards are going to come from the efforts you make now, since in the case of disaster prep, an ounce of prevention is worth far more than a pound of cure. Regarding the repair and rebuilding part of a disaster's aftermath, one method of "self insurance" is to have as many assets at your disposable as possible. A savings account and maybe an empty credit card or two would be really useful, as well as any skills you've developed that lend themselves to repair and rebuilding. The networking and connections you make as discussed in "**Cooperation**" will also help, as there's nothing more effective than neighbor helping neighbor.

> **Cash and account management**: Another form of "self insurance" is the way you manage some of your savings and income, or some of your regular accounts. The suggestion is for you to do one of two things, or a combo of the two. One is for you to open a separate interest-bearing bank account in which you keep at least one (preferably two) months worth of payments for all of your bills from mortgage or rent all the way down through average grocery expenses. Naturally this would include car payments, all your utility bills, etc. The second option would be to prepay some of your more important accounts during periods of heightened terror alert, or during your area's destructive weather season if any. We recommend the account option. The last thing you need to worry about in the middle of a disaster is whether or not your bills are paid. (**Note**: Do not dip into any of your emergency cash to pay bills. Save the cash for the unexpected needs and consider your bills as important, but secondary to your more immediate emergency assets.)

> **Barterable goods and skills**. There's no guarantee that any of the above mentioned policies, programs, etc. will be around after a major attack or disaster having severe economic repercussions. Therefore, to serve you in the short run, you need to have as many other "assets" at your disposal as possible. Chief among these, in an immediate post-disaster setting, are barterable goods and skills. The more you have to trade in the way of items or abilities, the better off you'll be. Cash may still be valuable, checks and credit cards may be useless, and current "investment" items such as gold and jewelry are actually luxury items that depend on functional banks and an operational economy to have any sort of true value. In a disaster, gold may only be as valuable in the short run as whatever you can melt it down and make out of it. Barterable *items* will be covered under "**Evacuation**." As for barterable skills, read back through our "Basic Training" section. Any skill or ability from that list would be useful. Heading this list would be medical or dental skills, and following those would be automotive repair, electrical work, small engine repair (generators), etc.

Services and Agencies

The following "**Services and Agencies**" list is extremely important, for after a disaster, these agencies and organizations will provide you with both immediate and long term relief assistance to include emergency money, supplies, and goods to start rebuilding.

> **FEMA:** In a federally declared disaster area, FEMA will be in charge. The personal information you gather for your "**Info Pack**" will be *extremely* useful in dealing with FEMA. Their hotline is 1-800-462-9029. General information can be obtained from: FEMA, 500 C Street SW, Washington, D.C. 20472, Phone: (202) 566-1600, 202-646-2500, or their website at http://www.fema.gov. The disaster assistance number is 1-800-621-FEMA (3362). When contacting FEMA, especially when planning your evac destinations and routes, ask where they might set up DRCs, or Disaster Recovery Centers, in your area in reaction to different scenarios.

> **_EMA:** Your state Emergency Management Agency. In a any emergency, most states' emergency management agencies will be on scene in some capacity or another. Look up your state's emergency management office information, either through your phone book's "Government Blue Pages," or the online listing at http://www.fema.gov/fema/statedr.shtm. You can also find more information about your state office through the National Emergency Management Association. NEMA is the professional association of state emergency management directors, c/o Council of State Governments, P.O. Box 11910, Lexington, KY 40578, (859) 244-8000, (859) 244-8239, http://www.nemaweb.org. Find your local office or representative, and record both the state and local contact info along with all the other requested information on your "**Important Contacts**" form located in the "**Appendix**."

> **Red Cross:** Know and record the number to your local Red Cross office, learn who in that office is in charge of what, and put their contact info on your "Important Contacts" form. Go to http://www.redcross.org. Their toll free number is 1-800-529-5558.

> **The United Way**: The United Way will certainly be one of the entities putting money into rebuilding efforts following in the wake of a disaster or terrorist incident. More on The United Way can be found at: http://national.unitedway.org/myuw, 95 M Street, Suite 304-A, SW Washington, DC 20024, Phone: 202-488-2070 (in most areas you can call by dialing 211), Fax: 202-488-2099. One of their programs is the Emergency Food and Shelter Program, as seen at http://www.efsp.unitedway.org.

> **Salvation Army:** In many cases, the Salvation Army will offer assistance. If nothing else, they are a *very* good source of low price pre-owned clothing and other goods (as is **Goodwill**) to give you a little economical resupply while rebuilding efforts are under way (don't forget thrift stores). Contact them at: http://www.salvationarmyusa.org, or The Salvation Army National Headquarters, 615 Slaters Lane, P.O. Box 269, Alexandria, VA 22313. Your phone book will have their local office.

> **Other National Relief Organizations:** A really good listing can be found on the "**Disaster Center**" website at: http://www.disastercenter.com/agency.htm, and a few more found at: http://www.disasterrelief.org.

> **The National Guard:** The National Guard, and in some areas **legitimate, official state militias** (see http://www.sgaus.org) will be on hand to keep order and bring in relief supplies. Find out who in your area is associated with either. Look up National Guard in your phone book or online at http://www.arng.army.mil, or see http://www.washingtonguard.com/statgard.html. To see if your state has an official militia (the state-appointed ones, not the fringe groups calling themselves militias) go to the **SGAUS** site or contact the State Guard Association of the United States at, P. O. Box 1416, Fayetteville, GA 30214-1416, Phone: 770-460-1215. (To see if a militia group in your area is on a watch list, go to http://www.militia-watchdog.org/m1.htm.)

> **Your local fire department:** Call them now and ask about post-disaster programs they offer and what kind of state, local, and private agencies provide help in your area. Really want to make a good investment? Volunteer some time to help either the Fire Department or one of these other agencies (and get to know the

people). Also, if you or someone you know has special medical or physical needs and may need assistance in an emergency, call your local emergency first responders now and register this person.

> **Internal Revenue Service:** In the event of property losses you will be entitled to take certain deductions. Your local IRS office can give you the up-to-date information concerning the declaration of losses. Find them at http://www.irs.gov, or call 800-829-1040.

> **Health and Human Services (local):** Call your local health department or **at least record their contact information**. In the event of a quarantine, or mass disaster, they'll be one of the key players in providing public medical care and addressing health concerns. They'll also be a point of contact for knowing when you can come out of isolation, or return to an area you may have evacuated. Additionally, there's no way to know exactly how intense stress will affect you or your family members. Your local health services will also be your contact point for mental health issues. Contact info for the federal **Health and Human Services,** is site at: http://www.hhs.gov. Direct contact is at: The **U.S. Department of Health and Human Services**, 200 Independence Avenue, S.W., Washington, D.C. 20201, Telephone: 202-619-0257, Toll Free: 1-877-696-6775. Additional contact information for state and local health services can be found through the website for "State Public Health" at: http://www.statepublichealth.org/index.php.

> **Local Food Bank:** Know the contact information for the food bank and other local relief organizations not run by any particular government entity. The best way to find such local offices and services is to call a local homeless mission and ask what programs and offices would come to their aid in the wake of a disaster. Though food banks are usually behind-the-scenes players in emergencies, it's best you know all your local options and network with the people involved.

> **Humane Society:** Relief and rebuilding affects more than humans. See which services will be available for your pets or livestock. Plan now by contacting your local **Department of Agriculture office, Farm Bureau or County Extension office, 4-H Club, FFA club, your vet, local animal shelter, or major pet store**. More on this under "**Evacuation.**" You can contact the **Humane Society** at: http://www.hsus.org/ace/352, The Humane Society of the United States, 2100 L Street, NW, Washington DC 20037, Phone: 202-452-1100, or find their local office in your phone book. List these numbers in your "**Important Contacts**" list.

> **National Volunteer Groups:** Some nationally networked volunteer civil-defense organizations may have an office in your area. You may be able to find contact information through "**Citizen Corps**" at: http://www.citizencorps.gov. (Citizen Corps is coordinated through **FEMA** at: FEMA, 500 C Street, SW Washington, D.C. 20472, Phone: 202-566-1600, and 202-646-2500 .)

> **Local Groups:** This will be covered much more extensively under "**Cooperation**." In fact, the concept of neighbors helping neighbors rebuild is what the "Cooperation" section is based on. You need to know (or better yet, become a part of) any local volunteer group that is organized for the purpose of disaster prep and/or aftermath relief.

> **Local Repair and Service companies.** Most disasters will have a destructive element to them, and you may need some sort of repair or cleanup, whether it's to your home, vehicle, or property. Though some insurance companies will have set providers, others won't. Be sure to record contact info for various companies in your area. If your yellow-pages is small enough, pack a copy in the "**Office Pack**" of your "**Bugout Kit**" (as discussed under "**Evacuation**"), or photocopy sections based on the checklist below. You can also use your address book or our "**Important Contacts**" forms. You'll want to do a little consumer research on various companies (making sure any potential choices are reputable, licensed, bonded, etc.) and record contact info now, since **the last thing you need after a disaster is the headache of figuring out who to call and whether or not they're qualified or capable of doing what you need. You'll also want to be able to contact these services early as there will be a waiting list.**

Local Repair and Service Companies

Put all contact info of selected companies on your "Important Contacts" form in your "Info Pack."

☐ Specific "Disaster Cleanup"	☐ Construction / Contracting	☐ General Carpentry
☐ Masonry / Concrete	☐ Plumbing	☐ Electrical
☐ Sheetrock / Plaster	☐ Appliance Repair	☐ Appliance Dealer
☐ Roofing	☐ Tree / limb removal	☐ Trash Service or Debris Hauling
☐ Landscaping	☐ Carpet Sales or Cleaning	☐ Auto Towing and Repair
☐ Equipment and Tool Rental	☐ Heavy equipment service or rental	☐ Hardware and Home Stores
☐ "Crime Scene Cleanup" or other BioHazard / HazMat cleanup specialists. (← VERY important.)		

Post-Disaster Grants

➢ **FEMA** may help you with disaster recovery loans and grants directly or through your state Emergency Management Agency. For grant information, contact FEMA directly by mail at FEMA, 500 C Street SW, Washington, D.C. 20472, or by calling the grant office at 202-646-3685. (Incidentally, a copy of the **post-disaster relief grant process** information can be downloaded at: http://www.fema.gov/pdf/rrr/dec_proc.pdf.)

➢ Other post-disaster financial aid programs might be found through the "**Catalog of Federal Domestic Assistance**" site. Info at: http://www.cfda.gov, with direct agency contact found in the extensive government directory "**FirstGov**" at http://www.firstgov.gov.

➢ The Dept. of Agriculture offers a **loan program for farm owners** to recover after a natural disaster. Contact them online at http://www.usda.gov, or directly at Loan Making Division, US Department of Agriculture, Farm Service Agency, AG Box 0520, Washington, DC 20250, 202-720-1632.

➢ The Dept. of Agriculture offers financial help for the **loss or damage to a home** after a natural disaster. Contact them at http://www.usda.gov, or for this program, at Director-Single Family Housing Processing Division, Rural Housing Service, Department of Agriculture, Washington, DC 20250, 202-720-1474.

➢ For disaster grants regarding crop loss, contact the Farm Service Agency of the USDA. Their online disaster page is http://disaster.fsa.usda.gov/fsa.asp, and their main office can be reached at U.S. Department of Agriculture, Farm Service Agency, 1400 Independence Ave., S.W., STOP 0506, Washington, DC 20250-0506, Telephone: (202) 720-7809.

➢ The **USDA** also offers a Rural Development disaster relief information page, online at http://www.rurdev.usda.gov/rd/disasters/disassistance.html.

➢ The government's new "Disaster Help" site offers a disaster grants listing. You can see it online at: https://disasterhelp.gov/portal/jhtml/index.jhtml. As this is a rather new site, you may have to go to the homepage at http://disasterhelp.gov.

➢ **HUD** offers loans to help disaster victims secure new homes. Contact information for your local office may be found in your phone book's blue government pages or by contacting Housing and Urban Development online at http://www.hud.gov, or directly at U.S. Department of Housing and Urban Development, 451 7th Street S.W., Washington, DC 20410, Phone: (202) 708-1112 TTY: (202) 708-1455.

➢ The **SBA** offers loans to certain disaster victims as well. Contact them online at http://www.sba.gov, the direct link at http://www.sba.gov/disaster_recov/index.html, or directly at Office of Disaster Assistance, Small Business Administration, 400 3d St., SW, Washington, DC 20416, 202-205-6734.

➢ In addition to these direct sources, you should contact all your **legislative representatives** at both the state and federal level to see what kinds of assistance programs are available. You'll find state information in your phone book's blue government pages, and you can find your federal reps by calling the congressional switchboard at 800-839-5276, or 202-224-3121. These sites will also help (visit each):

❑ http://www.congress.org   ❑ http://www.house.gov
❑ http://www.senate.gov   ❑ http://thomas.loc.gov
❑ http://www.vote-smart.org   ❑ http://www.firstgov.gov/Contact/Elected.shtml

While we're on the subject of seeking and **collecting specific information**...

Supporting Documentation – The "Info Pack"

We all know the old saying, "no job is complete until the paperwork is done." Our whole society revolves around paper and records, and relief and rebuilding efforts that follow a disaster will be an extreme exercise in red tape and bureaucracy. Gathering and protecting the below items in advance can make you or break you in dealing with the procedures that will be inflicted upon you by the above agencies.

The "**Info Pack**" checklist below concerns itself with organizing and storing the documentation necessary to post-disaster procedures. And, as redundancy is our friend in emergency planning, the checklist will also suggest various locations to keep duplicate "Info Pack" copies.

Once you've gathered all your important documents and paperwork, list all items about to be stored. You can make a simple list, or use our "**Stored Documents List**" found in the "**Appendix**." You'll want to have a list with you (in your "Info Pack") showing where all originals are stored, but more importantly, their **valid dates**. This way, you can check and update your documents periodically, making sure that each one is current and valid where applicable. You'll want to make sure all are updated during higher terror alerts or at the beginning of your annual severe weather season.

WARNING: BE VERY CAREFUL HOW YOU STORE YOUR SENSITIVE PERSONAL INFORMATION!

Supporting Documentation – The "Info Pack"

("←" Items have discussion following the checklist. ** Items have forms located in the **"Appendix."**)
(For items marked **(b)** make sure you include certified copies in your **Bugout Kit**.)

Primary Legal Documents

❑ Property Deeds	❑ Wills and/or Living Wills **(b)**	❑ Birth Certificates **(b)**
❑ Marriage Certificates	❑ Death Certificates	❑ Power of Attorney(s) ← **(b)**
❑ Original Social Security Cards	❑ Contracts and/or Decrees ←	❑ Vehicle Titles / Tags ←
❑ Stocks and Bonds	❑ Insurance Policies	❑ Latest Tax Return

Accounts and Services

❑ "Important Contacts" ** **(b)**	❑ Bank Statements	❑ Service Agreements
❑ Credit card statements	❑ Important Warranties	❑ Duplicate Credit Cards ← **(b)**

Property and Possessions

❑ Property photos / videos ←	❑ Household Inventory Form**	❑ Important Receipts
❑ Backup computer disks ← **(b)**	❑ Family Photo Album Copies ←	❑ Family Member Photos ← **(b)**

Personal Data

❑ Passport	❑ Photocopy Address Book ←**(b)**	❑ Photocopy wallet items ← **(b)**
❑ Secondary ID ← **(b)**	❑ Household Data Form** **(b)**	❑ Family Member Data ** **(b)**
❑ Pet Data Forms ** **(b)**	❑ Medical Records Copies ← **(b)**	❑ Emergency Prescriptions ←**(b)**

Documentation Storage Locations

Consider storing original copies at:

❑ Safe Deposit Box at Bank ←	❑ Fireproof Safe in Home ←	❑ With a trusted attorney

Store "Certified as True" copies at: ←

❑ The location not used above	❑ Bugout kit – Home ←	❑ Bugout kit – Vehicle ←

Store electronic copies on disk at: (Note: Electronic storage should <u>always</u> have paper backup.)

❑ Each of the above	❑ Secure location at work	❑ Storage trunk ←

Discussion Items (The "←" items from above)

Supporting Documentation

> ➢ **Powers of Attorney**: Since we're prepping you for the worst, we'll suggest that you have mutual Powers of Attorney and Medical Powers of Attorney in place for the heads of household should something happen. This will make it far easier for the other partner or other family member to secure proper medical attention and/or access necessary family assets.

> ➢ **Contracts and/or Decrees**: The more important the document, the more you'll want to safely store it and have a certified-as-true copy in your possession. Use your judgement as to which are more important. An example of some documents that might prove necessary in an evacuation are **divorce decrees, custody papers, guardianship documents**, etc.

> ➢ **Vehicle Title and Tags**: Naturally, you'd keep your <u>original</u> vehicle title in a safe place. You should get a **"certified as true"** copy of your title and tag info should someone steal your tag, or you need to verify ownership. It's always best to have as much documentation as you can.

> ➢ **Duplicate Credit Cards**: Though not a direct "documentation" item, duplicate credit cards will be an indispensable item to keep in your protected "Info Pack." Not only should you have duplicates of your regular credit cards, you should have one or two original cards (stored here and not carried by you) with a zero balance that you've set aside for such emergencies. Along with these credit cards, try to store as much **cash** as you can, as well as an **extra checkbook**. More on this under **"Evacuation."** While we're discussing emergency assets, we'll remind you to keep a copy of your automotive **roadside assistance club** paperwork in your Info Pack and make sure this paperwork lists all benefits of membership as well as club contact info. **Hint:** You should get in the habit of saving your current pay stubs and store them where

they would be accessible in an evacuation. If you're on the road, strapped for cash, and need to negotiate a loan or other cash source, having your proof of current employment on hand may help. The problem with this idea is that you'd have to constantly change your paystubs in order to have the current ones on hand. Old stubs are not likely to do you much good in this situation.

➢ **Property photos / videos**: Very necessary to have even if your insurance agent tells you they're not. When you take photos and videos, make sure to have **duplicates** of both and store them in separate locations. You don't want to hand over your one and only original set to an insurance company who's only got *their* best interests at heart. Make **2 sets of photos** and keep the negatives stored in your Safe Deposit Box. Also make 2 copies of any video tapes you've made of your property and keep them in different locations. **Note**: ALWAYS have **hardcopy** photos, even if you also shoot video, and make sure they're done with <u>film</u> and not digital photos, as sometimes digital pictures are called into question since they're easy to alter. Having original negatives gives you better proof that your pictures are genuine and accurate. Using digital pics as a backup is fine. We also recommend you include a "**personal identifier**," such as a driver's license, in some pictures to help prove the photographed items are yours. A good example is jewelry. When taking pictures of your jewelry spread out on a table, make sure your driver's license is in the picture too (show your photo on the license, but cover sensitive info). With other items, show the serial number or engraved identification number on the object. (You'll also want to carry copies of any **appraisals** you had made.)

➢ **Backup computer disks**: Intellectual property and other creations you've made on computer (both at home and at work) are just as valuable as anything else. Back up your computer at least monthly and change out disk copies in your "Info Pack" quarterly. **Important note**: All other items listed here should be primarily stored in <u>hardcopy</u> form as there's no guarantee you'll have access to a computer or even electricity. Having *secondary* backups on disk is fine.

➢ **Family photo album copies**: Old family photos are priceless and they're usually the first things destroyed in a disaster, and the last things grabbed in a hurried evacuation. Though not a documentation item, we realize the sentimental value of such heirloom items. While you're scanning, copying, and storing other documents we suggest you have all your old photos copied and printed, as well as scanned and placed on a CD. Lots of places will do it for you; office supply stores, copy centers, camera and photo centers, etc. You can also make hardcopy duplicates and leave those out for viewing while you safely and securely store the originals and their negatives. Speaking of **family photos**, you'll want at least **4 detailed photos of each family member** (hardcopy and not electronic) sealed, protected, and stored in your "Bugout kit." Should you become separated, you'll want to have the photos on hand to give to Law Enforcement and others to help you locate the missing. See our "**Family Member Data**" forms in the "**Appendix.**"

➢ **Photocopy address book**: During or after a disaster, it'll be extremely important to keep in touch with a variety of people. Since you can't count on being able to remember all necessary contact info once the adrenaline and stress hit, you should make copies of all important phone numbers, addresses, etc. You'll want to be able to stay in touch with your immediate family members, your rendezvous/contact people, as well as all the services and agencies listed earlier. You can photocopy your address book or use our "**Important Contacts**" forms in the "**Appendix.**"

➢ **Photocopy wallet items**: Find a **Notary Public**. Take all the IDs, credit cards, and everything else out of your wallet and have them photocopied front and back (as many as you can get together on each sheet of paper), and have the sheets notarized as being "true copies of the original." Having this is a good way to track account and document numbers, customer service numbers, etc., and in *some* cases *may* be acceptable as ID in case you lost your wallet. Keep these sheets in your "Bugout Kit." **Hint**: If you have too many cards or wallet items to fit on one page, make sure your driver's license or other official ID is **copied again** on the other sheets so it will help identify those items as being yours. **Hint #2:** Also have your original **Social Security card** copied along with these items even if you don't normally carry it in your wallet. In a situation where you'd need these copies, you'll need all the personal ID info you can get your hands on. Also, BE VERY CAREFUL HOW YOU STORE THIS INFO AS IT WOULD BE AN ID THIEF'S DREAM.

➢ **Medical Records**: Make sure you have copies of available medical records, especially those pertaining to conditions, allergies, current medications, etc. Also be sure these records have your current **immunization** records. We'll create our own duplicate records later.

➢ **Secondary ID**: ID is absolutely critical. Relief organizations will want you to prove you're who you say you are and that you live in the disaster area, Law Enforcement will want you to **prove you're a resident** and not a looter, and so on. Therefore, everyone should get a **passport** even if they don't intend to travel. It's a universally accepted photo ID. Also, many states will issue a "**non-driver's license**" ID card for the purpose of citizens having an official photo ID. Call your local law enforcement or Department of Motor Vehicles to see what's available. As a third option, if you are involved in a profession that either requires or offers **state licensing**, you should be sure to carry your license. Though not officially an "ID" it will be a government issued document with your name, and possibly your photograph on it. You might also pack away old college IDs, expired military IDs, or other cards. Similarly you should safely store your original **Social Security card** or an official (or "certified as true") copy. **Hint**: As an aid to ID, **label all clothing items**, especially children's clothing, with full name. Jewelry engraved with a full name is also useful in helping identify a person who may not be able to communicate. Here are some sources for passport info:

- **State Department's Passport** site: http://travel.state.gov/passport_services.html. U.S. Department of State, 2201 C Street NW, Washington, DC 20520, Main #: 202-647-4000, Travelers' Hotline: 202-647-522. An alternate contact point is: http://travel.state.gov/npicinfo.html, or at 1-877-4USA-PPT (877-487-2778), and TDD/TTY of 1-888-874-7793.
- **Passport office locator**: http://iafdb.travel.state.gov.
- **Passport Agency listing**: http://travel.state.gov/agencies_list.html.
- **"24 Hour Passports"** (private service): http://www.passportsandvisas.com, 1-800-860-8610.
- Another private service can be found at: http://www.americanpassport.com, 1-800-841-6778.
- Many **post offices** will carry passport applications and associated services.

➢ **Emergency Prescriptions**: Ideally, you'd be able to stock and pack enough of your current medications to get you through a crisis. As this is not always the case, and as many meds may be quantity limited and/or readily perishable, see if your doctor can write you an **emergency prescription to keep on hand**. You wouldn't fill the prescription, you'd keep it stored and fill it when necessary. List your doctor and pharmacist on your "**Important Contacts**" sheet.

➢ **"Certified as True" Copies**: We can't say it enough; in emergency preparations, redundancy is our friend and this certainly applies to important documentation. Trouble is, you only get one original. First thing to do is ask the document's provider for certified copies. Next, gather your **original documents**, and take them to a **Notary Public** (when you take your wallet contents). Have them **photocopied** and **notarized** as being true copies of legitimate originals. While this does not guarantee anything, in some cases it will carry more weight than a photocopy alone.

Document Storage Locations

➢ **Safe Deposit Box**: All in all, the safe deposit box is probably the best little cache you could ever have. It's secure, climate controlled, and relatively accessible. Banks may burn, get flooded, or damaged in storms, but that vault's going to be intact (get a bank on high ground if you're in a flood prone area). Besides, and this is VERY IMPORTANT, the information you gather for your various forms is VERY sensitive. Guard it well. (See the "Communication" section for suggestions on how to **code** sensitive information.) Another suggestion is, if your primary evacuation destination happens to be a location you visit often, get a safe deposit box in that town. Half of evacuation is coming home to rebuild, and having pertinent documentation at your evacuation destination will let you get your recoup efforts started that much sooner.

➢ **Fire-Proof Safe**: Most office supply stores will carry small, inexpensive, fire-proof and lockable storage safes for home and office use. These make good secondary storage containers for important documentation and other keepsake items. Be sure to keep these relatively hidden in case of burglary (maybe even chained to a wall stud in back of a closet), high off the floor if you live in a flood-prone area, and keep in mind that even though fire-proof, things inside may melt or bake due to the heat of a fire.

➢ **The "Bugout" Kit - Home**: Under "**Evacuation**" we'll discuss creating a "Bugout Kit," and within it, an "**Office Pack**" which will contain your "Info Pack," or at least copies. Your supporting documentation is important enough to have a copy with you in almost any situation.

➢ **Bug Out Kit – Vehicle**: Be careful how much personal or sensitive information you store in your vehicle as it could be stolen (your kit or the whole vehicle), and the recorded information used in an identity theft scheme against you. Remember to write sensitive info in code. One reason you DO want to have as much info as possible stored in your vehicle is the fact that there is no guarantee when or where an emergency will hit, and you may not be able to go home or to the bank to pick up your important documentation.

➢ **Storage Trunk**: We'll discuss this in more detail under "**Evacuation**," but one of the best things you could do is have emergency and secondary supplies stored in a number of locations. This can be either primary emergency reaction equipment, or simply a storage trunk full of extra food, clothing, and comfort items stored at your planned evacuation destination. In any event, a copy of your "Info Pack" should be among the items you store, **provided the location is secure**.

Filling out our Data Forms

Now that we've mentioned them a few times, we might as well discuss, and have you complete, the various **data forms** in the "**Appendix.**"

Before you go through these and start filling them out, take a quick look through the "Communication" section and find the part about **codes** you can use to protect some of the **sensitive information** you'll be putting on these forms. Remember, these pages are going with you in an evacuation and there's no guarantee they'll be secure. You'll want to do all you can to make sure things like Social Security Numbers, birthdates, mother's maiden name, etc., don't get into the wrong hands. The good news is, you don't have to code everything, just the **sensitive** stuff.

1. "**Important Contacts**": We suggested you photocopy the contents of your address book, or print a copy of your computer address book, and to pack a copy in your "Bugout Kit" and keep one in your vehicle. It's always great to have a backup of your contacts. However, address books are usually arranged **alphabetically**, and you have *everybody* listed. With our "Important Contacts" pages, we've organized them by **need** or **category**. This allows you to **isolate** your **emergency contacts** from your everyday. We've organized these in the following categories:

   ➤ **Quick List:** This is a single-page, easily readable listing of your most important emergency numbers. Fill this out and **post a copy** by your **phone**, one in your **safe room**, give one to each family member, and use another as the **cover sheet** for the rest of the Important Contact pages.

   ➤ **One - Emergency Contacts**: These are the people who would act as your **emergency message relays**. Pick both local and out-of-town friends, relatives, or volunteer group members, that you can call to relay messages and gather information in case there's no local direct contact between family or group members. (You may get a long distance line out even while local lines are overloaded, and your out-of-town contacts are less likely to be affected by your local emergency.) Make sure everyone has a copy of this page.

   ➤ **Two – Rendezvous**: In the "Evacuation" section we'll ask you to pick **rendezvous points** where your family could meet in an emergency. This list records the contact information for the location. If commo lines are open you might find out if the location is open, what the conditions are, and in *some* cases you might be able to ask if your family member is already there.

   ➤ **Three – Medical**: Record contact information for all your doctors, dentists, veterinarians, pharmacists, etc. As with any of these sheets, photocopy extras as you need them, or call up the wordprocessor files on the enclosed **CD** and fill them in on screen to print. **Hint**: Though you may save the files in softcopy format, you should always have a **hardcopy** printout as there's no guarantee you'll have access to a computer or that the electricity will be on.

   ➤ **Four – Local Officials**: This page is important for a variety of reasons. **One**, a few areas of the country still aren't on the 911 system. **Two**, these will be some of your info sources for gathering news on the extent and after-effects of an emergency or disaster hitting your home area, and you may need to call them from out of town. **Three**, if you're part of a volunteer group, you'll want close contact with these offices. Along with Law Enforcement, Hospitals, etc., list local **radio and television** stations. You might need clarification on news items you missed, and in some cases, you might be giving updates to them.

   ➤ **Five – Relief Services**: We asked you to become familiar with everyone from FEMA all the way down through local charity organizations. If the emergency is large enough, they'll all be playing a role, and they'll all have info you'll need at some time or other, and vice versa.

   ➤ **Six – Family and Friends**: Though all family and friends are important, some will play a greater role in an emergency than others. List them here to keep them separate from your general address book. Make a note as to why they were listed in case it's something you might forget.

   ➤ **Seven – Repair and Rebuilding**: Above we asked you to research repair and rebuilding services that might come in handy when cleaning up after a disaster. Record their info here.

   ➤ **Professional Contact**: Just as you may need doctors and other medical contacts, you'll also need your **attorney, insurance agent**, and other **professionals**. This page is also used as an attachment for your "Household Data" and other "Family Member Data" forms.

   ➤ **General Contact**: In case you didn't want to photocopy your whole address book, this page was included for general friends, acquaintances, and other assorted contacts.

2. "**Find Me**": While you're busy gathering contact data for everyone else in your life, remember that other people will need to know how to find <u>you</u> in an emergency. This sheet will allow you to give these folks numerous contact sources for you. You can also use this page as a blank to give out to others if you really need to find them. We suggest you swap "Find Me" pages with each of your emergency contacts. **Hint**: Fill one of these out and leave it in one of your brightly colored emergency envelopes and prominently displayed by your phone or on your refrigerator with the words "How to find us in an emergency" clearly written across the front. Make a note of this envelope's location on your "Dear First Responder" cards (found in the "**Appendix**"). **Hint # 2**: When a family

member is temporarily away, such as on a business trip, on military duty or TDY, in the hospital, kids on a field trip, whatever, be sure to **record their current location** and the date on your "Daily Activity Journal" that we'll discuss later under "Orange Alert." **Hint # 3**: Anytime you take a family member somewhere new, **collect a business card** if the location has one. For example, did you just drop the dog off at a new groomer's? Did you leave a child at a friend's birthday party at a restaurant? What would you do if an emergency hit a few minutes later? How would you get in touch with the establishment you left? You might get only one call and you don't want to waste it calling 411.

3. "**Notify in Case of Emergency**": Expanding on the everyday wallet-card, this is a four-panel, fold-up card, about the size of a business card, that will hold much more information. Make sure each member of your family carries one. On this card, you'll list two contacts. One should be an immediate family member such as a parent or spouse. The other should be someone **not living with you**, and preferably your main contact person from your "Important Contacts – One" sheet. Also, since different people have different emergency needs, we've given you two versions of this card to choose from. The second is an eight-panel card. Pick the one that serves you best.

4. "**Open Only in Case of Emergency**": Since your non-resident "Notify in Case of Emergency" person will be helping you in an emergency, they need to **know what to do**. This is the cover sheet for their instructions. Fill it out now, add instructions of your own, attach the recommended forms, and seal it in one of your brightly colored envelopes (that we'll discuss under "Communication"). Be sure you know **where they put the envelope for storage** because they may forget. It happens. Once you know where they've put the envelope, make a note on your "Notify in Case of Emergency" card. You'll see a blank saying "*Remind this person their instructions are located:___*". If you're incapacitated, the First Responder that calls your emergency contact can use the note to remind them where they put their special instructions. Forgetting happens a lot. When in doubt, write it out.

5. "**Household Data**": The purpose of this multi-page document is to record the information you may need to put your life back together after a disaster, or access services or assets in an emergency. This set will ask you for property info, household financial info, and services and people that are associated with the family as a whole. People and services associated only with *individual* family members will be listed on the "Family Member" pages.

6. "**Family Member Data**": These pages serve a variety of purposes. Some are useful if your family is separated in an emergency and you have to file a "missing persons" report. Others record pertinent information that may not only be important *about* a family member, but *to* them as well. When an emergency hits and adrenaline surges, you can't remember everything, and some information is too vital to forget. When in doubt, write it out.

> **One – Description**: Emergencies, especially those causing an evacuation, may mean the separation of family units. More severe emergencies may see injury or worse to some family members and this may prevent them from communicating or being identified. This "description" page is like a "Child ID Kit" for <u>all</u> family members. It provides a fairly thorough physical description and ties into the next page of "Identifiers."

> **Two – Identifiers**: This carries the "Description" page to the next level. The Description page can be copied and handed off to as many people as you'd like, as it will help searchers, and doesn't contain sensitive info. This *Identifiers* page confirms the identity of the person you're looking for as it will be carrying fingerprint, dental record, and DNA information. To make your own "Dental Fingerprint," as this page will ask for, find a piece of styrofoam that's smooth on all surfaces and is about the size of an audio cassette tape. Put it into your mouth and bite down on it just hard enough to leave a full bite-mark impression of all upper and lower teeth. That's all there is to it. Wrap it up so the impressions won't be damaged.

> **Three – Sensitive Info**: Each family member is going to have personal and important information associated with them that may be crucial at some point or another in an emergency. This info might be important *to* the family member, or to others *about* this person. For example, both parents may be ill or injured, and an older child might need to access important information to help them or the rest of the family.

> **Four – Medical**: Where under "Important Contacts" we had medical *professionals*, here we have medical *information*. This is essentially your own medical records that you create yourself. This kind of information might be extremely important if you have to seek treatment from a new doctor in an emergency situation where you were incapacitated and unable to communicate, or if communication lines were down and your regular physician and/or full medical records were inaccessible.

> **Five – Financial**: Just as the "Household Data" form recorded pertinent account and banking information for the *household*, this page does it for the *individual*. Here you'd list accounts and other info associated with just this individual family member.

> **Six – Student**: If you have a minor child in school, this may well become one of your most important forms as it will hold all manner of contact info for your child's school, teachers, and for the school's chosen emergency evacuation destination.

> **Seven – Pets**: Pet's are family members too, right? You'll need this to secure boarding or care for your pet in an evacuation, and you may need it to help find your pet if lost. You should also type out all this info on a "micro data card," that we'll describe below, and tape it around your pet's collar or leash, and on their carrier. Also, ask your vet to keep a copy of this on file in case you need it and can't access any of your copies.

7. **"Household Inventory"**: We asked you to photo and video all your possessions and to engrave an ID number on them if there wasn't one already. This form will allow you to list your more important possessions for purposes of insurance claims after a loss. Keep this form along with receipts and photo / video copies in your safe deposit box, and a copy in your fireproof safe at home. **Hint**: Do some online shopping or refer to current catalogs, etc. to gauge current replacement value or worth of your belongings if you don't remember the cost.

8. **Update Cards**: These cards will be a part of the "Office Pack" of your "Bugout Kit," and will help you update others on your current status, location, etc. after a disaster or evacuation. They also make easy-to-fill-out and recognizable cards to leave on an emergency shelter's communication bulletin board in case you leave the shelter for whatever reason, yet you know there are others looking for you. Print a few pages of these cards, cut them out, and put them in your Bugout Kit. You might consider using brightly colored paper that matches your "emergency" envelopes so anyone seeing the note will know it's from you. Having these preprinted saves writing time and ensures detail.

9. **Change of Address Cards**: Many scenarios can see you evacuating and relocating for an indefinite period. Though things in general will be in a bit of turmoil, the US Mail will be back on track in fairly short order. If you're in a situation where you'll be at a new location for four weeks or longer, you may want to put in either a temporary or permanent Change of Address notice with the Post Office (a copy of that form is also in the "**Appendix**"). Along with notice to them, you'll want to send a direct notice to each of the companies, services, and accounts you deal with, as well as the important people in your life. Use these cards in addition to your official change of address notice. Print out enough cards to send to everyone listed on all your various data sheets and pack these in your "Bugout Kit." Be sure that among your "Important Contacts" pages and your various "Family Member" and "Household Data" sheets, that you list everyone to whom you would possibly send a change of address notice.

As an interesting **hint**, and we'll touch on this more later, you can create "**micro data cards**" by using your wordprocessor to print very small versions of the info requested on the data forms. You'd print out very simplistic versions of the forms (just data and not the form part), with everything in 6-point font. You could print a lot of info on **business-card size cards, or on 3 x 5 cards**, laminate them for protection, and keep them in your wallet and in your various kits. Having smaller pages makes it easier to carry a lot of useful information in your wallet where you wouldn't have room for folded sheets of paper. You could keep them in other more pertinent places as well. For instance if you did one for your **pets** you could laminate it and wrap it around your pet's collar or on their carrier, sealing it with scotch tape and mark "**Read Me**" on the outside in **indelible red ink**.

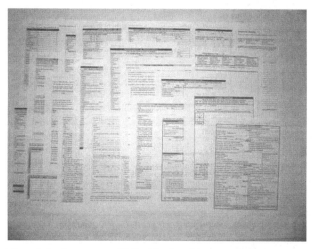

The data forms you'll find in the "Appendix" are an education in and of themselves, and are one of the best sets of tools you could hope for.

A ton of useful information can be printed on business card sized paper and kept in your wallet.

Notes:
_____
_____
_____
_____
_____

Emergencies and Disasters - An Overview

In a book about disaster prep, it would seem this section would come first. However, as we stated earlier, our first concern was getting <u>YOU</u> ready. Now that you have a better foundation to work from, we'll look at some common and not so common disasters and how these will affect sections to follow.

The one goal of this section is to review the emergencies and disasters that may affect your area. Some are obvious; California has earthquakes, Florida has hurricanes, etc. We want you to examine other situations that might have fallen off your radar, and remind you to be prepared for them as well.

Fill in the checklist below scoring each item as instructed. For example, if you've had 3 hurricanes in the specified time period, put a "3" in the blank next to "Hurricane." Add the numbers next to each emergency and put the section's total in the "Totals" column. Add everything in the "Totals" column for a Grand Total. This is your relative **"Emergency Risk Score."**

Basic Emergencies Risk Factors	
General Terrorism Risk Factors (Give a 1 for each of the following if your city is or has ....)	**Totals:**
___ Government Seat ___ Sea Port ___ Major Airport ___ Military Base ___ College	_____
Advanced Terrorism Risk Factors (Give a 1 for each of the following if your city is...)	
___ New York ___ Washington DC ___ Or is one of the top ten cities in population size.	
___ Listed on **Southern Poverty Law Center's** "Terror Group Intel" site (see below).	_____
Terror Targets (Give a 1 for each within 5 miles, a 2 if within 1 mile, and a 4 for ½ mile or less)	
___ Airport ___ Chemical plant / Refinery ___ Dam / hydroelectric facility ___ Nuclear power plant	
___ Other electric plant ___ Military facility ___ Government building ___ Railway terminal	
___ Mass transit station ___ Hospital with emergency room ___ Historical landmark	
___ Water treatment plant ___ Major bridge ___ College or University ___ Major tourist attraction	
___ Abortion clinic ___ Research facility with test animals on site ___ Large shopping mall	_____
Industrial / HazMat (Give a 1 for each one within 3 miles, a 2 for each one within 1 mile or less)	
___ Farm using insecticide ___ Interstate highway ___ Railway ___ Water treatment plant	
___ Manufacturing or processing plant ___ Chemical plant / refinery ___ Natural gas or propane facility	
___ Fuel storage site ___ Nuclear power plant ___ Trucking terminal ___ Railway terminal	_____
Violent Crime (Give a 1 for each within 5 miles, a 2 for each within 1 mile, in the past 4 years)	
___ Drive-by shooting ___ Drug seizure over 100 pounds ___ Armed bank robbery ___ Kidnapping	
___ Bombing ___ Riot / civil unrest ___ Armed home invasion ___ Carjacking ___ Pleasure Arson	_____
Severe Weather (Give a 1 for each within 50 miles in the past 10 years.)	
___ Hurricane ___ Tornado ___ Hail ___ Flood ___ Heavy snow / blizzard ___ Ice / freezing rain	_____
Other Natural Disasters (Give a 1 for each within 50 miles in the past 10 years.)	
___ Earthquake ___ Wildfire or Forest fire ___ Sink hole ___ Landslide ___ Mudslide	
___ Snow avalanche ___ Drought ___ Tsunami ___ Volcano	_____
Additional Personal Risk Factors (Give a 1 for each if you are a...)	
___ Hospital Worker ___ College or High School Student ___ College or High School Teacher	
___ First Responder ___ Mass Transit Rider ___ Employee at any other location on this checklist.	_____
(after totaling, see score ratings below)	**Grand Total:** _____
(Note that the above does not cover anything outside of 50 miles from you.)	

Emergency Risk Score

Let's look at your **score** from the checklist above and what it might mean:

> ➤ **0 to 5**: You're pretty **low as far as a direct incident** goes, but that probably means you're a **prime candidate to receive refugees** from situations in other locations.

- ➢ **6 to 10**: **Low average**. You run a risk of both **a probable natural disaster** and still a **higher than normal risk of being a refugee host**. Refugee cities may see a temporary drain on local supplies and services, a strain on infrastructure, and an increase in crime. Plan accordingly.

- ➢ **11 to 20**: **Getting into the "average" range**, you're still small and peaceful enough to attract refugees, and yet run an increasing risk of direct incident. Review all basic preparations.

- ➢ **21 to 30**: **High average**. You'd do well to pay close attention to the vast majority of the information contained in this book.

- ➢ **31 or more**: **High risk**. **Do everything this book asks**, then read it again to make sure you didn't miss anything, and that you're fully prepared.

(The risk factor checklist and these scores are ours and are used for illustration and comparison purposes. Your local officials will not be using these scores.)

Later on in the book, we're going to be helping you prepare for two types of ultimate reactions, "**Evacuation**" or "**Isolation**." Your checklist above, having reminded you of a few threats you may have forgotten about, should help you decide what you're most likely to experience. Again, some are obvious, others are not. For example, if you live in a beach house on the southeast coast, your most likely threat is a hurricane and your most likely reaction is evacuation. If you live in a heavily populated metro area, you face a high risk of biochemical terror attacks, and your likely reaction is going to be an extended quarantine or isolation. However, a terror attack on a bridge could leave you stranded or isolated at your beach house, or an earthquake that destroys part of your city could cause you to evacuate your metro area home for an indefinite period. So... you should plan for the obvious threats, but also keep the "other side of the coin" in mind. More on this later as we want to make sure you're fully prepared for _anything_.

Speaking of prepared, let's look at a few sources (aside from your regular news channels) that will give you more background information and keep you informed of a variety of developments. **It's extremely important that your "Foundation" continue with educational material from all of the below sources. To be prepared for a threat, you have to know as much about the threat as possible**. Visit each of the sites and sources below, see what they have to offer, print or download their educational materials, and bookmark the pages you'll want to visit again. You'll be needing this information to help you create your customized "**Threat Map**" in the section that follows.

Terrorism, Violent Crime, and Accidents

We'll start with a little background on what we call "manmade misfortunes." While going through each portion below, keep in mind that though you think one might not be as great a risk to you, that the purpose of this section is to point out the fact you may be at risk and to give you the information you need to determine just how at risk you are.

Terrorism

The question we hear most often in our work as security analysts is, "Well just _who_ is bound to attack us?" "_Who_ is our current threat, and would _they_ really want to attack us?" The true bottom-line conclusion is, "It doesn't matter."

It doesn't matter _who_ our "enemy" is this week, this month, or this year. What matters is the sad point that it will always be "_somebody_." As long as America is free and opulent, we'll always have detractors and numerous enemies in the worlds of the "have nots." That's what matters to those of us at the "end user" or "consumer" level of community and personal safety. The point is that in this day and age, with warfare as it's devolved to that of terror strikes vs direct military confrontation, there will always be "_somebody_" waiting in the wings to do us harm in our own backyard, and the next one that comes along will always try to outdo their predecessor.

The question of "who" is one for the federal government, the military, and for Homeland Security. What we need to know about is "how" and "when." _How_ might a group attack us, and if there is sufficient intelligence available, _when_ might it happen? Here are a few sources you should become familiar with in order to monitor the current war against terrorists, discussion of their methods, and information on their various ideologies.

- ➢ **Department of Homeland Security**: http://www.dhs.gov. This is your source to keep up with the current **Terror Threat Alert Level**. Mailing address: US Department of Homeland Security, Washington, DC 20528, fax number is 202-456-6337.

- ➢ For a current-events info on the war on terror, and related matters, see http://www.disaster-central.com.

- ➢ For emergency-related news, reaction analysis, and conflict analysis, see http://emernet.emergency.com.

- ➢ Homeland Security articles can be accessed at http://www.govexec.com/. Be sure to sign up for the free "**Homeland Security Newsletter**." Direct contact is Government Executive, 1501 M St. N.W., Suite 300,

Washington, D.C. 20005, 202-739-8500, fax: 202-739-8511. While on the site, visit their homeland security page at http://www.govexec.com/homeland/.

➢ You can sign up to receive the free "Civil Defense & Homeland Security Weekly" newsletter published by The American Civil Defense Association. Sign up online at http://www.tacda.org/defensealert/default.asp.

➢ Sign up for the Homeland Defense Journal newsletter online at http://www.homelanddefensejournal.com, or at the alternate signup site, http://www.surveysolutions.com/prs/hdj/hdj.htm.

➢ Some of the CIA's take on terrorism can be found at http://www.cia.gov/terrorism.

➢ Information and discussion on national security issues: http://www.nationalsecurity.org. This site is a project of The Heritage Foundation, 214 Massachusetts Ave NE, Washington DC, 20002-4999, ph 202.546.4400, fax 202.546.8328, http://www.heritage.org.

➢ Several New Counter-Terror Publications can be seen online at Rand: http://www.rand.org.

➢ Visit The Brookings Institution for terrorist info: http://www.brook.edu.

➢ The Cato Institute on Terrorism is at: http://www.cato.org/current/terrorism/index.html.

➢ Global Security provides good info and news coverage concerning conflicts and threats. You'll find news, maps, info sources, and more online at http://www.globalsecurity.org, or you can contact them directly at 300 Washington St., B-100, Alexandria, VA 22314, 703-548-2700, Fax: 703-548-2424.

➢ For a good collection of general information, be sure to visit http://www.disasterhelp.gov.

➢ A massive collection of terrorism info links is found at http://www.lib.msu.edu/harris23/crimjust/terror.htm.

➢ A good source covering numerous emergency categories is The Emergency Response & Research Institute, online at http://www.emergency.com, with direct contact at ERRI, 6348 N. Milwaukee Ave., Suite 312, Chicago, IL 60646, 773-631-ERRI (3774), Fax: 773-631-4703.

➢ For a variety of terrorism info and resources, visit the National Memorial Institute for the Prevention of Terrorism, online at http://www.mipt.org, or you can contact them at P.O. Box 889, Oklahoma City, OK 73101, Phone: 405-232-5121, Fax: 405-232-5132.

➢ Some good information on bioterrorism, as well as other disease or **epidemic** information can be obtained from the CDC at: http://www.cdc.gov, **Centers for Disease Control and Prevention**, 1600 Clifton Rd., Atlanta, GA 30333, 404-639-3311, 404-639-3312 (TTY), Public Inquiries 404-639-3534 or (800) 311-3435 .

➢ A good collection of **terror-related information** including NBC data, HazMat, and natural disasters, can be found at http://www.twotigersonline.com/resources.html. You should also sign up for their free email **newsletter** at http://www.twotigersonline.com/newsletter.html, or contact them directly at (910) 458-0690.

➢ The "**Terrorism Research Center**," at http://www.terrorism.com/index.php, contains good articles and other information. You can contact them at 877-635-0816, but only for questions not answered by the website.

➢ "**Jane's**" is a leading authority on the world's military, including military efforts against terrorism. The main page is at http://www.janes.com, and there's general info on the main page and more for subscribers. Jane's Information Group, Sentinel House, 163 Brighton Road, Coulsdon, Surrey, CR5 2YH, United Kingdom, Tel: +44 (0) 20 8700 3700, Fax: +44 (0) 20 8763 1006.

➢ **Terror groups and wanted terrorists**: http://www.fas.org/irp/news/2002/10/dos101102.html. For more information, including membership, contact: Federation of American Scientists, 1717 K Street, NW Suite 209, Washington, DC, 20036, (202) 546-3300. Visit their homepage at: http://www.fas.org/index.html, and their **threat page** at: http://www.fas.org/irp/threat/index.html.

➢ The CIA's "**World Factbook**" online: http://www.cia.gov/cia/publications/factbook/index.html.

➢ LA Public Health's "**Bioterrorism Preparedness and Response**" site: http://www.labt.org/links.asp.

➢ See the US General Accounting Office Terrorism Collection: http://www.gao.gov/terrorism.html.

➢ Defense Security Services' Security Awareness links: http://www.dss.mil/training/salinks.htm.

➢ Info and alerts on **domestic "Eco-Terrorism"**: http://www.stopecoviolence.com. "Stop Eco Violence," PO Box 2118, Wilsonville, OR 97070, ph: 503.570.2848.

➢ **Southern Poverty Law Center's** Terror Group Intelligence : http://www.splcenter.org/intelligenceproject/ip-index.html. 400 Washington Ave., Montgomery, Alabama 36104, 334-956-8200.

➢ See Officer.Com's "**Hate**" page at: http://www.officer.com/hate.htm.

➢ **The FBI's Terror Info page** http://www.fbi.gov/terrorinfo/terrorism.htm, or call (202) 324-3000.

➢ The **Center for Defense Information**, Policy Research Organization: http://www.cdi.org. You can contact them directly at The Center for Defense Information, 1779 Massachusetts Ave., NW, Washington, DC, 20036-2109, 202-332-0600, 202-462-4559 fax.

➢ **Terror related info** and current events: http://www.surviveanuclearattack.com/. This site carries a number of national security related news and info, stories, and links.

- Subscribe to the **Office for Domestic Preparedness email newsletter** by logging on to http://puborder.ncjrs.org/Listservs/subscribe_ODP.asp, and following the directions. An alternate site for the newsletter is http://osldps.ncjrs.org. Or, to subscribe directly to "**Daily Terrorism Update**," send an email to listproc@ncjrs.org with the words **SUBSCRIBE ODP** in the body of the email.
- Be sure to visit the ODP's main site at http://www.ojp.usdoj.gov/odp.
- Visit http://www.homelandsecuritygroup.org, and sign up for their free Homeland Security Report newsletter.
- If you ever get the chance, be sure to watch any rebroadcast of the PBS series "**Avoiding Armageddon.**" More info can be found at: http://www.pbs.org/avoidingarmageddon (which has some good informative links), or from **Public Broadcasting Service,** 1320 Braddock Place, Alexandria, VA 22314, or their local station locator at http://www.pbs.org/stationfinder/index.html.
- A great collection of bibliographical links related to terrorism can be found in the online Naval library at http://library.nps.navy.mil/home/terrorism.htm.
- Terrorism updates and news articles: http://dailynews.yahoo.com/fc/US/Terrorism/ .
- Counter-Terrorism Home Page: http://www.counterterrorism.com/.
- Terrorism Questions and Answers: http://www.terrorismanswers.com/home/.
- Terror Info Compendium: http://www.us-israel.org/jsource/Terrorism/terrortoc.html .
- Counter-Terror Issues: http://www.cdt.org/policy/terrorism.
- Department of State's "Patterns of Terrorism": http://www.state.gov/s/ct/rls/pgtrpt .
- A History of Terror Incidents: http://www.cdiss.org/terror.htm.
- General Terrorism Info: http://www.terrorism.net/, or a very similar page: http://www.terrorism.com.
- The Federation of American Scientists has good terror info pages at: http://fas.org/terrorism, and http://www.fas.org/irp/threat/terror.htm.
- If you're about to travel, you can call the DOT's Travel Advisory Hotline for Terror Activity at Foreign Destinations 800-221-0673, or 202-647-5225.
- Visit the Intel Center, online at http://www.intelcenter.com, for terror-related news updates. Some of the information is free, and some is not, but overall you can get a good picture as to what's going on in the world.
- Current terrorism-related news stories can be found online at http://terrorismdigest.com.

Violent Crime

Make no mistake about it. Violent crime can be just as devastating to you, the individual, as any organized terrorist attack carried out by any foreign interest. While various other parts of this book will touch on different aspects of your personal safety and security, these sources will give you a glimpse of the crime risk for your area, and the most common types of crime you might have to face.

- The Office of Victims of Crime offers numerous crime-victim related publications. Contact them online at http://virlib.ncjrs.org/VictimsOfCrime.asp, or directly at the US Dept. of Justice, Box 6000, Rockville, MD 20849, 800-627-6872, fax: 301-251-5212.
- For stats, look through the FBI's "**Uniform Crime Reports**" at http://www.fbi.gov/ucr/ucr.htm. It's also available in the research section of your local library.
- Also visit the FBI's "Most Wanted" page: http://www.fbi.gov/mostwant/seekinfo/temp021102.htm.
- Similarly, see the **Justice Department**'s site at: http://www.justice.gov.
- The best location for crime stats is the National Criminal Justice Reference Service at http://www.ncjrs.org. They can be contacted directly at the National Criminal Justice Reference Service (NCJRS), P.O. Box 6000, Rockville, MD 20849-6000, 800-851-3420, 301-519-5500, 301-519-5212 (fax), TTY: 1-877-712-9279, (local): 301-947-8374. You can also sign up for the free **NCJRS newsletter** at http://puborder.ncjrs.org/register.
- You'll find some really good crime stat info at the Bureau of Justice's Statistics Publications page at: http://www.ojp.usdoj.gov/bjs/pubalp2.htm.
- See the "Disaster Center's" **Crime Info**: http://www.disastercenter.com/crime.
- Your local **Police Department or Sheriff's Office,** and maybe your city or county website, will have more locally pertinent stats should you be interested.

HazMat Accidents

Last but not least in our "manmade misfortune" list is the threat of accident. As our factories and infrastructure age, this threat will increase. Know what's in your area, and the risks you may face.

- The National Institute of Occupational Safety and Health (NIOSH) book on **Chemical Hazards** is at: http://www.cdc.gov/niosh/npg/npg.html, or call 1-800-35-NIOSH (800-356-4674).

- Go to the Department of Transportation's Hazardous Materials info site at http://hazmat.dot.gov, and **download** their "**Emergency Response Guidebook**," or you can call them at 800-467-4922. Visit the download page at http://hazmat.dot.gov/gydebook.htm, and check out the various links on the same page. The direct link for the Adobe PDF file is http://hazmat.dot.gov/erg2000/erg2000.pdf. The book can also be ordered from the **US Government Printing Office** at http://www.gpoaccess.gov/, or by writing to Superintendent of Documents, 732 North Capitol Street NW., Washington, DC 20401, 202-512-0000, and 888-293-6498, or PO Box 371954, Pittsburgh, PA 15250-7954, 1-866-512-1800. (It's also on our **CD**.)

- **OSHA**, the Occupational Safety & Health Administration has some good information on hazardous materials at: http://www.osha.gov/SLTC/hazardoustoxicsubstances/index.html. Also, go to their home page at ohsa.gov and **click on "H"** to view links to other **hazardous material info**. You can contact OSHA directly at U.S. Department of Labor, Occupational Safety & Health Administration, 200 Constitution Avenue, Washington, D.C. 20210, 1-800-321-OSHA (6742).

- OSHA Fact Sheets covering various hazards, procedures, and safety topics, can be downloaded from http://www.osha.gov/OshDoc/toc_fact.html.

- Another source for Material Safety Data Sheets: http://hazard.com/msds.

- To learn more about **specific HazMat threats in your area**, check with the following sources:

  ❑ Call any nearby nuclear power plant, chemical plant, refinery, factory, etc. and ask their **public relations or public information** office about potential hazards and any **warning systems**. Most will not give you the worst-case-scenario, but it's good to get whatever information they'll provide.

  ❑ Next, call your local **Fire Department** and ask them about area hazards and warning systems.

  ❑ Call your local **Emergency Management Office**. They can be located through your state's Emergency Management agency, found in the government blue pages of your phone book, or you can locate them through NEMA, the National Emergency Management Association, c/o Council of State Governments, P.O. Box 11910, Lexington, KY 40578, (859) 244-8000, (859) 244-8239, http://www.nemaweb.org.

  ❑ Contact the **EPA** and ask them what hazards they have listed for your area. You'll find them online at http://www.epa.gov or contact them directly at: Environmental Protection Agency, Ariel Rios Building, 1200 Pennsylvania Avenue, N.W., Mail Code 3213A, Washington, DC, 20460, (202) 260-2090. The specific division of the EPA that will help you locate hazards in your area is the Chemical Emergency Preparedness and Prevention section. They're online at http://yosemite.epa.gov/oswer/ceppoweb.nsf/content/index.html, or you can contact them for more info at USEPA, Chemical Emergency Preparedness and Prevention Office (5104A), Ariel Rios Federal Building, 1200 Pennsylvania Avenue, NW, Washington, DC 20460, (800) 424-9346. While on the website, be sure to click on "**Chemicals in Your Community**."

  ❑ The CDC maintains a hazardous waste mapping site online at http://gis.cdc.gov/atsdr. Also visit the Agency for Toxic Substances and Disease Registry root page at http://www.atsdr.cdc.gov/atsdrhome.html.

  ❑ Additional information about emergencies and warnings at nuclear power plants can be found at http://www.pro-resources.net/signals.htm.

  ❑ To find nuclear power plants in your area, log on to http://www.nucleartourist.com/us/statemap.htm, and submit a request. As we don't like to provide the wrong kind of intel, we selected a mapping site that will ask who you are and why you want to know where the plants are.

A reminder about all these sources we give you. These are not bibliographies given to you for the purpose of verification if you're interested. These are sites we searched out for the **sole purpose of allowing you to continue your education outside the boundaries of this book**. We expect you to visit these sites and sources as you're able, and for you to learn from them.

Natural Disasters

As we leave the things man could do to us, either intentionally or by accident, we need to look at what nature can do all by herself. As we said at the outset, statistically speaking, you're more likely to fall victim to a natural disaster. The sources below will not only tell you more about the nature of the threat, but many will let you monitor current situations so you'll be forewarned and able to react quickly.

- Dams sort of bridge the gap between terrorist targets and natural disaster considerations. For more info on dam safety and alerts go see http://www.damsafety.org, or contact them directly at **Association of State Dam Safety Officials**, 450 Old Vine St., 2nd Floor, Lexington, KY 40507-1544, Phone: 859-257-5140, Fax: 859-323-1958, if you have any questions about the safety or alert status of any dam in your area.

- More data on dams can be found on the Army's "**National Inventory of Dams**" site at http://crunch.tec.army.mil/nid/webpages/nid.cfm. Contact them at U.S. Army Topographic Engineering Center, CEERD-TR-A, ATTN: National Inventory of Dams, 7701 Telegraph Rd., Alexandria, VA 22315.

- Go to either http://www.disasterlinks.com or http://www.disastercenter.com, to take a look at a variety of current conditions, locations, and general information.

- Multiple disaster information sites can be accessed from http://www.disaster.net/index.html.

- Also, go to http://www.esri.com/hazards/makemap.html, and enter your location info to create an **online map of hazards in your area**. (Very useful when creating your own "Threat Map.")

- For a variety of hazard info, log on to the **FEMA** site at: http://www.fema.gov. You'll find their national situation update page at: http://www.fema.gov/emanagers/natsitup.shtm.

- The **US Geologic Survey** Site at http://www.usgs.gov/, is a good place for information on many natural disasters. Direct contact is: USGS National Center, 509 National Center, Reston, VA 20192, 703-648-4000, 703-648-4748, or 1-888-ASK-USGS (1-888-275-8747), Fax: 703-648-4888.

- Browse all the info on http://www.disasterrelief.org/Library, and its homepage.

- Several **natural disaster newsletter and info sites** can be found at the: Natural Hazards Center at the University of Colorado, Boulder: University of Colorado, 482 UCB, Boulder, CO 80309-0482, 303-492-6818, fax: 303-492-2151, http://www.colorado.edu/hazards. While you're there, visit the links and info page at http://www.colorado.edu/hazards/resources/sites.html and the sub-pages. This is a great information source.

- Become a regular viewer of **The Weather Channel**. You should do this anyway as you need to know the **prevailing wind patterns for your area**, so you'll automatically know if you're in danger during a HazMat incident rather than waiting for officials to tell you. To keep up with weather information, weather patterns, and weather history, try: The Weather Channel, online at http://www.weather.com, or you can reach them at 310 Interstate North Parkway, SE, Atlanta, GA 39339, 770-226-9935 or 9338.

- Look at the **National Weather Service** at: http://www.nws.noaa.gov or at http://weather.gov.

- The National Geophysical Data Center, a division of NOAA, the National Oceanic and Atmospheric Administration, has a rather unique, interactive hazards map you can find online at http://map.ngdc.noaa.gov/website/seg/hazards/viewer.htm. The site allows you to search for historical occurrences of earthquakes, Tsunamis, and volcanoes. Be sure to visit their listing of other pertinent maps at http://map.ngdc.noaa.gov.

- For those of you with **scanners**, or if you just want to know, the National Oceanic and Atmospheric Administration's weather reports and updates are broadcast on: 162.400 MHz, 162.425 MHz, 162.450 MHz, 162.475 MHz, 162.500 MHz, 162.525 MHz, and 162.550 MHz. For the latest **list of frequencies and transmitter locations**, check the NOAA Weather Radio web site – http://www.nws.noaa.gov/nwr.

- Online **satellite weather tracking**: http://www.nhc.noaa.gov/graphics.shtml.

- Weather forecasts, information, and online camera links can be found at http://www.weatherimages.org.

- Anything Weather: http://www.anythingweather.com, AnythingWeather Communications, Inc., 2350 N Rocky View Rd, Castle Rock, CO 80104, Phone: 1.508.557.1555.

- Visit http://www.weatherpages.com, for regional forecasts, weather history archives, and educational info.

- For current weather, visit http://www.accuweather.com. You can learn more by contacting them at AccuWeather, Inc., 385 Science Park Road, State College, PA 16803, 814-235-8650.

- **Weather closings** of schools, etc. (in addition to local news) : http://www.weatherclosings.com. WeatherClosings, 6 Rock Ledge Road, Randolph, NJ 07869 Phone: 973-442-7668.

- **Hurricane tracking** can be seen at http://www.solar.ifa.hawaii.edu/Tropical/Gif/atl.latest.gif.

- "Hurricane Basics": http://www.nhc.noaa.gov/HAW/basics/hurricane_basics.htm.

- The "Hurricane Watch Net": Online only at: http://www.hwn.org. Be sure to visit their **links page** at http://www.hwn.org/hwn-links.shtml (lots of good general and related info links).

- Visit http://www.weatherpreparedness.com. It's home to the Hazardous Weather Preparedness Institute, LLC, 5107 Quaker Landing Court, Greensboro, NC 27455, 800-743-4989.

- We have quite a number of additional **weather information websites**. You'll find them listed in the "**Appendix**" and in softcopy format in the "Web Links" folder on your **CD**.

- Dartmouth maintains a good flood info and research site at http://www.dartmouth.edu/~floods.

- For **flood and flood plain information**, log on to http://water.usgs.gov/pubs/of/ofr93-641/.

- Current **flash flood warnings** can be seen at http://iwin.nws.noaa.gov/iwin/us/flashflood.html.

- Be sure to take a look at the "**Tornado Project**" at: http://www.tornadoproject.com, The Tornado Project, PO Box 302, St. Johnsbury, Vermont 05819.

- Another good tornado source is the **USA Today "Tornado Safety Page"** at: http://www.usatoday.com/weather/resources/safety/wtornado.htm.

- ➢ Just because you live outside the normal area for a disaster, don't cross it off your list. For example, there is a fault line in the east called the "New Madrid Fault." For more info on that go to http://quake.ualr.edu/public/nmfz.htm, or contact the **Arkansas Center for Earthquake Education**, College of Science and Engineering Technology, 2801 South University, Little Rock AR 72204, (501) 569-3085.
- ➢ More **earthquake** info can be found on the **National Geophysical Data Center's** website at: http://www.ngdc.noaa.gov/seg/hazard/earthqk.shtml.
- ➢ Visit the **National Earthquake Information Center** at: http://neic.usgs.gov, and their mapping page at: http://neic.usgs.gov/neis/eqlists/10maps_usa.html.
- ➢ Also be sure to check out the **USGS Seismic Hazard** page at: http://geohazards.cr.usgs.gov/eq. Information on earthquakes registered within the last 48 hours can be obtained through the USGS Earthquake Information Line, (303) 273-8516.
- ➢ Visit the WSSPC EPICenter, the Western States Seismic Policy Council Earthquake Program Information Center, online at http://www.wsspc.org, for a great collection of links, info, and warning sources.
- ➢ The site http://www.hbi.dmr.or.ir/hosting/disasters/links/Earthquake.html, is a great compendium of **earthquake** information sites.
- ➢ Good general information on a variety of **natural disasters and associated reactions** can be found at: http://www.abag.ca.gov/bayarea/eqmaps/eqmaps.html .
- ➢ You can download a weekly report from "**Earth Week**" detailing all manner of weather and geologic disasters as well as a host of other threats and conditions. Log on to http://www.earthweek.com.
- ➢ Good educational material on **wildfires** can be found at: http://www.nifc.gov, http://www.fireplan.gov, and at http://www.firewise.org. Contact Firewise directly at Firewise, 1 Batterymarch Park, Quincy, MA 02269.
- ➢ Check with the US Forest Service's **Fire and Aviation Management** page at: http://www.fs.fed.us/fire.
- ➢ Log on to http://geomac.gov, and click on their "**wildfire mapping**" link to track current wildfires in your area. For questions about **wildfire activity** call the **National Interagency Fire Center** at 208-387-5050.
- ➢ Visit the **National Interagency Fire Center** at http://www.nifc.gov for more wildfire data.
- ➢ You can learn about **volcanoes** through the Smithsonian Institution's site at: http://www.volcano.si.edu.
- ➢ More **volcano** info at the **National Geophysical Data Center's** website at: http://www.ngdc.noaa.gov/seg/hazard/volcano.shtml .
- ➢ Current **volcano** activity and mapping: http://volcanoes.usgs.gov/.
- ➢ **Landslide** hazard data: http://landslides.usgs.gov.
- ➢ **National Landslide Information Center**: http://landslides.usgs.gov/html_files/nlicsun.html, 1-800-654-4966.
- ➢ **American Association of Avalanche Professionals**: http://www.avalanche.org, and http://www.americanavalancheassociation.org, The Westwide Avalanche Network P.O. Box 8067, Alta, UT 84092, 801-694-9585.
- ➢ Forest Service National Avalanche page: http://www.avalanche.org/~nac.
- ➢ **Tsunami** info at the **National Geophysical Data Center**: http://www.ngdc.noaa.gov/seg/hazard/tsu.shtml.
- ➢ Be sure to visit the National Geophysical Data Center's **Natural Hazards Resource Directory** at: http://www.ngdc.noaa.gov/seg/hazard/resource.

Now that you've learned of a variety of threats, let's look at how they might affect your reactions, your emergency planning, and your geographic vicinity in general. Let's make a map, shall we?

The "Threat Map"

While you have these potential threats in mind, let's sit down and create a "**Threat Map**" customized to your personal area or situation. This map will come in handy under "**Evacuation**" where we'll help you plan evacuation destinations, rendezvous points, and travel routes, based on the potential dangers and safe zones you map here.

You'll need several maps:

1. **Vicinity street maps or aerial photos** of your **workplace**(s), children's **schools**, and your **home**.
2. **Your city map** (preferably a detailed city atlas for the larger locations).
3. **County map**.
4. **State map**, and map of adjacent states.
5. **A national atlas**.

Let's start with maps. Maps are extremely important tools regarding your disaster preparedness. You'll need them here to make your **Threat Map**, you'll need them later to make your "**Evac Atlas**," and you'll certainly need them during an actual **evacuation**. Therefore, start making your map collection now. In addition to the general road maps found at bookstores and gas stations, you can locate a lot of maps at the following sources:

➢ MapQuest at http://www.mapquest.com. (MapQuest used to have **aerial photographs**, and may again.)

➢ Maps.Com at http://www.maps.com. Lots of topographical and other types of maps.

➢ MapBlast at http://www.mapblast.com is a main competitor of the previous two sites.

➢ Topozone at http://www.topozone.com. All topographical maps.

➢ USGS maps at http://mapping.usgs.gov, http://ask.usgs.gov/products.html, or http://edcwww.cr.usgs.gov.

➢ Even more US Geologic Survey map resources are at http://erg.usgs.gov/isb/pubs/pubslists/index.html, their secondary mapping page at http://pubs.usgs.gov/products/maps, the national mapping service at http://ngmdb.usgs.gov, and the searchable maps at http://ask.usgs.gov/maps.html.

➢ See The Library of Congress map sites at http://www.loc.gov/rr/geogmap, http://www.loc.gov/rr/geogmap, http://lcweb2.loc.gov/ammem/gmdhtml/gmdhome.html, and at http://www.loc.gov/rr/geogmap/refweb.html.

➢ CIA Reference Maps at http://www.cia.gov/cia/publications/factbook/docs/refmaps.html.

➢ For satellite and aerial photos go to http://www.earthviewer.com, or http://www.globexplorer.com.

➢ Regional maps and satellite shots can be found at http://livingearth.com.

➢ Visit the "**National Atlas**" online at http://www.nationalatlas.gov, and at http://www.nationalatlas.org.

➢ See http://www.mapathon.com for printable road and city maps.

➢ Definitely visit the US Hazards Map page at http://www.hazardmaps.gov/atlas.php.

➢ See the University of Texas collection of map links at http://www.lib.utexas.edu/maps/index.html, and their other map page at http://www.lib.utexas.edu/maps/map_sites/map_sites.html.

➢ University of Texas has another rather decent map page and collection of outside links at: http://www.lib.utexas.edu/Libs/PCL/Map_collection/map_sites/map_sites.html.

➢ A Color Landform Atlas of the United States is at http://fermi.jhuapl.edu/states/states.html.

➢ A good collection of mapping links can be found at http://www.roadmaps.org/links/index.html.

➢ National Geographic's map page is at http://www.nationalgeographic.com/resources/ngo/maps, and their map making page at http://plasma.nationalgeographic.com/mapmachine.

➢ University of Iowa's extensive map collection at http://www.cgrer.uiowa.edu/servers/servers_references.html.

➢ Rand McNally offers both online maps as well as hardcopy maps for sale. They're online at http://www.randmcnally.com, or at P.O. Box 7600, Chicago, IL 60680-7600, 847-329-8100.

➢ Also visit DeLorme mapping software and GPS data, online at http://www.delorme.com, or contact them at 800.561.5105, or Fax: 207-846-7051.

➢ The folks at the Minnesota Map Store offer links to free maps from different states and also offer a number of map collections on CD. They're online at http://www.freemap.com/free.htm.

➢ A huge links collection of online maps and associated resources can be found online at http://www.greatdreams.com/maps.htm. Scroll all the way down to "Map Sources and Information."

➢ Though not free, you'll find a *ton* of useful maps in various formats at http://www.MapMart.com.

➢ For an extensive listing of all types of maps and hard-to-find sources, look through a copy of **The Map Catalog**, by Joel Makower and Laura Bergheim of Vintage Press. ISBN: 0-394-74614-7.

➢ The enclosed **CD** has a **links collection** with several more **mapping and aerial photo sites**.

➢ In addition to these sources, check your county Tax Assessor's office, County Extension office, or Agricultural Stabilization and Conservation Service for maps or aerial photos of your area.

➢ Additionally, your **local library, bookstore, newsstand, or gas station** will have maps and atlases.

Of greatest importance for your Threat Map, are potential threats that pose a danger to more than a single location. For example, a suicide bombing would be tragic, but its danger would be limited to the immediate area of the blast. Unless it triggered something else, there would be no danger to surrounding neighborhoods, etc. However, an industrial accident at a chemical plant could be a danger to an entire town.

On **each** of your maps, you'll want to mark the following things:

1. **Potential nuclear targets**. You can take an educated guess as to where a nuke would be detonated in your area. If you live in the suburbs outside a larger metropolitan area, you can bet that if a bomb was detonated in your vicinity, it would be in the heart of the city. Plot other potential targets in your city, county, state, and

bordering states. Plot targets within a 200-mile radius of your home area (to track fallout). With each potential target, draw a two-mile radius circle to plot potential devastation. Include **dams, government complexes, military bases, harbors, population centers,** and other specific and potential targets in your area.

2. **Prevailing wind directions**. Contact your local weather service, or other weather source listed earlier. On your maps, plot the prevailing wind directions and patterns for your area. You'll need this to project potential fallout if a nuclear detonation occurs, and you'll need to know upwind from downwind when locating chemical threats. One site that will help you plot fallout patterns is a Public Broadcasting Service page that can be found at http://www.pbs.org/wgbh/amex/bomb/sfeature/blastmap.html.

3. **Chemical and HazMat threats**. The EPA, local Emergency Management office, and your local Fire Department can help you note the location of your potential area HazMat threats. You'll want to know the locations of each nuclear power plant, factory, manufacturing plant, refinery, or any direct source or users of dangerous materials. You'll also want to know **transportation routes** on which hazardous materials are carried and loaded, so note the location of railway terminals, trucking terminals, pipelines, and temporary storage facilities. Also note your distance to any highways and railways, or to waterways that run past any of these facilities and near your house.

4. **High-crime areas**. Don't plan to evac through any of these areas, and certainly don't plan any rendezvous points there. (We'll be marking rendezvous points on this same map later on.)

5. **Potential terror targets**. These would be the ones listed above in the "Basic Emergencies Risk Factor" checklist. Mark a two-block area to show the places you would not want to chose as a rendezvous point, as these locations may be "ground zero" for a primary or secondary emergency (such as with a car bomb, etc.), or they may be an additional point of concern during an unrelated disaster. Best to leave them out of your planning.

6. **Transportation bottlenecks**. Mark the location of all **major bridges or tunnels**, or **physical barriers** such as mountains, lakes, or rivers. These will cause evacuation **bottlenecks** during any sort of mass traffic movement and should be avoided if at all possible. Bridges and tunnels may also be damaged in a massive natural disaster such as a hurricane or earthquake. The same bottleneck threat holds true for **toll booths**, and for any **railway** that crosses a road without benefit of an overpass. A stopped train could cut off a planned evacuation route. Bottlenecks will also figure prominently in your **evacuation destination and route** planning since you'll need to head in different directions depending on which side of these obstacles you're on when a disaster hits.

7. **Flood plains**. Using a **topographic map**, mark potential **flood zones** of rivers or waterways, or of the downstream area of any nearby **dams**. Also mark the location of **high ground** on your map. Is your home in harm's way? Are any of your planned evac routes or rendezvous points potentially compromised by flooding? In a flood, where could you go for the nearest high ground? Which roadways are likely to be open and could let you circumvent bridges or low-lying areas? Relatedly, mark the **flow direction** of streams and rivers on your map. Waterway flow directions will relate to HazMat releases in the same way your prevailing wind directions do.

Knowing all you can about threat locations and how they may affect you is extremely important. Under "**Evacuation**" we'll put together a completely different collection of information called the "**Evac Atlas**." This will be more a listing of services, sources, and other assets to be found locally and along potential evacuation routes. We mention this now as you'll be using your same threat map to mark the location of some of these local assets so you can plan rendezvous points and evacuation routes based on a comparison of threat locations vs asset locations.

Hint: Learning to use map and compass as we suggested under "Outdoor Survival" training will help you better prepare and use your Threat Map, as well as the maps you'll create later showing your evacuation routes.

Note: Your Threat Map is the first component of the various written plans you'll be creating throughout this book, and essentially, this is the first part of what others might call a "**Family Disaster Plan**," or a "**Family Reaction Plan**." We're not going to limit you to a set or simplistic checklist of things to put in your plan. In fact, we wish to reiterate that this entire book is your plan in rough draft format. However, it's also important to point out that it's the written portions of your disaster preparations that will be important to your family, as these plans will be their customized guidebook on what to do in the emergencies you're likely to face. Therefore, we wanted to take a second to remind you of how important it is to put all of your threats, assets, and plans in writing, and to make sure all family members had a copy of these written plans, or at least a copy of the parts that are applicable to them. In addition to this **Threat Map**, make sure the following items, which we'll cover in various sections of the book, are included as a part of your **written plan**:

❑ All data forms in the "**Appendix**" ❑ The "Safety Blueprint" ❑ Communication procedures
❑ Fire Drill and other drill procedures ❑ Applicable "Threat Reactions" ❑ Evacuation gear locations
❑ Rendezvous Points & procedures ❑ Your "Evac Atlas" ❑ Isolation procedures and info

Continuing with our preparations, let's keep our above emergencies in mind as we look at the preparations that two of our biggest assets, namely our **home** and our **transportation**, need in order to keep us safe and ready to react.

Basic Home Prep

Looking at your Emergency Risk Factors from above, keep in mind the types of events that might affect your home. Also keep in mind the fact that most homes are constructed to rest peacefully upright during the average windstorm, and not designed to resist much else. For this section, we'll provide the basics for improving the odds that your residence will survive the next event, and protect you in the process.

We also want to help factor out the unforeseen "nuisance" that can make an emergency worse. The last thing you need is to break a leg because of loose carpeting when hustling to the basement during a tornado, and you don't want to be incapacitated by some minor household mishap only to have a real emergency come along the next day.

Though much of this section will apply to privately owned houses, we haven't forgotten about our friends who live in apartments, condos, or mobile homes. There's plenty of useful information here for you too.

Goals of Basic Home Prep:

1. To make your home a **safer place** by avoiding common mishaps.
2. To better protect you against **destructive natural disasters, terrorist attacks, and fire**.
3. To better **protect you from crime**.
4. To make sure you have a **safe and solid structure** to come home to **after an evacuation**.
5. To give good **safety, functionality, and comfort** during an **isolation or quarantine**.

We've divided Basic Home Prep into the following categories:

1. **Miscellaneous Considerations.**
2. **General Home Safety**.
3. **Basic Security**.
4. **Structural Considerations**.
5. **Fire**.

Before you get started on any specific area of upgrading the safety, security, structural integrity, or functionality of your home, let's look at a few general concepts, and info sources.

We wish to point out that the projects in this section can be expensive, and that's one of the many reasons we included the "Frugality" section. Though some of these projects may be necessary, we want you to exercise thrift and good prioritization. Read through all sections first, decide what you truly need based on your particular risks, do as much as you can on your own (most people can do far more than they give themselves credit for), and budget your expenditures and planned projects wisely.

Miscellaneous Considerations

Additional home prep measures will be discussed under "**Isolation**" where we give you the specialized information you'll need to "feather your nest" for being cooped up for a while. Good home prep is just as important for an "**Evacuation**" as you'll want a safe, secure, and stocked home to come back to. Let's start with a few general points and concepts.

➤ Though *some* of the structural considerations listed below *may* be expensive, you should remember that they may **increase the value of your property** as well as help **save your life**.

➤ Some of you **rent your homes**. Some landlords will work with you on making the recommended upgrades (as it may increase their property value and reduce liability). Get permission from your landlord and ask about possible rent reduction, or repayment of costs of parts and materials.

➤ Other rental homes are **apartments**, and apartment complexes generally frown on any structural changes made to their property. Since there's always "more than one way to skin a cat," we've included ways to keep you physically safe with the creative use of your own **furnishings**. However, read everything below, as some of it may apply in ways you hadn't considered.

➤ In keeping with our previous section on "**Frugality**," we should remind you that many of the suggested structural changes can be accomplished with hardware store parts and tools, and a little "elbow grease." Also, you can probably barter a little help from some of your neighbors.

➤ All homes should be **inspected for termites** and other destructive infestations. If you rent, your landlord should take care of this. However, all households should *also* be aware of other pests such as **rodents, fleas, and mosquitoes**, as each of these can carry various **diseases**. Take precautionary steps against these pests even if you have to pay for the exterminator yourself. Advice is as close as your nearest hardware or lawn and garden store, or exterminator.

- **Orkin** has a really good education page on their site at http://www.orkin.com/edu.asp, with a great "Pest Identifier" page at http://www.orkin.com/identifier/identifier.asp.

- The National Pest Management Organization has a great "Homeowner's Info" site at http://www.pestworld.org/homeowners, with a downloadable "**Everything You Want To Know About Household Pests**" brochure. (You'll also find it on our **CD**.) You can find more information, including a pest control referral service, on their main page at http://www.pestworld.org, or you can contact them directly at National Pest Management Association Inc., 8100 Oak Street, Dunn Loring, VA 22027, (800) 678-6722, (703) 573-8330, Fax: (703) 573-4116.

- "ePest Supply," a pest control product retailer, has a really good pest information site at http://www.epestsupply.com/pest_information.htm. Contact them via their main page or at Corporate Offices, 10875 Plano Rd. #105, Dallas, TX 75238, 877-500-0011, Fax: 214-341-4105.

- See: http://www.cdc.gov/ncidod/diseases/hanta/hps/noframes/generalinfoindex.htm for some basic information on "Hanta Virus," which is spread through rodent feces.

- The site http://www.bugrunners.com/pests.html, has good general info on common household pests.

- More can be found at: http://www.allcountypest.com/pest_page.htm.

- See Google links at: http://directory.google.com/Top/Home/Gardens/Pest_and_Disease_Control.

- Your County Extension Office, Departments of Agriculture, or Health and Human Services should have extensive information on pests common to your area, as well as information on how to fight them and guard against them. You'll find contact info in your phone book's blue government pages.

➢ Another reason to control your pest problem is that after any major disaster, numerous services will not be available, including garbage pickup. Keeping pest populations in check now will help prevent them from getting a foothold later. The goal here is to limit your exposure to disease.

➢ While we're on the subject of pest and things that bite, we might as well give you some sources for information on **snakes**. You may run into the occasional snake, plus in many disasters, especially floods, snakes are driven from their normal habitats and may decide to visit yours.

- Your local info sources include pest control companies, municipal animal control services, county agricultural department, county health department, and county extension service.

- For specific info on poisonous snakes go to http://www.venomousreptiles.org/links.

➢ **Radon**, a toxic gas, should be included in your home safety list of "things to watch out for." One way to locate free test kits in your area is to call your County Health Dept., or County Extension Office and ask them. Also, you might want to log on to a search engine and enter the search string, +"free radon test kit"+"*your state*." We've found a couple of sites and sources for more information and low-price testing kits:

- For background info, visit the EPA's site at http://www.epa.gov/iaq/radon/pubs.

- Take a look at Air Check's Radon Information Center, at http://www.radon.com, or at 1936 Butler Bridge Rd., Fletcher, NC 28732-9365, 800-AIR-CHEK (800-247-2435), 828-684-0893, Fax: 828-684-8498.

- See National Safety Products, Inc., online at: http://www.radon.info, or call 877-412-3600.

- Visit Radonzone.com, online at http://www.radonzone.com, or call 540-772-2600.

➢ Learn how to **shut off and restart all utilities**. Keep the necessary tools in their own little spot. We recommend a **red** plastic toolbox (maybe marked with some reflector tape or glow-in-the-dark paint) that each family member knows the location of and can get to. **Don't use these tools for anything else since if you need them, it's going to be urgent**.

➢ Mark all safety gear, emergency evac or isolation equipment, fire extinguishers, fire exits, fire escape equipment, flashlight locations, etc., with **glow-in-the-dark paint or tape**, and/or reflectors or **reflector tape**.

➢ If you have children, have them help you while you make all these safety changes and home improvements. Not only is it educational, it makes them feel involved in their own safety, and it's a great opportunity to teach them about all kinds of safety from injury through fire prevention, in a non-threatening way.

➢ For **general home repair** and maintenance info go to: http://www.doityourself.com.

➢ Consider joining the **Handyman Club of America** for their magazine and discounts. Info can be found at: http://www.handymanclub.com/home.asp, 12301 Whitewater Dr., Minnetonka, MN 55343, 612-988-7403.

➢ Read **Home Maintenance for Dummies**, by James & Morris Carey, and **Household Hints for Dummies**, by Janet Sobesky, both books are available at http://www.dummies.com and http://www.amazon.com .

➢ If you're new to anything electromechanical, such appliances, tools, or home improvement, you can find some basics at the "**How Stuff Works**" site: http://www.howstuffworks.com/index.htm.

➢ Most of us are "average" tool users. We might work on one or two major projects over the years, and therefore really don't put our tools to the test. Invest well in your lifesaving gear, but go for economy in tools you're only going to use once or twice around the house. Since we always try to help you save money, take a look at some sources for economical tools:

- **"Dollar Stores"** (where everything is literally a dollar or less). Simple hand tools can be found here. To find a store, check your phone book or go to: http://www.dollarstore.com, or http://www.dollartree.com.

- **Big Lots:** Big Lots stores usually have a really good selection of economically priced tools. Check your phone book for the nearest location. More info and email sales notifications are at: http://www.biglots.com.

- **Harbor Freight Tools**: http://www.harborfreight.com, 3491 Mission Oaks Blvd. Camarillo, CA 93011, 1-800-444-3353.

- One of the discount sites we managed to find is http://www.hand-tools-power-tools.com, **Hand Tools Power Tools**, 16210 Gundry Avenue, Paramount, CA 90723.

- Another promising tool site was found at: http://www.maxtool.com, with direct contact at Maxtool, Inc., 834 West Cienega Ave, San Dimas, CA 91773, 1-800-MAX-DEAL.

- Be sure to check the universal gear page at http://www.equipment.net, to search for related tools and supplies. You can contact them directly at EQUIPMENT.NET, INC., PO Box 810461, Boca Raton, FL 33481, FAX: 305-675-5896.

- A great source of tools, surplus, and odds and ends is at **American Science and Surplus** at: http://www.sciplus.com, P.O. Box 1030, Skokie, IL 60076, 847-647-0011.

- For the serious electronics hobbyist who'll wind up wanting to make some of the below-mentioned alarms and other devices, be sure to sample "Nuts & Volts" magazine. It's online at http://www.nutsvolts.com, with direct contact at Nuts & Volts Magazine, 430 Princeland Court, Corona, CA 92879, Voice: 909-371-8497, Fax: 909-371-3052.

➢ Speaking of tools, here's a **checklist of basic and advanced tools**, as well as some "parts and materials" you should have on hand as part of your general household equipment. All of these items can be found at your local hardware store. You'll notice some of these will be redundant with tools suggested under "**Basic Vehicle Prep**." Their uses will be discussed along and along throughout the book.

Tools You Should Own		
Basic Household Tools		
❏ Large claw hammer	❏ Assorted screwdriver set	❏ Carpenter's handsaw
❏ Hacksaw	❏ Regular pliers	❏ Vise-grip pliers.
❏ Channellock pliers	❏ Needlenose/wirecutter pliers	❏ Basic socket set
❏ "Monkey wrench"	❏ Razor knife / box cutter	❏ Utility scissors / "EMT Shears"
❏ Heavy duty stapler	❏ Crowbar or prybar	❏ Hatchet or machete
❏ Caulking gun	❏ Tape measure	❏ Electrical extension cords
❏ Punch or carpenter's awl	❏ Sewing kit	❏ Tool belt or toolbox
❏ Work gloves	❏ Protective eye wear	❏ 5-gallon metal bucket
❏ 5-gallon plastic buckets	❏ Broom, wisk broom, dust pan	❏ Cloth or "string" mop
Basic Parts and Materials		
❏ Duct tape – large roll	❏ Masking tape (3" wide)	❏ Heavy plastic sheeting
❏ 5-gallon plastic buckets	❏ Plastic bins and tubs (new)	❏ Plastic 30-gallon trash cans
❏ Heavy and light utility wire	❏ Assorted screws and nails	❏ Assorted nuts and bolts
❏ 50 feet of ½" nylon rope	❏ 100 feet of ¼ " nylon rope	❏ 100-foot roll nylon string
❏ Spackling paste	❏ Old newspapers	❏ Assorted angle brackets & straps
Advanced Household Tools		
❏ 8" Circular saw	❏ Electric drill w/ assorted bits	❏ Screwdriver set for drill
❏ Socket set for drill	❏ Small propane torch	❏ Sewing machine
❏ Small camping shovel	❏ Fire extinguishers	❏ Smoke detectors
❏ Carbon monoxide detectors	❏ Garden sprayer(s)	❏ Garden hoses (for fire and water)
❏ Spare hose nozzle (for fire)	❏ Spare lawn sprinkler (for fire)	❏ Five-gallon gas can(s)
Landscaping or Outdoor Tools (especially if you have a yard)		
❏ Shovel	❏ Axe	❏ Sledge hammer
❏ Extendable limb saw	❏ Heavy rake	❏ Chainsaw (optional)
❏ Hedge clippers	❏ Wheelbarrow (optional)	❏ Heavy broom

➢ A word about the extension cords and garden hoses. You'll notice we emphasized plural in the checklist. One thing neighbors should do to help each other after a disaster is to share utilities from functional houses to deprived ones. You might help a neighbor by running an extension cord from your house over to theirs to power their freezer. They might run a hose from their water to yours after the main in your front yard burst, or to help you fight a fire. We'll cover other ideas later, but this concept is important since one of the best things you can do is help those around you.

➢ For plastic tubs and things like that, check out Rubbermaid products, online at http://www.rubbermaid.com, or you can get more info from Rubbermaid Home Products Division, ATTN: Consumer Services, 1147 Akron Road, Wooster, Ohio 44691-6000, 1-888-895-2110.

General Home Safety

Much of what follows is considered "pedestrian safety" as it deals with preventing the more common or simple household accidents. We want to keep you safe from the petty things so you're fit to handle the big stuff. Let's look at general safety issues:

Overview

➢ In the "**Appendix**" you'll find the "**Safety Blueprint**" template that will allow you to draw a floorplan of your house and note the location of each safety feature you employ. Doing this will help you and your family memorize the location of safety equipment, plan for good equipment distribution, and make sure you haven't left any areas unattended. We suggest that you have this template out as you go through this section so you can show where you already have safety gear, and note the locations you want to add equipment. After completion, put a copy of this form in your safe room.

➢ Be sure to take a look at the **Institute for Business and Home Safety**: http://www.ibhs.org, 4775 E. Fowler Avenue, Tampa, FL 33617, Voice: (813) 286-3400, Fax: (813) 286-9960.

➢ Also look at the **US Consumer Product and Safety Commission** site at: http://www.cpsc.gov, or write to U.S. Consumer Product Safety Commission, Washington, D.C. 20207-0001, 800-638-2772.

➢ You'll also want to see the **National Safety Council** site: http://www.nsc.org, 1121 Spring Lake Drive, Itasca, IL 60143-3201, 1-800-621-7619, (630) 285-1121; Fax: (630) 285-1315. While on the National Safety Council page, be sure to look through their **extensive fact sheet page** at http://www.nsc.org/library/facts.htm.

➢ **Home Safety Council**, PO Box 1111, North Wilkesboro, NC 28656 (or at 1605 Curtis Bridge Road, Wilkesboro, NC 28697) Phone: 1-336-658-5634, http://www.homesafetycouncil.org.

➢ Though primarily intended for the workplace, good general safety information can be downloaded for free from **OSHA's** website at http://www.osha-slc.gov/OshDoc/toc_fact.html.

➢ General home safety for kids: http://www.stayingalive.mb.ca/kids_roomsafety.html.

➢ Home safety in general: http://www.rtc4safety.com/safetyworks.htm.

➢ More at: http://www.lifeessentialsbyzee.com/zee/zLifeE_safety.html.

➢ Definitely read the articles and follow the links found on the **Injury Control Resource Information Network** page at: http://www.injurycontrol.com/icrin.

➢ Be sure to visit http://www.efwhomesafety.com. Sign up for their free **email newsletter**. They can be contacted directly at EFW Home Safety Products, PO Box 1050, Fillmore, CA 93016, Phone: 1-805-524-7826, Fax: Toll-free: 1-888-830-9247 or 1-978-418-5673.

Landscaping:

➢ Take care of all dead tree limbs, or dead trees, as they pose both a fire risk and a falling debris risk. For info on tree health, visit http://www.arborday.org.

➢ Take care in your planting of trees or large bushes as large roots under concrete walkways can eventually cause them to buckle and create a trip-and-fall risk.

➢ Don't allow vegetation to create sight-picture blocks for walkways or driveways. Motorists on the street need to see you coming out of your driveway and vice-versa. When you're pulling in to the driveway, you need to be able to see if your path is clear.

➢ Check for hornets nests, wasp nests, scorpion nests, and other **stinging insects**. If you have cluttered storage areas, be careful of **spiders** such as Black Widows, or Brown Recluses.

➢ While on the subject of insects, let's mention mosquitoes. **West Nile** virus will be with us for a long time, and one of two ways to reduce the threat is to reduce the mosquito population. Keep your yard free of any **standing water**, even very small amounts. Rain water gathering tubs are okay provided you have a screen covering so mosquitoes can't get to the water to lay their eggs. For ponds and open bodies of still water, you can get **larvicide** tablets at your county extension office or at most lawn and garden stores. Depending on your mosquito problem, you might look into getting a "**mosquito zapper**." You can get these at most any lawn and garden store.

➢ Similarly, make sure your outdoor torches and candle holders use Citronella candles and fluid.

Walkways:

➢ Make sure all floor and walkway tiles, inside and out, are securely fastened in place.

➢ Any outside wooden stairs should be secured with wood screws and not just nails. The same goes for interior wooden stairs. Make sure any nails are hammered down and not protruding.

➢ Exterior stairs and walkways should have lights at night.

➢ Interior stairs and walkways should have lights. Use overhead lights as needed and nightlights after dark. Some nightlights are rechargeable flashlights that not only act as nightlights, but they come on in case of a power outage. These are available at department and hardware stores.

➢ Numerous companies make battery operated walkway lights marketed as "Tap Lights" or "Touch Lights" because you push on the cover to turn them on. Other models have a small pull chain or switch. You'll find these at most department and hardware stores.

➢ If you have small children, make sure you have a baby gate at the top or bottom of staircases.

➢ Have outside wooden porches, stoops, decks, staircases, and all railings routinely checked for termites and rotten wood.

➢ Outdoor stair steps should have traction strips.

➢ All banisters and handrails inside and out should checked for support.

➢ All removable carpeting on walkway areas should have a non-slip backing.

➢ Make sure permanent carpeting, especially that on stairs, is properly tacked down and secured.

➢ All floorboards should be checked to see if any are weak, or are loose and need to be secured.

➢ Make children pick up their toys when done, especially from driveways, and walkway areas inside and out.

➢ Don't allow water to puddle in any area of the house. Slip-and-fall reasons notwithstanding, you don't want a breeding ground for molds or fungi.

Family Areas

➢ Make sure all windows are secured and functional. Falling and breaking glass is a danger in a destructive event, and a permanently sealed window may prevent your escape in a fire. Similarly, an unsecurable window is a security risk, and a poorly fitted window is more difficult to seal in a shelter-in-place scenario.

➢ Have at least two regular corded phones along with any cordless phones you may have. The cordless units won't work in a power outage, unless the base unit has a battery backup.

➢ Check the stability and anchoring of any ceiling fans and overhead light fixtures.

➢ If you have small children, be sure to put protective covers over electrical outlets.

➢ You'll also want to teach children to stay away from such things as the fireplace, candles, radiators, space heaters, and other potentially harmful items.

➢ Even though some gas space heaters say "no venting necessary" you'll still need to make sure the room it's in has plenty of fresh air for you to breathe.

➢ Secure freestanding bookshelves to the wall, and place heavier items on the lower shelves. If you have wall-to-wall carpeting you also might want to put rubber "feet" under the front of the bookshelf to keep it from leaning forward into the soft carpet.

➢ Secure any other heavy items that might topple in a destructive weather event or earthquake, such as the television, stereo cabinets, window-unit air conditioners, aquariums, wall mirrors, or heavy framed pictures. Basically, if it's something you wouldn't drop on a baby, secure it.

➢ Have a flashlight accessible in every room.

Kitchen

➢ Make sure your stove is anchored to the wall. Children might try to climb it and tip it over, or it may shift in a destructive weather event or earthquake. The same goes for all other heavy appliances such as your hot water heater, refrigerator, dishwasher, and washer and dryer. Do what you can to bolt them in place and/or secure them with heavy strapping.

➢ When cooking on the stovetop, turn pot and pan handles inward so they don't hang over the edge of the stovetop where you might bump them or kids might grab them.

➢ Use oven mitts or an oven stick to handle or move pots, or to slide oven items or hot oven racks in and out.

- Establish a "no kids or pets" rule for the kitchen when food prep or cooking is going on. Lots of sharp items and hot items pose a danger.

- Make sure your childproofing includes locks on the refrigerator, washer, dryer, oven, and microwave.

- If you have a gas stove, hot water heater, or dryer, make sure you use flexible gas fittings in case of appliance shifting during a destructive event.

- Have your gas company come check all the fittings between the street and your appliances (including the meter) for leaks. This will not only improve your safety, but may save on your monthly bill if a leak is detected and fixed.

- Keep all medicines and household chemicals secured and out of the reach of children.

- Separate chemicals that you know to be reactive with each other. For example, you wouldn't use bleach or abrasive powder products together as they'd release chlorine. For that same reason, you should keep them in different cabinets in case of unforeseen spills or situations that might cause a rupture of the containers. If you're trapped in a damaged house after a tornado or earthquake, the last thing you want to worry about is your own little HazMat incident or a fire being started by reactive household chemicals.

  - See "Oregon Toxics Alliance" at http://www.oregontoxics.org/chart.html. Contact them directly at 1192 Lawrence St, Eugene, OR 97401 · or at P.O. Box 1106, Eugene OR 97440, (541)465-8860.
  - Visit the NIH's "Household Products Database" at http://householdproducts.nlm.nih.gov, or at 8600 Rockville Pike, Bethesda, MD 20894.
  - The EPA's guide to the disposal of common household chemicals can be found online at http://www.epa.gov/glnpo/p2/Lkwatchc.html.

- To dispose of potentially hazardous household chemicals, save them until you use them up, give them to neighbors, sell the remaining amount at a yard sale, or call your local fire department or recycling center for disposal instructions. Never dump them down drains, on the ground, or into the sewer system. Also, don't burn them or dump them in the trash. They may mix with other chemicals discarded by your neighbors and react in the garbage truck or at the dump.

- For reasons of accidental poisoning, you should store consumables and non-consumables separately.

- If you have an older, or well-used microwave, have it checked for microwave leakage. You can get detector cards at most appliance stores or department stores.

- Be very careful with plugged-in electrical appliances near the sink when it's full of water.

- Don't "modify" any kitchen appliances by cutting off the plug's ground wire prong. If your kitchen outlets don't have a ground wire, have an electrician take care of the problem. Electric shock is a very common household accident.

Bathrooms

- Be careful with children and hot water. Scalds are far too common. Turn your water heater down to about 130 degrees, and color code your hot and cold faucets red and blue so children can learn which is which.

- Keep hot items such as curlers, curling irons, or hair dryers, out of the reach of children.

- Put a non-slip shower mat in your tub or install non-slip appliques. You might also take a look at a new tub treatment at http://www.stopslipping.com, or call 310-428-4567.

- Put a non-slip bathroom rug outside your tub if you have a tile floor.

- Consider adding a sturdy support rail on your shower wall and next to the toilet. This isn't something just for infirm individuals. It could be a well-anchored, heavy duty towel rack that will support you if you fall.

- If you have an older shower with glass doors, replace them with a non-breakable variety. Many people huddle in the bathroom in an emergency, and glass may only add to the problem in a destructive event. Another problem is slipping in the shower and falling against breakable glass.

- If you have children, you've probably already child-proofed the medicine chest and bathroom cabinets. You should also **adult-proof** them by keeping dangerous medications away from common ones. Hospitals see numerous cases each year where a sleepy adult thought they were taking one medication, but had accidentally grabbed the bottle next to it. If space is limited, try putting the more dangerous medications in a Ziploc bag or wrapping rubber bands around the bottle so they can be **identified by feel**.

- You should also separate consumables from non-consumables. (Ever notice how some popular antacid bottles are shaped just like fabric dye bottles?)

- Don't keep any plugged-in electrical appliances near the tub, especially if you have children. This includes hair dryers, curler sets, space heaters, televisions, radios, etc. Battery operated items are okay if UL listed for in-shower or in-bath use. See: "**Underwriter's Laboratory**" at: http://www.ul.com, 1285 Walt Whitman Road, Melville, NY 11747-3081 USA, phone: 1-877-ULHELPS (1-877-854-3577), Fax: 1-631-439-6464.

Bedrooms

➤ Install a **carbon monoxide alarm** in the hallway near the bedrooms, following the alarm's installation instructions. You can get good carbon monoxide alarms at most department or hardware stores. These are a good backup to smoke detectors for a number of reasons. One is that some fires may burn and produce carbon monoxide, but very little smoke. Another reason is that some gas appliances will produce carbon monoxide which will kill as quickly and quietly as smoke or other noxious fumes.

➤ The discussion of window safety and functionality holds doubly true for bedrooms. Windows need to be properly installed and insulated, easy to use as an exit in a fire, and yet secured to keep out an intruder.

➤ Secure loose furnishings. Anchor tall bookcases, televisions, wall shelves, etc. In addition to securing these items, be sure to decorate accordingly. Place heavier objects on bottom shelves, give extra support to heavier hanging picture frames, etc. If you live in an earthquake-prone area, consider anchoring knick-knacks and other shelf items with "museum wax," modeling clay, or other sticky substance made to temporarily attach items to surfaces.

➤ Pay close attention to the support of any large pictures, mirrors, or shelves on the wall over your bed. You don't want broken glass or heavy objects falling down on you in your sleep because of an earthquake or tornado, or because something simply broke. Hanging items should be hung from closed eye-hooks so they don't bounce off the hook.

➤ Also in earthquake areas, think twice about putting your bed (especially baby cribs!) underneath regular glass windows that could easily shatter and fall onto the bed's occupants.

➤ If you keep any kind of defensive weapon in the house, make sure you keep it out of the reach of children, and that it's kept in as safe a mode as possible.

Work Areas

➤ Many home-related trips to the emergency room come from weekend hobbyists working in their shops or other work areas within the home. Though we could go for pages and pages on individual "shop safety" suggestions, they'd pretty much boil down to two things: Know and follow the **safety recommendations** for the equipment you're using, and **take your time and pay attention** to what you're doing.

➤ Don't cut the ground wire prong off of your electrical plugs. Always be safe with electricity.

➤ Basements can be considered "work areas" though many are often family rooms, and some just storage areas. Make sure your basement is kept clean and dry as molds and fungi are a cause of many illnesses. Seal the walls with water sealant, and make sure you have good ventilation. If you have problems with excess moisture, you might consider buying a dehumidifier or using desiccant packs. Charcoal briquettes and/or uncooked white rice inside an old nylon stocking make excellent expedient desiccants. Change them out as often as needed.

➤ Never allow water to puddle anywhere in your house. Not only is it a slip-and-fall waiting to happen, it could make a mild electrical shock deadly if you're standing in the water.

➤ Make sure all your breakers or fuses are properly labeled as to what they control. Remember to mark the breaker box with glow-in-the-dark paint and/or reflector strips.

Security

A criminal act perpetrated against you, your family, or your property, would be a disaster in and of itself and is something best prevented. It would be bad enough to lose belongings, but truly unfortunate if you or a loved one were injured or worse. Pay close attention to your security needs so you can protect your newly gathered **emergency gear** from burglary, and have protection in place in case of **temporary social unrest** after a major disaster. Too, if you or someone in your family is a pilot, government employee, first responder, doctor, or a key employee of any infrastructure system, <u>you</u> may be the target of a **terrorist kidnapping or assassination**.

Landscaping

➤ To help first responders find your house in an emergency, make sure your house number is prominently displayed on your mailbox, by your door, and along the curb if possible. Also make sure at least one of these numbers is properly illuminated at night.

➤ Good outdoor and entranceway lighting is one of the best security measures you can take. Make sure you have floodlights that can be automatically <u>and</u> manually operated and that they cover all areas of your property. Similarly, if your street does not have streetlights, ask your local municipal government what can be done to remedy the situation.

➤ Have at least one outside light be photoelectrically controlled to turn on at night and off in the daytime.

- Add automatic and manually controlled lights on the entrances and far side of any outbuildings where tools and other equipment are stored. You don't want break-ins or hiding spots.
- Don't provide hiding spots for intruders. Hedges should be thinned, bushes trimmed, and lighting strategically placed. Also, follow the "6 and 2" rule. No tree limbs lower than 6 feet and no hedges taller than 2 feet.
- If you don't have a fence, or sections of fence to cut off hidden areas, consider adding one.
- Repair any holes in any fences, and consider putting a lock on your gates.
- Put up all toys, lawn and garden tools, and ladders when done. Some items are targets for theft, and in the case of ladders, are tools that will help a burglar. You can install anchor points (more under "Structural Considerations") that can be used to padlock things like bikes, lawn equipment, boats and trailers, etc.
- One old trick is to place gravel in flower beds underneath windows. It helps keep weeds at bay plus it makes a loud crunching sound if a would-be intruder steps on it.
- Another old trick for creating barriers is to plant hedges of Holly bushes. They're decorative yet thick with pointy leaves and therefore tough to get through. Plant these under windows or other hidden access points. The same holds true for Spanish Bayonets, Hicks Yews, Washington Hawthornes, Red Barberry, and Pyracanthas to name but a few. Don't forget Rose bushes.
- Look into getting a remote-control porch light. You can turn it on as you pull into your driveway. One model can be found through "Smart Home" online at http://www.smarthome.com/2005.html, or at 16542 Millikan Avenue, Irvine, CA 92606-5027, Office: (949) 221-9200, Toll Free: 800-762-7846, Fax: 949-221-9240.

Access

- A good hardware store can help you with all the following items regarding locks, hinges, door plates, etc.
- All exterior doors should have a deadbolt lock in addition to the knob lock, and the door frame should be securely anchored to the supporting studs with long wood screws.
- All exterior doors should be metal, or at the very least, solid-core wood doors. Doors with glass windows (unless barred) are too easy to breech. If you have a door with a window in it or next to it, reinforce the window with a layer of plexiglass on the inside. Otherwise it's too easy for an intruder to make a hole in the window and reach in to unlock your door. Some people simply have deadbolts that require keys on the inside, but that could be a problem in a fire if you couldn't get to the key. **Hint**: If you do have a deadbolt requiring a key on the inside, keep an extra key near the door where all family members can reach it, but not in range of a burglar reaching through a window.
- If any exterior door has the hinges exposed to the outside, make sure they're "pinned" hinges that can't be removed or disassembled from the outside.
- Entry doors should have a secondary stop such as a chain, floor stob, or floor mounted brace rod. Your local hardware store or department store should have everything you need.
- Though they provide little protection, many people use chains on their door. If you use a chain, get the type you can lock and unlock from outside. This allows you to add the little bit of extra protection the chain provides, while you're out of the house entirely. Also, when you mount it, use longer screws (2") that will penetrate into the door **frame** rather than the short screws that will only attach to the molding. This will make the chain <u>much</u> more effective! To be even more effective, mount 2 chains in this manner with one near eye level and the other nearer the floor.
- Reinforce door frames with a mending plate held in place with long screws underneath the molding. Further reinforce the frame by using longer screws to hold the lock's strike plate in place on the door frame. These steps make it more difficult for an intruder to kick in the door.
- Reinforce the knob and lock area of wooden doors with a reinforcement plate. This protects the weakest part of the door if someone tries to kick it in.
- The lock area of the outside of the door should be protected by a cover plate to prevent someone getting a pry-bar between the door and the frame.
- Install wide-angle-view peepholes in all solid exterior doors. Make a cover for the inside lens of your peephole so that no one can come to the door and use a small "pocket scope" to see inside (a more common trick than you might think).
- If your entrance areas have hidden spots that can't be seen through the peephole or nearby window, consider installing a small, curved, "security mirror" so you can see who might be hiding around the corner.
- Also, learn to use natural reflective surfaces to see who's hiding out of view of the peephole. The windows on your car, or the neighbor's window across the street might provide all the "security mirror" you need.
- Your main entrance way should have a light source close to the door so that visitors will have their faces illuminated and they won't be silhouetted when viewed through the peephole.

- Numerous companies sell video surveillance systems. Many of the newer models will let you log on to the internet from any location to see what's going on in and around your home. You can find basic systems at SAM's Club all the way through a couple of sources we found for you:

  - **ADT** home security has such a system. Contact them at ADT Security, http://www.adt.com, PO Box 5035, Boca Raton, FL 33431-0835, Tel: 561-988-3600, 800-238-4459.

  - "**All Phase Video Security**" at http://www.allphasevideo.com, 70 Cain Drive, Brentwood, NY 11717-1265, 800-945-0909.

  - **CCTV Wholesalers** at http://www.ccTVwholesalers.com, 1308 Dealers Avenue, New Orleans, LA 70123, 800-291-0523.

  - The "**X-10**" **Camera System** at http://www.x10.com, or call direct at 800-675-3044 .

  - **123 Security Products** at http://www.123securityproducts.com, 387 Canal Street, New York, NY 10013, 1-866-440-CCTV (2288), FAX: 631 451-2408.

  - **Keepsafer Home Security** at http://www.easyhomesecurity.com, or call 800-528-4454.

  - You'll find a home monitoring camera setup at: http://www.save-on-security.com/xcam2.html. Be sure to visit their main page at http://www.save-on-security.com, and contact them directly at Save on Security Systems, Inc., 108 – 2750 Faithfull Ave., Saskatoon, Saskatchewan, Canada S7K 6M6, Ph. (306) 931-3707, Toll Free: 1-800-931-3707, Fax (306) 931-4455.

- Sliding glass doors, or doors with plain glass windows, should be reinforced against breakage with plastic film. This film can be found at most auto-supply stores (mild tinting) or get a really tough film through **BlastGARD** at: http://www.blastgard.com/home.html, 1.888.306.7998, Fax – 888.646.8913. Some film can be mirrored so you can see out but "they" can't see in, or it may be tinted for both privacy and sunlight reduction in hotter climates (to save energy).

- Sliding glass doors should be reinforced with a secondary sliding bolt near the base, a cross-bar lock, or by dropping a broom handle down in the track so the door can't slide open.

- Garage doors should have key locks as a backup to any remote control opening device. Lock these at night and on vacation, since the remote control devices are so easy to clone. Having locks will also let you disconnect your automatic openers during an extended power failure and open the doors manually while still having a way to lock them.

- Ground floor windows should also get bars, reinforcement film, an internal plexiglass cover, or at least wooden dowels cut to fit their tracks to keep them from being opened all the way. Let your local crime rate determine what's right for you, but remember you may have to use one of these windows as a **fire escape**.

- Window unit airconditioners should be securely bolted in place, not only for safety, but to prevent their removal which would allow a burglar access.

- If other windows are routinely left open for ventilation, be sure to have pins or broom handles in the tracks that will prevent them from being opened far enough to allow a person in.

- Make sure any skylights are securely fastened in place.

- Make sure you have a doorbell. You can get a simple battery-operated wireless doorbell at most hardware and department stores.

- Make sure all outbuildings and garage doors are lockable. Similarly you might want to chain and lock all lawn and garden gear, or mobile recreational equipment such as bikes, or boat or camper trailers.

- If you're going to hide a house key somewhere, be original. DON'T hide one above the door, on the porch light, around the mailbox, under the mat, under a flower pot, or in one of those fake rocks. Those are the first places a burglar will look. Pick a unique spot, but make it one that's easily describable to family members or first responders (You may get a call at work from your alarm company saying responders are on the way to your house because the intruder alarm or fire alarm went off. You'll want them to get in without breaking in the door.) **Hint**: If you have a trusted neighbor, hide your key in *their* yard, and their key in *yours*. **Hint #2**: If you think you'll buy several combination locks, get the kind that can be set to the same combination. This way everything's locked, but your family only needs to learn one combination.

Alarms

- Alarms come in many shapes and sizes. Choosing one for your particular location and needs will be tough. Sizes range from simple Radio Shack single-location entry alarms, through professional alarm service companies that monitor your system for you and can detect fire, smoke, light, sound, motion, entrance openings... the whole nine yards.   Others still, will let you monitor room cameras in your house via the Internet. You can get more information on individual alarms and components from the following sources:

  - Start your info gathering at "**Smart Home,**" online at http://www.smarthome.com, or at 16542 Millikan Avenue, Irvine, CA 92606-5027, Office: (949) 221-9200, Toll Free: 800-762-7846, Fax: 949-221-9240.

- See several alarms and other systems at http://www.safetyandsecuritycenter.com. Direct contact is through: Safety and Security Center, P.O. Box 699, Carlsbad, CA 92018, Telephone: 760-929-9650, FAX: 760-929-9601, Sales: 800-378-2957 .
- A wide variety of home and personal alarm devices can be found online at: http://www.womensdefensecenter.com/selfdefense.htm. See "Home Protection," and "Personal Alarms."
- **ADT** Security, http://www.adt.com, PO Box 5035, Boca Raton, FL 33431-0835, Tel: 561-988-3600, 800238-4459.
- **Brinks Home Security**, http://www.brinkshomesecurity.com/index_hi.asp, 8880 Esters Blvd, Irving, TX 75063, 972-871-3500, 800-725-3537.

➤ When choosing an alarm **system**, some of the more desirable components are:
- ❑ Entry sensors on all exterior doors and windows (at least ground level windows).
- ❑ Fire (heat) and smoke detection.
- ❑ Glass-break detectors (usually an acoustical sensor).
- ❑ Central control panel with panic button (preferably to include an additional panic-button necklace).
- ❑ An external audible alarm and an automatic phone dialer that calls 911.
- ❑ Battery backup.
- ❑ Interior video surveillance system tied in to alarm system, with the video system accessible online so you can remote view from an off-site computer.
- ❑ Insurance policy to cover system failure.
- ❑ Choose a system recommended by your insurance company so that it may reduce your premiums.

➤ When choosing an alarm **company** to monitor your system, some desirable services are:
- ❑ Codeword identification for residents.
- ❑ 24/7 Monitoring.
- ❑ Service will call 911 redundantly with system.
- ❑ Service will call your residence to check alarm activation with residents.
- ❑ Insurance policy to cover system failure.

➤ If your alarm system/company does not offer the panic button on a necklace, you might consider a "Life Alert" or "Life Call" system or similar.
- Find "Life Alert" online at http://www.lifealert.net/home.htm, or contact them at 800-815-5922.
- The "Guardian" system http://www.safetyandsecuritycenter.com/peremalsys.html, or call 760-929-9650.
- The "Life Fone" system at http://www.lifefone.com, or call 800-882-2280.

➤ Some experts feel that posting signs or decals stating that your home is alarm protected, will help deter some crime, whether your home is actually alarmed or not.

➤ Consider a proximity light on your garage or by your front door. The proximity switch will turn on the light when anyone comes near, and then resets after they leave. The assembly is inexpensive and you can get it at almost any department or hardware store.

➤ You can also find driveway alarms that let you know if anyone's pulling into your driveway.

➤ Pets make great alarms. Small, nervous dogs will bark readily, birds will squawk, big dogs will growl and offer some real protection, and cats will alert when someone's approaching.

➤ Learn to be creative with the gear you have. One cute trick we've seen uses the simple Radio Shack entry alarm coupled with an old baby monitor. The monitor's transmitter was placed by the alarmed door, and the receiver was placed in the master bedroom. There was no doorbell so the baby monitor served double duty. Anyone knocking could be heard in the master bedroom, and so could the entry alarm if it were activated.

➤ Don't forget the simple but useful intrusion alert tricks such as a soda can with some marbles balanced on the doorknob, a tripwire that knocks over a lamp, etc. However, we recommend you use these only if nothing else is available.

➤ **Word of caution**: We've seen a lot of material on various ways to "boobytrap" a home when people are away on vacation, etc. We do not recommend boobytrapping and would like to remind you that most jurisdictions frown on any type of automatic device that might injure or kill an intruder, especially when no resident life is in danger.

Communication

➤ Almost every room should have some sort of phone in it. We recommend at least one regular corded phone at each end of the house as the cordless units won't work during a power outage.

➤ Prepaid, and 911 cell phones are plentiful and cheap. Consider getting one to keep in the house as a backup, even if you already have a regular cell phone. You never know if a storm is going to blow down a phone pole, if a burglar is going to cut your phone lines before breaking in, or in the middle of everything, you find you left your regular cell phone somewhere else (or its battery runs out in the middle of a call).

- Speaking of phone lines, make sure yours is protected by metal conduit where it runs alongside the house if the incoming phone line is exposed and reachable from ground level. You don't want to make it easy for a burglar, or flying debris in a storm, to cut your lines.

- Communication also means from room to room. Consider an intercom system or baby monitor if you have family members such as children or the elderly, who may need help but may not be able to come ask for it. If this is the case, especially with the elderly, be sure to put an intercom in the bathroom, possibly with the panel near the floor or close to the tub since a collapse or fall is the most likely scenario. Make sure the system has a battery backup for use in power outages. You can find intercoms and baby monitors at most department or electronics stores.

- As a really inexpensive alternative, try a battery-operated wireless doorbell in the same locations you'd put an intercom, such as the bathroom. Put the button in the bathroom (again, at a lower level) and the receiver in a central area. Wireless doorbells can be found at most department stores. You could also put the button on a necklace worn by an elderly or sickly person. This would allow them to signal from wherever they might be. Make sure the "alert" doorbell has a different tone than your regular doorbell.

- You might also consider an intercom system between the front door and remote areas of the house such as the garage, and your safe room. Again, make sure this system has a battery backup.

Safe Room

Many people have seen the movie "Panic Room" and have also seen the marketing of extremely expensive panic-rooms that followed. This kind of effort and expense is not high on our list of recommendations, though the concept of a safe room does have validity. Every family needs a secure area in which to gather during an emergency. The structural considerations for making such a safe room are in the next section, "Structural Considerations." Here, we'll discuss what your safe room is, and some details that will make it functional.

- There are 2 concepts for a safe room, and you may wind up with both, or one may serve both purposes. Make your decision based on your home's layout and on your "Emergency Risk Score" from earlier. If you're high on the crime scale and low on the natural disaster scale, you'll want a safe room closer to your family and bedroom areas. If you're high on the natural disaster scale and low on the crime scale, you'll want your safe room to be on the lowest floor possible near the center of the house (except in flood areas). Of course, depending on your home's size and construction, these two locations may well be one in the same.

- Keep in mind though, that one of your "safe areas" may be a room or hallway that you intend to seal in reaction to a biochemical event. In this case you'll want a central room with as few doors and windows as possible, and as high off the ground as possible. Most chemical compounds and biological weapons will be heavier than air and will settle to the ground. This is one instance you do **not** want to be in your basement! (Of course, in a small house or apartment, the response to all of these "potential" safe room locations might well be, "the main closet." Go with what you have.)

- You'll want to make sure any safe rooms each have the following:
  - ❏ First aid kit (or main Bugout Kit containing your main first aid kit).
  - ❏ A corded telephone and a copy of your "Important Contacts" lists.
  - ❏ Prepaid or 911 cell phone (in addition to your regular cell phone).
  - ❏ Extra battery or battery-operated power adapter for your cell phone(s).
  - ❏ An automotive "antennae relay" or "booster" if your cell phone gets poor reception in the safe room.
  - ❏ Air horn <u>and</u> whistle for signaling rescuers or alerting household members (airhorns are louder, but may leak all their air before use).
  - ❏ If you have an intercom system with the front door, put an extra intercom panel in your safe area to talk with first responders. Again, make sure the system has a battery backup.
  - ❏ Electrical outlet.
  - ❏ Battery operated radio/TV combo (with extra batteries) or a hand-crank powered radio.
  - ❏ Electric light fixture or lamp.
  - ❏ Flashlight with extra batteries (no flammable light sources!), or a batteryless flashlight.
  - ❏ Fire extinguisher.
  - ❏ Personal defensive equipment. (A more extensive discussion of defense and personal security will be found in the "**Isolation**" section.)
  - ❏ At least one day's worth of water and ready-to-eat food for the family and pets.
  - ❏ A manual can opener if you have canned goods in your safe room.
  - ❏ Protective and regular clothing for each family member.
  - ❏ Blankets (and maybe a couple of inflatable mattresses and pillows if you have sleeping room).

❏ An extra set of all your important keys.

❏ Your "little red toolbox" that contains the tools needed to shut off and restart all your utilities and associated appliances (such as long matches to relight some pilot lights).

❏ Possibly your "Bugout Kit" (as described under "**Evacuation**") and some of your other emergency goods.

❏ A set of "**Bolt**" clothing (described under "**Evacuation**"). Use this clothing to house the keys, extra cell phone, etc. that you keep stored in your safe area. That way it would be in the safe area, yet already loaded in the pockets of these clothes in case you needed to "bolt."

❏ Depending on how much disposable income you have to develop your safe room, you could also have your surveillance camera monitor there.

➢ **Note**: Some of these items such as your cell phone and keys need not be duplicates <u>IF</u> you get in the habit of keeping your purse handy, or your pocket items together in a small basket that can be easily grabbed to carry to the safe room with you. However, we do recommend the duplicates. Make sure you practice grabbing your purse or basket during any **reaction drills**.

➢ If you live on a flood plain, downstream from ANY dam, or in an area that has a chance of flooding, include some extra supplies in an **upper floor or attic safe room** or area. Since your situation may be that of a minor isolation, you'll also want to have, in addition to the above items:

❏ Three days worth of food and water for family members (in addition to that in your pantry).

❏ Extra signal gear such as aerial flares, highway flares, chemical light sticks, colored smoke, a signal mirror (or the shiny side of a CD), powerful flashlight, brightly colored towels or sheets (dye or spray-paint some old ones bright orange), airhorn, or aerial fireworks (if legal in your area).

Keeping your pocket goods in one spot makes it easier to grab everything and run.

❏ If you don't own a boat that you keep near the house, you might want to consider keeping an inflatable raft in your flood safe-room. Check your local department stores and sporting goods stores.

➢ **Special note**: Unless your safe room is a full-sized room or hallway, do **<u>NOT</u>** use this as the area you would seal with duct tape and plastic sheeting in the case of a biochemical or HazMat incident, as the amount of breathable air would be a concern. More on this under "**Isolation**."

Procedurals and Tricks

➢ Make it a bedtime ritual to check the locks on all doors and windows. Do the same after any stranger has been in your house unattended. A small percentage of burglaries are committed by service or maintenance people who will unlock a back door or window while they're in your home and then come back later to commit the crime.

➢ Suspend all mail and newspaper delivery when going on a planned trip over 3 days in length, or ask a trusted neighbor to pick them up for you.

➢ For longer trips, arrange to have your grass mowed, and be sure to notify the local police or your landlord (and a trusted neighbor) so that your property can be watched in your absence.

➢ You might also ask a trusted neighbor to park one of their cars in your driveway.

➢ Be sure to note tag number and description of any suspicious cars in your neighborhood (see our "**Suspicious Activity**" forms in the "**Appendix**").

➢ "Light timers" that will turn certain lights on and off while you're away, are cheap and readily available at most department stores. They make your house look occupied.

➢ Also, turn down the ringers on your phone, and the volume on your answering machine, so they can't be heard from outside. You don't want to broadcast the fact that no one's home to answer.

➢ If you live alone, **especially if you're a female living alone**, make it look like you don't.

❏ Have a male voice on your answering machine.

❏ Yell "I'll get it!" when someone knocks on the door.

❏ Yell "It's okay, I've got it!...," or play a tape of a large dog barking in the background if you've received a suspicious phone call.

- ❏ Consider leaving a heavy chain and food bowls outside to make it look like you have a large dog. You might even put up a "Beware of Dog" sign.
- ❏ Have alarm signs or decals even if you don't have an alarm.
- ❏ If you're required to put a name on your mailbox, don't use "Miss" or other titles that would denote an individual living alone.

➢ Always carry a "call in case of emergency" card in your purse or wallet, and make sure that whoever's listed on the card is versed in how to help you in such a situation. For example, if you live alone and list your neighbor on your card, does your neighbor know how to get in your house to feed your pets, and do they know who in your family to call? Use our "**In Case of Emergency Please Notify**" card<u>s</u> and "**Open only in Case of Emergency**" sheet in the "**Appendix**."

➢ Have an innocent code word that you could use over the phone in case you were in trouble. Suppose you were the victim of a home invasion and the burglar let you pick up that incoming call so as not to arouse suspicion from the caller. If the caller was a close friend who knew your code word, you could relay that you were in trouble. This technique is used regularly by battered women reporting in to case workers, and stalking victims checking in with friends and family. A common code word is to say "adios" instead of a normal "see ya" or "goodbye."

➢ If you come home late one night, or unexpectedly after a trip, and can't get your key in the lock or can't get the lock open, back off a safe distance or go to a neighbor's house. A common trick used by burglars is to jam the door lock so home owners can't come in and surprise them.

➢ Similarly, when coming home, try to take stock of your home as you're approaching. Scan for doors that are ajar, open or broken windows, house pets being out in the yard, or lights that are on or off that shouldn't be.

➢ One trick used by forcible-entry burglars or "home invaders" is to smudge the peephole with something like chapstick. It blocks your view without looking intentional, so most people open the door. If you can't see out, don't open the door. If you don't have a camera system, try looking through a secured window or call a neighbor to see if they can tell who's at your door.

➢ As you're making all these Disaster Prep supply purchases, stocking your pantry, etc., don't mention too much to outsiders. You don't know where this information might go. It might go to someone who would use you as their own personal supply depot in the wake of a major disaster.

Miscellaneous

➢ Good security may mean Law Enforcement or Fire Department involvement at one time or another. Make it easy for them to find you by making sure your address is prominently displayed on the front of your house, on your mailbox, and/or painted on your curb. Also, be sure that at least one of these addresses normally has a light on it at night.

➢ Earlier we mentioned a remote-control porch light. You can find the same setup in an interior light configuration. This would allow you to turn on your inside lights on arriving home.

➢ If you don't have caller ID you should consider getting it. It's a great way to screen unwanted calls plus identify those that might be lying about who they are. Burglars sometimes call to see who's home, and phone scams cost people billions per year either through fraudulent sales, or through identity theft. Don't buy or pledge to anything that doesn't offer a local contact, and don't give out your Social Security Number, date of birth, or any other personal information that could be used to access your money or create false accounts in your name.

➢ Have "**intruder drills**" (see "**Tornado**" drills in the "**Reaction**" section) just like you would a fire drill so that all family members, especially children, know what to do, what not to do, where to go, and who to call. Intruders and fire are both good reasons that children should have airhorns, whistles, and other alarm devices in their room (along with 911 phones), just as adults do.

➢ For general information on several home security topics, be sure to visit the "How Stuff Works" page at http://home.howstuffworks.com/channel.htm?ch=home&sub=sub-home-security, for some surprisingly detailed and sophisticated educational tidbits.

➢ General crime-prevention suggestions can be found at http://survivaltoolkit.com/crime-safety/crime.html.

Structural Considerations

The focus of this section is the structural integrity of your residence, since the place you live could be subjected to any number of destructive disasters or terrorist attacks. We'll take an outside-in approach to looking at ways to make your home a safer place to be in an earthquake, tornado, or maybe even a nearby nuclear detonation.

Landscaping

➢ As far as structural considerations go, the landscaping points will mostly cover a reduction in potentially dangerous debris. Be sure to trim dead limbs from trees, have dead trees cut down, and clean up loose yard debris that could become projectiles in violent weather.

➢ Speaking of projectiles in violent weather, strong winds or rushing flood waters can easily pick up the average automobile and flip it or toss it several houses away. Mobile homes can be picked up and tossed the same way. As both homes and cars are vitally important to our emergency reactions, you need to consider using anchor points, especially if you live in areas frequently visited by destructive weather events such as hurricanes, tornadoes, or floods. Though not an entirely foolproof system, yours would certainly be the last car in the neighborhood to be lost.

➢ Anchor points can be decoratively installed, or installed and hidden, in different places in your yard. These anchor points can be as simple as long spikes with a ring attached, or heavy concrete bases with a chain. They can be used to tie down everything from lawn furniture, to tool sheds and outbuildings, to vehicles such as boats and cars, a tree that might fall on your house, or even your mobile home. For commercially manufactured models, go to http://www.google.com and enter the search string "earth anchor."

● To make your own concrete anchor points, take a 5-gallon plastic bucket, hang the two loose ends of a loop of heavy chain down in it, and fill it with concrete mix (found at any hardware store for next to nothing). The free ends of the chain should hang down in the bucket with only a loop of about 4 inches or so coming up out of the top. Bury these completely so that only a small part of the chain is sticking up out of the ground (so you won't hit it with the lawnmower). Or you can skip the bucket part and just dig a hole to hang the chain and pour the concrete mix.

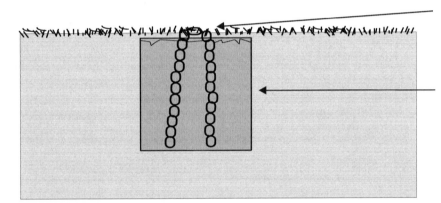

Keep the chain accessible but low so you don't hit it with the lawn mower.

Use a 5-gallon bucket filled with concrete to anchor the chain.

These anchor points are cheap and easy to make. They could save your vehicle or mobile home from high winds.

➢ If you live in a tornado-prone, or hurricane area, we highly recommend having a couple of anchor points to secure your car if it's parked outside (it's safe in a garage or well shielded carport as the moving wind or water can't get underneath it). Though you wouldn't anchor it all the time, you should do so on approach of inclement weather, such as during a tornado watch (NOT a tornado *warning*!). Simply pull your car over the anchors, and loop a heavy chain over your axles through the chain loops of the anchors using a simple hook rather than any lock or link that may jam after being tugged on. Anchors should prevent high winds from getting too good a foothold under your vehicle. Afterwards, don't forget about it being anchored when you go to drive off! (Leave a sticky-note inside your windshield.)

● Two anchor points, one at each end, may suffice for the average car (though we recommend four). You might want two for the average boat and trailer, and one for all your yard furniture that can be tied together with rope. For the average mobile home, we recommend eight large concrete anchor points connected to the frame with heavy chain or heavy cable. This may save your mobile home from being flipped, but we still don't recommend you be <u>in</u> a mobile home during weather that violent, as it can still be destroyed in place.

➢ One way to protect mobile homes from wind is to add siding or other structural material around the base of your home to block wind from blowing underneath. This may help prevent lesser winds from being able to lift your home off its foundation.

➢ Your heavier anchor points can be used to prevent theft. For example, if you have a small electric generator, the only place you can run it is outside, and you may need a place to chain and lock it so it won't be stolen.

➢ Another "landscaping" consideration is flooding. Parts of a town, neighborhoods, and even small sections of a neighborhood may flood depending on terrain, water sources, and drainage. Flooding can occur from the collapse of a dam, heavy or sustained rains, or from storm surge produced by hurricanes along coastal areas. To prep your "landscape" for potential flooding, the first thing you should do is have your public works department check the storm drains in your area. If they drag their feet on the project, and inclement weather is predicted, get together with neighbors and take a look at the storm drains yourself. Better a little elbow grease now than flood mud later. Take the matter up with officials after the storm has passed.

➢ Check with your appropriate municipal departments to see if all area drainage systems, ditches, levees, etc., are routinely checked and cleared. While in contact with these folks, check local regulations on what kind of drainage-oriented "landscaping" you can do on your property.

➢ Also, this seems as good a place as any to suggest you stock up on empty sandbags, heavy plastic sheeting and tarps, and a small sump pump or two, if you live in an area prone to flooding. Better yet, this is an effort and expense best shared with neighbors.

The Structure

As we stated earlier, most homes are designed to resist only gravity and the average storm, and not much else. Though building codes in some of the more earthquake or tornado prone areas have improved recently, many homes are not as structurally sound as they could so easily be.

The types of destructive force homes are most likely to face are high winds from hurricane or tornado, rising and rushing waters of a flood, or the shaking and shifting of an earthquake (distant nuclear blast or industrial plant explosion also being a possibility). During these events, occupants may be in danger of either roof and/or wall collapse, or flying debris. The problem lies in the fact that the vast majority of homes, regardless of brick or wood facade, are simple wooden frames, held together with nails, and merely resting on a foundation. Subsequently, during a destructive event, wood can shift, nails can pull loose, and structures become unstable. Simple changes can be made using cheap and common hardware store parts. In fact, with just these simple parts and simple labor, the structural integrity of your home can be increased dramatically.

If you're in the process of having a home built, or in the midst of heavy remodeling, incorporating these little structural augmentations will be easy. Simply show this list of considerations to your contractor and they'll know exactly what to do. For the rest of us though, let's look at the following:

➢ **Helpful hint**: We mentioned that making these structural upgrades may increase the value of your home. We suggest, as you make these upgrades, that you **photograph or video the changes** and create a home improvement scrapbook. This documentation will be helpful should you decide to sell your home, or if required by your insurance agent for premium reduction.

➢ If you live in known earthquake, hurricane, or flood-prone areas, have your home inspected to make sure it is constructed according to local building codes. If not, some of the changes you may need might be the responsibility of the home's original builder.

➢ You'll definitely want to visit the Disaster Center's page on disaster building and retrofitting for the homeowner. Log on to http://www.disastercenter.com/build.htm.

➢ For a list of earthquake-proofing parts and hardware, log on to http://www.qsafety.com/index_products.html. You can also call them at 800-997-2338.

➢ Also visit the Seismic Safety page at http://www.seismicsafety.com/Basics.htm, for more construction and reinforcement info.

➢ Survey your home and see where you can **access the frame without removing wall boards**, sheetrock, or attic floor boards. Can you get to the studs making up the outer walls? Can you get to the tops of the walls where the roof joists and trusses (or rafters) are attached in your attic? Can you get to the frame where it rests on the foundation? If you can't readily get to these points, that's okay. You can probably access your frame/wall studs by cutting through the walls in inconspicuous spots (such as inside closets). Besides, sheetrock is easy to repair.

➢ After seeing how many access points you have to the frame, go to the hardware store and get the following:
   ❑ A large box of 2" wood screws.
   ❑ Another box of 3" wood screws and ¾" diameter washers to go with them (100 count).
   ❑ Angle brackets, mending plates, or "strapping ties."

➢ If you have a **solid concrete slab** as your **foundation**, be sure to get:
   ❑ A masonry bit for your drill (and you may have to rent a hammer drill).
   ❑ As many concrete anchor screws or studs (and an equal number of 2" diameter washers) as you have access points to the frame where it meets the foundation.

➢ If you have a **wood frame resting on a block foundation**, and you can access where the wood meets the concrete, you'll need either:
   ❑ Corner brackets or "T" mending plates.

❏ Strapping ties.
❏ As much concrete mix as you think you can get into the open blocks.

➢ Our goal with these parts is going to be to:
❏ Make the roof truss system a little more stable.
❏ Secure the roof structure more solidly to the frame.
❏ Add structural support to the frame to keep the walls from shifting.
❏ Anchor the frame to the foundation.

➢ Though this sounds like a difficult goal, it's not. For the "average" size house, you'll probably use around 40 angle brackets or mending plates, about a dozen concrete anchor screws and washers, and just under 200 wood screws. You'd be surprised at how much stronger your home will be with just the addition of a few pieces of hardware. **Hint**: One of the things that will help the most is the use of screws instead of nails. Nails pull loose, but screws will hold, especially when they're long enough to sink into the wood, and you use washers to keep the screw from being pulled through the piece of wood it's holding.

Angle brackets and mending plates are inexpensive, and will add a surprising amount of stability to your home's frame if strategically placed.

➢ **For the roof joists**: Make sure the truss units themselves are structurally sound, make sure they're anchored well to the frame of your house, and add a little support against each other. After all, the only thing keeping the truss units from shifting side to side and falling is the thin plywood of the roof held in place with only a few nails. While in the attic, do the following:

❏ Dress to protect yourself against insulation and spiders. Wear a cap, long sleeves, long pants, gloves, and most importantly, goggles (wrap-arounds with no openings) and a dust mask. Insulation not only itches when it gets on your skin or in your eyes, it can cause lung problems.

❏ Make sure the truss units are held together with nailing plates on both sides and that the nailing plates are further attached with nails or screws. If not, add screws and mending plates on at least every other one.

❏ Add angle brackets in various spots around the attic where the joists rest on a part of the frame. Don't worry about anchoring each and every joist. Anchoring every other joist in one or two spots will make a tremendous difference. Plan on anchoring in about a dozen spots if you can access that many.

❏ In addition to angle brackets, you can use lengths of galvanized strapping.

❏ Along with the internal frame attachment points, see if you can the locate points where the truss units rest on top of your exterior walls. Anchor the trusses there as well.

❏ If the truss units are a simple series of parallel units (most common) and are only attached to the frame and the roof, you'll want to brace them against each other so they lend each other some support. Use 2x4 lumber connected horizontally to the joists much in the same way a rail supports fence posts, or a hand-rail supports banister uprights. Use screws and small angle brackets instead of nails (always!). You can even use this opportunity to create a small shelving system in your attic to store your extra junk or extra emergency supplies. Not only will this increase your home's ability to withstand high winds, it will help protect against such things as falling trees which might otherwise crash through your roof. In fact, if you have a tree that might fall on the house, add extra support inside the attic where the tree might hit.

❏ You might also see if you can access your roof studs underneath the eaves of your house at the soffets. These are the areas your roof extends over your outside walls. If you can get underneath any of the soffets and reach the roofing studs, you can attach these studs to the outside wall frame with small to medium-sized angle brackets or strapping. This won't work on all houses, so if you can't get to this area, don't worry. Your earlier efforts to anchor the trusses to your house's frame in the attic should be fine.

❏ In addition to using common angle brackets, take a look at the engineered braces and brackets offered by Simpson. They're online at http://www.strongtie.com, or reach them at Barclay Simpson, Simpson Manufacturing Co., Inc., 4120 Dublin Blvd., Suite 400, Dublin, CA 94568, Tel: (800) 999-5099, or (925) 560-9032. While on the website, be sure to click on "Literature" and download some of their info and plans.

➢ While you're in the attic, see if you can add extra support to overhead light fixtures and ceiling fans. Seal around them for better insulation and a better seal against a biochemical attack. Also, they're nothing more than falling hazards in a destructive event so anchor them well.

➢ **For supporting the frame**: As mentioned earlier, all it takes to add a great deal of support to the basic frame of a house is a few angle brackets or mending plates, and some wood screws:

❏ While still in the attic, see if you can access the wall headers. These are the boards that form the top edge of your walls. Add a 3-inch screw here and there (avoiding electrical wires or plumbing) where you know you'll hit a vertical wall stud (especially at the corners of interior support walls). "Spread the wealth" a little bit and don't worry about anchoring each and every stud. About two dozen screws, evenly spaced around the attic can be enough to make a difference.

❏ If there are spots in the attic with wide gaps between trusses, you can add a little support to the frame and protection for the rooms below by adding a few 1x4s or 2x4s at right angles to the ceiling studs over the room or hallway below. Install them as if you were putting a floor in your attic. As with all other pieces, make sure these boards are held in place with screws and washers.

❏ Wherever you can access the frame, whether in an open section of unfinished wall, or in an inconspicuous spot in a closet, etc, add an angle bracket or mending plate. This is especially important regarding the central "core" of the house, or the interior supporting walls (and certainly where they meet exterior walls). However, any bracket you can add is a good thing! The best way to accomplish this is by creating safe rooms out of closets as we'll discuss below.

❏ If you ever remodel or change out old window frames, be sure to augment your window frame's structure with a few screws or some inlaid angle brackets.

➢ This extra internal structural support will help reinforce central hallways, closets, and bathrooms. In lieu of a specifically augmented safe room, these will probably be the areas your family would go in the event of severe weather or earthquake. The more you can do, the stronger your home.

➢ For anchoring to the **foundation**: Other structural support does you no good if high winds or flood waters push your house right off its foundation. Consider doing the following:

❏ If you're on a solid slab foundation, use your concrete anchor screws to anchor the frame to the foundation anywhere you can find a spot.

❏ If your foundation is a footer of concrete block, try supporting the blocks by dropping rebar, metal piping, or old metal rods down in the holes and filling the holes with concrete. Concrete mix is really cheap and you can complete this task in small increments as time, money, or energy allow. While working the block, use mending plates or metal strapping to anchor the wooden frame to the blocks, taking advantage of the wet concrete you're pouring. Just attach it with screws to the wood frame and let the straps or plates hang down into the wet concrete inside the blocks. Use the longest straps you can.

Use heavy screws to hold the strapping to your house frame.

Use heavy concrete anchor screws to attach your frame to your foundation. Use washers on all screws.

Dig a three-foot hole and fill it with concrete to hold your anchor strap.

A little anchoring can make all the difference in the world in keeping your home on its foundation.

❏ If you can't access your foundation from inside the house at all, you can still use "strapping" (a 6-ft heavy-metal strip or length of galvanized pipe) to anchor your home more thoroughly. If there are hidden spots outside the house (at least 4 spots) where you can anchor a strap directly to the frame, you can dig at least a two-foot deep hole and anchor the strap with concrete as shown above, using wood screws or bolts long and heavy enough to bite into the frame's wall studs. You can also drive a 6-foot length of 1-inch galvanized steel pipe into the ground and attach it to the house using half-moon clips and heavy screws.

❏ As an expansion on the idea of external strapping, some homeowners have added heavy strapping in *numerous* places around their house and then sealed and decorated all the way around the house with a three-foot high brick veneer and/or shrubs.

➢ **Note**: This will not keep your structure perfectly intact during a direct hit from a tornado, or the likes of hurricane Andrew, but it should provide you much, much better protection than without it. Like we said earlier, it'll make the difference between a damaged home and a destroyed one and that can make the difference between you walking away, or receiving severe injuries or worse. Or, you may not be home at all. If that's the case, these simple structural enhancements will protect the property that's inside your house. Better a partially collapsed home containing all your worldly possessions that a house and its contents strewn down the block, or floating down a new river.

➤ Two more places to check: One is your **exterior doors**, and the other is your **garage doors**. If either give way during heavy winds, that will allow the wind to get into the house and under your roof. Make sure you follow all door and deadbolt, and chain and doorstop recommendations under "Security" above. Not only will this help keep a person from kicking in your doors, it'll keep the wind from doing the same. As far as garage doors go, make sure the tracks are securely anchored with heavy screws and that the door material itself is strong and in good condition. If your garage doors are primarily wood, you may run a 1x4 piece of lumber across panel sections to reinforce the door. You might also cut a protective sheet of plywood to cover your garage doors for hurricane protection, just as you do your windows.

➤ While still under the concept of violent weather, most of us are aware of how destructive lightning is. If your house is a little taller than the rest, sits on top of a hill, has taller trees nearby, or if you have a lot of electronics you wish to protect, you might want to look at having lightning rods installed. Not only is it a physical safety measure, it's a fire prevention measure as well (especially if you have aluminum siding). You can get more information from your local hardware store, or from the "Lightning Rod Parts" store at http://www.lightningrodparts.com, and at 6217 S. Candice Path, Homosassa, FL 34448, 877-866-3189.

➤ Moving along to other structural considerations, we'd like to reiterate the importance of good insulation, and sealing open seams. Not only will doing this help cut your utility bills, it'll make your home a safer place in the event of a biochemical attack or HazMat accident. Be sure to check the fit of your doors and windows, caulking around the outside, or adding weather stripping and threshold seals as needed. Also, if you were ever considering having new siding added to your frame home, new siding will help cut your utility bills, add a small layer of projectile protection, and will add a tiny bit of sealant against hazardous materials.

➤ You can also increase your home's insulation either by adding insulation sheets or blankets to exposed framing and attic areas, or by introducing "expanding foam" into hollow walls (be sure to follow manufacturer's instructions on the foam). Good insulation not only saves on the bills, but makes it easier to regulate your home's temperature during a utility outage.

➤ Speaking of temperature regulation, if you live in warmer areas, you're sure to want roof vents to help keep your attic cool during the summer. If you haven't yet gotten roof vents, consider this hidden benefit: One of the ways tornadoes destroy houses is by creating a sudden vacuum outside. Normal pressure inside a house can't vent quickly enough and the house literally explodes. Having roof vents, in either a ridge line, or turbine configuration, may help pressure equalize more easily and may help spare your house.

Windows

Windows have their own unique needs, so we'll give them their own section. You'll want to save on utility bills, provide good security, and have your windows help seal your home during a terrorist biochemical attack or local HazMat accident. During any kind of destructive event, windows and glass will become nothing short of deadly razor-sharp shrapnel. Also, broken windows during a violent storm will allow destructive winds to get underneath your roof and lift it off your house. Let's look at a few things to do to your windows.

➤ As stated earlier, be sure windows are caulked around the frame on the outside, that they fit snuggly in their tracks, that the window frame is firmly attached to the house's frame, that they have good locks, and are of tempered, double-paned glass to resist breakage and provide good insulation.

➤ You can add to the strength of the glass and reduce its shrapnel effects by applying a permanent plastic film, either automotive window tinting (which will also help cut down on summer sun heat) or a thicker film from http://www.blastgard.com/home.html which will hold broken window glass in place after an explosion. Contact BlastGARD at 1.888.306.7998, Fax – 888.646.8913. Also look at the "Roll-on™ Shrink Film Window Insulator Kit" produced by http://www.duckproducts.com. Contact Duck Products at 1-800-321-0253.

➤ In earthquake areas, beds, and especially cribs, should not be placed underneath regular windows.

➤ To even further reduce the potential destruction of flying glass, make sure you have both blinds and heavy curtains that can be drawn over windows, especially your larger windows or windows near your family's safe room or safe area.

➤ If you live in hurricane or tornado prone areas, you may want to invest in storm windows <u>and</u> storm shutters. In other areas where destructive weather is possible but not so prevalent, you might consider having your decorative louvered shutters (if any) actually be functional. These louvered shutters can be inside and/or outside and can be structurally augmented by using heavier hinges, longer screws, and heavier hasps. (Keep this in mind when reading the section on "Nuclear Detonation.")

➤ For areas with frequent hurricanes, make a set of reusable window covers out of ¾" marine plywood (if you have room to store them). Cut the plywood to overlap each window 4 to 6 inches each direction, paint it to help protect it and make it reusable, label it as to which window it fits, and go ahead and mount handles and maybe heavy sliding bolts (to hold the panels up while you screw them in place) so you can put the covers up and take them down with ease. If it takes two sheets or pieces of plywood to cover a window, you should also cut a length of 2x4 to go across it as a brace. If your home is brick, you'll need to go ahead and drill holes to install anchor spots for the screws that will hold the plywood in place during a storm. Waiting until the last minute to make a frantic trip to a hardware store as a storm approaches is *not* the thing to do.

➢ The closer you live to the coast, the more you should consider having an additional set of heavy plywood storm panels to mount *inside* your house to cover all windows.

➢ The Department of Energy has a grant program to help those on lower incomes improve their insulation and storm windows. Contact them at Director – Office of State and Community Programs, Mail Stop EE-44, Office of Energy Efficiency and Renewable Energy, US Department of Energy, Forrestal Building, Washington, DC 20585, 202-586-4074, or at http://www.eren.doe.gov/buildings/weatherization_assistance.

➢ If your windows have burglar bars, make sure they can be opened with a latch from the inside. Otherwise, they'll only trap people inside during a fire. Also, make sure your windows aren't painted or nailed shut.

Utilities

Some utilities, such as gas or electricity, can be destructive in and of themselves if safety precautions aren't taken. They can also be easily lost during a destructive event unless preventative measures are considered ahead of time.

➢ As mentioned before, know how to shut off and restart all utilities, and make sure all family members and older children know how to do it, and where the tools are kept.

➢ To help locate necessary switches, valves, and other important locations during a power outage, give them a coat of **glow-in-the-dark paint, and apply reflector strips**. You could even label your individual breakers using glow paint. We've found you some sources for glow-in-the-dark products to include paint, tape, signs, and stickers. If your local art store, hardware store, or department store is out of stock, try these locations:

● **eGlow** products at http://www.eglow.com, and at eGlow, 1601 N.W. State St. Bldg 1A, Lehi, UT 84043, Fax: 801-437-2654, toll-free: 866-999-4569.

● "**Evac Map**" online at http://www.evacmap.com, with direct contact at PictoGraphix Designs – 5191 Doherty Ave. Montreal QC Canada H4V 2B4, 1-800-504-3822.

● **Glowinc** at http://www.glowinc.com/Default.htm, and at Glow Inc., Suite 100, 1539 Florida Ave., Severn, MD 21144, 410-551-4874.

● Visit http://www.chemlite.com, and http://www.chemlite.com/glo/safety.html. For Chemical Light, Inc., 595 N. Lakeview Pkwy, Vernon Hills, IL 60061, 800-FOR-GLOW (367-4569), 800-446-3200, Fax: 847-680-9250, International: 847-403-0100.

➢ Automatic gas shutoff valves will automatically cut off your gas in the event of earthquake, building destruction, or fire. Contact your local Fire Department or fire equipment distributor.

● Once source of earthquake triggered gas valves is **Safety Central**. Locate them online at http://www.safetycentral.com/safetycentral/autgasshutof.html, or contact them directly at 1370 Highway 10 West, Livingston, MT, 59047, Phone # 406-222-3171.

● The Earthquake Store, online at http://www.earthquakestore.com, offers a wide variety of gear to include gas shutoff valves. You can also reach them at Earthquakestore.com, 1300 Gardena Avenue, Glendale, CA 91204, Toll Free: (888) 442-2220, (323) 245-1111, Fax: (818) 240-1492.

➢ In high-risk destruction areas, invest in flexible fittings for your gas and water connections.

➢ In "Isolation" we'll discuss water storage considerations. One technique concerns structural modifications so we'll go ahead and mention it here. The suggestion is to have in-line water tanks along your incoming water line somewhere in the house (if you have room). These tanks can be specific cistern tanks such as those found at the sources below, or you can use old and inoperable, but clean and non-leaking, hot water tanks hooking one or more in series. A plumber can help you further. You could even install a regular water storage tank along your pipeline in the attic, provided it had proper support. This attic tank is a good idea if you have poor water pressure as this could act as your own private water tower. As this is an optional consideration, balance the cost against your probable threat, and your other disaster prep needs.

● **The Tank Depot** at: http://www.tank-depot.com, 641 S.W. 7th Street, Pompano Beach, FL 33060, 727-546-1790, or 954-783-0115, Fax: 954-783-9420.

● **Essential Water Storage** at: http://essential-water-storage.com/, Essential Water Storage, PO Box 70, Winfall, NC 27985, 1-252-426-9953.

● **Complete Water Storage and Disposal**: http://www.watertanks.com/, Water Tanks.com, P. O. Box 340, Windsor, CA 95492, Toll Free (877) 655-1100, Local (707) 535-1400, Fax (707) 535-1450.

➢ You can also set up a few rain collection barrels in your yard. Just make sure they're sealed with screen to prevent mosquito breeding so you don't create your own West Nile production facility. Some barrels have a funnel-top and simply sit in the yard. Others utilize your rain gutters (roof runoff cannot be used for drinking, but is great for washing or watering the garden). We found some examples at:

- **"The Green Culture,"** http://www.watersavers.com/docs/rainbarrel_main.shtml. Contact these folks at 23192 Verdugo Dr., Suite D, Laguna Hills, CA 92653, Phone 800-233-8438, Fax 949-360-7864.

- See the **rainbarrel guide** at http://rainbarrelguide.com, (has good sources and links).

➤ Make sure any **propane or fuel-oil tanks** are securely anchored lest they float in a flood, or shift in a violent storm or earthquake. Similarly, make sure external lines and connections are protected from damage.

➤ With propane or fuel oil tanks, like your car's gas tank, **never let them get below ½ full**.

➤ Also in flood prone areas, consider moving all external utility connections such as electric, phone, etc., to an upper portion of the house, or up near the roof on a one-story house, rather than at ground level. Also, make sure lower floor outlets are on different circuits from upper floors so you can cut off lower floor electricity at the breaker box to help prevent an electrical short during a flood, which could burn your house down.

➤ As time and money allow, you might want to consider the same moves for your central heating/air-conditioning unit(s), furnace, and any home electric generator you may have. Move them to the roof or attic if you live in an area likely to flood. Check with your flood insurance provider to see if any of these moves will result in a premium reduction. Also check with your local building and zoning departments to see if these modifications should have been incorporated in your home by the original builder.

➤ Ask your electric company to let you put a lock on the box around your electric meter. Meter removal is not only a breaking and entering trick, it's big business in some urban areas. Meters are stolen and then illegally sold to people who've had their power permanently turned off for non-payment. Keep the key handy.

➤ Speaking of securing your meters, most people will put boards over windows as a hurricane approaches, but won't protect their utility connections. If you ever do hurricane prep work at your house, don't forget to make a plywood cover for your electric meter, gas meter, water meter if it's exposed, and any exposed connections to an outdoor fuel tank. Power and gas to the neighborhood might be restored quickly, but it would be useless to you if your meter was impaled by a flying 2x4.

➤ If you're connected to a municipal sewage system, look into getting an anti-backflow valve (or sewer plugs) to put into the sewage line between your house and the street connection. Also install float-plugs to sink and floor drains. This will prevent a backflow of raw sewage into your house; a common occurrence in floods. (The flood doesn't even have to be in your yard, only somewhere along the sewage line.) Your water / sewage company or local plumber can tell you more. While you're gathering info, ask if your water line also has an anti-backflow valve that would allow water to stay in your pipes should the municipal main be broken.

➤ Insulate all exposed piping. You'll want to prevent lines from freezing, and hot water lines from wasting energy. Also look into getting little styrofoam covers for your outdoor faucets in the winter. It would be a shame to have a fire that you couldn't help fight because your outside faucet was frozen solid.

Safe Room

Previously, we discussed the "safe room" as a security concept and listed some of the equipment items one should contain. Here we'll cover some of a safe room's *structural* considerations.

➤ If you choose not to structurally augment an area to be a safe room, you should at least know what part of your house is the strongest so you can gather the family there in a destructive event. Choose an area of the lowest portion of your home where you have additional structure such as in a central hallway, central bathroom, a closet near the center of the house, a portion of your basement near a supporting wall (or under a heavy workbench), or in the case of floods, a place in your attic. You can also find safe spots in an area of a room that has sturdy or well-anchored furniture, such as under a heavy wood dining table, or behind a heavy sofa next to an interior supporting wall (never near an exterior wall).

➤ Your augmented safe room or area may also be a reinforced closet, a modified central bathroom, a structurally enhanced hallway, a secure area of your garage, a reinforced portion of your basement, an outdoor underground storm cellar, or even a specially constructed bed frame.

➤ Our favorite is the reinforced closet. This will work for a standard frame and sheetrock house in a non-flooding severe weather or distant-nuke scenario. Pick a walk-in closet or small hallway near the center of your house on the lowest floor (a centrally located bathroom will work, but there's more work involved due to all the fixtures). We're going to reinforce this area to the point that if the rest of the house were blown away by a tornado, this closet might be the one little "box" remaining on your foundation. Let's do the following:

❏ Empty the closet of all contents, shelves, and remove any wood molding.

❏ Remove all sheetrock from the walls and ceiling, and remove all sheetrock nails and screws. Wear a dust mask while you do this to protect your lungs from sheetrock dust.

❏ Install angle brackets as needed to secure the upright studs, ceiling studs, and corners.

❏ Install additional ceiling studs and vertical wall studs (with screws) if the spacing between existing ones is greater than 16". If you can afford it, add extra studs anyway, regardless of current spacing, for the extra strength that the studs will provide.

❏ Install random horizontal cross-pieces of 2x4 between every other stud set, anchoring them with screws and angle brackets.

❏ If your lowest floor is your only floor and/or you can access this closet's ceiling from the attic, add some 2x4 cross members to reinforce the ceiling. Attach them at right-angles to the existing boards and studs.

❏ If the closet rests on the foundation, take this opportunity to help anchor the frame to the foundation by using concrete anchor screws and washers.

❏ Have an electrical outlet and overhead light fixture installed if one does not exist already. If you already have an overhead light fixture and don't want to spend the money to install an extra electrical outlet, buy one of the extension sockets that screws into your light fixture and has a plug socket on the side, or at least snake a permanent extension cord into the closet from a nearby outlet. In any event, you want to have power available in the closet when the door is closed.

❏ Install a phone jack, or at least run an extra line off a splitter from a nearby phone jack. (Easy to do with department store parts.)

❏ If you have the materials, paint all studs, supports, cross-pieces and the backside of adjoining rooms' sheetrock with a fire-retardant paint.

❏ Consider installing a layer of fireproof insulation, and/or galvanized sheet metal. Both will give extra fire protection, and the sheet metal will give a little extra protection against projectiles. For fireproof insulation, look at Dow's new "Great Stuff Pro, Gaps and Cracks Foam Sealant." It's been recently approved for use as a fire-block sealant. You can learn more at http://www.dow.com/greatstuff/pro, or by calling 1-800-668-3801.

❏ For the walls and ceiling, use ¾" plywood (not fiberboard, not particle board, not chip board – plywood.) instead of sheetrock, and mount it to your studs using generous amounts of Liquid Nails ®, and numerous 2" wood screws and washers, anchoring them with a screw at least every 6 inches along the edges, and at least four around the middle of the plywood sheet. Also, if you're using the flame-retardant paint, be sure to paint the underside of the sheets before mounting them in place (you can paint the exposed surface later).

❏ If you want extra strength, such as in a situation where one of your safe room walls has to be an exterior wall, use two layers of the ¾" plywood against that wall. You might want to add a layer of inexpensive galvanized sheet metal between the plywood layers and/or a layer of chicken wire between the boards and the wall studs. This inexpensive "armor" will help protect against projectiles common in high winds. (A hurricane or tornado can drive a loose piece of lumber through a concrete block wall with no problem!) Consider the same for the closet floor and ceiling.

❏ If this safe room is to be part of the area you'd seal in a biochemical incident, be sure to seal each and every seam with exterior acrylic caulking as you're putting your plywood sheeting in place. The Liquid Nails will help a little, and the caulking will finish the job.

❏ Paint the plywood as the closet was originally painted, and remount your shelves and reload your closet (to include all recommended emergency supplies). **Note**: If you're doing this to a bathroom, small hallway, or section of hallway, you can resurface the plywood using sheetrock mud, and then sand it, and you'll never know an alteration was made! If you have a smaller house, making these kinds of reinforcements to a central hallway could actually help strengthen the whole structure.

❏ For added structure and protection, if you have room in your closet, bring in a dresser, maybe a tall and sturdy book shelf, or build a sturdy wall shelf unit inside the closet. Make sure each is mounted securely to the walls. (In fact, if you absolutely can't do any of the other steps, adding this sturdy furniture will greatly increase the safety offered by your interior closet.)

➢ If this area is also intended for safe-room protection from crime, you should install a solid-core wood door (or metal door), reinforce the door frame as you would an exterior door, and use a deadbolt lock. Whether a closet or a hallway, you can close yourself off from the rest of the house. You can make your interior walls stronger by using two layers of plywood, and a layer of galvanized sheet metal as mentioned above. Again, balance the cost against your true need.

➢ Speaking of crime, while you're doing all this work demolishing your closet, you can mount an inexpensive wall safe, or mount a heavy chain to a stud to secure your fireproof safe.

➢ If your home is constructed in such as way that you can't make these kinds of alterations, such as with a pre-fab or mobile home, or an older home with more delicate plaster and lattice construction, you can go into a larger closet and basically construct a box within the closet leaving its original walls intact.

❏ Create a framework of 2x4 studs anchored to each other using angle brackets, corner plates, and 1.5" wood screws. These frame pieces are anchored to each other and not to your existing walls. Also, to save space, you'll want to turn the 2x4s so that they're "flat" against the wall (with the 4" side up against the wall).

❑ Once the frame is in place use ¾" plywood sheeting anchored to the frame using generous numbers of 1.5" wood screws and washers. Using screws is stronger, plus it makes it easier to remove the setup if you want to make changes or completely disassemble it later. The washers keep the screw heads from being pulled through the plywood if the structure is placed under stress. You don't want to use Liquid Nails ® here since you might not want this box to be a permanent fixture.

❑ Paint the plywood walls as desired and reinstall shelving and hanger rods.

❑ If no overhead light fixture or wall outlet was previously installed, try running a small extension cord along a baseboard from a nearby outlet so you can have power in your safe room when needed.

➢ If your safe area is part of the basement or garage, reinforce the area much as you would the closet above.
❑ Use angle brackets and support pieces on accessible studs.
❑ Add ¾" plywood walls for both structure and protection from flying debris.
❑ Don't forget all your safety gear.
❑ Remember you can also have a heavy workbench as part of your protection.

➢ Consider this; if you did this kind of remodeling to all the closets in your house, you'll have effectively strengthened the entire structure.

The average frame house is held in position with just enough nails to hold the lumber together. An example is on the left. Adding angle brackets and mending plates in key junctures, and holding them in place with long screws, will add a tremendous amount of structure to the average house. See how many places you can reinforce your home's frame.

➢ In earthquake prone areas, you might want to do each closet in your house like this, and certainly follow all recommendations for the overall structure we mentioned earlier. To this, we'd add the suggestion of reinforcing all doorways where you could access the frame.

➢ In flood areas, you'll want a safe room as high in the house as possible. Make all the same reinforcements as above since high and moving flood waters will try to push your house off its foundation and shift the frame structure in the process. You need to stay safe until you can get out. The only difference with this type of safe room is that you'll want a second way out so you can get to the roof. If it's an attic safe room, make sure it's adjoining a gable or window, or make sure you have the tools necessary to cut through the roof so you can get out if waters continue to rise. If it's an upper floor room with no space in the attic, have a ladder, strong trellis, or other setup that will allow you to climb out a window and onto the roof.

➢ From the late 40's through the late 70's the "Cold War" was in full swing and many families built fallout shelters. Though shelters are temporarily out of vogue, "storm cellars" are as useful as ever in some areas. Many companies offer commercially manufactured and prefabricated underground shelters. For a rather high price, they'll come to your property, dig the hole and drop your new shelter right in place. (More in **"Advanced Home Prep"** under **"Isolation."**)

➢ For a more economical in-ground storm cellar (depending on the size of your family), call your local septic tank installation company. Since there would be no plumbing involved, they can probably give you a good price on installing a <u>new</u> concrete tank with cover. All you'd need to do is fabricate two entrance hatches, make sure water and vermin didn't accumulate inside, and make sure the entrances were located so that you couldn't be trapped inside the cellar. In fact, your storm cellar, doesn't even have to be completely buried. It could be partially buried with a recreational deck built over it.

➢ You could also purchase a length of 6-foot diameter corrugated metal culvert pipe to create a small cave in a hillside, or weld on ends and use as it an underground shelter.

Furnishings

➢ We mentioned that your safe area could be a specially built bed frame. As that's more a furniture item than a structural modification, we'll start our "Furnishings" portion with it. If absolutely nothing else, make an extra sturdy bed frame that's high enough off the floor for the bed's occupants to roll off the bed and underneath for protection. This is a great concept for studio apartment dwellers in California as earthquakes can happen in the middle of the night without warning. The same can be said for tornadoes. If you choose the bed frame route though, make sure you don't use the space for extra storage. It won't do you any good if you can't dive underneath. However, you're good if the bed is near an interior supporting wall and there's room for the bed's occupants to roll of the bed between it and the wall.

➢ Though your bed frame support could be made out of anything that gave both good support and good clearance, like a couple of rows of milk crates, it really needs to be constructed as one strong, cohesive, unit. We recommend a box frame made out of 4x4 and 2x4 lumber, which should be easy enough to make if you're handy with tools and wood. This bed frame should be anchored to both the floor and to the wall.

➢ We also recommend the same kind of safety gear for under the bed as was mentioned for a closet safe room. At the very least, you'll want an AM/FM radio and a phone already underneath your bed (or close enough to reach out and grab) so you'll know what's going on and you can call the appropriate authorities if you wind up trapped in your room.

➢ As stated at the beginning, the greatest threat to a house's occupants is from collapse of the structure. Sturdy furnishings, such as the bed frame mentioned above, can offer expedient safe areas. You can buy sturdier furniture such as heavy hardwood tables, dressers, book cases, and entertainment centers; you can modify some of what you've already got; or you can learn what's already sturdy, such as your refrigerator, stove, heavy metal bathtub, or washer and dryer, and make a mental note to use these to your protective advantage in a house-threatening situation. Make sure everything's secured in position.

➢ If you live in an earthquake zone, see if your wall cabinets, or any other wall-mounted furnishings are properly anchored. Also see if they can be retrofitted with knobs or childproof latches that have to be turned to open. This will help prevent contents from spilling out during an earthquake or other destructive event.

Miscellaneous Sources

➢ This site offers an extensive list of links to free construction magazines, tool catalogs, etc. Very detailed and extensive, it's worth seeing: http://www.freeconstructionmagazines.com/Links.aspx.

➢ See the "Storm Survival Page" at http://www.stormsurvival.homestead.com/Pre_Hurricane_Season.html.

➢ Be sure to read "A Residential Construction Checklist to Help Reduce Earthquake Damage," online at: http://www.seismo.nrcan.gc.ca/hazards/prepare/eqresist_e.php.

➢ Remember to visit the Institute for Business and Home Safety, online at http://www.ibhs.org, or contact them directly at 4775 E. Fowler Avenue - Tampa, FL 33617, Voice:(813) 286-3400 · Fax:(813) 286-9960.

Fire

Though listed here as a subsection, fire is a disaster and unforgiving enemy unto itself. It's a tool of terrorists and criminals, it's a secondary effect of other disasters, and it's a costly result of ours or a neighbor's carelessness.

Since it's such a huge topic, you'll find separate fire sections in this book. Under "Basic Training" we recommended you find as much local firefighting training and background information as you could. Here we'll cover some fire safety changes and equipment purchases you should make to keep your home as protected as possible. Under "**Reaction**" we'll discuss some situational scenarios, and under "**Isolation**" we'll cover some useful techniques on how to safely keep a fire in check until the pros can arrive.

Good fire protection depends heavily on three key elements:
1. **Thorough prevention.**
2. **Early detection.**
3. **Rapid reaction.**

For now, let's start with a little safety background.

Overview

Safety always begins with you. Under "Prevention" below we'll cover some of the specific things you need to do to your home to reduce the risk of fire. In the meantime, here are a few independent safety information sources. (Don't forget all the sources you reviewed under "**Basic Training**" earlier.)

- In addition to these websites, your **local Fire Department** can help you with any fire safety questions this section doesn't answer.
- Definitely visit Google's Fire Safety links collection at http://directory.google.com/Top/Kids_and_Teens/Health/Safety/Fire_Prevention_and_Safety.
- More good info can be found on the Elbert, CO Fire Department website. Visit them online at http://www.elbertfire.com/ElbertFireRescue/EducationAndResources/FireSafetyAndPrevention.htm, or contact them at P.O. Box 98, 24310 Main Street, Elbert, Colorado 80106, (303) 648-3000.
- Another good fire information links collection can be found at "Teach-nology's" page. Visit them online at http://www.teach-nology.com/edleadership/disaster_preparedness/fire.
- The LA Fire Department has a number of safety info downloads at http://www.lafd.org/factsheets.htm.
- Basic fire safety and prevention is at http://www.flagleremergency.com/fire/default.asp.
- And a little more can be found at http://firesafety.buffnet.net.
- Fire safety and product recall site http://www.firehouse.com/infozone/firesafety.
- Fire-safety and coloring book pages for kids http://www.stayingalive.mb.ca/kids_colour_pages.html.
- More fire safety for children http://www.sparky.org.
- Some fire extinguisher and fire safety info http://www.stayingalive.mb.ca/fire_extinguisher.html.
- Fire safety info http://www.stayingalive.mb.ca/fire_checklist.html.
- More fire and home safety http://www.stayingalive.mb.ca/fire_safety.html.
- Wildfire protection tips http://www.firewise.org/fw99/home.html.

Prevention

For general preventative measures, let's take a look at the same areas we did earlier under "General Home Safety":

Landscaping

- Your earlier landscaping efforts will help with fire prevention. Making sure that all dead limbs, overgrown hedges or grasses, are cleared away will make it more difficult for a neighboring brush fire or house fire to spread to your property. Also keep dead leaves raked, and roof and ground gutters clean.
- Definitely keep the area around your outdoor air-conditioning compressor clean, and make sure to clean the unit's filter seasonally. Dried grass and other debris can collect inside the unit where a spark can set it off creating your own little brush fire in the backyard.
- If your area is at a higher than normal risk of wildfire, keep a clear area around your house of at least 30 feet (including the firewood for your fireplace), and make sure grass in the area is kept nice and green. More on this subject can be obtained through "Firewise" at http://www.firewise.org, and their "Communities" page at http://www.firewise.org/usa. Contact them directly at Firewise, 1 Batterymarch Park, Quincy, MA 02269.
- Similarly, make a serious "burn proof" zone around any outside propane or fuel oil tanks, and add a sprinkler head to cover your outdoor tank if you have an in-ground lawn sprinkler system. You could also come out of your attic with a branch of either your automatic or "Poor Man's" sprinkler system we'll cover later.
- Make sure your chimney has a wire-mesh spark arrestor on top with a cover.
- If you're in a wildfire area and can afford it, have an in-ground lawn sprinkler system installed in your yard. Also, even though there may be very little grass there, have the system also cover the small gap between you and any house closer than 30 feet.
- Consider getting an auxiliary pump and water holding tank for the sprinkler system, because in a wildfire situation, everybody on the block is going to be trying the same sprinkler trick, and that will leave very low municipal water pressure. Make sure your lawn sprinkler system can tie into this auxiliary water system.
- If you live in an area with potential wildfire or forest fire, and you have a swimming pool, get a pump that can use the swimming pool water for your sprinkler system if your regular water source is out of commission.
- Get insulated covers to winterize your outside faucets. If you have to fight a fire, you don't want to be hampered by a frozen faucet.
- Make sure there are no flammable materials underneath any outdoor decking attached to the house. In the case of extremely low decking, seal around the edges so that burning embers cannot blow underneath, and/or include the deck area as part of your lawn sprinkler coverage for the purposes of fire prevention.
- Also in wildfire areas, make sure all your window and porch screens are metal rather than plastic or other meltable materials. In a wildfire, the wind will carry burning embers to your area well in advance of the flames. For this same reason, you should keep your gutters clean and make sure that any openings in the roof where embers could settle and catch the roof on fire are sealed with screen or metal flashing, and that they're regularly cleaned.

➢ Metal storm shutters coupled with storm windows are highly useful in a wildfire situation. Also, cut and prep plywood boards to mount over each window as you would for a hurricane. Treat these boards with a flame retardant mixture and paint them with a flame retardant paint. Include storm shutters or boards to cover any attic or basement openings or vents. We've found a few sources of flame retardant paints and treatments at:

- Flamestop at http://www.flamestop.com, PHONE: 817-306-1222, FAX: 817-306-1733, Toll Free: 1-877-FYRSTOP or 1-877- 397-7867.

- Flamebusters at http://www.flamebusters.com, Flamebusters, Inc., 70 Plain St., Pawtucket, RI 02860, 1 (401) 729-53221, (866) 729-5322.

- Unishield LLC, http://www.universalfireshield.com, Corporate Office and Chemical Manufacturing, 6311 Washington, Unit N, Denver, CO 80216, 866-860-6584, 303-853-0204.

- Firetect, Inc., http://www.firetect.com, 26951 Ruether Ave., Unit D, Canyon Country, CA 91351, (800) 380-8801, (661) 298-8801, Fax (661) 298-8851.

- National Fireproofing Co., http://www.natfire.com, 427 Anne St. Verona, IL 60479, Phone: (815) 287-3473, Fax: (888) 202-7801.

➢ Know your local burning laws before burning any trash or leaves. When burning, make sure you're always present and that you have a heavy rake, garden hose, and phone nearby.

➢ If you live in an apartment or condo where your unit is attached to someone else's, be sure that the property is up to fire code. Call your local Fire Inspector or Fire Marshal for more information. You'll want to check on attic sprinklers, fire alarms, fire walls between units, smoke detectors in your unit and in common areas such as hallways, laundry rooms, etc.

Walkways

➢ When going back through all your walkway safety concerns from earlier, remember that these are your avenues of escape during a fire, making walkway safety that much more important.

➢ Never run permanent extension cords under walkway rugs. Temporarily is okay, but if you need power for an area, device, or permanent appliance, have an outlet installed where needed.

Family Area

➢ Smoking is still a leading cause of home fires. If you smoke, use deep, stable, ashtrays and make sure all cigarettes are out before emptying them. Also, never smoke in bed or when drowsy. If that's a hard habit to break, the least you can do is put a smoke detector over your bed, your favorite chair, or wherever it is you might fall asleep while smoking.

➢ Make sure your fireplace has a glass cover, or at least a screen, and that the chimney is regularly cleaned and inspected. One company sells a "chimney sweep" log (check your local hardware or home-supply store). According to the manufacturer, you burn the log in your fireplace and it helps keep the flu and chimney clean. It won't replace professional cleaning, but it may help in the interim. The Chimney Sweeping Log can be found online at http://www.chimneysweepinglog.com, or through Joseph Enterprises, Inc. 425 California St, Ste 1300 San Francisco, CA 94104, Tel.: (415) 397-6992, Fax: (415) 397-0103.

➢ Never leave a fire going in the fireplace regardless how much screening or protection you have in place. If you go to bed or have to leave the house, put out the fire by separating remaining logs and embers and dousing with water. Do not pull any logs out of the fireplace as they'll still be hot and can reignite.

➢ Also, never close your fireplace damper with hot ashes in the fireplace, and never scoop hot ashes out of the fireplace, even if your ash bucket is metal (and it should be).

➢ Make sure outdoor charcoal grills are out completely before leaving them unattended.

➢ Inspect your gas or electric furnace regularly and make sure the filter is changed or cleaned every 3 months. Some of the more devastating house fires have been caused by gas furnaces with a heavy dust buildup in the main duct. If ducts full of dust catch fire, it can turn every heater vent in the house into a blow-torch. Have your ducts cleaned at least every other year (or as your furnace manufacturer recommends).

➢ Put surge protectors on all your electronics such as your TV, computer, stereo, etc. Not only will it prolong the life of your electronics, it can help prevent a short that may cause a fire.

➢ Also make sure your electronics are dusted regularly, that no dust has accumulated inside cabinets or frames, and that they all have ventilation room around them. TVs, computers, stereos, etc. get hot and being dusty and confined is a great way to start a fire.

➢ Keep drapes and curtains away from anything electrical including space heaters, lamps (especially halogen lamps), wall sockets in use, etc. If you go to replace your curtains, see if any of your choices have a fire retardant rating. Also, one problem created in nuclear detonations is the blast and heat waves shattering windows and catching curtains (and other nearby flammables) on fire.

- During religious holidays, make sure decorations are properly utilized, candles are not near anything flammable, and that real Christmas trees are properly watered and kept hydrated (for a good tree hydration formula, see: http://www.meds.com/archive/mol-cancer/2000/12/msg00364.html).

- During power outages, it's far safer to use flashlights and battery operated lights than it is to use flame. Candles are a common recommendation in emergencies, and they're on our list too, but we recommend batteries before flame for both fire safety and breathable air reasons.

- One major cause of fire is children playing with matches. Kids are curious, so it's up to you to teach them what a danger fire is. One of the sources above has a downloadable fire safety coloring book for kids.

- Accordingly, this seems as good a place as any to tell you two things about children's' clothing. One, make sure pajamas and sleep wear has a UL fire retardant rating. Two, some popular fabric softeners can neutralize the fire retardant treatment of these clothes. Check your labels.

Kitchen

- Never leave food cooking on the stove top unattended.

- Never leave the house when anything is cooking.

- Don't use small pans on large burners or vice versa.

- If you have something non-edible on a low simmer such as potpourri, set a timer to make sure you remember to turn it off. Since you're not going to eat it, it's easy to forget.

- Clean the hood, filter, and fan over your stovetop regularly. Do it either quarterly or twice a year (at daylight savings time). Be sure to do the same underneath your stove top where grease and crumbs can build up.

- If you have a gas stove, don't wear loose-fitting clothes or long sleeves while cooking. They can drag the flames and catch fire. For the same reason, be sure you're careful about tossing oven mitts or dish towels on the stove. The pilot light can ignite items even if the burners aren't on.

- Check the vent pipe coming off your stove and/or other appliances all the way from source to roof exhaust. Check for any cracks, openings, or loss of any insulation around the pipe. (Gas water heaters and gas furnaces will use a vent, and sometimes it will tie in to the stove's vent.)

- Make sure all family members know what natural gas smells like by turning off a pilot light. It's also a good opportunity to show everyone how to relight the pilot. BE CAREFUL WHEN DOING THIS. Follow the appliance's instructions and vent the room before relighting.

- While doing the gas drill above, teach everyone that if they smell gas they should not turn on any lights, they should open the nearest door or window, and then go outside, turn off the gas, and wait for about twenty minutes before coming back in to open more doors and windows to finish venting any built up gas.

- **Hint**: Natural gas is lighter than air, and petroleum based gas is heavier than air. Vent accordingly.

- When cooking with oil or grease, **keep the pan's lid nearby**. If you have a tiny grease fire in the pan, the easiest way to put it out is to put the lid on the pan and let it smother. You should also keep baking soda near the stove for the same reason. Baking soda is a great little fire extinguisher for grease, and since you'd sprinkle it on, it's easier to clean up than if you'd used a pressurized fire extinguisher (which may actually spread the flame by literally blowing the flaming grease out of the pan if you hold the extinguisher too close).

Utilities

- Have a certified electrician inspect the condition of all wiring and electrical connections.

- Have technicians inspect your HVAC (Heating, Ventilation, and Air Conditioning) system.

- Be sure you're not overloading your wall outlets. Though you may have 3 or 4 things plugged into each of them, make sure no two of them are appliances or lights that need to run at the same time. Even then, limit what you keep plugged in. Also, dust can easily build up behind "octopus" outlets and start a fire when you're not there. So clean regularly.

- If you do have to plug numerous items into relatively few wall outlets, and you've made sure you've "divided the duties" so that the most-used items aren't all plugged into the same outlet, use a "power strip" that has its own breaker or fuse.

- Power strips that have built-in surge protectors are extremely inexpensive. You should have all your valuable electronics such as televisions, stereos, and computers, protected by them. For computers, buy the surge protector that includes a phone jack protector as well. Most department stores and office supply stores will have these. (This is also a money saving consideration. Don't want to lose valuable electronics to a power spike, do we?)

- Keep your breaker or fuse box free of obstructive and/or flammable clutter, easy to access, and mark it with glow-in-the-dark paint and/or reflectors.

- If you have an older style fuse box, never use anything but approved fuses in it.

- Periodically check all the gas line fittings from your meter through to each gas appliance. You can check the fittings by pouring on a small amount of soapy liquid made from 2 parts liquid dishwashing soap to one part water, or you can use the kids' "bubble toy" liquid. If you see bubbles at a fitting, you have a leak. Leaks will not only increase your bill, they can build up gas pockets that can be flammable and maybe explosive.

Bathrooms

- When taking your romantic candle-lit bath, be sure you didn't put the candles underneath a towel, a flammable artificial flower arrangement, or the toilet paper roll.
- The same goes for space heaters. Don't place them under flammables either.
- If you smoke in the bathroom, flush the leftover cigarette butt and don't toss it into the wastepaper basket.
- Rubbing alcohol is a common bathroom staple. Remember that not only is it highly flammable, but the flame produced by alcohol is virtually invisible in the light. Be careful using alcohol around open flame such as candles or heat sources such as your hair dryer or space heater.
- Have an electrician check the condition of the motor and wiring of your bathrooms' exhaust fans, and clean out accumulated dust once a year.
- Never use a space heater to dry clothes.

Bedrooms

- Don't smoke in bed.
- If you have a smoker in your family and they smoke inside the house, make sure you put an extra smoke detector in their bedroom and other rooms in which they smoke.
- Don't cook anywhere but in the kitchen or outside in a safe grill area.
- Make sure any and all candles are out before going to sleep.
- Scented candles really aren't all that effective. Try using other non-flammable air fresheners.
- Make sure nightlights or plug-in air fresheners are in outlets away from drapes, bedding, or other flammable materials.
- Similarly, make sure space heaters are not near anything flammable.
- Check electric blankets, electric alarm clocks, radios, stereos, etc. for any frayed or damaged wiring, too many items plugged into one outlet, and for excessive dust around any outlets. Also check for excessive dust in or around your TV, stereo, etc.
- In mobile homes, make sure your windows can function as escape routes. Mobile homes burn too quickly to not have as many escape routes as possible. If you can afford it, look into having skylight escape hatches.

Laundry, Work, and Storage Areas

- Apply all the overloaded electrical outlet rules to your workshop area as well.
- If you have appliances that are supposed to turn themselves off after use, don't trust them. Get into the habit of making sure the unit is turned off rather than relying on devices. Make this a part of your bedtime ritual as you're going around checking door and window locks.
- Keep all gasoline and fuel containers tightly capped, in a cool place, on the floor or lowest shelf, and as far away from the living and bedroom areas of your house as possible, preferably in a metal outbuilding. Also, be sure to keep all flammables, especially the liquids, away from electrical outlets. (**Hint**: Though secured and safe, keep gas cans handy as you may want them to top off your car's tank before an evacuation.)
- Clean up flammable material spills promptly and keep all cleanup rags in a metal bucket with a lid.
- When working with any flammable liquid or gas, make sure there's plenty of ventilation. This is for your breathing safety as well as fire safety.
- In any gas appliance with a pilot light, the appliance should be situated so the pilot light is at least eighteen inches off the floor. This is to allow a safety margin between the pilot light and any gas fumes that may waft through the area (sometimes the gas dryer and gas hot water heater are in the garage or basement area where gasoline is stored for yard tools).
- Empty the dryer after it finishes cycling.
- Clean out your clothes dryer's exhaust system at least twice a year.
- Keep potentially reactive household chemicals away from each other.
- The average family accumulates a lot of mementos over the years, usually storing them in cardboard boxes stacked in the garage or attic. As plastic tubs are inexpensive, you should start swapping out old cardboard boxes for the plastic tubs as time and money allow. Though the plastic is probably just as flammable in the long run, it doesn't catch fire as quickly as cardboard, and you can use them to store water.

> Similarly, check to see how you have things stored in your attic. Make sure nothing is stacked on top of wiring junctions or ceiling fixtures, and the same rule applies about cardboard boxes. Many household fires have burned straight up from the point of origin until they hit the bare wood and stored flammables of the attic where the fire then fans outward to finish consuming the rest of the house. You don't want anything in the attic to help the process.

Miscellaneous

> If you decide to remodel your home or put up new siding, look into having firewall or fireboard put up underneath your siding or facade (unless you're using brick) in key fire break locations. Among these areas would be the exterior wall between you and a house that's closer than 30 feet, between the house and the chimney, over the kitchen between it and the attic, or over any indoor garage or workshop areas (especially if your workshop/garage is under the house). Your local hardware or home supply store can tell you more.

> Install wire mesh over attic vents and inside the attic under any turbine or ridge vents. This will help keep sparks or small airborne embers from nearby fires entering your attic. (Mesh will keep out bats too.)

Detection

> Buy smoke detectors and install them according to the manufacturer's instructions. We recommend **at least three smoke detectors**. Priority goes to the hallway near the bedrooms. The others should be placed near potential fire areas such as your furnace area, and the kitchen area. As cheap and plentiful as smoke detectors are, you should have one anywhere you feel there might be a risk of fire. In addition to the locations mentioned, have a detector over your washer and dryer, in your workshop area, in the basement, the attic, and in the garage. Smoke detectors can be tied in to a central alarm, or you can use the baby monitor trick we mentioned earlier and change its batteries when you change those in your smoke detectors.

> Another reason to have extra smoke detectors is the fact that though they have a sterling track record, there is no guarantee that every smoke detector is going to work as planned. As with any piece of crucial emergency equipment, always have a backup. Use battery-powered units as a backup to electrical units.

> The carbon monoxide detector we recommended earlier may also help detect a fire as some fires may burn "clean" producing very little smoke, yet they'll still produce carbon monoxide.

> Make sure you've swapped phone numbers with all your neighbors, especially in apartment complexes where everyone's home is attached to everyone else's. Many home fires are actually discovered by a neighbor. You help watch their house or apartment, and they should help watch yours. (We'll mention this again, but you should know which of your neighbors might be mobility-challenged, so they can receive the proper help and attention in an emergency.)

> Keep an **airhorn** and **whistle** (air horn is louder, but may leak its air over time) or maybe a hand-held **bell** (which you can use while covering your mouth due to smoke), as a fire alarm. Have something like this in each bedroom and near each phone or flashlight in select areas of the house.

Commercial Fire Extinguishers

Now that we've covered ways to prevent fires from occurring, and ways to detect one that's started, let's talk about the equipment you'll need to have on hand to help you fight a fire.

Under "Basic Training" we asked you to learn how fire works, and how fire extinguishers put them out. Hopefully you learned about the three elements necessary to fire; oxygen, fuel, and heat, and that taking away any one (or better yet two) of these elements will put out a fire. Fire extinguishers are designed to remove or neutralize one or more of those elements. Let's take a look at some fire extinguishing equipment that should be part of your fire safety arsenal:

> There are three common classifications of fire:
> **Class "A"** is flammable "**solids**" such as paper, wood, fabric, etc.
> **Class "B"** is flammable **liquids** such as gasoline, grease, oil based paint, etc.
> **Class "C"** is **electrically** started fire, or fire with an electrical equipment component to it.
> (The less common Class "D" pertains to flammable metals such as magnesium.)

> There are two common types of household fire extinguisher: **Water** units, and "**dry chemical**" units. (CO_2 units are common in industrial settings, but a bit too expensive, and limited in application, for home use.)

> The water unit is called such as it's a pressurized unit filled with water. It's good on class "A" fires of flammable solids as water both smothers the flame and reduces the kindling temperature. It cannot be used on flammable liquids (Class "B") as most flammable liquids will float on top of the water and therefore the pressurized stream will only help spread the fire. It also cannot be used on electrical fires as electricity may follow the stream of water back to the user and electrocute them. Water units typically come in a 2.5 gallon size, and are rather expensive. We'll discuss alternatives below under "**Additional Equipment**."

- "Dry chemical units" are so named because they are pressurized units full of powder specially intended for certain types of fire. Dry chemical units rated "B and C" (usually a sodium bicarbonate base similar to baking soda) can only put out Class B flammable liquid, or Class C electrical, fires. They can't put out "Class A" flammable solid fires. Dry chemical units rated "A, B, and C" can put out all three classes of fire.
- As a caveat, the ABC dry-chemical units do not have all that high a rating on putting out class A fires as they do not lower the temperature of the burning materials (a major component of fire).
- Dry chemical units are inexpensive and can usually be found at hardware or department stores in the following sizes: 2 ½ lb. units (commonly found in kitchens, boats, or cars), 5 lb., and 10 lb. sizes. 20 lb. units and above can be purchased through fire equipment distributors found in your yellow pages.
- We recommend **water units** and **class ABC dry-chemical units**. We don't recommend the BC-only units as most household fires will usually have a flammable solid element to them regardless how they start. For example, a grease fire on the stove might catch the kitchen window drapes, or wooden cabinets, on fire. An extinguisher rated BC (commonly sold as "kitchen" fire extinguishers) can put out the grease fire, but will probably not put out the drapes.
- We also recommend that you have **several** of the smaller fire extinguishers rather than one of the larger ones. Having only one is "putting all your eggs in one basket," so again, redundancy is our friend. Fire extinguishers can lose their pressure, and they can only be used once before needing recharging, so having only one could be a serious handicap, especially in a situation where you could not leave your home to get your spent units recharged.
- Having several smaller units will also allow you to strategically locate several units in key locations rather than keeping your one large unit in a central location. Keep a fire extinguisher near each area where you have a smoke detector.
- **Hint**: Never store a fire extinguisher where you'd have to go *towards* a fire to get to it. For example, in the kitchen, the most likely spot for a fire is the stove. Therefore you wouldn't put your fire extinguisher in the cabinets over or near the stove. You'd want to put it closer to the doorway. Consider this when locating units in other parts of the house. This includes keeping extra units (including water units) in bedrooms.
- **Hint #2:** Have your fire extinguishers serviced annually. Pressure can leak, powder can settle, and any number of little things could go wrong that only a trained technician can spot.

Automatic Fire Systems

Speaking of "over the stove" as we did above, let's talk a bit about fire extinguishers or fire systems that you <u>do</u> want to have where the fire will be.

- The most commonly recognized, but the most expensive, is the overhead sprinkler system. Unfortunately, most homes do not have sprinkler systems. If you opt to have one installed, make sure it covers your attic <u>and</u> your basement, that it has a water-flow alarm, and an accessible yet unobtrusive shutoff valve. We'll discuss a "poor man's" alternative below.
- Other automatic systems come in individual units or "pods" that allow you to mount them yourself over potential hot spots. One example is the "**Dryer Watchguard**" available through http://www.aspensafety.com, or ASPEN MANUFACTURING INC 712 Broad St., Riverton, NJ 08077-1154, Phone: 1-800-327-1794, Fax: (856) 786-7210. It's primarily intended to mount over your clothes dryer though it has a sister model called simply "The Watchguard" that could be used to cover a small workbench or workshop area, gasoline powered yard tools, stored gasoline, etc.
- Another great product that we <u>highly</u> recommend is called the "**Range Queen**." It's a small unit, about the size of a tuna can. It has a magnet on its back to let you easily mount it underneath the fan hood over your stove. Any unnoticed stovetop fire will cause the unit to automatically discharge, smothering the fire. You can get **Range Queens** (among other fire fighting supplies) through **Fire Safety Equipment**, P.O. Box 2067, Valdosta, GA 31604, Voice: 800-228-9839 or 229-242-3567, Fax: 229-242-3568.

Other Fire Extinguishing Equipment

- Since we highly recommend you have a water unit fire extinguisher, and since the commercial ones are so expensive, we thought we'd offer a simple (and only slightly less effective) alternative: "**Garden sprayers**." Go to your local department store, hardware store, or garden store and pick up one (or preferably two or three) of these little plastic, two or three-gallon capacity, hand pump sprayers with an adjustable nozzle.
- They're made for water, fertilizer, insecticide, etc. but they make great little water unit fire extinguishers. They're cheap enough to have several, they have a range of about 30 feet with the nozzle adjusted for "stream," you can refill and repump them on the fly as many times as needed, you can use them to water your plants and still have them as fire protection, and in a water emergency you can drink the water replacing

it with dirty water from another source. Store these units in various places throughout the house. Keep one near the kitchen, one in the garage, and maybe one in your safe room (for both water and firefighting needs.)

➤ You can also fill your garden sprayer with your own choice of fire extinguishing or fire retardant mixes. (If you do that, be sure to take a Sharpie and mark the sprayer as being filled with non-drinkable contents.)

➤ Also, you can use your sprayer to apply a **fire retardant mix for fabrics**, that can be easily made at home. This formula was pulled from an old Popular Mechanics publication, and we've not made any laboratory tests or comparisons to see how well it works. Here's the formula in case you want to make it and test it (might come in handy if you choose to make the "fire blanket" below). "To 7 ounces of Borax, add 3 ounces of Boric Acid. Have a 2-quart pan of hot water ready. Add a small amount of water to the powder, enough to make a paste. Mix the paste thoroughly before adding the paste to the water. Wet fabric thoroughly and let dry. Apply to fabric again after laundering." We recommend that if you want to try this, that you use a test batch on the type of fabric you're wanting to treat to not only see if it works, but if it stains. Again, mark your sprayer unit is being non-drinkable water if you use it to apply this mix. If you decide it works on clothing, consider trying it on wood and apply the solution to exposed wood areas that may be exposed to flame. An example would be bare garage or workshop walls.

➤ A commercially available **fabric fire retardant**, UNIVERSAL FIRE SHIELD, can be found online at http://www.firechemicals.com, or at Corporate Office and Chemical Manufacturing Plant, 6311 Washington, Unit N, Denver, CO 80216, Toll Free 1 866-860-6584, OFFICE: 303-853-0204. You can also find other fire retardant chemicals for fabrics at the sources listed above for fire retardant paints.

➤ Another fire-fighting substance touted to do a tremendous job of fighting liquid fuel fires is "Cold Fire ®." You can find out more at http://www.firefreeze.com.

➤ Some of these fire retardant compounds are useful on unfinished wood. If you have window storm covers cut from sheets of plywood, treat the plywood. While the shutters are in storage they're a potential fire hazard, and if you put them up to help protect against an approaching fire instead of a storm, you'll definitely need them to be fire retardant.

➤ While still on the subject of water, we highly recommend you attach a garden hose with an adjustable spray nozzle behind your washing machine and leave it coiled up but accessible. Use a "Y" attachment on the cold-water faucet, and make sure your hose is long enough to reach from the washing machine to the farthest point in the house. A moderate to large Class A fire, such as a mattress or a couch on fire, will quickly outpace the capabilities of your dry chemical extinguishers and garden sprayers. You'll need access to a garden hose, and running outside to find yours, hook it up, and run back inside, will take way too long.

➤ You can also do the same from the drain faucet of your hot water heater. This is especially useful if your hot water heater is in one part of the house and your washing machine in another. This gives you extra coverage. Again, the advantage is that the hose is right there, right then. The only drawback is that cold water is slightly preferable over hot water for its heat-reducing value in quenching a fire. However, hot water is far, far better than your little dry chemical extinguisher or garden sprayer. We suggest you have hoses hooked to both your washing machine and your hot water heater. This gives you two sources, possibly two locations for hoses, and having two hoses may also allow **two people** to help fight a growing fire at once.

➤ Another neat little extinguisher we've found, but haven't had the chance to test, is the "Chimney Flare." It looks like a highway safety flare, but it produces smoke. It's made to use in case of a chimney fire. All you do is ignite the flare, toss it in your fireplace, and the smoke and gasses produced are supposed to rise into your chimney and smother any fire. The original model was called "Chimfex" and was manufactured by the Orion flare company. Currently it's out of production, but tons were bought up by Wal-Mart so some might still be available there.

➤ Under landscaping, we mentioned the suggestion of having an in-ground lawn sprinkler system installed if you live in a wildfire risk area. If you do that, have an extra line run to the top of your roof. You'd want a separate valve since you wouldn't want to water your roof every time you watered your lawn, but you'd want to have the sprinkler there in case a wildfire or brushfire threatened your yard and home.

➤ Let's discuss the "**Poor Man's Sprinkler System**" as mentioned earlier. Imagine your manually-controlled, in-ground lawn sprinkling system mounted in your attic and/or basement. Such a system is far less expensive than a standard fire sprinkler system, and is much easier to install. In fact, your yard sprinkler company could do it for you. However, if they want to charge too much, you could do it yourself with simple hardware store parts. Let's look at the details:

❏ Draw a floor plan of your attic and basement and decide which areas are potential hotspots.

❏ For the "average" house, you'd probably divide the attic into 4 quadrants, with one over the kitchen, one over the family areas or fireplace, one over the master bedroom, and the fourth over the other bedrooms.

❏ For the basement, you'd probably have one or two locations, either over a workshop area, garage, a laundry area, over a den or fireplace if any, or over the furnace or other appliances.

❑ Each of your quadrants or hotspot areas above will have its own **separate line and sprinkler head** run from the main "**valve box**." Don't forget to include an extra line for your rooftop if it's not covered by your lawn sprinkler system already.

❑ You can use any piping material you want. Standard systems use copper pipe or galvanized pipe since the standard systems are always under pressure. This "poor man's" system is **dry** unless **manually activated**, so use what you're familiar with using and/or what you can afford. You could even use a series of garden hoses if you like. We recommend simple PVC.

❑ If you have great water pressure, use ¾" pipe. If not, use ½" pipe. However, don't use anything bigger than ¾" or smaller than ½.". You'll want to have enough pressure to get the water where it needs to go, and yet enough volume to do the job. (Also, don't count on water pressure always being high. It'll drop considerably if several houses are fighting a wildfire or brush fire.)

❑ The **valve box** you'll need for this system should be located outside, in an unobtrusive location, and the box itself should have a **lockable cover**. You want your box to be accessible from the outside (since you'd only use it if the fire got bad enough to drive you out of the house), yet you don't want vandals or pranksters turning on your system in the middle of the night. Locate your box a few inches from an **outside faucet**.

❑ To make your valve system, you'll need the following: one in-line valve (may also be called a stop-cock valve or butterfly valve), and a "T" connector *for each line you'll be running*. You'll need the box itself, a short section of garden hose with connectors at each end, and a "Y" connector with "leg" valves for your outside faucet. You'll also need two female garden hose connectors (made for whatever piping you're using) to connect your water input sources to the valve system, and another in-line valve to close off the secondary input connector until it's used. To connect valves to Ts, and Ts to each other, use small sections of your piping. Assemble your valve box as shown below, and make sure each line is labeled as to where it goes.

❑ Once you understand how the valve box regulates the flow of water, you can construct it out of any number different types of valves and connectors. Search through your hardware store or garden center and let your imagination or your wallet dictate what you make. (In fact, you might find an economically priced, manually operated lawn sprinkler valve system that will do the trick.)

❑ The "secondary input" connector is to allow a neighbor to hook their hose into your system, or for you to connect some other outside water source (such as from a water pump connected to a swimming pool). This allows you to increase the incoming volume of water if you have more than one fire spot break out.

❑ The amount of piping, "elbows," and other connectors you'll need will depend entirely on how far you need to run the pipe, and where you're taking it. Once you draw it out on paper, your hardware store rep can help you with what you need.

❑ Since you may be using PVC, you'll want to run your pipe in such a way that the pipe leading to a potential hotspot doesn't travel <u>over</u> that spot. For example, the line you run to cover the attic area over the kitchen should not cross over the kitchen, and should in fact spray the kitchen area from the other side of the attic. The reason is, the fire may get out of control and hit the attic before you can react. Therefore you don't want the fire melting the PVC that *would've* carried water to that area.

❑ If you <u>have</u> to run pipe across its own hotspot, you can run PVC from your valve box up to the attic, but use copper pipe inside the attic. There are lots of options that your hardware store rep or a plumber can help you with. If nothing else, if you follow the "no crossover" rule, you could even use sections of garden hose inside the attic if running pipe is too big a pain.

❑ Once you get the pipe run, you'll need to have some sort of sprayer head. You can use an old shower head, a twist-adjustable hose nozzle, a lawn sprinkler, or even a pipe cap with holes drilled into it. Regardless of what you use, test it first to see what kind of range and spray pattern it has so you'll know about where to mount it to cover the desired area. You can aim the spray to fall down on the target room, or you can "bounce" the spray off the underside of your roof to rain down on the target room.

❑ **Hint**: Though this system is intended for the attic, there's no reason some sprayer heads couldn't poke through the ceiling to protect rooms below. Sprayer heads are easy to decoratively camouflage.

❑ For the roof, you may want to use a length of PVC running along the roof's ridge. Drill small 1/8" or 1/16" holes down each side of the pipe at 6-inch intervals so water will spray down each side of the roof evenly. You can also use one of those flat "watering hoses" with perforations already made down its length, or you can even use several small yard sprinklers. Whichever you choose, make sure to test the spray pattern before you mount it in place. Be sure to mount it securely. Afterwards, you can paint it to match the roof.

❑ Once your system is in place and set, be sure the water is off, the valves are in the "off" position, lock your valve box, and have plenty of the keys around so you can get into it in a fire. Be sure to put one key in your fire trunk, and you may want to hide one near the valve box like you'd hide a house key. That being said, let's talk about how to use this system.

❏ You really don't want to turn this system on unless a fire has gotten out of hand. This means your dry chemical unit or garden sprayer aren't doing the trick, and neither is the garden hose you have hooked to the washing machine. Either that, or the fire got too big too quickly to even try anything else and you did the right thing by getting everyone out of the house and calling 911. Should a fire start to engulf a room, should smoke start coming out of the attic, or should a moderate fire in the house or basement start to seriously "lick the ceiling," that's when you'll want to go to your valve box and turn on the water over the room the fire is in (more detail under "**Isolation**"). In any event, it's far better to have a gutted room than to lose a house. And again... all at the low, low price of a few hardware store parts. The only drawback to this is that it's not an automatic system. You'd have to be there to turn it on, or you'd have to let a trusted neighbor know where the valve box key is hidden in case a fire broke out while you were gone.

❏ **Hint**: Your valve box should be okay since this is a dry system, meaning no water is in the pipes until the unit's in use. However, you should **winterize** the setup by draining any accumulated water in the faucet or hose leading to the valve box, and you should cover your faucet with an insulating foam cover. You might temporarily disconnect the hose running from the faucet to the valve box. You run a greater risk of fire in the winter due to space heater usage, etc. and you don't want your sprinkler compromised by a frozen faucet.

❏ **Hint #2**: Though this poor man's system is economical and extremely workable, an automatic fire sprinkler system installed by a professional fire equipment company is preferable if you can afford it. We mention our system here as a very desirable alternative to having nothing at all.

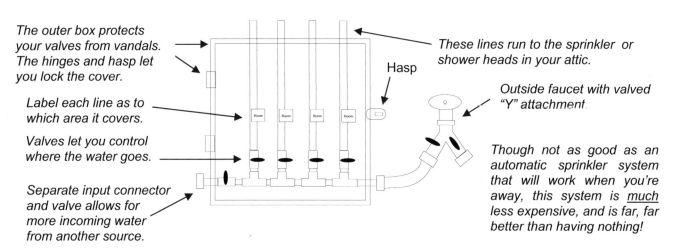

The outer box protects your valves from vandals. The hinges and hasp let you lock the cover.

Label each line as to which area it covers.

Valves let you control where the water goes.

Separate input connector and valve allows for more incoming water from another source.

Hasp

These lines run to the sprinkler or shower heads in your attic.

Outside faucet with valved "Y" attachment.

Though not as good as an automatic sprinkler system that will work when you're away, this system is _much_ less expensive, and is far, far better than having nothing!

➢ Another "poor man's" hint is to store some of your water supply jugs over potential fire locations, in roughly the same places you'd aim your nozzles with the fire system mentioned above. The thinking here is that if an undetected fire were to burn through into the attic that these jugs of water would melt through or at least fall into the fire effectively stalling it or diminishing its size for a few crucial moments. This is not high on our list of recommendations since its effectiveness is minimal. However, it is an option, and options are our friends. A related idea concerns the plastic water storage tanks we'll discuss under "Isolation" later on. If you were to put one in your attic, you might get double benefit from it if you were to locate it over your number one fire risk area just as you would the water jugs or system nozzles.

➢ Also, you might try installing one or two "Dryer Watchguards" over the same areas you'd aim your shower heads. The dry chemical isn't as good as water, but the Watchguards are automatic and they will stall a fire for a few crucial moments, maybe even minutes.

➢ While we're on the subject of alternate equipment and fire, be sure that if you have an electric generator and your only water source is a well with an electric pump, that you're able to get the generator into operation and hooked to your pump in short order. In fact, we suggest you practice this a few times. You'll want to be able to do this if a fire should break out when you have no utilities such as municipal water or power. This is when you'd be most at risk since, with no power, it's likely you'd be cooking or reading by some sort of flame. Always be ready to put your alternative emergency equipment to work.

Additional Fire Safety Equipment

Let's start with "**The Fire Trunk**." You'll want to create a "fire trunk" for a couple of reasons. Number one, it's always good insurance to have redundant gear accessible in a secondary location away from the primary danger. Number two, since you're reading this book, it's obvious you're the responsible one in your neighborhood so you'll want to have some fire fighting tools you can grab on the fly to go help the neighbors (or extra gear to help them help you). Simply put, the "fire trunk" is literally a trunk or box that you keep in an outbuilding or in your garage near the doors where you can still get to it even if most of your house is involved in fire. Let's discuss contents and considerations:

➢ You'll want redundant gear in one spot because the minute you need a hose to fight a fire, either a neighbor's borrowed it or you just ran over it yesterday with the lawnmower. Your fire trunk should contain **equipment that is <u>never</u> used except in an emergency**.

➢ If you keep your fire trunk in an outbuilding or garage, make sure you **keep a key to its location hidden somewhere outside**. You may be forced out of the house wearing next to nothing and have precious little time to grab anything at all on the way out, much less your keys.

➢ Your fire trunk should be marked with reflector tape, and should contain the following: (Some uses will be explained under "**Isolation**.")

❑ A garden hose long enough to reach from any outside faucet to more than half-way to any other outside faucet, and/or to the top of the highest window of your house. Also make sure it's long enough to reach any outbuildings or other flammable locations on your property, or whichever distance is greater.

❑ An adjustable spray nozzle for the hose. If you use the squeeze-type make

❑ A metal yard sprinkler attachment for the hose. One that sprays all directions at once.

❑ An extra 2 ½ lb. ABC dry chemical fire extinguisher (or a couple of them).

❑ The following tools: Hatchet, full-sized axe, full-sized sledge hammer, crowbar or prybar, gas shutoff wrench, water meter valve wrench, wire, duct tape.

❑ A brick or baseball-sized rock (or something to break a window from a safe distance).

❑ 50' length of ½" nylon rope with grappling hook (to drag things from a safe distance).

❑ Two or three old bath towels to wet and cover your head, breathe through, or to smother flames in secondary fires.

❑ A gallon jug of water (to wet the bath towels or mop if there is no other water).

❑ Fire blanket (see below).

❑ A bucket (regular or collapsible).

❑ A set of coveralls, shoes, and heavy work gloves (you may be forced out of the house wearing next to nothing).

❑ A NIOSH smoke-rated respirator, or an "Evac-U-8" smoke hood.

❑ Hardhat and safety goggles, or better yet, a full face shield.

❑ Keys to your manually operated attic sprinkler system valve box, if any. (See above.)

❑ Extra house and car keys (you may have had to escape through a window and you'll need keys to get back in, or you'll need to move your car or get into it for a variety of reasons).

❑ A key to your electric meter if you have it locked. You may need to yank the meter to turn off the HVAC system to keep from feeding air to the fire, or your fire may have an electrical equipment component to it.

❑ Air horn <u>and</u> whistle (and other signal items) to alert the neighbors.

❑ Flashlight (either get a non-battery generator type or remember to change the batteries).

➢ Along with the items actually inside the trunk, make sure you have a **ladder** tall enough to reach the higher windows of your house (don't count on borrowing the neighbor's), a **heavy garden rake**, and a **mop** with a cloth-strand head. A wet mop is great for smothering small fires, such as a grass fire.

➢ A **fire blanket** may be the cool tool to have. If you're forced out of the house by flame and you <u>HAVE</u> to go back in yet all you're wearing is undies, the blanket in the fire trunk will help. It may also help if you're fighting a fire from outside the house, but you're close enough to feel the heat. Buy a commercially produced and rated blanket if possible. To make a fire blanket, take a twin or queen-sized wool, nylon, or rayon blanket, and stitch it to a rubberized canvas tarp or thermal tarp the same size. The blanket will help hold the water you use to wet it down before use (**key step**) and will help provide some of the insulation against the fire. Make sure your blanket and tarp already have a fire retardant or flame-resistant rating, and you may also want to apply one of the previously mentioned fire retardant treatments. **Remember**: This will give you a little insulation for a few critical seconds. It will **<u>NOT</u>** make you fireproof.

➢ Speaking of higher windows, you might want to have **fire escape ladders** in select rooms of your house. You can find a discussion of escape ladders at the following locations:

● We found a lot of good products at "**Fire Evac**." They're online at http://firevac.com, or contact them at Ready Prep, P.O. Box 261639, Littleton, CO 80163-1639, 303-346-4498, Toll Free: 866-723-2562.

● More equipment can be found on http://www.preparedness.com/homsafemprod.html, and they can be contacted directly at the Preparedness Center at 1370 Highway 10, West Livingston, MT 59047, Phone # 406 222-3077, Fax # 406-222-3309, Toll Free: 800-547-4223.

● Fire Escape Systems, http://www.fireescapesystems.com, 1-888-347-3727, Fax: 303-790-2942.

- Bold Industries at http://boldindustries.com, 1069 Connecticut Ave., Bridgeport, CT 06607, 203-334-5237, Fax: 203-362-1644.
  - **Hint**: For able children and athletic teenagers, a heavy, **knotted rope** will work just as well.
- If you have escape ladders in the children's rooms you can make a small "window seat" type box to house the ladder. Remember our little Radio Shack entry alarms from earlier? These make great "tamper" alarms for these ladder boxes. If the box is opened for a real fire, the alarm will help alert others. Or, it can be a great deterrent to keep the teenagers from using the ladder to sneak out in the middle of the night.
- If you opt for this "window seat" box, or "**Safety Box**" as it's called on the "**Safety Blueprint**" page of the "**Appendix**," you should keep other safety-related equipment in it as well. Consider the following (which should be in each bedroom anyway):
  - ❑ A respirator or Evac-U-8 smoke hood.
  - ❑ An air horn, whistle, bell, or other alarm device.
  - ❑ Flashlight.
  - ❑ A "911-only" cell phone and/or walkie talkie.
  - ❑ A small backpack or duffel containing the room occupant's share of "Bugout" gear or special-needs equipment, or at least their "Safety Necklace." (Or alternate gear in addition to the family kit.)
  - ❑ Enough duct tape and pre-cut and pre-labeled plastic sheeting to seal the room.
  - ❑ A quart (or more) of water as either part of the household's overall stored water, or enough to wet clothing or other items for evacuation or sheltering during a fire.
  - ❑ Though the room's occupant might not be able to use the ladder themselves, such as with the mobility challenged, the ladder should be there for someone else to use to help them.

Miscellaneous Sources

- Visit the Residential Fire Safety Institute at http://www.firesafehome.org, or contact them directly at 712-829-2734 or 763-416-0527.
- Fire extinguisher selection info, and a quiz can be found at http://www.fireextinguisher.com/.
- Be sure to visit the fire safety page at http://www.safehaven365.com/fire-safety.htm, and their main page at http://www.safehaven365.com.
- See the National Library of Medicine's http://www.nlm.nih.gov/medlineplus/firesafety.html. While you're there, search the other safety links and visit the home page.
- You'll find home safety products online at http://www.firstalert.com, or contact them at 1-800-323-9005.
- Flame resistant fabrics can be found at Westex Inc., online at http://www.westexinc.com or contact them at Westex Inc., 2845 West 48th Place; Chicago, IL 60632, Phone (773) 523-7000 : Fax (773) 523-0965.
- Flame Stop provides spray-on fire inhibitors. See them online at http://www.flamestop.com, or contact them at Flame Stop, Inc., 924 Bluemound Rd. Fort Worth, TX 76131, Tel. 817-306-1222, Fax. 817-306-1733, Toll Free: 1-800-397-7867.
- Firetect also offers fire inhibitors online at http://www.firetect.com, with direct contact at 26951 Ruether Ave, Unit D, Canyon Country, CA 91351, 800-380-8801, Tel: (661) 298-8801, Fax (661) 298-8851.
- Chemco offers fire retardant for wood. See them at http://chemcoproducts.com, or contact them at CHEMCO Inc. • P.O. Box 875 • Ferndale, WA 98248 • Phone - 360.366.3500 • Fax - 360.366.3831.
- More fire inhibitors can be see at National Fireproofing Co.'s http://www.natfire.com site, or reach them at 427 Anne St., Verona, IL 60479, Phone: (815) 287-3473, Fax: (888) 202-7801.
- See the links at http://www.equipment.net/list/fireprotectionequipment.htm.

Notes:

_____
_____
_____
_____
_____
_____
_____
_____

Basic Vehicle Prep

Many of us spend a great deal of time in our vehicles. Not only that, but in the case of an emergency, our vehicles are our life savers. We'll rely on them to get us out of harm's way, to get us from point A to point B, and protect us in the process. They may also be our family's ambulance, and in some cases a temporary shelter.

Goals of Basic Vehicle Prep:

1. To ready your **primary vehicle** so you can **react to numerous scenarios**.
2. To continue with "**Frugality**" in making sure such a large an investment as a car is **well maintained**.
3. To keep your vehicle **safe and secure, and prevent theft,** so it will be there when you need it.
4. To provide you **operational considerations** that will make the difference in a sudden emergency.
5. To give you a little **advance preparation for an evacuation**.

We've divided "Basic Vehicle Prep" into the following categories:

1. **General Considerations**.
2. **Basic Vehicle Prep Checklist** .

General Considerations

> The most important piece of equipment in a vehicle is the driver. We've found a site with a few general motor-safety considerations and links. Go to http://www.cheap-auto-car-insurance-quotes.com/safety.htm.

> More detail concerning your vehicle and its role in your safety and protection will appear under the "**Reaction**," "**Evacuation**," and "**Isolation**" sections. Your vehicle is that important.

> Perform **maintenance checks** not only on their scheduled dates or mileage, but each time the Federal **Terror Threat Alert is raised** to either an "Orange / High" or "Red / Severe" alert. Be sure you're ready to hit the road at any time. Too, have your vehicle checked when your area enters a **potentially dangerous weather season,** such as hurricane season in Florida, etc. For your convenience, we've included an "**Auto Maintenance Record**" form in the "**Appendix**."

> If you've read this far, you're sure to have seen our suggestion under "Basic Training," that you **take classes on vehicle maintenance** and simple repair.

> **Never let your gas tank get below half-full**. You might even put a tiny piece of tape on your fuel gauge to remind you to refill the tank once it gets down to ½. If you have a metal outbuilding and use gas-powered lawn equipment, consider always having at least **two 5-gallon containers of gasoline** on hand (keeping one always full), provided you can store it safely. You'll make fewer trips to the gas station during mowing season, and you'll have enough gas to top off your car's tank before hitting the road in an emergency.

> **Security** is a major component of the preparation measures listed below. You don't want to lose your car or any stored equipment due to a theft that could've been easily prevented. Also, if your car has an **entry sticker for any secure facility** such as a military base, your car might become a more likely target for theft.

> Many car makers offer a "**valet key**" that will operate the door and ignition, but not your glove compartment, trunk, etc. It's called a valet key as that's the one you'd give to a valet or parking lot attendant to allow them to park and retrieve your car, but not rifle through secure areas.

> Some cars use keys with an ID chip installed to prevent unauthorized keys from being used to crank the car. Since these are so expensive to duplicate, you might want to make the cheap key copies for your "hide-a-key," and keep the ignition duplicate, with the chip, **well-hidden** inside the vehicle somewhere.

> **Never put personal information on your keys**. Some travel clubs will give you a tag for your keys that has a code number so that they can be returned to you if someone finds them and drops them in a mailbox. Use those since they don't give away sensitive data.

> Be VERY careful with any **remote, keyless entry device** you have for your car. Drop that, and anyone who picks it up can not only *find* your car, but they *have* your car. Also, don't use these "convenient" key-chain credit cards. You lose your keys, you lose a credit card too.

> **Pay attention to where you park**. This applies to home, work, errands, leisure, and travel. Be sure you're not in a situation where your vehicle could be easily blocked in or accidentally damaged. For instance if the huge tree in your front yard is old and diseased, don't park under it until you get the tree taken care of. Also, when parking at public locations or gatherings, make sure you can get out as easily as everyone else, should something happen.

➤ **Crime awareness** is also important. Be sure to park in well-lit, public areas, keep valuables out of sight, don't approach your vehicle if strangers are milling about closely, and keep your doors locked and windows up when driving.

➤ Though we'll go into much more detail under "**Evacuation**" on what to keep pre-packed in your car, in this section we're giving you what we feel to be the minimum as you never know where you'll be or what you'll be doing when an emergency arises. As far as emergency gear goes, **it's always better to have it and not need it than to need it and not have it**. And... redundancy is what?... It's our friend.

➤ Regarding all the equipment listed below, you never know when you'll be **using your gear to help someone else**. We'll say this constantly throughout this book, that sometimes our true "*first*" responders are neighbors and bystanders. Though you should be crime conscious since there are millions of people on this earth who are either terrorists or criminals, there are *billions* who are not. We need to learn to help each other more.

➤ Pack as much of the recommended equipment as possible into **removable packs**. It'll make it easier to stay organized and save space in your vehicle, plus it'll make it easier to remove your gear to change vehicles, or to simply take the equipment with you if the situation so dictates.

➤ Naturally, the checklist below will not apply to all vehicles. It should, however, be applied as well as possible to your **primary vehicle**, and especially to the vehicle you would choose to use **during an evacuation**. For example, your primary vehicle for daily travel to and from work might be a motorcycle, and your primary evacuation vehicle might be a boat or private aircraft. For the purposes of this book though, we'll focus on automobiles for two main reasons. One, autos are more common and plentiful. Two, land vehicles are the easiest to use and therefore the easiest to take for granted. Aircraft and boat owners can't operate their craft without being highly knowledgeable about them, and therefore, are usually way ahead of the game as regards maintenance and operational preparation. However, since we want everyone to be as prepared as possible for an emergency, regardless of what kind of vehicle they'll be using, you should **apply the checklist below as well as you can to as many of your vehicles as you can**.

➤ Since quite a number of people use a variety of transportation methods, we'll provide a few information sources for our boating and flying friends.

➤ For prep and safety information on **boats**, try these links:

● US Coastguard's Boating Safety: http://www.uscgboating.org/links/safety.aspx, or 1-800-368-5647.

● "Nautical Know-How Boating Tips Notebook" : http://boatsafe.com/nauticalknowhow/safetips.htm.

● Boating information: http://www.commanderbob.com .

● See the online magazine ,"By the Sea" at http://www.by-the-sea.com.

● Google's boating links: http://directory.google.com/Top/Recreation/Boating/Resources, and http://directory.google.com/Top/Recreation/Boating/Resources/Safety/Boating_Laws_and_Regulations.

● John Keyes has a good boating info page at: http://johnkeyes.com/boating.

● Be sure to visit the Boating Safety page at: http://www.boatingsafety.com.

● Definitely bookmark the mass links page at http://www.cbel.com/boating_resources.

● And the sites at http://100topboatsites.searchwho.com.

● With even more links at http://www.boatinglinks.com.

➤ Here are a few sources for private **aircraft**.

● Aviation begins and ends with the FAA. You'll find them online at http://www.faa.gov, or you can contact them at 800 Independence Ave, SW, Washington, DC 20591, 202-366-4542.

● Visit the Aircraft Owners and Pilots Association online at: http://www.aopa.org, or at AOPA, 421 Aviation Way, Frederick, MD 21701, Phone: 301/695-2000, Fax: 301/695-2375.

● See Sporty's Pilot Shop at http://www.sportys.com, or by contacting Sporty's Clermont County Airport, Batavia, OH 45103, Phone: 1.800.SPORTYS, Fax: 1.800.359.7794, Customer Service: 1.513.735.9000.

● Pilot's Friend has some good online info sources at http://www.pilotfriend.com.

● Go to AVWeb, the "Internet's Aviation Magazine," online only at http://www.avweb.com.

➤ Much of the automotive equipment suggested on the checklist below can be obtained through:

● The "**Safety Center**" at http://www.safetycentral.com/travelsafety.html, 1370 Highway 10 West, Livingston, MT, 59047, Phone # 406-222-3171.

● The **JC Whitney** company at http://www.jcwhitney.com, or contact them at JC Whitney, 1 JC Whitney Way, LaSalle, IL 61301, (800) 469-3894, or fax at (800) 537-2700.

● **AutoZone**: http://www.autozone.com, 1-800-AUTOZONE (1-800-288-6966), or find a local store in your phonebook.

● **NAPA** Autoparts: http://www.napaonline.com, 1-877-805-6272, or find a local store in your phonebook.

- **Pep Boys at:** http://www.pepboys.com, 215-430-9000, or again, look for a local store.
- **Advance Auto Parts**, online-only at http://www.advanceautoparts.com. While online, visit their "How-To's & Tips" page.
- Be sure to check the universal gear page at http://www.equipment.net, to search for related tools and supplies. You can contact them directly at EQUIPMENT.NET, INC., PO Box 810461, Boca Raton, FL 33481, FAX: 305-675-5896.

Vehicle Prep Checklist

Let's take a look at specific maintenance and equipment items you should consider:

Basic Vehicle Prep Checklist

("←" Items have discussion after the checklist. ** Items have forms located in the "Appendix.")

(Items marked "(RP)" are to be kept in a Removable Pack so they can be easily taken elsewhere.)

Maintenance: Check the current condition of each of the following:

☐ Battery and Alternator	☐ Electrical System & Starter	☐ Bulbs and Fuses
☐ Engine Tune-up & Timing Belt	☐ Transmission	☐ Radiator and Cooling System
☐ Brake System	☐ Differential	☐ Joints and Bearings
☐ Steering System	☐ Belts and Pumps	☐ AC and Heating
☐ Oil & Fluid Change	☐ Jack and Spare Tire	☐ Maintenance Record **

Service and Security:

☐ Roadside Emergency Club ←	☐ Alarm System	☐ Locking Gas Cap ←
☐ Steering Wheel Lock Tool ←	☐ Ignition Kill Switch ←	☐ "LoJack" ←
☐ All Parts Engraved ←	☐ Accident Form ** ←	☐ Disposable Camera ←
☐ Inside Trunk Release ←	☐ Check the Radio ←	☐ Window Tinting ←
☐ Other safety gear pertinent to your vehicle ←	☐ Double-check Vehicle Insurance Coverage ←	
☐ Bike Rack ←	☐ Luggage Rack ←	☐ Trailer Hitch ←

Equipment:

"Parts"	General Tools	"People Care" (All RP)
☐ Hidden Keys ←	☐ Basic Tool Kit ←	☐ Vehicular **Info Pack** ←
☐ Extra Oil	☐ Map Atlas (**RP**)	☐ First Aid / Trauma Kit ←
☐ Jumper Cables	☐ Flashlights (**RP**) ←	☐ Spare Clothes and Shoes ←
☐ Solar Recharger ←	☐ Spotlight ←	☐ Toiletries ←
☐ Battery Pack ←	☐ Small Shovel	☐ Gas Mask or "Evac-U-8" ←
☐ Tire Repair Kit	☐ Hatchet or Axe	☐ Food & Water ←
☐ Two Cans "Fix a Flat"	☐ Rope (**RP**)	☐ Collapsible Water Bucket
☐ Extra Wheel (Lug) Nuts	☐ Tow Straps	☐ Blanket(s)
☐ Snow Chains	☐ "Come Along" or Winch ←	☐ Heating Pad or Seat ←
☐ Radiator Patch Kit	☐ Roadside Safety Signals ←	☐ Large Trash Bags ←
☐ Spare Fan Belt(s) ←	☐ Window Breaking Tool ←	☐ Cash & Credit Cards ←
☐ Spare Bulbs & Fuses	☐ Air Pump ←	☐ AM/FM Radio and/or TV ←
☐ Second Spare Tire ←	☐ Gas Can and Siphon Hose ←	☐ Commo Gear ←
☐ Extra Transmission Fluid	☐ Two Fire Extinguishers ←	☐ Signal Kit ←
☐ Extra Brake Fluid ←	☐ Bag of Salt or Sand ←	☐ Car Cover, Tarp, or Tent ←
☐ Assorted Hardware ←	☐ Crowbar or Pry Bar	☐ Defensive Weapon ←
☐ Extra Gas Cap	☐ Windshield Ice Scraper	☐ Cloth Towels & Paper Towels
☐ Extra Radiator Cap	☐ Duct Tape and Masking Tape ←	☐ Air Filter ←
☐ Fuel Filter	☐ Binoculars ←	☐ Umbrella

Discussion Items (The "←" Items from above)

Service and Security:

➢ **Roadside Emergency Club**: Check with your insurance company or gas credit cards to see if they offer a roadside assistance program. Though you couldn't count on them during a mass evacuation in a regional disaster, you may need them for those little "nuisance" occurrences that will happen at the worst possible moment. Also check with **AAA auto club**: http://www.aaa.com/scripts/WebObjects.dll/ZipCode. **American Automobile Association**, 8111 Gatehouse Road, Falls Church, VA 22047, 1-800-AAA-HELP (800-222-4357). Whichever group you choose, make sure one copy of your **paperwork** (with all benefits and services listed) lives permanently in the vehicle, and another is included in your Info Pack.

➢ **Locking Gas Cap**: Get a locking gas cap even if your gas cap cover locks from inside the car. Some emergencies could see civil unrest and economic repercussions. This means that gasoline theft will be much more prevalent.

➢ **Steering Wheel Lock Tool**: We've all seen "The Club ®." We suggest you get not only The Club ® or similar, we also suggest you add another lock that goes from the steering wheel to the brake pedal. This tool prevents the brake pedal from being used and also helps prevent the steering wheel from being removed which is a method of defeating steering wheel locking tools.

➢ **Ignition Kill Switch**: This can be an automatic function of an anti-theft alarm, or a simple switch that you have installed in a hidden location so you can manually disengage the ignition after you've parked. Works great for preventing theft if done right.

➢ **"LoJack"**: You can use the brand LoJack ®, On-Star ®, or other tracking service that will trace your vehicle if stolen. Tracking services have a good record of recoveries. Base your decision to purchase such an alarm on the value of your vehicle and how likely it is to be stolen.   On-Star ® services also include other emergency services. Find out more about them at your local autoparts store or car dealership.

➢ **All Parts Engraved**: One theft-prevention method involves engraving a serial number (though *not* the VIN number) on all resalable parts, especially the windows. The more traceable parts are, the less useful they are to thieves. We've included these security suggestions because of the value your vehicle holds as a asset in an emergency.   Speaking of the VIN number, if you drive a newer vehicle, be sure to cover the VIN number on the dash when you park. Simply slide a piece of paper over it to prevent anyone from reading it. At the time of this writing, it's easy in some states for thieves to get a VIN (Vehicle ID Number) and use it to forge documents making it appear the thief is the vehicle's owner. Play it safe and cover your VIN.

➢ **Accident Form**: In the "**Appendix**" you'll find a form for your use if you're involved in an accident. We've done this for **three reasons**. **One**, there's certain info you'll want to record immediately. **Two**, it's always best to have a detailed report independent and redundant to the initial Police report. **Three**, you may be in an evacuation situation where officials tell you to "swap info and move on" after a minor accident with no injuries, where all vehicles are mobile.

➢ **Disposable Camera**: This ties in with the accident form above. You'll need photos to document your accident report and to record damage to your vehicle. Photographic documentation is extremely valuable in such cases. We've seen situations where cars were towed off only to be stolen by the towing company who denied ever having the car, we've seen theft and damage committed by some towing companies, and we've seen passengers omitted from Police reports and thereby causing a denial of insurance coverage, etc.

➢ **Inside Trunk Release:** If someone locks you in your trunk, which is common in some carjackings, you need to have a way out. Have a trunk release installed or fabricate your own.

➢ **Check the Radio:** Make sure the AM/FM radio in your car works. It may be the first place you hear an emergency warning.

➢ **Window Tinting:** Actually, your windows don't have to be tinted much at all. The idea is to apply a layer of window tint film to reinforce your vehicle's glass making it a tiny bit safer in an accident and a little tougher for criminals to commit a "smash and grab" robbery of your gear. Window tint films come in shades of "almost unnoticeable" to "can't see through." (Our descriptions, not a manufacturer's.)

➢ **Other Safety Gear Pertinent to Your Vehicle:** As we said earlier, this list can't be applied to every vehicle, and not all emergency vehicles are going to be cars or trucks. Other types of vehicles, such as boats or private aircraft, will have their own safety equipment needs.

➢ **Double-check Vehicle Insurance Coverage**: Make sure your coverage is up to date, and meets realistic coverage needs. Also, see if your insurance covers things like lodging, car rental, roadside assistance, emergency cash, etc.

➢ **Bike Rack**: If you have a bike, get a rack. If you don't have a bike, but are able to ride one, get a bike and then get a rack. Some situations may arise where you can't use your car. Though you'd have to consider numerous details of the situation, you could probably use your bike to get out of the immediate area to a secondary source of transportation.

- ➤ **Luggage Rack**: Though you might not transport anything on your roof with any regularity, consider having a rack since you'll need all the carrying capacity you can get during an evacuation. Several companies make detachable racks that you don't have to put on until you actually need it. You can also purchase an enclosed roof-top cargo carrier which is much more secure than a simple rack. You could leave your cargo carrier pre-packed with some of your evacuation gear, and ready to mount in case of an emergency. **Hint**: If all you have is the luggage rack, put your least valuable or useful gear up there. Items are too easy to steal from the luggage rack and you don't want to lose your most important equipment.

- ➤ **Trailer Hitch**: Similarly, even though you don't regularly tow a trailer, you might want to have a small hitch installed, or at least purchase an economical detachable hitch. One of the concepts we'll discuss later is buying a small trailer (such as a small U-Haul trailer) and keeping it prepacked with some of your gear. More on that later. For now, just consider getting the hitch.

Car "Parts":

- ➤ **Hidden Keys**: It's always good to have a hidden key. We recommend hiding 2 sets of keys (if your door key is different from ignition) in 2 different locations (and not obvious locations) on the exterior of the vehicle. We suggest 2 keys in case one works its way loose and is lost. While you're at it, you might as well hide a house key or other important key in the same spot(s). Also, we don't recommend the simple magnetic holders. They'll get lost. Whatever you use, be sure you also have it tied in place with wire. One unique way to carry a spare car door key is to put one in the cover of your cell phone or pager, if your cell phone or pager has a removable cover such as the more popular leather carrying cases with belt clips. We also suggest having a door key in your wallet. With spare keys and potential emergencies, it really is "the more the merrier."

- ➤ **Solar Recharger**: Several companies make a small recharger that uses solar cells to generate the current to charge your battery. This isn't a "jump start" system as it charges slowly (trickle charge), and of course, it won't work at night. However, it's highly useful if caught out in the middle of nowhere with a weak battery.

- ➤ **Battery Pack**: Several companies make a battery pack that remains connected to your car's battery and alternator system. This is like having a second battery for your car. It can be used to jump-start the main battery, or it can be disconnected and used as portable battery power for spotlights, radios, air pumps, etc. Check JC Whitney, and most automotive supply stores for these and for power inverters that allow your battery to power 110 volt AC household electronics.

- ➤ **Fan Belt**: The best thing to do is carry spare(s) of the belt(s) your specific vehicle uses. In addition, you'll want to carry some spares of other parts that you can replace, that break easily, and that your vehicle can't function without. Emergency fan belt kits (that fit numerous vehicles) can be found at the "**Safety Center**" mentioned above under "General Considerations," and at http://www.preparedness.com/fanbelrepkit.html.

- ➤ **Second Spare Tire**: Though you wouldn't want to carry one in your car all the time, you might want to have another one available in case of evacuation. Always remember, in an emergency, redundancy is our friend. If you can afford it, you might want to consider self-sealing, or "run flat" tires. Ask your local mechanic, car dealer, or auto supply store about them.

- ➤ **Extra Brake Fluid**: This ties in to those "tricks" we asked you to learn about "Auto Repair" under "Basic Training." Different fluids are interchangeable on some vehicles. Sometimes brake fluid can be used as clutch fluid. See what's interchangeable on your car in an emergency and carry extras of each.

- ➤ **Assorted Hardware**: There's nothing more frustrating, or dangerous in the case of an emergency, than to be broken down because of something incredibly simple. Be sure to keep a small assortment of **hose clamps, nuts and bolts, and machine screws**. An auto parts dealer can recommend some types to keep based on your type of vehicle.

General Tools:

- ➤ **Basic Tool Kit**: You should carry the tools you'll need to perform the maintenance and emergency roadside repairs you're capable of. Maybe even more as you may be supplying tools to someone with more mechanical experience than you. We recommend your kit carry at least:

❏ Assorted screwdriver set	❏ Socket set (English & Metric)	❏ Spark plug socket
❏ Needle-nose pliers	❏ Wire-cutter pliers	❏ Vise-Grip pliers
❏ Wire (electrical and bailing)	❏ Duct tape	❏ Electrical tape
❏ Masking tape	❏ Can of WD-40	❏ Electrical circuit tester
❏ Hammer	❏ Tire pressure gauge	❏ Wire brush
❏ Spark plug gapper and file set	❏ Super glue	❏ Windex ®
❏ Leatherman Tool ® or similar	❏ Hacksaw	❏ Crescent wrench
❏ GoJo ® or other waterless hand cleaner		❏ Large and small Channel-Lock pliers
❏ "Water Cooler" Paper Cups (conical cups for drinking or use as a funnel)		

- ➤ **Flashlights**: We certainly recommend having more than one. The small, 2-AA battery size flashlights can be found at most "Dollar" stores or elsewhere for next to nothing. Get several of those and you're sure to always have light when you need it. For logistics' sake, make sure your flashlights (all of them, not just the ones in the car) use the same kinds of batteries as other equipment such as your walkie talkies, AM/FM radio etc. Interchangeability is a great thing.

- ➤ **Spotlight**: Spotlights are indispensable for signaling, searching, repairs after dark, and for safety – especially at an accident scene. Get the type that plugs into your cigarette lighter. The rechargeable units we've tried all drained easily and weren't designed to work while plugged in. If you find a unit that does both, great. Get a spotlight with a million candlepower (or more). You can make a spotlight using the same type **headlight lamp** your vehicle takes, mounting it on a stick, and attaching an electrical cord with a cigarette lighter adapter on the end. It makes a great spotlight, plus you have a spare headlight should you need it.

- ➤ **"Come Along"**: This is basically the "poor man's winch." It's a lever-operated winch that can help get you out of a ditch or unstuck from ice or mud if you have a nearby stationary object to anchor your rope or tow straps. If so inclined, get a real winch that mounts to a bumper. Though many people might think, "I don't need this, I don't do any off-road driving," remember that it only takes a small patch of mud or ice to get stuck. How valuable would such a small tool be in a mass evacuation when you drove on the shoulder of the road or across a rural intersection corner and got stuck? What if this happened during a family emergency?

- ➤ **Roadside Safety Signals**: Among these are ❑ Extra Flashlights ❑ Flares ❑ Chemical Light Sticks, and ❑ Reflectors. You'll want these for safety (along with your spotlight) if stalled, broken down, or at an accident scene (if at an accident, always check for spilled gasoline before lighting flammable signals).

- ➤ **Window Breaking Tool**: You might need to get someone (or you) out of a car in a hurry when the doors are jammed or locked. These tools are available at most any automotive or department store, and look like little plastic hammers with pointy ends on the hammer. Get one that has a seatbelt cutter on the other end, and a built-in flashlight if possible.

- ➤ **Air Pump:** You may need to reinflate a tire after you've patched it. Some models are electric and can be plugged into your cigarette lighter, whereas other models are manual. We recommend the manual models as they're more economical, less mechanically complex, and you never have to worry about your car's electrical system being operational. You can also use the manual pump for other things in other locations.

- ➤ **Gas Can and Siphon Hose**: You'll need an empty gas can in case you run out of gas, and you may need a siphon hose to get to the gas (though some car models have siphon blocks), though we don't suggest doing anything dishonest. A fellow evacuee may offer you gas out of his tank, or you may arrive at a gas station where the power is out (you'll need cash to pay the attendant). In that instance, we suggest your siphon hose be at least 8 feet long and that you carry a small plastic "squeeze pump." These pumps and hoses are available at most automotive supply and department stores. Also, you can find a "Gas-O-Haul" collapsible gas "can" at http://www.safetycentral.com/gasohaul.html, or at the above stores. Due to vapor and spillage concerns, you shouldn't carry full fuel cans inside your car (temporarily is okay). If you can carry external gas cans, or "Jerry cans," do. You can use the setup for gas or water, just make sure each is labeled.

- ➤ **Two Fire Extinguishers**: Car fires are a lot tougher to extinguish than most folks think, and the small 2.5-lb. dry-chemical units that most recommend just won't do the job. We suggest you carry **two 5-lb., ABC rated dry-chemical units**. Also, you might need to help another person with their car fire, and since dry-chemical extinguishers are no good after use, you'll want the second unit as a spare for your needs. You learned about all this under the **"Firefighting Training"** in the **"Basic Training"** section, right?

- ➤ **Bag of Salt or Sand**: Great to have in your trunk to sprinkle under your tires during the winter if you live in an area prone to road icing.

- ➤ **Duct Tape and Masking Tape**: Though already mentioned under the tool kit, we felt this important enough to mention again. You may find yourself caught in a contaminated area, or having to evacuate during a HazMat incident or biochemical attack. Having tape in your car, you can seal the door and window seams, and cover your AC vents, helping to seal the car and protect its occupants from outside substances. If you carry some spare trash bags, you can also seal any incidental broken windows or other large openings.

- ➤ **Binoculars**: Though seemingly a leisure item, binoculars are useful in figuring out if that wreck a half-mile down the road is something that should be avoided. Binoculars also make it easier to read those hand-written notes displayed by other cars in your evacuation convoy. Though optional, you should consider having a good pair of 10-power or better binoculars.

"People Care":

- ➤ **Vehicular "Info Pack"**: Earlier, under **"Planning Ahead for Phase II,"** we discussed gathering all your pertinent information together to create an "Info Pack." We also recommended that you keep a full or partial set of this "Info Pack" in your vehicle. **At the very least**, the "Info Pack" you keep in your car should contain:

❑ Copy of your title	❑ Copy of auto insurance policy	❑ Vehicle Registration papers
❑ Tag receipt	❑ Tag related documents	❑ Safety inspection certificates
❑ Emissions inspection papers	❑ Maintenance records	❑ Car repair warranties

➢ **First Aid / Trauma Kit**: We'll go into greater detail on contents under "**Evacuation**."

➢ **Spare Clothes and Shoes**: Carry clothing that will allow you to dress for all weather in all seasons, provided you have room in your vehicle. If you pack for *all* types of weather, you can leave your pack in there and not have to worry about constantly changing things in and out. Be sure your clothes and shoes are durable and you consider each of the following:

❑ Ponchos and umbrellas	❑ Work gloves	❑ Work boots or hiking shoes
❑ Hat and bandanas	❑ Heavy coat	❑ Thermal underwear
❑ *Layers* of clothing so you can change and adjust to the weather		❑ 2 changes socks and undies
❑ Eye protection such as extra sunglasses or work goggles.		❑ Coveralls for heavy labor

Keep all this packed in a **removable pack** in case you need to grab it and run.

➢ **Toiletries**: See both the "**Basic Toiletries**" checklist and the "**Non-Prescription Medications**" checklist under "**Prepping the Gear**" in the "**Evacuation**" section. Carry a smaller version in your car. A good way to gauge the amount of toiletries that you keep packed in the car, is to keep enough for an overnight stay at a friend's. As a bare minimum we recommend you carry:

❑ Soap and washcloth   ❑ Toothbrush and toothpaste   ❑ Comb or brush   ❑ Toilet paper   ❑ Towel

➢ **Gas Mask or "Evac-U-8"**: All terrorist acts aside, you never know when you'll be sitting in traffic when up ahead, a tanker truck full of hazardous materials is involved in a wreck. We'll discuss masks in more detail under "**Reaction**." If you decide to carry respirators, be sure to carry yours where you can reach them from inside the vehicle, and make sure you have one for each of your regular passengers or family members.

➢ **Food & Water**: Any number of scenarios could see you stranded in or near your vehicle for an extended time period. Having extra food and water on hand is always good. We suggest you carry at least a quart (preferably a gallon) of water, and make sure the food does not need water or cooking. We recommend military style MREs (discussed under "**Evacuation**") though their shelf-life will be shortened somewhat by the extreme temperatures of being stored in a vehicle (soft-sided insulated lunch boxes may help and they're inexpensive and readily available at most any grocery or department store). You also might want to include some gear to acquire food and water. Along with your collapsible water bucket we recommend:

❑ One-quart canteen	❑ Canteen cup	❑ Water purification tablets
❑ Small hand-pump water filter	❑ Waterproof matches	❑ Small roll aluminum foil

We also recommend that you keep some food and water (at least a one-quart canteen and some meal replacement bars) **accessible from inside the vehicle**, as some scenarios could see you unable to exit your vehicle to even get to your trunk.

➢ **Heating Pad or Seat**: Numerous companies make blankets or car seats that plug into your cigarette lighter. Some also make space heaters that run on your car's battery. You should have something like this if you live in an area where you could be stranded on the road due to snow or ice and wind up with no other way to stay warm. Another little useful gizmo is an immersion "coffee heater." It's a small handle with a small coil on the end that's made to immerse in a cup of water while the other end plugs into your cigarette lighter. It'll let you heat liquids, which could be a lifesaver in a below-freezing situation.

➢ **Large Trash Bags**: Trash bags are like duct tape in that they have universal applications. You could make an expedient shelter, make a poncho, use them as an outer shell for your blanket as you stay warm, or you can use them to clean out your car after an immobilizing accident. Carry at least 4 of the 8-mil thickness or better. As all plastic bags may prove useful, carry a few plastic **grocery bags** in your car as well.

➢ **Cash and Credit Cards**: Keep as much cash as you can in the form of **small bills and coins**. Also carry **duplicate credit cards** for both gas and general purchases. In addition, be sure to have an **extra checkbook**, and maybe some **traveler's checks**. Cash and coins will be the most important since you'll need it for **vending machines, toll booths**, and situations where you can't use anything else. Besides, you never know if the power's going to be on to process a credit card. Also, be sure you have a **prepaid calling card** (get one with NO expiration date) or a **calling card account** that charges back to your **home phone**.

➢ **AM/FM Radio or TV**: You'll want a small battery-operated radio in addition to your car's radio. That's in case you need a radio after you've left your vehicle, your vehicle's radio isn't working, or you want to monitor two stations. Additionally, small battery-operated televisions are inexpensive. Though we don't recommend you watch TV while driving, consider having one in your car for times you need the extra news source.

➢ **Commo Gear**: Even if you have a cell phone, look into getting a **CB or FRS radio**. Radio works anywhere, even where cell phones don't. Your radio could help you keep in touch with associates or with strangers.

Speaking of cell phones, you can use almost any working cell phone as a "911" phone as most locations consider 911 a free call. You don't have to have an account for the phone, it just has to function. You also might consider a **pre-paid cell phone** that lives permanently in the car. Be sure to check any "expiration dates" of the phone's service. More on this in the "Communication" section below. As for your regular cell phone, make sure you have a power adapter that plugs into your cigarette lighter, and that this power adapter lives permanently in the vehicle. As a minimum, we recommend you have **both** a **CB radio** that lives permanently in your vehicle, and at least a **pair of FRS radios** that lives in your "Bugout Kit."

➢ **Signal Kit**: In addition to the safety signals mentioned earlier, you may want other items that will help you get noticed in case you're stranded. You might also want to carry:

❑ **Aerial flares or smoke signals**. Both are available at most marine shops, camping supply stores, or military surplus stores. The flares are valuable at night, and smoke during the day.

❑ **Fireworks such as Roman Candles** (depending on local laws) if you can't find aerial flares.

❑ **Signal mirror**. You can also use the shiny side of a CD.

❑ **Whistle or Air horn**.

❑ **Colored flag** for antennae (important). This will help get you noticed in a blizzard situation, as will the colored smoke signals from above. It also might get you noticed at a crowded rendezvous point. More later.

❑ **Sun visor with "HELP" written on one side**. These sun visors are also highly useful in creating shade in situations where you might be stuck in your vehicle in the hot sun for hours on end. You can get these shade visors at any automotive store, or any store's automotive department. Most are made of cardboard with a decorative scene on one side and the other side is usually printed with something useful, such as the recommended "HELP" request.

❑ Also, make sure your **blankets, towels, and trash bags are brightly colored** (preferably bright orange), or dye or spray-paint an old sheet. More extensive discussions on signaling and communication will be found below in the "Communication" section.

➢ **Car Cover, Tarp, or Tent**: Though usually a vanity item, a car cover might come in handy in a survival situation. You could make a tent out of it, or an awning for shade in a high heat situation. Also, **keeping your car covered during a biochemical attack or HazMat situation** will help protect its interior from contamination so your car will be ready to use when it's safe to leave the house. Similarly, if you drive an open-bed pickup truck, you'll at least want to get a shell cover for it. Such a simple accessory will provide shelter for people and will protect loaded gear from easy theft during a panicked evacuation.

➢ **Defensive Weapon**: We'll discuss this concept in detail in the "**Security and Defense**" portion of the "**Isolation**" section. It's a serious enough concern to get its own section.

➢ **Air Filter**: If you're stuck in a car for a long time with a lot of people or, or when airborne contaminants may be in the car, an air filter might come in handy. Though we haven't tested it, one interesting model that caught our eye was the Ionic Breeze car model by Sharper Image. It's a tiny little unit that plugs into the cigarette lighter, but it's supposed to filter just as well as its big brother. Contact Sharper Image at: http://www.sharperimage.com online, or directly at Sharper Image, 650 Davis St., San Francisco, CA 94111, Customer Service: 800-344-555, or order toll free at 800-344-444.

Notes:

_____

_____

_____

_____

_____

_____

_____

_____

_____

_____

_____

_____

_____

The EAS Alerts

The "**Foundation**" section is arranged in a logical flow taking you progressively closer to an event reaction through a planned succession of preparedness activities and education. We started with you and your health, added to your knowledge base, helped you become more aware of possible dangers, and then helped you prep your recovery information, home, and vehicle. Next in the grand scheme of things is the approaching emergency itself and making sure you get as much warning as possible. Therefore, let's talk a bit about warnings, the **E**mergency **A**lert **S**ystem, and the steps you can take to make sure you get the word that the potential for a disaster is high.

Goals of The EAS Alerts:

1. To **identify the current systems** that exist or are available in your area.
2. To **find additional notification services** to help keep you informed of emergencies.
3. To **augment both the local services and private services** to make sure you'll be warned in time.

We've divided this section into:

1. **Standard Alerts.**
2. **Alternative Alerts.**
3. **Getting the Word.**

Standard Alerts

> To many, the "Emergency Alert System" is that annoying "pop-up ad" with the red screen and obnoxious tone that comes on in the middle of our favorite late night movie. Learn not to turn it off though, as one day it may be for real, and it may save your life. To find out more about the EAS system itself contact the **FCC's National Call Center** at 1-888-CALL-FCC, or visit www.fcc.gov/cib/eas. Also visit the FCC's emergency communications discussion at: http://www.fcc.gov/cgb/consumerfacts/emergencies.html, and be sure to see the rules page at http://www.fcc.gov/cgb.

> Go see the **National Communications System** site at http://www.ncs.gov and their **EAS** discussion at: http://www.ncs.gov/n5_hp/Customer_Service/Xaffairs/NewService/NCS9833.htm.

> Many communities have a **siren** system set up to warn of tornadoes, or nuclear attack. Ask when the **next scheduled test** of the local siren system is, to find out if you can hear it and what it sounds like.

> Some plants or factories have their own alert system, either operated independently or tied into a community's alert system, to warn of accidental chemical releases, or possible fire/explosion. For sites on your threat map, find out what kind of alerts or alarms would be activated at the site or by the community.

> Additional information about emergencies and warnings at nuclear power plants can be found at http://www.pro-resources.net/signals.htm.

> Depending on the nature of the incident, Law Enforcement, Fire, or Emergency Management personnel may drive through neighborhoods making announcements via **loudspeaker**. They may also call certain area businesses or industries to pass the word.

> "**Weather Alert**" radios are constantly being improved. When they first came out, the National Weather Service could send only one warning signal to a region. That means you'd get weather warnings for locations halfway across the state. Now, these radios can be set to your own and surrounding zip codes. If you don't yet have a "Weather Alert" radio, get one.

> Contact your local Emergency Management director or officer and ask them if your area has additional forms of alert. Some communities have a "**Reverse 911**" system, which is an automated dialer capable of calling mass numbers of residents simultaneously and telling them to turn on their TV or radio, or relaying a short message. By the way, if your area does not use 911 as its emergency number, log on to the National Emergency Number Association site at http://www.nena.org, or contact them at NENA, 350 North Fairfax Drive, Suite 750, Arlington, VA 22203-1695, 800-332-3911, Fax: 703-812-4675, for more info.

> Similarly, check the institutions you frequent most often to see if they have any in-house EAS monitoring, or if they utilize any of the "Alternative Alerts" listed below. You'll want to know what your children's' school has arranged, what your workplace has set up, etc.

> In the "**Cooperation**" section we'll discuss ways for your community to expand their Emergency Alert System. This is extremely important as not all communities (even some states) are not on board with the Emergency Alert System. See more about this on the FCC's Media Security and Reliability Council at: http://www.mediasecurity.org, Tribune Company, 1501 K Street, N.W., Suite 550, Washington, DC 20005, Voice: (202) 775-7750, Fax: (202) 223-3844.

Alternative Alerts

This is yet another of the many ways in which redundancy is our friend. We may miss the municipal alert for any number of reasons, so therefore, we need to set up alternate methods of making sure we get the word if an emergency occurs.

> As current EAS methods are still rather antiquated, several companies have started providing their own EAS relay services using the electronic devices many of us carry. Different companies will notify you of an EAS activation in your area by way of your phone, cell phone, pager, or email.

- One is "**AlertsUSA**" at http://www.alertsusa.com, which will notify you by cell phone or email of Department of Homeland Security alert changes. Contact them through their website or by Telephone: 910.458.0690, Fax: 910.458.0279.

- **Emergency Alert Email**: Online only: http://www.emergencye.com/.

- **The Weather Channel** offers an emergency weather alert service called "**Notify**!" Find them online at http://www.weather.com/notify, or contact them at 310 Interstate North Parkway, SE, Atlanta, GA 39339, 770-226-9935 or 9338.

- **AlertAmerica** at http://www.aaofcc.com, with direct contact at Telecom Ventures, LLC, 7520 East Independence Blvd., Suite 430, Charlotte, NC 28227, PHONE: (704) 567-3058, FAX: (704) 567-3056.

- Visit the **Emergency Email** service online at http://www.emergencyemail.org.

- It's rather pricey at a hundred dollars, but Ambient has a satellite activated EAS warning device for "all hazards." See: http://www.weatherconnect.com/ExtDesc.asp?id=105, or call them at 877-413-8800.

> Some companies are thinking of designing "Weather Alert" **televisions** with an instant-on feature. Also, the National Weather Service is working to incorporate all forms of EAS into their weather alert system so you'll get more than just severe weather warnings through your "weather alert" radio. (It's a pet project of ours to help make this idea a reality. Help us by writing various television manufacturers, the FCC, and your congressional reps in Washington, DC.)

> In keeping with TV, this company, Information Station Specialists, Inc., provides good information on the upgrades available to "weather alert" radio listeners. See them online at http://www.issinfosite.com/the_source_news_eas_codes.htm, or contact them at PO Box 51, 3368 88th Avenue, Zeeland, MI 49464-0052, Phone 616.772.2300 • Fax 616.772.2966.

> A new group that you'll be hearing a lot about in the future is the **US Security Council**, an initiative of the **Georgia Security Council**. When you visit the site, be sure to click on "Forums" which is one of the vehicles for passing along emergency notifications to members and communities. You'll find the USSN online at http://www.ussn.org (or http://www.safertogether.org), or you can reach them through the GA Security Council, 250 Williams St., Suite 1001, Atlanta, GA 30303, Phone: 404-525-9991, Fax: 404-525-8977.

Getting the Word

Until the day comes that the EAS is all it's capable of being, and until all states and communities decide to participate, we may have to rely on our own efforts to receive prompt EAS activation, or other emergency notice. Aside from signing up for the commercial alternative alerts listed above, we'll have to regularly seek emergency information directly or set up our own alternative source "networks."

> Find a source to monitor the current Office of Homeland Security's **Terror Threat Alert** Level. You can monitor it directly on the **Homeland Security** website at http://www.dhs.gov. Other sources include **Fox News Channel**, or their site at http://www.foxnews.com.

> One group helping to expand public warnings is the **Partnership for Public Warning**. Visit their website at http://www.partnershipforpublicwarning.org, or contact them at Partnership for Public Warning, 7515 Colshire Drive, Mail Stop N655, McLean, VA 22102, Phone: 703-883-2745, Fax: 703-883-3689.

> EAS won't come through stereos or TVs that are playing tapes or CDs, nor will it come through while you're watching a subscription cable channel or "pay-per-view." During periods that may see emergency activity, take occasional news breaks to listen to open radio or TV stations.

> Know which of your TV and radio stations are your all-news stations. We've found several news sites and sources, several of which will let you set up your own "news alert" email. You'll find some below and others in the "**DP General Links**" collection on the enclosed **CD**.

> Buy a "**Weather Alert**" radio and make sure it's a type that has a battery backup. You can get them at most department stores or electronics stores such as Radio Shack or Circuit City.

➢ **Have alternate inputs for your television.** Your cable company may be knocked out during a large-scale emergency, so it's a good idea to have a second signal source such as satellite dish, or regular antennae. Another idea is to keep a small battery-operated TV handy for use in such situations. This will also let you monitor two separate stations, or maybe three if your main set has the "picture in picture" function.

➢ If you live near a factory, plant, or refinery that uses hazardous materials or potentially dangerous substances, ask what kind of **public alert or alarm system** they employ and how to recognize it. Also ask to be included in any emergency notification **phone-tree** they may activate in the wake of an accident.

➢ If your business is large enough, ask local Law Enforcement or Emergency Management to make sure your company is on their "emergency alert phone tree" and that your location will be called in case of local emergency. To help matters, make sure your building has a PA system that can reach everyone at once to broadcast the alert.

➢ Sometimes, the only way you're going to find out about an alert is if **someone tells you**. The **best** way to make that happen is to be proactive with friends and coworkers to set up an informal network or "**phone tree**." Anyone that hears of an alert or pertinent news, will notify the others. This happens on its own anyway, but go ahead and do the **network** thing to hedge your bets.

➢ Later, under "Communication," we'll tell you how to set up a Yahoo Groups email group. Though not as rapid as a phone warning, email is still too useful to ignore as far as EAS goes.

In addition to the alert sources and hints above, one of the best ways to receive an "alert" is to **stay current on all forms of news** so that in some cases you won't even need an alert; you'll already know something's up. Here are some general news sources and others with a specific focus:

➢ Aside from your own newspaper, radio, and television, you can find a variety of current newspapers and magazines at your **local library and bookstore**, depending on how often you visit. Though you won't be able to keep abreast of *breaking* news, you can certainly follow trends in the weather, and with political climates that might indicate potential terrorist activity.

➢ Many of the major newspapers and television news providers will also have a website, and some will offer a newsletter. Though we'd normally give you phone numbers and contact info, our purpose here is to put the speed and power of the internet to work for you. Some of our favorite direct sites are:
   ● CNN has one of the best at http://www.cnn.com.
   ● Another great source is Fox News at http://www.foxnews.com.
   ● BBC News: http://www.bbc.co.uk.
   ● ABC News: http://www.abcnews.com.
   ● LA Times: http://www.latimes.com.
   ● New York Times: http://www.nytimes.com.
   ● Reuters: http://www.reuters.com.
   ● USA Today: http://usatoday.com.
   ● The New York Post: http://www.nypost.com.
   ● Chicago Tribune: http://www.chicago.tribune.com.
   ● See the newslinks at http://newslink.org.
   ● Also visit http://www.newsdirectory.com.
   ● http://www.rapidtree.com has links to newspaper, radio, TV stations and more.

➢ The following are some online news source compendiums. Some of these will offer **newsletters**, some will email you their general daily news **digest**, and others still will send you an email alert on news stories pertaining to subject categories you select. Visit each of these and set up any automatic notification services or newsletters that any of them might offer. In addition to these, remember the search sites we gave you earlier under "Information Research."
   ● Reuter's "Alertnet" at http://www.alertnet.org.
   ● "Online Newspapers," at http://www.onlinenewspapers.com.
   ● More online newspapers can be found at http://www.newspapers.com.
   ● Another major online newspaper link source http://www.newspaperlinks.com.
   ● "Ether Zone" Alternative News Source, at http://etherzone.com.
   ● "Planet Word's" World Headline Search http://planetwords.triqi.com.
   ● Yahoo News http://news.yahoo.com.
   ● "Newslinx, All of Today's Web News" http://newslinx.com.

- News Index will allow you to search and set up email delivery http://www.newsindex.com.
- "Total News" news search http://totalnews.com.
- RocketInfo's News Site Search Engine http://www.rocketnews.com/2corporate/searchengine.html.
- JournalismNet News and Search Site http://www.journalismnet.com.
- Newsdirectory's News and Media http://www.ecola.com.
- News Service Search Engine http://www.pandia.com/news/index.html .
- "First" Headline News http://www.1stheadlines.com.
- News from the United Nations organization http://www.un.org/news.
- See the "I Want Media" site's massive links collection: http://www.iwantmedia.com/resources/index.html. While there, be sure to visit the home page.
- If you want to keep up with what's going on in DC, log on to the Federal News Service page at http://www.fnsg.com, for transcripts and other Washington news.

➢ Below, under "Communication," we'll discuss internet radio, XM radio, shortwave, and other ways to keep abreast of what's going on in the world around you.

Okay... Spot check. Have you been following our web links by using the "DP1 Book Links" file? Have you been collecting all the pages you chose to print from websites, your data forms, and your notes that you've made so far and put them in your 3-ring binder? We hope you have, because now we're going to look at some situations where you'll need to put that information to use.

Now that we have a respectable list of warning sources, let's look at how these warnings might come your way, and how you might pass the word to others. Let's look at "**Communication**."

Notes:

_____

_____

_____

_____

_____

_____

_____

_____

_____

_____

_____

_____

_____

_____

_____

_____

_____

_____

_____

_____

_____

_____

_____

_____

Communication

Communication in an emergency situation is an absolute must. You and everyone involved will need to know what's happening, when and where it's happening, how everyone else is, and where they are. The hard part is doing all this if standard forms of communication aren't available.

The problem with good communication is that we've become so heavily dependent on electronic communication that we tend to overlook some of the more simple and subtle ways we can get messages across. Too, there's a lot more to consider with electronic communication than most people tend to think about. So, for this section, we'll look at a little bit more detail concerning the electronics, and at a few other ways people can pass the word along. Rather than a technical education section on how things work, this is an introduction to practical application so you can take what's out there, decide what works for you, and then put it to good use. It's also a major **imagination** primer.

Goals of Communication:

1. To keep you **in touch with the rest of the world**.
2. To help you **stay informed during an emergency**.
3. To **increase your ability to communicate** with family members and others to coordinate your safety.
4. To introduce you to **alternative forms of communication** and signaling.
5. To provide a brief **overview of equipment you should own**.
6. To help you **develop codes** to use for verbal messages, and in securing sensitive written information.

We've divided communication into

1. **Suggested Equipment Checklist.**
2. **Telephone and Email.**
3. **Radio and Television.**
4. **Visual and Audible Signaling.**
5. **Text, Symbols, and Codes.**

The detail we've gone into in this section, the amount of left-field "Rube Goldberg" ideas we've presented, and the long lists of options that have been included might make this subject look like one you'd want to come back to later. However, being able to communicate is so important that you should become familiar with all these concepts and choose what works for you *now*, since you never know exactly when you're going to need them. Let's start with a few quick scenario questions to show you why you might want to expand your communications capabilities.

1. A disaster occurs, local phone and cell lines are overloaded from everyone trying to use them, and the Police just pulled through your neighborhood announcing over loudspeaker that everyone must evacuate. How do you get in touch with a part of your family that lives over a mile away?
2. Your neighbor is stranded 50 yards away on the other side of a bridge that was just washed out by flood waters. He has no cell phone, and the rushing water is too loud to yell over. How do you tell him you're going to get help? How would he tell you that actually you're the one in danger?
3. Your rural home is surrounded by calm flood waters and you're stranded on your roof. There's no phone service and your civilian walkie talkie can't communicate to rescue helicopters. How many ways do you know to tell them you need rescuing? Or medical attention? Or that you're okay and they can take more extreme cases first?
4. Your neighborhood is quiet, peaceful, and wooded with very few houses able to see many other houses. You're the neighborhood watch commander and the terror threat alert has just been raised to Red Alert. Besides calling each person on an increasingly busy phone system, how do you warn your neighbors?
5. Your daughter is at work one night and someone she knows to be stalking her abruptly approaches as she's on the phone with you. What is her codeword to discreetly let you know she's in trouble?
6. A terrorist group has taken advantage of the aftermath of an earthquake and released a biological weapon in your area. A massive quarantine has been ordered, power is out, all phone systems are down, and after 10 days all the batteries in your radios are dead. How do you relay messages in and out to get help if you need it, or to simply find out what's going on?
7. Your volunteer group is tasked with assisting first responders and local officials at disaster scenes. Your city was just rocked by 6 simultaneous incidents requiring several command post locations. Power is out, phone systems are down, traffic is gridlocked, and first responder radio units are spread too thin and not everyone can talk to everyone else. How could you and your group help first responders communicate?

8. A massive evacuation has been ordered, cell towers are down, and the authorities just told you to leave your home. How do tell other family members to meet you at the primary rendezvous point?

Let's start with a short checklist of communication items we feel you should own, learn, or do. These are the basics. The text below will probably add things to your list, and maybe give you some options that will let you subtract a few items. Gear mentioned in the checklist below will be explained in more detail in the portions that follow.

Suggested Communication Equipment		
Telephone and Email		
❏ At least 2 corded phones	❏ At least 1 cordless phone	❏ 1 fully activated cell phone
❏ 1 "911-only" cell phone	❏ Full cell phone accessory kits	❏ Dry-cell battery connectors
Radio and Television		
❏ Alternate TV antennae	❏ 1 small portable television	❏ 1 "Weather Alert" radio
❏ 1 pair FRS or GMRS radios	❏ 1 handheld CB radio	❏ 1 vehicle-mounted CB radio
Visual and Audible Signaling		
❏ Handheld spotlight	❏ Camper's aerial flares	❏ Colored smoke signal
❏ Chemlight sticks	❏ Small dry-erase board	❏ Binoculars
❏ Sports whistle or air horn	❏ Learn simple hand signals	❏ PA speaker for your CB radio
Text, Symbols, and Codes		
❏ Colored envelopes and paper	❏ Flourescent orange spraypaint	❏ Chalk
❏ Grease pencil	❏ Indelible markers	❏ Memorize a code system

Let's move on to some of the detail.

Telephone and Email

The two most common forms of electronic communication in use by the civilian population today are the telephone and email. Instead of trying to go into unnecessary discussions of how the telephone was invented, or which politician invented the internet, let's talk a bit about general hints and pointers on how to put these tools to better use. We'll break these down into Telephone, Cell Phone, and Email.

Telephone

Your telephone will do you absolutely no good if you can't hear it, can't get to it, or can't use it in general. Here are a few things to keep in mind to make yours a little more useful.

> Location is important with all your communication devices. You should have at least one phone in your home's kitchen, living areas, each adult's bedroom, garage, and safe room. You can do this with regular phones, or have a couple of regular phones and one cordless you carry with you.

> Regular land line telephones are cheap and plentiful. As we discussed in other areas of the book you should have at least 2 regular corded phones, one at each end of the house, and at least one on each floor, with another one in any safe room or safe area. The cordless phones are great, but they will not work in a power outage unless you bought one of the more expensive models that has a battery backup in the base unit. Even then, the battery will last only so long.

> However, we do recommend that you have at least one cordless phone, preferably two. This way you can have your phone with you while out in the yard, over at a neighbor's, etc., and having two cordless units allows you to swap phones while the battery on the other is recharging. **Hint**: When you buy a cordless phone, go ahead and buy a **replacement battery**. Most batteries will take only so many charges before needing replacement regardless of how careful you are with recharging.

> Under "Basic Home Prep" we mentioned the importance of protecting your outdoor utility connections in anticipation of a storm and to protect your communications from being cut by a burglar. This suggestion applies to your phone line and junction boxes as well. At the very least, protect your phone line with metal conduit to prevent cutting by burglars or damage from storm debris.

> **Program all important phone numbers into each programmable phone you own.** It will be much easier for you to call everyone on your "Important Contacts" list, since you won't have lost anyone's number, and regardless of which phone you have in your hands, you can call anyone.

> Though it's not free, we recommend that you get certain "Touch-Star" features on your phone if they're available in your area. Chief among these are *69 or the last-call-in feature, and the various call forwarding

and select-call forwarding features.   If you have to leave in a hurry, you can **forward some or all of your calls** to your cell phone, or to your family's contact person. The *69 feature will let you find out who called while you were on the other line (if you don't have call-waiting Caller ID), if you're in an emergency situation with several people calling in and out.

➤ We also recommend you have call waiting, caller ID, and call waiting caller ID. In an emergency, certain people and groups in your life will take extreme precedence over others and you'll need to know which caller is which. Relatedly, some internet service providers provide an on-screen caller-ID to let you know who's trying to call in while you're online. Do anything you can to keep your lines as open as possible and to be able to prioritize your incoming calls.

➤ Another concept in direct contradiction to our "Frugality" section is that we suggest you have voice mail as a part of your phone service, in addition to any answering machine you may have at home. If there's an emergency and you have your phones thoroughly tied up (or if you're online), callers will still have a way to leave or retrieve a message. It's also a good way to leave rendezvous instructions and updates for family or group members who have the voice mail retrieval access codes.   Your house and regular answering machine may be destroyed, but voice mail will probably be intact and can be retrieved from anywhere.

➤ Speaking of frugality, we thought we'd pass along a link to the FCC's "Afford-A-Phone" program, which is designed to help lower income families afford a phone.   The FCC's program can be seen online at http://www.fcc.gov/cgb/getconnected, or by contacting them at Federal Communications Commission, 445 12th Street SW, Washington, DC 20554, Phone: 1-888-CALL-FCC (1-888-225-5322), TTY: 1-888-TELL-FCC (1-888-835-5322), Fax: 1-866-418-0232.

➤ Answering machines physically located at your house can act as a redundant backup to voice mail, and they offer one more little trick. With an answering machine, you can call your house after an evacuation to see how things are. If the machine picks up, then the power is on, phone lines are up, and your house is probably intact. We suggest you have both voice mail and an answering machine. You can leave one off most of the time, or coordinate the use of both together. **Hint**: Make sure you record your access and control codes for both voice mail and your answering machine on your "Household Data" form.

➤ **Get calling cards**. Get a calling card number that will allow you to charge calls back to your home number account (save your payphone coins for vending machines). If you get *prepaid* calling cards make sure they have no expiration date. In addition to these are the various "call collect" services such as 1-800-CALL-ATT, and 1-800-COLLECT which are supposed to let you place lower-rate collect calls.

➤ Speaking of payphones, they may be operable while the local cell tower is down or residential phone service is out.   Additionally, when calling your out-of-town commo / rendezvous contact, you might get a long distance line through a payphone more easily than you could a local line from your home phone since sometimes the two use different wiring trunks.

➤ Also, and we'll cover this more extensively in other parts of the book, you need to be a part of a **phone tree** set up by your neighbors or your volunteer group. If the phone systems are up, this is one of the quickest and best ways to get an emergency warning in or out.

Cell Phones

Cell phones are miraculous little gizmos and we're constantly amazed that we were ever able to function without them (though we did, and that's what this commo section is all about). Of course on other days we have to restrain ourselves from hurling the poor little things against the nearest brick wall when we can't get a signal or when the battery runs out at the worst possible moment. So what are some of the things we can do to make our love relationship with our cellular devices grow stronger?

➤ For background, we found a surprisingly good source of introductory material that answers not only basic questions, but helps with advanced information and with shopping. Go see the "How Stuff Works" page on cell phone details, usage, and comparisons at http://electronics.howstuffworks.com/cell-phone.htm.

➤ The first order of business is to suggest that you <u>have</u> a cell phone, or at least a "911 phone." In most areas of the country, 911 is specifically legislated as a free call. That means you don't need a cell phone *account* in order to call 911. In fact, you could probably walk into any thrift store, pick up a cell phone, and if the battery worked you could immediately call 911. Some cell providers will even allow you to place collect calls with a "non-account" cell phone. So, it doesn't matter if you can't afford the monthly service fee of a cell phone. Get one for a 911 phone. Later on, if you can afford the service, you can probably get a newer phone for free from your cellular provider, or you can have your 911 phone activated for full service.

➤ A better suggestion is that you have both a regular cell phone with a service account, and have a separate 911 phone. Keep the 911 phone(s) as a part of your vehicle's safety equipment, your safe room's safety gear, or as part of your "Bugout Kit."   We found a 911-phone source for you at

● See "911 Phone" at http://www.911phone.net, or contact them at AA Communications, 291 Watershed, Noblesville, IN 46060, Phone: 1-800-484-8034 security code 9160, or 317 877-2038.

- Also try http://www.emergencycellphones.com, or call 1-866-WANT-911 (926-8911).

➢ A compromise between the two are the "**disposable" cell phones**. You buy the phone and it comes with a certain amount of prepaid minutes, or "talk time." After you're done, you can "recharge" the minutes, discard the phone, or keep it as a 911 phone. Most of the cellular service providers listed below offer some sort of prepaid deal. You should also check with any local office these providers may have. To find more about the prepaid offers, we've located a couple of sites:

- Prepaid cell phones, at http://www.pre-paid-cell-phones.net.
- First in Cell Phones, online-only at http://www.1st-in-cell-phones.com.

➢ **Fully accessorize**. Cell phones are only useful if they work. Simple enough, huh? Make sure you have **recharge adapters** for both your **car** and your **home**, and if your phone model offers it, make sure to get a **battery discharge or "refresh" unit**. Discharging the battery fully before any recharge keeps it from developing a "memory," which will shorten a battery's useful lifespan. Relatedly, get the **highest capacity battery** you can for your model and get a **backup** as well. Charge the backup and keep it with you. Last but not least, you'll want to get an "**ear bud**" or "**hands-free device**" for your phone whether it's required by local law or not. Having your hands free in an emergency is a must. **Hint**: For your adapters, use a labelmaker and print a **label for each adapter** showing what it belongs to. (You might be incapacitated and a family member may have to use the gear.) Here are a couple of good phone accessory sources for you:

- Cell Phone Shop online at http://www.cellphoneshop.net, 806 Buchanan Blvd, # 115-041, Boulder City, NV 89005-2144.
- And more, online only, at: http://www.cellphone-accessories-4u.com.

➢ We found a **hand-crank cell phone recharger / generator**, called the SideWinder ®, at iPrepare.Com, online at http://www.iprepare.com/sicephch.html. You can reach them at iPrepare.com, P.O. Box 344, Roseville, CA 95678-0344, Toll free: 800-219-9696, Phone/Fax (707) 982-7292.

- Freecharge has also produced a dual hand-crank and outlet-chargeable battery backup for emergency cell phone power. See it online at http://www.freeplay.net/website/product/charger.php.
- Even more of the same, including the hand-crank cell phone charger, can be found through Innovative Technologies, online at http://windupradio.com, or call them at 1-250-386-2556.

➢ Improvise. There may be situations where you're without any sort of power to recharge your phone, and you've used up both your charged batteries. You can **make your own battery pack** that uses the same **standard dry-cell batteries** you'd put in your flashlight or AM / FM radio. If nothing else, take your phone and its owner's manual into pretty much any Radio Shack or other electronics hobby store and ask the people there to help you put one together. You can find a commercial version of this idea at: "**9 Volt Power**" online at http://www.9voltpower.com, or by phone at 215-513-9572.

➢ Speaking of **batteries**, we've listed a few battery sources in the "**Evacuation**" section. However, since we've already recommended you get a backup phone battery, here are a few sites and sources for now:
- A **disposable** backup cell phone battery can be found through **CellBoost**, online at http://www.cellboost.com, or you can call them at 800-833-1070.
- "**Batteries.Com**," online at http://www.batteries.com, or at Batteries.com, 6024 West 79th Street, Indianapolis, IN 46278-1727, 888-288-6500.
- **Battery Mart**, online at http://www.batterymart.com, or at Battery Mart, 1 Battery Drive, Winchester, VA 22601-3673, Tel: 540-665-0065, Fax: 540-665-9623, Toll-Free 800-405-2121.
- **Battery World**, online at http://www.batteryworld.com, or at Battery World, Inc., 23164 Ventura Blvd, Woodland Hills, CA 91364, Phone: 818-225-0478, Fax: 818-337-7544.
- **Phone Batteries**, online only at http://www.phonebatteries.com.
- **E-Batteries**, online at http://www.ebatts.com, or at eBatts.com, 703 Rancho Conejo Blvd., Newbury Park, CA 91320, Toll Free 800.300.1540, International 805.499.4332.

➢ Protect your phone. In an emergency, your phone may be your lifeline. So, all accessories and batteries aside, you'll want items that will physically protect your phone from damage. Many phones have belt clips, but you should look into **belt clips that are part of cases and covers**, and you should also have a separate cord running from the case to your belt in case the regular clip fails. You should also look into protective covers that have a neckstrap that will allow you to carry your phone under your shirt for extra protection. Similarly, many companies make waterproof pouches with belt loops or neck straps that are specifically made to protect small personal items such as phones, wallets, keys, etc. **Hint**: If you get a cell phone cover, you can hide a spare car key and a small amount of cash inside. **Hint #2**: For lack of anything else, if you're in a situation where you need to keep your phone protected from the elements (chemical, weather, etc.) just put it in a resealable plastic bag. You can even use your phone without taking it out of the bag.

➢ Different areas and different cellular services will offer various "**touch star calling**" services that use the star (*) key on your cell phone. For example, through Cingular in Georgia, dialing "*DOT" will connect you to the

Department of Transportation so you can ask about road conditions. "*GSP" will connect you with the Georgia State Patrol. Find out what touch star services are offered in your area by your cellular provider, and record them on your "**Important Contacts**" pages and your "**Evac Atlas**" sheets.

- ➤ **Coverage area** is another frustration you'll experience, especially during an evacuation since you may be travelling from one coverage area to another with major gaps in between. The best way to solve this problem is to **diversify**. This is much more easily done with a family than with an individual. If your family has several cell phone users you should get **each phone from a different provider** (at least 2 different providers). Even though it may be more economical to get the phones from the same service, you're limited in coverage area. Having different providers will mean that no matter where you go, **someone in your family or group should be able to get a cellular signal**. You can find local offices in the phone book, but in the meantime, here's the contact information for the most popular cellular service providers:

  - AT&T Wireless: http://www.attws.com, 800-888-7600.
  - Cingular: http://www.cingular.com, 866-CINGULAR (246-4852).
  - Sprint PCS: http://www.sprintpcs.com, 800-480-4727.
  - Verizon Wireless: http://www.verizonwireless.com, 800-466-4646.
  - T-Mobile: http://www.tmobile.com, 800-T-MOBILE (866-2543).
  - Nextel: http://www.nextel.com, 800-639-6111.
  - Southern Linc: http://www.southernlinc.com, or 800-818-LINC (5462).

- ➤ Speaking of Nextel and Southern Linc, they have nationwide "**push to talk**" **radio coverage** where, to put it in simplest terms, your cell phone doubles as a walkie-talkie. **Verizon** Wireless is also starting to enter this arena. This feature is useful if evacuating as part of a group having the same phone system, or if working with a volunteer group. The drawback is that the walkie-talkie coverage area is dependent on cell towers. If the tower's out of range or out of commission, the push-to-talk feature won't work.

- ➤ Another useful feature we've seen is the **photo-phone feature** with the built-in digital camera. This could prove highly useful in neighborhood watch settings. Sure would be convenient to take a picture of a crime in progress and have it immediately sent to the Police Department, wouldn't it? It may also help in sending a claim in to your insurance company, sending accident scene details to the Police, or **sending injury photos to a doctor** who might be coaching your distant first aid and treatment efforts over the phone.

- ➤ A close cousin to the cell phone is the rather expensive **satellite phone**. As its name implies, it uses satellite signal relays to communicate rather than local cell towers. Therefore, you'll still get service even if all the area cell towers are out of commission or out of range. Satellite phones are definitely useful, but currently too pricey for anyone but the well-to-do, or organized groups (though some companies will lease phones). However, in the near future, the technology may drive down the price. To see more sources of satellite phones than you could ever use, go to http://www.google.com and enter the search phrase "**satellite phones**," just like that, quotes and all. Here are a few sources we found:

  - Iridium Satellite Solutions, online at http://www.iridium.com, or at 1600 Wilson Boulevard, Suite 1000, Arlington, VA 22209, 1-703-465-1000.

  - GlobalCom, online at http://www.globalcomsatphone.com, or at GlobalCom Satellite Communication, 2709 Compton Dr. SW, Decatur, AL 35603, 256-432-2685, 888-636-0707, Fax: 256-350-7859.

  - Globalstar, online at http://www.globalstar.com, or at Globalstar LP, 3200 Zanker Road, Building 260, San Jose, CA 95134, , Tel: +1.408.933.4000, Fax: +1.408.933.4100.

  - Outfitter, online at http://www.outfittersatellite.com, or at Outfitter Satellite, Inc., 2727 Old Elm Hill Pike, Nashville, TN 37214, 877-436-2255, 615-889-8833, Fax: 615-627-0580.

- ➤ Another feature you should consider is being able to send and receive **email messages from your phone**. Though still dependent on cellular signal and area coverage, email is another message avenue, and having a cell phone that can access it will give you one more commo method at your disposal. Email can be used when your voice mail is full or inaccessible, it can tie you in to "Yahoo Groups" type communications that we'll discuss below, it may carry larger messages than your voice mail could hold, and having an email capable phone will allow you to receive messages that were sent by someone who may have had email access through computer, but had no phone. More on that under "Email" below.

- ➤ An important consideration with cell phones is the question, "who gets one?" The adults in the family should have one, but what about children in school? Many schools that had once banned cell phone possession changed their views and policies after September 11th. What about those schools that still ban cell phones? One thought is that it's easy for your child to keep a cell phone in their locker and keep it turned off except for emergency use. The problem with that is that during a hurried evacuation of the school, when you'd want to hear from your child the most, children will not be allowed to retrieve anything from their lockers. Another option is to give your child the smallest cell phone available and that they keep it on them, but turned off except for emergencies. Other options are to lobby your PTA to change the school's policy, and/or for you to insist that your child's teacher own a cell phone and give you the number. (**Hint**: You'll be prompted to gather teacher data on a "**Family Member Data**" form in the "**Appendix**.")

➢ Finally, in order to get the most use from your cell phone, ask your cellular provider for accessories or programs that will allow you to hook your **standard laptop modem** to your cellular phone.

Email

Email is certainly no less fascinating than cell phones, and can be just as annoying when problems crop up at the most inopportune times. The key to email, as was with telephones and cell phones, is in how you set it up for success and how you put it to use.

➢ Email's role in disaster prep is two-fold. First, it's useful to keep family or group members in contact, and second, it's one of the ways mentioned earlier to get EAS warnings.

➢ If you'll look back through the "Frugality" section, you'll notice we gave you a couple of **free or low cost internet service and email providers**. One other free email service to try is Hotmail. Go to http://www.hotmail.com for more details.

➢ For a family or group to stay in full contact with all other members through email, each person would have to have everyone else in their "address book." An easier way to do this is through "**Yahoo Groups**." The way this works, is that everyone in your family or group would log on to YahooGroups, register as a new user, and sign in as a member of your group. Any email message sent by any member to the group would automatically be sent to <u>all</u> other members. This way you'd only have to remember one email address in order to message everyone. The setup process is simple too. Go to http://www.yahoogroups.com, register as a new user, and then click on "Start a New Group." By the way, YahooGroups is a great place to become a member of a huge variety of **educational, hobby, and disaster-prep groups**. Be sure to visit.

➢ Similarly, consider having a **family website** for posting information for those rare instances someone might have internet access without email capability.

➢ A major drawback to email in an emergency situation is going to be accessibility. That's why above, we mentioned the cell phone email capability. To help you find even more access, we'll be asking you to create an "Evac Atlas" under "**Evacuation**." One of the things you'll be asked to list about your destination and stops along the way, is the location of Public Libraries, Internet Cafes, or other locations that might offer internet and email access should you find yourself without computer, email phone, or other access.

➢ While we're on the accessibility issue, if you can afford a laptop with a cellular modem, get one, and make sure it's listed on your Last Minute List. Not only is this a great commo tool, you can add a web-cam that can help you with anything from emailing Neighborhood Watch suspicious activity video to Law Enforcement, or with sending injury pictures to physicians who may be directing your first aid efforts from far away.

➢ In addition to computers and cell phones, **PDA devices** (such as Palm Pilot, etc.) have found a definite place in our society, and many offer email access. Though too pricey an option for us to recommend that everyone immediately go out and get a PDA, we do suggest that if you're considering buying one that you research which have the better deal on full email access. We suggest you check Consumer Reports and other consumer sites we gave under "Frugality."

➢ PDAs aside, if you have a laptop computer and can afford a cellular modem, then by all means get one, and make sure you list your laptop on your Bugout Kit's "Last Minute List."

➢ While we're on the subject of text messaging, this seems as good a place as any to give honorable mention to **pagers**. We don't put pagers high on the list of recommended communication equipment since their use is usually one-way only (depending on the model) and the service is dependent on phones being operational. Also, the vast majority of pager services are limited to local area unless the more expensive larger coverage areas are part of the contract. However, if you or someone in your family already has a pager, there are ways to put it to work. First, several of the private EAS message relay companies mentioned earlier will forward warning messages to your pager, even if it's only a numeric pager (additionally, some of the text messaging pagers can be set to receive news updates). Secondly, you can create your own **emergency code phrase or number** to alert the pager wearer should they be without a cell phone or other communication. Thirdly, though this same person may have a cell phone, they may be in a limited coverage area affecting the phone and not the pager. Therefore, the pager would act as redundant communication.

➢ You can do your part to help keep commo lines open by not allowing your computer to be knocked out by a virus, or to pass a virus along. Log on to the US Computer Emergency Readiness Team at http://www.us-cert.gov, to sign up for their virus alert notification program. Many Internet Service Providers (ISPs) will also offer some sort of virus protection and/or update information, as will Microsoft at http://www.microsoft.com.

Actually, we're going to cover television first as it has fewer considerations as regards your disaster prep needs. Then we'll cover radio which may be much more useful to you than previously thought.

Television

TV is our prime information source. We watch it in the mornings before we go to work, at work if we can, at lunch if there's one around, in the evenings after work, and a lot of times it's what lulls us to sleep. As with any other piece of equipment, a television's worth, as regards disaster prep, is all in how it's used.

➤ Most televisions have four main weaknesses; the station you need to watch, the cable or satellite system linking you to it, the power supply for your TV, and that most pay channels and pay-per-view events are not subject to interruption by the Emergency Alert System.

➤ Though you can't control operations at the station you're trying to watch, you can locate alternatives. For example, when it storms in our area, the better of the local weather stations seems to always get knocked off the air. Fortunately, we have alternative stations we tune to when this happens. You should always know your alternate stations. In fact you should try to **know alternate stations** wherever you might travel with a portable television. We'll ask you to list stations when you complete your "Evac Atlas" under "**Evacuation**."

➤ Other times when it storms around here, the entire cable system goes out. One option if this happens to you is to immediately turn on your radio. Another option for your television is to have an **alternate antennae**. If you have the old pole antennae up against the house, great. If you don't, that's okay since many sets still come with the indoor "rabbit ear" antennas that will work in a pinch. You can also get all manner of electronic and/or amplified antennas at almost any electronics hobby store, or the electronics departments of most department stores. Make sure you have **both options** at your disposal.

➤ You should also consider the interior antennae if you're on a satellite system and if your reception is regularly compromised for whatever reason. One thing to try in that situation, if your home is already wired for cable, is to **see if any free signal at all is coming in through your cable**. We certainly are not recommending any sort of cable or signal theft here, but some local companies allow a couple of local stations to run through the cable system at all times for free. It's the other stations and pay channels that are filtered out. (This is common at many apartment and condo complexes).

➤ To make it easy to switch from one signal source to the other, several companies manufacture simple **switches** just for this purpose. You can find several makes and models at most electronics and department stores. Be sure to get one for your main television.

➤ As for the power supply, the only thing that can help you with the average set is an auxiliary generator or battery backup which we'll discuss later under "**Isolation**." One tidbit to give you here regarding electricity is to make sure all your communications electronics (TV, radio, cordless phones, cell phone rechargers...) are protected by **surge and spike protectors**. These protectors are inexpensive and can be found at any department, electronics, or computer supply store.

➤ To keep you informed of news and events while out and about (especially during higher terror alerts), or if you're forced to evacuate, you should look into getting a **portable battery-operated TV**. Technology has allowed the price on these to drop dramatically, so keep your eyes open for sales. You can readily pick up a small black and white model for next to nothing, and even color sets are extremely reasonable. If you get one, make sure you have all the proper accessories:

❑ Make sure the set uses standard batteries and not some expensive proprietary type.

❑ AC power adapter.   ❑ Cable connector.

❑ Cigarette lighter power adapter.   ❑ Protective case.

❑ External antennae with adapter.

➤ General criteria for choosing a portable television:

❑ It has to have all the accessories listed above.

❑ A nice optional feature is an earphone jack.

❑ Go as small as you can afford, but where you can make out **important detail** on the screen. (Either that, or make sure you pack a magnifying glass.)

❑ Research the reputation of the product through Consumer Reports, especially if going with an unknown brand or unfamiliar dealer. The set's **durability** rating is a crucial factor.

❑ Go with as many added features as you can while keeping the set's size and price within reason. Include AM / FM, weather alert radio, and/or short-wave reception capabilities.

❑ If you're buying a new vehicle and a built-in television is offered as an accessory, by all means get it if it meets the above criteria. Many people might discount this accessory if they don't spend a lot of time on the road, don't have kids, etc., but since you might have to make an emergency evac in this new vehicle, you'll want all the communications gear you can get. (Relatedly, one of the things we're pushing for is to get all automobile manufacturers to include a mounted 911 cell phone as part of every vehicle's accessories.)

➤ One of the bigger drawbacks to television is that the set in your home doesn't have any sort of instant-on or weather alert capabilities. It also won't show you any EAS warnings if you're watching a pay channel, pay-per-view, or watching a video tape or DVD. One of the things we're pushing for is to have an instant-on warning light flash on the set if it's already on and an EAS signal has been sent. This warning light will tell the viewer to switch stations or turn off the video and tune into a broadcast station. **Help us by writing to the FCC**, your Congressional Representatives in Washington DC, and to various television manufacturers to have such a feature added to all television sets. Early warning is very necessary in an emergency situation.

Radio

By far, the most commonly known, and the most useful of the electronic communications devices is the radio. Though older than television, it's still the preferred tool for emergency communication. Radio doesn't require wires, cables, cell towers, or bulky antennas; most are far cheaper than any TV on the market, and almost anyone can own and operate certain broadcast units. We're going to break this Radio portion down into two sections. One is **Receive-Only Radio**, and the other is **Two-Way Radio**.

Receive-Only Radio

➤ **AM / FM radios** are probably the most plentiful entertainment and news device in existence. Every household is likely to have one in one form or another, either as part of a stereo, a clock-radio, or as portable entertainment. The first order of business is to make sure you have more than one, and at least one as a backup to the others. In fact, be sure to keep a spare in your safe room and another in your vehicle as we've suggested on previous checklists.

➤ The second thing you want to do is make sure the radio you have is durable and protected. Numerous companies make "sports" radios that are specially made to go camping, jogging, even swimming with you. Get one of these models, and make sure that in addition to batteries, it takes an AC adapter, and that it will accept an earphone or headphones. As redundant protection, make sure you have the same kind of waterproof sports bag (or at least a resealable plastic bag) we recommended for your cell phone earlier.

➤ Third, as regards AM / FM units, try to find one that uses "**triple power**." Several makes are manufactured to use not only batteries and AC, but solar power, and some have a built-in hand-crank generator. These are important enough for us to locate a few examples for you:

● iPrepare.Com has one at http://www.iprepare.com/amsoldynradl.html. You can also reach them at iPrepare.com, P.O. Box 344, Roseville, CA 95678-0344, Phone/Fax (707) 982-7292.

● Crank 'n Go has one at http://www.crankngo.com/intro.htm, or call 877-9-ARDENT (877-927-3368).

● Oreck has a wind-up radio with AM/FM, Weather, an emergency light, and it will power your cell phone. Go to http://www.oreck.com and do a search for "radio," or call 800-289-5888.

● Even more of the same can be found through Innovative Technologies, online at http://windupradio.com, or call them at 1-250-386-2556.

● This solar unit also picks up Short-wave, Weather alerts, TV audio. It's online at http://www.radios4you.com/solar-dynamo.html, with direct contact at 954-925-8788.

➤ To find local stations, you usually scan until you pick up a station. If you're on the road during an evacuation, you'll want to know immediately where to tune without wasting time. We'll cover this more extensively in the "**Evacuation**" section's "Evac Atlas." For now though, here's a couple of internet sources for finding public radio stations: http://publicradiofan.com, and http://www.radio-locator.com.

➤ Stepping up a notch from simple AM/FM, you should make sure at least one radio in your house is an **NOAA Weather Alert radio** that will come on automatically when it gets a weather warning signal for your area. You can find these radios at pretty much any place where AM/FM radios are sold. In addition, it's also important enough for us to locate a few sources:

● Radio Shack will have several models. They're online at: http://www.radioshack.com, with direct contact at Radio Shack, 200 Taylor Street, Suite 600, Ft. Worth, TX 76102, 817-415-3200, 817-415-3240 (FAX).

● Weather Radio Store, online at http://www.weatherradiostore.com, with direct contact at Weather Radio Store, 1315 S. Hemlock, PO Box 1190, Cannon Beach, OR 97110, Information: (503)436-1844, Fax: (503)436-1426, 1-800-414-8655.

● Weather Alert Store, online at http://www.weatheralertstore.com, with direct contact at M/A Digital Services, 106 West Main Street, Enterprise, OR 97828, Phone: 541-426-3643, Phone Orders 866-237-2373, Fax 443-346-0452.

- Safety Central carries several. They're online at http://www.safetycentral.com/communications.html, with direct contact at Preparedness Industries, Inc, 311 E. Perkins St., Ukiah CA 95482, Phone # 707-472-0288, Fax # 707-472-0228.

- Weather Connection, online at http://weatherconnection.com/weatherradios.asp?page=1, with direct contact at WeatherConnection.com, 171 N. Larch St, .P.O. Box 1190, Cannon Beach, OR 97110, Orders: (800)414-8655, Information: (503)436-1844, Fax: (503)436-1426.

- Weather Products, online at http://www.radiowarning.com/new2083.html, has a couple of interesting accessories. One is a **weather alert extension kit for the deaf** and hard of hearing, and the other is an extension siren for a remote location in your house. The remote will go off when the weather alert radio activates. Contact Weather Products directly at Regional Weather Alert Radios, 2338 Pinecroft Dr., Burlington, NC 27217, Phone: (800) 484-1087 ext. 7714, Fax: (775) 923-6894.

- With this, as with source info for many necessary goods or gear, don't forget we'll have a larger directory of miscellaneous sources in the links collection on the enclosed **CD**.

➢ As we mentioned with television, most stereo systems, especially the one in your vehicle, do not have any sort of instant-on warning system. The same considerations mentioned about helping us push for upgrades to television sets should be applied to radio. All car stereo systems should have weather alert, instant-on features, and a warning light in case the listener is playing a tape or CD. EAS has to get through, and a driver on the road is more vulnerable than a person sheltered in their home.

➢ Let's chat a second about another form of local emergency info, and that's to hear it via Police, Fire, and Emergency radio channels picked up through a **scanner**. If you decide to get one, make sure it's a programmable scanner (some models still use crystals which limits your channels) with the widest range of frequencies you can get for the money (**Caveat**: Many Police frequencies are inaccessible by anything but the more expensive scanners). Also, make sure you can use regular batteries and that you have both a home and cigarette lighter power adapter. Here are a few **sources of equipment** and more information:

- Radio Shack: http://www.radioshack.com, 200 Taylor Street, Suite 600, Ft. Worth, TX 76102, 817-415-3200, 817-415-3240 (FAX). (Radio Shack also carries books listing scanner channels.)

- Universal Radio: http://www.universal-radio.com, Universal Radio, Inc., 6830 Americana Pkwy, Reynoldsburg, OH 43068, 800 431-3939, Info: 614-866-4267, Fax: 614-866-2339.

- Radios 4 U: http://www.radios4you.com, 954-925-8788.

➢ More information about scanners, and **frequencies** that should be monitored, can be found through:

- Monitoring Times magazine claims to be "the leader in scanner and shortwave communications." Online at http://www.monitoringtimes.com, or contact them at Monitoring Times, 7540 Hwy 64 West, Brasstown, NC 28902,1-800-438-8155 or fax 1-828-837-2216.

- Check the links collection on scanners at http://www.spiesonline.net/scanner-resources.shtml.

- "How Stuff Works" at http://electronics.howstuffworks.com/radio-scanner.htm.

- Bearcat's Scanner Frequencies and Codes. Online at http://www.bearcat1.com/free.htm,

- Scanner Pro, Scanners and Frequencies, online at http://www.911scanner.8m.com, or contact directly at P.O. Box 241, Elwood, Indiana 46036, 800-551-3152, Fax (765)552-9826.

- "CityFreq" at http://www.cityfreq.com, an online-only state by state scanner frequency listing.

- General Scanner Frequencies, online only at www.angelfire.com/wi/scanner/frequency.html.

- Black Cat Systems' Radio Links: http://www.blackcatsystems.com/radio/shortwave.html.

- For a little alternative detail concerning scanners, visit: http://www.stupidscannertricks.com (previously http://shell.exo.com/~rbarron).

- If you're planning on carrying a mobile scanner as part of your evacuation gear, be sure you know the laws in areas you may travel. See http://www.afn.org/~afn09444/scanlaws.

➢ Let's take a quick look at **short-wave radio**. Though not quite as important for immediate emergency info as the radios previously mentioned, short-wave is a good way to keep up with news from out-of-area sources. The hidden benefit is that short-wave will pick up news sources should your local, area, or regional radio stations be out of commission. Most of the above sources will have a variety of short-wave radios that are all reasonably priced. Though not really high on the list of suggested family equipment, if you're part of a volunteer group, you should have at least one short-wave radio for the group. Here are a few more sources:

- Short-wave radio on the web: http://www.chilton.com/scripts/radio/R8-receiver.

- Short-wave Radio and International Broadcasters: http://www.shortwave.be.

- Worldwide Short-wave Listening Guide: http://www.anarc.org/naswa/swlguide, Listening Guide, 45 Algonquin road, Clifton Park, NY 12065-7703, fax: 518-383-0796 (no phone listed).

➤ A newer cousin to short-wave radio is **satellite radio**. Similar to having a satellite service for your TV (though you don't need a dish for the radio), you can subscribe to a satellite radio network. The advantage this offers in disaster prep is that you can get news feed from satellite and you don't have to worry about local or area stations being operational. While this isn't high on our list of recommended gear for the individual or family (it's a little pricey), it should be considered by volunteer groups. For more on what satellite radio is, how it works, and who has it, visit these sources:

● For an explanation of the satellite radio system, go see http://www.howstuffworks.com/satellite-radio.htm.

● One provider is **XM Radio**, online only at http://www.xmradio.com.

● Another is **Sirius Radio**, online at http://www.siriusradio.com, with direct contact at SIRIUS, 1221 Avenue of the Americas, New York, NY 10020, 888-539-SIRIUS (7474).

➤ A close cousin to satellite radio, but one that requires more hardware and connections, is "**Internet Radio.**" Numerous stations around the country allow listeners to tune in to their broadcast via different websites. While email is a faster way get internet news and information, there are times it's better to hear the original broadcast. This is last on our list, but we felt it should be mentioned since this is a good way to keep up with current news and events that might indicate an emergency situation is brewing, or will let you tune in to stations around the country while you're at your evacuation destination. Some of these sources will charge a subscription fee, while others won't:

● For info, sources, and a subscription, go to http://www.radiotower.com.

● http://www.radio-locator.com will give you regular stations as well as internet feeds.

Two-Way Radio

Often called the "poor people's cell phone," we like to refer to two-way radio as the "smart people's backup." Nothing beats two-way radio for easy, rapid, and economical communication. As we'll see below there are so many types of radios, frequency ranges, and price ranges, that you'll need this primer in order to make an educated decision about what you need, what you don't, and how best to overlap the gear you have with what you need to add. We'll look at GMRS, FRS, FM transceivers, the venerable CB, and finally, HAM radio.

➤ Occasionally we'll use the term "walkie-talkie." We use this to describe any hand-held, push-to-talk, two-way radio. We'll use the rest of this section to hopefully give you enough of an education to make a good decision on what kind of two-way radios you might need.

➤ **GMRS** stands for **General Mobile Radio Service**, and is a band of frequencies set aside for personal communications. The system has repeaters (sort of like cell towers for radios) operating on this frequency and the handsets are usually more powerful, so this system gives you a greater range of operation than many hand-helds. Because of the range and power of this system, licensing is required (but may soon not be), but it's a simple process, and one license will cover a whole family. For licensing info contact the FCC at 1-888-CALL-FCC. You can also obtain a form from the FCC's Web site: www.fcc.gov/formpage.html .

➤ **FRS** stands for **Family Radio Service** and is a different band of frequencies also set aside for personal and family communications (businesses and commercial broadcasters should not use these frequencies). FRS radios do not require a license and they can be purchased at most department stores, camping supply stores, hobby electronics locations, and even office supply stores. The average 14-channel FRS radio has a respectable range of about 2 miles (depending on terrain and conditions) making it pretty good for commo from neighborhood house to house, or from vehicle to vehicle on the road.

➤ **CB** is the old **Citizen's Band** radio used by truckers for years and made famous by such movies as "Smokey and the Bandit." CB is still heavily in use and it's 40-channel spread occupies a different range of frequencies from either the GMRS or FRS bands. CBs, like many of its cousins, are found in vehicle mounted, or hand-held configurations, and do not require a license.

➤ Some **FM transceivers**, as they've been called, are still in use though that set of frequencies is used mostly as in-house communications in business settings. The most commonly seen application would be in large restaurants where the hosts and hostesses are seen with little headsets used to keep them informed of table openings, or building maintenance crews needing close-range communications. As emergency communication, these little 4-channel units are not high on the list of recommendations. **Note**: Don't get these units confused with FRS radios since the FRS units also use part of the FM spectrum. FRS has more channels and more range.

➤ Before we continue your radio education with more sources, let us leave you with our **general recommendations**. As always, we try to balance our suggestions between need, performance, and economy. As far as two-way radio goes, we recommend your family own at least **a pair of FRS radios** and least **one CB radio** (even though the two won't work with each other). The FRS units will allow you to communicate person to person or vehicle to vehicle, and the CB will allow you to contact others on the road who still use CB. (**Note**: The FRS radios will communicate with other FRS radios on the same channel. The pairs of FRS radios are not frequency-matched.) Both types of radios can be found for extremely low prices and from a variety of manufacturers and retailers. Base your actual purchases on need, number of users, projected range of use (how far away will users be from each other?), and on economy. Once you've decided all this and the price range, go for the units that have the best reputation for service and durability.

➤ **Hint**: You might give your child an FRS or GMRS radio to keep with them, even at school, during higher alerts. These would work when cell phones would not, and would allow you to talk with your child when you got in range of their school, the school's evacuation destination, your rendezvous point, etc.

➤ The best way to **continue your education about the wide world of two-way radio** is to provide you several sources of background information:

• A good list of radio terms and definitions is on http://www.gmrsweb.com/gmrswords.html.

• The GMRS, or General Mobile Radio Service, information portal can be found online only at http://www.gmrsweb.com/gmrs.html. Be sure to visit this page and read the **articles**.

• The same company hosts a CB, or Citizen's Band radio, portal with a respectable number of articles and links at http://www.gmrsweb.com/cbweb/index.html.

• Good information on the licensing and regulations pertaining to GMRS radios and others can be found through the "Personal Radio Steering Group" at http://www.provide.net/~prsg, with direct contact at PO Box 2851, Ann Arbor MI 48106, (734) 734.662.4533 voice.

• The Amateur Radio and DX Reference Guide, online at http://www.ac6v.com, has a great multi-topic radio subject search page at http://www.ac6v.com/swl.htm#AM. Visit both pages.

• "National Communications" magazine: http://www.nat-com.org, National Communications, P.O. Box 291918, Kettering, OH 45429, 800-423-1331.

• "Popular Communications" magazine: http://www.popular-communications.com, CQ Communications, Inc., 25 Newbridge Road, Hicksville, NY 11801-2953, Office: 516-681-2922, Fax: 516-681-2926.

• Visit "Radio Resource" magazine at: Communications Specialists, Inc. 426 West Taft Ave., Orange, CA 92865-4296, Phone: 800-854-0547, Fax: 714-974-3420, http://www.com-spec.com.

• "Amateur Radio" magazine: http://www.cq-amateur-radio.com, CQ Communications, Inc., 25 Newbridge Road, Hicksville, NY 11801-2953, Office: 516-681-2922, Fax: 516-681-2926.

• "Mission Critical Communications" (formerly Radio Resource Magazine): http://www.radioresourcemag.com, Pandata Corp., 7108 S. Alton Way, Building H, Centennial CO 80112, TEL: +1 303.792.2390, FAX: +1 303.792.2391.

• Additionally, some questions might be answered by the Federal Communications Commission, or FCC. You'll find them online at http://www.fcc.gov, or you can contact them directly at 1-888-CALL-FCC.

➤ Before leaving the general two-way radio discussion, we thought we'd provide you a few sources to let you start shopping, do some price comparisons, and learn more about the kinds of equipment on the market:

• C. Crane Company. For a wide variety of radios and other communication gear. http://www.ccrane.com, 1001 Main Street, Fortuna, CA 95540-2008, 800-522-8863.

• Both Office Max and Office Depot will carry a limited line of FRS and GMRS hand-held units.

• Visit the Radio Accessory Headquarters at: http://www.rahq.com .

• Discount Radio at: http://www.discounttwo-wayradio.com.

• RMI Radio at: http://www.rmiradio.com.

➤ Moving up the radio food chain, we come to **HAM radio**. Ham radio, or "Amateur Radio," is a world unto itself. Due to the high cost of equipment, the extensive licensing, and the experience necessary to become an able operator, we do not suggest you immediately jump into Ham radio as a means of emergency communication. However... you, and especially any volunteer group you may be part of, need to know who in your area is a Ham operator. **DURING A LARGE-SCALE EMERGENCY, HAM RADIO WILL PLAY A MUCH LARGER ROLE IN PUBLIC SAFETY THAN THE AVERAGE PERSON WILL EVER KNOW.**

➤ You can find more info on Ham Radio, clubs, and organizations at:

• The "How Stuff Works" page at http://electronics.howstuffworks.com/ham-radio.htm.

• **ARL**, the Amateur Radio Relay League: http://www.arrl.org, 225 Main Street, Newington, CT 06111-1494, Tel: +1-860-594-0200, Fax: +1-860-594-0259. Be sure to check out their "Emergency Backgrounder" page at http://www.arrl.org/pio/emergen1.html.

• **RACES** – Radio Amateur Civil Emergency Services: http://www.races.net/ (click on the "Links" button for state by state Ham organizations), or see the FEMA site at http://www.fema.gov/library/civilpg.shtm.

• HamRad Amateur Radio Resource: http://www.hamrad.com .

• **ARES**, or Amateur Radio Emergency Services: http://www.ares.org .

- More info on RACES, the Radio Amateur Civil Emergency Service, and how to become a part of the system, can be found through.
- A similar communications group is **REACT**, or Radio Emergency-Associated Communications Teams. You'll find them online at http://www.reactintl.org, or you can contact them directly at React International, Inc., 5210 Auth Road, Suite 403, Suitland, MD 20746, Phone: 301-316-2900, Fax: 301-316-2903.
- You'll also want to contact the Salvation Army Team Emergency Radio Network, or **SATERN**, at http://satern.org. Contact this office directly at The Salvation Army, 5040 North Pulaski, Chicago, IL 60630, Or call 773 725 1100.
- To this list, add the US Army's "Military Affiliate Radio System," or **MARS**. You can find them online at http://www.asc.army.mil/mars, or contact them directly at ATTN NETC OPE MA, US ARMY NETCOM/9TH ASC, 2133 CUSHING STREET, SUITE 3102, FORT HUACHUCA, AZ 85613-7070, 1-800-633-1128.
- Amateur Radio Community site: Online at http://www.eham.net, or contact directly at eHam.net, LLC, 1268 Old Alpharetta Road, Alpharetta, GA, USA 30005, 678-513-1535. Be sure to visit their links page at http://www.eham.net/links (lots of good info and sites).
- Links collection of Ham Radio information http://www.qth.com/#ham.
- Yahoo's Ham Radio links collection:
http://dir.yahoo.com/News_and_Media/Radio/Amateur_and_Ham_Radio/Clubs_and_Organizations.
- Keyworld's Ham Radio: www.keyworlds.com/a/amateur_and_ham_radio_organizations.htm.

➢ And not to forget our boating friends, here are a couple of links to information on marine radio:
- See BoatSafe's radio page at: http://boatsafe.com/nauticalknowhow/radio.htm.
- Visit the Coast Guard's page at: http://www.uscgboating.org/safety/metlife/radio.htm.
- Another lesser-known radio relay service is the Maritime Mobile Service Network. You'll find them online at http://www.mmsn.org, and you'll find a really good links page (to include radio manufacturers) at http://www.mmsn.org/hamradiolinks.html.
- And the FCC's frequency page at: http://wireless.fcc.gov/marine/vhfchanl.html.

Visual and Audible Signaling

Just how did people communicate before the days of cell phones, and email? Some of the things we'll mention below are interesting and should be stored in the back of your mind in case you ever need it, but others discussed should be made an important part of your training and foundation. For the most part, this portion on Visual and Audible Signaling, and the following portion on Text, Symbols, and Codes is meant to be a brainstorm section on backup methods of communication should your electronics be down, or should you need a redundant messaging system.

Visual Signaling
Your visual signals can be anything from flashlights to flares, with quite a few things in between. Some of the things we cover here can be useful in group communications, especially to help relay EAS warnings.

➢ One of the older forms of visual signaling is **smoke signals**. The modern day equivalent to this comes to us in the form of colored smoke distress signals and military surplus smoke grenades. Though of lesser value for relaying warnings since the duration of most smoke signals is relatively short, colored smoke is extremely useful for daytime signaling if you're stranded in a blizzard, lost in the woods, or at sea. Be sure to pack a couple in your outdoor survival kit. Two companies (among many) that sell smoke signals are:
- Orion Safety, at http://www.orionsignals.com, Orion Safety Products, Customer Service, RR. 6 Box 542, Peru, IN 46970, 1-800-851-5260, Voice (765) 472-4375, Fax (765) 473-3254.
- Shomertec, online at http://www.shomertec.com, with direct contact at Box 28070, Bellingham, WA 98228, Phone: 360.733.6214, Fax: 360.676.5248.

➢ What's a close cousin to smoke? **Dust**. Suppose you're trapped in a building awaiting rescue and don't have any communications equipment or anything to wave as a flag, but you do have a bottle of talcum powder, a jar of powdered coffee creamer, etc. If it's daylight outside, you could dump the powder out the window, which would act almost like a little smoke signal. The uses are limited, but if it's all you have...

➢ Where there's smoke there's **fire**, and fire is the oldest visual signal. Aside from building three evenly-spaced fires if you're lost or stranded, you can use torches for signaling, or more commonly, you can use highway safety flares. You can get both the flammable type, and the battery-operated "road flare" at pretty much any store that sells automotive parts.

➢ **Candles and lanterns** are also useful as we hear from the days of Paul Revere. Candle or lantern signals could be useful in such situations as signaling during a forced quarantine while the power is out.

➢ A less flammable option is **chemical light sticks**, sometimes called "Cyalume ®" or chemlight sticks. You can find them at many "dollar stores," party and novelty stores, and most camping or automotive outlets.

➢ A close cousin to the road flare is the **aerial flare**. Aerial flares are highly recommended for your outdoor survival kit. They're also useful to relay emergency warnings (especially the flares with an audible whistle or bang). Orion Signals listed above carries a good selection of both hand-launched and pistol-launched aerial flares, as do most camping supply, marine or boating supply, and military surplus stores. In fact, the military surplus stores may sell some hand-launched parachute flares. The only drawback to aerial flares is that people have to be looking for your signal and know what it means. The good news is that many flares come in a variety of colors, and using different colors could be a way to send different messages.

➢ Speaking of colors, **skyrockets** and other **aerial fireworks** are a great way to signal. For instance in one of our studies ("Enhancing the EAS" found on the enclosed **CD**) we mention red starburst fireworks as a way to alert a town that an Emergency Alert has been issued and for citizens to turn on their radio or TV. Check your local laws before investing in any fireworks. Note: With any flare, skyrocket, chemlite, or any other colored signal used for EAS alerts, be sure to use ONLY the color <u>RED</u>, and don't mix colors.

➢ A variation on this theme, depending on local laws and population levels, would be **tracer ammunition**. Tracer rounds glow as they fly making them visible for miles. These are recommended only for vertical firing while signaling in an unpopulated area. An example can be found at Gold City Gun & Cartridge Co., online at http://www.22ammo.com/tracer.html, with direct contact at Gold City Gun & Cartridge, Co., LLC ~ 385 Lumpkin County Pkwy, Suite E ~ Dahlonega, Georgia ~ 30533 ~ (706) 864-1205 ~ Fax (706) 867-8313.

➢ A cheap, low-altitude substitute for an aerial flare, if you have a **bow and arrow**, is to tie a chemlight stick to an arrow and shoot it into the air. If you can retrieve the arrow you can use the signal over and over until the light stick wears out. You can do the same thing using a slingshot. In fact, you could attach a small toy parachute to the chemlight stick so it would descend more slowly. Also, chemlight sticks come in a variety of colors, so certain colors or combinations could mean different things. If your hobby is model rocketry, you're probably already aware you could launch a chemlight stick as a signal. Always go with what you have.

➢ Staying with **light**, these items can be used to relay a signal using light. However, as we'll discuss later, you'll need something like Morse Code to realize the full potential of your signals.

● **Flashlight or hand-held spotlight**. For the spotlight we recommend the corded kind as most of the rechargeables we've tried have faired rather poorly. Get one with more than one million candlepower, and for both the spotlight and flashlight, get **colored lenses** to signal with different colors. Spotlights can be found at most camping or automotive supply stores.

● Your **porch light** could be used as either a Morse code type signal, or as a simple on = yes and off = no kind of thing. This would be useful in a forced quarantine situation where communications were down and watch groups or authorities did limited patrolling and needed a way to see if residents required assistance.

● Metal **signal mirrors** are good to reflect sunlight. You can use these mirrors in your toiletries kit so you don't have to carry two mirrors. You can also use the **shiny side of a CD** for signaling. CDs weigh next to nothing and are virtually unbreakable.

● **Chemlight sticks** also come in a variety of **colors** and last several hours after activation. You can signal with them by putting them in your window, on your roof, up a flagpole, etc. Messages can be sent by various colors, or by numbers in a "one if by land, two if by sea" kind of thing. Or, a combination of certain colors might have meaning just as variously colored nautical flags have meaning. Make color combos that are easy for others to remember the meaning. (This is useful for neighbors signaling each other during a forced quarantine, as a signal that you've evacuate, etc.)

● **Laser pointers**, sold at many department or novelty stores, as well as office supply stores, are great for long distance signaling. They're also good for precision signaling as you could focus a laser pointer on a particular window of a specific house as far away as you could see.

● We'll cover Morse Code a little under "Audible Signaling" below, but we wanted to point out that Morse Code and any of its cousins and derivatives, could also be used with signal lights. You can use **light pulses as the dots and dashes**, or you could have two different types or **colors** of light, one meaning dot and the other meaning dash. For example, if you had two or three flashlights, one light could mean dot, and two lights together mean dash (same with chemlight sticks). If you had colored cones or colored lenses, one color could mean dot and the other dash. The list is endless, and if this section made you think of something new entirely then it will have served its purpose.

➢ Next in the visual signal lineup is **flags**. Flags and banners came along shortly after smoke and fire and have been used to convey rather intricate messages. Here are some improvised flags:

● If you carry **towels or blankets** in your car, make them **brightly colored** so you can signal if stranded.

● The mylar "**space blankets**" you might pack in your survival kit are silver for a reason. They're meant to help you signal if lost.

- Small pieces of **brightly colored cloth**, to include such things as bandanas and neckties, are useful for putting on your car's antennae if you're stuck in the snow. This is also useful to ID your car to a family member waiting for you at a rendezvous point.

- **Ping-Pong paddles**, or similar, can be brightly painted, or you can tape several chemlight sticks to them for use at night.

- Depending on how your neighborhood is physically arranged (if homes are in a position to see), your Neighborhood Watch chairman might want to have a **tall flagpole** at their house. Colored flags could be used to pass messages. Red might mean an EAS alert has been issued, gray for inclement weather, yellow means "call in for message," etc. Your group should come up with its own series of colors or patterns.

➢ Flags can also be used as redundant backup to assistance requests, especially in such scenarios as neighborhood damage assessments conducted by Watch Groups after a disaster. For example, using the little utility or construction flags (2-inch by 2-inch colored fabric square on a 12-inch stiff-wire post), you can signal different situations or needs at various locations. After a destructive event in the community, you could place different flags in front of homes to indicate needs or situations to make it easier for first responders or repair crews to respond. Consider the following color suggestions:

❑ **White**: Family is at home, all is okay.

❑ **Green**: Family known to have evacuated safely.

❑ **Red**: Need onsite medical assistance.

❑ **Yellow**: Need utility repairs or assistance.

❑ **Blue**: Resident / family status unknown.

❑ **Black**: Death at this location.

➢ Related to the concept of colored flags having different meanings, you should review and consider the existing signals and meanings of the variously colored and patterned **Nautical Flags**. In the "International Code of Signals," different flags not only correspond to different meanings and messages, but flags exist for each letter of the alphabet. You can see some examples at the following locations:

- US Navy Signal Flags http://www.chinfo.navy.mil/navpalib/communications/flags/flags.html.

- The Boat Safe page at http://www.boatsafe.com/nauticalknowhow/flags.htm.

- International Code of Signals at http://www.geographic.org/flags/code_of_signals_flags.html.

- An Australian government site at http://www.anbg.gov.au/flags/signal-flags.html.

➢ The subject of flags cannot be considered completely covered until we mention **Semaphore**. Semaphore is the use of two hand-held flags, held in a unique position (similar to hands on a clock) for each letter of the alphabet. Though Semaphore might be a little tough to learn for simple interpersonal communication, it would be an extremely useful asset to a volunteer group who might be trying to assist local first responders while other communication options were extremely limited. If you're part of a group, we highly recommend that at least three people learn Semaphore. With three two-person flag teams (one sender/reader and one person to write the letters called out to them), and some binoculars, you could cover over two miles in distance. In addition to the charts you'll find on our **CD**, take a look at the following sources:

- The last Boy Scout Fieldbook and Handbook we saw had a Semaphore Alphabet chart in it.

- See the international scouting page at: http://inter.scoutnet.org/semaphore/semaphore.html.

- Visit http://www.cubsrus.com/Semephore.htm, and its subpages.

- San Jaquin Delta has a semaphore page at: http://www.sacdelta.com/semaphore.htm.

- An Australian government site at http://www.anbg.gov.au/flags/semaphore.html.

➢ Your Semaphore flags really don't need to be flags at all. Flags are good for daytime, but what about night? Let's toss out a few ideas:

- Flashlights with colored plastic cones on the ends. (Or kid's "Star Wars Light Saber" toys.)

- Flaming torches.

- Sticks or ping pong paddles with chemlight sticks taped to them.

- Lanterns.

➢ By the way, did you know that flags could be used to signal in Morse Code? Swinging a flag in a quick figure-eight to the right hand side of your body means dot, and a quick figure-eight to the left-hand side of your body means dash.

➢ The subject of flags leads us to **printed signs** and their close cousins. This isn't a complicated concept, so we'll just throw out a few clever examples we've seen:

- During 9-11, Police, Fire, and Port Authority boats helped evacuate New Yorkers from the disaster area. They took old **bed sheets and used spray paint** to make signs showing the boats' destinations. Simple, and highly effective. **Hint**: If you live in a flood prone area, enough so to have taken our advice earlier about stocking some goods in your attic or top floor, you might want to include a can or two of bright, flourescent-orange or yellow spraypaint. This may help relay messages to rescue helicopters, etc., even if you don't have a sheet and you have to paint directly on your roof.

- Similarly, after Hurricane Andrew, since all the street signs and house numbers were blown away, residents **painted their address and cell phone numbers** on handmade signs, or whatever wall of their house was left standing so insurance adjusters could contact them.

- Another trick we've seen is the use of **dry-erase boards**. As long as you have line of sight, and maybe a pair of binoculars at each end, you could write notes all day long. This is good for house to house or car to car when other systems are down, and it saves paper. One nice kit we've seen is actually a kid's toy, but it's a zippered notebook with a dry-erase board inside and it comes with markers, erasers, and some game ideas for the kids. Most office supply stores and toy stores will have all you could ever want in this area.

- And what about billboards? We'll discuss the concept of CCAPs, or Corporate / Civilian Aid Pledges under "**Cooperation**" later. CCAPs are pledges of donations that aren't given until an actual emergency occurs. One CCAP might be arranged with the owner of a billboard in your area to allow your volunteer group to **hang a bed sheet on the billboard with a message painted on it,** in event of certain disasters, Red Alert, EAS alerts, or other critical situations.

- We'll cover notes, symbols, and printed codes that could go on other signs in the next portion on "Text, Symbols, and Codes."

➤ While we're on the subject of things to wave in order to get people's attention, we'll toss in the left-field idea of **balloons**. Regular party balloons could be used in place of semaphore flags, a red balloon in the window might mean one thing, while a blue one on your car's antennae means something else. You get the idea. In addition to regular balloons, hundreds of places sell little "spray cans" of helium meant to fill a couple dozen balloons at home. Helium balloons on a long string can be seen from quite a distance.

➤ Also, several companies sell rather large **weather balloons** (about 8 feet in diameter) that can either be filled with warm air from your vacuum cleaner's exhaust, or with helium. These larger balloons can *really* be seen, and you might be able to float a signal flag or two or maybe a couple of chemlight sticks, in a, "One if by land... two if by sea" kind of thing. While this doesn't occupy a high position on our list, it's certainly worth considering, especially for neighborhood watch and volunteer groups. Suppose you were driving home one night, didn't have your radio on, and for some reason couldn't get a signal on your cell phone, but you round the corner and see a tethered weather balloon suspended over your neighborhood and trailing a few red chemlight sticks. It's a highly visible signal that would certainly mean something's wrong. To help, we found a few places that carry the larger balloons:

- You definitely have to visit http://www.overflite.com, for a discussion of several types of homemade balloons and links for more information.

- Be sure to visit their British sister site at http://www.ifo.co.uk/index.html.

- How to make a tissue-paper hot air balloon: http://www.explorium.org/tissue_balloons.htm.

- See the Arizona Balloon Company, online at http://www.arizonaballoon.com, or contact them directly at 4333 W. Paradise Lane, Glendale, AZ 85306, 602-938-3550 or 1-800-791-1445.

- You'll find a few models, including blimps at: http://www.advertisingballoons.com. Contact Above & Beyond Balloons, 800-564-2234 - USA Phone, 800-481-3299 - USA Fax, 949-586-8470 - Intl. Phone, 949-586-5489 - Intl. Fax.

- Another source of balloons and all sorts of useful odds and ends is **American Science and Surplus** at: http://www.sciplus.com, P.O. Box 1030, Skokie, IL 60076, 847-647-0011.

- Relatedly, check **Edmund Scientifics** online at: http://www.scientificsonline.com, 60 Pearce Ave. • Tonawanda, NY 14150-6711, TEL: 800-728-6999 • FAX: 800-828-3299.

➤ If you have the time, wind, and ability, you can also do a great deal of signaling with a **kite**. In fact if the flags were made of lightweight material, you could hoist a few of your nautical signal flags or similar with a kite and actually send intricate messages. Don't forget you could also **fly a few chemlight sticks with your kite** as well. Kites have come a long way from the simple diamond shape of years ago. We found some interesting sites and sources:

- In addition to your local toy or department store, you should check out the above links for American Science and Surplus, and for Edmund Scientifics. Both have kites on occasion.

- Links for kites, patterns, and more can be found at: http://sewing.about.com/cs/kitemaking.

- Clem's Homemade Newspaper Kite: http://www.clem.freeserve.co.uk.

- Uncle Johnathan's Easiest Classroom Kites: http://www.aloha.net/~bigwind/20kidskites.html.

- Dragon kite pattern made from plastic trash bags: http://familyfun.go.com/crafts/buildmodel/feature/famf199703_kite1/famf199703_kite4.html.

➢ Next, let's look at simple **hand signals**. American Indians circumvented tribal language differences with an elaborate system of signing. Almost every military unit in history has used some sort of silent hand gestures, and the extensive sign language system in use today by the deaf is a wonder to behold. Even the least educated among us use some sort of hand signals or other bodily gestures. We nod our head yes, shake our head no, and shrug our shoulders when we don't know. We all know gestures for "call me," or things to eat or drink, and every baseball team on the planet has some sort of hand signals. The suggestion here is to come up with a simple set of hand signals for your family and/or volunteer group. **This is highly recommended.** In fact, we suggest you not only create a few "secret" signals that only group or family members know, but that you learn standard signing as it will give you an additional way to communicate with others. Let's look at a few types of signing and signaling and some educational sources.

➢ **Indian Sign Language.** If you're going to learn a complete sign language, we suggest you go with the American Sign Language for the deaf. However, for a little food for thought on creating some privately understood signs we suggest you look at Native American sign language. Most bookstores and libraries will have some pretty good books on the subject. For now, here are a couple of online sources:

● An extensive dictionary and discussion of Indian Sign Language can be found online at http://www.inquiry.net/outdoor/native/sign/index.htm.

➢ **Alphabetical sign language.** This is where you use just your hands to convey letters of the alphabet.

● American Sign Language Dictionary, online at: http://www.where.com/scott.net/asl/abc.html .

➢ **Signing.** True signing uses hand and arm positions and motions for whole words and concepts, and it's not only a method of communication, it's an art. If you have the time and money, you should take a good course on signing to not only be more able to talk with our hearing impaired friends, but to have a backup method of communication when sound is absolutely unavailable. This is highly recommended. Your best education will come through a course although books and videos are a good start. Check your **library, bookstore, and video rental outlet** to see what they've got. Some online sources of info include:

● Definitely see the links at http://www.fcps.k12.va.us/DIS/OHSICS/forlang/amslan.htm.

● And another links collection at http://babel.uoregon.edu/yamada/guides/asl.html.

● Visit "Your Dictionary" at http://www.yourdictionary.com/languages/sign.html.

● Another can be found at http://www.masterstech-home.com/ASLDict.html.

● Michigan State has an ASL site at http://commtechlab.msu.edu/sites/aslweb/index.htm.

● "Signing Online" offers a fairly reasonably priced online course in American Sign Language. You can find them online at http://www.signingonline.com.

● See http://www.theinterpretersfriend.com/indj/dcoew.html for a huge collection of sign language links.

➢ **Military patrol hand signals.** This is a fairly easy to learn short series of simple hand signals.

● On our enclosed **CD** you'll find Field Manual FM 21-60 on **visual signaling**.

● A few hand signals can be seen at http://www.tpub.com/seabee/6-49.htm.

➢ **Helicopter signals.** This is another item on our list of recommended education. Let's say you're in an emergency situation, such as being stranded on your roof during a flood, and you've run out of spray paint. How do you communicate with the helicopter? Well, turns out there is a set of hand signals specifically designed for use by ground personnel who need to communicate with a pilot. Here are a couple of sources:

● A few can be seen at http://www.tpub.com/content/combat/14234/css/14234_309.htm.

● Another source is http://www.tempe.gov/fire/docs/210.05.htm.

● On our enclosed **CD** you'll find Field Manual FM 21-60 on **visual signaling**.

Audible Signaling

The first audible signal to bridge the distance gap that voice couldn't cover, was from sticks beaten against logs, and that eventually graduated to drums. The first electronic form of audible signal was in the form of Morse Code. Here we'll discuss some forms of audible signaling that may be necessary to communicate in situations where all electronics are down and you don't have line of sight necessary for visual communication.

➢ Before we get too complicated with this subject, don't forget good ol' yelling. Remember cheerleaders and their "**megaphones**?" Megaphones are nothing more than cones you yell through. The cone increases the range of your voice, just as a longer gun barrel increases the range of a bullet. So if you're trying to yell at the neighbor across the street to tell them a tornado is coming, remember you can yell through something **conical** (like an orange traffic safety cone) and be heard from farther away. Ready for the other half of this coin? **Listening** through the same cone to someone yelling at you from far away will help you hear better. It's an ultra-simple trick with limited applications, but now that we've mentioned it, you won't forget it.

➤ Adding electronics to the above, don't forget "**bullhorns**," which are nothing more than electronic megaphones. About the only time you'd want to go buy one of these is if you were part of a volunteer group who was tasked with some sort of safety or security. **Hint**: Most **CB radios** meant to be vehicle mounted have a "PA" or **Public Address** connector that connects to a **loudspeaker** mounted behind your vehicle's grill. Though inexpensive and simple, this makes an effective address system for relaying a message.

➤ Relatedly, any neighborhood group or small community should have its own **alert siren** with a non-electronic backup alert. These sirens should be in addition to any municipal alert sirens.

➤ Before we get into codes or sound patterns, let's list a few simple things that you can use for audible signaling above and beyond the electronic sirens, etc.

❑ Whistling (by mouth) ❑ Sports whistles ❑ Compressed-air horns
❑ Car horns ❑ Clapping or tapping ❑ Hand or cow bells
❑ 2 pieces of metal pipe ❑ Metal pot and heavy spoon ❑ Whistling fireworks
❑ Musical instruments ❑ Firearms ❑ Regular fireworks

➤ The first of your signal codes could **simply be whether or not you heard the sound**. For example, a cow bell (or a church bell, ship's bell, etc.) could be a prearranged alert signal for your neighborhood watch. The same could be said for the whistling fireworks (usually skyrockets of some sort), or a long blast on one or more car horns or air horns. This way you don't have to worry about learning or teaching any complicated codes or patterns. **Hint**: Use simple alarms and alerts for the general public or group members at large, and message patterns and codes for your core group or family members.

➤ Other signals using the simple checklist above, could use simple **patterns** to differentiate you from others, or for simple messages. For example one long burst on an air horn might mean one thing, and three short blasts another. In close quarters, simple claps or tapping noises could relay entire messages using Morse Code or the POW Code we'll see below. With musical instruments, especially with a trumpet or bugle, you could play "Charge," "Reville," "Taps" or other recognizable tunes that convey a message or alert. You could also do three short blasts on a sports whistle (to make it different from kids playing.) This is one area that's limited only by your imagination, and the reason for including this section was to get your imagination going.

➤ Speaking of sound codes, you should have a few code words that can be slipped into conversation so you can securely convey messages in front of people you don't want privy to what you're saying. Earlier under the "Security" portion of the "Basic Home Prep" section me mentioned saying "adios" instead of the usual "goodbye" to indicate the person on the phone was in some sort of trouble.

➤ The most universally recognized sound code is the Morse Code for "**SOS**." SOS was chosen because the **international distress number is *three***. (If you're lost while hunting in the woods, you fire three quick shots. A ship at sea would fire three rockets. Three parallel lines dug in the dirt would tell a rescue pilot that someone was lost and needed help.) The Morse Code pattern that makes up SOS is ...---..., or "**dot, dot, dot, dash, dash, dash, dot, dot, dot**." This was done to create a group of three threes. As this was started as a distress code for ships at sea, the reason the code isn't three dots, or "SSS" is because most ships started with "SS." They needed something different. (SOS is not an acronym for anything though some think it means "Save Our Ship.") So... whether you have flashlight and you're signaling, or you're trapped in a damaged building and tapping for help, remember dot, dot, dot, dash, dash, dash, dot, dot, dot.

● If really interested in learning Morse Code, you can download shareware lesson software from http://www.downloadfreetrial.com/hobbies/hobb3371.html or http://www.net-magic.net/users/w4fok/.

Basic Morse Code Chart

A .-	K -.-	U ..-	1 .----	, --..-- comma
B -...	L .-..	V ...-	2 ..---	. .-.-.- period
C -.-.	M --	W .--	3 ...--	? ..--.. question mark
D -..	N -.	X -..-	4 ....-	; -.-.- semicolon
E .	O ---	Y -.--	5	: ---... colon
F ..-.	P .--.	Z --..	6 -....	/ -..-. slash
G --.	Q --.-		7 --...	- -....- dash
H	R .-.		8 ---..	' .----. apostrophe
I ..	S ...		9 ----.	( ) -.--.- parenthesis
J .---	T -		0 -----	_ ..--.- underline

> A simpler cousin to Morse is a **tap code** created by American **POWs** in Vietnam to communicate with each other using taps or clicks since they weren't allowed to speak. Not everyone in the service knew Morse so a simpler system had to be devised. They came up with a matrix for the English Alphabet but shared the use of C and K to use only 25 slots. The matrix looks like this:

Taps	1	2	3	4	5
1	A	B	C, K	D	E
2	F	G	H	I	J
3	L	M	N	O	P
4	Q	R	S	T	U
5	V	W	X	Y	Z

In this matrix, each row and each column has a number. To tap out the letter you'd give the **number of taps** for the **row**, pause, and then the number of taps for the **column**. For example, the letter "H" would be tap, tap, pause, tap, tap, tap. This is a rather simple system to learn, and one with some potential uses. Imagine a neighborhood with most houses out of line of sight of each other. One person with a pair of metal pipes, or a metal pot and a heavy spoon could tap out a message loud enough for the whole neighborhood to hear. **We recommend you and yours learn this code**. You could also use this code as a *written* form of secret code. The letter "H" would simply be 2-3, the letter "P" would be 3-5, and so on.

Speaking of written codes...

Text, Symbols, and Codes

Sometimes the written word is the best way to get a message across. After all, you're sitting here reading a book, right? However, going beyond simple note and letter writing, to make sure your message is distinct, received, and **secure**, you might have to consider a few new techniques.

> The most common written message is likely to be a note or letter left behind for someone in case of an emergency. A couple of examples can be see in the "Update Cards" and "Open Only in Case of Emergency" forms in the "**Appendix**," but for now we'll discuss the most important part, and this is making sure your message is read. For that to happen, the note has to be **found**. The two best ways to do this are to make sure the note is very **distinctive**, or that it's left in a **prearranged or designated spot**.

> To make sure your messages are seen, consider using **bright and distinctively colored envelopes**. Make them a color you would not use for anything else but emergency communication. Flourescent colors, especially orange, yellow, or green are good. Keep them in the "Office Pack" of the "Bugout Kit" we'll discuss later. This way, if your house is destroyed in a tornado, you had to evacuate for whatever reason, or you were at one emergency shelter and left to go somewhere else, colored and distinctive envelopes will let you leave a recognizable note. If a family member that's searching for you sees one of those envelopes on a stick next to the rubble of your former home, taped to the refrigerator, or pinned on a bulletin board at a shelter, they'll know that it's an emergency message from you. To make these envelopes even more distinctive, add a couple of tiny stickers or decals (that you can get at any toy, department, art, or office supply store) and make sure all the important players in your life know what these envelopes look like and what they're for. Also carry plastic bags to protect the notes from weather and water. **Hint**: Write your notes with a ballpoint pen or Sharpie ® (pencil will fade), so it will still be legible if weather damaged.

> **Location** will also play an important role in leaving notes. Keep the number of potential locations low so your note will be easy to find. In addition to the locations mentioned above of nailed to a stick or on the refrigerator, you might want to consider other specific locations such as the place you hide your house key outside, behind or underneath the mailbox (sad fact of life: it's illegal for anything other than postage-paid US mail to go in mailboxes, even your private mailbox), or similar agreed-upon locations.

> You should also agree to **preplanned message drop locations** at each of your rendezvous points in case other forms of communication are unavailable and you're trying to get everyone together. Make the note location one that no one else would bother except for the person searching for it. For example if your rendezvous point is a convenience store, and they have an ice or vending machine out front, your message drop location might be behind the machine, underneath its front edge, etc.

> Now let's move on to **symbols** because they're a second choice in this situation if you don't have your envelopes, notepaper, etc., or you want to **convey a simple message *quickly***. Using the above example of leaving a note at a convenience store rendezvous point, you might consider a unique yet simple symbol that's recognizable by all group or family members, and one that you can quickly and easily leave in chalk or other removable writing. For example, you might take a piece of chalk or a grease pencil, and put your

symbol on a wall, in a phone booth, or other **specified location** to let another person know you were there. Such symbols could be accompanied by a short note, or the symbol itself could mean something. Keep your symbol vocabulary simple. You don't want a long message that takes time to write, or that might be noticed and wiped away by someone else. Use simple signals for simple messages like "stay here," or "go to rendezvous point number two," and so on. Here are some examples.

Possible Symbol	Possible Message
(↓)	**Stay here.** You might want one person in your family to stay put while you go round up others.
←(R2)	**Proceed to Rendezvous point # 2.** The person heading for this rendezvous point might be more mobile than others and you can send them to where the rest of the family or group is gathering.
(M↓)	**Message (note) for you at this location.** See above about picking a specific spot to leave a note.

As another suggestion, you might want something, such as **your initials, to personalize your symbol** on the off-chance somebody else might be using the same thing. Remember, these are just examples of symbols you might use when <u>nothing</u> else is available and you're frantically trying to regroup with loved ones in the midst of a chaotic situation. These symbols might seem like a dumb idea now, but in a massive emergency evacuation, you'll be glad you have them.

➤ As an associated side note, we've suggested to several municipalities that they adopt a **city-wide "okay / not okay" symbol system** for use if communications were to go down during any sort of mass quarantine. We suggested a simple "✔" or "X" placed in a front window or door of a residence using tape. A checkmark would mean everything's okay, and an "X" would mean "we need assistance." A lack of any mark at all might mean the occupant couldn't even make it to the window and might need help. Keep this in mind when formulating alternate communication methods for your neighborhood watch or civic volunteer group.

➤ Let's get back to **security** for a second. You might leave notes with information that you don't want getting out. One simple way for some light security with voice or written communication is to **talk in references** that only your intended recipient will understand. For example you might say "we're meeting at the place you and John went fishing last summer" rather than specifics.

➤ A second option is to use **written code**. Whatever you come up with, make it **simple** to understand, **easy** to read and write, and easy to **decipher**. The most common written code uses the position numbers of the letters of the alphabet. A=1, B=2, C=3, etc. Therefore a message such as 8, 15, 23, 4, 25 would say "howdy." You could also use the reverse, such as Z=1 and A=26, etc. Conversely, you could use an alphabet code to record sensitive numbers such as unlisted phone numbers, social security numbers, and other sensitive info that might be placed on your "Family Member Data," "Household Data" and other forms. To code numbers, you'd only need **A** (=1) through **I** (=9), and **O** would stand for zero. For example, the phone number 770-555-1212, would be written as G,G,O,E,E,E,A,B,A,B.

➤ A cousin to this code is the **telephone keypad code**. Your phone keypad will have numbers *and* letters:

1	ABC 2	DEF 3
GHI 4	JKL 5	MNO 6
PRS 7	TUV 8	WXY 9
*	OPER 0	#

This code is pretty simple and can be used to communicate both text and numerical messages.

❏ Let's start with coding **numerical messages** since that's easier. First, you'll notice there's no "Q" or "Z" on most keypads, and that the zero is the OPER or operator key. Keep those in mind. To write an encoded **number**, simply **use any letter that appears on that number's key**. For example if you wanted to write the phone number 555-1212, you could write it as JKLQAQB. Notice that J, K, and L each appear on the "5" key. You could've just as easily used LLL. Notice something else too. We used "Q" for the one. Why? Because there is no "Q" on most keypads, and the number 1 has no letters. Use either the "star sign" or the "pound sign" to represent zero (don't use the zero itself because it looks like the letter "O"). So for another example, the number 21079 would be BQ#RX. Now why is this code important? Here's why: When you go to fill out all the **sensitive information** that will appear on your various "data forms" in the "**Appendix,**" you'll want a way to safeguard the info, such as your Social Security Number, so that if someone steals your gear, they can't steal your identity.

❏ Now let's use the keypad to encode **text messages**. This is simple to remember though it takes longer to write. Remember, there is no "Q" or "Z" so we'll be using our 1, *, or # keys again. For the letter "Z" use the other symbol key you didn't use for zero. To write the code, give the **key number and then the letter's position**. For example, the letter "**D**." The code for "**D**" is **3-1** since it's on the # 3 key, and it's the **first** letter. The letter "U" would be 8-2, as it's on the # 8 key and it's the *second* letter.. For the letter "Q" use 1-1. So, the word "quickly" would be 1-1, 8-2, 4-3, 5-2, 5-3, 9-3. Actually, if all your recipients know you're using the keypad code, you wouldn't even have to put dashes, you could simply write it as 11, 82, 43, 52, 53, 93. This is another useful way to encode sensitive text information such as mother's maiden name, etc.

❏ **Hint**: Some phone manufacturers put the letter "Q" in with PRS on the # 7 key and the "Z" on the # 9 key with WXY. However, since many phones don't do that, it's best to plan around it.

➢ **Important hint**: If you're the only one who knows these codes, be sure to let someone else in your family know how to decipher the messages should something happen to you. You'll notice on some of the forms in the "**Appendix,**" we left a line asking you for a "code key" so you can remember how the information was encoded. Use simple statements to jog your memory. For example, if you used the above phone-pad code, your reminder key could be "phone me and I'll tell you." You used the word "phone" and yet most would-be identity thieves would not know this to be the answer. You could also make it look like a sorority thing by saying something like, "ZTA forever!" (for a Z to A, or Z=1 to A=26 code). Whatever code you choose, don't make it so complicated that it can't be broken by a family member. Oh yeah, and don't use invisible inks either. Many will fade, and there's no guarantee you'll have the substance or heat source to make the inks appear when you need to read your documents.

➢ A final point regarding written communication: **The mail will probably be running.** Though we plan for worst case scenarios and expect the shutdown of all utilities and services, we should keep in mind that government functions will try to return to normal as quickly as possible. They may be up and running before you are able to return home. So, don't forget the US mail and other private carriers as a text communication option. In the "Office Pack" of your "Bugout Kit" we'll ask you to include stationery, stamps, and more.

➢ Regarding the mail, you should consider having a **Post Office Box** (provided you can afford it after all other needs are met). Should something happen to your home, you have an instant location for **forwarding your mail**. Also, it provides a secure address to give out should you want to keep your home location less public.

By the way, all those questions we asked at the beginning of this section? There are no specific answers. The answer would be for you to use whatever you had available that would work for you under those or similar situations. Use your imagination, and then go out and test whatever you come up with. We cannot say it enough; **experience and practice are the best teachers, and your reaction in an emergency situation is only going to be as good as the training you've allowed yourself.**

Now that you've learned ways to get the warning and how to communicate it to others, let's look at how you're actually going to **React**.

Notes:

III. REACTION

"There is no terror in the bang, only in the anticipation of it." –Alfred Hitchcock

✓	Goals of the Reaction Section		
	Activity	**Date Completed**	**Pg.**
	Homeland Security Terror Alert Levels:		130
❑	We've read and understand the section on **The Lower Alerts**.	___/___/___	
❑	We've read and understand the section on **Elevated / Yellow Alert**.	___/___/___	
❑	Made plans and reactions based on the section **High / Orange Alert**.	___/___/___	
❑	Made plans and reactions based on the section **Severe / Red Alert**.	___/___/___	
❑	Have located a source to monitor the **current terror alert threat level**.	___/___/___	
❑	Each family member has started a **Daily Activity Journal**.	___/___/___	
	Basic Personal Safety:		136
❑	A "**Get You There Pack**" has been made for each family member.	___/___/___	
❑	Made subtle changes in our **daily wardrobe** to increase our protection.	___/___/___	
❑	We've read and understand the section on "**Gas Masks and Moon Suits**."	___/___/___	
❑	Each family member has a "**Safety Necklace**."	___/___/___	
❑	Our family has practiced all at-home **safety and reaction drills**.	___/___/___	
❑	We know the basic steps of **personal decontamination**.	___/___/___	
	Manmade Misfortunes:		161
❑	We've memorized the immediate reactions to a **Nuclear Detonation**.	___/___/___	
❑	We've read and understand the portion on "**Dirty Bombs**."	___/___/___	
❑	We know immediate reactions to a **chemical attack or HazMat incident**.	___/___/___	
❑	Studied the symptoms of the most common **biological terrorist weapons**.	___/___/___	
❑	We've participated a **Fire Drill at our workplaces** or requested one.	___/___/___	
	Weather and Natural Disasters:		198
❑	We've memorized the various immediate reactions to all **Severe Weather**.	___/___/___	
❑	Our area's **destructive weather season** is from _____ to _____.	___/___/___	
❑	We know immediate reactions for **earthquakes, landslides, or avalanche**.	___/___/___	
❑	We know all **safe areas**, and **safety equipment locations** at home.	___/___/___	
❑	We know all **safe areas, exits, and safety equipment locations** at **work**.	___/___/___	
❑	We've learned to look for all **safety features** of buildings we **visit**.	___/___/___	
	Goal Setting:		A-1
❑	Entered **start dates and goal dates** onto **calendar** for all the above.	___/___/___	

As we move from the "**Foundation**" section to the "**Reaction**" section, we'd like to congratulate you on making it this far, and hope that you've found useful information, both here and through the outside sources, that has either helped directly, or sparked your imagination for creative alternatives.   So congratulations, and welcome to "**Reaction**."

Though some things in here *might* be simple, and some things *might* look like they won't apply to you or your area, we caution you to read everything and learn as much as you can, since there are no absolutes in life, and you never know where you'll be or what you'll be doing.   For example, you may live in an earthquake-free area but visit California only to wind up in an earthquake. As the Boy Scout motto says, "Be Prepared."

You'll also find that many of the incidents listed in this section will have similar or overlapping reactions, and that some precautions or advance preparations will apply in several circumstances.   Also, each immediate reaction discussed here will have an inherent "step two" which will either be leaving immediately or sheltering-in-place, so in addition to specific measures mentioned in this "**Reaction**" section, be sure you give equal attention to the "**Evacuation**" and "**Isolation**" sections that follow, as there's a definite overlap and interdependence of all three sections.

The actual reaction-to-incident suggestions, guidelines, and pointers we'll present in this section have either been recommended by other experts, used successfully by victims, or are just plain common sense, and they're presented in as much detail as possible.   However, and we can't stress this enough, **no two situations are going to be identical**. No two tornadoes are going to be the same, no two earthquakes will be alike, no two people in the same event will have the exact same experience.   The best thing you can do is treat these reaction instructions as general guidelines and be prepared to handle your unique situation, as it unfolds, and to the best of your ability.

Similarly, you'll want to remember that Murphy's Law will prevail and that misery always loves company. Don't expect one emergency to be only one emergency – always be ready for the other shoe to drop. An earthquake might trigger a HazMat incident, terrorists might have instructions to launch their next attack following a natural disaster, you may experience a medical emergency during an evacuation following another emergency...   Any number of things can crop up together. Be ready.

Let's begin the "**Reaction**" section by discussing the **Homeland Security Terror Threat Alert Levels**, what they mean to you, and how you should "react."

Homeland Security Terror Threat Alert Levels

There are good things and bad things about the color-coded Homeland Security Terror Threat Alert system. One really good thing is that at least we have some sort of barometer with which to gauge potential terror activity. Another good thing is that it's primarily intended for Fire Departments, Law Enforcement, and other first responders, so we don't really have to worry *quite* as much as some people have, over an increase in alert status. To us civilians, a change in alert status is almost like the "two-minute warning" in football. Rather interesting, and we should pay attention, but it means a little more to the players on the field than to us in the stands. (Though don't ever forget, we're definitely involved.)

The bad things about the alert system are that it's only a generalized gauge, it's certainly not foolproof, and it still can't warn us about sudden natural disasters, accidents, or personal emergencies.

We're starting this "**Reaction**" section with the alert system because most people tend to "react" to a change in the level as they would to an actual event. It's good to know *how* to react. Regarding these alerts, let's cover a few critical points basic to any level. For this reminder, we'll use the acronym **A.L.E.R.T.S.**:

Act	**Act** on the alert, but otherwise act normal. Alerts are a heads-up, but they're not an actual emergency.
Listen	**Listen** to the news and stay aware of developing situations.
Evaluate	**Evaluate** your daily travel and family activities based on the specific nature of the current alert or threat.
Review	**Review** any and all family reaction plans and information forms to make sure everything is current.
Train	**Train** and practice the things you learned in classes, and the drills we'll cover below.
Safety	**Safety** is key in whatever you do. You don't need a minor mishap just before an actual emergency.

The Lower Alerts

The lower alert levels of Green / Low, and Blue / Guarded simply mean that there <u>may</u> be a little less on our plates for us civilian types to worry about regarding potential terrorist activity. It's almost like driving on the freeway with 10% fewer cars. Easier to do, a little less stress, but still a lot out there you have to be careful of. So, lowered alerts do not excuse us from our disaster prep needs or duties.

However, a little less to worry about is still a little less to worry about, so put this "down time" to good use. Continue working on the foundation items you may have had to skip, your reaction items from this section, and your other planning and equipment needs from the sections that follow. If you're ahead of the game and done with everything in here, you may want to spend a little time with family members, friends and neighbors, coworkers, or members of a volunteer group, and help bring them up to speed.

Ready for a "fly in the ointment?" What if lowering the Terror Alert Level to Green / Low is the signal for some sleeper cell to launch the next major attack?

Let's look at the other levels and some suggestions on what to do. **Note: Most of the items presented for the alerts are useful suggestions when your area is in its severe weather season**.

Yellow Alert

A Yellow / Elevated alert is a time for a little caution. Just like the yellow light of a traffic signal, it's our "heads up" signal that in this case means something may be brewing in the terrorist world. Let's take a look at a few things you as an individual or a family might want to do.

> **Act normal**: The alerts are more for those protecting our borders, watching our backs at home, and checking out the bad guys overseas. However, it is a signal for us to keep an extra eye on the terror threat just as we would watch The Weather Channel when bad weather is brewing. Other than that, go about your business, and relax knowing that you've done a good job with all the preparations you've made so far.

> **Monitor:** Keep an eye on the news, double-check the dates of your Bugout Kit's next "refresh," check the stocks in your pantry, and run the family members through the drills, programs, and emergency responses you've set up. It's a short hop from Yellow to Orange, so you'll want to know if the alert level changes. You can monitor the current alert level online at the **Homeland Security website** http://www.dhs.gov, or by watching the **Fox News Channel** or logging on to their website at http://www.foxnews.com.

> **Counter-Intel Vigilance**: As Yellow Alert means something might be brewing, pay a little extra attention to possible "suspicious activity." Most terrorists will perform a great deal of **intel gathering and surveillance** on a target before launching an attack, and this is an area of opportunity to stop them. Watch for anyone "casing a target," especially sensitive areas such as government buildings, utility or service facilities, malls, mass transit, mass public gatherings, water towers, fire hydrants, etc. If you see anyone taking unusually detailed videotapes or photos, taking notes, pacing off distances, counting surveillance cameras, etc., report them. Make your report first to **local authorities**. If you get no response from them, call both the **FBI** (find your nearest office on http://www.fbi.gov or in the blue government pages in your phone book), and the **US Rewards hotline** at 1-800-USREWARDS or online at www.rewardsforjustice.net. To give them the information they'll need, see the "**Appendix**" for the "**Suspicious Activity**" forms.

> **"Pad Your Work Account"**: Some of us have jobs that allow a certain number of sick days or personal days on top of vacation time. During higher alerts, don't use up any of your days off for anything frivolous, as you may need to use them to lay low during Red Alert or as an extra safety precaution after an actual attack.

> **Investigate Telecommuting**: With similar regard to work, if your job is a type that might allow you to "telecommute," you should see what is needed to actually do that in the event it's not safe to go to an office. As business-continuity is a major concern of any company, you might want to suggest to your employer that they consider all employee safety options that would still allow the company to function. You don't need to actually *start* telecommuting, just set it up.

> **Update your tape collection**: As any number of scenarios could see you stuck at home for days on end, you'll want to make sure you have some entertainment in order to fight isolation's number one enemy – boredom. Buy some new books, or videotape a few shows or movies and set everything aside for a rainy day. In fact, this is where you'd review everything you set up under "**Isolation**."

Orange Alert steps up the urgency for us just a bit, and we now see several things we need to do beyond simply keeping an eye on the news or on suspicious individuals. Everything from Yellow Alert still applies. To those, let's add the following checklist of things to do followed by their explanation:

Orange Alert Checklist		
❏ Be alert but not alarmed.	❏ Weigh your activity options.	❏ Keep a Daily Journal.
❏ Dress for chemicals.	❏ Reaction prep review.	❏ Review your plans.
❏ Update documents.	❏ Wear your communications.	❏ Vehicle maintenance.
❏ Prep the evac vehicle.	❏ Prep your home.	❏ Top off all fuels.
❏ Check utilities.	❏ Bring your bike.	❏ Medical checkups.
❏ Check your medications.	❏ Pet checkups.	❏ Prepay bills.

> **Be alert but not alarmed**: Keep your eyes open, listen to the news, and keep your cell phone close, but go about your daily business. For most of you, this will be as easy as watching out for potential accidents while driving. It's just something you do and you don't get stressed about it.

> **Weigh your activity options**: Though you want to live your life as normally as possible, you might want to rethink some of your plans that may put you at unnecessary risk. For example, if the upcoming sporting event is not extremely important to you, you might want to consider doing something with a smaller crowd rather than going to a crowded stadium. If you do choose to go out, expect **tighter security** and therefore a **longer delay** when visiting any government building, transit facility, or public function. Before going to any of these, call to make sure they have not been temporarily closed or the function postponed.

> **Keep a Daily Journal**: <u>THIS IS EXTREMELY IMPORTANT.</u> **For the duration of the alert, keep a daily activity journal of where you go and what you do for the purposes of bioweapons vigilance**. Keep a **small notepad** with you and write when you went to work, the route you took, your mode of transportation, when and where you ate lunch, what you ate, who you came in contact with, when and where your family ate dinner, which movie you went to see... <u>Everything</u>. Here's why: Though much has been said and done about protecting yourself against biological attacks, victims find only out *after the fact* that they've been exposed to a bioweapon (due to the incubation period). If you fall ill, your journal might **help authorities figure out when and where you were infected** (especially if you can't speak). This information, if gathered quickly enough, can help save others by allowing authorities to find those in the community who may have been in the same place at the same time. It will also prove invaluable in a "ring vaccination" where doctors will need to know who you've been in contact with, so they may be notified and vaccinated, or if *you* were the one in contact with the victim and *you* need to be vaccinated. Be sure to keep a daily journal.

> **Dress for chemicals**: As we'll discuss below under "**Basic Personal Safety**," there's much that can be done to protect yourself against some chemical exposure by subtle modifications in your daily wardrobe. For the duration of the higher alerts, be sure to follow all recommendations made regarding your wardrobe and the "Get You There Pack" that will be explained in detail.

> **Reaction prep review**: Take serious stock of your preparations and any remaining needs concerning a potential "**Evacuation**" or "**Isolation**" outlined in the corresponding sections of this book. Check your "Bugout Kit," its "Last Minute List," your pantry levels at home, and double-check to make sure all items in your "Important Contacts" list(s) and your "Notify in Case of Emergency" card are current, and that all family members have copies.

> **Review your plans**: Double-check the reaction plans for your work, children's schools, etc. Also review your family's procedures that will be followed in response to various scenarios.

> **Update documents**: check all your **documents** to make sure all pertinent legal documents, contracts, accounts, policies, or license paperwork are all **up to date** (see the "**Stored Documents**" list in the "**Appendix**"), and appropriate copies are stored in your Bugout Kit's "Info Pack." Also, make it a habit to back up your computer or at least copy pertinent files to disk and store the disks in a safe place. **Hint**: You might also email important files to yourself for later retrieval.

> **Wear your communications:** If you have a pager or cell phone, or cordless phone at home, always have them with you, preferably <u>on</u> you. (In fact, some of the newer FRS radios are the size of a golf ball and come on a neckstrap.) Rapid communication is vital, and it would be a shame to be out of touch simply because you were out working in the yard, or stepped out of your office for a minute. For this reason, we've included spaces on your "**Find Me**" page (see "**Appendix**") for at least 3 of your neighbors, and for at least 3 contacts at each person's workplace, etc. so wherever you go, someone can reach you.

➤ **Vehicle maintenance**: Double-check all vehicular maintenance needs, and make sure any and all vehicles you own are ready to do anything you need them to. Keep all vehicles, tuned, gassed, accessible, protected, and equipped and stocked as indicated by the checklist under "Basic Vehicle Prep." Also, cover your car when not in use so as to protect the car's interior against a biochemical attack or HazMat incident. (If you live near a HazMat source, you should always keep your car covered in case of accidents anyway.) The same goes for any camper, RV, pop-up camper, or other vehicle you might use for evacuation or secondary shelter. Keep it covered.

➤ **Prep the evac vehicle**: As we'll mention under "Evacuation," sometimes the vehicle you use every day isn't the one you'd prefer to evacuate in. For example, you might own an RV that you keep parked in a lot when not in use, or you might have a boat that you'd use if you live along a waterway and roadways were gridlocked. Under Orange Alert, make sure any such vehicles are prepped and ready, and in fact, you should relocate them if possible to a better and more accessible location, such as in the case of the RV. The same goes for any camper shell for your truck. Go ahead and put it on, and make sure it's ready to go.

➤ **Prep your home**: Just as you want to prep your vehicle for an evacuation, you want to make sure your home is ready should you have to shelter-in-place. Our "**Isolation**" section will give you everything you need. The emphasis here is on the structure itself, and on food and water. We've mentioned utilities below.

➤ **Top off all fuels**: (Don't forget oil for generators, chainsaws, and other machinery that needs it.)

- ❏ Fill your **car's tank** (and never let it get below 1/2 full). Evacuation notwithstanding, you'll need a full tank of gas after an isolation event as there's no guarantee stations will be open, operable, or even stocked.
- ❏ Fill the tank on **any transportation vehicle** you own (some vehicles in long-term storage may have fuel removed). **Hint**: Having fuel in all your vehicles gives you some to siphon later if you need it.
- ❏ Fill the **five-gallon gas can(s)** you keep on hand for **lawn equipment**.
- ❏ If you have an **electric generator** that uses **diesel**, or other type of fuel, stock up on that.
- ❏ Have your on-site **propane** tank serviced and filled.
- ❏ Swap any **small propane tanks** for full ones, getting an extra or two if you can afford them.
- ❏ Top off any **fuel oil** tank you might have, or **coal** if your furnace uses it, or *can* use it.
- ❏ Refill any **kerosene** containers you may have to fuel space heaters.
- ❏ Stock up on **firewood** if the alert is near any sort of cool weather.
- ❏ If you have a **charcoal** grill, make sure you have plenty of charcoal and **lighter fluid** for cooking.
- ❏ Make sure you have plenty of **fuel/oil** mix for power tools such as **chain saws**.

➤ **Check utilities**: If you have **utilities that need servicing**, like new coolant for the AC unit, etc., have it taken care of now. If you're suddenly forced to shelter-in-place, you'll need everything to be in working order. (Don't wait. If the status is raised to Red Alert, service techs might not come.)

➤ **Bring your bike**: As we explained under "Basic Vehicle Prep," if you work in a populated metro area, any number of scenarios could see traffic gridlocked to the point that you'd never get out of there in a car. If you drive to work and you have a bike and a bike rack, keep the bike with you. Similarly, if you work in an area with good waterways, carry a canoe, kayak, or even a small inflatable raft. Be prepared to put any of your escape avenues to use. **These are not recommendations**, only options and food for thought. You always have to think of the alternatives available, and options in an emergency, like redundancy, are our friends.

➤ **Medical checkups**: Get a medical and dental checkup. If an emergency occurs, the last thing you'll want is to be hampered by some minor ailment, like a toothache, that could've been discovered and easily taken care of beforehand. Too, you want your health to be all it can be in case you fall direct victim to an attack. Lastly, a current physical is useful to establish an up-to-date medical baseline and current history should it be needed to compare to your condition if you're directly affected by an event. While at your checkup, make sure all your **immunizations** are up to date as emergencies may place you in close proximity to others.

➤ **Check your meds**: Refill any important prescriptions, update any emergency backup paper prescriptions you might have, and check your stock (and expiration dates) of all necessary non-prescription medications.

➤ **Pet checkups**: Make sure all your **pets' immunizations** are up to date, and maybe get them a checkup as well. If you're forced to evacuate, you may have to board your pet somewhere and that usually means you'll need **up to date veterinary records**. (See the "**Family Member Data - Pets**" form in the "**Appendix**.")

➤ **Prepay bills**: Prepay one extra month of some of your utility accounts, or your rent or mortgage if you're able, and if prepayment is allowed. If you have to hit the road for a while or be quarantined for a bit, or if something happens to your source of income, this will take one less worry off your plate. Consider it a tiny and temporary investment in peace of mind. In addition, **draw some extra cash** from the bank or ATM to have on hand. If phone lines are knocked out or in heavy use during an emergency, ATMs might not work and credit cards cannot be processed.

Red Alert

Yellow Alert involved a little planning and a few activities. Orange Alert added even more activities and a request for greater caution. **Red Alert** is more a list of what <u>NOT</u> to do, since such a high alert would not be issued on a simple whim. If this were equated to a weather condition, it would be a close cousin to a tornado warning; best to stay in and lay low. Let's look over some considerations we would add to all our previously listed items in a Red Alert.

Red Alert Checklist

❑ Relay the alert.	❑ Check your commo & rendezvous.	❑ Contact volunteer group members.
❑ Revisit the Orange Alert list.	❑ Check work status.	❑ Check school status.
❑ Rethink leisure.	❑ Preseal the house.	❑ Sleep semi-prepped.
❑ Keep communications open.	❑ Establish check-in times.	❑ FILL your car's tank.
❑ "Basket" your pocket items.	❑ Keep your tub scrubbed.	❑ Turn your refrigerator down.
❑ Protect heirlooms and valuables.	❑ Backup your computer.	❑ Relax.

➤ **Relay the alert:** Call the rest of your family, your volunteer group's phone tree, and the rendezvous/contact people on your "Important Contacts" list.

➤ **Check your commo & rendezvous:** See if your in-town and out-of-town rendezvous points and commo people are ready and able. Is that store you're supposed to meet at still the same store or has it changed names? Is "Uncle Al" at home this week or is he out of town? Has your child changed to a new school that has a completely different evacuation plan than the old one? During a Red Alert, double-check your info.

➤ **Contact volunteer group members:** If you're part of a **volunteer group** (and you should be) contact your fellow members to make sure everyone's personal needs are met, their families secure, and that they are ready to respond if needed. Then try to **stay off the phone** in case of emergency calls to you, or in case something happens locally and officials need the lines.

➤ **Revisit the Orange Alert list**: If we went straight from a lower alert to Red, go back through Orange to take what steps you can, especially those relating to **supplies**, **communications**, and **utilities**.

➤ **Check work status**: Don't skip work unless instructed to do so. However, if you're able to telecommute, do. Also, if you have vacation time, accumulated sick leave, etc., and can use them without putting your job or company in a bind, then consider staying close to home, especially if you live or work in a potential target city or area. This is why we asked you to save all your sick leave and personal days. However, go about your daily business only <u>IF</u> you <u>AND</u> your local officials feel it's safe.

➤ **Check your school's status**: Many school systems will consider short-duration closings during such a high alert. If schools remain open, have your child bring home all books every day so that if the school closes, they can study at home. (You might want to contact your school and ask them to adopt that as a universal policy, and for them to create take-home study packets in the event of prolonged closings.) For information on **School and Event Closings** call your local radio and TV stations, your local Law Enforcement non-emergency lines, or go to: http://www.cancellations.com. "Cancellations.com," c/o Lazerpro Digital Media Group, 220 Regent Court, Suite B, State College, PA 16801, 814-238-6201.

➤ **Rethink leisure:** Under previous alerts we advocated going about your normal life but keeping both eyes open. At this level, you might want to cut back, spend more time at home, and err to the side of caution. Don't go **out to eat** as much (especially at restaurants with accessible buffet lines), stay away from **crowded venues** such as malls, concerts, or sporting events (if they occur at all), **rethink air travel** or other mass transit, and postpone family vacations (especially to **metro areas**). And, the most difficult of all, **make the kids stay at home**.

➤ **Pre-seal the house:** If you sleep with your **windows open**, consider **not doing so** for a while. Go ahead and keep your home a little bit sealed against a possible biochemical strike. You might also pre-seal a bit, especially in unused rooms or attic vents, but don't seal the whole house unless something actually happens and you're instructed to shelter-in-place by officials.

➤ **Sleep semi-prepped**: Sleep with **durable, easy-on clothing** handy. By easy-on clothing we mean something you can get into quickly. An example would be a set of coveralls just inside the closet door with slip-on shoes in the pockets; something you can grab easily and get into in a flash. We call this "**bolt clothing**" in case you need to "bolt" out the door. More on this under "**Evacuation**." At the very least, sleep with good pajamas, or something to keep you covered if you literally have to run out the door. You might even keep your car's door key on a necklace or pinned to your PJs so you can access your supplies there.

- ➤ **Keep communications open:** Keep a radio or television on as much as possible. If anything happens, the sooner you know about it the better. Keep your phone with you as you did under Orange Alert. In fact, for a Red Alert, the phone tree you're a member of should have already notified the community of the new alert.

- ➤ **Establish check-in times**: As communication is much more important under Red Alert than under Orange, we remind you to always <u>wear</u> **some sort of communications device**, and set a **regular check-in time** for each family member to call in if anyone has to venture out. If anyone does venture out, they should tell you *exactly* where they'll be.

- ➤ <u>Fill</u> **your car's gas tank:** You should never let it get below ½ anyway, and you should make sure all tools and supplies listed under "**Basic Vehicle Prep**" are kept loaded in your vehicle. If you have a camper shell for your open-bed pickup truck, put it on. You might also consider bringing home any camper trailer, RV, or other useful transportation / shelter items you might have parked elsewhere, and make sure each is serviced, stocked, and ready to use if needed.

- ➤ **"Basket" your pocket items**: You probably do this anyway. Most people are creatures of habit and leave their purses, keys, wallets, cell phones, etc. in the same place each time. To make these items easier to retrieve, make sure you keep them in the same place <u>all</u> the time and that the spot you keep them is near your safe room. Also, use a small basket or other container to keep these items in since it's faster to grab the whole basket and run than to stand there and fill your pockets.

- ➤ **Keep your tub scrubbed:** In an attack or disaster, you'll want to fill your bathtub with water for drinking or washing in anticipation of loss of water. The cleaner your tub stays, the quicker you can give it that final bleach-wipe before filling it in an emergency. Also, **remove any automatic bowl cleaner** from your toilet's tank. The water in the tank is useful for washing, etc. provided it doesn't have harsh chemicals in it. Additionally, we suggested keeping extra plastic tubs and trash cans to store last-minute water in. You can store these tubs and trash cans by double-stacking them with the ones in use. In fact, every trash can or plastic storage bin you own should be doubled. It's conceivable you could use them in any type of emergency whether it's gathering your things in an evacuation, or storing drinking water, and then later, garbage and refuse during an isolation.

- ➤ **Turn your refrigerator down:** Even if you have an electric generator, you'll want your refrigerated food at its coldest so it will last longer if the power is lost. Also be sure to follow our suggestions in the "Water" section under "**Isolation**" and put water bottles in any empty spots in your refrigerator and freezer. In fact, double-check <u>ALL</u> your **evacuation and isolation** supplies, and be ready to react accordingly.

- ➤ **Protect heirlooms and valuables:** Just as you would in preparation for a hurricane or potential flood, you should round up your valuables and irreplaceables, put them in resealable plastic bags and inside plastic tubs sealed with duct tape, and put them up off the floor in a secure area such as your safe room. This way they'll be redundantly protected against physical harm or biochemical contamination, a little better hidden from looters, and a little better secured against being strewn about the neighborhood after a destructive event. **Hint**: Since fireproof safes are just that, and not water or chemical-proof, you should do more to protect the contents. Keep something like a computer monitor dust cover (or at least a plastic bag) over yours to keep out hazardous materials should your home fall victim to an attack or accident. You don't want to wrap the *contents* in plastic as it will melt all over your documents and heirlooms in a fire.

- ➤ **Back up your computer:** Store the disks in a safe place, and **email pertinent files to yourself** for later retrieval in the event your home and safe location of your backup disks are destroyed. A handy tool to help you do this is "WinZip." This will help you compress the size of your files for easier mailing, and fewer disks necessary for downloads. You'll find WinZip for free at http://www.winzip.com.

- ➤ **Relax:** After you've finished all this, rest as much as possible since there's really not much else you can do. Just be confident in your own preparations, in the capabilities of your local first responders, and spend your time doing relaxing things close to home. When's the last time you read a good book?

Though we're now in the "**Reaction**" section, we're still dealing with things that *might* happen. Let's go to a section that will prepare you to handle an actual event. Let's talk about your **personal safety gear**.

Notes:
_____
_____
_____
_____
_____
_____

Basic Personal Safety

Since we just left terror alerts and we're about to go to actual reactions, it's time to introduce some of the personal safety gear and procedures that will help you with many of the immediate reactions to follow.

Goals of Basic Personal Safety:

1. To make you aware of how **subtle wardrobe changes** can equal greater personal protection.
2. To discuss the **everyday items**, that if carried with you, **could save your life**.
3. To open your eyes to **improvising effective protective equipment** and learning to use what's around you.
4. To **explain commercial equipment,** what's good and what's not, and what's needed or not.
5. To teach you **how to organize and execute various drills** in preparation for actual reactions.
6. To teach you **basic decontamination procedures** should you be on your own.

We've divided Basic Personal Safety into the following subsections:

1. **Dress for Success.**
2. **Gas Masks and Moon Suits.**
3. **The Drill to Survive.**
4. **Basic Personal Decontamination.**

Dress for Success

Think of first responders arriving on the scene of an accident or other urgent situation. Do they show up, look around, and then put on their equipment, or do they have their gear on before they get there?

For you, your "scene" is going to be wherever you are at any given moment, since you don't know what's going to happen, when, or where. Therefore, we need to discuss your "uniform and equipment" which happens to be your day to day clothing and the things you carry with you. Subtle modifications can be made to these items to help protect you before, during, and after an unforeseen event. So for "uniform," we'll look at "**Dressing for Chemicals**," and for "equipment," we'll look at the "**Get You There**" **pack**.

We've included this section on clothing modification since it's extremely unlikely that you'd carry a gas mask and full protective clothing with you at <u>all</u> times. It's also unlikely that you'd be warned of an incident far enough in advance to put your protective gear on. Most incidents will come as a complete surprise, so unless you plan to wear a protective suit as your daily wardrobe, you need to consider ways your everyday wear can be made to help protect you.

Dressing for Chemicals
During heightened terror alerts, you can do a little advance protection against chemicals by learning to "**Dress for Chemicals.**" <u>**This will not make you impervious to chemicals**</u>, but it will go a long way toward limiting the actual quantity you may be exposed to, which with some chemicals can make all the difference in the world.

Let's cover the general points:

> The three most important things to remember here are, **wear extra layers, make the layers count,** and **carry useful accessories**.

> Wear **long sleeve** shirts, even in the summer. Long sleeves can be rolled up in heat, but short sleeves can't be rolled down to protect skin.

> Wear an **undershirt** whenever possible. This gives you an extra layer of protection.

> Office workers and other professionals, male and female, should wear outfits that include a **suit jacket**. These jackets should be made of tighter-weave, less absorbent materials. Avoid wool or cotton unless treated with something like **Scotchguard** ®. With these jackets, consider adding a small Velcro tab or button under any collar, at the neckline, so that if the collar was turned up and wrapped around the neck, it could be held in place. Turn your collars up to see if this will work with your suit jackets. When done, your coat should look almost like the old Neru jackets. This will be helpful in sealing around any sort of hood you may put on. Show the same consideration to overcoats.

> **Pants** are preferred over skirts where possible.

> Other outer layers, such as your **blouse, or dress shirt**, should be of a **non-absorbent, tighter weave material** such as silk, nylon, etc. Fashion should be a secondary consideration during a potential terror threat. If you prefer cotton or can't wear synthetics, the tighter the weave the better. Nonabsorbent clothing

will shed a certain amount of some chemicals so when you evacuate you carry less contaminant with you. Absorbent *outer* layers will *cause* you to carry some chemical with you. Conversely, absorbency is preferred for your under layers due to perspiration.

➤ **Scotchguard** ® or other stain protection may help make some clothing less absorbent to some chemicals. For non-office work clothing, less expensive options than Scotchguard can be considered. **Word of caution**: Some Scotchguard substitutes will reduce the flame retardant capabilities of some fabrics. For example, some silicone sprays sold in camping stores as "waterproofing compounds" really do work well, and can be used on quite a number of fabrics as additional repellant protection from some chemicals. However, make sure you read the label carefully to see what it says about flame and fire retardants.

➤ Office workers who wear starched shirts should have their shirts pressed with **extra starch**. This will help make them a little less permeable.

➤ In cooler weather, opt for **turtlenecks**. The neck can be pulled up over your mouth and nose to help hold your breathing filter (or even a simple handkerchief) in place, and/or the base of your smoke hood. If you wear the more permeable or absorbent sweater-type clothing, make sure you have adequate outer and/or under layers for protection.

➤ In colder weather, consider **Gore-Tex** ® fabric coats and jackets. Though not rated as being chemical-proof, the fabric repels most substances well while allowing clothing to "breathe" and vent trapped moisture.

➤ Also in cooler weather, opt for **dress gloves**. Leather is preferable to cloth. If yours are cloth, get the Scotchguard ® treatment or similar.

➤ Some **hats** may not only be fashionable, but protective. This applies both to formal office dress, and to more casual dressers who would do well to wear ball caps, etc.

➤ For contact wearers, consider wearing **glasses** during higher alerts. They cover more area of the eye, which may block a mist or dust from blowing into your eye, and they don't trap substances in place against the eye like contacts may. Consider changing your casual eyewear, such as sunglasses, to a more protective wrap-around style. Glasses are easy to clean and decontaminate whereas contacts would have to be disposed of.

➤ Carry useful accessories such as cloth **handkerchiefs** (dust mask and washcloth), scarves or plastic **rain caps** (additional head cover, or tourniquet in the case of a strong scarf), a sturdy **belt** (tourniquet or sling), and carry a small **umbrella** as most chemicals are heavier than air and may settle on you if released from altitude, or they may act as a mist if released at ground level. Also, slip-on shoes may be faster and more comfortable, but **shoe laces** have a considerable number of uses. **Neckties** have numerous uses as well.

➤ Casual dressers should still follow the long sleeve and extra layers rule, and should wear (or at least carry) an additional outer layer such as a **light nylon jacket** or windbreaker, preferably with a hood. This holds especially true for service technicians who spend most of their day on the road. See if these jackets respond well to Scotchguard ® treatments.

➤ Wear clothing that carries a UL **flame retardant** rating. Wearing layers will not only help protect you from chemicals, but layers of flame retardant clothing could also offer a great deal of protection from **fire** that may result from some of the more severe destructive scenarios. Of natural fibers, **wool** is the most fire-retardant.

➤ For more information on **fireproof textiles** and fireproofing compounds, go back to "Basic Home Prep's" portion on "Fire" and look through the various outside sources.

➤ Wear **athletic socks** under dress socks if wearing pants (and if shoes will fit). At least wear longer dress socks so your pants legs can be tucked in preventing a draft from allowing chemicals up your pants legs (during a known chemical release).

➤ Apply **aloe vera gel** or other skin lotion when you're about to head out in public. Lotions may help protect your skin from absorbing certain chemicals. For a look at an interesting new product, see the **new lotion called RSDL.** It's just been FDA approved for use as a topical protectant against chemicals. We've found a few info sites for you:

- http://www.nbcteam.com/personal.shtml.
- http://www.ezem.com/corporate/rsdl2.asp?action=corporate.
- http://home.golden.net/~flatcat/ODEL/faqs.html.
- http://www.applesforhealth.com/HealthySurvival/fdalotbloc4.html.
- http://www.fda.gov/bbs/topics/NEWS/2003/NEW00888.html.

➤ Among the "useful accessories" we mentioned that you can carry, the section below concerning the "**Get You There**" pack tops the list. The clothing items we've listed here will help protect your skin and thereby some absorption of chemical hazards, but will do nothing to protect your face, eyes, or lungs.

Again, these measures will not make you chemical proof, and are not even as good as a simple plastic rain suit. However, these steps are things you can do all the time without much notice, and they'll certainly limit your exposure to a certain degree which is always a good thing if you're in a situation where you're caught completely unaware.

There are a couple of hidden benefits to the extra layers and tighter weaves of materials. **One** is they'll help protect you against projectiles to a certain degree. Though they certainly won't make you bullet-proof, more layers of tougher fabrics will protect you against some smaller and lighter shrapnel, such as glass fragments, etc., that you might be exposed to in an explosion. Also, battlefield history is full of stories of how watches, medallions, books, coins, etc. deflected bullets and saved lives. Even Teddy Roosevelt survived an attempted assassination because the speech he had folded up and put in his coat's breast pocket was thick enough to slow the bullet.

Two, think back to the World Trade Center '93 bombing where thousands of the buildings' occupants were forced out of the building through smoke and fire and into cold weather. The wardrobe modifications mentioned above would be helpful in this fire and weather scenario. Keep these things in mind when going though all these suggestions, since chemicals aren't your only day-to-day threat.

As we've already mentioned the "**Get You There Pack**" a few times already, let's see what it's all about.

The "Get You There" Pack ───

This is the smaller sibling of the "**Bugout Kit**" you'll see under "**Evacuation**." The difference is that the "Get You There Pack" is so small you can have it with you at all times, and economical enough to have several in different locations. It's meant to *get* you out of harm's way, and help *get* you from wherever you are to wherever you want to be. From here on, we'll simply call this your "**GYT Pack**" (pronounced "**get**" as in "**get** you there").

Practice accessing your pack and any and all of its contents blindfolded. You may in a smoke-filled room and need the items to help you escape. You might also be in a power outage and need to access other items. **Practice!**

The "Get You There" Pack	
"The Smoke Hood" ←	
❑ Large gauze pads	❑ Three-foot length of heavy-duty aluminum foil, folded
❑ Triangular bandage	❑ Pill or squeeze bottle of baking soda solution
❑ Reynold's ® 12-lb. baking bag	❑ Gym socks
❑ 2 Reynold's ® 6-lb. baking bags	❑ 2 small rubber bands, and 2 large
Exposure and Shelter	
❑ Plastic poncho ←	❑ 2 large garbage bags ←
❑ "Space Blanket"	❑ 4 medium safety pins
❑ Lighter or waterproof matches	❑ Tire-strip tinder ←
❑ Small pack dental floss ←	❑ Small roll duct tape or individual strips
❑ Sewing needle taped to dental floss pack	❑ Small bottle insect repellant
Communication and Signaling	
❑ Small flashlight ←	❑ Whistle
❑ Extra battery for pager or flashlight	❑ Extra cell phone battery pack ←
❑ Small chemlight sticks ←	❑ Sharpie ® marker and chalk ←
❑ Small notepad and pencil	❑ "Micro" AM/FM Radio ←
Tools	
❑ Leatherman Tool ® or equivalent ←	❑ P-38 can opener ←
❑ Pepper spray ←	❑ "EMT Shears" ←
First Aid	
❑ Latex gloves ←	❑ Backup of crucial medications
❑ Individual alcohol or antiseptic wipes	❑ Waterless hand sanitizer
❑ Assorted Band-Aids and Butterfly Band-Aids	❑ Small roll gauze and Maxi Pads ←
❑ Tweezers ←	❑ Backup eyewear ←
❑ Saline Solution ←	❑ Aspirin or substitute

Decontamination		
☐ Comb ←		☐ Small bar of soap
☐ Toilet paper or small pack facial tissue		☐ "Corn starch" ←

Transportation		
☐ Cash and coins		☐ Mass transit pass or tokens, car discounts ←
☐ Spare car and house keys		☐ Personal data and map photocopy ←
☐ Credit card(s) and blank checks		☐ Bike lock key ←

Sustenance and Comfort		
☐ Protein food bar		☐ MRE accessory pack with spoon ←
☐ Iodine tablets (water purification)		☐ Small packs over-the-counter meds ←
☐ 4 quart-size resealable plastic bags ←		☐ Caffeine tablets ←
☐ 2 plastic drinking straws ←		☐ Small packs "comfort" food ←

Pack the above items in as small a container as possible. We suggest a "fanny pack" you always carry with you. Periodically go through it so you'll remember what you have, where it is, and what it's for.

Discussion Items (The "←" items from above.)

"The Smoke Hood"

As the "Smoke Hood" is such a detailed and useful concept, we'll discuss each item under it. The purpose of the "Smoke Hood" section is to create a piece of improvised protective gear that can keep you relatively or momentarily safe from fire, smoke, and some forms of toxic substances. Some of the more debilitating fire injuries are smoke inhalation, burns to the lungs, face, and hands, and some of the more lethal chemicals and bioweapons attack by way of being inhaled. Let's look at our components, and then we'll look at how to assemble and wear the hood later:

➢ The **large gauze pads** have several uses. Here they're part of an improvised "dust mask" or temporary breathing filter. The type we carry is an 8 inch by 7.5 inch, "Surgipad" combine dressing. These pads can also be used as gauze in a first aid situation, or washcloths in a decontamination. In place of gauze pads, you can use any densely woven or multi-layer fibrous cloth or pad (preferably cotton) that you can breathe through when wet. Other examples are washcloths, cloth diapers, "Baby Wipes," a clean pair of gym socks, ACE bandages, etc. (Experiment with other fabric items too.) **Note: None of these products are designed, sold, rated, or intended to be a dust mask or breathing filter.** They're listed here as examples of commonly found items that <u>might</u> be used as an expedient filter. We'd prefer you to carry a **NIOSH** rated gas mask, or at the very least an "**Evac-U-8**" smoke filter (discussed below in "Gas Masks and Moon Suits"). However, we realize that price and size are obstacles, so we're giving you these minimal substitutes in hopes that you'll at least carry *something* with you at *all* times.

➢ The **Triangular bandage** makes up some of the insulation of the smoke hood and can also be used in numerous first aid situations. Additionally, it makes a great scarf in cold weather.

➢ The **Reynold's 12-lb. baking bag** (again, a product **NOT** marketed or intended for this use) is meant to go over your head to give some protection against direct contact with flame. However, just as with the turkey in the oven, it will <u>not</u> protect you from being "cooked." The bag should be **pre-punctured with a small mouth hole**, and marked with a Sharpie marker across the "<u>back</u> of the head" part so you don't put it on backwards. You want that mouth hole in the right place in case you pass out. Don't want to suffocate now do we? We'll explain this a bit more under "**Using the Smoke Hood**." **Note**: If you live near facilities that pose an extra chemical threat, you may want to include a small pair of **swimmer's goggles** in your kit. They cost pennies, weigh next to nothing, and they seal around each eye rather well.

➢ **Hand protection** is the intent of the two 6-lb. baking bags. In a fire, you don't want anything meltable under the bags. For instance, don't wear the listed latex gloves under the bags in a fire escape situation. Wearing the latex underneath the bags as insulation against cold is fine.

➢ **Aluminum foil** has a million and one uses that could fill another book. Here, it's a part of the smoke hood's insulation. In other scenarios, the foil could make utensils to cook food or boil water.

➢ The average child-proof, amber-colored **pill bottle** you get from the pharmacist will hold water quite well without leaking. You can also find little **travel-size "squeeze bottles"** at almost any drug or department store. These little containers can carry just enough water to wet down your breathing filter to make it a bit more effective. Adding **baking soda** to the mix may help neutralize some slightly acidic particles, mists, or gasses. For example, some sick prankster might think that hosing down your subway car with pepper spray is funny. Baking soda may help neutralize the pepper spray better than plain water. To make a **saturated solution** of baking soda, bring two cups of water to boil and stir in baking soda until no more will dissolve, and you start getting a little residual powder lying on the bottom of your pan. Allow the solution to cool and fill your bottle with the liquid (but not the residue).

➤ **Gym socks**, or heavy cotton athletic socks, have several uses. They can add an additional layer to your **breathing filter**, you can wear them as **mittens** (or stitched together using your floss and needle to make a headband or earmuffs) in severe cold, as **heat insulation** under your hand's baking bags in a fire, or office workers can **swap out thin dress socks** for the gym socks if having to walk a long distance to evacuate a disaster scene. They'll also come in handy as **washcloths** during any decontamination.

➤ The **rubber bands** are meant to hold your **baking bags in place**. The 2 large ones are for your head, and the 2 small ones for your hands.

Exposure and Shelter

Think back to "World Trade Center '93" where thousands of people were evacuated from the building out into the cold. Not only will the images of their soot-blackened faces bring to mind the importance of the "Smoke Hood," it will support the need for items to protect you against outside weather or worse.

➤ **Plastic ponchos** can be found at most any "Dollar Store" or department store. We're talking about the kind that folds up to about the **size of a small wallet**. A poncho can protect you from the rain, the cold, or if you're evacuating a known chemical attack or HazMat incident, a poncho can help **protect your skin and clothing from further contamination**. **Hint**: You might want to carry 2 in case you have to decontaminate and you don't have clothes to change into. You'll also want them to be brightly colored for use in signaling and being seen in general.

➤ **Dental floss** has more uses than just flossing. It can be used as twine, or even thread (hence the **needle**) to sew ripped clothing, etc., to protect you from the elements. In an outdoor survival situation, it can be used to make small snare traps. Carry the waxed, unscented variety.

➤ Your two **large garbage bags** can be used to augment your poncho, to give to others who might need a poncho, to help gather and transport miscellaneous items or contaminated goods for later cleaning, or you can slit them open to make plastic sheeting for expedient temporary shelter or to seal yourself in a small room. Using the bags, some dental floss, and some duct tape, you could actually make a small tent. Or, you can use the plastic to seal a broken window in your car. Make sure your trash bags are brightly colored as you may need their help in signaling. An additional use (along with your resealable plastic bags) is in decontamination situations. You'll need something to hold personal items like wallets, watches, jewelry, glasses, etc., and you'll need them to seal and dispose of any contaminated clothing articles.

➤ **Tire-strip tinder** is simply small strips of rubber, about 3 inches long and ¼ inch wide, cut from an old tire. These make great little fire starters as they don't stay wet, they're easy to light, and they burn well making it easy to light fire wood. Make this a part of your pack if you frequent outdoor areas in which you may become stranded in the cold and/or forced to stay overnight.

Communication and Signaling

Whether you're trapped in a building, or stranded in a blizzard, you'll want to let people know you're there.

➤ You can find little **flashlights** at any "Dollar Store" or department store. Get one that uses only **one battery** (carry the two-battery pen-light type if you have room), and make it a battery size that you frequently use. For example, if you wear a pager, get a flashlight that uses the same size battery. Depending on the current need, you can rob one to feed the other. This way, the "extra battery" listed can be used in either or both.

➤ Usually sold as party favors, little **inch-long chemlight sticks** can be useful as a signal in the dark (or carry the full-size ones if you have room). Using these instead of your flashlight can help save your battery. You can find these at most toy stores, party supply stores, or through:

● eGlow products at http://www.eglow.com, and at eGlow, 1601 N.W. State St. Bldg 1A, Lehi, UT 84043, Fax: 801-437-2654, toll-free: 866-999-4569.

● Brigade Quartermasters at http://www.actiongear.com, 1025 Cobb International Blvd., Kennesaw, GA 30152-4300, 800-338-4327.

➤ Cell phones are notorious for giving out of power in the wrong place, at the wrong time. When it happens to you, you might not be in a place that has power, or you might not have your adapter with you. Keep an **additional charged battery** in your pack (swapping it out to recharge), or for other battery and accessory options, revisit the cell phone portion of "Communication."

➤ **Sharpie** ® markers are inexpensive, plentiful, and more importantly, **indelible**. Keep one in your pack to leave notes that need to be written on any number of surfaces, or read after being wet. We also mention **chalk** as you may want to write **temporary messages** on walls, etc. (good tool for commo at your rendezvous points). Carry a short piece of chalk inside an old chapstick tube.

➤ Several companies make miniature **AM/FM radios** that we like to call "micro radios" as they're so tiny. Some are about the size of a 9-volt battery, making it easy to keep one in your pack.

Tools

There's no guarantee that any emergency will be free of obstacles, or gear that needs repair. Make sure you have the minimum tools necessary to take care of the annoyances that could turn into major trouble.

> Our favorite multi-tool is the **Leatherman Tool** ®. Its compact collection of blades, pliers, screwdrivers, and more, just can't be beat. Depending on the size of your GYT Pack container, you might want to consider other options such as Swiss Army knives, or the combo tools that are made out of a single piece of flat steel about the size of a credit card as shown to the right. Whatever you get, make sure you can cut with it, and turn screws and bolts. You might even want to include a short length of **hacksaw blade**.

> The venerable **P-38 military surplus can opener** is worth its weight in gold. You can get these through many of the equipment suppliers listed under "**Evacuation**." Get several as you'll want one in your pack, one on your key chain, one in your safe room, one in your car, one in the kitchen, and certainly a couple in your main Bugout Kit (to use or trade).

> **Pepper spray is optional.** Carry it if you feel you may be involved in a panicked evacuation or other situation where someone might think they need your gear more than you.

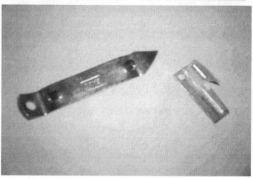

> Not only is a good pair of **EMT shears** a great tool, it'll come in handy during decontamination since you'll want to cut off some clothing rather than pull it over your head. Pulling clothing over your head will wipe the chemical across your eyes, mouth, nose, and exposed facial skin.

On top we have a smaller Swiss Army knife to the left, the Leatherman Tool ® in the center, and to the right, a compact multi-tool.

Above, the standard bottle/can opener on the left, and the military P-38 on the right, are also highly useful tools.

First Aid

> As you may be giving first aid to others, you'll want the **latex gloves for your own protection**. They'll also provide some protection from certain chemicals for a short time. Pack **Nitrile** gloves if you can, since it's turning out that more and more people are allergic to latex.

> **Tweezers** are important since most emergencies involving buildings result in breaking glass. This means **glass shards and splinters**, in either you or someone else.

> Not only is **saline solution** good for dry, tired eyes, it's good for **irrigating the eyes in a decontamination** situation. To make your own, take a quart of distilled water, bring it to a boil, add one level tablespoon of **non-iodized** salt (and some formulas recommend a literal pinch of baking soda), stir until dissolved, and package in a sterile plastic squeeze bottle with a cap until needed. Replace the homemade stuff at least monthly. Commercial saline is probably more sterile, but making your own will allow you to carry more. (**Hint**: Put this in a different style bottle than your baking soda solution so you don't get the two mixed up.)

> A **small roll of gauze** is always useful in a first aid situation. Surprisingly, so are **Maxi Pads** (plain, unscented). Not only are they a useful backup for their intended purpose, but they make wonderful gauze pads for fairly large wounds.

> **Backup eyewear** covers more than just eyeglasses, and ties in with saline solution. Carrying something as simple as sunglasses will give you a little extra eye protection. Also, if you wear contacts, you might want to carry a pair of your prescription glasses in your pack or another set of contacts, as you may have to change them out after a chemical exposure. **Hint**: If you want something to really protect your eyes, remember what we said about **swimmer's goggles** earlier.

Decontamination

In addition to some of the antiseptic wipes mentioned under First Aid, you'll need some items to help clean you up if exposed to either a known hazard or unknown substance.

> Depending on your hair style, you may want a **comb** in case you have to comb **dust or other particulate matter** out of your hair until you can get to a location you can wash it.

> We mention **corn starch** as a generic item under decontamination as its a great absorber. If you're exposed to a liquid that gets on your skin or in your hair, a little corn starch will help absorb it until you can find a

source of enough water to put your soap and gauze pads to work. Keep a few tablespoons in a small baggie or in a travel-size talcum powder bottle. (It's also recommended by vets as an emergency first aid treatment for pet wounds.) In addition to, or instead of, corn starch, take a look at some of the items and substances mentioned under "**Personal Decon Alternatives**" in the "**Basic Personal Decontamination**" section below.

Transportation

Once you survive the event, you may need to find an alternate means of getting out of the area you're in and back to an area you want to be. The checklist items are pretty much self-explanatory except for:

➢ Though you may always drive to work, something might happen to deny you access to your vehicle. Always carry a **mass transit pass or tokens** in your pack. Additionally, many roadside assistance clubs offer **car rental** discounts to members. Keep a copy of your discount card in your GYT Pack, and note the location of local rental offices on your Threat Map.

➢ Your personal data may be a copy of your "**Family Member Data**" forms, a copy of a pertinent page from your "**Important Contacts**" sheets, or it might simply be your "**Call in Case of Emergency**" info. Your **Threat Map photocopy** will help you recall the locations you may have stashed goods or gear, and will remind you of alternate routes out of your frequented areas, and locations of alternate transportation. It should also show your **family's emergency rendezvous points**. (In a stressful situation, it's best to have things written down.) Also, remember what we said earlier under "Info Pack" about creating "**micro data cards**" and printing all this info on small business cards to keep in your wallet. If you had only *one* micro data card, it should contain at least the following:

❑ Your emergency contacts' names, phone numbers, and email addresses.

❑ Your spouse's or significant other's "Find Me" contact info.

❑ Insurance policy numbers and agents' phone numbers.

❑ Answering machine and/or voice mail retrieval codes.

❑ A list of your rendezvous points in preferred order, along with written reaction plan copies.

❑ A listing of your cache locations.

➢ We suggested under "Homeland Security Terror Alert Levels" that if you drive your own vehicle to work and you also ride a bike, that you carry your bike to work with you. If you need it to get out of a gridlocked area, you'll want to be sure and have a duplicate of your **bike lock key** in your pack. (Or use a combination lock and write the combo down, keeping it in your pack.) **Note**: If you wind up putting a lock on numerous items in different places, spend a little extra and get combination locks that can be set to the **same combination**.

Sustenance and Comfort

As evidenced by other items we've mentioned, we see scenarios where you might be stranded for a little while. You could be huddled in your car in a blizzard, caught in extreme traffic gridlock for hours, or in some other situation where you're simply stuck. Your cash and coins will help you with vending machines, but sometimes all you have is all you have *on* you.

➢ For immediate comfort, carry a few **medicinal items** in addition to any prescription meds you may be on. Some examples are aspirin, a mild "calmative" or **sedative**, allergy medicine, etc. For additional suggestions, read ahead to the "**Evacuation**" section's Bugout Kit checklists and look for the "Non-Prescription Medications" checklist.

➢ Your **resealable plastic baggies** can be used as **food serving utensils**. You can make instant coffee in them (after heating the water in a pot you made from tin foil), you can **gather drinking water** from a faucet, use them as a collapsible bucket to fill your radiator, or you may need them to hold your wallet and other personal items during a **decontamination procedure**. In fact, we recommend using a gallon-size Resealable plastic bag as a liner for your whole GYT Pack for later use, and to protect your pack's contents. You can also inflate your plastic bag to use as a **pillow**, or inflate all your bags and stuff them under clothing to use as **flotation devices**.

➢ **Caffeine tablets**. Adrenaline will certainly keep you going for a while, but we have no idea what's going to come next. If you have to evacuate an area and get in your car and keep going for hours and hours, you might need a little boost to keep you awake behind the wheel.

➢ To drink out of your baggies, you'll need **the straws**. Interestingly enough, a couple of straws connected end to end might help you **access breathable air** in some situations.

➢ The **MRE** (military acronym for "Meals Ready to Eat") accessory pack is a gem unto itself. They usually contain salt and pepper, a beverage powder, napkins, toilet paper, gum, and a spoon. We'll give you a couple of sources for MREs under "**Evacuation**."

➤ Nothing will help you handle the stress of an emergency like a little of your favorite **comfort foods**. We suggest carrying a little hard candy, chewing gum, maybe some tropical chocolate (doesn't melt in heat), a teabag or two, or maybe some small packets of instant coffee, cream, and sugar.

Miscellaneous Considerations

Many of us live in or frequent more rural areas, and we also realize that our GYT Pack might be needed to sustain us in an outdoor situation, or after a natural disaster in less populated or civilized areas. In addition to the above items you might consider adding:

➤ A compass for navigation.

➤ Signaling items such as a mirror or small aerial flare.

➤ A small fishing kit including line, hooks, and sinkers.

➤ Other items you learned of while studying the "Outdoor Survival" portion of "Basic Training."

Now that we have the gear listed and explained, let's learn how to use part of it. Let's learn how to put on your "Smoke Hood."

Using the Smoke Hood

Again, we wish to point out that the "smoke hood" we've mentioned is <u>NOT</u> a NIOSH rated protective system, but is rather an extremely economical and easily obtainable set of common components that may offer a certain degree of protection. The most important aspect of our little smoke hood is that due to its size and economy, you're more likely to carry it with you at *all* times. The best protective gear in the world is useless if it's inaccessible when you need it.

This smoke hood can help protect you from smoke and some flame, and will also offer a surprising amount of protection against low levels of chemical exposure. Let's look at the steps involved in putting it to use. (**Always practice putting on your gear!**)

Putting on the smoke hood as an emergency reaction:

1. If you've made the decision to put on your smoke hood, it might be in a rapid emergency reaction to a fire or chemical incident. As your first order of business is to protect your lungs and eyes, you should immediately close your eyes and **stop** breathing. Don't *take* a breath in order to hold it, as that will cause you to inhale what you're trying to filter out. Just simply *stop* breathing.

2. Since you've practiced this as we've suggested and you know your pack's contents by feel, pull out your gauze pads and baking soda solution, and saturate the pad. If you *have* to breathe and clothing allows, you may pull your shirt collar up over your nose and mouth to take slow and shallow breaths for a second or two while you prep your gauze pad. Tilt your head down as if you were breathing the air off the surface of your chest. If you followed our advice under "Dressing for Chemicals" you'll be wearing a couple of layers of clothing which will do a *little* filtering while you're prepping your pad or respirator. However, you should be able to prep your pad in only a couple of seconds.

3. Hold the wet pad over your nose and mouth and begin to breathe slowly. Hold it in place firmly enough so that you're breathing through the pad and no air gets in around the edges.

4. With your free hand, locate your larger baking bag, open it up, and pull it over your head.

5. Using the same free hand, align the mouth perforations over the wet gauze pad, and hold everything in place while you slide the first hand out from under your hood.

6. Use your large rubber bands to go around your head in such a position that they're holding the baking bag down on the gauze pads, and holding both the bag and gauze in place over your nose and mouth. (The rubber bands should be positioned under your nose, and just above your chin, but not around your neck. You want good blood-flow to the brain.) You can now open your eyes.

7. If clothing allows, flip your collar up over the base of the bag to help seal against heat, smoke, or chemical contaminants.

8. When you can, you'll want to help hold the gauze against your face to keep a good seal. Re-wet the gauze as necessary and be sure to breathe slowly so the pad has a chance to filter the air. The only time you should move the pad away from your face is if you need to vomit (and then you reposition the mask immediately), or you've escaped the area and have been given the all-clear.

9. You should be able to do all the above in **15 seconds** or less. Be sure to practice.

10. If this is in response to a chemical event, put your poncho on too. It'll help protect your clothing from additional contamination. However, once out of the immediate danger area, take your poncho *off* (but not your smoke hood) as it may be trapping some chemicals on your clothing and preventing them from evaporating.

(**Reminder**: <u>None</u> of this will make you <u>impervious</u> to fire or chemicals. It's only meant to give you those few precious moments of protection that will either reduce your exposure, or allow you time to escape. Also, with chemicals, some will start to eat right through the plastic poncho. BUT, better it eat the poncho for a few minutes than your skin. Also, don't sit around putting on your smoke hood if your evacuation needs to be rapid. Balance your needs. If you're under localized chemical exposure, just stop breathing and run. If you're in a safe spot but about to have to evacuate through smoke and fire, then take a couple of seconds to put on all your protection. Don't forget to practice.)

If you had a little more time and were putting on your smoke hood while a nearby fire was increasing in intensity, or in anticipation of having to evacuate through a heavy smoke or flame (or biochemical) area, you'd do the following:

Putting on the smoke hood with more advance notice:
Some scenarios might see you with advance notice of approaching flame, an impending chemical incident, etc. An example may be that you're trapped in a burning building waiting for rescuers to get to you. In this instance, you'll have more time to prep your complete hood, and more reason to do so.

1. Drape your triangular bandage over your head like a scarf. It should cover the top, sides, and back of your head. You can tie it under your chin, or wrap the ends across your breathing filter for extra filter protection. This bandage is for insulation against heat.
2. Unwrap the tin foil and wear it in scarf fashion over the triangular bandage. This is extra protection against both heat and flame.
3. Put on your wet gauze pad and large baking bag as described above.
4. Pull any shirt or coat collar up over the bottom edge of the bag to keep smoke or gasses from entering your smoke hood. If your jacket has its own hood, pull it up over your smoke hood. The more insulation the better.
5. To protect your hands, wear your gym socks as mittens and cover them with the smaller baking bags, using the smaller rubber bands to hold them in place.
6. Roll your shirt sleeves or jacket sleeves down over the edge of the bags.
7. If you have access to any water at all, wet your clothing.
8. If evacuating in a fire situation, do <u>NOT</u> wear your poncho. It will hamper your movements, and most of the inexpensive ponchos are made of flammable material.

Ultimately, your GYT Pack will be customized by you as you're the only one who knows exactly what your needs are. For example, you might live and work in a rural area and escaping a building fire would be low on your list of probable emergencies. You may want more in the way of exposure and shelter items, signal items, or food and water.

The pack we've described above is a general kit that will help protect the user from fire and smoke, some hazardous chemicals, and from the elements for a little while; the basic results of a terrorist attack. **Its importance lies in its economy, simplicity, ease of use, and extremely portable size**. There is no reason at all to not make this kit or to not have it with you at all times. Customize yours accordingly.

The kit we created lives in a small "fanny pack." Depending on its final size and the places you plan on carrying it, yours could be put in a briefcase, a small backpack, ladies could carry a slightly larger purse, or you could use an old military ammo pouch. You should keep one on you, and you can also leave one in the office, one in the car, etc. **Make sure your protective gear is there when you need it**.

Speaking of protective gear, let's move on to some commercially available protective gear, and how to judge its capabilities so you can choose what you need.

Gas Masks and "Moon Suits"

The threat of a terrorist-instigated biochemical attack has spawned the sales of thousands of gas masks and related PPE, or "Personal Protective Equipment," products. Most of these were inferior equipment that was given an inflated price tag and advertised to do more than the item was able. The purpose of this section is to teach you how to distinguish the good equipment from the bad, how to decide what you truly do or do not need, and to help you understand some of the expedient items you can use in a pinch.

Before we even get into describing form, function, and sources of protective equipment, let's help you answer the biggest question of all, and that is, **should you buy any**? Well, the best answer we can give to that is, "it depends." This is yet another area of preparedness that hinges on your unique situation, and there is no definitive yes or no answer. Let's look at a few considerations.

Should you *consider* buying a "gas mask" or protective clothing?	
Yes	**No**
➢ If you live within 2 miles and downwind of a potential HazMat source.	➢ If you live in a rural area far away from potential HazMat sources.
➢ If you live in a populated metro area that is considered a likely terrorist target.	➢ If you do not plan to research the gear's uses and limitations.
➢ If you can and will carry this equipment with you at all times.	➢ If you could not, or would not, carry this gear with you or near you at all times.
➢ If you can easily afford it and having it will make you feel more secure.	➢ If you scored low on the "Emergency Risk Score" and can't really afford the gear.
➢ If you're certain you will receive training and learn and respect the gear's uses and limitation.	➢ If you're not going to be trained or fitted.

Keep these considerations in mind, and let's proceed, even if you're undecided about the equipment, and even if you've decided against it. We'll not only give you some more information that will allow you to make a more informed decision, we'll also give you a few suggestions for some alternative gear.

Gas Masks

Bones will heal, skin can be grafted, but eyes and lungs are virtually irreplaceable. Here we'll cover the basics about gas masks and what makes a good one.

The best equipment in the world would be a Level-A self-contained unit with its own air supply. But that's neither practical nor affordable. Something more obtainable would be the current military issue MOPP gear (**MOPP** stands for **M**ission **O**riented **P**rotective **P**osture), but those too can be rather expensive and are too cumbersome to have with you everywhere you go.

The first thing we should cover is **what masks do**, and more importantly, what they **DON'T do**. Let's consider the following general information about masks, or to use the more appropriate term, **respirators**:

➢ No one respirator will do everything you need. Despite what the ads say, there are different types of masks and filters designed to perform different functions. So the question, "Should I buy a gas mask?" is answered with, "What specific substance will you be exposed to?" The answer to that is, "Who knows?"

➢ No respirator will help you in an oxygen-deprived environment unless they have their own air supply. Therefore if you're in a situation where there's so much chemical that it's crowding out breathable air, the only way a mask is going to help you is if it's attached to an air tank.

➢ Respirators do not work well unless fitted properly to ensure an adequate seal against drawing in outside air. One size does not fit all. If a person cannot get a mask to fit properly, such as men with beards, they might want to get a respirator with a hood system as part of the mask.

➢ All respirators, military types included, fair poorly against ammonia, a common industrial hazard.

➢ Any type of respirator system requires that you be fitted for it and trained in its use.

Now let's cover how respirators are described or categorized.

There are **two** basic types of respirators. One is **Air-Supplying**, and the other is **Air-Purifying**.

1. **Air-Supplying** respirators are similar to scuba diving setups in that they **supply air from a reservoir** either through a long hose (like the deep sea divers of old), or through a tank carried by the individual wearing the mask (as with modern scuba gear). If you're a scuba diver, your tanks are an option, but don't mess around with this type of equipment unless you've been trained to use it. Compressed-air equipment and its use require experience, caution, and safe handling.

   The units using a **hose** are called **SARs** or **Supplied Air Respirators**, and the ones with **attached tanks** are called **SCBAs** or **Self Contained Breathing Apparatus**. Sometimes a combination of the two are used. An operator will carry a backup tank of air while trailing a hose from an outside source. One is there in case the other fails or runs out.

2. **Air-Purifying** respirators are divided into three categories, though most utilize filter cartridges of one type or another to do their job. Let's look at these categories:

> One type of air-purifying respirator is called a **particulate respirator** as it filters particles from the air. Simple particulate respirators are commonly called "dust masks." These masks will not help filter gases or vapors and will only filter chemicals if the droplets or particles are large enough to be physically trapped in the filter.

> The second type of air-purifying respirator is the **gas and vapor respirator** which uses a chemically treated filter to either absorb or neutralize a specific chemical or specific type of chemicals. Though rather effective, the gas and vapor respirators are limited to a certain concentration of chemical (and again, matched to the type of chemical – one type does not do all). Too much chemical, and they just can't handle the volume.

> The third type of air-purifying respirator is a **combination particulate filter and gas and vapor type**. This is the preferred type of mask to have if you don't know what you'll be running into.

Next, let's take a look at how these masks are **rated for effectiveness** so you'll know how to chose the right one if you decide to purchase a commercial model. (The "Sources" listed below will also give you more information.)

> <u>THE</u> authority on gas masks, respirators, protective gear and the like is the National Institute for Occupational Safety and Health, or **NIOSH**, and their National Personal Protective Technology Laboratory or **NPPTL**. See the NIOSH data on masks and respirators at http://www.cdc.gov/niosh/respinfo.html, or contact them at NIOSH, P.O. Box 18070, 626 Cochrans Mill Road, Pittsburgh, PA 15236, 800-356-4674, Fax: 513-533-8573. The alternate address for the NPPTL is National Personal Protective Technology Laboratory, 1095 Willowdale Road, Morgantown, WV 26505-2888.

> **OSHA**, the Occupational Safety and Health Administration also has some pertinent info on respirators. Go see http://www.osha.gov/SLTC/etools/respiratory/respirator_selection.html, or you can contact them directly at U.S. Department of Labor, Occupational Safety & Health Administration, 200 Constitution Avenue, Washington, D.C. 20210, 1-800-321-OSHA (6742). Also se their FAQ page at: http://www.osha.gov/SLTC/etools/respiratory/oshafiles/faq.html.

> Any respirator you choose should be listed as "**Approved under 42 CFR part 84" by NIOSH**."

> Some systems are rated "**evacuation only**." This is what you'd probably want in a disaster prep situation. This notation is to warn users who may be purchasing the equipment for professional, or prolonged exposure, HazMat use, that the gear is only rated for limited exposure.

Let's take a look at a list of **items you should look for** if considering a mask system. We'll start with a short list:

❏ Seal and fit ❏ Adjustability ❏ Anti-fog faceplate or goggles
❏ Cartridge selection ❏ Rating ❏ Protective hood
❏ Will accept eyeglasses ❏ Drinking adapter ❏ Cartridge and part availability

> **Seal and fit**: Make sure your mask properly fits your face and that it will hold to your face as you inhale and create suction. If the mask won't seal, you could draw in contaminants through gaps.

> **Adjustability**: One size does not fit all. Even if the face area seals properly, the setup will be useless if it falls off you. Make sure the unit you choose comes with suitable and adjustable head straps.

> **Anti-fog faceplate**: You can't evac if you can't see. Cheaper respirator systems do not have anti-fog capabilities. Though this feature might cost a tiny bit more, it's well worth it.

> **Cartridge selection**: If you don't know the specific substance you might be exposed to, the safest type mask would be a **combination** particulate filter and gas and vapor filter, with a **type "P" filter** that is impervious to **oils** and other **chemicals** that might adversely affect the filter itself.

> **Rating**: You'll notice different **"class" or "rating" numbers** for different masks. Different manufacturers use different terms, but these numbers refer to the **percentage of particulate matter** the mask filters out. These are usually 90%, 95%, and 99.7% (which is referred to as 100). Make sure the respirator you're considering has a rating, and that you get the highest possible. Any brand you consider should be listed with **NIOSH**, and you should verify the mask's rating through their website or by phone.

> **Protective hood**: The unit should have **full face and eye protection** as a minimum. Respirators that cover only the mouth and nose may protect your breathing, but leave your eyes and much of your facial skin vulnerable. Even better are the full face masks that come with an integral **hood**. The better units will have the mask held in place over your face with head straps and the hood will simply drape over everything. This is the best setup for men with beards.

> **Will accept eyeglasses**: As we said above, you can't evacuate if you can't see. If you wear glasses (even if you normally wear contacts), make sure the face plate is large enough to accommodate your frames. Some of the rather expensive models will mount your prescription lenses into the face plate portion.

- ➢ **Drinking adapter**: Most military MOPP systems have an adapter to allow you to insert a sealed drinking tube from your mouth, through the mask, into your canteen. This allows you to rehydrate without taking off your mask. It's a feature we recommend.
- ➢ **Cartridge and part availability:** Don't take advantage of a "sale" if a year down the road you can't find parts or additional cartridges. Similarly, get all the masks you'll ever buy at one time. This way, everyone's mask parts and cartridges are interchangeable.

As a minimal suggestion, take a look at some of the commercial "escape hoods." Two that come to mind (and we have not verified their NIOSH rating, we've simply located them as examples) are the "Evac-U8" hood, and the "Quick 4," both listed below under "Sources and More Information."

Some Alternatives

As we pointed out with our "GYT Pack," there are common items that can make fairly decent improvised dust masks. By now you should realize that these expedient items that make dust masks **do not make true respirators** and that anything improvised should be used only when there's **nothing else**. However, as we'll state over and over, improvised sure beats nothing at all.

If you're bored one day, you should go around the house and take note of basic household items that might work as an impromptu dust mask. Try some of the following:

- ❑ Oven mitts
- ❑ Towels and washcloths
- ❑ Cloth diapers
- ❑ Gauze pads
- ❑ Folded clothing

- ❑ Throw pillows or bed pillows
- ❑ Layers of paper towels
- ❑ "Baby wipes"
- ❑ Neck ties
- ❑ Comforters or quilts

To test these items, pick them up and breathe through them. They should provide a little resistance to air flow otherwise they're not filtering. If you can, try wetting them to make sure you can still breathe through them while wet. Some items might surprise you. Also, stay away from anything with chemicals of its own. This would include some pretreated "dust mop" covers, etc. You don't want to have an unexpected chemical reaction with the substance you're trying to filter, or an allergic reaction.

Sources and More Information

Each of the sites and sources listed below will have a mix of additional information and equipment for sale. As always, we include these so you can not only continue your education outside the boundaries of this book, but you can shop around for the best prices should you decide to purchase any of this gear.

- ➢ The Mine Safety Appliances Company, or MSA, line of protective products can be found online at http://www.msanet.com, or by contacting them at P.O. Box 426, Pittsburgh, PA 15230-0426, 1-800-672-2222, Fax: 412-967-3521.
- ➢ For Scott protection products, log on to http://www.scotthealthsafety.com, or contact them at P.O. Box 569, Monroe, NC 28111, 1-800-247-7275, 1-704-291-8300.
- ➢ ApprovedGasMasks.com, online at http://www.approvedgasmasks.com, seems to offer a complete line of retail protective masks, respirators, and suits. You can also reach them at Box 9509, San Diego, CA 92169, 1-877-AGM-1010 (877-246-1010).
- ➢ A commercial **fire-escape smoke hood, "Evac-U8"**: http://www.evac-u8.com, Brookdale International Systems Inc., 1-8755 Ash Street, Vancouver, B.C., Canada, V6P 6T3, 1-604-324-3822, 1-800-459-3822, Fax: 1-604-324-3821.
- ➢ Another escape hood is the **Quick4** mask found at http://www.airsecurity.com/quickmask4.htm.
- ➢ See the CDC discussion of respirators and its NIOSH links at http://www.cdc.gov/niosh/respinfo.html, http://www.cdc.gov/niosh/npptl/cbrncheck.html, http://www.cdc.gov/niosh/npptl/scbasite.html, and http://www.cdc.gov/niosh/npptl/respstdpg.html. Be sure to visit all four pages.
- ➢ Interactive Learning Paradigms, Inc. has a site at http://www.ilpi.com/terrorism/index.html that's worth seeing. Also visit the other terror site at http://www.ilpi.com/msds/ref/respirator.html, or you can contact them directly at ILPI, 4905 Waynes Blvd., Lexington, KY 40513-1469, Phone: 859-396-5218, Fax: 859-523-0606. Visit their sister site at http://www.safetyemporium.com.
- ➢ 3M, online at http://www.3m.com/market/safety/ohes2/html/respirators.html, has an extensive lineup of respirators and other safety equipment. Contact them by phone at 1-888-3M-HELPS (364-3577).

- American Airworks, online at http://www.americanairworks.com/scba.html, specializes in SCBA gear. Contact them directly at American Airworks, 209 East Main St., Sophia, WV 25921-1000, 800-523-7222, ext. 999, Fax: 304-683-3257.
- Neoterik, online at http://www.neoterik.com, boasts a good line of general first responder gear. Contact them directly at Neoterik Health Technologies, Inc., 401 S. Main St., PO Box 128, Woodsboro, MD 21798, P1: 301.845.2777, P2: 877.840.1469, FX: 301.845.2213.
- "Safety Square" Gas Hoods, Emergency Equipment and FAQs can be seen online at: http://store.yahoo.com/safetyproducts-store/index.html. While you're at this site, follow the links to a variety of other preparedness info.
- I.B.N. Protection Products at http://www.ibnprotection.com, offers a unique line of German manufacture respirators. Contact them directly at 77 Quaker Ridge Road, Suite 217, New Rochelle, New York 10804, Phone: 914-738-0400, Fax: 914-738-4474.
- Safety.com offers equipment as well as rating and selection criteria. See them at: http://www.safety.com, or contact them at SafetyHQ, Inc., 670 North Commercial Street, Manchester, NH 03101, Tel: 1-603-226-7233, Fax: 1-603-223-5003.
- Leader Safety, online at http://www.leadersafety.com, has a good lineup of a variety of safety gear including respiratory protection gear. Contact them at **Leader Safety Inc.,** 61 Victoria Road, Guelph, Ontario, Canada, N1E 5P7, Toll Free: 866-343-3389, Fax: (519)-763-1284.
- Safety Supplies, online-only at http://safetysupplies.com, has an offering of PPE as well as first aid gear.
- The Industrial Safety Company, at http://www.indlsafety.com, has a good line of respirators as well as some information about choosing the right type. They also have a good selection of other safety equipment. You can reach them at Industrial Safety Company, 1390 Neubrecht Rd., Lima, OH 45801-3196, 1-800-303-5995.
- Go to http://www.sefsc.noaa.gov/HTMLdocs/protectionequip.htm for good reference charts on eye protection and glove selection.
- A sort of respirator "consumer reports" is found at http://www.bobandjulie.com/gasmask.html.
- Even more of the same can be found at http://www.approvedgasmasks.com. You'll also find equipment for sale and a "recalled filter" consumer info link at the bottom of the page. Contact them at Approved Gas Masks, Box 9509, San Diego, CA 92169, Toll Free: 877-AGM-1010, Tel: 301-931-6700, Fax: 301-931-6655.
- Survive America has a good selection of "Gas Masks" and NBC products. They're online at http://www.surviveamerica.com, or contact them at Survive America, 4100 US Hwy 45 N, Bruce Crossing, MI 49912, (906)827-3600 phone, (906)827-3601 fax.
- The Earthquake Store, online at http://www.earthquakestore.com, offers a wide variety of gear to include respirators. You can reach them at Earthquakestore.com, 1300 Gardena Avenue, Glendale, CA 91204, Toll Free: (888) 442-2220, (323) 245-1111, Fax: (818) 240-1492.
- Visit the Oak Ridge National Laboratory's "Emergency Management Center" site at http://emc.ornl.gov, and click on "Publications." This will take you to a page offering free downloads on a variety of chemical info and personal protection information.
- A variety of masks and hoods can be found online at: http://www.ilcdover.com/Homeland/EscapeHoods.htm, and http://www.ilcdover.com/Homeland/PetShield.htm.
- Millennium Ark has an online comparison chart of a few gas mask models at http://www.millennium-ark.net/News_Files/NBC/Mask.Chart.html. They also have a listing of manufacturers and suppliers links at: http://www.millennium-ark.net/News_Files/NBC/Mask.dealers.html.
- For a variety of industrial masks and protective gear, log on to Web Soft Safety Solutions, at http://www.websoft-solutions.net, or call them at 866-272-1334.

"Moon Suits"

As an accessory to a gas mask and as an expansion of the previously mentioned "Dressing for Chemicals" section, let's discuss protective clothing specifically manufactured for the purpose of protection against biochemical substances. The official term for protective mask and clothing systems is "PPE" or "Personal Protective Equipment." Let's cover some of their pros and cons.

For the most part, *most* specially designed protective suits are impractical, both because of the high price of those that offer full protection, and the limited number of scenarios in which you'd need one. This is not to dissuade you from protection, but rather to make you think twice about spending top dollar for a top of the line suit that you might not carry with you. At least that's our opinion. We feel that a surprising amount of protection can be gained from simple solutions mentioned earlier. However, we'll let you decide. Go with what you feel you need, what you can afford, what you'll keep with you, and what makes you and yours feel secure.

How Suit Systems are Rated

Personal Protection Equipment is classified into levels **A, B, C, and D**:

> **Level A**: Is a fully enclosed, liquid and vapor protective unit providing the highest level of skin, respiratory and eye protection. This is similar to what you'd expect to see in a bio research lab or as part of a HazMat team's gear at a chemical manufacturing plant.

> **Level B**: A splash-resistant suit unit combined with the highest level of respiratory protection. Military MOPP gear would fall in at the bottom of this level.

> **Level C**: A splash-resistant unit with the same level of skin protection as Level B, used when the concentration(s) and type(s) of airborne substances(s) are known and the respirator matched to the specific hazard as a minimum. You might expect this level of protection in a hospital operating room, or in the paint room of an auto body shop.

> **Level D**: Used with no respiratory protection and where minimal skin protection is required. Recommended only when the atmosphere contains no known hazard and work functions preclude splashes, immersion, or the potential for unexpected inhalation of, or contact with, hazardous levels of any chemical. This would encompass our recommendations found under "Dressing for Chemicals."

Some Alternatives

If you want a more protective set of clothing that's extremely economical and easy to assemble (though slightly bulky to carry), consider the following:

1. Get a full plastic rain suit, one with a jacket and pants. Make sure the jacket has a hood. You can find them at pretty much any department store.

2. If the suit does not come so equipped, get a pair of rubber galoshes or rain boots. Though even bulkier, you could use a set of fishing waders instead of the pants and galoshes.

3. Get a pair of rubber household cleaning gloves and a pair of latex gloves.

4. Keep a full-face gas mask with these items.

5. Also keep a large, 3"-width, roll of duct tape.

6. The procedure is simple. Put on your gas mask, then put on the rain jacket pulling the hood up over your head. Put on the pants and then the galoshes. Put on the latex gloves and then the rubber gloves pulling the gloves over the ends of the jacket sleeves. Now seal each and every seam with the duct tape. This includes the rain suit hood where it meets the gas mask, even if the mask comes with its own hood. Cover all skin. Lastly, take the poncho that you should have in your GYT Pack, and pull it on over everything. Now... you'll find this **VERY** hot and stifling and you'll only be able to wear it for a short time, probably just long enough to get out and do what's needed. But... that's the way it's going to be with any full-body protective gear. Another **word of caution**; the plastic will only last so long against some chemicals, but better they eat at plastic for a bit than at your skin.

Shall we leave you with one more "left field" idea? If absolutely nothing else, and if you scuba dive, you can achieve some protection by wearing your diving mask, air tanks, and covering yourself with a clear plastic bag (so you can see). Again, this is not a rated system or officially recommended gear, it's just an option and food for thought.

Sources and More Information

> A good sampling of protective suits, and informative links can be found at: http://www.nbcprotect.com/.

> DuPont is a world leader in protective gear. See them online at http://personalprotection.dupont.com, or call them at 800-931-3456.

> Survive America has an equally varied collection of NBC protection products. They're online at http://www.surviveamerica.com, or you can call them at 906-827-3600.

> A short list of protective clothing goals, and an index of the chemicals it should protect you against are online at: http://www.cdc.gov/niosh/ncpc/ncpc1.html.

> OSHA's indexed discussion of protective clothing and its selection factors can be found online at http://www.osha-slc.gov/dts/osta/otm/otm_viii/otm_viii_1.html.

> The Mine Safety Appliances Company, has a great online PPE selector site at: http://www.msanet.com/response. You can also contact them directly at Mine Safety Appliances Company, PO Box 426, Pittsburgh, PA 15230, 1-800-MSA-2222.

> The OSHA discussion of protective clothing and decontamination, along with other useful info, can be found online at http://www.osha-slc.gov/dts/osta/otm/otm_viii/otm_viii_1.html.

The Safety Necklace

Finishing out our section on PPE, and bridging the gap between equipment and reactions, let's look at a piece of gear that's **particularly important to children** who aren't in the habit of carrying wallets, purses, or GYT Packs, and who might be involved in emergencies where the family is separated. Actually, each family member should have one of these. The "Safety Necklace" is an extension of the "Dog Tags" concept, or an expansion of the old "my name is ___, if lost, please call __" note pinned to your child's shirt. It's a simple necklace on a **breakaway lanyard** (for safety) containing personal identification info, as well as a couple of items that may prove useful in an emergency. Make sure yours contains the following:

> Whistle for signaling.

> Small chemlight stick and/or battery-operated "blinkie" light, for signaling and safety.

> Squeeze activated LCD light, or small flashlight, to see in the dark.

> Copy of personal identification and "Notify in Case of Emergency" card (see "**Appendix**").

> A "Micro Data Card" (see "Info Pack" under "Planning Ahead for Phase II").

> Pertinent medical information (see "Family Member Data" forms in the "**Appendix**").

A child's Safety Necklace should provide safety and signaling devices along with info and ID on a breakaway neck strap.

Kids should have one necklace for school and one for home. The one at home should be kept near any bugout gear they keep, or simply accessible on a door knob or coat hook. They should be taught to grab it and put it on during any and all drills. For adults, keep yours in places it would be accessible should you be separated from your wallet, purse, or GYT Pack (such as with your "bolt clothes" that we'll discuss under "**Evacuation**").

Hint: The micro data card for kids to carry with them should at least contain the following:

> A reminder that if they're at school when an emergency hits to do what their teacher tells them and to stick with teachers and classmates.

> A list of their rendezvous points should they be away from school.

> A reminder of how, when, and why to call 911.

> A short list of "Do's" and "Don'ts" based on your family's reaction plan for different emergencies.

> "Find Me" information for parents and older siblings.

> Contact information for the family's "Notify in Case of Emergency" contact people.

> A list of people authorized to come get them in an emergency (your "Minor Child Release" form recipients.)

Hint #2: Use a "Find Me" card from the "**Appendix**" to fill out with your contact and evac destination info and put it on your pet's collar. Should your pet become lost somewhere along your trip, people finding the animal can find you.

Let's move our discussion to the drills you need to practice, where Safety Necklaces might be put to use.

The Drill to Survive

Do not underestimate the effectiveness of drills. Drills are extremely important, and participating in them will give family members, especially children, a sense of safety and well-being through preparedness and inclusion. The same holds true for coworkers, students, volunteer group members, and anyone else that participates in regular safety drills as a part of school or job. This sense of safety and well-being will do worlds of good in mitigating some of the stress and trauma *after* a disaster as well, since some of this stress is brought on *during* an emergency by the adrenaline released when you realize you really don't know what to do. You should practice each of these drills now to work out any "kinks" and you should repeat them each time the terror threat alert changes, each time the season changes, whenever your area is in its severe weather season, or if everyone happens to be home and is bored one night.

Essentially, there are two basic types of drills. There are the **evacuation drills**, the most common of which is the fire drill, and there are the **shelter drills** or "duck and cover" drills, such as a tornado drill. As we'll see below, each is useful for preparedness against other eventualities.

The Evacuation Drills

The Fire Drill will suffice for our evacuation drills as it contains elements of gathering the "three Ps," People, Pets, and Property, and gets everyone out of the house in a rapid and safe manner. This type of drill will help prepare you for other rapid evacuations such as a pending chemical accident or attack scenario, a breach in a dam, a bomb scare, approaching wildfire, a hurricane that suddenly changes course, etc.

Fire Drill

The fire drill is an evacuation-type drill, in that you're practicing to flee an immediate area to avoid a danger. This type of drill is extremely useful in practicing evacuation in general. Conduct the drill in a "fun but serious" mood, especially if children are involved. They'll need to know the drill is important, but performing it should never instill fear or dread of either the drill itself or the event it's representing. Children should respect the danger, but not be a victim of fear or anxiety. Some people recommend simply "talking through" the drill, but as we believe experience is always the best teacher, we recommend actually performing the items listed below.

Setup: Before performing any actual drill, complete the following:

1. Before springing all this on the whole family, the adult heads of household should lay out the entire plan in detail and go through it a few times before involving others. You want to make sure you're in agreement with the whole procedure, and that you have the process worked out well enough so that you present a unified and confident image to the rest of the family, and so that smaller children, who will record their first lessons most strongly, are taught the correct procedure the first time around.

2. Speaking of younger children, when you're going through any drill or set of instructions, teach them to respond to **simple one-word or two-word commands**. More than that and you may confuse them. Similarly, for the elderly or forgetful, you might want to create a simple, large-print, instruction sheet for what to do in an emergency, and keep it prominently displayed in their room, in similar fashion to the safety info you'd find in a hotel room.

3. Have everyone sit down together and fill out the **"Safety Blueprint"** found in the **"Appendix."** If you have a regular babysitter, nurse, hospice, etc., invite them over to participate in filling out the blueprint and to participate in a fire drill. It's important that they be familiar with your emergency equipment and procedures.

4. Make sure everyone knows the **difference** between the way an **automatic alarm** and a **manually activated alarm** sound. The difference is important since an airhorn or whistle means someone else knows about the fire, but a smoke detector or automatic fire alarm going off means the family was caught off guard.

5. Make sure all children know the "symptoms" of fire; smoke, the smell of something burning, heat, etc., and let them know that it's okay to wake up a grown-up and have them check out the threat and that they won't get in trouble for doing so.

6. Go room to room and have children identify **escape routes** and **demonstrate** that they can actually use them. If the escape route involves a window and an escape ladder, have the child open the window, lower the ladder and climb down (under supervision of course). Some people recommend drawing this or going over it verbally, but experience is <u>always</u> the best teacher.

7. Have everyone demonstrate where they would **hide for protection** if they could not get out of the room. If someone can't get out of their room, they should stay on the floor nearest the exit that's *away* from the fire. For instance, if the fire is in the center of the house, the trapped person should stay on the floor of their room near an exterior window. If the fire is outside the house (such as in an approaching wildfire) and starting to burn in, the trapped person should stay on the floor near the closed but unlocked interior door. These positions put them away from the fire, and nearest the entry point rescue personnel would use to come get them. Point out that hiding under the bed or in a closet only places them near flammable materials and hides them from rescuers.

8. Teach everyone what you learned under "Gas Masks" about which household items make **expedient dust masks**. All family members should be familiar with all available methods of protecting their face and lungs.

9. Teach everyone how to **seal around doors** using tape on the seams and a towel or other cloth along the bottom, in case the fire is burning from the inside out. With that, point out to not break windows unless absolutely necessary. Broken glass is dangerous, and depending on the smoke flow, you may need to close the window.

10. Have all family members go room to room and point out other **safety equipment** such as telephones, flashlights, and all equipment listed on the "Safety Blueprint." In addition to the safety gear, have them point out any secondary **sources of water** (in case the main is shut off) such as "snow globes," flower vases, soft drinks, toilet tanks, fish tanks, etc. This water may be necessary to wet down clothing, towels to seal under doors, or to wet cloths to breath through in a heavy smoke situation.

11. Have all family members demonstrate how they would **alert the rest of the family** in case of fire. This should include using the whistles, bells, or airhorns while maintaining good personal safety.

12. Make sure children know how, why, and when (and when <u>not</u>) to **call 911**, and where the family's "**Important Contacts – Quick List**" sheet is posted showing alternate emergency numbers and family contact info.

13. Have children repeat to you, your home phone number, street address, household rendezvous point, and other basic information from memory. Turn it into a fun game and see if you can rattle them in a good-natured way to see if they forget anything. The better they remember this info, the better they'll be able to **repeat it to authorities** if they need to call 911.

14. Make sure all family members know how, why, and when to **activate emergency equipment** such as your alarm system's panic button, or your manually operated fire sprinkler system. Discuss different scenarios in which each might be needed.

15. Similarly, make sure each family member knows how to **shut off the gas, turn off the HVAC system, and turn on the water** (if it's off for whatever reason), or at least how to quickly trouble-shoot the water if there seems to not be any. Also, if you have alternate water pumping, or electrical generators, make sure all able family members know how to safely operate those.

16. Pick a **rendezvous point** outside your home where all family members should meet if they evacuated separately. Pick a spot that's easily recognizable to all members of the household and that's easy to describe to authorities. This point should be far enough away for safety if the house is burning, and free of obstacles or safety hazards.

17. Before starting actual drills, **talk through different scenarios** with the family to get everyone's input and to make sure everyone understands the importance of the drills. Discuss how to react if fires started at different times or in different areas. Point out that pets and some safety gear might be gathered but nothing else. Also point out that in some cases, a fire may be so far along that only people can be saved. Discuss all scenarios and associated options and ramifications.

Execution: When performing the actual drills, consider and include the following:

1. The most dangerous time for fires to occur is at night when everyone's asleep, so start your drill with everyone in their bedrooms since you always want to practice for the worst case scenario. Some of the dangers of nighttime fires are that everyone is in separate rooms, they're all asleep and may wake disoriented, it's dark, and everyone will be wearing less clothing than normal.

2. Use the actual device you'd use to warn of fire. Don't pretend. You want everyone to learn the sound of the smoke alarm, airhorn, whistle, or bell and associate it with emergency action. Add a verbal command as well, such as, "fire drill!" In fact, you might consider using the same "drill" command during an actual fire to get the same reactions from family members but without stress. (In a real fire, use the less threatening command with small children and the real thing with teenagers.) Also, if doing this drill at night, use your alarm quietly or at least tell the neighbors it's only a drill.

3. Regardless of the scenario, some factors will remain the same. Therefore, **in each drill, include the following**:

    ❑ On hearing an automatic fire alarm, do your part to **warn others**. When hearing a human-activated alarm, call out that you heard the alarm and ask for further instructions. (This is where parents would tell children to either go out their window, gather in the hallway immediately, or buddy up, get the pets, and grab their individual "Bugout Kit.")

    ❑ Check doors and door knobs for heat before opening them. **Use the back of your hand** for this and **never the palm**. Burns to palms and fingers can be debilitating and hamper further escape efforts.

    ❑ Even if you don't feel heat, you should stay low and brace against the door while opening it slowly.

    ❑ As you open the door, close your eyes and hold your breath even if you didn't feel heat or smell smoke. Fire can be tricky, and you don't want a sudden back-draft or flash, burning your eyes or lungs.

    ❑ If you smell smoke, low crawl along the floor to stay under the smoke. Smoke rises and breathable air will be close to the floor. Stay about 12 inches or lower to the floor.

    ❑ Use the buddy system, especially with larger families, but don't pair young children together. Pair younger children with adults or at least older siblings. Anyone paired with an elderly person should be physically capable of carrying them out of harm's way if necessary.

    ❑ **Each pet** should be assigned to a specific person. It will be that person's job to bring that pet to safety (exercising reasonable caution and time limits) *after* pairing up with their "buddy."

    ❑ Consider how to handle each pet if alternate escape routes such as windows have to be used. (For example, one person may go down the escape ladder and the family dog wrapped in a bed sheet and lowered down by the remaining buddy before that person evacs.)

    ❑ Each able person should have a role to play, but the heads of household have the final responsibility of making sure everyone's safe and that everything's been done in a safe and timely fashion.

- ❏ When people leave a room they should shut the door behind them to cut off a fire's air supply, but they should NOT *lock* the door. Fire fighters will need full access to the house.
- ❏ Your HVAC system should also be turned off if possible to prevent air circulation that feeds the flame.
- ❏ The family "Bugout Kit" should be retrieved during each drill and carried outside to the rendezvous point, since in a real fire, or any other household evacuation, you'd want your kit with you if it was at all accessible.

4. For the drill itself, make sure everyone knows it's about to commence. You don't want to unduly frighten anyone or "cry wolf." You do, however, want to change the scenario for each drill so the reaction will be slightly different. In any event, alter the drill and alter the fire location scenario.

5. For one drill, have everyone check their bedroom door for heat (always with the back of the hand), and then low crawl into the hallway where the family will meet and walk orderly outside to the rendezvous point.

6. For another drill, have everyone go out their windows (one at a time for drill safety), or at least open their windows and set up any escape equipment, such as a rolled-up escape ladder.

7. Have enough drills to where each family member gets to be the one who discovers there's a fire and sounds the alarm for the others. Some drills will have parents calling out commands, while in others the children should react on their own and alert the adults (save these for last until children get a better handle on what's going on).

8. Have each person practice being "trapped," and actually demonstrate how they would position themselves while waiting for rescue. (Teenagers may rebel at this but could probably be made to participate if they believed *they* were doing the teaching. It's also a good idea to remind them they could be a true hero in an emergency by saving the lives of family members.)

9. Make each person do at least one fire drill blindfolded as in a real fire they might not be able to see because of smoke and/or because the power might be out.

10. Have each person practice carrying their buddy. Make sure everyone knows the different ways to carry someone whether it's a simple "pick them up and go," or a "Fireman's Carry" while low-crawling.

11. Have each able person demonstrate how their respirator works. If they don't have a respirator, have them demonstrate which common household items would make a suitable alternative.

12. If the scenario allows, adults should grab their "bolt clothes", keys, wallet, etc., and children should put on their "**Safety Necklaces**" (as discussed above under "Gas Masks and Moon Suits"). Each should grab any "Level 1 Bugout Kit" stored in their room. If accessible on the way out, the family's main "Bugout Kit" should be taken as well. Reiterate the fact that personal safety takes priority over gear. Let children know not to go for their Bugout Kits unless instructed by parents. (More on all these items under "**Evacuation**.")

13. The most important consideration of all in these drills is **safe speed**. Everyone should be gathered and out of the house, or vice versa, **in ONE minute or less** while ensuring safety. It would be a shame to *almost* escape a fire but fall and break a leg while running. This time limit includes the time necessary to round up pets, for everyone to grab their individual Bugout Kits, and for someone to grab the main family Bugout Kit. (**WARNING**: **A fire will double in size every 30 seconds!**)

14. Once everyone is out safely and within time limit goals, discuss each of the following steps that would take place during a real fire:
- ❏ Alert any neighbors whose home is attached to, or near yours, and if 911 has not already been called, you would do so from the neighbor's. Do NOT go back inside the house to call.
- ❏ Decide if the fire is small enough to fight or if you can safely impede its progress from outside the house while the Fire Department is enroute. Otherwise, do NOT go back inside.
- ❏ If the valve is safely accessible and you have a separate gas line running to your home (from pipe or tank) remind everyone that it might be necessary to shut off the supply. If you never got a chance to turn off the HVAC system, you should know how to yank the electric meter from outside the house to shut off power. Have your electric company's meter reader show you how to do this safely.
- ❏ The last thing to do after people are safe is to move your car away from your burning home if it's in danger of burning, and provided you can reach it safely.

15. Have each family member show where the rendezvous point is outside the home, recite why that spot was chosen and why a rendezvous point is important. (This also helps make sure the spot is accessible to everyone.)

16. After everyone is thoroughly familiar with the different exit scenarios and knows how to respond to a fire, start teaching two advanced concepts. One is how and when to operate the household fire extinguishers and the second is the "stop drop and roll" reactions to clothing on fire.

17. On occasion you should extend the fire drill to be a **full evacuation readiness drill**. In this instance you should go through the whole process of looking over your "Last Minute List," gathering your Bugout Kits, and actually load them, people, and pets into the car. (More on this under "**Evacuation**.") This way you'll not only conduct a fire drill, but loading the car will give you a little practice for the sake of speed and may uncover some unforeseen stumbling blocks to your evacuation procedures or plans. **Very important!**

The Shelter Drills ——

Shelter drills are the opposite of evacuation drills in that their objective is to get the family **back inside** the house and huddled together in a safe area as safely and quickly as possible. Though more applicable to natural disasters than terrorist attacks, there still exists some counter-terror benefit to these drills in such instances as a nuclear attack with advance warning, some chemical accident or attack scenarios, prowler in the neighborhood, etc. For the purposes of "planning without panic" though, we'll focus on the natural disaster applications of these drills. We'll look at the two types of "gather and shelter" which involve heading for "hardened low ground" such as in a tornado (or nuclear detonation), and their distant cousin, "duck and cover," such as in an earthquake (or again, a nuclear detonation).

Tornado Drill

The Tornado Drill is a "**reverse fire drill**" aimed at gathering family members together and heading to your safe room or safe area. Though a tornado is just as destructive as fire, it's simpler in a few regards as it starts outside the house, it's usually preceded by some sort of warning, and these safety measures are proactive, before-the-fact actions, rather than any sort of hurried reaction (hopefully). Tornado drills serve a couple of other useful purposes as well. As we stated before, they're an innocuous way to practice one possible reaction to the advance warning of a nuclear detonation, etc. Also, it's also your chance to teach children how to tell the difference between average storms and dangerous ones, so they won't be afraid of normal weather. Consider the following:

Setup: Before performing any actual drill, cover the following:

1. Hopefully, you followed our advice earlier under "Basic Home Prep" and included the whole family in preparing any safe rooms or safe areas. Not only will that give them an important role in helping with the family's safety preparations, it will stress the importance of the safe room and will increase their confidence in that spot's ability to protect them in an emergency.

2. Teach children the possible warning signs of tornadoes and teach them to report in during foul weather if they're playing at a friend's. Given our current state of affairs, it's more important than ever to always know where your children are anyway.

3. Pets can be trained to head to the safe room on command. You'll have too much to do during an emergency to go round up the animals. The training is fairly simple. Hide a generous amount of their favorite treats in the safe room. Go find your pet and lead them back to the safe room with a single treat, while repeating the command, "Tornado!" (or "Apocalypse!," or "Safe Room!" or whatever command you want to use) over and over as you go. Do this several days in a row, then drop it back to once every few weeks and also throw in a few instances where you give the command but no treats are provided. Once in a while you may want to give them a toy and play a little instead of providing food treats. You want to train your pet but not burn it out on the trick. Your local pet store or pet trainer might have some additional ideas, and yes, this will even work for cats. By the way, you should consider leash training your cats. Get a little harness for them and get them used to it.

4. If training won't work and you have one or more small children, and/or one or more untrained pets, get an inexpensive baby gate for your safe room or area. If you get the kind that expands against the door frame for support, it won't need to be permanently mounted. This way, you can grab the kids, put them in the safe room, and if you have time, round up the pets and drop them behind the gate so no one wanders off again. You might consider getting two gates for height since most cats and some dogs can jump a single gate.

5. Decide on a distinct alarm, such as quick blasts on an airhorn or whistle, and voice commands for a tornado, so everyone will know to react differently than they would for a fire. Teach the kids that any time they hear this signal, even while out playing, it means to come home <u>immediately</u>.

6. Remind everyone that not all steps will be followed, only those dictated by an actual event. For example, during the drill you would not include closing storm shutters because that should have been done as weather got worse, or when your weather-alert radio or weather channel issued a tornado watch.

7. Think of another excuse for having this drill if your area rarely has tornadoes.

Execution: Execution is simple since most of it only involves going from one room to another with few deviations. Include the following:

1. For the drill itself, you'll want to practice the second-worst-case-scenario which would be an incident while everyone was sleeping, just as with the fire drill. (The worst case scenario would see children playing with friends away from the home, or family members separated and away from shelter.)

2. Similar to the fire drill, take turns so each family member can be "it" and notify the others. You never know who will be the one to see the alert on TV, spot the funnel cloud, or hear the local sirens.

3. On command, everyone should pair up with their "buddy," grab their personal safety kit, any pet that's their assigned responsibility (unless it's trained), close (but don't lock) their room doors behind them, and gather in the safe room or area.

4. Heads of household should close and lock exterior doors and windows, close storm shutters if you have them, and check behind other family members.

5. Once it's clear that everyone is aware of the "emergency" and reacting accordingly, parents should gather their important personal goods such as wallets, purses, car keys, and cell phones, etc. (which should be kept in a basket or at least together in the same spot).

6. Once everyone's in the safe room and accounted for, sit down and huddle in the center of the area and close any doors. If there are any windows in your safe area, react as you would in a real tornado and cover them.

7. If your safe area is an outside storm cellar, make sure everyone is able to open and close the shelter.

8. Place pets on a leash or in a carrier (which should either be stored in the safe room or gathered with the pet). This makes nervous pets easier to manage, and all pets easier to rescue or evacuate.

9. All of these actions should be completed in 30 seconds or less. (Alter the scenario from time to time and act as if a tornado had touched down with <u>no</u> warning and everyone had to be in the safe area in **UNDER <u>five</u> seconds.**)

10. While still huddled in the safe room during the drill, use the time together to discuss what to do next had this been an actual emergency. This would also be a good opportunity to subtly mention that this type of reaction would be useful in other scenarios besides violent weather. This is a good way to plan without panic.

11. Reiterate the fact that the kids might hear the "tornado" signal while out playing and that it means to come inside immediately and tell others. Also discuss communication and signaling options and techniques should anyone be trapped in another room, or in the safe room, should part of the house be destroyed. (See "Communication.")

12. To expand the drill, locate places in the yard, neighborhood, or while out traveling, that would offer protection from tornadoes. Look for places such as culverts, ditches, etc. This will be explained in more detail in the "Tornado" portion of "Weather Disasters" below.

13. Similarly, go from room to room like you did in the fire drill and have each family member show where they would go during a tornado if they were trapped in the room or if access to the family safe area was blocked for some reason. Also, have them choose the best alternate room if the safe room is blocked, and describe why they chose that particular room.

14. As a final expansion of the drill, discuss performing all the same steps except moving to a safe spot on the *highest* floor of the house in case you were reacting to a flash flood or chemical release. (If you missed it, see the safe room discussions under "Basic Home Prep.")

Earthquake Drill

The Earthquake drill is a close cousin to the tornado drill but much more rapid in its reaction and less dependent on any gathering of the family, or taking time to head to any one location. The earthquake "drill" is less drill and more education, planning, and brainstorming.

Setup and Execution are the same here. Consider the following:

1. Again, we hope that you included all family members in any earthquake-proofing work you did around the house. It helps drive home the danger of natural disasters and the importance of preplanning.

2. Have each family member go room to room and demonstrate both the hazards they may encounter in that room, and the safe spots in each room that would offer them protection from falling objects. Also, have them demonstrate "duck and cover" by lying low and using their arms to cover their head and neck in order to protect themselves from falling debris.

3. Discuss how to spot safe areas and potential earthquake dangers while at work, school, or traveling.

4. Drive home the point that reactions in an earthquake must be immediate.

5. Point out that if caught in the open near buildings, they should look up to keep an eye out for falling debris. In earthquakes most threats come from above, others from the side. Teach everyone that avoidance is just as beneficial, if not more so, than a universal "duck and cover" reaction. Use this as a reminder; you always look both ways before crossing the street. It's better to avoid the car than to rely on a "duck and cover" move to protect you from impact.

6. Make sure each able family member knows how to shut off the utilities. (Don't actually do it unless you're proficient in relighting pilot lights.)

7. Make sure children know how, when, and why to call 911.

8. Make sure all family members are schooled in various communication and signaling methods should they become trapped after an earthquake. (See "Communication.")

These drills are here to set the stage for the specific reaction items to follow. We've given them to you early because the things you'll learn while doing drills will apply in a number of situations. Under the remaining portions of this "**Reaction**" section we'll go into more detail about reacting to specific events.

Next, let's cover another educational tidbit that may be required in a number of different situations as well. Let's talk about the basics of **decontamination**.

Basic Personal Decontamination

Let's assume you were exposed to an unknown chemical substance, that you were able to put on your protective gear and get out of the immediate vicinity, and that you suffered no immediate ill effects. Once out of the danger area, you may be directed to an official decontamination site. If that's the case, you'll be told what to do. If the situation is too large, officials too far away, or other circumstances dictate that help won't be coming, you may have to start decontamination procedures on your own. If on your own, make sure you follow these simple procedures:

Simple Personal Decontamination

1. The most important point to make is the warning that **you probably won't know what you were exposed to** until well after time to begin decontamination. The steps below should prove sufficient to help clean you up after exposure to pretty much anything. However, since you won't know what you're dealing with, err to the side of caution. Again, we're assuming you're on your own for whatever reason and you <u>have</u> to decontaminate without the aid of officials.

2. The second most important point to make is to not panic. The vast majority of biochemical weapons or other harmful substances can be tackled by simple means. So if you get a mystery substance on you, don't worry about what it might be. Focus instead on the procedures involved in safely getting it off you and neutralizing it to a certain degree.

3. As soon as you're able, you'll want to start precleaning. If you're exposed to a liquid, start blotting off excess from any exposed <u>skin</u> with a cloth, or one of your extra gauze pads from your GYT Pack. You can also use your handkerchief, gym socks, moist towelettes, "baby wipes," dirt, crumbled up soda crackers, or any of the other "absorbing agent" items we've listed below under "Personal Decon Alternatives." Clean your hands first, put on your latex gloves and then clean other exposed skin areas. Don't worry so much about excess liquid on clothing until your skin is taken care of. Remember, **blot, don't rub**, and be sure you're wearing your latex gloves. **Note**: If your eyes were exposed, or your eyes are starting to hurt, rinse them before other exposed skin areas.

4. To start full decon, stay away from areas you do not wish to contaminate (such as other people, or inside your car or home), but find an area protected from further contamination, and away from the danger area. Try to at least stay upwind and uphill from an affected area. Otherwise you'll just contaminate others or be recontaminated after cleaning up. You'll also want adequate ventilation since some chemical agents may be aerosolized by the cleanup process and remain airborne in the immediate vicinity until carried off by fresh air.

5. For adequate decontamination you'll need access to water, soap, good drainage, or preferably, something to collect the dirty water in (such as a small plastic "kiddie" pool). You'll also need at least three large garbage bags, duct tape to seal the bags, resealable plastic bags for personal belongings, towels to dry off with (optional), and fresh clothing to change into. The garbage bags, duct tape, resealable plastic bags, soap, and poncho to change into later should all be in your GYT Pack.

6. In addition to those items, keep your simple sources in mind. If you have an uncontaminated person to help you and you're in an urban environment, remember that you can get soap from most public restrooms, clothes detergent from laundromats, all manner of soaps and bleach from nearby grocery stores, convenience stores, etc., and pool chlorine from pool and spa supply stores or hardware stores. You can also make a great setup at home by using liquid dishwashing soap mixed with water in one of your clean garden sprayers (the ones you bought to use as fire extinguishers).

7. Speed is preferable to equipment, so if all you have is water, go ahead and start. (If you don't have water, don't rule out simple things like soft drinks, water-unit fire extinguishers, etc.)

8. If you're wearing your smoke hood and poncho, keep them in place and have someone hose you off from as great a distance as possible (and in fresh air) using a standard garden hose with sprayer. (**Note**: At this point, if the chemical starts to **react with the water**, such as emitting smoke, bubbling, "fizzing," or the victim complains of burning, stop what you're doing and use an absorbing agent listed below.) Be sure you're standing in a good drainage area such as near a storm or sewer drain, or in a reservoir that will collect the contaminants for later disposal by officials (the kiddie pool).

9. Regarding **water**, if you have a choice, don't use water so cold that it may cause its own problems (such as with an outdoor decon in the middle of winter) and don't make it so hot that it could either scald the victim, or that it

could cause the skin's pores to open and allow more chemical in through the skin. Tepid or luke-warm water is perfect; warm enough to be comfortable and to dissolve substances, but not too cold and not too hot for people.

10. Regarding **soap**, if you have a choice, use a brand that is hypoallergenic and as generic as possible. It's unknown how fragrances, color dyes, or other additives will react with the substance you were exposed to. Also, try to stay away from brands with abrasives or exfolients mixed in. You want to clean your skin, not abrade the skin and allow chemicals to absorb into your system.

11. If contaminated by a **dry substance**, and you have limited water, keep your respirator on and start carefully removing outer clothing and folding it outside-in to contain the dust, and then place it in a plastic garbage bag. You can then use a household vacuum cleaner to remove any dust from your skin, hair, or remaining clothing, provided you then double-bag the whole vacuum cleaner for later disposal. (Don't let an unprotected person use a vacuum cleaner on you as it will kick up the dust and may contaminate them.) You can also start wiping dust off yourself using a damp cloth or towel, or duct or masking tape reversed as you would when cleaning lint off a suit. Bag the towels or tape just as you would the vacuum cleaner and outer clothing.

12. After initial cleaning, remove your poncho and outer layer of clothing being careful not to drag clothing over your face. You might cut off some garments to avoid the chance of getting dust or other contaminants in your eyes, ears, nose, or mouth. Do as much clothing removal as you can with your respirator or breathing filter in place.

13. Remove personal effects from your pockets and put them in a separate resealable plastic bag. Place your removed outer layer of clothing in a garbage bag and seal the bag with duct tape.

14. Remove the rest of your clothing. Always start from top to bottom when removing any clothing, and don't pull anything over your head.

15. Place the remainder of your clothing in another garbage bag but leave it open. More will go in it.

16. If your hair was exposed to the chemical but wasn't hosed or vacuumed, you should now comb out any dust or blot any liquid, and then start direct decontamination by washing and rinsing your hair. Lean back while combing and rinsing so excess water runs down your back and not your face. If your respirator can stay in place as you wash your hair, all the better. Rinse your hair well when done.

17. You can probably take off your mask or respirator now.

18. Next, clean your face, eyes, and ears. Wash your face but be sure to wash away from your eyes, nose, and mouth. Rinse the same way. Alcohol prep pads are great for this as alcohol helps kill biological materials and acts as a solvent to many chemicals. However, the water rule still applies. If a reaction to the alcohol takes place, stop using it.

19. If you wear contacts and your eyes were exposed, you'll want to discard those lenses. Rinse your eyes with plain water or simple saline solution for at least 10 minutes. Rinse each eye by leaning your head to the side and irrigate from the nose side of the eye outward, where the waste water drips off your face and not down your face or into the other eye. You can also lean your head straight back and irrigate both eyes from the nose area outward (after you've cleaned the bridge of your nose).

20. If your ears were exposed, gently clean them with clean cotton swabs using a fresh one for each ear.

21. Using generous amounts of soap and water, wash, rinse, and repeat over every square inch of your body starting from the top down and from inside to outside. For instance you'd wash your arm from the shoulder outward and downward to the fingers and not the other way around. Scrub in one direction only (outward and downward) and not back and forth. Scrub hard enough to clean, but not hard enough to abrade the skin which could allow residual chemical a way to cause further damage.

22. If you wind up receiving information about the chemical or biological agent you were exposed to, you may be able to take appropriate steps to help neutralize the substance. This should be secondary to the soap and water treatment. If the chemical is acidic in nature, baking soda may help neutralize it. If it's base or alkaline in nature, vinegar may help. If it's a biological agent, alcohol, peroxide, or a 10-percent bleach solution (**1 part unscented non-soap household bleach to 10 parts water – you do NOT want pure bleach on your skin, and you don't want to use bleach unless in you're in a well ventilated area**), or even a topical application of Listerine may help kill some germs. DON'T play chemist or doctor unless you know for sure and certain what the substance is and how it will react with household remedies. See "**Personal Decon Alternatives**" below.

23. If you're using a water collection setup such as the kiddie pool suggestion, step out of the dirty water, on the opposite side from where you got in, washing and rinsing your feet as you do. Wash and rinse your feet again outside the pool. Dry off from top to bottom and from inside outward.

24. Put the wet towels and any washcloths in the bag with your clothing.

25. You can now close the bag and seal with duct tape. Put this garbage bag and the first one together in a third bag and also seal that bag with duct tape.

26. You should be able to get dressed now. **Hint:** You may want to carry a second poncho in your GYT Pack in case you have to decontaminate and you don't have spare clothing to change into.

27. If you decontaminated near a storm or sewer drain, it's okay. More contaminant will be flowing into the sewage system from general environmental runoff after an attack than would come off of you. Even the EPA expects this and has allowed municipalities temporary dispensation from pollution standards in the wake of certain hazardous materials incidents or attacks. The only reason you would want to use the kiddie pool setup is if you had no access to any sort of drainage, and you needed officials to come remove the contaminated water. Cover the pool to keep children or animals from getting into the waste water. **Hint**: Your friendly neighborhood open-bay car wash (the kind with the pressure hose, not the brushes) can be used as an expedient decontamination station. Just be sure not to get too close to skin with the pressure hose, and only use soap and water (not wax).

28. Leave your bagged clothing, (the vacuum cleaner if you used one) and the dirty water together, and call local officials to advise them of your location and situation, and your mode of travel, if any, since that vehicle may also need to be decontaminated.

29. Now that you're cleaned up enough to travel, ask officials where they'd like you to go for any testing to detect residual contaminants or for medical observation. In any event, definitely see a medical professional as soon as possible, even if you think the decon was successful and you don't feel any ill effects. Also, don't travel in the same vehicle that got you out of harm's way as it will more than likely recontaminate you. **Note**: The jury is out on exactly how to handle private transportation out of a contaminated area. One school of thought says you should roll down the windows once you're out of the danger area, turn the fan on high, and set the climate controls to "fresh air," to give you a lesser concentration of residual contaminant in the vehicle posing less danger to you while driving. The rationale is that there would not be enough chemical coming off the average person to seriously contaminate other areas. The other school of thought is that if you have plenty of personal protection in place, you should roll your windows up, and turn your HVAC system off, in order to prevent the transmission of any contaminant from your vehicle to bystanders. Base your decision on how much personal protection you have in place, the type of area and how many bystanders might be present while you drive away, on how you may be feeling and whether or not the chemical seems to be affecting you, and on what the substance might be (if you can even take a guess). Focus on your safety first, but the safety of others should come in a very close second.

Personal Decon Alternatives

The hose, garbage bags, soap, and all the things listed above are the basics. Let's cover some alternate components to remember when self-decontamination is absolutely necessary.

Let's look at **simple cleansers, advanced cleansers, absorbing agents, and disinfectants**:

➢ **Simple cleansers** are things like soap or bleach for skin, and simple saline solution for eyes. They're common items that will help wash off whatever it is that got on you. The only drawback is that you won't know exactly what it is that you were exposed to and you won't know how your cleanser is going to react with the agent until you try it. Go ahead and try *something* since even a small adverse reaction is better than waiting too long. Time is of the essence, so don't forget simple things like pre-moistened towelettes or "baby wipes" that can be used to decontaminate until better stuff is found. Similarly, since the best decontamination usually involves water, let's discuss some **alternative sources of water** if the nearest faucet isn't cooperating:

- Bottled water or soft drinks.

- Drinking fountains or water cooler jugs.

- Aquariums.

- Water-unit fire extinguishers. (Provided there's absolutely NO threat of fire in this incident.)

- Toilet tanks.

- Hallway fire hoses. (**Warning**: This should be used for group decon only and the hose should be handled by at least **two** strong people as it will tend to sling the average person around like it was a runaway horse. Spray a wide-angle spray upwards and let the water fall like a shower. Don't aim a stream or a wide-angle spray at anyone or you'll only knock them down and injure them.)

- Overhead fire sprinklers (see "Fire" below). Only do this in a group or severe emergency decon where there is NO alternative. Activating one sprinkler will not automatically activate the others. That's a movie myth. However, the water can only be turned off at the main valve.

- Swimming pools, provided they were not contaminated by the event.

- Creeks and rivers provided they were not contaminated by the event.

- Public water fountains – the decorative kind – provided they were not contaminated by the event or by other people already decontaminating. Most fountains are sealed systems that simply recirculate the same water. If you use fountain water, dip it out for use and don't' contaminate the water by getting *in* the fountain. If others are decontaminating with you make sure they do the same.

➤ **Advanced cleansers** are items that will not only clean, but may neutralize the agent you were exposed to. Interestingly enough, some of the more useful advanced cleansers are actually household chemicals or combinations thereof. Let's look at a few:

● In addition to being a good disinfectant, **household bleach** works well as a solvent for some chemical compounds. Essentially alkaline in nature (or a base, the opposite of an acid), it will also help neutralize some acidic compounds. This is why you'll see bleach listed as a recommended cleaner for either Sarin or VX gas as it helps break down the chemical components. **Note**: Bleach should only be used in a well-ventilated area. For use on skin it should be diluted ten to one. This means **one cup of bleach to ten cups of water**. Pure bleach will eat your skin. That's actually the reason bleach feels slippery to the touch. It's instantly dissolving your epidermis.

● Just as bases will neutralize acids, acids will neutralize bases. One common household acid is white vinegar. Though not high on the list of decontamination substances, vinegar can be used to offset a substance if you know it to be a base. This is useful if you're exposed to something such as sodium hydroxide (lye) during an industrial chemical accident.

● One concoction the military is working with is a simple mixture of household **baking soda, alcohol, and peroxide.** The baking soda neutralizes most chemicals, the alcohol and peroxide are both natural solvents, plus both the alcohol and peroxide have disinfectant properties. This mixture can be used equally well against chemical or biological substances. As it's also harmless as a topical application and it's relatively environmentally friendly, it should be included as part of any home decontamination arsenal. Leave it as separate substances until ready to use, but when needed, mix equal parts of alcohol and peroxide, and then mix in as much baking soda as will dissolve. Do not heat. Vapor is minimal, but it's still advisable to mix these ingredients in a well-ventilated area. To use, spray this mixture (using one of your garden sprayers) over whatever needs to be decontaminated, let it sit for a few seconds and then wash off with some fresh solution. You can also apply it with a washcloth. To be on the safe side, go for *three* applications of this stuff, rinse well, and finish up with soap and water.

● On our CD, you'll find a file named **Cwirp_guidelines_mass.pdf,** pertaining to mass decontamination. The text claims that a fire-fighting additive called "Cold Fire ®" (see http://www.firefreeze.com) has a dual use as a decontamination agent.

● If you can't find any of the listed household chemicals, and the liquid substance you're trying to wash off isn't cooperating, you can try either hair conditioner, Dawn™ dishwashing liquid, or any number of clothing detergents. Also, the powdered detergent for your automatic dishwasher may work rather nicely as most contain a fairly decent concentration of both bleach and soap.

➤ **Absorbing agents** are essentially dry cleansers. These substances will absorb liquid chemical agent off your skin. This list includes such simple items as corn starch, baking soda, baking powder, flour, talcum powder, sawdust, powdered charcoal, unscented clay kitty litter, instant mashed potatoes, and even oatmeal. Other absorbing agents easily obtained are various types of diatomaceous earth (Zeolyte) or "Fuller's Earth." It's a close cousin to natural chalk and talc except much more absorbent. You can find it at most hardware or janitorial supply stores. If *absolutely* nothing else, you can crush a few handfuls of soda crackers and use them to soak up some liquid chemical. If you want to make a good absorbent "cocktail," take a quarter-cup each of corn starch, baking soda, instant mashed potatoes, unscented clay kitty litter, uncooked white rice, and regular dry oatmeal, and run them together through your kitchen blender until they make a relatively fine powder. Keep these as separate items until needed. To use, heap it generously on the spots of liquid you wish to clean up. **Hint**: If you decide to make up an absorbent "cocktail," keep it tightly sealed in an old talcum powder or other shaker bottle. If you don't keep it tightly sealed, it will absorb moisture out of the air.

➤ **Disinfectants** pretty much apply only to biological agents. These are simple and common substances that will kill germs. Alcohol, peroxide, and bleach are the most common, though iodine or Listerine could be used if nothing else is available. However, soap and water still rule. Whatever you clean up with, don't think you can skip a medical checkup.

*Note: Even though these substances are widely used in decontamination procedures, you need to remember three things. **One**, don't play chemist. If soap and water is working well for you, stick with that. **Two**, don't skip medical attention just because you feel okay and your decontamination efforts seemed to work. **Three**, these measures are advisable only if official decontamination is absolutely unavailable.*

To help you remember the most important points of decontamination, we have another acronym for you; **W.A.S.H.**:

Water   Start with **water**, or a suitable substitute mentioned above, to start washing the chemical off of you.

Air   Don't start a detailed or thorough decontamination unless you're in the middle of plenty of fresh **air**.

Soap   **Soap** is the king of decon cleansers. Use other items if you have to, but remember, don't play chemist.

Hospital   Go to the **hospital** and get checked out even if you feel fine.

Sources and More Information

Don't forget all the military NBC manuals on our **CD**, specifically **FM 3-5**, and all the previous material from the "NBC" portion of "Basic Training." In addition to those, you can find some more decontamination info at these sites:

➤ Field decontamination tidbits can be found at: http://www.jrhenterprises.com/report/chem1.htm. Other goods and info can be found at the home page, or by contacting JRH Enterprises, PO Box 317, West Green, GA 31567, (912) 379-9441.

➤ One of our sources of first aid information has a really good online discussion of Chemical Decontamination Info: http://www.emedicine.com/emerg/topic893.htm.

➤ The Federation of American Scientists has a decontamination discussion available online at: http://www.fas.org/nuke/guide/usa/doctrine/army/mmcch/Decontam.htm.

➤ A variety of additional protection and decontamination info can be found online at the Sierra Times' site http://www.sierratimes.com/archive/files/oct/09/armk100901.htm.

➤ Decon Solutions has quite a variety of equipment at http://www.deconsolutions.com, which might be useful to businesses or groups.

Notes:

_____

_____

_____

_____

_____

_____

_____

_____

_____

_____

_____

_____

_____

_____

_____

_____

_____

_____

_____

_____

_____

_____

_____

_____

_____

_____

_____

_____

_____

_____

Manmade Misfortunes

Short of a nuclear detonation, or wildly successful biological attack, no single terrorist action will produce the kinds of destruction we've seen in severe hurricanes, F-4 or F-5 tornadoes, or in some of our larger earthquakes. So why are potential terror strikes so "terrifying" and so much in the forefront of everyone's worries? It's because of the unknown factor inherent in human malice. It's in the cunning that is unquantifiable and unpredictable in scope, nature, or target. It's because of the "not knowing" and the fact that despite the terror alert levels, there is no truly accurate warning system. It's also in the fact that the cost in human lives will be so much higher in a terror strike than in a natural disaster. Natural disasters tend to take their toll in property more than human life, and with terror it's the opposite.

Goals of Manmade Misfortunes:

1. To prepare you for the **emergencies the weatherman can't predict**.
2. To point out the fact that **accidents are just as deadly as intentional attacks**.
3. To reiterate the fact that **actual reaction is more important than an academic study** of a few weapons.
4. To provide you the "meat and potatoes" information that will **save the lives of you and yours**.

We've divided Manmade Misfortunes into the following sections:

1. **Nuclear Detonation**.
2. **"Dirty" Bomb**.
3. **Chemical Attack or HazMat Incident**.
4. **Biological Attack or Natural Epidemic**.
5. **Fire**.

Note: As you go through these reaction sections, you'll notice that in many we've included an acronym to help you remember some of the more important points. To get started, we'll give you one that covers all your reactions and planning in general. Appropriately enough, it goes to the word **R.E.A.C.T.**:

Research	**Research** the threats you may face, the assets at your disposal, and the options you have available.
Equip	**Equip** yourself with the goods and gear you need to adequately supply yourself in any reaction.
Awareness	Stay **aware** of developing situations, potential threats, and various dangers as emergencies unfold.
Coordinate	**Coordinate** and communicate your efforts with your family and friends. Life involves others.
Train	No plan will work unless you've practiced the important parts. **Train** and practice whenever you can.

Nuclear Detonation

General Discussion

If there's one comforting thought regarding nuclear weapons, it's that the only country capable of inflicting a mass nuclear attack on us is Russia, and at the time of this writing, we're on pretty good terms with our old cold-war enemy. Other nuclear powers currently lack either the number of bombs or the means to deliver. North Korea may be able to reach us with one of their few ICBMs (Inter-Continental Ballistic Missile), but they're rather inaccurate. India and Pakistan are more interested in their conflict with each other, and China, overall, has been on a pretty good footing with the US as their interests lie in becoming more an economic power than a power based solely on military threat.

The other comforting thought with nukes is that even with perfect delivery of a large weapon, there will be survivors.

At this point in time, the highest probability of a nuclear strike comes to us in the form of a device detonated in a ship in a harbor, or a smuggled "suitcase bomb" or other tactical nuke stolen from an unsecured stockpile of one of the nuclear powers. Though the tactical devices are slightly less effective than a full warhead, any nuclear weapon is something to be both feared and respected.

Primary Danger

The primary danger from a nuclear detonation comes from the intense blast and heat waves, and the secondary return wave or "backlash" wave. The backlash wave occurs as air comes rushing back in to fill the vacuum left by the initial blast wave. The secondary threat is radiation in the forms of direct radiation emissions, and later fallout of irradiated dust. To make matters slightly worse, there will be an Electro-Magnetic Pulse or "EMP" generated by the nuclear reaction, that will wreak havoc on all electrical circuits and anything with a transistor or microchip that happens to be in the area. This will effectively knock out communications, many forms of transportation, and many appliances to include your personal computer.

The overall destructive effect a device may have is dependent on several different contributing factors. Among these are size or "yield" (in kilotons or megatons) of the weapon, its placement within a particular area, its elevation relative to the ground at detonation, terrain features of the target area, and to a limited degree, the weather.

For ideal performance, a nuclear weapon has to be detonated at a certain altitude above a target for an "air burst" effect, which allows the bomb to reach its full effective power. Detonation at ground level will greatly reduce the weapon's effectiveness, as will sloping ground, surrounding hills, etc., but it will increase contaminants and fallout.

Advance Preparations

This leads us to a question we hear from time to time, and that is, "What ever happened to all the civil-defense fallout shelters?" The short answer is, "times change." From the early '50s through the late '70s, America became a much more mobile society. In the early '50s we were just starting the Interstate Highway project (which was built for nuclear evacuation and military purposes), a large percentage of America's roads weren't paved, and if a family had a car, it was just <u>one</u> car. Also at the time, the threat was a completely devastating nationwide nuclear strike from the Soviets, so there was no place to run to, even if you could get there. Cities and private citizens did everything they could to get ready for mass sheltering in place. We had civil-defense shelters, some people had fallout shelters in their basements or dug into their backyards, and all of us knew that "duck and cover" was NOT a menu item.

Since we're a much more mobile society, and since our current nuclear threat is more along the lines of a single bomb detonated without warning, reaction is basically "if you can still walk, <u>run</u>." Evacuation is the order of the day with the plan of some emergency management agencies being to offer temporary lodging and care in evacuation destinations. The hidden advantage to this is that the infrastructure strain is shared since not everyone will head to the same destination (see "**Evacuation**"). Hence, official fallout shelters have fallen out of vogue in most areas, and their supplies redistributed long ago to hospitals and Emergency Management offices for use in natural disasters. (See the "Civil Defense Museum" at http://www.civildefensemuseum.com.)

However...On the flipside of the coin, there are so many more people with so many more cars these days, that on attempting evacuation, you could be gridlocked in traffic and exposed to fallout before getting far enough from the area of a detonation to be safe. You may be faced with a tough choice between evacuating and sheltering in place. Base your "should I stay or should I go?" decision on:

> - How far you are from potential targets.
> - Prevailing wind direction. Are you in the main fallout zone?
> - How mobile you are in general.
> - How quickly you can grab your gear and get to a major transportation artery.
> - If you have an alternate source of transportation (such as a boat near a waterway).
> - If you have immediate access to an adequate fallout shelter. Some communities still maintain their shelters. Call your local Law Enforcement, Fire Department, or Emergency Management office to ask about your city.
> - Whether or not you were injured in the attack.

As far as your advance preparations, there are three main factors that determine immediate protection:

> - Your **distance** from ground zero.
> - The amount of both radiation and blast **shielding**.
> - The amount of **time** you're exposed to radiation.

The less protection you have from one of these, the more you need the other two.

The advance preparations you can make are actually rather simple. If your biggest fear, or location's primary danger, is nuclear attack, you should do the following:

1. **Plot potential destruction**: As discussed with the Threat Map, you can probably take an educated guess as to where a nuke would be detonated in your area. If there's a positive side to a suitcase nuke, it's that its placement will be more precise than with a foreign ICBM. Knowing where a nuclear weapon *might* be detonated, you can plot your relative distance to target, between ground zero and your home, school, or workplace, and you can determine if you will have any "flash to crash" lag time. This "flash to crash" is the lag time between seeing the flash and feeling the shock wave, such as with the time lag between seeing lightning and hearing thunder. Interestingly enough, a nuclear flash to crash is roughly the same as lightning to thunder; just under 5 seconds

per mile. Use your city and area maps to plot potential target sites. One site may be in the center of a downtown area, and others may be important government sites, military bases, sports arenas, harbors, or large dams. Your Threat Map should already show target sites, destruction zones, and fallout patterns (based on prevailing winds).

2. Follow all **structural reinforcement and safe-room** advice given under "**Basic Home Prep**," as it will not only help you with the average natural disaster, but it could also make the difference in protecting your structure against the blast wave of a relatively nearby nuke. (**Note**: this will only help you if you are 3 miles or more from a probable ground zero based on an "average" size nuclear weapon.)

3. Certainly a number of factors come into play, but you may also want to consider making some advanced preparations at your **workplace** as you may be far enough away from a blast to survive, but close enough to sustain some building damage. Among these preparations would be:
   ❏ Reinforcement film for windows.
   ❏ Increased first aid supplies.
   ❏ Additional fire safety equipment.
   ❏ Stored hard hats, goggles, work gloves, and protective clothing.
   ❏ Simple structural reinforcements to buildings or rooms where feasible.
   ❏ Heavier office furniture or reinforced workbenches, etc. (as applicable).
   ❏ Reaction drills for employees.
   ❏ Secondary communications (in addition to phone system).
   ❏ Augmented Emergency Alert notification.
   ❏ Stored food and water in anticipation of a short period of isolation.

4. IF you feel your area is at high risk for nuclear attack, IF you DO feel at risk of a larger weapon, IF you feel that there will be adequate warning, IF you have sufficient yard space, IF you have the money, and IF you know you CAN'T evacuate after a nuclear incident, you might want to consider having your own in-ground shelter. See "**Advanced Home Prep**" in the "**Isolation**" section.

5. You did follow all our advice under both "**Basic Home Prep**," and "**Basic Vehicle Prep**," right?

6. At home you should have a variety of reaction drills to include Fire Drills and Tornado Drills. The Tornado Drills are a great way to practice safety measures that will apply to both tornadoes and nuclear detonations. Calling it a tornado drill may scare the kids less. Practice both **duck and cover measures**, and getting from your bedrooms or family areas **to your safe area in under 5 seconds**.

7. Carry your GYT Pack with you at all times and place a heavier emphasis on its **fire, first aid, transportation,** and **exposure and shelter** items if the nuclear threat is high on your probability list. (**Note**: Nothing in this kit will protect you against radiation, but the little things you carry could help get you out of a damaged building and on the road faster if you're on the outer fringes of destruction.)

8. Continually take a good visual inventory (and then make a map) of all the **potential "safe spots" that exist in areas you frequent**. The areas you frequent would be your **home and neighborhood, your route to work or school, areas at work or school, and your most common leisure areas**. The "safe spots" would include any location or object that would offer shielding from the heat and blast effects of a nuclear detonation. Examples are ditches, culverts, heavy concrete walls, river or creek beds, in-ground swimming pools, underground basements, etc. Pretty much any spot that's at or below ground level with heavy earth, concrete, or steel between you and a probable ground zero will work. Remember, you'll need protection from the heat, and both the blast and the return waves. Stay away from *seemingly* safe spots that could trap you or expose you to secondary hazards. You wouldn't want to duck and cover near a glass exterior office building that may collapse on you, nor would you want to run *towards* a potential hazard such as a nearby propane tank that may explode in the heat wave.

9. If any locations in your area are still maintained as fallout shelters, know their location as well.

10. You should already know the fire escape and fire and safety equipment locations at work, school, or other locations you frequent. Get in the habit of noting the same info in any new location you visit.

11. Know what your children's schools would do in the event of both a nearby and non-local attack. Would you be called to the school to pick up your kids, or would they be bussed to a safer, more accessible area to rendezvous with parents? Would they automatically evacuate to a predetermined site? Would they shelter in place? How would they let you know what they did? (You'll be asked to record this information on the "**Family Data - Student**" sheet in the "**Appendix**.")

12. What if you have a family member in the hospital or a nursing home? What is the hospital's reaction plan?

13. You might consider a supply of Potassium Iodide tablets. Potassium Iodide helps protect the thyroid from radiation damage. You'll find some information sources at the end of this section.

14. If we're ever again faced with a nuclear standoff, as we were with the Cuban Missile Crisis, you might want to go ahead and pre-seal parts of your house (see "Advanced Home Prep" under "**Isolation**") as this will give you a jump on sealing your house against fallout, which you'll want to do whether you evacuate or shelter in place. You'll want a contaminant-free house when the all-clear is given to return home or come out of your shelter.

Warning Time

If the weapon is an ICBM (InterContinental Ballistic Missile) we *may* have up to 30 minutes warning.

For the smuggled suitcase nuke, or the ship in the harbor scenarios there is no warning.

Forms of Alert

In the event of a detected ICBM, the Emergency Alert System *may* be activated on television and radio, and community sirens may be sounded. This is another important reason to make sure you've augmented your own Emergency Alert System warning sources.

Again, there will be no alert for a surprise detonation of a smuggled nuclear weapon.

Warning Signs

Aside from the alerts activated after the detection of an inbound ICBM, your only warning may be a brilliant flash.

Reactions

Naturally, all reactions below are based on the assumption you're outside the immediate blast area or "instantaneous kill zone" (the two-mile circles you drew on your Threat Map). The time within in which you can react is based entirely on your distance from "Ground Zero." Where you are when a detonation occurs will dictate some of the immediate reaction procedures you follow:

The basic immediate reaction to a nuclear detonation has always been and will always be "**duck and cover**." You'll want to immediately dive for one of the "safe spots" you chose earlier under number 8 above. If you've plotted your relative distance-from-target and know your "flash to crash" lag time, you may have a couple of seconds to make it to a safer spot before the blast wave hits. Once covered, you'll want to **wait two full minutes** to make sure the heat and blast waves completely pass.

We've put together an acronym to help you remember your immediate reactions. Since most of us remember the old "duck and cover" drills, this one goes to the word **D.U.C.K.**:

Don't   **Don't** look at the flash, and **don't** waste time! And actually yell "duck!" to alert others.

Under   **Under** something now! Protect yourself from blast, heat, and debris.

Count   **Count** a full **120 seconds** (two minutes) to allow the blast and heat waves to pass and then backlash.

Know   **Know** your situation and intended reaction. Evacuate or go to shelter immediately.

Part of "*knowing your situation*" will be to watch where the mushroom cloud starts to drift (after waiting your full two minutes). It will probably follow your area's prevailing wind patterns, but it's important to watch since the direction it heads is the area of heavier radioactive fallout.

Your basic immediate reaction will have some additional elements to it depending on your location. Let's take a look at some of the places you may frequent.

Home

> ➤ IF the scenario is that of the incoming ICBM and IF you have the 30-minute warning, react as you did in your Tornado Drills and call in the children and pets, close your storm shutters (won't have time to board windows), deadbolt and brace your exterior doors, close interior doors, close your fireplace flue, close the windows, blinds, and curtains, and then head for your safe room or safe area. You should practice this procedure so you can have it done in **10 minutes or less**.

> ➤ During this ten minutes, you should also prep for fire and make sure all your extinguishers are handy, sprayer units nearby and full, and you might even want to fill a couple of buckets of water to have handy (in addition to the tub and plastic bins you filled in anticipation of being without water).

> ➤ **Special Note**: In any emergency where your outdoor pets, such as dogs, are kept in an outside area separated from your house, you should lead them back in on a leash even if they normally run straight to the house when told. In an emergency, people will be acting a little differently. Animals can sense this, and can also sense that something's wrong in some natural disasters. For these reasons, don't expect your pets to act as they normally do. Lead them on leashes, and if already in the house, put them on leads or in carriers to take to your safe area. This will also make it easier to rescue the animal should your home be damaged and first responders have to extricate you and your pets.

- Depending on your distance-from-target, lying in a low ditch or creek bed in your backyard may be safer than sitting in your above-ground safe room inside a frame house. Always go for the lowest area with the greatest shielding. A ditch at the bottom of a hill is better than a house with no basement on the top of the hill. Preferred would be a below-ground-level basement, followed by a partially buried storm cellar, or even an empty in-ground swimming pool.

- If there is no warning at all, and you suddenly see the bright flash of detonation, the old rule of "duck and cover" comes immediately into play. Drop to the floor and dive underneath a heavy piece of furniture, or to the other side of *something*, anything. You'll want protection from flying debris, such as glass, and from the heat wave. (Remember our discussion of stronger furnishings earlier?) Keep in mind that a shock wave will be coming from two directions. One goes out, then the other comes back. Stay under cover for a full two minutes. (This is another reason we included the "Earthquake Drill" earlier.)

- Regarding the flash, keep in mind that it's not like a camera flash. It may stay blindingly visible for several seconds. Don't look up thinking it'll be over "in a flash." Don't look at anything at all until your full two minutes are up.

- Hopefully, you alerted others in the house as you were diving for cover. There may be those who didn't see the flash.

- If you determined earlier that you had a good "flash to crash" lag time, you may have time to scurry to your safe area. Stay low. Your "Tornado Drill" mentioned earlier, should be accomplished faster than your estimated lag time.

- Though we asked you to plot one and two-mile circles around potential nuke targets in your area, keep in mind that the destruction from some of the larger nukes can extend beyond that, and that secondary effects can be felt up to twenty miles away. If you live anywhere within this twenty-mile range, you'll still want to do all you can regarding structural and safety enhancements for your home and workplace.

"Work"

- In this section and each subsequent section, "Work" will refer to your workplace, school, or any indoor public venue. The common denominators are that each location is away from home, around other people, an area you might frequent, and for our intents and purposes, near a potential hazard area.

- You should know your distance-from-target as it relates to work or school, just as you do for your home. Know the lag time from "flash to crash."

- Take stock of your workplace (or school if that's where you spend most of your day) as directed in number 8 of "Advance Preparations." Know the areas that will afford immediate protection. Know where to duck and cover and where not to, and be sure to know the location of all emergency exits and safety equipment.

- Large-span, open buildings such as aircraft hangars, warehouses, school gyms, etc. do not offer good protection and will collapse easily under stress. Taller buildings (especially nearer ground zero) have more surface area to catch the blast waves and are therefore in more danger of collapse than smaller structures nearby. Plan to vacate these types of buildings immediately (if you get warning) and do not include them on your list of safe areas.

- With most buildings, the safest areas, structurally speaking, are against interior walls. This places you near structural supports and away from outer walls or windows that may collapse, give way to projectiles, or themselves become projectiles.

- Should you be caught in the debris of a collapsed building, do the following:
  - ❏ Above all else, maintain your composure, and be prepared to wait.
  - ❏ If you're able to move, check yourself for injuries and tend to them as best you can with the supplies in your GYT Pack. Then help others as much as you're able.
  - ❏ Put on your basic breathing filter as there will be a lot of dust (and possible smoke from fire) associated with a structural collapse.
  - ❏ Though prepared to wait, you should not sit there and do nothing. Try your cell phone to see if lines are still up, or see if any land line within reach is still operable. Just because a building is damaged doesn't mean everything's out of commission.
  - ❏ Start signaling that you're trapped. You may have several options at your disposal. First of all, don't yell, as you'll only inhale dust. You should have packed a whistle in your GYT Pack, you can activate your pager if you have one, and you can keep ring-testing your cell phone. In addition to sound, you should also have a small flashlight on you and maybe a chemical light stick. You might want to save the lights and electronics until you hear rescuers searching for you. You can also tap with a heavy object.

- Even if your building seems undamaged, be careful. Shock waves can do odd things to architecture. Watch for falling objects, keep an eye out for possible hidden damage (windows about to fall, etc.), and don't use the elevator. Even a small shift in the building's structure can jam the elevator in its tracks, and power outages can occur suddenly.

- ➤ Carry your **evac map** in your GYT Pack as it will show all your primary and secondary evacuation routes as well as locations of alternate transportation. We'll discuss the map in more detail under "**Evacuation**."

Car

- ➤ Driving a moving vehicle during a nuclear explosion carries with it some unique challenges, the most immediate of which is potential wrecks on the roadway. The effects of the blast and heat waves notwithstanding, the flash alone is liable to cause some serious distraction for most drivers on the road. Your first concern is to pay attention to the cars around you (as best you can without looking at the flash). Maintain control of your car, avoid others, and get out of harm's way as quickly and safely as possible.

- ➤ Next, since you're supposed to know not only your distance-from-probable-target, but also the possible protective locations along any regular routes, immediately look for a protective spot you can pull your car within two seconds. If no protective parking site is near, stop as best you can, exit your vehicle, and dive for immediate cover.

- ➤ This cover may be a depression such a ditch, a creek or river, a culvert, behind a small hill or earthen rise, or behind a solid structural wall. Avoid buildings as walls may collapse. Also avoid getting underneath an overpass since you're still exposed along the sides to the heat wave. Once in a safe spot, lie face down with your hands under your body for protection from the heat and stay there a full two minutes. If you have your GYT Pack on you (as you should) try to put on your smoke hood to help protect your head from heat.

- ➤ Do not stay in your car unless you've parked in a protective spot large enough to protect the whole car. The blast wave may toss your car more easily than a tornado would, or it may shatter the windows injuring you.

- ➤ If you know, based on distance-to-target, that there is absolutely no time to exit your vehicle, stop the car and apply the parking brake (if possible), keep your seatbelts on, draw your elbows in tight against your body, lay down as best you can, and cover your face with your hands since you may be about to go for a ride!

- ➤ Afterwards, be prepared for the fact that your vehicle might not be able to start again due to the effects of EMP on automotive circuits.
  - A good discussion of what EMP is, as well as several sources for **shielding gear** and outside links, go to: http://www.unitedstatesaction.com/emp_and_faraday_cages.htm.
  - For a good discussion of the EMP threat, see: http://superconductors.org/emp-BOMB.htm.
  - Also see: http://popularmechanics.com/science/military/2001/9/e-bomb/print.phtml.
  - A list of UK EMP shielding suppliers: http://www.applegate.co.uk/elec/pselect/ps_2575.htm.

- ➤ While under the heading "Car," we'll remind you that you may be on public transportation when something happens. Duck and cover as best you can, hang on to whatever is available, and know your options for step two. Step two will involve getting out of the vehicle you're in if need be. Therefore, always note the location of escape windows, hatches, and doors of any bus, train, ferry, or commuter plane you're on.

Outdoors

- ➤ If caught outdoors and relatively unprotected, follow the advice above about diving for the nearest depression and protecting your face, hands, and exposed skin. The deeper the depression the greater the protection from blast, wind, debris, and heat.

- ➤ If a detonation happens during any sort of higher terror alert, you may be partially protected if you chose to follow our advice regarding "Dressing for Chemicals." Any insulation against heat and some of the milder forms of radiation is a good thing.

- ➤ Try to avoid things that may collapse or become debris themselves. Among these are large trees, deep construction ditches (may cave in on top of you), glass or other flimsy buildings, or lightweight objects such as patio furniture, etc.

Specific Follow-Up

1. After the detonation and the two minutes have passed, check for serious personal injury to yourself first, then to those around you.
2. Scan your immediate vicinity for imminent safety hazards such as a ceiling or wall about to collapse, downed power lines, broken gas mains, etc.
3. Check for fires caused by the heat wave. It's common for the blast wave to shatter windows and the heat wave to cause curtains or other room items to ignite. Also check outside your building for fire.
4. Check for minor injuries now that you've taken stock of the first three items.

5. Shut off the ventilation system of the building or vehicle you're in, or have others do it. Fallout notwithstanding, the air will be heavy with dust from the blast wave, smoke from fires, and there may be a chemical element involved if any area factories or plants experienced damage.

6. Do the same at home, and also shut off your gas and water until told it's safe to turn them back on.

7. See if your TV or radio are operable, and while taking preparatory evac or shelter steps, listen for word of secondary damage (such as a dam about to rupture, blocked evac routes, etc.) and for instructions from officials.

8. If communication systems are operable, try to contact your immediate local family members. Though cell phones may certainly be out, you may be able to make a long distance call on a regular phone. If that's the case, call the people you've listed as your rendezvous contacts on your "Important Contacts" list.

9. Prepare to shelter or to evacuate, based on the planning dictated by your individual situation. (After the blast wave clears, you have roughly **20 minutes** to react before fallout starts becoming a threat. Eighty percent of all fallout is down in the first 24 hours following a detonation.)

10. If your home is undamaged and you're away from the danger area and not in the fallout zone, you still may want to evacuate, at least temporarily. If that's the case, you'll want to do a **rapid** preliminary sealing of your house before leaving since you'll want a clean environment to return to. (See "Advanced Home Prep" under "**Isolation**.") If you're **well outside** the blast-effect area of a local attack, but **in** the **projected fallout area**, you'll want to seal your home, and yet still be prepared for an evacuation order. If evacuation is your preferred response, go ahead and leave after sealing. **Whatever you do, be practiced enough to where you can have everything done and be on the road in less than half an hour** (less time than that if you live closer to the blast area). Don't forget to seal your vehicle from the inside once you're underway.

11. If your planned reaction to a nuclear detonation is to evacuate, you'll want to evacuate either upwind or at a 90 degree angle to incoming fallout clouds. Don't head in the direction the prevailing winds blow because fallout will follow. If the winds are blowing east or west, you go north or south, etc.

12. Watch for the symptoms of **radiation sickness** which include (and may include more):
    ➢ Noticeable burns on skin, possibly with blisters.
    ➢ Red, itchy skin.
    ➢ Headache.
    ➢ Nausea.
    ➢ Vomiting.
    ➢ Diarrhea.
    ➢ Hair loss.
    ➢ Bleeding.
    ➢ Sore mouth or throat.

13. Don't come out of your shelter or back to the area until officials give the all-clear.

14. Once back home, be sure to follow instructions for decontaminating your home as outlined in "Coming Home" under "**Evacuation**" provided authorities say radiation levels are low enough for you to return.

Miscellaneous Points

➢ A nuke in another city will prompt precautionary actions in your area just as 9-11 saw the grounding of all commercial flights. Be prepared for an immediate threat level of Red Alert and the closure of all schools, public venues, and all non-emergency, non-essential businesses and services. In this case, you'll want to gather the family at home and await further information.

➢ Regardless of alert status, you'll probably want to gather your family at home and stay put a few days (unless your town was the one hit) as wide-spread social unrest is an extremely likely result of such an attack. We'll discuss home security measures a bit more under "**Isolation**."

➢ If you come into contact with victims of a nuclear attack, do not be afraid of them. You cannot "catch" radiation poisoning from them provided they've been decontaminated by bathing and changing clothes.

Finally, this is a difficult section to write because of the vast number of variables that can readily change the effectiveness of an attack, the effectiveness of your defenses, and your follow-up or secondary reaction procedures. With a worst-case-scenario such as nuclear attack, people want solid, unchanging answers as to what to do and how they're going to be affected. Unfortunately, no one can give you such specific answers. No one. These variables make a nuclear attack a close cousin to a tornado in that you don't know exactly when or where it will hit, how it will act, or whether or not you'll be a victim or an untouched witness. The best you can do is evaluate your unique situation based on the guidelines we've given, and adjust your preparations and planned reactions accordingly.

Overall, the threat of a nuclear detonation reminds us that all emergencies are going to be different and have their own unique situations. Every situation is going to be fluid and dynamic with a life of it's own and you need to be able to react accordingly and instantaneously.

Sources and More Information

➤ Remember the sources listed in the "**NBC**" portion of "**Basic Training,**" and all the manuals on our **CD**.

➤ A great source of nuclear explosion technical data and related planning and expectations can be found in **Strategy for Survival** by Thomas Martin and Donald Latham (Library of Congress Catalog # 63-17720). Long out of print, you can still find this book through used-book sources.

➤ Nukefix, online at http://www.nukefix.org, offers an interesting collection of nuclear weapon information. The main purpose of the site is to foster an anti-proliferation movement through **downloadable software** showing the statistical probabilities of a nuclear attack given variable input. It's interesting and worth a look. Also, their links page, at http://www.nukefix.org/link.html, and other info on the site offers an extensive collection of related information to include blast and fallout protection info. We also suggest you read their blast information page at http://www.nukefix.org/weapon.html.

➤ For numerous CERT manuals on this and other "Reaction" subjects, visit the Los Angeles CERT site at http://www.cert-la.com/manuals/instman.htm.

➤ The Trinity Atomic Website: http://nuketesting.enviroweb.org/.

➤ FAQs on Nuclear Weapons: http://nuketesting.enviroweb.org/hew/Nwfaq/Nfaq0.html.

➤ The Nuclear info links page: http://nuketesting.enviroweb.org/hew/News/Bigbig.html.

➤ Nuclear Weapons Effects: http://www.fas.org/nuke/intro/nuke/effects.htm.

➤ NATO HANDBOOK ON NBC DEFENSIVE OPERATIONS AmedP-6(B) PART I – NUCLEAR: http://www.fas.org/nuke/guide/usa/doctrine/dod/fm8-9/1toc.htm. (Also on our **CD**.)

➤ The Effects of Nuclear War: http://www.fas.org/nuke/intro/nuke/7906/index.html.

➤ Multiple links including fallout shelter plans: http://members.aol.com/rafleet.

➤ International Nuclear Stockpiles: http://www.nrdc.org/nuclear//nudb/datainx.asp.

➤ See the CIA's "Electronic Reading Room" at http://www.foia.cia.gov, and enter the search word "nuclear."

➤ Accounts from Hiroshima survivors: http://nuketesting.enviroweb.org/hew/Japan/Testimon.

➤ Nuclear technical info: http://nuketesting.enviroweb.org/hew/Nwfaq/Nfaq5.html#nfaq5.4.

➤ Visit NukePills.com, online at http://www.nukepills.com/pages/749857/index.htm, or contact them at 631-407 Brawley School Road #165, Mooresville, NC , 28117, 1-866-283-3986.

➤ For **Potassium Iodide**, log on to http://www.ki4u.com, or contact KI4U, Inc., 212 Oil Patch Lane, Gonzales, Texas, 78629, (830) 672-8734. Be sure to visit their subpage http://www.ki4u.com/free_book/s73p916.htm, which contains a detailed look at nuclear war and domestic preparedness.

➤ Another source of Potassium Iodide: http://www.pro-resources.net.

➤ The Nuclear Regulatory Commission discussion on potassium iodide and its usage: http://www.nrc.gov/what-we-do/regulatory/emer-resp/emer-prep/potassium-iodide.html.

➤ New counter-radiation drug called "HE2100" http://news.bbc.co.uk/1/hi/health/2924197.stm.

➤ For a good discussion of nuclear war and its associated affects, go online to visit: http://www.surviveanuclearattack.com/NuclearWeaponsFactoids.html, and be sure to visit the homepage at http://www.surviveanuclearattack.com.

➤ Read the tables at http://www.surviveanuclearattack.com/FlashtoBangTime.html.

➤ Read the online version of "Nuclear War Survival Skills" by Cresson H. Kearny. You can find it online at either http://www.oism.org/nwss, or at http://www.homelandcivildefense.org.

➤ Also be sure to visit: http://nuketesting.enviroweb.org/hew.

➤ For the history and science of the Atomic Bomb, see: http://www.atomicarchive.com.

➤ See the "Wrongdiagnosis" site concerning radiation sickness, complications, and treatment, at: http://www.wrongdiagnosis.com/r/radiation_sickness/intro.htm.

➤ Visit http://www.webpal.org, for a collection of independent opinions regarding post-nuclear rebuilding.

➤ Also for radiation sickness, visit: http://www.atomicarchive.com/Effects/radeffects.shtml.

➤ See http://www.healthcentral.com/peds/top/000026.cfm for radiation sickness symptoms and treatment.

➤ Visit the Nuclear Radiation links at: http://www.nacworldwide.com/Links/Nuclear-Radiation.htm.

➤ Remember our PBS "Blastmap" site at http://www.pbs.org/wgbh/amex/bomb/sfeature/blastmap.html.

General Discussion

Many people get "dirty" bombs and nuclear devices confused due to media hype and street rumors. The bottom line is this: a dirty bomb is not a nuclear bomb as there is no nuclear *reaction*. A dirty bomb is nothing more than a "standard" bomb designed to disperse radioactive materials in the same way a regular bomb would spread shrapnel.

Let's use this analogy. Let's say you wanted to get talcum powder over as much of a downtown area as possible. You'd make a pipe bomb, or you use regular explosives, you'd pack as much talcum powder around it as you possibly could, and you'd put the whole the whole thing on top of a tall building to give it some height. When the bomb blows, there goes the talcum powder, floating through the breeze to eventually settle over the downtown area. A dirty bomb is exactly the same except a terrorist would use some sort of radioactive material instead of the talcum powder.

The dispersal goal is the same though. The bomb does not use the radioactive material as a source of its power, it's simply a conventional explosive that spreads the radioactive material (usually some sort of low-grade waste product) over as wide an area as possible.

Interestingly enough, the precautions and reactions listed below will also help you react to a radiological accident at a nearby nuclear power plant as the end result will be virtually the same. It will be a release of radioactive materials spread over a fairly local area and without the devastation and destruction caused by a nuclear detonation.

Primary Danger

Actually, the primary danger is from the explosive itself, but again, it is no more than a conventional explosive. The damage from the blast might not be much more than that of a car-bomb or similar device.

The effectiveness of the radioactive threat depends on how much material the terrorist was able to get their hands on, the purity and level of radioactivity of the material (the radioactive material will most likely be some sort of leftover or partially depleted nuclear fuel or medical equipment waste acquired from an unsecured dump site), its form (powders drift farther than "chunks"), elevation of the device on detonation, how long you're exposed to any radiation, and on wind and weather.

The good news is that radioactive material may be relatively easy to clean up, as the material will tell you exactly where it is by its radiation emissions. These emissions can be picked up on any Geiger counter very easily. Also, the less material there is, the less of a threat it poses, since a smaller amount means less radiation (generally speaking). That means that if a downtown area were the scene of a dirty bomb attack, the first good rain that came along would reduce the amount of material present on exterior surfaces. Lastly, most waste material, though radioactive, is not as radioactive as "fresh" materials.

The direct physical threat to people comes from residual radiation that may be emitted by remaining materials. Usually, the radioactive material is nowhere near strong enough to cause immediate radiation sickness. The only danger is in long term exposure and its effect on later health. Again, this danger is reduced with time, and cleanup of the radioactive material. In most scenarios you'll be fine after simple decontamination, provided you protected your lungs during the event.

All in all, as compared to a true nuclear detonation, a biological attack, or a chemical attack, the dirty bomb is the easiest to clean up after.

As a caveat to our discussion of the threat posed by dirty bombs, it's fair to say that experts disagree widely on the dangers and end results. Under "Additional Information" below, we'll give you a couple of links that will let you see contrasting opinions.

Advance Preparations

As the immediate physically destructive force of a dirty bomb is limited to its core explosive, there is no need for any other preparations than you'd take in anticipation of a suicide bomber or car bomb.

However, the breathing filters mentioned earlier will help keep any radioactive dust out of your lungs. Again, this is a good reason to know your area's prevailing wind patterns, and to acquire a stock of **Potassium Iodide** pills as mentioned earlier. (**Note**: Taking Potassium Iodide should **not** be an immediate or automatic reaction. Seek medical opinion as soon as possible.)

Another advance precaution you can take, as we mentioned earlier, is to know the location of nearby nuclear power plants, and to also know what kind of public warning system they'll be using in the event an incident occurs.

The only other advance preparations necessary would be those found under "Advanced Home Prep" in the "**Isolation**" section concerning sealing your home, as those caught in the downwind area of the explosion may be asked to either shelter-in-place or evacuate. If you evacuate, you'd want to do a little bit of quick sealing first to give yourself a clean environment to come back to.

Warning Time
The only way there would be any warning is if the explosive device were discovered before detonation and a radiation scan were conducted. Otherwise, there will be no warning of the explosion itself, but there may be a rather rapid ascertation that radioactive materials were involved. The testing involved may be completed as quickly as within the hour and the information made public within two hours.

The warning time for a nuclear power plant incident would depend on how early a problem was discovered, or on how rapidly an attack on the plant was reacted to.

Forms of Alert
Television and radio news sources would be the primary forms of alert, though some communities may activate their sirens and local EAS to warn residential areas downwind of the blast of potential radiation. For a power plant situation, you might also have alarms and sirens activated at the site itself in addition to the above warnings.

Warning Signs
Unless you have your own Geiger counter, your only warning is going to be word from officials.

Home
> If you're at home and your home is downwind from ground zero, you'll want to listen to the media for instruction from officials and be ready to react. Go ahead and start some **pre-sealing** measures. **Don't wait for official word**. Go ahead and shut down your HVAC (Heating, Ventilation, and Air Conditioning), bring outdoor pets and their food inside, close all the doors and windows, close the fireplace damper, and tape over any permanent openings such as the exhaust vent over your stove, and any window unit air conditioners, or in-floor convection heaters. See "Advanced Home Prep" in the "**Isolation**" section for a more extensive discussion of sealing.

> If you're told to **shelter in place**, you'll want to go ahead and **finish sealing**.

> If you're told to **evacuate**, you'll want to grab your Bugout Kit and be on the road within 30 minutes, preferably with some sort of breathing protection in place. (Lungs are far more vulnerable than skin in a dirty bomb scenario.)

> Before leaving the house, be sure to **complete your preliminary sealing** as mentioned above. You'll want a clean home when you return.

"Work"
> Treat this as you would any other HazMat incident. Make sure your building's HVAC system is shut down. Close all doors and windows and move to the center of the building to await instructions. Do not go into stair wells of taller buildings as stairwells are usually pressurized by outside air via a system separate from the main ventilation system. Don't take chances.

> Your workplace should keep materials on hand to seal rooms and sections of the building.

> Call and alert family members.

> Turn on a radio or TV to await official instruction. (This scenario is one of the few times you'd wait around for instruction. You'll want to wait because you'll need to know the extent of the radiation danger, and that's something only officials will know.)

> Have your GYT Pack's breathing filter and protective cover (such as the poncho) ready to put on in case you're told to evacuate. You'll want to be protected while outside.

> If you deal with livestock, bring them into a sheltered area or locate evacuation transportation for them. (Livestock will probably not be an issue since a "Dirty" bomb would more than likely be detonated in a metro area unless it's detonated in an agro-terrorism attack.)

Car
These apply only if you're near a known dirty bomb and an official announcement has been made. Otherwise, you'd never know.
> If stuck in your car downwind from a dirty bomb, immediately roll up your windows, turn off your airconditioner or heater, and set circulation controls to "recirculate." You don't want any outside air drawn into the car. You may also want to go ahead and put on any dust mask or respirator you carry in the car.

➤ If you can, use the masking tape we listed under "Basic Vehicle Prep" to seal around door and window seams, and cover AC vents on the dash and floorboards.

➤ If this happens in a hot climate, put on your respirator, but not any other protective clothing. You'll want to stay as cool as you can inside a sealed vehicle with no AC. In this instance, skin protection is not as necessary since radioactive debris from a dirty bomb is not high on the skin absorption list and it'll wash off. Your lungs are what should be protected.

➤ Once official instruction is given, you may be told to proceed to an official decontamination site.

➤ The HVAC rule applies to public transportation just as it would yours. If you're in a public vehicle such as a bus or train, make sure the operator has closed any windows or outside vents and has turned off the AC.

Outdoors

➤ If you're outdoors and well away from the effects of the explosive charge, your only concern will be finding protection from the radioactive contaminant (provided your EAS enhancements have alerted you to the threat). The items kept in your GYT Pack will go a long way toward protecting you until you can get to a place to decontaminate. Put on your "smoke hood" or other breathing filter, and your poncho. You'll want to keep as much dust off you as possible.

➤ If you feel you've been exposed, or think you have some of the radioactive material on you, do not go home, into a building, or into any mass transit system or other public place. You don't want to contaminate others. Call local authorities for assistance and instruction. They may be on the way to set up a decontamination site within walking distance of where you are. Otherwise, follow the "Personal Decontamination" steps outlined earlier.

➤ If you're already protected, don't worry too much about beginning decontamination procedures. As we said, the radiation level will probably not be intense enough for immediate reaction, and besides, dust may be settling for some time and you'd only be recontaminated after cleanup anyway. Wait until you get to a sheltered area that's set up for decon. However, continue to protect your lungs at all times.

Miscellaneous Points

➤ The same precautions you take here can also protect you from an incident at a nuclear power plant.

Specific Follow-Up

1. Listen for official word on the strength, concentration, and dispersal of radioactive materials so you'll know whether or not to take your Potassium Iodide pills. With a dirty bomb, the answer will probably be "not," since levels will probably be low and a widespread contamination would lead to an evacuation which would reduce exposure time.

2. Call all family members to see where they are, if they'll be heading home or to a rendezvous point, or if they're involved in reactive measures being initiated by local officials. Alert them to hot zones and other travel and reaction advisories.

3. Listen for official instructions before changing your status. If you've evacuated, stay gone until given the all-clear, and if you've sheltered-in-place, wait for the all-clear before coming out.

4. If returning home from an evacuation, if you have any doubts about the interior of your home, do not enter until officials have had a chance to test for radioactive material. Otherwise, follow instructions for decontaminating your home as outlined under the "Coming Home" portion of the "Life on the Road" section of **"Evacuation."**

5. Some people recommend that you purchase radiation detection equipment. There's nothing wrong with that except that it's an expensive investment on gear you (hopefully) will never use. In the case of a radiological incident FEMA, the EPA, NRC, and a host of others will show up in sufficient numbers to ascertain which areas are hot or not, and most agencies will tend to err towards the safe side. However, if having detection equipment makes you feel better, and cost is not an obstacle, then by all means get some equipment. To help you, we've listed a couple of sources below.

Sources and More Information

➤ PBS has a program and an informative links page about the Dirty Bomb subject. See them online at http://www.pbs.org/wgbh/nova/dirtybomb, or contact them at **Public Broadcasting Service,** 1320 Braddock Place, Alexandria, VA 22314, or their local station locator at http://www.pbs.org/stationfinder/index.html, to see when this topic is to be aired again.

➤ For some good radiation and health information, visit the Health Physics Society, online at http://www.hps.org, or contact them at Health Physics Society, 1313 Dolley Madison Boulevard, Suite 402 , McLean, Virginia 22101, Phone: (703) 790-1745, Fax: (703) 790-2672.

➤ A general dirty bomb description, plus some Q&A links can be found online at: http://www.terrorismanswers.com/weapons/dirtybomb.html.

- ➢ "Guidance for Radiation Accident Management": http://www.orau.gov/reacts/guidance.htm.

- ➢ The Nuclear Regulatory Commission discusses dirty bombs online at http://www.nrc.gov/reading-rm/doc-collections/fact-sheets/dirty-bombs.html. For more information on this subject, nuclear reactor safety, or potassium iodide, you can contact them directly at U.S. Nuclear Regulatory Commission, Office of Public Affairs (OPA), Washington, D.C. 20555, Toll-free: 800-368-5642, Local: 301-415-8200, TDD: 301-415-5575.

- ➢ For varied news story opinions, see the "Wired" Dirty Bomb story online at http://www.wired.com/news/conflict/0,2100,53110,00.html.

- ➢ Read the FAS "Dirty Bomb" page at http://www.fas.org/faspir/2002/v55n2/dirtybomb.htm .

- ➢ See the Center for Defense Information's Dirty Bomb discussion at http://www.cdi.org/terrorism/nuclear.cfm.

- ➢ Visit the "How Stuff Works" page at http://www.howstuffworks.com/dirty-bomb.htm.

- ➢ For **THE** discussion, source, and extensive links collection regarding **Potassium Iodide pills, radiation detection equipment, and radiation info** in general, see http://www.ki4u.com. You can contact them at KI4U, Inc., 212 Oil Patch Lane, Gonzales, Texas, 78629, (830) 672-8734. Be sure to see their detector page at http://www.radmeters4u.com.

- ➢ For radiation detectors and other gear visit http://www.saferamerica.com, and click on "radiation detectors." You can also reach them at: Safer America, 226 East 54th Street, Suite #308, New York, NY 10022, Tel: 866-SAFER-99 (723-3799).

- ➢ See "Dosimeter.com" at: http://www.dosimeter.com, or contact them directly at 5 EASTMANS ROAD, PARSIPPANY, NJ 07054, 800-322-8258, 973-887-7100, Fax: 973-887-4732.

- ➢ Visit "Radmeter" at http://www.radmeter.com, or at 1-877-RAD-METER (723-6383). While you're at their website, click on "News" and subscribe to their free email **newsletter**.

Chemical Attack or HazMat Incident

General Discussion

Now it's time to upset the apple cart just a bit. We wish to apologize in advance to our friends and colleagues who've invested time and money putting all manner of biochemical threat books on the market. We've seen work after work of bioweapon this, and chemical weapon that, each one going into extreme but useless detail on chemical properties, biological development, weapon history, instructions for using unobtainable military equipment, explanations of how antidotes and vaccines work, etc., etc.

Truth is, as regards individual civilian protection, NONE OF THAT MATTERS! Yep, you read correctly, **IT... DOESN'T... MATTER!** An academic discussion of the incidental details of a limited number of substances will not help us in the middle of a release. What's important is that we know **_how_** to react, **_regardless_** of the threat or **substance involved**.

Let's look at it this way. If a Police Officer pulls a car over and the driver jumps out with a gun, is the officer thinking: A) "How do I protect myself and neutralize the situation?" or B) "Hmmm... now is that a Taurus 9mm or a Colt .45 he's holding?" Of course, the answer is "A." The type of weapon is a subject for secondary discussion well after the threat has been taken care of.

So it is with chemical weapons. In simplest terms, if you think you're being exposed to a chemical, you'll want to **reduce further exposure, stay protected as you evacuate the area, decontaminate as soon as possible, and seek attention from a medical professional**. The chemical makeup of the substance is for first responders and doctors to worry about. Not us. If you see a crop duster coming at you laying a fog over a city street, you're going to want to know how best to protect yourself with whatever it is you may have on you. Any discussion of how VX gas was accidentally discovered is absolutely pointless.

Many of the books we've reviewed have focused on discussing all manner of superfluous background information associated with the "*most common chemical weapons*" or "Weapons of Mass Destruction" that may have been produced by ours or a foreign military. What they fail to mention is that there are so many more chemical compounds you may be in danger of facing. There are any number of industrial chemicals that could be released intentionally or by accident, and there are just as many noxious or lethal compounds that can be easily produced by terrorists using locally and readily available chemicals.

In this section we'll provide you with reaction information that will help protect you in a number of situations, regardless of the substance used. That's why "IT DOESN'T MATTER" as regards specific chemicals. **It's the core reaction that counts**. Let's proceed.

Primary Danger

There is no way a paragraph or two can categorize the specific threats posed by countless chemical compounds. That's one of the reasons we gave you so many sources of basic information for "**NBC Data**" (to include hazardous materials) under "Basic Training." Some chemicals will absorb through your skin and affect certain organs, some will attack the skin itself, some won't touch the skin at all, but will destroy your lungs. For a listing of the specific ways in which some of the known chemical weapons can affect you, look at the chemical weapons charts on the websites we gave you under "NBC Data," and the applicable **military manuals** we've included on our **CD**. As for this section, we'll focus on the reactions that should be taken when faced with any chemical threat.

Advance Preparations

Most advance preparations have already been covered under "Basic Home Prep," "Gas Masks and Moon Suits," "Dressing for Chemicals," and what will be covered under "**Isolation**." In addition to those, consider these items:

1. We've mentioned the importance of making a Threat Map. One of the things this map should show is the location of factories and plants near you that may be host to a potential HazMat incident. It should also note the location of highways, railways, railway and trucking terminals, and their **relationship to your home, workplace, and prevailing wind patterns**.

2. You'll want to be able to **react as quickly as possible** to protect your own physical person, and your home or immediate environment against chemical exposure, contamination, and its negative effects. Therefore, **practice** putting on your **protective gear** (with your eyes closed) and have a few dry runs **sealing the house**.

3. Regarding a chemical incident, it, like a tornado, is one of a few things that will necessitate a rapid return to shelter if you're outside. Therefore, at home, work, and school (especially school), you should conduct "**Reverse Fire Drills**" (which we called "Tornado Drills" earlier).

4. Make sure you have your "**Daily Activity Journal**" with you so you can jot down notes (after you're safe) concerning details relating to the chemical you may have been exposed to. You'll want to record whether or not you saw someone using the chemical, what form it was in (liquid, gas, powder, etc.), if there was a color or odor to the substance, and symptoms noticed by you and others affected. **Medical officials will need this information** if you're able to record it. **Don't** take time for notes until you and yours are safe and clear of the affected area. (More on this below, under "Specific Follow-Up.")

5. **Install filters** where you can. They may help protect you somewhat in the event a chemical incident has unfolded and caught you completely by surprise. Install a simple counter-top **water** filter connected to your kitchen sink water tap. Filter all water that will be consumed, even for cooking. Even if you already have a good **HEPA** filter on your HVAC system, buying a **room air filter** will not only help reduce allergens in the house, it may trap hazardous vapor droplets or particulate matter.

6. Some HazMat incidents happen within the home due to the increasing number of complex household chemicals in use these days. Call the poison control center, the EPA, and the Red Cross to ask about treatment methods and antidotes for injuries resulting from pesticides, detergents, various medications, bleach, gasoline, and workshop solvents. Pay attention to what the label says regarding the disposal of various types of household chemicals. For some more information on dealing with household chemicals, see:
   - Child-safety tips as well as first aid steps for poisoning appear on http://homesafety.ws/poison.htm.
   - The National Library of Medicine maintains a comprehensive online database of household chemicals and related information at http://householdproducts.nlm.nih.gov/products.htm.
   - We've also located a few information and product sources regarding non-toxic household cleaners:
     - http://www.ecos.com
     - http://www.naturallyyours.com
     - http://www.mrsmeyers.com
     - http://www.safehomeproducts.com

7. The other preparation you can make in advance is the very first thing we suggested you work on, and that's your **health**. Have you been **exercising** and watching your **diet**?

Warning Time

Your only real warnings are going to come in the event of a local industrial accident, or highway or railway accident that involves hazardous materials. Even then, you'll only have a few minutes to react.

As far as a terrorist attack goes, your only warning *may* be to see someone tossing a suspicious canister, seeing a crop duster buzz a non-agricultural area, etc. Actually, you'll be lucky if you see anything at all.

Otherwise, you, or others around you, will start feeling symptoms that say "something's wrong!" In any of these instances, you'll have to react immediately.

Forms of Alert

In the case of an industrial accident, some plants or factories near you may have a warning siren or system. Earlier, under "**The EAS Alerts**," we asked you to find out what kind of alerts or alarms would be activated at the site or by the community in such a situation, and whether they would differentiate between a chemical release or a possible fire/explosion situation. If you consider the 1984 industrial accident that claimed 5,000 lives and inflicted 60,000 injuries in Bhopal, India, you'll see the importance of knowing what's going on around you.

In the case of other forms of accident, your only warning may come from law enforcement or other first responders cruising your neighborhood using their loudspeakers to announce the situation and that you should either evacuate or shelter-in-place. (Your phone tree should help relay such a warning.)

A third possibility is that the news will be announced on radio or television.

Warning Signs

In the absence of a warning, or your witnessing a chemical event unfold, there are certain environmental and physical symptoms you should commit to memory. If enough symptoms are present, you should react accordingly and not wait for official word. It may be too late by then. It's better to "seal first and ask questions later." Most of these will be obvious, others will not:

> Unusual amounts or colors of smoke are seen coming from a plant or factory, a railway area, or from the direction of the highway.

> There is an unusual haze and/or odor (such as bitter almond, freshly mown hay or grass) in the air, you notice a funny taste in your mouth, or you notice an oily residue or droplets on surfaces. (Remember, though, that many dangerous chemicals are colorless, tasteless, and odorless.)

> You see someone spraying where they shouldn't, or people wearing an unusual amount of clothing or breathing protection.

> You notice that animals are acting strangely, or that some are dead or dying. You may notice this in birds, squirrels, or in your outside pets.

> Similarly, a chemical attack could come to us through our water supply. Keep an eye on how your pets and house plants act as they may be more sensitive to the chemical than you are.

> Outside plants and vegetation suddenly appear wilted or dying.

> You start feeling one or more of the following symptoms:
  - Dizziness, light-headedness, or vertigo.
  - Blurred vision or sudden loss of vision.
  - A sudden or unusual headache.
  - A sudden feeling of anxiety, depression, or disorientation.
  - Sudden, uncontrollable laughter.
  - An intense ringing in the ears.
  - A burning sensation in your eyes, nose, mouth, throat, or skin.
  - Sudden and excessive salivation or runny nose.
  - Localized sweating or twitching of skin.
  - Unusual muscle twitching.
  - Unexplainable slurred speech.
  - A sudden cough, choking, trouble breathing.
  - Tightness of the chest (which may also indicate heart attack), or you feel as if you're suffocating.
  - Sudden and unexplained rapid breathing.
  - Stomach cramps, nausea, or you suddenly vomit, or lose control of bowels or bladder.
  - You're smoking, or drinking or eating something that suddenly becomes tasteless or offensive tasting.

➢ Where would our manners be if we didn't give you another acronym to help remember some of these? Since people with these symptoms would be in a lot of trouble, this one goes to the word **D.I.S.T.R.E.S.S.**:

Dizziness or disorientation.

Irritation of skin, eyes, nose, or throat.

Salivation and/or runny nose to excess.

Twitching of skin or muscles.

Respiration abnormalities such as rapid breathing, difficult breathing, or a tightness in the chest.

Elimination; sudden vomiting, or loss of bowel or bladder control.

Stress, sense of pending doom, or feeling of anxiety.

Suddenness with which symptoms appear. The more sudden, the more likely it's a chemical incident.

➢ You may also notice one or more of these symptoms in people around you. You may be in an office watching people on the street, or in a subway car watching the people in the car up ahead, when suddenly people start falling or acting strangely. Always be aware of your surroundings.

Reaction

The most important aspect to reacting to a chemical contamination is **speed**. You have to react quickly because the chemicals certainly will. In this case it's better to err to the side of caution and enact your protective measures even if you have a tiny doubt or two that a chemical incident is really unfolding.

Your first order of business is to protect yourself from exposure or from *further* exposure. Once you're protected and if you feel okay, you'll want to protect your immediate environment in the same vein, either from exposure or from further exposure.

Since many people envision **chemicals** as something being **sprayed**, we've put together a reaction reminder, using the acronym **S.P.R.A.Y.** :

Survey   **Survey** people and animals around you to see if they're being affected.

Protect   **Protect** your lungs, eyes, and skin as rapidly as possible.

Retreat   **Retreat** from the affected area, preferably heading uphill and upwind to a clean area and fresh air.

Alert   **Alert** others around you, **Assist** them, if you're able, and **Alert** the authorities.

Yell   **Yell** "Medic!" If you were near an incident, go see a doctor even if you don't feel ill.

Home

Let's assume the worst, that these reactions are in response to a detected release and not a protective response to a warning. In an advance warning scenario you'd start by sealing the house. Here we're going on the notion a chemical release caught you by surprise.

➢ Protect you first. Grab something close by to use as a breathing filter to protect you while you access your respirator or gas mask. You can grab an oven mitt, a wet towel, or even a small pillow off the couch. Anything is better than nothing, so protect you first.

➢ Alert any others in the house as to what's going on and what to do.

➢ Call 911 unless they're the ones that notified you. Definitely call 911 if you feel at all ill.

➢ This will sound cruel and will be extremely difficult to do, but if you have outside pets, and they appear to be **already affected** by whatever's going on, think twice about bringing them inside or you'll be introducing extra contaminants into your immediate environment. **Otherwise**, bring them in **immediately**. Pets are family too, and worrying or grieving will only add to the problems you're facing. Having your pets safe will bring comfort.

➢ Since this might be a shelter-in-place scenario, do the following as quickly as possible, and in this order:
- ❏ Close all doors and windows.
- ❏ Shut down your HVAC system.
- ❏ Turn off any exhaust fans or interior fans.
- ❏ Close the fireplace damper.

➢ Be sure to turn off your HVAC system in any NBC situation even if the power's off. You don't want power coming back on and the air conditioner sucking in contaminants.

➤ Call and alert family members not at home, and then call and alert your neighbors.

➤ Once done with the calls, be sure to **wear your communications**. Keep a phone <u>on</u> you.

➤ When you're certain your primary sealing is complete, turn on any room air filters you have. Some may filter some types of contaminants out of the air. (Exercise caution when cleaning or changing filters later.)

➤ If time allows, tape over any permanent openings such as the exhaust vent over your stove, window unit air conditioner, or in-floor vent heaters. Though there may be a little contamination in your house, it's like it would be with a flood. An inch of water in the house is bad, but a foot is worse. Do what you can to reduce any additional contamination. (See "Advanced Home Prep" in the "**Isolation**" section for a more extensive discussion of sealing.)

➤ Turn on your radio or TV for news and instructions. Better yet, turn on both. Watch the TV with the sound down and turn the radio up. Switch when it looks like a news report is coming on TV.

➤ Put your Bugout Kit by the door, and continue sealing your house if necessary. Be ready to shelter-in-place or evacuate. Though you may evacuate, try to pre-seal the house if you have time since you'll want a safe home to come back to.

➤ In a chemical event, your safe area should somewhere high in your structure. Stay on an upper floor in a closed-off central room rather than heading for a basement or other lower level as you would during a destructive event. Most chemical compounds are heavier than air and will settle.

"Work"

➤ This is one of the places your GYT Pack really pays for itself. Though nowhere near a NIOSH rated system, your simple homemade smoke hood and poncho will go a long way toward protecting you. At the very least, put on your smoke hood, with breathing filter, to protect your lungs, eyes, and facial skin.

➤ If this incident is occurring during a heightened terror alert, and you followed our earlier advice about "dressing for chemicals," then you're already wearing a few extra layers of protection. You should have your poncho with you to provide an extra outer layer of protection.

➤ Once you're protected, do what you can to help others. Make sure everyone knows there's a situation, make sure all windows are closed and that someone is shutting off the ventilation system. Shutting down the HVAC system should be a standard reaction to any chemical event whether the chemical was introduced inside or outside the building, or whether or not your primary reaction will be to shelter-in-place or evacuate.

➤ Move toward the interior of the building, away from windows and entrance ways, but do not go into central stairways that may be ventilated with outside air by a separate ventilation system. (Find out how your building's stairwells are ventilated.)

➤ Make sure the authorities have been alerted unless they're the ones who alerted you. Definitely call 911 if someone is ill or injured.

➤ Turn on a radio and TV, or call authorities for further instruction.

➤ If it's obvious the building you're in is ground zero, you'll want to evacuate as quickly as possible. As you should be wearing some sort of protection, you may have a little bit more time to evacuate than others. Put this time to good use by not being part of any panicked dash for an exit. It would be a shame to be protected against chemicals only to be trampled to death. Besides, you took our advice and you know where all the alternate exits are, right?

➤ <u>IF</u> the interior of your building was definitely affected by a chemical, and <u>IF</u> it's obvious some measure of decontamination is needed right away, you might consider intentionally activating one or more overhead fire sprinklers to act as an emergency shower. (They won't all go off if one is activated, that's only in the movies.) Check with authorities first, before doing this.

➤ If you're in a situation where you have a liquid come in contact with your skin, immediately wash it with soap and water. If you don't have access to water, blot it off with cloth, or use cornstarch or some other absorptive substance, even crumbled up soda crackers. You could even use a business card or something similar to gently scrape it off your skin like you'd remove excess water from a window with a squeegee. (If you do that, make sure your gently scraping with something that won't scratch or abrade your skin.)

➤ Once you're safe, call and alert family members, then proceed to a decontamination site, or start your own preliminary decon steps.

Car

➤ If stuck in your car during a HazMat incident, roll up your windows, turn off your airconditioner or heater, and set any circulation controls to "recirculate." You don't want any outside air drawn into the car. You may also want to go ahead and put on any respirator you carry in the car.

- Use your clothing or poncho to protect any exposed skin.
- Use masking tape or duct tape to seal around door and window seams, and cover AC vents on the dash and floorboards.
- Listen to your radio for official instruction. Once official instruction is given, you may be told to proceed to a decontamination site.
- Remain aware of your surroundings (be ready to react to any secondary situations caused by panic or traffic problems caused by evacuation efforts) and keep an eye out for emergency vehicles and responders.
- Let's discuss **public transportation** for a second, as we may see another scenario similar to the Tokyo subway attack. Being a passenger of public transportation has the disadvantage of placing you in a potential terrorist target. However, it gives you the advantage of being around other people who may show warning signs of chemical exposure before you're affected.
- In any event, react like you would anywhere else and protect you first. Then help anyone you may be travelling with. As always, you should know the location of any emergency exit, emergency alert devices (alarm switches), or the safety equipment locations in any place you frequent, which includes public transportation vehicles and facilities. Also be sure to ask bus or train personnel to turn off the ventilation system, and close windows or vents *unless the vehicle is ground zero for the incident*.
- Knowing all available exits in a public transportation facility will also help protect you from the physical injuries of being caught up in any crowded mad dash for the *main* exit.

Outdoors

- If you're outdoors during a major industrial accident, you're more likely to be exposed to a greater amount of contaminant than those inside their residences. Yet another good reason to know your local potential threats and your prevailing wind patterns.
- Conversely, if you're outdoors during an isolated terrorist attack, you're likely to get more fresh air and less chemical than those at ground zero (provided you're not AT ground zero).
- If you're out in the open and an alert is given, or you notice that suddenly birds are dropping out of the trees, or people around you start showing symptoms, hold your breath, protect yourself with clothing and any gear you have on you and immediately head away from the location, preferably moving both uphill and upwind. Some manmade chemicals, and most chemical weapons are heavier than air. If there is a visible haze or smoke to the chemicals, try to see how it behaves and where it blows. You go the other direction. Stay away from creeks or rivers, as they may be carrying large amounts of the chemical threat.
- Head for what cover you can. If this cover offers full protection, begin preliminary decontamination. Shed what clothing you can as quickly as possible since you can probably get rid of about 80% of your previous surface area that way. Turn the clothing inside out as you take it off, drop it in a small pile and leave it alone. As soon as you've done that, call 911 and notify them of the situation and that you may need treatment. You memorized the decontamination steps listed under "**Basic Personal Decontamination**," didn't you?
- In any event, utilize what personal protection gear you may have with you, even if it's taking off your shirt to cover your face and breathe through. Have we mentioned yet how much we encourage you to keep your GYT Pack with you at all times?

Miscellaneous Points

- Though we keep mentioning our GYT Pack, we fully recommend you acquire and carry the best chemical protection you can **find**, **afford**, and will **keep with you**. Our section "Gas Masks and Moon Suits" should have provided you enough information to make a good decision regarding protective gear. The reason we push the GYT Pack is because it's small and economical enough for everyone to have one or more, and *something* is always better than *nothing*!
- One minor reassuring thought is that gas dissipates and most chemicals wash away or eventually evaporate.
- Another reassuring thought regarding a terrorist application of chemical weapons is that though some substances are "officially" classified as "Weapons of Mass Destruction," true <u>mass</u> destruction depends on mass deployment of mass quantities under ideal conditions, and a massive lack of preparation on our part.

Specific Follow-Up

1. Once you KNOW you're out of an affected area, and authorities have been notified as to what's going on, proceed to a decontamination site or go ahead with preliminary personal decontamination measures as you're able. See "Basic Personal Decontamination" above.

2. Depending on the severity of the situation and the presence of officials on-scene, you may be immediately *taken* to an official decontamination and medical screening site.

3. Afterwards, even if you never felt ill during the incident, get checked out by a medical professional.

4. If you've been instructed to shelter-in-place, don't come out until given the all clear.

5. On returning home after an all-clear, be sure to follow the home-cleaning instructions given under the "**Evacuation**" section's "Coming Home" portion.

6. Call local officials to see if there are any specific instructions on dealing with a cleanup following a disaster with a specific chemical. For example some chemicals may be acid base and neutralize easily with a heavy application of a simple baking soda and water solution. Ask before playing chemist on your own.

7. Though many people recommend that you stockpile various antidotes and medications for self-treatment in chemical exposure cases, we recommend against it. The reason for our position are the "two Ds," or **Diagnosis and Dosing**. Making an incorrect assumption about the type of chemical involved, and administering the incorrect dosage of medication for your incorrect diagnosis, can be just as dangerous as the chemicals themselves. Besides that, many of these medications are only available to civilians "under the counter." In that case, you have no idea about the quality or source of the antidotes, or even if it's that drug at all. We stand by our position that you should see a doctor. **However**, if you can acquire the proper medications, store them safely, and only have them to ensure they're available when your **doctor** makes the diagnosis and sets the dosage, then there's absolutely nothing wrong with that. Also, there's certainly nothing wrong with receiving the proper training in diagnosis or dosing if you're educated by qualified personnel. Extra education is always a good thing.

8. Speaking of doctors, they and first responders will want to know everything you can tell them about the chemical you were exposed to. We made another acronym to the word **D.O.C.T.O.R.S.** to help you remember:

Delivery   Did you see the attack? How was the chemical **dispersed**? Liquid? Powder? Gas?

Odor   What did you **smell** if anything?

Color   If you saw the chemical, what **color** was it? Did you see a colored haze or smoke?

Taste   Did you develop a strange **taste** in your mouth? What did it taste like?

Oily   Was there a **feel** to the substance? Was it dry, wet, grainy, slippery, or oily?

Reaction   How did the chemical **affect** you or others? What were the noticeable **symptoms**?

Speed   How **quickly** did the chemical act or spread? How **rapidly** did symptoms appear?

Sources and More Information

➤ See all sources from "**NBC Data**" under "**Basic Training**" in the "**Foundation**" section.

➤ The EPA's Chemical Fact Sheet: http://www.epa.gov/chemfact, Environmental Protection Agency, Ariel Rios Building, 1200 Pennsylvania Avenue, N.W., Mail Code 3213A, Washington, DC, 20460, (202) 260-2090.

➤ Visit the EPA's "Pollutants/Toxics" page: http://www.epa.gov/ebtpages/pollutants.html.

➤ Agency for Toxic Substances and Disease Registry: http://www.atsdr.cdc.gov/atsdrhome.html, The ATSDR Information Center, 1600 Clifton Rd., Atlanta, GA 30333, Fax: 404-498-0057, Toll-free Phone: 1-888-42-ATSDR or 1-888-422-8737.

➤ Go to the **NIOSH** (National Institute of Occupational Safety and Health) site, and download the "**Pocket Guide to Chemical Hazards**": http://www.cdc.gov/niosh/npg/npg.html. (By the way, it's also on our **CD**.)

➤ A great links page to find all manner of MSDS or "Material Safety Data Sheets" is located at http://www.ilpi.com/msds/index.html.

➤ The CDC has an Emergency Response Hotline (24 hours) at 770-488-7100.

➤ Read Field Management of Chemical Casualties at: http://www.vnh.org/FieldManChemCasu/ or try: http://ccc.apgea.army.mil/reference_materials/handbooks/fmcc/ncohandbook2000.htm.

➤ The NATO Manual, FM 8-285 on Chemical Agent Casualties can be found online (along with other titles on the homepage) at http://www.nbc-med.org/SiteContent/MedRef/OnlineRef/FieldManuals/fm8_285/toc.htm.

➤ Read the FM 3-7 NBC Field Handbook at: http://155.217.58.58/cgi-bin/atdl.dll/fm/3-7/toc.htm, and you'll also find it on our enclosed **CD**.

General Discussion

As we mentioned with chemicals, there are so many books on store shelves right now full of detailed information on a limited number of potential biological weapons, and very little of it matters. The bottom line with biological events is, always has been, and always will be, "**If you feel sick go see the doctor!**"

If there's one important point we can make in this section, it's **don't <u>wait</u>** to go see the doctor. Some people may think, "Oh, it's only the flu." as a disease progresses, and then it's too late. Furthermore, going to the doctor early may have you treated and released early, thereby avoiding increasing numbers of other people seeking help. Being around such large numbers of people may expose you to other illnesses, or to a *real* illness you only *thought* you had.

Also, if you really are infected by a bioweapon and you go to the doctor early, you might save others by helping authorities reach a more rapid conclusion that a biological event had occurred. Even an extra two days could save thousands of people.

By the way, have we mentioned how important your overall health and immune system function is?

Primary Danger

Generally speaking, the primary danger is the disease itself and any lingering effects it has on your body or your health. For example, though there's currently no cure for SARS, most victims survive. One thing the media is not publicizing as it should is the fact that many of these survivors have a certain amount of permanent lung damage.

Specifically, the primary danger of disease comes from the microorganism causing the illness. Simply put, if you don't get any on you or in you, you'll be fine. We'll continue along that vein as **prevention** is the one area under our control, and disease is certainly one area where prevention is far preferable to cure.

That leads us to the fact that biological considerations are a yes or no. There are no partials. You're either infected or you're not. There's no such thing as "slightly" infected. With a tornado or earthquake, your house can sustain minor damage or major damage, with a broad range of possibilities in between, depending on what kind of preemptive mitigation was performed on your part. With disease, it's like a light switch; on or off. Therefore, your only "reaction" other than going to the doctor, is to do everything you can to keep from being exposed in the first place.

Advance Preparations

Regarding a biological threat, your advance preparations are pretty much the only "Reaction" available to you. Anything after the fact will be up to medical professionals and not us, so instead of a "reaction" we'll focus on "preaction." Therefore, our main discussion on defense against a biological threat will be prevention instead of location reactions such as home, work, car, etc. The factors covered below should help protect you regardless if the biological event is an intentional attack or a natural epidemic.

You can best protect yourself from biological weapons or epidemics by doing the following:

1. **Avoiding exposure during higher alerts and known epidemics.**
   - ➤ Act as you would during cold and flu season and keep your physical distance from others as best you can. Avoid people who seem to be symptomatic, even if it looks to be only a common cold. Even a simple ailment can lower your resistance making you more vulnerable to a more dangerous germ.
   - ➤ If you normally use public transportation, see if you can find a less populated form of travel such as a carpool. If not, keep to yourself as much as you can. (It's likely that our mass transit system will be ground zero for an intentional bioweapon attack.)
   - ➤ If you're stuck near someone who seems symptomatic (of anything), don't be afraid to use your GYT Pack's breathing filter as an impromptu mask.
   - ➤ If your job allows you to work from home through telecommuting, or some other arrangement, do it. Do anything that will reduce your contact with other humans. If an outbreak reaches pandemic proportions and no official quarantine or cessation of services has been instituted, you might want to use a little vacation time and stay home. Don't wait for official word.
   - ➤ Do a little more shopping and bill paying online and over the phone, thereby reducing the time you spend shopping in public, and the amount of mail you have delivered. Though our last Anthrax attacks were a while back, the stage has been set and there may be more. If you're <u>really</u> paranoid about your mail, wear surgical

gloves and a respirator while opening it outside, or while opening it with a clear plastic bag over your hands and mail. We have some mail handling info sites under "sources" below.

➢ Do not eat at open-buffet restaurants (cafeteria-style serving lines are okay). Publicly accessible food arrays should be avoided as they're easy to contaminate, and the utensils are handled by too many people. During higher alerts or epidemics you should eat at home for those reasons, and to avoid contact with other people.

➢ As mentioned in our discussion of the higher terror alerts, it might be a good idea to keep your home slightly sealed. Don't go the full duct tape and plastic sheeting route, but if you live in a more heavily populated area that might be ground zero for a bioweapons release you might want to sleep with windows closed, etc.

2. **Being current on all immunizations and medical checkups.**

➢ As listed under heightened terror alerts, be sure to be up to date on all checkups and immunizations. Simply put, you want to keep your slate clean in order to deal with the bad stuff that might come along.

➢ Always use your "**Daily Activity Journal**" as it will be extremely important with "ring vaccination" even if you haven't been actually infected. A "ring vaccination" is where an infected person's family, friends, acquaintances, and their circle or "ring" of contacts are vaccinated against a particular disease in anticipation of the friends and acquaintances being exposed or infected. This is a common reaction to a Smallpox outbreak. Your journal can tell you if you were ever in contact with anyone who got vaccinated as part of the ring, and it can tell physicians who to vaccinate if you're the infected victim.

➢ If you ever have the opportunity to be vaccinated against Smallpox, do it, provided your personal physician agrees. However, think twice about taking it if the vaccination is being given in a **mass vaccination program** in **response** to an outbreak. It is our non-medical lay opinion that the crowds at inoculation facilities will either help spread the very disease being treated, or it will allow for mass transmission of some other ailment. It would be a shame to show up for a Smallpox vaccination, only to catch SARS. Not only that, but *large* crowds under stress are a riot waiting to happen. It's our suggestion you keep your distance.

3. **Practicing increased personal hygiene.**

➢ Wash your hands more often. This is by far the simplest and best way to prevent illness. Get in the habit of carrying a waterless hand sanitizer or antiseptic moist towelettes. This is particularly important after visiting public gatherings or for school children who are in closer proximity and physical contact with others.

➢ Wash any fruits or vegetables that are to be eaten raw, even if they came in some sort of wrap from the grocery store. Someone had to handle them to put them in the wrap, didn't they?

➢ Hygiene also includes keeping your immediate area clean. For office workers, this will include your air. Factory workers usually don't have as much to worry about in this regard as most factories are keen on ventilation, and workers generally have more space between each other. For the office workers among us, try an ionizing air purifier in your office or cubicle. Though we have no test data that shows the ionizers to be a recommended anti-biological device, it stands to reason that anything rated to drastically reduce airborne particles, would help reduce any sort of droplet or airborne organism that might be floating around in your area. Also, cleaner air will protect you from dust mites, molds, fungi, or other bits that might cause you a minor ailment and increase your susceptibility to something worse.

➢ Similarly, keep your home clean. Many common ailments are caused by the things listed above like dust mites, molds, etc. You should also be concerned about mosquitoes, fleas, rats (in fact, Hanta Virus is particularly nasty and is spread by rat feces), and other common disease carriers. Your lawn and garden or hardware store will carry chemicals you can spread around your house to ward off any number of pests. If nothing else, spreading moth crystals around your home's foundation will help keep most pests at bay, and you should always wear repellant during mosquito season. (Don't forget the pest info sources we gave you under "Basic Home Prep.")

4. **Being prepared for a quarantine, whether self-imposed or officially instituted, if a disease is particularly dangerous or gets out of hand.**

➢ That's what our "**Isolation**" section is about.

To help you remember these, we have another acronym for you: **G.E. Rx. M.**:

Germs	**G**o wash your hands!
Exposure	**E**at at home and generally reduce your contact with others.
Rx	**R**est, diet, and exercise. Remember to maintain your health.
Medic	**M**edical attention is your only option if you think you're sick.

Warning Time

There are only two instances in which there would ever be a warning. One is if an outbreak occurred elsewhere, as we've seen with SARS, or if a local outbreak was announced on the news. The other is if solid information were given that a terrorist was in the process of releasing a bioweapon. In these cases, your warning time would depend entirely on how long it took officials and the media to get the word out.

Other than those two instances, **your only warning will be <u>after</u> the fact**, when you or those around you start showing symptoms.

Forms of Alert

Once an epidemic has been reported or a bioweapon released, radio and television news will be your primary source of up-to-the-minute information. You may also receive a call from your doctor or notice from your health department.

Warning Signs

In lieu of anything obvious such as a crop-duster scenario, someone tossing a canister, or using a sprayer (you'd be lucky to see anything), your only warning signs are going to be either news stories or symptoms.

Some common symptoms of more dangerous diseases (including biological weapons) include, but are not limited to, the following:

- Fever and muscle aches as commonly associated with the flu.
- Heavy congestion of sinuses, throat, and chest.
- Unusually sore throat.
- Unusually severe headaches.
- Severe neck stiffness.
- Bloodied whites of the eyes, or inflamed, infected, or swollen eyelids.
- Stomach cramps, nausea, vomiting, or diarrhea (with blood possibly present in vomit or stool).
- A localized or whole-body rash.
- Odd skin coloration; a yellowish hue or ash-gray tone.
- Severe swelling of glands or lymph nodes.
- Odd sores, boils, or pustules, especially new ones that develop quickly, or ones that won't heal.
- Asthma-like symptoms.
- Tremors of head, hands, or other extremities.
- Delirium, delusions, or hallucinations.

To help you remember some of these symptoms that may indicate a more serious ailment, we have an acronym to the word **A.I.L.M.E.N.T.S.**:

Asthma-like symptoms.

Intestinal tract problems such as nausea, vomiting, or diarrhea.

Lymph nodes may be swollen.

Mucus buildup in chest or sinuses.

Eyes are bloodied or lids swollen.

Neck is unusually stiff or sore.

Temperature is at fever level.

Sores, boils, or rashes that appear suddenly or won't heal.

Diagnosis of what these symptoms may mean is up to your doctor. If you have any of these symptoms, or others not listed here, go see your doctor! We repeat; **"If you feel sick, go see the doctor!"**

Miscellaneous Points

➤ One form of bio attack may take the form of a "dirty" bomb except using biologically contaminated waste instead of radioactive material. Such a device may be planted in high-traffic areas such as a shopping mall where lots of people may be in close proximity. A similar scenario would see an intentionally infected suicide bomber choosing a crowd as his target so his infected blood would expose passersby. If you're a close witness to any sort of bombing, begin personal decontamination measures immediately and then seek a medical checkup soon thereafter even if you don't think you were directly affected. It could be a "bio" bomb or a radioactive "dirty" bomb. You just never know.

Specific Follow-Up

> **If you're sick, go see the doctor!**
> If your doctor is unavailable, call your largest area hospital <u>and</u> your county health department to ask them where you should go for treatment.

Sources and More Information

> In addition to the biological weapons information given in the "NBC Data" portion of "Basic Training," you'll find a surprising amount of information in the sources we listed under "**Health**."
> The USPS mail security page is at http://www.usps.com/news/2001/press/pr01_1022gsa.htm.
> FP Mailing has a mail security page at http://www.fpusa.net/mailroomsands.cfm.
> Remember the CDC Emergency Response Hotline at: **770-488-7100,** http://www.cdc.gov.

Fire

General Discussion

As promised, here is our **second** of three parts regarding fire.

Earlier under "**Basic Home Prep**," we discussed home fire prevention measures and fire equipment needs. In this section, we'll discuss your **personal safety reactions** to a fire, and how to get you and yours out of harm's way. As with everything else in this portion of the book, once you've reacted and protected yourself, you can figure out what step two should be. In this case, step two *may* be fighting the fire. This we'll discuss in the "**Isolation**" section under "**Security and Defense**."

We can't stress strongly enough to the need to have **good background information on fire**. If you missed any information source recommended earlier under "**Basic Training**," go back and look at it now.

Not only is fire an enemy itself, it's a common terrorist weapon, it's more sudden and less forgiving than other threats, and it's an all too common after-effect of other major disasters.

As with many threats, the key to your safety lies in early detection and rapid reaction. Let's talk about the subject some more.

Primary Danger

The primary danger from fire is smoke. More people who die <u>in</u> fires die from smoke inhalation than from burns. A component of smoke inhalation is the asphyxiation that occurs as flame consumes all available breathable air and replaces it with toxins, and unbreathable gasses. Next in frequency are those who die as a result of burns. A smaller percentage die from injuries received either in a panicked evacuation of a burning building or from debris as the burning structure collapses on top of them.

Advance Preparations

Most of your reactions to a fire are going to center on early detection and rapid reaction. Knowing a fire has started, reactions include getting out of its path, protecting yourself in the process, or protecting yourself if you *can't* leave. Let's take a look at a few things you'll want to have done beforehand:

1. Make sure you've completed all the prevention steps, and the detection and extinguishing equipment recommendations made under "**Basic Home Prep**." You should also apply those suggestions to your **workplace**. Especially important is increasing the number of water-unit fire extinguishers or their equivalent.

2. Anything critical to safety is worth practicing, and **fire drills** top the list. Conduct fire drills at home and ask that they be conducted at work. Make it a habit to learn all the fire exit locations and fire and safety equipment locations of <u>any</u> building, structure, or vehicle you frequent. If you skipped it (and you shouldn't be skipping parts, should you?) go back and read the section, "The Drill to Survive."

3. Also make it a habit to notice the fire exits and fire safety equipment locations of **any building you enter**, and not just the ones you frequent. Consider this: Not too long ago, there was a fire in a nightclub that was started when stage pyrotechnics ignited ceiling materials. Over 100 people were killed trying to get out the main exit when other exits were open. Additionally, the film footage of the ceiling catching fire was broadcast for weeks, and the one thing that was *never* seen on the footage was *any staff member heading to the stage with a fire extinguisher!* YOU need to know where fire and safety equipment is. Don't expect anyone else to react.

4. Know and memorize all the **"hidden" water** sources of the places you frequent. You may need them to wet you down to protect against flame. Among these would be flower vases, snow globes, etc. (We have a longer list below under "**Work**.")

5. Practice assembling your GYT Pack's smoke hood with your eyes closed. You'll always want to be able to put it on quickly, and you may have to put it on in total darkness or in heavy smoke.

6. One of our recommendations under "Dressing for Chemicals" was that you check your clothing for a UL fire retardant rating. Clothing layers will not only protect you from chemicals, but could also offer a great deal of protection from flame and heat.

7. If you live or work in a tall building above the tenth floor, examine alternate escape methods in case all exits are blocked by fire. Fire truck ladders are limited to about the tenth floor. Getting down from anything higher will be up to you. That's why two options discussed below will mention escape parachutes or rappelling. IF you decide to investigate either, one of your "Advance Preparations" will be to make sure you keep this equipment accessible, and that you're trained to use it.

8. If your home is in a potential **wildfire** area, have tools and materials ready to perform all the steps we'll list below regarding wildfire.

9. You'll also want to keep a small Bugout Kit **away from home** during fire season in case something should happen to your home while you're away. In this situation, the most important component of your Bugout Kit that you should keep with you is your Info Pack with all your insurance information.

10. If you live in a wildfire area, you should have **alternate sources of water**. The municipal main may be low on pressure and volume due to firefighting efforts nearby. Also, having alternate water sources will free up the municipal water desperately needed by others. For alternate sources of water you might consider the following:

    ● If your water table is accessible, you should have a well. Having an additional water source is good advice regardless of wildfire threat.

    ● Any swimming pool can be used as a water reservoir. So can any nearby creeks, rivers, or lakes, provided you have access to a water pump and hoses.

    ● Hoses and water pumps should be considered standard equipment for any home in a potential wildfire area. (Hoses and pumps are also useful in either flood or drought situations.)

    ● Electric water pumps are the most common. Since you can't count on the power being on the whole time, you should consider running the pumps off your generator. However, be careful in your use and placement of your gas-powered generator since you don't want it being a contributing factor to the fire. The same holds true if you're using a gas-powered water pump.

    ● Go back to "Basic Home Prep," and revisit our list of yard tools to own. Make sure you have a shovel, heavy rake, axe, sledge hammer, chain saw, and enough garden hose to reach from two separate water sources to the farthest side of your house or yard.

Warning Time

The longest warning time you'll ever have for fire is in a scenario where a wildfire is approaching your house and it's still a few miles away. In this case you might have the better part of a day to prep your house and to gather your gear and goods to evacuate.

If you live in an apartment complex where another apartment in your building has caught fire, you may have up to **two minutes** (absolute **max**!) to grab your stuff and run, but that's about it. On average, a fire will double in size every 30 seconds. Try to be out and gone in less than one minute.

Your next longest warning would come when your smoke detector alarmed from a smoke source that had not yet become a fire. You may have a minute or two to find the source of the smoke and prevent a blaze from erupting.

Other than situations like these, you'll know there's a fire when a fire alarm goes off, or if you see the flames or feel the heat. At that point, all reactions have to be immediate and properly executed.

Forms of Alert

The wildfire or brushfire scenario is the only one you'll see on the media. Other fire warnings are going to come to you through smoke detectors, fire alarms in your building, or someone pounding on your door yelling "fire!"

Warning Signs

Aside from alarms, the warning "signs" of fire are rather simple:

➢ The smell of smoke or of something burning.

➢ Seeing smoke.

➢ Animals running from a wilderness area.

➢ Seeing the flames or hearing the crackle of flame.

➢ Feeling the heat from the flames.

➢ Hearing the commotion of people reacting.

Reactions

We have another acronym for you. Actually, there's already one in use now called **R.A.C.E.** which stands for **Rescue, Alert, Contain, Extinguish**. It's not a bad acronym really, but here's one that goes to the word **F.I.R.E.**:

Find   **Find** the others you're with, family members or coworkers, and get them to safety.

Invite   **Invite** the Fire Department. **Call 911**.

Restrict   **Restrict** the flames by closing doors, closing windows, and cutting off the gas.

Evaluate   **Evaluate** quickly as to whether you can **Extinguish** or you should **Evacuate**.

Home

Most of the information you'll need to react to a fire in your home was given in the "Fire Drill" portion under "The Drill to Survive" earlier. The intent of this reaction section is to **get you out of harm's way**. Under "**Isolation**" we'll talk more about actually fighting a fire. For now let's look at some additional considerations for getting you to safety and away from a fire **inside** your home.

➢ The Fire Drill information earlier covers most of the information you needed for the very basics. If the house is on fire, let everyone know, and leave in a safe, orderly, but rapid fashion. Most fire deaths occur when people were never able to get out, either because they didn't have adequate detection, or they had not fully studied their exit options.

➢ Other problems occur when people try to go back inside the house thinking they can gather valuables. DO NOT GO BACK INSIDE THE HOUSE. If the fire was severe enough to drive you out, then don't go back in unless it's a calculated risk you're willing to take in order to rescue another human. We'll give you more info on this under "**Isolation**."

➢ Scenarios in which you could immediately fight a fire will also be covered under "**Isolation**."

➢ The most important point to reiterate is, **stay with breathable air**. When exiting a building on fire, you should expect a lot of smoke. Stay low, no higher than a foot off the floor. Protect your lungs if you can, with, if nothing else, a wet cloth over your nose and mouth.

➢ The second point to drive home is the importance of the "**Stop, Drop, and Roll**" reaction to clothing on fire. Here's *why* this reaction is important, or rather, why it works. If your clothing is on fire, moving or running only fans the flames and feeds the fire, making it worse. That's why you need to "**Stop**." As fire burns, it burns *upward* toward your face where it can cause debilitating burn injuries to sensitive facial skin and eyes. Also, being excited, you could inhale flame and burn your lungs causing irreparable damage. Therefore you need to "**Drop**" which essentially pulls your head and face out of the way of the flames and also places you on the ground where you can smother the fire. Rolling is what smothers the flame. "**Roll**" slowly so as to put the fire out. If you roll too quickly, you'll only fan the flames again.

Now let's discuss a few considerations to keep in mind if a fire is **outside** your home and threatening your home and property. Here we'll use the scenario of an approaching wildfire with **several hours** notice. More threatening scenarios would simply see you leave immediately.

You should do things in this order:

1. Prep the family to evacuate.

2. Set up any water pumping and spraying gear you might have.

3. Shutdown and seal the house.

4. Perform any last minute protective landscaping.
5. Load any valuables or heirlooms that will fit in your vehicle.
6. Protect valuables and heirlooms that have to be left behind.
7. Evacuate.

Note: These steps are all **time sensitive**. Any fire in your area should be monitored so you'll know exactly how much time you have. Anything **less than two hours**, perform only numbers **1 and 7**.

Another quick way to remember what to do in advance of a <u>wild</u>fire uses the acronym **W.I.L.D.**:

Water   Set up your sprinklers and any **water** pumps you may have.

Interior   Prep the **interior** of your house as outlined below. Help protect your valuables.

Landscaping   Perform some last minute **landscaping** to help slow the spread of fire to your home.

Depart   Time to **depart**. Get out as soon as you can, and don't stick around attempting to fight the fire.

Let's look at some things to keep in mind while doing the above.

➢ At the first hint of approaching wildfire, round up the family. Make sure children are home and/or accounted for, and all pets rounded up with carriers and leashes ready.

➢ If you can, get most of your family out of the area well in advance of any approaching fire. Have them take your main Bugout Kit and other items they're able to load based on the hierarchy we'll explain under "**Evacuation**." If the fire is too close for comfort, don't worry about the house, just leave! For the rest of the steps that follow, we're assuming you could not start these measures while everyone else was packing to evacuate. (Otherwise you should complete these steps and leave with the rest of your family.)

➢ If you decide to stay behind to make final fire prep arrangements, try not to stay alone. There's safety in numbers so keep an able family member with you, or work as a team with a neighbor.

➢ Whatever you do, do things in advance of the fire to protect your home and do NOT stick around to fight a wildfire. Small grass fire maybe, but NOT a full wildfire or forest fire.

➢ Hopefully you followed all our landscaping advice earlier under the "Fire" section of "Basic Home Prep." All you should have left to do now is to close down the house, some very quick work in the yard, and setting up sprinklers (if not already installed).

➢ Before you do anything at all, make sure <u>your</u> evacuation vehicle is set and ready to go. Have it prepped and ready so that all you have to do is hop in and go! **Word of caution**: Don't keep the keys in it. You never know who might panic and be in desperate search of a way out of the area. Don't want them to wind up taking your ride do you? Keep your car keys secured to a belt or on a neck chain. You don't want them to fall out of your pocket somewhere.

➢ Constantly keep an ear to the news to make sure your planned escape route isn't compromised. We suggest keeping a portable AM/FM radio <u>on</u> you. Similarly, pay close attention to smoke and animals. Animals will be running away from the fire and will not only be a good indicator of how close the fire may be, but some wild animals may pose a danger to you while you're out prepping the yard. Pay attention!

➢ Similarly, this is dangerous situation, so be sure to **wear your communications**. Keep a cell phone and/or two-way radio on you while completing all these steps.

➢ Start inside the house and work outward. Close all outside storm shutters if any, and close the windows and any interior shutters. If you have absolutely no other window protection you should tape your windows with generous strips of duct tape up and down, across, and diagonally over each window as you might for a hurricane. Add a layer of aluminum foil if you have it. Anything's better than nothing when you're dealing with the winds, debris, and embers stirred up by a massive fire. If you have metallic window blinds, leave them in place and close them.

➢ Move all flammables away from the windows. This includes drapes, plastic blinds, and any nearby furniture.

➢ Close any screens in front of the fireplace and leave the damper *open*. (If a spark gets in, you want it to drop into the fireplace and not smolder inside the creosote-coated chimney.)

➢ Turn off your HVAC system, close all interior doors, and turn off all interior lights.

➢ Outside the house, shut off the gas at the meter, or at the valve if you have an outside tank.

➢ If you live in a hurricane area and/or have boards pre-cut to cover your windows (instead of storm shutters), put them up now. (Earlier we suggested you treat exterior wooden shutters with fire retardant. This is why.)

➢ Move anything flammable away from the house. This includes outside trashcans, wood or plastic patio furniture, firewood piles (which shouldn't be next to the house anyway), etc.

> Set a lawn sprinkler on your roof and fasten it in place (though you should've already installed a roof system if you live in a wildfire prone area). While on the roof, wrap any roof turbines, vents, and accessible chimney spark arrestors with aluminum foil and duct tape. Work safely.

> Cover any other attic vents or gables with aluminum foil and duct tape unless you have storm shutters or pre-cut boards for them.

> Once back down on the ground, cover any basement windows or low-level vents with aluminum foil and duct tape, again, unless otherwise protected.

> Set sprinklers around the house to cover as much exposed wall as possible. Make sure they're as secure as everything else since large fires will create a windstorm and will blow loose items around as well as blow burning debris into your yard. Along with dousing flaming debris and embers, the water will help keep the walls cool. Some fire is transferred by sheer heat alone without the aid of direct contact with flame.

> If you have an outside gas tank, cover it with sheets or old towels and make sure you have a sprinkler covering it. The sprinkler will keep it cool and wet, and the fabric will hold water longer.

> IF time allows, cut down any tree branches near the house, and all bushes next to the house. Also cut larger limbs that offer a connection from treetop to treetop. This is one of the many ways fires spread.

> Turn on all the sprinklers and LEAVE. (**Note**: Don't use municipal water if you have alternate sources. Municipal water will be depleted or suffer pressure reduction due to area firefighting efforts. This will reduce the effectiveness of your sprinklers, and massive sprinkler use may hamper local firefighting efforts.)

> Never stick around to fight such a large fire on your own or even with other people.

> When you leave, call your fire department's non-emergency number and let them know your address and your status. If your house winds up burning in your absence, that's one less house authorities will have to search for victims.

> Also when you leave, turn your **outside lights on**. Fire crews may be going house to house through increasingly heavy smoke looking for stragglers. Make it easy for them to find your house and the **note you left on the front door** telling them everyone is gone.

> If you're working on the above preparations and the fire suddenly changes but there are no officials around to notice, call 911 and let them know of the fire change.

> At any time during these preparations you feel threatened, leave! Insurance will pick up where you left off.

> However, once you've completed these protective steps and you're CERTAIN you have a few minutes (and your escape route is sure to be open) you might want to consider a little extra protection for some valuables.

  - If your safe room is on your lowest floor and you gave it a lining of sheet metal as suggested earlier (or flame-retardant insulation) you might want to place a few valuables here as this part of the house may be the last to burn. You can hedge your bet by placing a couple of your "Dryer Watchguard" or "Range Queens" in your safe room to retard any encroaching flame. (Neither product is rated for such use, but it stands to reason either would hamper flame a little, and a little can be a lot in some instances.)

  - One trick is to fill your tub and your empty trash cans and plastic bins with water, and drop in any waterproof valuables you want to protect.

  - **Hint**: You may have "waterproofed" some valuables earlier in the way you prepped them for severe weather storage or for long-term storage. To do this, use a plastic bin (with no holes in any of the sides) with a solid plastic lid. Put your valuables inside resealable plastic bags or vacuum pack them in plastic, and seal any seams with duct tape. Fill your bin about 2/3 full of valuables and fill the last 1/3 of the bin with bubble wrap, or wadded up newspaper sealed inside plastic bags. This will act as waterproof padding. Next, seal your lid in place with a bead of glue or silicone caulking around the rim of the bin, placing your lid on and sealing the seam with duct tape. To "use" this container in the tub full of water, simply turn it upside down and place a heavy object on top to keep it submerged. Being upside down, the bin will trap air inside and the valuables will now be resting on top of the bubble wrap or padding, keeping everything out of any water that might seep in. (**Note**: This is a lengthy process and _not_ something you want to be sitting around doing as a wildfire approaches. We mention this here as a good way to protect some things **IF** you have them **already sealed** in this manner, IF you have time to fill the tub, and IF these items couldn't go with your family.)

  - If you have a swimming pool you're not using as a source of sprinkler water, you could also drop a few water-proof valuables in there.

  - If nothing else, move as much as you can to the center of the lowest floor of the house, cover everything with a waterproof tarp and then with towels or bed sheets, and hose the whole thing down to keep the cover wet. If the house burns, this will probably be the last stuff to go.

  - However, we wish to remind you: don't mess around with possessions if time is at all questionable! Insurance will pick up where you left off.

> Even though a wildfire might seem small enough to handle, don't take chances. Out of numerous successful "home fire fighters" interviewed, most said if they had to do it all over again they'd leave. Many have reported lung damage from smoke and fumes, some experienced serious burns despite safety measures, and all reported the actual fear factor was much higher than anticipated.

"Work"

Again, we expand our definition of "work" to include the places you spend most of your time away from home, and where you go for leisure activities. The steps below apply to your job, school, etc. For these steps, we're assuming a building fire in which you may be trapped. If you're not trapped, the only thing to mention would be to walk safely to the nearest exit, protect your breathing, and low-crawl if you have to. Let's look at some more serious considerations:

> As stated, the basic reaction to a building fire is to exercise what you've practiced in fire drills and just leave in an orderly manner. The most important thing to remember is to **not use elevators** since the power may fail and leave you trapped.

> Also remember how important it is to know all **alternate fire exits** so you can avoid crowds at the main exit.

For the remaining steps, we'll assume the worst case scenario of a high-rise-building fire that may have you trapped above the tenth floor. Let's assume you're in a room or office, the fire alarm has been sounded, you've been alerted by fellow tenants, or you see smoke that's obviously from a major fire.

> Remember the **F.I.R.E.** acronym : **F**ind, **I**nvite (call 911), **R**estrict, and in this case, **E**vacuate.

> Gather everyone that's in your room or office, check the exit door (surface and knob) for heat with the back of your hand, and close, but DO NOT LOCK your door behind you after you leave. Firefighters will need full access, but more importantly, you may have to turn back and you'll need the safety of that room. If you're in a hotel, leave something to prevent the door from closing all the way as many hotels use electronic keys which may not work in a power failure. **Note**: Even if a door is cold to the **touch**, brace against it as you open it slowly. Air interchanges can cause a backflash, which is essentially a small explosion. Once you have the door open a bit, **look** for smoke, **smell** for smoke, **listen** for other people and the crackle of flames, and see if you **feel** any hot air before proceeding out into the hallway.

> Proceed in an orderly fashion to the nearest exit. (For the rest of these steps, let's assume your exit was blocked by fire or heavy smoke and you had to return to your original location.)

> Activate a wall alarm, especially if no other alarm was given. This will at least let authorities know that someone was on your floor.

> Take a last look around the hallway noting the location of fire hoses and fire equipment. Don't retrieve it unless flame is present and you have to fight the flame for immediate survival.

> In most office buildings, restrooms are in the hallways. If this is the case and you feel you have a second, use a couple of the resealable plastic bags from your GYT Pack to fill with water to take back to the office.

> Make sure the room you return to has an exterior window.

> Stay calm. Though fire claims many lives each year, there are always more survivors than victims due to the professionalism and capabilities of most Fire Departments.

> Call authorities to let them know you're trapped. Try your regular phone first and then your cell phone. Using the regular land-line phone will verify your location to 911 dispatchers.

> Next, seal yourself in. Use wet cloth along the bottom of doors and tape to seal around the seams. The less smoke that gets in to contaminate your air the better.

> Start to signal from the window. Open the window if you can, but DO NOT BREAK THE WINDOW. Smoke or flame may billow up from the outside and you need to be able to close the window to protect yourself. If possible, open the window slightly at the top and at the bottom. The top opening will allow interior smoke to leave and the bottom will allow access to fresh air provided the floor below you isn't engulfed in flames.

> If the window is a solid pane of glass and you can't open it, try signaling anyway. Use a lamp or flashlight if it's dark outside and wave a brightly colored object during the day. Some windows have a reflective film which will make it difficult for anyone outside to see in. If your windows are like this, see if the film is on the *inside* of the window and if it can be scraped off. This will let you signal without breaking the glass.

> However, as a last resort, break the glass. Stand back and throw something through the glass to break it rather than risk getting cut or falling out the window. Break out all the glass so no shards will be left to fall on you as you lean out the window to signal.

> Hang a bed sheet, towel, or other brightly colored object out the window to draw attention. Remember our mention of "dust" or "powder" as a signal in the "Communication" section? This is a situation where you'd need all the attention-getters you could find.

➢ As smoke increases, put on your smoke hood or evacuation respirator. If you're using the GYT Pack components, this is the situation in which you'd want to use all your layers of insulation, both with the hood and with protective clothing. (Remember, in a fire, <u>don't</u> wear the poncho.)

➢ As heat increases, you'll know fire is close. You'll want to have sources of liquid in order to keep you wet and cool, and your breathing filters wet. You should have gathered some water in anticipation of being trapped. Other less-obvious sources of water and liquid include, but aren't limited to, the following:

❏ Soft drinks and bottled water
❏ "Snow globes" (not for breathing filter)
❏ Water-unit fire extinguishers
❏ Ice machines
❏ Drinking fountains
❏ Janitorial sinks
❏ Plant watering cans (not for breathing filter)

❏ Aquariums (not for breathing filter)
❏ Fire sprinkler heads
❏ Cold coffee
❏ Water cooler jugs
❏ Toilet tanks (not for breathing filter)
❏ Shampoo or conditioner (not for breathing filter)
❏ Flower vases (not for breathing filter)

➢ Let's talk about fire system sprinkler heads for a second. If you're in a smoke-filled room and **really** feeling the heat from flame, and there's a sprinkler head in the room you're in, go ahead and manually activate it. Sprinkler heads will only go off on their own in the presence of direct and relatively intense flame, and it's a movie myth that when one sprays they all spray. Each sprinkler head will have a little vial of liquid in it, usually red in color, or will have a small metal rod in the center. The vial or metal are set to melt at a certain temperature and thereby release the plug that holds back the water. To activate a sprinkler, chip out that vial or metal rod. You can pull it out with pliers or pry it out with a letter opener or screwdriver. Don't do this unless you're <u>really</u> feeling the heat though as it may draw off water needed on other floors to put out the flames that are heading your way.

Break the vial or post to activate the sprinkler.

➢ If there is fabric in your room that you can use as an additional protective layer, and as a way to keep water on you, use it. Look around for towels, throw rugs, bedspreads and sheets, curtains, etc.

➢ All you can do at this point is hope your constant signaling is seen and that help is on the way. Keep calling 911 if no one has noticed your location.

➢ Let's discuss a couple of seemingly uncommon options that have been researched, and in some cases successfully used, in situations where help did not arrive in time. The ONLY time any of these options should be actually utilized is if you find yourself in a "burn or jump to your death" situation. The two concepts are escape parachutes and rappelling.

● We've listed a couple of companies below, that manufacture parachutes designed for "base jumping" which is jumping from a stationary object such as a building. While not really an option if you're travelling and stuck in a burning hotel (since you probably wouldn't carry one on vacation), it could be an option for your office if you work on a floor higher than fire ladders can reach. If you entertain this option, be sure to receive the proper training. We do not recommend this as your first escape option. However, if you're pushed into a burn or jump situation, you should have every option at your disposal that modern technology and ingenuity will allow.

● Slightly less risky, but more demanding of skill, is the fine art of rappelling. Again, the equipment is not something you'd lug around on vacation, but it would be easy enough to keep the gear locked away in a filing cabinet drawer at work. If you consider this option and keep the equipment, make sure you're trained to use it. Also, when planning this escape, make sure you have enough rope to get you where you want to be! Too, since you're the responsible one, make sure to include a seat harness with your gear so you can lower others to safety before you yourself rappel down.

● In a hotel situation, you *may* tie bed sheets together. The drawback to this is that you'll only have a few sheets in the average room, probably just enough to lower you one or two floors. However, if you REALLY need to get out, and a couple of floors is <u>all</u> you need to drop, then by all means, go for bed sheets! The best way to tie bed sheets together is with a "Fisherman's Knot," as shown to the right.

- As a really remote idea, we heard of one guy who works on the fiftieth floor of an office building who's an experienced hang glider operator. His boss lets him keep a fold-up hang glider in a storage closet. This guy's claim is that if he's ever in a "burn or jump" situation and can get to the roof, he'll take his chances with his glider. Recommended? No. But it is a good example of creative thinking in the search for alternatives.

➢ As promised, we've found a few sources of parachutes made expressly for the purposes of escaping a fire in a high-rise building if you have absolutely no other way out and are about to die from fire or smoke:

- One escape parachute system is produced by Aerial Egress Inc. They're info is online at http://www.aerialegress.com, and you can contact them at AERIAL EGRESS, Inc., 236 E. 3rd Street, Unit C, Perris, CA 92570 Voice (909) 940-1324, Fax (909) 940-1326.
- The Evacuchute, online at http://www.evacuchute.com, or call 1-866-E-CHUTES (866-324-8837).
- Go see The Executive Chute, online at http://www.executivechute.com, or call 866-393-2888.

Car

There are multiple facets and therefore different reactions to situations regarding fire and vehicles. Since there are different places and ways a vehicle could be involved with fire, we'll break our discussion into a few different subcategories.

Fire in your car. A fire in your car's interior could be caused by an occupant smoking, a passerby flipping a cigarette that comes in through an open window, or by faulty wiring. Here's what to do for an interior fire.

➢ Cut off your AC or heat fan, and turn off any electrical accessory.

➢ Pull over as quickly as possible (since your car may be filling with smoke), turn on your emergency flashers, and get everyone out of the vehicle making sure doors and windows are closed behind you.

➢ Grab your fire extinguisher and only stay around your car long enough to discharge the extinguishers (hopefully you took our advice and carry two).

➢ Call 911 as car fires are not easy to extinguish.

➢ Close the car doors and don't open them again until fire units arrive, or another motorist has stopped with additional fire extinguishers. Even then, don't go back to the car if flames have reappeared or smoke is still present.

Fire under the hood. One of the more common types of vehicle fire, a fire under the hood can be caused by electrical short or overheating of flammable debris that may have accumulated on the engine. In turn, a fire will ignite various plastics and other materials that are part of a car's engine equipment.

➢ If you notice smoke or flame coming from under the hood, if your hood's paint starts to bubble, or the hood starts to buckle (you might not see smoke or flame yet), pull over as quickly as possible, turn off the engine, and any electronics that operate independently of the ignition (with the exception of emergency flashers).

➢ Get everyone out of the car. A fire under the hood could spread to the rest of the car by way of the fuel lines or by igniting flammable materials you may have inadvertently parked on.

➢ Don't open the hood since flame may be building just underneath.

➢ If fire is obvious and flame is present, try to discharge your extinguisher into the engine compartment by way of an opening in a wheel well or from underneath. Don't GET underneath the car, just discharge your extinguisher from that angle, or through the front grill if you can bypass the radiator. Then call 911.

➢ If that seems to have worked, wait a few minutes to see if the flames come back. If the flames stay out for a while, you can consider slowly lifting the hood. Feel it for heat first using the **back** of your hand. The metal should be cool enough to maintain contact. Step well back as you <u>slowly</u> lift the hood.

➢ If the first extinguisher obviously didn't work, spray your second one and back away.

➢ If the extinguisher worked, leave your car off and let everything continue to cool. Call a tow-truck.

Fire and fire safety at a gas station.

➢ Never smoke while filling your gas tank or even a small gas can.

➢ Always put gas cans on the ground before and during fill-up to discharge any static electricity. Though more important with the older metal gas cans, the plastic ones can still carry a charge.

➢ Don't get back in your car while filling the tank as it can create a static-electricity charge. Several incidents of static electric discharge have caused fires at stations. It's not a myth.

➢ Don't use your cell phone near your open gas tank or while filling. Cell phones are pieces of electronic equipment and that means they can spark.

➤ Make it a habit to notice the location of fire safety equipment at gas stations you frequent. You may need to put out a fire at your own car, or help a fellow driver. Most gas stations will have accessible fire extinguishers and emergency fuel shut-off switches.

➤ If a fire breaks out at another car, You have two choices. The first is to grab the nearest fire extinguisher and help put out their fire thereby helping them and reducing the threat to you. If their fire is larger than you think you can handle and no human life is in danger, then your second choice is to protect you and your property, by getting your car out of the way. Stop filling your tank, put the nozzle back in its bracket, put your gas cap back in place, and pull your car a safe distance away. Dropping the gas nozzle or leaving your gas cap off could cause a fuel spill and could allow the fire to spread or could allow a fire to follow your car.

Your car starting a fire.

➤ Once in a while your car can start a fire. If it does, the reaction would be to move your car and treat the fire it started as you would any other. Notify the authorities if need be, or fight the fire if it's small enough. A common example is when highway travelers pull off the road into moderate grass and the hot exhaust pipes of the car catch the grass on fire. If you ever pull off into grass, be careful because you don't want a small grass fire igniting your gas tank.

Caught in your car near fire. Common scenarios involve motorists stuck in traffic either near an accident that has resulted in flame, in the middle of a pileup where fire is starting, or near a wildfire that is threatening the roadway.

➤ If you see heavy smoke or fire up ahead, go ahead and take an exit, or get off the road for a minute. Tune in to your local emergency-only, or news-only radio station, or call any DOT hotline to find out what's going on. **Avoidance is always the first order of business**.

➤ If you're in committed traffic and still moving, the smoke will be a visibility hazard just as heavy rain or fog would be. Turn on your low-beam headlights (never high beams) and slow down.

➤ Do not slam on brakes unless you know you're about to hit something. Many accidents become pileups when drivers become stationary obstacles rather than maneuver out of the way. In a smoke situation, put on your emergency flashers and try to pull off the road, but don't stop in the middle of the road.

➤ If no official vehicles are on scene, call 911 to make sure the situation has been reported.

➤ After calling 911, call your local **radio station** so they can warn other motorists to stay clear of the area. If your area DOT has a motorist's input line, warn them too. (Numbers for both these groups should be recorded on your "Important Contacts" sheets and a copy kept in your car.)

➤ If you have a two-way radio to communicate with fellow motorists, try to find out what's going on. If you don't have a radio and the situation is looking serious, call 911 for advice and instruction. You may also call a friend or family member and ask them to help you monitor the news.

➤ Keep your car running, keep your seatbelts fastened, roll up the windows, and set your heat or AC to recirculate. If you don't have the recirculate option, turn your system off altogether and close your vents. You don't want to draw in outside fumes or smoke. Stay in your vehicle and keep your respirator handy. If you happen to have an air filter for your car, it could help reduce remaining smoke inside your vehicle.

➤ Look around for an escape route. If you can maneuver your car out the area and to an escape route, do so even if you have to cut across non-roadway areas. Be careful you don't get stuck.

➤ If you can't move your car and the situation is one in which fire will reach you, look to see if there's a visible area that will provide definite physical safety or another form of evacuation. If you can move you but not the car, then leave the car for greater safety. However, DON'T do this unless you're either guaranteed better safety, or that you'll burn if you stay in your car. When getting out of your vehicle, look for other dangers and oncoming traffic. Don't cross the road unless you absolutely have to.

Outdoors

Fire tops the list of tools for the agro-terrorist, since unchecked burning of fields and forests can wreak havoc on our food supplies. The danger to homes notwithstanding (since we discussed that earlier), we'll discuss fire as a threat to people caught in the open, such as when hiking or camping, near a wildfire or forest fire. If this happens to you, keep the following things in mind.

➤ Regardless where you go, you should always know where you are, and your position relative to evacuation routes or areas of safety.

➤ Several factors will determine how you'd escape an outdoor fire. Among these factors are:
  ● Your distance to the fire.
  ● Terrain features.
  ● Wind speed and direction, weather, and moisture of vegetation.
  ● Your distance to areas of safety.

- ➤ Areas of safety include:
    - Rivers, lakes, or other sources of water.
    - Areas of civilization that include transportation and communication.
    - Major paved roads that can lead you away from the fire.
- ➤ If you're caught in a fire situation such as this, try calling authorities for as much information and advice as possible regarding the fire's location, movement, rescue or firefighting operations underway, and the location of safe areas you might not know about. This is yet another reason in a long list of many why you should always carry some sort of communications device on you. If there's no way to call out, tune in to news radio to see if there are any broadcast reports that describe the details of the fire.
- ➤ In a wilderness setting, the safer areas are larger bodies of water that are both downhill and upwind from fire.
- ➤ If you make it to a creek or river, try to follow it downstream where it should become larger as it goes along.
- ➤ Don't try to outrun a fire unless there is absolutely no other direction to travel.
- ➤ Don't run towards a fire unless you know for sure you're 5 times closer than it is to an area of safety such as a large lake or reservoir.
- ➤ Be aware of wild animals that are also trying to escape a fire. They're most likely in a panicked state and may be a danger to you.
- ➤ <u>If absolutely no other safety option exists</u>, and you successfully avoid the smoke and flame, and can definitely access an area that's already burned, go there. Watch for falling trees, etc., and be aware of heat, embers still burning, and carbon monoxide.

Miscellaneous Points

The information we've given here is both situational and general. Fire is more variable than any other emergency, and situations and dangers will change by the second. The only hard and fast rules for dealing with fire are **protect yourself as best you can, stay with breathable air, and get out of the area as quickly as possible using the safest route available**. As you're reading the suggested reactions above (suggestions that should be tempered with your own common sense during an actual reaction), constantly be looking for additional safety information, new and innovative safety equipment, and first-hand accounts of survivors and what they experienced.

Specific Follow-Up

- ➤ You should seek medical attention for any burns or exposure to smoke, even though you feel fine. Doctors can easily measure oxygen intake by way of a small device that clamps on to the end of your finger. Oxygen levels can indicate whether or not your lungs received any damage.
- ➤ Report any property damage to your insurance agency immediately. The information and documentation you collected earlier under "Planning Ahead for Phase II" will be worth more than its weight in gold here.

Sources and More Information

- ➤ The bulk of your outside sources of information on fire were given earlier under "Basic Training." Many of these sources will have wildfire information as well.
- ➤ Technical and professional information on the subject of wildfires can be found online at: http://www.nwcg.gov/teams/wfewt/biblio/index.htm, through the National Wildfire Coordinating Group, and at http://www.fs.fed.us/fire/news_info/index.html, through the US Forest Service.

Bombing or Explosion

General Discussion

Bombings and explosions are areas in which the reaction is actually simpler than most think. Since most bombings come as a complete surprise, you're either affected or you're not. There's little in the way of additional information that has not been covered in other subjects.

Let's cover a few peculiarities to bombings and explosions, and cover some of the protective measures hidden in other sections.

Primary Danger

The primary danger from an explosion lies in the effects created by the explosive force and fire. Among the effects created by the explosion are the concussion or shock wave, and flying debris commonly called "shrapnel." The concussive force can cause damage to people and structures, depending on the size of the explosion and the range, and shrapnel can cause bodily harm or property damage.

Advance Preparations
Any and all advanced preparations you as an individual could make against an unexpected explosion or bombing have been mentioned in other sections. Let's go back over a few:

1. Go back over all suggestions made for increasing the structural integrity of your home in anticipation of destructive weather. Do as much as you can afford to strengthen your home.

2. Fire drills are next as that's how you'd start an evacuation if told to leave your residence or workplace during a bomb threat or potential industrial explosion.

3. Under "Dressing for Chemicals" we pointed out how wearing layers of tougher fabrics will not only help protect your skin from some chemical exposure, but it will help protect you against a certain amount of debris or shrapnel in an explosion.

4. Any area you frequent, such as your workplace, that you feel to be at above-normal risk for a bombing, remember what we said about heavier furnishings and about window protections such as the film produced by BlastGard ®.

Warning Time
The only time there would ever be an advance warning would be in either a fire in which an explosion was a distinct possibility or in the case of a "bomb scare" that is actually a genuine warning.

Forms of Alert
In the case of a bomb scare, it would be the authorities, or building security or administration, doing the immediate notification and call for evacuation. In the case of a factory fire, there might be an on-site or community siren, an alert made through local radio or television, or authorities may make announcements via loudspeaker.

Warning Signs
Other than officials relaying message in the above fire or bomb scenarios, there are no general warning "signs."

The only other indicators may be suspicious behavior associated with someone planting a bomb or with an approaching suicide bomber.

Reaction
Reactions will generally be the same regardless of location, as explosions are usually an "on" or "off" thing like biologicals. You're either warned of a pending explosion, or one happens and you follow up accordingly. There is very little "during," and therefore relatively few variables. So for the purpose of this section, we'll divide our reactions into "**Before**" and "**After**."

Before
> Treat all bomb scares and warning of pending explosions seriously. Don't stick around thinking it's a hoax or a mistake. Authorities wouldn't issue a warning lightly as they don't want to create crowds they might have to control, anymore than you want to be uprooted from what you're doing and become part of this crowd. To help you remember this general reaction, we have an acronym to the word **B.O.M.B.**:

Believe   In this day and age, it's best to **believe** the threat. Evacuate when warned.

Others   Encourage **others** around you to believe the threat too, and **help** them to safety.

Move   **Move** away from the threatened area to a safe distance and/or protective area.

Beware   **Beware** of locations that might hide a secondary device or sniper (in some scenarios).

> The most difficult warning/evacuation would be from your home in the middle of the night. This is one of the many reasons we suggested you practice some of your fire drills at night, so you can react properly and handle any sort of surprise evacuation that catches you asleep, disoriented, and partially dressed. React exactly as you would for a family fire drill, except go ahead and get in your car to leave the area.

> Any warning during the day, say at work, would be handled in the same fashion as a simple fire drill. Everyone should calmly file out of the building and await further instructions. Don't forget your GYT Pack.

> Regardless of the location or time, if the situation involves a fire and pending explosion at a factory, or other unintentional situation, it will more than likely be your only threat (no secondary devices, etc.) and authorities will tell you which is your best route for evacuation and protection.

> If the situation is an intentional or planted bomb, you have to consider the human factor and the possibility of unknown elements. Therefore when evacuating due to a threat or scare, be sure you have your GYT Pack on you and that you're wearing your extra layers of protective clothing. Once out of the building or immediate area, pay close attention to spots where a secondary device could be hidden, or open areas

where a sniper might be waiting for all his targets to come filing out of the building. This scenario is more likely in a metro or business vicinity and less likely in a residential area.

➢ If circumstances prevent you from exiting, react as you would in anticipation of a tornado and move to a more structurally secure interior area away from windows. You'll also want to stay low and situate yourself so as to place as much sturdy furniture between you and a potential explosion as possible (but not furnishings that may fall over on you).

➢ In the rare case where you know a bomb's about to go off and you know you don't have time to evacuate, such as in a suicide-bomber scenario, you should drop as low as possible, preferably behind something solid, cover the back of your head with your hands, and turn your face away from the bomber. Lay as flat as possible and close your eyes. You should learn to immediately spot good cover for bomb scenarios, just as you learned to spot safe areas for nuclear detonation protection.

After
➢ If indoors when a major explosion occurs, in your building or nearby, quickly move towards an interior wall or central hallway, if you're able, in anticipation of secondary effects. These secondary effects may include such things as windows shattering and falling after cracking, walls collapsing, etc. In the Oklahoma City bombing of 1995, there were actually several seconds between the explosion and the building collapse.

➢ Don't stay inside for more than a few seconds if you feel the building you're in is in danger of collapse. If collapse is not a consideration, then you may want to stay where you are provided no fire was started. Moving outside may expose you to debris, outside fires, or secondary devices if the incident was intentional. You may want to wait for word from authorities before you evacuate the building. Use your best judgement until instructed by authorities.

➢ Even if your building seems undamaged, don't use the elevators. Even a slight building shift can cause elevators to jam in their tracks, or power outages can occur suddenly. Use the stairs.

➢ While waiting, or as you evacuate, check yourself for injury and then help those around you. Keep an eye on objects that may fall, items on fire, or injured people that are incapacitated.

➢ If there's a fire, proceed as you practiced in your fire drills and don't forget your GYT Pack.

➢ Should you be caught in the debris of a collapsed building, the same things you learned under "Nuclear Detonation" apply. Remain calm and try to signal for help.

Miscellaneous Points
➢ One opportunity to thwart intentional bombings is to keep an eye on suspicious individuals and any packages they may carry. Watch for anyone leaving a box, backpack, or other container behind as they leave an area.

➢ In a similar vein, though suicide bombing hasn't, as of the time of this writing, been a problem in this country, you still need to know what to watch out for when the news and political climate might indicate we're at risk for such a thing. Watch for suspicious or nervous people wearing unusually bulky clothing (such as a coat in the summer), or backpacks or shoulder bags, and loitering in areas likely to be a target. An example target area is dense crowds in such areas as public transportation or shopping malls.

➢ Some explosions resulting from industrial accidents can be particularly devastating. In Texas City, near Brownsville, Texas, in 1947, a ship was being loaded with ammonium nitrate fertilizer when a fire caused the ship to explode, leveling most of the town.

Specific Follow-Up
➢ The first order of business following any explosion for which you were a bystander is medical attention. Get checked even if you feel fine. Overpressures and shock waves can cause injury to ears, eyes, pulmonary cavities, and your heart, not to mention possible concussion.

➢ Once you're checked out, write down anything you can remember about the incident. Authorities will need your help in putting clues together whether the explosion was intentional or not.

➢ The only other follow-up to an explosion would be damage assessments and contacting your insurance agent (if any of your property were damaged) just as you would for a fire or natural disaster. **One word of caution:** As we mentioned under "Planning Ahead for Phase II," many insurance companies are quietly rewriting their policies to categorize terrorist attacks as "act of war" and therefore not a coverable incident. Make sure you know your policy coverage.

Sources and More Information
➢ Many of the terrorism info sources we listed under "Emergency Risk Score" in the "**Foundation**" section will have information on bombings and their associated risks and safety suggestions.

➢ See the Marine Corps guide for understanding terrorism at http://www.tpub.com/content/USMC/mcrp302e.

➢ Be sure to visit Bomb Security, online at: http://www.bombsecurity.com.

General Discussion

Sniper attacks have been with us for a while and will continue to be a potential threat. We've seen Charles Whitman, the "Bell tower sniper." Most recently we've had the "DC Snipers" and a couple of Columbine-type school shootings have also been sniper scenarios.

Additionally, sniper tactics have been studied extensively by al Qaeda and other terror groups, so we know it's a likely terror tactic.

Primary Danger

Naturally, the primary danger in a sniper attack is in being shot.

Advance Preparations

Just as with biological attacks, your only real opportunities lie in your advance preparations since there is no reaction other than first aid. You're either shot or you're not, and our advance preparations in this section will hopefully provide you a few ways to not get shot.

Let's look at a few things people can do to protect themselves should a sniper be on the loose:

1. Dress in more conservative colors. Don't draw attention to yourself. Keeping a low profile is good advice all the way around in this kind of situation. Noise is another consideration. Don't allow excess noise to draw attention your way.

2. Park as close as you can to wherever it is you're going. At least park as far away from thoroughfares and sniper vantage points (such as wooded lots) as possible.

3. When fueling your car, start the gas then go inside to pay, or position yourself where there are obstacles between you and outside areas. (Don't get back in your car while fueling as static electric spark is a very real fire hazard.)

4. When walking, don't stroll or dawdle. Walk with a purpose and don't walk in a straight line for long.

5. When eating in a restaurant, don't sit near big picture windows.

6. In your home, arrange your lighting and activities so that you don't throw a silhouette. Keep your blinds closed and curtains pulled. Same goes for work.

7. Shop online more, and if you're able at all to telecommute for work, do that for a few days.

8. Lay off golf or other outdoor recreation (as participant or spectator) until the situation is resolved.

9. If you're a witness to a sniper attack, get down and stay low when you hear the shot. Other shots may be fired. IF you can find a safe spot and observe, then notice what you can so you can report to authorities. Your safety comes first though. One safe way to observe your surroundings is through reflections. Rather than standing up and looking around, see what you can observe through reflections in store windows, car windows, etc.

10. Know what to tell police. Record any pertinent details relating to people or vehicles. Be ready to convey number of people, sexes, ages, descriptions of face, hair, skin, clothing, marks or tattoos. Also report vehicle types, makes, model, year, color(s), and tag number if you can get it. Also keep an ear out for the number of shots. Again, don't jeopardize your safety to try and get details. (See the "Suspicious Activity" forms in the "**Appendix**.")

11. If this happens again and a particular group seems to be targeted, offer to run errands for any of your neighbors that might fit the target description. Do what you can to help them stay out of harm's way.

12. Neighborhood Watch groups can extend their "patrols" to neighboring businesses, schools, or other likely and nearby target spots. Don't patrol visibly, but pick a secure vantage point from which to observe and report.

13. Business owners in a target area who happen to have video surveillance cameras could probably do a public service by training these cameras outward in an effort to catch the perpetrators red-handed.

14. Businesses may also hang decorative signage, large tarps, or erect temporary landscaping which will offer concealment to approaching customers. Schools may park school busses alongside playgrounds or common areas to provide cover for students. Many things can be done to reduce risk.

15. We do not suggest Kevlar vests simply because of the astronomical numerical odds against you actually being a victim, and because most snipers will try for head shots. However, the risk is there or else we would not have included this section. Therefore, if you feel yourself to be at greater risk than most, and the price of Kevlar is no obstacle, by all means take those protective steps that will give you peace of mind. Peace of mind is one of the main goals of this book.

Warning Time

Your only warning time will be in a scenario similar to the DC Snipers where there has been a previous incident in your area and a reasonable expectation of another.

Forms of Alert

Your only forms of alert are going to be announcements on the local news.

Warning Signs

In addition to the news items mentioned above, your only warning may be to see a sniper preparing to act. Several people saw Whitman on his way to the bell tower.

Specific Follow-Up

Other than first aid for the injured or reporting the event to the Police, there is no specific follow-up to an incident occurring near you. The only thing we would suggest is for you to monitor your stress levels and be aware of any personal or emotional effect an incident may have.

Source and More Information

No additional sources have been chosen for this section. The only applicable information can be found in the "First Aid" section of "Basic Training."

Held Hostage

General Discussion

Hostage taking is being mentioned as it's a definite possibility when considering terrorist actions. While not a "disaster" in the classic sense, being taken in a hostage situation is every bit as dangerous.

Essentially kidnapping, hostage situations can be perpetrated on or by individuals or groups, and can take place in many locations or venues. We've seen lone gunmen hold hostages intentionally or as the result of a botched robbery. We've seen terrorist groups take control of passenger aircraft, cruise ships, hotels, and the like. The scenario is all too common and the threat all too real.

Primary Danger

There are **three** danger periods in any hostage situation. The **first** is when hostages are taken. Terrorists will use force to herd people where they want them, and a hostage or two may be killed outright so terrorists can show they mean business. The **second** danger is during the ordeal when terrorists may feel the need to act on threats or to kill more hostages to force others to meet their demands. The **third** danger area is during any attempted rescue by authorities. Hostages may be killed by terrorists during the rescue, or hostages may be caught in the crossfire.

Warning Time

There is no warning unless you see suspicious behavior that tells you something's wrong and that it's best to leave. Even then you'll only have a few seconds to act.

Warning Signs

The suspicious behavior you should look for includes but isn't limited to:

➢ "Suspicious" individuals attempting to communicate and coordinate their positions "secretively." In other words, they're using slight hand signals or gestures, and eye movements or head nods, instead of talking or waving like everyone else around them might.

➢ The presence of a weapon means little in and of itself. Watch the behavior of its owner. It's not difficult to see if someone's angry, nervous, distracted, or simply seems "out of place."

➢ Someone who seems to be "studying" or casing an area and then comes back a few minutes later carrying unusual packages.

Advance Preparations

There's really not much you can do in the way of any material advance preparations.

Mental preparation is a different story. The best way to prepare is to learn the suspicious behavior that may allow you to avoid being taken, and what to expect and how to handle it if you wind up caught in such a situation. For this discussion, let's divide our attention into avoidance, and life as a hostage.

Avoidance

Avoidance of a potential hostage situation depends on the same factors as successfully dealing with a fire; early detection and rapid reaction. Let's consider the following:

- ➢ Memorizing the little suspicious behavior pointers we gave you is easy. Applying them will take some practice. Whenever you're out and about, spend a little time people watching. See how normal people act and try to pinpoint traits that make you feel a certain person or group may look "suspicious" or "out of place." Is it their behavior? Their dress? Their physical appearance? A combination of these? People watching is one of the best educations you could ever have.

- ➢ If you notice extremely suspicious behavior and see warning signs that some sort of violent incident is about to unfold, the best thing to do is leave through the front door and behave normally. For example, if you see suspicious individuals in a bank lobby and you feel a robbery's about to commence, it's easy to pat your pockets and mumble something like "left the checkbook in the car..." Call authorities once you're safe.

- ➢ This is another area of many where it's important to know all exits and alternate exits of any building you enter. Though there aren't really any exits in a passenger aircraft at 30,000 feet in the air, or a cruise ship in the middle of the ocean, buildings will have several. Knowing them may allow you to slip out if you notice a situation brewing and you didn't leave soon enough through the front door.

- ➢ Of course, the advice we gave earlier regarding the higher alerts might reduce your risk of being involved in a hostage scenario. Do more online and over the phone, eat at home more often, travel a little less, and telecommute if work allows.

Life as a Hostage

Let's assume the worst, and actually the more probable, that a situation unfolded more quickly than you could react. There are several things you'll want to do, and not do, if you're taken hostage.

- ➢ Stay calm. People who take hostages usually have an agenda, and killing all the hostages is rarely the true agenda. Terrorists will hang on to their bargaining chips as long as possible.

- ➢ Give the hostage takers time to calm down. The whole situation will be an extremely tense one for the first few hours, and it's then the hostage takers are the most dangerous. They're nervous and uncertain about what's going to happen next (just as you are).

- ➢ Don't react. Don't speak, don't argue, don't even look at your captors. Simply do what they say.

- ➢ Don't try to call anyone or reach for an alarm switch unless that's your specific job at the location where the incident is taking place. Hostage takers may alert authorities on their own in order to reach their objective.

- ➢ Unless there is serious reason to do otherwise, try to draw ZERO attention to yourself. Don't look your captors in the eye, don't move around, don't ask to go to the restroom, don't try to become a spokesperson for the other hostages, and don't try to play hostage negotiator.

- ➢ If spoken to, answer in a calm, half friendly, half respectful tone. Be neutral. Don't show anger, don't be sarcastic, don't try to make a new friend, but don't grovel or demean yourself.

- ➢ If your captors do talk to you, answer in ways that humanize you. Give them your name, mention family if you have the chance, etc. Don't force this information on them, but slip it in if you have the opportunity. If they view you as a person rather than a bargaining chip, it's more difficult to harm you.

- ➢ Listen to everything that's going on. Try to remember the number of perpetrators, names and any other bits of information the terrorists may inadvertently leak. Authorities may need this information later. While we're on the subject, let's give you a quick list of things to notice:

  - ❑ For a list of things to remember, study the "**Suspicious Activity**" forms in the "**Appendix.**"

  - ❑ Otherwise, you'll want to remember as much of the following as you can: ____ Number of perpetrators, ____ Names, ____ Sexes, ____ Physical descriptions, ____ Accents, ____ Languages spoken, ____ Conversation items, ____ Odors, ____ Weapons, ____ Vehicles.

- ➢ Unless you're extremely capable, unless multiple deaths (including you and yours) are imminent, <u>and</u> a golden opportunity presents itself, DON'T try to fight the hostage taker(s). Adrenaline and emotional commitment are on their side, if you fail you risk not only your life but the lives of the other hostages, and you don't want to confuse the situation should a rescue operation be launched just as you're fighting an armed terrorist. Police might shoot you by mistake. However, keep this in mind: In New York city in 2002, a bar was taken over by a man carrying a large knife, a pistol, a can of gasoline, and a lighter. He was subdued by two of the bar's unarmed female patrons before he could hurt anyone!

- ➢ Do pay attention to what's going on. Instead of fighting a hostage taker, you may have an opportunity to slip out through an alternate exit. If you do this, proceed cautiously with your hands in plain site as authorities will certainly be watching all exits. You don't want them mistaking you for a suspect. Once outside, provide authorities all the intel you can.

Miscellaneous Points

➢ If you're a government worker, or an employee at a potential terror target, tool source, or at a bank, *you* might be the target of a hostage taking or kidnapping. It's best you take these points to heart.

➢ One danger sign to remember is if your captors were masked and suddenly take off their masks or start divulging information. This may be a sign they don't care about security as they won't be leaving witnesses.

➢ As we discussed earlier, key employees, or their family members, may be the specific targets of kidnapping.

Specific Follow-Up

1. If you escape before a hostage situation unfolds, go to a safe place but remain in contact with authorities to give them as much information as you can about the number of perpetrators and their descriptions, information about actual hostages, and detail about the rooms or areas where everyone is held.

2. Any time you escape or are rescued from a hostage situation, expect to be handcuffed and questioned. It's a generally accepted procedure to make sure there are no perpetrators trying to escape claiming to be a hostage.

3. If you're a victim of such a situation, try to talk with friends or professionals soon after. This is one of the most stressful situations to be in for anyone. Don't let the stress get to you. Talk it out.

Additional Sources

➢ Some information about foreign soil kidnappings can be found on Robert Young Pelton's "Come Back Alive" site at http://www.comebackalive.com/df/kidnapp/goodhand.htm.

Notes:

Weather Disasters

Moving from the area of man-made catastrophe into the realm of nature, we get into nature's answer to "mass destruction." Nature usually wins.

The weather sections presented here and the other natural disaster discussions that follow should give you a good cross-section of information to keep you safe in a number of scenarios. As you read through these sections, borrow from them and see if anything will apply to unrelated emergencies. For example, we suggested under Orange and Red Alert that a lot of the preparatory steps suggested would also do you well in preparing for your region's destructive weather season. See if anything presented here strikes a chord and would also apply to your protection against possible terrorist attacks.

Not to be doomsayers, but we wish to point out that a natural disaster could trigger an industrial accident, or a terrorist sleeper cell could have been given instructions to use a natural disaster as their "launch signal" and their orders are to make matters worse. In any event, always remember there's always another shoe that could drop at any moment. Emergencies and disasters are seldomly one-dimensional.

Read through these disaster sections detail by detail as it's the little things that make the big difference. Though we can't cover every nuance that may present itself, if we teach you something new, or something we've presented causes you to think of something entirely different that helps protect you, then this book will have served its purpose.

Goals of Weather Disasters
1. To reiterate the point that **we face more dangers than just terrorist attack**.
2. To give you **additional reasons to plan ahead** for emergencies.
3. To remind you that if you're fully prepared for one type of disaster **you're protected against others**.
4. To give you a **non-alarmist "cover story"** for your family's terrorism protection planning.

We've divided Weather Disasters into the following sections:
1. **Hurricane.**
2. **Tornado.**
3. **Lightning.**
4. **Flood.**
5. **Blizzard.**

Hurricane

General Discussion

Hurricanes are among the most devastating forces on the planet. If they had a radiation component, or an equally nasty after-effect, they'd be more feared than nuclear weapons.

Fortunately, modern technology has allowed us to more accurately track these storms and predict their path. The downside is that too many people think that technology and tracking equal personal protection. They don't. Hurricanes are just as dangerous as ever and should be feared and respected on the same level as a nuclear device.

Primary Danger

The primary danger from hurricanes comes from their extremely destructive winds. These winds tear down structures, turn debris into deadly projectiles, and in coastal areas the wind literally pushes the sea up on to land in a phenomenon known as "storm surge." A secondary danger comes from tornadoes that are commonly created along the outer fringes of the hurricane.

The Saffir/Simpson Hurricane Scale

Category	Winds	Storm Surge	Associated Damage
1	74-95 mph	4-5 ft	**Minimal**. Primarily to trees, foliage, and unattached mobile homes. Some small boats may be pulled from moorings.
2	96-110 mph	6-8 ft	**Moderate**. Some trees blown over, some damage to doors, windows, and roofs. Some evacuation of shoreline areas and low lying islands.
3	111-130 mph	9-12 ft	**Extensive**. Large trees blown down, some structural damage to small buildings. Mobile homes destroyed, serious coastal flooding. Many small structures near coast destroyed by wind and waves. Almost all small boats torn from moorings.
4	131-155 mph	13-18 ft	**Extreme**. Extensive damage to roofs on many small structures. Terrain 10 feet or less above sea level flooded. Escape routes cut by rising water 3 to 5 hours before center arrives. Massive coastal evacuation required.
5	156 mph or more	18 ft or higher	**Catastrophic**. Complete failure of roofs on residences and many commercial structures. Small buildings overturned or blown away. Massive evacuation of low ground within 5 to 10 miles of the coast.

While we're discussing wind speeds, let's take a look at the "Beaufort" wind-speed estimation scale.

The Beaufort Wind-Speed Estimation Scale

Wind Speed in MPH	Common Indicators
Under 1 MPH	Calm; smoke goes straight up.
1 to 3	Direction of wind shown by smoke, but not by wind vane.
4 to 7	Wind felt on face; leaves rustle; wind vane moves.
8 to 12	Leaves and small twigs move steadily; small flags will be held outward.
13 to 18	Dust and loose paper raised; small branches are moved.
19 to 24	Small trees sway; waves form on lakes and ponds.
25 to 31	Large branches move; wires whistle; umbrellas hard to use.
32 to 38	Whole trees in motion; hard to walk against the wind.
39 to 46	Twigs break from trees; very hard to walk against the wind.
47 to 54	Small damage to buildings.
55 to 63	Much damage to buildings; trees uprooted.
64 to 72	Widespread damage from wind.
73 and up	Violence and destruction from wind. Beginning of hurricane strength.

Advance Preparations

Advance preparations for the destructive power of a hurricane lie in two distinct areas. One is in **structurally enhancing your home** as discussed under "Basic Home Prep," and the other in your **ability to evacuate quickly and efficiently**. Let's look at a few considerations that cover both bases:

1. Go back through the considerations listed earlier concerning **Orange** and **Red Alert**. You'll find that many of those suggestions apply rather nicely during hurricane season.

2. Also make sure you've completed all our landscaping and structural suggestions mentioned under "**Basic Home Prep**" as destructive events, such as hurricanes, are the main reason those suggestions are a part of this book.

3. In addition to boarding your windows and bracing your doors, you should **construct a protective box around your electric meter, gas meter, or the valve and line system from any outdoor fuel or gas tank**. Projectiles can damage them as easily as anything else, and you don't want any utility outage to be longer than it has to.

4. **Note**: Depending on how close you live to the coast, you're going to have to balance your protection against both high winds and flood waters. The best protection from destructive wind is found in the basement. However, the closer you live to the coast the more in danger you are of storm surge which will cause flooding, and therefore you wouldn't want to shelter below ground. Balance your protection as best you can staying aware of both dangers.

5. Don't mess around with hurricanes; be fully prepared to **evacuate**. If officials say evacuate, you should go. Even if they don't say you should go, if you have doubts, go.

6. Should you decide to stay and ride out the storm, be sure you've completed all the supply checklists provided under "**Isolation**" and that you've completed them as early as possible. Making an emergency run to the store as a storm approaches is not the way to do things. You'll have other things you need to do and most stores will be stripped of goods by the time you get there.

7. Check your neighborhood **storm drains**. If blocked, clear them yourself if the city can't get there in time.

8. Look ahead to some of the other suggestions mentioned under "**Flood**."

9. Double-check the water, food, first aid kit, tool kit, and battery supplies in your safe room.

Warning Time

Hurricane paths can be projected up to three days out with relative accuracy. However, they can change paths suddenly and leave you with less than 12 hours warning, and they can spawn tornadoes miles and miles away.

Forms of Alert

The Weather Channel and your local television and radio stations are your best source of advance notice. This is also another reason we suggested you buy a NOAA "Weather Alert" radio with battery backup.

Warning Signs

The main warning sign that the hurricane is right up on you will be the heavy rains and squalls. One other interesting phenomenon is "green lightning" which is produced when lightning flashes behind clouds heavily laden with water.

With any approaching storm, if there's less than 30 seconds between lighting and thunder, it's time to go back inside.

Reaction

To remember the basic reactions as storms approach, memorize the acronym **S.T.O.R.M.S.**:

Shelter	As severe weather threatens, keep your family close to **shelter**.
Time	Pay attention to storm trackers or the **time** between lightning and thunder. Know where the storm is.
Organize	**Organize** your goods and gear for a shelter-in-place, or an evacuation; whichever seems more likely.
Reinforce	**Reinforce** your doors and windows in anticipation of heavy winds. Prep for some flooding as well.
Move	**Move** to an evacuation destination if the storm's severity dictates and it's early enough to leave safely.
Safe room	Gather your family in your **safe room** or area if you intend to shelter-in-place during the storm.

Home

As there's so much advance warning of a hurricane and only two real choices – **shelter** or **evacuate** –most of the reaction points here will cover your efforts at home. It's highly unlikely that a hurricane would "catch" you at work or in your car. It shouldn't anyway. Let's start with a couple of evacuation suggestions and then cover what to do at home.

➢ Know the **official hurricane evacuation routes** out of your area. You should stick to them and do not deviate. Side roads or other routes may prove to hold hidden dangers and they will not be manned by authorities and associated services like the main evac routes will (unless you left early).

➢ If evacuation looks like it might be your response, **leave as soon as you can**. The storm should be easy enough to beat, but what you want to avoid is the traffic crunch and the drain on basic local services and supplies both at home and your destination. It's best to leave early.

The points that follow apply to additional advance preparations you can take in anticipation of the storm. Once a hurricane hits, you should either be gone or sitting in your safe room listening to TV or radio. You shouldn't be out doing anything else.

➢ Perform last minute outdoor preparations at <u>least</u> **12 hours** before the storm is due. Close storm shutters or put up plywood, strap down lawn furniture, and tie down trailers or boats. This includes chaining your car down to any concrete anchors if you decide to stay, and anchoring any trees that might fall on your house.

➢ If the storm is projected to be particularly severe, or if you live close to the coast, you should put up a second set of plywood reinforcements on the **inside** of your windows, and bolt them to the wall studs. Add additional bracing as you're able. Remember that 3/4 inch marine plywood, held in place with 2-inch screws (never nails) **with washers**, is the best for hurricane reinforcement. Screws hold much better than nails, and washers keep the boards from pulling off of the screws.

> Your **food, water, and other supplies** should be kept at sufficient levels so you don't waste time dealing with crowds and bare shelves as the folks that don't plan ahead go on panicked last-minute shopping sprees. Do this even if you evacuate as you'll want plenty of supplies on hand when you return to clean up.

> If severe weather approaches before you've had a chance to act on any of these suggestions, place a big "X" across the **glass of all windows** with duct tape, inside and out (and maybe add a layer of corrugated cardboard inside, if you have some), close all blinds, and draw all curtains. If you have solid furniture such as wooden bookcases that can be pushed up against windows and braced in place, do that. The tape won't keep a window from breaking, but it may help reduce the amount of broken glass in your house making walking around after the storm safer, and cleanup a bit easier. However, remember this; the biggest reason for boarding over windows is to keep wind from getting under your roof and lifting it off your house.

> If there is a **window in your safe area**, give it a little extra protection (or give *you* some extra protection from it). You should have an interior storm shutter made for it. If not, duct tape it heavily, put up some cardboard, nail a blanket in place over it, and/or throw a mattress up against the window and brace the mattress in place with furniture, even if you've boarded the window from the outside. Redundancy is our friend.

> Deadbolt and brace **exterior doors** so they don't blow open allowing wind in the house. Close interior doors for the same reason and to protect from flying debris. Make sure the garage door is closed and braced.

> Don't forget your protective box over your **electric and gas meters,** or any other exposed utility connection.

> **Protect your heirlooms and valuables**. Just as you would during a Red Alert, you should round up your valuables and irreplaceables, seal them in plastic tubs and put them up off the floor in a secure area such as your safe room. They'll be redundantly protected against physical harm, a little better hidden from looters, and a little better secured against being strewn about the neighborhood as the result of a destructive event. For additional protection, read the plastic tub setup we described in the "Fire" portion above.

> **Cover your interior furniture**. If you live close to the coast and you're worried that even with all your structural enhancements that your roof might blow off, wrap plastic sheeting around your more valuable furniture and seal the plastic in place with duct tape to protect it against water damage. If flooding is possible, put your furniture on blocks (or expensive furniture on top of cheap furniture) to keep it off the floor.

> **Keep your tub scrubbed.** As a hurricane approaches, fill your bathtub with water for drinking or washing in anticipation of a loss of municipal water. The cleaner your tub stays, the quicker you can give it that final bleach-wipe before filling it in an emergency. Also, **remove any automatic bowl cleaner** from your toilet's tank. The water in the tank is useful for washing, etc. provided it doesn't have harsh chemicals in it.

> In addition to **filling your tub with water**, use the empty plastic tubs and trash cans we asked you to buy earlier, and fill them with water. The loss of all utilities is a common after-effect of any natural disaster.

> You should also fill any empty spot in your freezer(s) and refrigerator with **bottles and jugs of water**. Not only do you need the water, but a freezer or refrigerator works better and stays cold longer when full. The bottles you put in the freezer should be **plastic** and filled about ¾ full to allow for expansion of ice.

> Even if you have an electric generator, turn your **refrigerator to its coldest setting** so the food inside will be as cold as possible thereby lasting longer if the power is lost.

> Do the same with your **thermostat**. Hurricanes usually come in warm weather, so use your AC to make your house cooler than normal. This will give you a comfortable temperature inside the house for a longer period of time should the power go out (assuming no structural damage).

> Unplug any and all unnecessary electrical devices that are not protected by surge protectors. Lighting is a big factor in any storm and there's no sense loosing valuables over a simple electrical spike.

> As the storm arrives, react as you would for any destructive event and gather in your safe room or area, and stay low. Make sure you have your safety gear, communication equipment, keys, wallets, and everything else from your "Last Minute List." Listen to your radio and TV for instructions, updates, and other reports.

> Beware of the eye of the hurricane. It will be calm if it passes directly over you, but all that means is that the other half of the hurricane is about to hit.

Miscellaneous Points

> Some severe weather can damage locations housing hazardous materials adding a second danger to the situation. Always be ready for the hidden danger and for situations to change.

> Other hidden dangers include flooding, downed electrical wires that are still live, trees or other structures that may fall later, weakened bridges and roadways, etc.

Specific Follow-Up

1. Stay inside as much as possible afterward.
2. Stay away from damaged areas, or flooded areas.
3. Stay off the phone unless you need help. Leave the lines open for official use or for those actually injured and needing medical attention.
4. Though we suggest staying off the phones, you should report any utility outage. Just because your utilities are out doesn't mean anyone else's are, or if they are, that anyone else has called it in.
5. Once your personal situation is taken care of, see what you can do to help your neighbors and then check in with any emergency volunteer group you're part of.
6. More aftermath considerations will be discussed in "**Coming Home**" under "**Evacuation**."

Sources and More Information

➢ Visit http://www.weatherpreparedness.com. It's home to the Hazardous Weather Preparedness Institute, LLC, 5107 Quaker Landing Court, Greensboro, NC 27455, 800-743-4989.

➢ Other useful sources can be found earlier under "Emergencies and Disasters. An Overview."

Tornado

General Discussion

Though less devastating as far as "square mile coverage" and "total cost in property loss," tornadoes are more feared than hurricanes because the immediate destruction is more severe and tornadoes appear with less warning.

The US has more tornadoes per year than all other countries combined. Though some areas of the US see more tornadic activity than others, no location is immune.

Primary Danger

The primary danger from tornadoes lies in their incredibly destructive winds, the suddenness with which they can appear, and their unpredictable path. A smaller danger presents itself in the hail that sometimes accompanies a tornado, and the fact that the winds can hurl debris at incredible velocities.

The Fujita Scale

F – Scale #	Winds	Associated Damage
F 0	40-72 mph	**Light**. Some damage to chimneys; breaks branches off trees; pushes over shallow-rooted trees; damages sign boards.
F 1	73-112 mph	**Moderate**. The lower limit is the beginning of hurricane wind speed; peels surface off roofs; mobile homes pushed off foundations or overturned; moving autos pushed off the roads; attached garages may be destroyed.
F 2	113-157 mph	**Considerable**. Roofs torn off frame houses; mobile homes demolished; boxcars pushed over; large trees snapped or uprooted; light object missiles generated.
F 3	158-206 mph	**Severe**. Roof and some walls torn off well constructed houses; trains overturned; most trees in forest uprooted. This wind speed is the upper limit of hurricanes.
F 4	207-260 mph	**Devastating**. Well-constructed houses leveled; structures with weak foundations blown off some distance; cars thrown and large missiles generated.
F 5	261-318 mph	**Incredible**. Strong frame houses lifted off foundations and carried considerable distances to disintegrate; automobile sized missiles fly through the air in excess of 100 meters; trees debarked; steel reinforced concrete structures badly damaged.
F 6 (never used)	319-379 mph	**Inconceivable**. These winds are very unlikely. The small area of damage they might produce would probably not be recognizable along with the mess produced by F4 and F5 wind that would surround the F6 winds. Missiles, such as cars and refrigerators would do serious secondary damage that could not be directly identified as F6 damage. If this level is ever achieved, evidence for it might only be found in some manner of ground swirl pattern, for it may never be identifiable through engineering studies.

Advance Preparations

1. Everything from "Basic Home Prep" earlier, and "Hurricane" above, apply here.
2. You should have already had "Tornado Drills" at home.
3. As we caution you to always know the fire escape and fire and safety equipment locations of places you frequent, you should also know the protective areas in each. Where is the safest place to go in case of tornado?
4. Similarly, as we pointed out in number 8 of the "Advance Preparations" under "Nuclear Detonation," know all locations that would offer you protection against a tornado. The more practice you have in looking for such spots, the easier it will become to pick them out if you're caught in unfamiliar areas.
5. Make sure you're part of a neighborhood or community emergency phone tree. Sometimes it's the only way to get the warning.

Warning Time

Tornado *Watches* are routinely issued when conditions are right for tornado formation. However, there is rarely more than 15 minutes advance warning for an actual tornado. Tornado Warnings are usually sounded only when a tornado has formed or touched down.

Forms of Alert

In a Tornado *Warning* (which means one has actually touched down), community sirens may sound, a National Weather Service Warning will be issued over your Weather Alert radio, and local TV and radio stations may have their own emergency alerts.

However, as we pointed out under "EAS Alerts," you can't count on always getting the word. What if you think it's merely a thunderstorm outside and you're at home watching a movie on tape with the sound turned up? You won't see any EAS alerts on TV, you probably won't hear the community sirens, and it's doubtful you'd hear your Weather Alert radio. That's why we continually suggest that you be part of an emergency notification phone tree.

Warning Signs

 ➢ Dark green or black clouds, maybe with a "thick pea soup" consistency.
 ➢ Green lightning. This occurs when lightning flashes behind clouds that are heavily laden with water. (You might also see this with hurricanes.)
 ➢ Hail.
 ➢ "Horizontal rain." This is one of many indicators of strong winds.
 ➢ Swirling clouds even if no funnel has formed.
 ➢ You may feel sudden air pressure changes, both high to low. Your ears may pop.
 ➢ You may hear a rushing roar similar to a freight train passing by, or a jet engine.
 ➢ A very sudden and very eerie stillness in the middle of what was a violent storm seconds earlier.

Reaction

Your biggest threat from a tornado, as opposed to other storms, is its incredible winds. Our reminder for your reactions here uses the acronym **W.I.N.D.S.**:

<u>W</u>arnings Pay attention to watches and **warnings**, and be sure to **warn** others. Get people and pets inside.

<u>I</u>nterior Prep the **interior** of your home by closing windows, and exterior and interior doors.

<u>N</u>ow Don't waste time! Get to safety **now**! Don't watch the storm, and don't try to videotape the tornado!

<u>D</u>uck Get in your safe room and get down. Even in the safe room **duck** under heavy items and cover yourself.

<u>S</u>ignal **Signal** for help if you are in any way injured or if your area sustained damage.

Home

Tornado Watch: (When conditions are right for a tornado but one has not touched down.)
 ➢ Call the kids home if they're out playing. Use the signal you agreed on during your drills earlier.
 ➢ During a watch, you may want to crack <u>ONE</u> window on the same side of the house as the approaching storm. Also open your fireplace damper. Do NOT open any doors and do not open that window more than an inch. Sometimes structures are destroyed by the sudden vacuum created by a tornado. Normal air

pressure inside a building can cause the building to "explode" if it has no way to vent. Cracking a window will allow pressure to vent, and leaving doors closed will prevent strong winds from "getting under" your roof.

➢ Close all other windows, close storm shutters if you have them, pull the blinds, and close the drapes. Not only do you not want strong winds inside your house, closing the drapes and shutters will catch a small amount of any broken glass resulting from damaged windows. Though only a small amount of glass will be deflected, it might that one shard that would've gotten in your eye.

➢ Deadbolt and brace exterior doors so they don't blow open allowing wind in the house. Close interior doors for the same reason and to protect from flying debris. Make sure the garage door is closed and braced.

➢ If you decided you needed one, put your baby gate up across the door of your safe room to help you corral toddlers and pets should the watch turn to a warning. In fact, you could probably go ahead and put up your pets for the duration of the watch (in their carriers if they have one). That'll be one less worry in a warning.

➢ Get out of a mobile home and get into any prepared external safe room, or head for natural shelter or a sturdier building. Don't wait for an actual warning if you may be caught in the open and between locations.

➢ If a tornado watch is issued around bedtime, the best suggestion is for someone in the family to stay up and monitor the news and weather. If that's not possible, try to sleep as dressed as possible with your car keys, wallet, cell phone etc., either on you or stored in your safe room.

Tornado Warning: (When one has touched down.)

➢ All points from your earlier "Tornado Drill" come into play here.

➢ Sound your home tornado warning alarm if the tornado has caught you by surprise and you need to warn everyone in the household.

➢ Get the family to the safe area and put your pets on leads or in their carriers.

➢ If absolutely nothing else, immediately head for the lowest and most central point of your home. Close doors behind you and duck and cover. Though some people recommend heading to a bathroom to hide in the tub, we don't recommend it unless the tub is the old metal style. Modern fiberglass tubs offer little protection.

➢ Call 911 to report the funnel cloud if you've seen one before any warnings were sounded.

➢ Do you have a hardhat or helmet of some sort? Is it in your safe room? If so, put it on. You'll also want to pull clothes on top of everyone, pull an old mattress over you, or drag a table or other piece of furniture over you to protect everyone from glass shards and other flying debris (yep, even in your safe room).

Work

➢ Your reaction at your workplace should be essentially the same as at home. During a watch, you'd want to keep an extra eye on conditions, close doors and windows, and keep your personal effects and your GYT Pack close at hand.

➢ During a warning or sighting, you'd want to head to the most structurally sound section of the building you're in. For example, restaurant workers routinely seek shelter inside their indoor walk-in coolers or freezers. These have extra wall supports and heavy-duty shelving inside. In the average office building, the stairwell area is the most secure. Learn which areas in your workplace offer the most protection.

➢ Avoid wide-span buildings like gymnasiums, airplane hangars, etc.

Car

➢ Do not try to outrun a tornado. Though some people have accidentally succeeded in doing this, the odds are not in your favor.

➢ Immediately look for natural shelter or a safe location. You should already know these on your normal travel routes and should have learned to look for them when travelling new areas.

➢ During a tornado warning, even if you don't see a funnel cloud, pull off the road as soon as you can upon seeing adequate shelter, since winds aside, you'll want protection from hail and debris. **Note**: Highway overpasses are NOT adequate shelter though they've worked in a few instances. There is no protection from debris, and the wind can come through and blow you out.

➢ Get out of your car and into a sturdy building or down in a ditch or other depression that keeps you below wind and debris. If you wind up in a ditch or other semi-exposed area, cover your head and neck with your arms. Ditches offer better protection than overpasses, because you're <u>underneath</u> the threat in a ditch. Debris will pass over you, and wind can't get underneath you to lift you off the ground. Cars are dangerous because the wind can get underneath and turn your car into a toy airplane.

Outdoors

➤ Remember that hail is commonly associated with tornadoes. Seek shelter if any is available but avoid things that might become debris in a high wind. For example, a tree might be good protection from the hail but large limbs may fall on you, and it may be a target for lighting.

➤ As with all other locations, seek low ground that will protect you from wind and debris.

Miscellaneous Points

➤ You may have to divide your attention and reactions in a heavy storm situation. On one hand you'll want to find low ground for protection against tornadoes. On the other hand, a heavy storm may bring flash flooding. The good news is that, though severely destructive, tornadoes usually last only a few minutes. Our suggestion here is to not stay tucked away in your below-ground area too long if flooding is a possibility in your area.

➤ Many storms carry destructive winds without forming an actual tornado. Also, some destructive winds in the vicinity of a tornado may come in the form of "microbursts" which are just as destructive. Just because the sirens haven't sounded, don't think you don't need to play it safe.

➤ Aside from protection during an emergency, one of the reasons we urge you to keep durable clothing stored in your home, vehicle, and workplace, is that you may become part of an impromptu search and rescue group digging through rubble after a tornado. The more gear you have, the more able you are to help.

Specific Follow-Up

1. Stay inside as much as possible afterward, unless you're part of your Neighborhood Watch Group and it's your duty to help the injured.

2. Stay away from damaged, or flooded areas.

3. Stay off the phone unless you need help. Leave the lines open for official use or for those actually injured and needing medical attention.

4. Once your personal situation is taken care of, see what you can do to help your neighbors and then check in with any emergency volunteer group you're part of.

5. One way to help neighbors, if you can do so safely, is to go to heavily damaged houses near you and shut off both gas and water at the meters, and then help arrange utility sharing from neighbors with functional utilities.

Sources and More Information

In addition to the weather sources listed under "Emergencies and Disasters. An Overview" earlier, try the following for tornado-specific information.

➤ Tornado Safety: http://www.tornadoproject.com/safety/safety.htm.

➤ Also visit the F-Scale page at http://www.tornadoproject.com/fscale/fscale.htm, and the home page.

➤ USA Today's Tornado Safety Page: http://www.usatoday.com/weather/resources/safety/wtornado.htm.

➤ Virginia's Tornado Preparedness Page: http://www.vaemergency.com/03torn.

➤ Tornado Safety Tips: http://www.tornadoproject.com/safety/safety.htm.

➤ Weather Alert Service: http://www.weather.com/notify.

➤ http://www.weatherpreparedness.com, Hazardous Weather Preparedness Institute, LLC, 5107 Quaker Landing Court, Greensboro, NC 27455, 800-743-4989.

Lightning

General Discussion

Though not a specific disaster in and of itself, lighting is associated with all forms of violent weather and does present its own unique hazards.

Primary Danger

The primary danger from lighting is from electrocution. A more common effect is fire. Quite a number of house fires, and many wildfire and forest fires, are started by lightning.

Advance Preparations

1. Install lightning rods at home, especially if your home stands taller than most surrounding objects.

2. Buy surge protector / power strips for all your expensive electronics such as your computer, television, stereo, etc.
3. Change your outdoor activity plans if a thunderstorm is predicted.

Warning Time

As with any approaching storm, anytime there's less than 30 seconds between lighting and thunder, it's time to get back inside. Lightning can precede a storm by 10 miles, and strike another 10 miles away from its cloud.

Forms of Alert

There is no direct or official alert for lightning alone, only for severe weather. Your only lightning warning would be the flash or thunder.

Warning Signs

> An approaching storm and thunder.
> Warning signs a split second before you're about to be struck include a tingling feeling of building static electricity, and sometimes the smell of electrical burning.

Home

> React as you would in any storm and do not bathe, shower, or use corded electrical appliances.
> Similarly, during a storm, unplug any expensive electrical appliances not protected by surge protectors. This includes your computer, stereo, and unused televisions.

Work

> Most workplaces will have invested in protection for their wired electronics and thereby protect the user. Still, react as you would at home and stay under shelter and away from conductive objects as much as possible.

Car

> The biggest thing to focus on in your car is to maintain control of it in case your car's struck, or if lightning hits nearby. You'll also have to watch for the reactions of other drivers around you.
> Though it offers some protection, it's a myth that you're fully protected inside a car.
> If driving through a heavy storm, it's advisable to pull off the road for a little while anyway for the sake of simple traffic safety.

Outdoors

> As lightning can hit as far as ten miles away from a storm front, you should head for cover if any lightning is flashing and you can't count past 30 before the thunder is heard.
> The number of seconds you count is not the distance in miles. To get the actual distance, divide the number of seconds by **five**. If you can count to five between a flash and thunder, the strike is one mile away.
> Go to the lowest ground possible without putting yourself in danger of flash floods.
> Avoid hilltops or elevated areas.
> Stay away from anything that might be a known target of lightning. This includes metal light poles, a solitary tall tree, any sort of metal tower, etc.
> Stay away from objects that might conduct a lighting strike, such as wire or metal fences or railings, metallic vehicles, bicycles, etc. However, you are safer inside a hard-topped car than out in the open, though it's an old wive's tale that the rubber tires offer complete protection from electrocution.
> If you're out on the water, get to shore as soon as possible. On the water you are the tallest object around.
> If you feel a static charge, smell electricity, or feel your hair stand on end, these are all signs of an impending strike, so you should immediately reduce both your height and your contact with the ground. Crouch down but stay on the balls of your feet. Lower your head and cover your ears and back of your neck. Do NOT lie down since that puts you in too close contact with the ground.
> Be aware of your surroundings and watch for fires started by lightning.

Miscellaneous Points

> If you happen upon a lightning strike victim, go ahead and start CPR. Lightning victims do not carry any sort of residual electrical charge.

Specific Follow-Up

> ➢ If there is no fire, damage, or injury, there is no follow-up.

Sources and More Information

> ➢ You can get a little information from the National Lightning Safety Institute, 891 N. Hoover Ave., Louisville, CO 80027, http://www.lightningsafety.com.

Flood

General Discussion

No one can fully appreciate the destructive force or the extensive damage a flood causes until they've been through one. A flood can be either a direct or indirect result of severe weather or heavy rains, it can also be the end result of an earthquake through either Tsunamis or levee breaks, or the result of a terrorist attack against a dam.

Few areas are truly safe against floods. Any area that forms a natural pocket or "bowl" can be flooded. This means individual neighborhoods can flood even if the city has never experienced one as a whole.

Primary Danger

The primary danger is from the water itself, and this danger changes with the rate at which the flood occurred. Slowly rising waters are one thing, a raging torrent from a ruptured dam is another. Heavy debris carried by strong currents only adds to the problem.

A secondary danger from floods is the contamination carried by flood waters when sewage treatment plants, chemical production plants, etc. are overrun by flood waters.

Advance Preparations

1. All suggestions from "Basic Home Prep" apply here. If your area is prone to flooding, you should have already done what you could to anchor your home to its foundation, enhance your home's structure, safeguard your utility connections and functions, and to create a safe room as high within your home as possible.

2. Also in flood areas, it's advisable to own sump pumps for basements, water pumps for greater volumes of water, and necessary hoses and power sources, as well as empty sand bags, shovels, and heavy plastic sheeting. You may choose to acquire this gear as part of a community protection effort, or as part of an emergency volunteer group. It's also advisable to own a pair of fishing waders to help protect you from contaminants commonly found in flood water.

3. Get together with neighbors, your volunteer group, community leaders, or municipal planners to decide what part you would play during a flood. Find out who would be in charge of sandbag filling details, etc.

4. Also in flood-prone areas, you should own a boat or an inflatable raft. A canoe or small "John boat" is preferable, and should be equipped with a small trolling motor, oars, and life vests. You never really know how a flood is going to develop, and you should have all safety and transportation options open to you.

5. Check the stock of food, water, first aid supplies, batteries, and other safety equipment of your upstairs safe room.

6. If portions of your neighborhood are partial to flooding, you should install or rig certain flood level indicators to let you know if roads are still passable, etc. One way to do this is paint water level warning marks at the bottom of sign posts or telephone poles (after checking with local emergency management or city engineers). Use colored paint for different levels. Use green, yellow, and red. Green is at the bottom. Any water lower that that, it's okay to "**go**." If it's over the green and under the yellow, go, but use **caution**. If it's over the yellow and about to reach the red, and certainly if the water is up over the red, **STOP**. As the saying goes, "Turn around, don't drown.

7. One common cause of localized flooding is backed up storm drains. Have the city or county check yours on a regular basis to make sure all drains are clean and open.

8. You should show on your Threat Map which areas in your immediate vicinity are subject to flooding.

9. In anticipation of the dam burst scenario, if you live downstream of one, get in the habit of looking for high ground and protected areas with solid structures just as you looked for low and solid structures in preparation for other destructive events mentioned earlier. You should also know any official warning signals the dam uses.

10. Know which nearby areas of your community are usually above flood level. Not only would you choose a rendezvous point there for some of your flood-related scenarios, but it would be a good place to park your vehicle if flood conditions were right and you were waiting around to see if the threat actually developed.

Warning Time

The warning time for a flood is directly dependent on the event causing it. Hurricanes will come with a few days advance warning. Heavy rains will see a day or less. A terrorist attack on a dam will carry no warning, though some dams have been equipped with breach alarms tied in to community alert sirens.

Forms of Alert

Radio and television news sources are going to be your best source of continually updated weather info. The Weather Channel and the National Weather Service (through your NOAA Weather Alert radio) will also provide up-to-the-minute information concerning flash floods and developing flood situations.

Warning Signs

Your only warning signs will be from the weather itself, or from news announcements. You may also hear the warning sirens from an area dam if they're so equipped. Very few are so equipped.

To help you remember your general flood reactions, we've created a reminder using the word **F.L.O.O.D.**:

Fixtures   If flooding is about to occur, turn off the appropriate utilities and prep your home for protection.

Levees   Do what you can to block the water and help neighbors with walls, sandbags, and pumps.

Options   Objectively weigh your "stay or go" options, and set yourself up to react accordingly and quickly.

Observe   Keep an eye on the weather and water levels so you're not caught off-guard.

Depart   People are more valuable than property. If you and yours can *safely* leave the area, do so.

Home

- If flooding is a seasonal possibility, do the plastic tub thing to protect heirlooms and valuables.
- You should have already made your primary "stay or go" decision to a flood and be prepared to react accordingly. Should you decide to evacuate in advance of severe weather, shut off gas, water, and power, open all interior doors (to let water drain if flooded), and leave early. (However, we recommend evacuation.)
- Should you decide to stay, be sure to fill all available containers (including that bathtub we keep telling you to clean) with drinking water in anticipation of a loss of utilities. The loss of local drinking water during a flood is a certainty. Once you've filled your water containers, shut off your water at the meter and at any other valve between the meter and the house. This will help prevent contamination of the water currently in your pipes, making startup easier later on. Still, it's a good idea to boil municipally supplied drinking water for a while after a flood. Better safe than sorry.
- As a flood develops, use cheap furniture, milk crates, cinder blocks, etc., to raise expensive furniture off the floor. Shut off your gas and turn off breakers to lower floors if water starts to rise near your house.
- Watch your sink and tub drains. You may want to tape over them and sandbag them to prevent raw sewage from backing up into your house if you don't have an anti-backflow valve installed.
- If there's a section of your neighborhood or town you know will stay dry, and that you can get to by boat and then on foot if your flood waters rise, go ahead and park your vehicle there if you're in a situation where you're facing a possible flood but nothing's developing yet. This way your vehicle is protected and accessible should waters start to rise and you decide you need to go. Don't try to drive through any water.
- If the water level starts to rise near any electrical appliances, make sure you cut off the power. If the water reaches the appliance, a family member might suffer electrocution, or a short circuit could start a fire that would burn the rest of your house down to the water line.
- You should already know what part you might play in any community protective reaction. The use of sandbags, pumps, and other flood-fighters will be decided upon by local leaders and residents based on the unique situation of local geography, and the event causing the flooding.
- Assuming nothing was effective in fighting the flood, and assuming your house is now flooding, you should move everyone to your flood safe room and prepare to leave as soon as possible. If weather and conditions allow, you should load the family in your boat and get them to safety.
- If the flood developed too quickly, you should gather your family in your upstairs safe room (where your flood reaction gear should be). As conditions allow, signal from your roof that you need assistance.
- In more rural areas, wild animals displaced by flooding can be a danger. Keep your eyes open since the last thing you want is a pack of Armadillos, or an angry Aardvark taking up residence in your safe room with you.

Work

- Your reaction at work is going to depend heavily on the event causing the flood, and on whether or not you're trapped at work or able to leave.

- In a slowly developing situation, non-essential businesses may be closed, and that might mean you will be allowed to leave work and return home.

- In a more rapidly developing situation, you may become isolated at work in which case any personal supplies you stashed at work will come in handy in keeping you comfortable while you wait for an opportunity to leave.

- If the situation is that of a dam burst, you'll either be a direct victim or you won't. If a victim, react as you would to any previously covered destructive events and get to safety, treat your injuries, help those around you, and then protect yourself as you proceed with whatever secondary option is available to you.

Car

- The biggest danger to drivers is from being swept away by flood waters. Most drivers that are victims in flood situations were doing non-essential driving and taking chances with water-covered roadways that should've been avoided.

- Do NOT drive in any water too deep to see the road. As little as 6" of rushing water can move a car. Besides, if you can't see the road through the water, you don't know if the road is even there! If weather is that bad you shouldn't be driving. Remember, "Turn around, don't drown."

- While we're on the subject, let's look at what to do if you're in your car and in the water:

  - If there's no way the car will respond and the car is being carried away by running water, the first thing you have to figure out is where you're heading. If you're just moving slowly in a small area and there's no danger of submersion or of being carried into a river or deeper body of water, then just sit in your car and signal for help. Don't try to walk away. Stay in your car.

  - If your car is moving towards a more rapid current or more dangerous body of water, you have to try to get out IF there is a stationary object you can get to. You won't be able to open your door so you'll have to roll down your window, or break the window if it's a power window and your electrical system has shorted out due to water. That's one reason we recommend you carry a window-break tool with you, and keep it accessible from inside the vehicle.

  - If for some reason you find your car in water deep enough to be submerged, you'll have to get out. Most modern cars will float for a few moments but that's about it. Don't try to open the car door as the water pressure will be too great to get it open. Roll down your window to crawl out while you're still floating.

  - If you start to submerge, don't panic. You'll have some breathable air trapped in the top of the car for a few moments. Once the car is under water and is rather full of water, the pressure will be equal inside and out so you may be able to open the door. If not, you can probably roll down or break the window to get out. **Caution**: Don't break a window if the interior of the car is still mostly full of air. Water will rush in with enough force to injure you. It'll be tough to do, but wait until there's enough water inside the car to offset any incoming water. A cool head is key.

Outdoors

- Campers and hikers are the ones most at danger from floods, and more from flash floods than anything else. As a general rule, you should not camp in dry river beds, etc., and should instead choose moderately high ground in your immediate area. Pay close attention to what's going on around you if storms with heavy rains develop. In a flood situation, the storm doesn't even have to be in your area. Storms can be miles away but the rains can cause rivers to flood numerous downstream areas.

- As a general rule, don't walk into moving water or water of unknown depth. If you **have** to move through flood waters, tie a strong rope around you and to something stationary. You learned about knots and rope harnesses during your "Outdoor Survival" training, right?

Specific Follow-Up

- Follow general "returning to a damaged home" instructions you'll find in the "Coming Home" portion of the "**Evacuation**" section.

- Flood waters should be treated as a biohazard and anything or anyone exposed to the water should be decontaminated as soon as possible with a mild bleach solution (1 part household bleach to 10 parts water).

- **Hint**: If you were unable to evacuate before the flood developed and you have sick or injured people to get out of the area, you'll have to make some sort of improvised flotation device if you don't have a boat and if (and only if) flood waters are shallow enough and still or calm enough to wade through. Flotation devices include life vests, swimming pool floats, air mattresses, two hollow-core doors lashed together, and even water bed mattresses that have been drained and inflated with air. You can also use any of your large plastic tubs (as long as they don't have holes) as small barges to carry your children, pets, or gear

Sources and More Information
- ➢ Most information on floods can be found earlier under "Emergencies and Disasters. An Overview."
- ➢ **Hint**: Heavy plastic trash bags can be used to make emergency sand bags.

Blizzard

General Discussion
Snow's a lot of fun to play in. That is until it starts blowing in sideways so hard you can't see, and piling up so deep you have to dig to find your car.

When considering the following points and procedures, keep in mind that though many areas of the country won't experience blizzard conditions, many will experience an "**ice storm**," and the same considerations will apply.

Primary Danger
Most people think the primary danger associated with blizzards is the temperature. Sometimes it is. However, what's more dangerous is the effect blizzards have on mobility and on utilities. When a blizzard hits, no matter where you are, there you stay. If you're stranded at home, stuck in your car, shut off from the world, and you haven't planned ahead, you're in trouble.

A second threat is the temperature. Many of the suggestions and points below will apply equally well to subfreezing temperatures that hit without snow.

Advance Preparations
1. Go back through "Basic Home Prep" and make sure you have all insulation and weatherproofing needs taken care of, especially those pertaining to insulating your pipes.
2. As winter approaches, have all your home heating equipment and supplies checked and filled.
3. Do the same with your food and water supplies. More on this under "**Isolation**."
4. During winter, cover any roof vents or turbines to add insulation and retain heat in the attic.
5. As extra insulation may become necessary, make sure you have a supply of plastic sheeting and painter's tape to cover windows and unused exterior doors. This supply should not come from your emergency chemical attack room sealing supplies.
6. Winterize your car to include minute details as WD40 in car locks. Do the same for your house locks.
7. During periods of potential snowfall, move all of your warmth and comfort items out of your vehicle's trunk to inside the vehicle where you can get to them without having to get out of the vehicle.
8. Add an extra pair of gym socks, extra garbage bags, and maybe a small tube tent to your GYT Pack.

Warning Time
Heavy snowfall and freezing temperatures can be accurately forecast several days in advance.

Forms of Alert
Notification will come through radio and television reports, with some communities making public service announcements on their own. These announcements will more than likely be based on the projected severity of the weather. So, if you hear a police car coming though your neighborhood announcing a pending blizzard over its loudspeaker, you'd best pay attention.

Warning Signs
Other than professional forecasts, your only warning will come when it's too late, after you're snowed in.

Home
- ➢ For the average stand-alone home, one of the dangers a blizzard poses is a potential collapse from too much snow and weight on the roof. You should have already sealed normal roof vents. In a projected blizzard, you should also seal your gables and any other attic vents. This will allow the attic to retain more heat. If heavy snow should start accumulating on your roof, you may have to open the access door to let your house heat rise into the attic to help melt the snow on the roof. Be careful if you use any kind of

additional heater inside the attic itself. Follow all fire safety rules from earlier, and any manufacturer's instructions for the heater.

- ➢ To keep your pipes from bursting, be sure to keep all faucets dripping, and all under-sink cabinet doors open. One trick that may help is to keep some sort of heat source near the incoming water supply pipe inside your house. If you have metal piping, you can place a clothing iron or a 100-watt light bulb against the pipe to keep it warm. This will warm the water somewhat as it trickles through the pipes toward all the dripping faucets. If you have plastic pipes, you'll need to be more careful as to what actually touches them. In any event, make sure your "heater" is safe and that no flammables are in the area. Only keep this active while you're around to monitor its safety and don't leave it on while you sleep.

- ➢ Another hint to help with pipes and with keeping the house warm applies to houses with unheated basements or crawl spaces. Make sure all openings, windows, and vents to any under-house area are covered with plastic and sealed. Leave one or two 100-watt light bulbs burning underneath the house to warm exposed plumbing. In some basement situations, you could leave an electric or kerosene space heater operating for short intervals. Again, exercise good fire safety.

- ➢ In anticipation of a loss of utilities, set your refrigerator and freezer(s) to their coldest setting and your heat to its highest (unless you're on a limited fuel supply). Doing these will allow for a longer "staying time" of both fresh food and interior house warmth if you should lose power. **Note**: Though you may have gas heat, your thermostat and blower fan are electric.

- ➢ In extended periods of sub-zero weather without power, your interior heat may begin to warm the foods in the fridge. In those temperatures, you can leave a few food items outside the house to stay frozen.

- ➢ The "Climate Control" portion of the "**Isolation**" section covers what to do if utilities are out and the temperature becomes a problem.

Work

- ➢ Should heavy snowfall, freezing rain, or an ice storm be a distinct possibility while you're at work, do what you can to leave before conditions leave you stranded. As with any evacuation, the key is to leave early.

- ➢ Should you be stranded at work, how this experience will affect you is directly proportional to the amount of preplanning you do and how much personal equipment you keep stored at work (or in your vehicle).

Car

- ➢ If you have to drive in wintry conditions, especially at night, let someone know where you're going, the route you're taking, and when you'll be back. Also set up prearranged times to call them and let them know you're okay. This is sort of like a pilot filing a flight plan, and is necessary since heavy snow could knock out phone lines and cell towers preventing you from calling for help.

- ➢ Make sure your car is fully checked, prepped, and stocked, and that you're dressed warmly. You might also want to carry a thermos of warm broth or hot chocolate with you.

- ➢ If you can't see due to the heavy snowfall, don't drive.

- ➢ If the road is iced over, don't drive.

- ➢ If snow conditions get to the point you can no longer drive, and no shelter option is available (such as open restaurants, gas stations, etc.), pull off the road a safe enough distance to avoid other cars pulling or sliding off the road, but not so far that rescuers can't find you. Turn on your emergency flashers. If you can pull off the road near a business establishment, do that even if the place is closed. Someone will be along sooner or later to check on the building.

- ➢ Stay in your car unless you see a nearby building that's obviously occupied. If you vacate your car, leave a note for rescuers telling where you've gone. (Authorities will be out checking for stranded motorists.)

- ➢ If you accept a ride in a more capable vehicle be sure to leave a note telling rescuers where you've gone.

- ➢ If you stay with your vehicle, run your engine and heat for ten minutes out of every hour. More than that and your run the risk of excessive carbon monoxide buildup. Less than that and you or your engine could freeze or your battery could drain. Use your cell phone to call 911 (if it's working) and notify authorities of your situation and to get an update on what to expect.

- ➢ Listen to your battery-powered radio for updates as well. Using your battery-powered radio keeps you from using your car's radio which could drain your battery.

- ➢ Once every two hours, get out (if you can) to make sure snow has not blocked your car's exhaust. Exercise and stretch for a minute or two while you're out. When you get out of the car, close the door to retain heat, but DON'T LOCK YOURSELF OUT OF YOUR CAR.

- ➢ If you have no source of heat, do not sleep! Sleeping will slow your metabolism and may allow you to freeze to death. This is why we suggest you carry heating pads or blankets that plug in to your cigarette lighter.

Being stranded is also one of many reasons we recommend keeping food and water in your car. You might also stock a few pocket hand warmers, battery-heated socks or mittens, an immersion heater for making hot beverages, and a space heater that plugs in to your cigarette lighter. If you have several items running off your car's battery, you'll want to run your car's engine 20 minutes out of each hour. However, be careful about carbon monoxide buildup inside the car.

➢ If you have a passenger with you and you <u>DO</u> have heat, you can take turns sleeping. One person should stand watch looking for rescue crews. Four-hour sleep shifts should work well.

➢ Signal. Keep your interior light on, hang a chemical light stick from your antennae, and during daylight hours, use a bright orange cloth to mark your location in the snow if the snow is so deep you can't walk away. At night while running your engine, turn on your headlights for a couple of minutes at a time and blow your horn. Signaling will help rescuers find you, or you may wind up offering safe haven for someone who took a chance on walking.

➢ Once daylight comes and the blizzard is over, you can leave your car if civilization is nearby. Once the snow starts to melt, the sand and snow chains you should have in your trunk will help get your car unstuck.

Outdoors

➢ Other than the driving suggestions above, the only people likely to be caught outdoors by a blizzard are hikers and campers. If you're out and about in potential snow weather, be sure you're carrying the proper gear and are trained to use it.

Miscellaneous Points

➢ Know symptoms of frostbite. They include numb and/or very pale fingertips, toes, earlobes, or the tip of the nose. These points are usually the first to be effected. Treatment involves an immediate warming of these areas and protection against additional exposure. But you learned all this while doing your first aid background, right?

➢ Know the symptoms of hypothermia which include: drowsiness, disorientation, uncontrollable shivering, slurred speech, and severe fatigue.

➢ Hot fluids like broth, hot chocolate, and decaf coffee are good warmers. Avoid caffeine or alcohol which may ultimately drain your energy or alter your metabolism and cost you body heat.

Specific Follow-Up

For most people, a snow storm will be more a memorable inconvenience than a disaster. If that's the case for you, the situation is over once the snow starts to melt. If the snow or low temperatures have caused any damage or injuries, tend to them accordingly. This is yet another situation where the "cure" lies mostly in prevention.

Sources and More Information

➢ All weather related sources can be found under "Emergencies and Disasters. An Overview."

Notes:

_____

_____

_____

_____

_____

_____

_____

_____

_____

_____

_____

_____

_____

Other Natural Disasters

Sad to think that our list of potential mayhem is not yet complete. We only have a few more to go though, as we discuss some of the more geologically-originated emergencies you may face. Though some of these sound like remote possibilities, you never know where you'll be or what you'll be doing. Besides, the reaction steps for some of the below will also apply to other types of emergencies.

Goals of Other Natural Disasters:

1. To complete the **very basic list of disasters and emergencies** you may experience.
2. To have you consider **situations you might not normally think about** but may have to face.
3. To provide **additional reaction scenarios** that may apply to numerous other emergencies.
4. To give you additional **"non-panic" reasons for your family to plan ahead**.

We've divided this section into:

1. **Earthquake.**
2. **Landslide, Mudslide, Avalanche.**
3. **Volcano.**
4. **Tsunamis.**

Earthquake

General Discussion

Fortunately, earthquakes don't happen as often as other disasters. However, when larger ones hit, they hit hard and many times carry secondary dangers with them. As with many of the destructive events already discussed, an earthquake can cause a hazardous materials release from an area chemical plant or factory. To make matters worse, they can cause the collapse of area dams and levees. As if that wasn't enough, building destruction, the rupture of gas and water lines, and the downing of electrical lines not only knock out utilities for some time to come, but they routinely start fires.

Most people look to California every time they hear the word earthquake, but as we pointed out under "Emergency Risk Score," no area is truly safe. Not the south, not the northeast, and certainly not the southwest. All areas of the United States have had earthquakes and will continue to do so.

Primary Danger

The primary danger from earthquakes comes from falling debris, as most injuries and deaths are caused by crushing injuries and blunt trauma. A secondary danger comes from problems on the road and subsequent car wrecks. Add to this the fire danger mentioned above and you can see the true danger of a more intensive earthquake.

There are two different scales used to report an earthquake's intensity. The most widely known in the US is the "Richter Scale," which measures the magnitude or inherent strength of a quake. The other is the "Mercalli Intensity Scale" which uses a general description of the damage produced by an earthquake to describe the intensity or degree to which the quake is felt in a specific location.

Interestingly enough, for each number increase on the Richter Scale, there is an increase in actual force by a factor of times ten! For example, a 3 on the Richter Scale is 10-times more powerful than a 2. On the next page, you'll find a comparison chart showing the relative relationship of the Richter Scale to the Mercalli Intensity Scale with a corresponding description of the damage that can be expected.

The Richter and Mercalli Intensity Scales:

Richter Scale	Mercalli Intensity	Observable Damage
1 & 2	I & II	Usually not felt, but may be felt by people on the upper floors of tall buildings.
3	III & IV	Slight vibrations noticeable inside buildings as some loose fixtures slightly shake as if a heavy vehicle or low jet were passing by.
4	V	This level will be felt by most people. Some objects on shelves may fall, plaster may crack, and a window or two might break.
5 & 6	VI	Minor to moderate damage. This level will be felt by everyone in the area, even those in automobiles. Weak buildings will see heaviest damage, brickwork may crack, some chimneys may fall.
6	VII	Damage is average to heavy. Furniture will topple, most brickwork will crack, and some walls will fall. Well water may be affected.
7	VIII & IX	All buildings will experience heavy damage. The ground will crack, underground pipes and foundations will shift.
8	X	Virtually all brickwork and plaster destroyed, railways will shift and bend, large ground cracks will appear, and bridges will collapse.
8 & Higher	XI & XII	Total destruction of man-made structures. Waves will be seen rippling through the ground with some objects being thrown into the air.

Advance Preparations

1. Revisit everything from "Basic Home Prep" that addressed structural enhancements, securing your utilities, and sturdy furniture.

2. Use "museum wax," poster clay, modeling clay, or double-sided tape to help hold your knick-knacks in place on shelves. Hanging pictures and the like should be hung from closed eye-hooks so they can't bounce off the hook.

3. If you live in an earthquake-prone area, learn to spot and avoid structures and items that may collapse on you, or in your path if traveling. Some examples are buildings with glass facades, old brick chimneys, power transformers on telephone poles, bridges, tunnels, etc.

4. Similarly, learn to look for potential safe areas that will keep you protected, or at least out from under such items as those just listed. Where in other sections we taught you to get under something, here we're asking you to pretty much stay out from under anything unless you know it's not likely to collapse, or that it offers full protection.

5. Know what to expect. People who've been in more violent earthquakes say it's like wearing roller skates and standing on a trampoline that has mushy springs, while people are shaking the trampoline back and forth.

Warning Time

There is no warning time with any sort of pinpoint accuracy for an earthquake.

Forms of Alert

The ground and everything on it starts shaking. That's about it.

Warning Signs

The only warning signs a major earthquake might have are smaller tremors in advance of the main one, or in the case of the Pacific Rim areas, a major earthquake or volcanic activity in another portion of the rim. Even then, no one can be 100% sure when a quake will hit or where, only that its likelihood is increased.

Reactions

Like the other emergencies, we've broken down your earthquake reactions by location. However, to help you remember your core reaction to everything starting to shake, we give you a reminder to the word **S.H.A.K.E.**:

Structure	Always know where the **solid structure** is around you, such as heavy doorways, and dive to it!
Hold	**Hold on** to whatever structure you're sheltered under, even if it's just a heavy table.
Aftershocks	Be ready for an immediate **aftershock** (which is really a delayed continuation of the main event).
Know	**Know** where your escape routes are and where debris or HazMat dangers might come from.
Evacuate	Safely **evacuate** the area you're in, especially if it has sustained appreciable damage.

Home

> If you're in the kitchen cooking, your first threat is the stove. Turn it off, back away, and duck and cover.

> If you're using any sort of temporary appliance such as a kerosene space heater, make sure you always keep it anchored. That's good safety whether you live in earthquake prone areas or not.

> You'll only have time for a few steps, so step towards an interior supporting wall, reinforced doorway, or piece of sturdy furniture. You should already know which walls in your home are supporting walls.

> Know which pieces of furniture in each room will give you adequate protection. The best case scenario would see an earthquake hit while you were a couple of feet away from your safe room, but that would be a rare occurrence.

> Don't run out of the house. Most injuries occur while people are exiting buildings and parts of the buildings fall on them. Besides, everything is shaking, not just the house, so you might run out the front door only to have a tree fall on you. When you leave a building, leave carefully and practice good vigilance.

> If your home is structurally damaged, turn your electricity off at the main breaker if you can get to it, and turn off your gas. If a gas leak is present, an electrical spark can start a fire. HOWEVER... if you get to your breaker box and notice a strong gas smell in the immediate area, don't do anything since throwing the breaker can also cause a spark. Just get out, make sure your gas is turned off, and let the area ventilate a while. If the gas smell dissipates, try turning off your electricity by yanking the meter. Only do this if you're sure of your safety.

> If trapped in your home, do what we mentioned under "Nuclear Detonation" regarding being trapped in a collapsed building. Stay calm, check for injuries, and try to signal for help. However, though we mentioned pyrotechnics earlier, do NOT use them for signaling after an earthquake since gas leaks are likely.

Work

> Stay put for a while if you can. Most injuries come from hasty evacuations while debris is still falling. However, if you work in a building with a wide roof span such as an aircraft hanger or gymnasium, try to get out or get under something really sturdy as this type of building will readily collapse.

> Get under a desk or heavy table and hang on to it to keep it over you.

> If there's no desk or table, move against an interior supporting wall away from windows and cover your head with your arms.

> Don't use the elevator. The shaft may be bent, or the power will go out, and you'll get stuck.

> If you can't get out of the building, follow our "trapped in a building" suggestions under "Nuclear Detonation."

> If you're in a leisure area, you're sure to have chosen your safe spots in advance like we recommended. Sometimes your only safe spot will be to duck down between stadium seats, etc.

Car

> Watch for wrecks in your immediate vicinity.

> Pull off the road as soon as possible, preferably in a safe area you've chosen if you're travelling one of your usual routes, and turn on your emergency flashers.

> If it's safe to do so, stay in your car and keep the engine running. You may have to move if you notice a secondary hazard or objects starting to fall your direction.

Outdoors

> If you're outside in an open area, stay there. The greatest dangers come from falling debris. You're safe where you are.

> If outside in an urban area, look up. Your danger will come from above. Hold your ground as best you can unless something is falling or rolling your direction.

Miscellaneous Points

> Don't forget that earthquakes can trigger secondary situations such as HazMat incidents, dam ruptures, landslides, or Tsunamis. Always be ready for the second punch.

> Earthquakes aside, a construction company may have broken a gas main. If a gas main is broken nearby, not only will you smell the gas, but you may hear a high-pitched whine or rushing sound similar to a jet engine. If you hear that, you'll DEFINITELY want to shut off all flame or spark producing appliances.

Specific Follow-Up

1. Check yourself for injuries, and then help those around you.
2. Be ready for aftershocks and fires.
3. Sniff the air for gas and listen for the hiss of leaks.
4. Listen to the radio for news of secondary dangers.
5. Pay attention to structures that may be damaged and about to fall, or that may present other hazards.
6. Shut off all utilities and leave them off until you've inspected for damage. Turning off the water is very important since a broken main may cause all your pipes to drain depriving you of accessible emergency drinking water.
7. Don't flush toilets until you've asked if sewer lines are functioning.
8. Open your cabinets carefully as items could be ready to fall out on you.
9. Watch for sudden behavioral changes in normally friendly animals.

Sources and More Information

➢ Remember all the sources listed under "Emergencies and Disasters. An Overview."
➢ See The Virtual Times: The New Madrid Earthquake at: http://hsv.com/genlintr/newmadrd.
➢ Take a look at the Southern California Earthquake Center http://www.scec.org.
➢ The Earthquake Store, online at http://www.earthquakestore.com, offers a wide variety of gear to include gas shutoff valves and earthquake alarms. You can also reach them at Earthquakestore.com, 1300 Gardena Avenue, Glendale, CA 91204, Toll Free: (888) 442-2220, (323) 245-1111, Fax: (818) 240-1492.

Landslide, Mudslide, Avalanche

General Discussion

If there is a "good side" to an avalanche or landslide, it's that it's usually a localized event. Landslides are not likely to wipe out a whole town though they may play havoc with a neighborhood. At least that's the way it is in most of the US where, for the most part, building and zoning codes are pretty good and take such potential into consideration. However... Mother Nature doesn't always pay attention to such codes, and things will happen. Besides, landslides can be a more likely danger in other countries that don't put such consideration into how and where they build. So, since you never know where you're going to be or what you're going to be doing, let's take a look at a few things to keep you safe regarding landslide.

Primary Danger

Landslides carry a couple of inherent dangers. One, they can cause a structure to collapse on you just like many other destructive events. Two, if out in the open, you can be swept away by debris and covered up or pushed into water. You may be crushed, or you may suffocate or drown. In the case of a snow avalanche, you may receive crushing injuries or suffocate.

Advance Preparations

1. Call your department of natural resources to have someone come check your property and surrounding areas for landslide risk. If their information is current, they might be able to tell you what you need over the phone.
2. On hillsides near your home, plant grasses and other fast-growing vegetation that have a good root system.
3. Follow all previously mentioned fire safety measures as the burning and loss of vegetation is a prime cause of some landslides. When vegetation is burned off and the root system dies, there's less holding the soil in place.
4. Look at natural diversion features such as creeks and see about building diversionary retaining walls to redirect any potential mudflow or landslide threatening your home. Check with community building and zoning or your department of interior representatives to make sure your changes are acceptable.
5. Look at anyone else's landscaping changes around you to see if they put you in danger.
6. Learn the difference between stable and seemingly stable objects and structures. Steel-reinforced concrete buildings resting on bedrock are generally stable. Trees, roads, sidewalks, and other seemingly solid structures are simply "resting" on the topsoil and will shift in a landslide.
7. Once again, all structural enhancements suggested under "Basic Home Prep" apply here.

Warning Time

Any warning time would be dictated by the event that allowed a landslide to occur, such as heavy rains after a fire burned off the vegetation, and how soon any of the warning signs from below were recognized.

Forms of Alert

If a potential for a landslide exists within a community or area, and officials are aware of it, there may be a news announcement, or police units may come through the area making loudspeaker announcements.

Neighbors should all know the warning signs and help keep each other aware of any potential danger.

Warning Signs

> Watch for changes in normally stationary objects:
>   - Trees, outbuildings, light poles, or fence posts that begin to lean.
>   - The ground has "shifted" or there's an unusual lump or crack along the side or at the base of a hill.
>   - Door or window frames shift, or interior or exterior walls show cracks or "signs of settling."
>   - Sidewalks, driveways, and chimneys crack or shift their location slightly.
>   - Floors are no longer level, or develop a noticeable "buckle."
>   - You notice a tear, gap, or buckle in the lawn.
> Water starts seeping up through the ground in new places.
> Utilities may be disrupted by ground shifting and breaking local connections.
> Signs of an imminent landslide may be a trickle of water, or the sounds of moving trees, creaking wood, shifting rock, or cracking concrete.
> Landslides, especially mudslides, occur most often after a brush fire has removed vegetation, and a heavy rain has loosened soil. They can also be a secondary result of an earthquake.

Home

> If told to evacuate, do so immediately. (This is a "Level 1" evac. More under "**Evacuation**.")
> If you feel threatened, move to a safe central area on an upper floor. Flowing mud or debris may break through walls or windows and pose a threat to anyone on a lower floor.
> If trapped, react as you would during a flood, and stay in an upper safe area and signal for help.
> If caught in a collapsed structure, react as you would in a nuclear detonation and remain calm, tend to injuries, and try to signal for help.
> Be aware of the possibility of broken gas lines and don't use anything flammable to signal. Even using a phone or cell phone is a slight risk because either can throw a spark.

Work

> Work in this case is essentially an extension of home, and all points from above apply.

Car

> How you would react to a landslide while in your car would depend entirely on the situation itself. Some people have equated a landslide or avalanche with a "stampede of cattle down a wide city street." They don't necessarily cover a wide path, but they're coming ahead full speed and you're not going to stop them.
> If you're in your car and you see a landslide coming, you're pretty much confined to where the road will take you. Keeping the "stampeding cattle" theory in mind, if you can move at right angles to the oncoming debris, try that. It may be easier to sidestep a landslide than outrun it.
> If your only choice is to try to outrun the landslide then while you're on the move, look for any solid structure that will offer protection from the debris or allow you an alternate escape direction.
> If you can't avoid the slide and you're about to be hit by debris, roll your windows up, lock your fingers behind your head with your arms covering the sides of your head and get ready to go for a ride! After you come to a stop, check your surroundings carefully before trying to get out of your car. If you find yourself in water, remember the points we gave you under "Flood."
> This category can also include rock fall on mountain roads. If you encounter a rock fall, drive around it carefully if you can and call 911 to notify them of the problem. If you run across this at night and have your spotlight, a highway safety flare, or chemical light stick accessible from inside the vehicle, then drop one out the window to warn other motorists of the debris. We do not recommend you stop and get out to do this as there may be other rocks about to fall. Stay in your car.

Outdoors

➢ If outdoors, you should pay more attention than ever to the possible warning signs of landslide. The best reaction is to not be in the area when one occurs.

➢ Stay out of ravines, riverbeds, and other natural channels that may become avenues for debris or mudflow.

➢ If a landslide or mudslide starts, move at right-angles rather than try to outrun it, unless there is something in the immediate area that will offer guaranteed protection. Remember, protection is on high ground with solid, stationary, objects. An example might be a rock facing on a hill, a concrete building, or other structure anchored to the bedrock.

➢ In a snow avalanche, try to move away at a right angle. If caught in it, most experts recommend trying to stay on top of the snow by doing something resembling the backstroke in swimming. This keeps you on top.

Miscellaneous Points

➢ The suggestions carried here and in the "Earthquake" section will also prepare you for a sinkhole.

Specific Follow-Up

After a landslide, mudslide, or avalanche has settled, you should go with the standard reactions of remaining calm, checking yourself for injuries, and trying to communicate if you need help.

Sources and More Information

➢ Landslide info can be found under "Emergencies and Disasters. An Overview."

Tsunamis

General Discussion

A true, natural Tsunami is sometimes mistakenly referred to as a tidal wave. Though essentially that's what it is, a tidal wave is usually just one big wave, while a Tsunami can be a series of waves just as an earthquake may be a series of shocks and aftershocks.

Though generally an after effect of an earthquake, a small Tsunami could be the after effect of a nuke detonated in a harbor, or just off shore. Also, at the time of this writing, there is a concern about a volcanic island called LaPalma, part of the Canary Islands, off the coast of West Africa. Fears are that should the island collapse, it would result in a Tsunami powerful enough to reach the east coast of the US. Relatedly, if anything were to ever happen to Hawaii or California, the destruction of one might cause a Tsunami that would damage the other.

Primary Danger

Interestingly enough, the primary danger of a Tsunami is not necessarily the water, but the force with which it hits. The amount of water, though, is a close second.

If the same amount of water carried in by a Tsunami were to trickle in to an area slowly, as in a flood slowly rising from heavy rains, all you'd have is a general flood situation. Once the water drained away, everything would be wet and water damaged, but otherwise intact. With a true Tsunami, whole towns could be literally wiped away leaving only foundations where buildings once stood. It would almost be like a nuclear shock wave only in water form.

Tsunamis can be over a hundred feet high, and travel at speeds up to 300 miles per hour.

Warning Time

Warning time depends on how far out the wave originates, and on detection and communication channels. Waves can be spotted by the Coast Guard, coastal observers, various forms of radar, and sometimes satellite.

Other warning could come from geologists monitoring, measuring, and pinpointing the epicenter of an earthquake. If an epicenter is out to sea, an alert may be sent to coastal authorities for a Tsunami watch.

In either situation, the warning time will be short, usually less than 30 minutes.

Warning Signs

> An earthquake is usually the cause of a Tsunami, so this is usually your first indicator.

> The sea or tide going out unusually far is usually another indicator, as water is drawn up by the waves moving in to shore.

Advance Preparations

1. If you live near a coastal area, you should know your area's elevation from sea level and distance from the shoreline. Officials may evacuate areas based on this data.

2. You should take preparations seriously if you live lower than 50 feet above sea level and within one mile of shore, and especially if you live in Hawaii or on the west coast.

3. The closer you live to shore, the more your reaction should focus on evacuation. There's very little you can do to augment the structure of your home that will withstand a Tsunami. If you live five miles or more from shore, your home will begin to stand a chance of offering you some protection.

4. If a Tsunami is high on your list, you should choose an evacuation destination that offers high ground and solid structure. Be advised, though, that in areas a Tsunami may hit, very few spots exist that will offer true protection. Check with your local officials to see if certain locations have been designated as Tsunami protection sites. Another point to keep in mind here is that the route you take to seek protection should be easy for you to access, and certain to get you out of harm's way quickly. A long, straight, stretch of highway moving directly inland to high ground is preferable. Remember, this is another scenario where you'd want to evacuate quickly to beat traffic.

Forms of Alert

As a Tsunami is detected, you can pretty much bank on all forms of alert being used. Community sirens may sound, a local EAS alert may be issued, and the National Weather Service may issue a redundant alert as well. Police officers will not have time to tour every neighborhood, but will probably make loudspeaker announcements at their current locations. This is another reason to be part of a phone tree.

Home

> Any reaction that you could take at home will depend entirely on how far inland you are. If you live along the coast, your only hope is to be able to evacuate quickly enough to make it to a protected area in time.

> Those living further inland and at the outer edges of a Tsunami's reach should react as they would for a landslide and head for an upper safe area. Basements may flood and lower floors may receive damage from water and/or impact of debris. By the time a Tsunami gets inland a ways, you can expect a great deal of debris to be carried by the water.

> Your other option is immediate evacuation. Get everyone into your vehicle and on the road as rapidly as possible. IF you can grab your Bugout Kit on the way out the door, great, but this is one situation where speed is truly of the essence.

> As you evacuate, head for your prechosen destination that offers high ground and/or a protective structure. How successful you are depends on how early the warning was given, how quickly you were able to react, and how far inland you already are. Another helpful item is to know the geography of your area. Do you know where the ground elevation increases and where it decreases? Be sure to keep this in mind when creating your Threat Map and planning your evacuation routes.

Work

> Again, work reactions are the same as those for home. Your only real options are to find immediate protection where you are or evacuate as rapidly as possible.

Car

> Finally we find a situation where it actually might be advantageous to be in your car. As running away is the order of the day, you're actually better off if you're already in your main mode of transportation. The only factor you have to consider is other family members and their location, and available routes out.

Outdoors

> If you're able to receive the warning, about the only thing you can do on foot is head for high ground and/or any naturally protective structure. Otherwise, you'd secure transportation and evacuate.

Miscellaneous Points

> As there would be little time to gather your gear before hitting the road, this is another scenario that illustrates the importance of having supplies stored at your intended destination. More on this under "**Evacuation**" below.

Specific Follow-Up

1. Just as there are aftershocks with earthquakes, there can be follow-up waves to a Tsunami.
2. If there's anything left to go back to, to clean up and rebuild, be extra careful with biohazards. Among the debris will be dead animals and the like, and therefore associated disease. Also, depending on the season, you may soon see an drastic increase in the mosquito population (just as you would after any flood) and that may see an increase in West Nile and other communicable ailments.
3. More on cleanup will be covered under the "**Evacuation**" section's "**Coming Home**" portion.

Sources and More Information

> Tsunami info can be found earlier under "Emergencies and Disasters. An Overview."

Volcano

General Discussion

We'll end the specific disasters list with the least probable of all, the volcano. We're including this section for our friends in Hawaii, the Pacific Northwest, and those who may travel.

Essentially, once you've mastered the reactions to a nuclear detonation, earthquake, landslide, and chemical incident, you're pretty much set to deal with a volcano. Simple, huh?

Primary Danger

Blast, lava flow, and mud flow are primary problems during an eruption, with ash and dust running a close second.

Advance Preparations

1. Prepare as you would for a hurricane or wildfire. Geologists may be able to provide advance warning of a potential eruption, so you can board up your windows (if there's time) and evacuate.
2. The landscaping steps you take to prevent wildfire may help protect your property from fires commonly started by eruption and lava flow.
3. Dust, ash, and toxic gasses from an eruption should be treated as a chemical attack or HazMat incident, and you should have protective equipment accessible, or remember that some gear can be improvised or fabricated.
4. On your Threat Map and evacuation route map, you should consider which of your routes may be compromised by any after effects of a volcanic eruption. Will your primary winds carry smoke and ash over your main route? Are there any bridges that may be damaged by earth tremors? Will any rivers be swollen by mud flow? Does the local topography show that lava may flow across any of your roadways out of the area?

Warning Time

The science of geology, as regards volcanic eruption, is actually rather accurate. The probability of an eruption can be estimated several days out.

Forms of Alert

If an eruption is likely, an evacuation order may given in much the same manner as with a hurricane. Word will probably be relayed through radio and television news, and by loudspeaker from patrolling Law Enforcement.

Warning Signs

> Earth tremors are usually felt as pressure builds within a volcano.
> Smoke activity is frequently seen at the top or mouth of the volcano.
> Sometimes a bulge may be seen in the side of the cone as internal pressures build.

Home

- Though we mentioned you should know which of your evacuation routes may be compromised during an eruption, we recommend you evacuate when ordered and that you not be in the area long enough to worry about any volcanic effects.

- However, your home may be well outside the danger area of eruption, and concerns regarding your home may include ash or dust. React to these as you would a chemical release. It's best to seal your home as volcanic dust is highly irritating to humans and animals.

- If no warning was given, and eruption suddenly occurs, react as you would for a nuclear blast. Run for your safe room if you're in the house, or duck and cover in a low protected area if outside. There will more than likely be a shock wave depending on the severity of the eruption even though you don't have to wait a full two minutes like you did for a nuclear blast. Once the blast is over, get out of the low areas since the secondary danger may be mud or lava flow, depending on your location.

- Be aware that there will likely be falling debris. Stay protected as you gather family members and your Bugout Kit to hit the road.

Work

- Again, work is essentially an extension of home. Your immediate concern would be evacuation in advance, or to duck and cover during an eruption.

- Considering the possibility of an unforeseen eruption, your evacuation map should also reflect all routes leading away from where you work, go to school, or other areas you frequent. Remember to keep a copy of this map in your GYT Pack.

Car

- If caught in your car during a surprise eruption, react as you would during a nuclear detonation. Your first concern is controlling your vehicle and avoiding wrecks. Pull of the road as soon as possible while you take stock of your situation and your location as regards evacuation routes open to you. However, keep moving if you're pelted by debris, or you see lava or mudflow.

- Smoke, dust, and ash will cause severe visibility problems on the roadways. Turn on your low beams and emergency flashers, and drive as you would through heavy fog.

- Drive through settled dust and ash carefully as they can cause roadways to be slippery (though dry) and they'll wreak havoc on your air intake filter and brake linings.

- Also while driving through dust and ash, react as you would to a chemical release and seal your car as well as possible. You might also want to wear a breathing filter as dust and ash are fine enough to slip through unsealed seams, and they're both harmful to lungs.

Outdoors

- If caught outdoors, react as you would during a nuclear detonation and duck and cover. Again, the blast won't be quite as strong so you won't have to stay down a full two minutes.

- Remember that some debris may come raining down on you, so pay attention overhead as you would during an earthquake. Some lighter debris can be deflected by hardhats or umbrellas if it's still falling on you and you have **no** other cover.

- After the initial blast of eruption, your main threat may be dust, ash, or toxic gasses. React as you would to a HazMat incident and protect your skin and breathing.

- Head for high ground and/or protected areas as landslide, mud flow, lava flow, and wildfire are all distinct possibilities.

- If you're camping or hiking, always have your cell phone (if you're in an area that gets a signal) and know the number of the nearest ranger station or county extension office. If not, simply call 911. Give them your current location and ask for advice regarding location of threats and possible routes out.

Specific Follow-Up

1. Keep an eye on fires that may have been started.
2. Clean accumulating ash off of roofs and out of rain gutters for the same reason you'd want to remove accumulated snow. Too much could cause structural damage due to the weight.

3. Deal with dust and ash as you would with fiberglass insulation. Wear full skin and eye protection as well as a dust mask, though ash can be so fine you'd want a full particulate-filter respirator. Both volcanic products can cause health problems to lungs and eyes.

Sources and More Information

> ➤ Volcano info sources can be found earlier under "Emergencies and Disasters. An Overview."

What if Nothing Happens?

As a final "Reaction," let's discuss what to do if something major happens near you but you're not directly affected. Sometimes the hardest thing of all to do is nothing. Let's look at a few considerations.

The rules of engagement for a civilian are much more limited than those of our first responders, but just as important. Let's look at a few things to do if your area is the victim of a small-scale disaster or emergency, yet you are not immediately or directly affected (and won't be). An example of this might be a conventional terrorist attack, such as a car bombing, on a local target. Another example might be a tornado where some parts of town were destroyed, but your area is untouched.

1. **Relay the emergency.** Call your local family members to make sure all are safe, especially if the event is in the midst of occurring and you need to relay the warning. If you're part of a **volunteer group** (and you should be) activate your phone tree.
2. **Stay off the phones.** After you've relayed the emergency info, if you're not directly involved in an incident or situation, and you don't have to communicate with others for help or to coordinate an evacuation, etc., stay off the phones. Leave the lines open for those needing emergency assistance, for official first responders, and to allow for financial processing for those needing credit card use. Out of town family members can contact you later.
3. **Stay off the roads.** If the incident is an isolated one, and you happen to be driving near the area but safe, the best thing you can do is leave the area, or find a parking lot and sit for a while. You'll want to help free up the roads so emergency vehicles have less traffic to deal with.
4. **Stay put.** One of the toughest things to do is sit back and watch. Granted, this may seem to conflict with all the encouragement we've given you to volunteer and help, but sometimes the best thing you can do is *not* show up. In actuality you may have skills and assets to offer, but though an asset, you'll still be something the incident commander on scene has to manage. If you're part of an organized volunteer group, the best thing to do is get ready, but wait to be called into action.
5. **Observe and report.** If you're helping by watching your own neighborhood, or you have information pertinent to unfolding events, you should call authorities to provide useful information, but do so without compromising your own safety.
6. **Help your neighbors.** Even though you're staying put and out of the way, you can still help by checking on your neighbors.
7. **Listen.** Keep your TV and/or radio on for pertinent information as it develops. If your home or neighborhood is close enough to an incident, you'll need to listen for further instruction.
8. **Assist locally.** One thing your volunteer group or watch group could do without showing up at an incident scene is to patrol your own area, or take over some other low-level function to free up first responders for the incident.
9. **Donate.** After an emergency, everyone wants to help. That's always a good thing. However, you might want to limit your donations of tangible goods, like clothing, until it's directly requested. Donated goods require people to handle them, store them, inventory them, and distribute them. This is something best left to a volunteer group or other charitable organization. The best donation is always money, or in the case of the Red Cross, blood.

Now that we've covered the basic reactions of the most common terror attack scenarios and natural disasters, let's start talking about the first of two major secondary reactions. Let's talk about the worst-case-scenario, "**Evacuation.**"

Notes: _____

_____

_____

_____

IV. EVACUATION

Let us not look back in anger or forward in fear, but around in awareness. –James Thurber

✓	Goals of the Evacuation Section		
	Activity	**Date Completed**	**Pg**
❑	Per our "**Emergency Risk Score**," we know why we might have to evacuate.	___/___/___	54
	Prepping the Gear:		224
❑	We understand the **urgency levels** of different evacuations.	___/___/___	
❑	We've prepared a primary "**Bugout Kit**" for the family.	___/___/___	
❑	Each family member has a set of **"bolt clothes"** prepped and ready.	___/___/___	
❑	**Support & Secondary packs** have been added to the main Bugout Kit.	___/___/___	
❑	Some **redundant evacuation gear** has been stored **away from home**.	___/___/___	
	Planning the Destinations:		275
❑	We have identified **three official shelters** in our area.	___/___/___	
❑	We have chosen at least **three out of town evacuation destinations**.	___/___/___	
❑	**Three different routes** to each destination have been identified.	___/___/___	
❑	**Alternate routes <u>out</u>** of **neighborhood**, **work**, and **school** areas are set.	___/___/___	
❑	Our "**Evac Atlas**" is filled out, covering each destination and all routes.	___/___/___	
	Getting There:		300
❑	In-town and out-of-town **rendezvous points** have been established.	___/___/___	
❑	In-town and out-of-town **contact persons** have been recruited.	___/___/___	
❑	Alternate **communication methods** are set and **equipment** purchased.	___/___/___	
❑	We're aware of all local and long distance **transportation alternatives**.	___/___/___	
❑	All portions of the "**Life on the Road**" section have been discussed.	___/___/___	
	Life at the Destination:		313
❑	We've read and understand the section on "**Shelter Dynamics**."	___/___/___	
❑	Our family understands the importance of "**rest and recharge**."	___/___/___	
❑	We've noted **communication methods** and **outside services** to get home.	___/___/___	
❑	We understand the precautions of **returning to a damaged home**.	___/___/___	
❑	We've recorded contact info for **commercial cleanup & repair services**.	___/___/___	
❑	We're aware of and ready to face **stress and other emotional issues**.	___/___/___	
	Goal Setting:		A-1
❑	Entered **start dates** and **goal dates** onto **calendar** for all the above.	___/___/___	

Despite occasional travel, most of us are home bodies. We go to work and out to play a little now and then, but home is always great to come back to. No one wants to leave it. Sometimes we may have to, though, whether the reason is a terrorist attack, natural disaster, or even a local industrial accident. So if you have to go, it's best to be prepared and able to leave quickly, make the trip as efficient, easy, safe, and comfortable as possible, and that you're also prepared to safely return home and deal with the aftermath of whatever it was that drove you out in the first place.

The first consideration of evacuation planning is the "should I stay or should I go?" decision. The answer is simple. **You should evacuate if officials tell you to, or if you feel you should**. Better safe than sorry. The second consideration is the fact that if you do evacuate, **you leave as soon as you possibly can**. This serves two purposes: **one**, you should outrun whatever's causing you to leave, and **two**, you should "beat the rush." You'll want to be on the roads before they're clogged with evacuees, and you'll want to be at a destination and settled before unprepared and under-supplied masses of people arrive and drain the local supplies and strain the infrastructure.

Your "Emergency Risk Score" from earlier should give you a glimpse of what might be in store for you, and you can probably figure whether your primary reaction will be an evacuation or a sheltering in place. With isolation vs. evacuation, you should plan first for whichever is most likely, but prep for the other as well. Divide your attention and investment 55/45 between the two, and certainly don't neglect either.

When deciding how much emphasis you should give evacuation planning, as opposed to isolation, one of the things you should consider is that state and federal emergency planners have based *some* of their assistance plans on the fact that America is a much more mobile society than in years past, and that, depending on the type of disaster, it may be easier to set up refugee and assistance centers in towns surrounding a disaster area rather than at "ground zero." Add to that the growing threat of a terrorist attack that will result in some sort of area denial, and you can see that evacuation is a very likely reaction.

Many people conjure up mental images of hurricanes and wildfires when the word "evacuation" is mentioned. Others know there are other incidents which could cause an evacuation, but they still consider it to be just an "annoying road trip" where they'd pack the car with a few things and head off to a friend or relative's house a few towns away. We wish it were always that simple.

What if a massive wildfire suddenly changes direction, forces you out of your home, and cuts off your main evacuation route? What if your main evac destination was destroyed by the very thing forcing you to flee? What if there's a nuclear detonation near your city and massive traffic hits all roads in all directions in a panicked evacuation? How would you gather your family together? Who in your family is in charge of picking up the kids? Who goes home to get the supplies? What if authorities have blockaded the main evacuation routes to direct evacuees to "tent cities" or other aid sites and you never make it to your destination where some of your supplies are stored? What if you're fleeing a biological attack and vigilante groups have blockaded the town you were heading to? Where would you go? What would you do? What if you were only evacuating a tornado-damaged house? How long could you stay in a motel? How long could you stay camped out in your yard? What if terrorists seize the dam just up from your town and are about to blow it up? How quickly could you load your gear and go?

In this section we're going to make sure that if you have to evacuate, that you know what to take, where to go, when to go, how to gather everyone together, how to get there safely, how to protect what you left behind, and how to deal with the aftermath. We're going to prep you to escape the worst case scenario disasters. That way you can rest assured that you're set, and if you experience anything less, it'll be a piece of cake. Considering all that and the old adage that, "no matter where you go, there you are," you'll need things to comfort and protect you along the way and when you get where you're going. As we always preach preparation before execution, we'll look at gear first.

Prepping the Gear

For "Prepping the Gear," we'll discuss the essential items that you <u>have</u> to take with you during a worst-case-scenario evacuation. We'll also look at items that would be *nice* to have with you but that you can survive without, since the more dire the emergency, the less time you have to grab things to take. That's why your evacuation gear will be organized by necessity. We'll divide your Bugout Kit into your **Primary kit, Support kit, Secondary kit,** and **"As Able"** items. Other interrelated conditions exist that will influence what gear you carry. We'll look at those in a bit.

Goals of Prepping the Gear:
1. To help you prep your gear for the **most dire evacuations**.
2. To help you understand the various **levels of urgency** with which you may have to evacuate.
3. To make sure you can perform the **most important task**, which is to evacuate <u>**quickly**</u> and <u>**early**</u>.

4. To prepare for **lesser known extenuating circumstances** surrounding an evacuation.
5. To make sure you have all the **goods and gear to keep you safe and comfortable**.
6. To make sure you **don't buy what you don't need**.
7. To **help you make some common equipment** at home.
8. To help you **organize your gear's locations** to make your evacuation smooth and efficient.

We've divided this section into:
1. **The "Bugout Kit" Overview.**
2. **Contents in Detail.**
3. **Packing and Storage.**

The "Bugout Kit" Overview

Let's look first at the **evacuation urgency levels**, and some examples of **what may cause them**:

Level 1: <u>Immediate</u> evacuation, as if the house was on fire, a breach in a nearby dam, a nuclear detonation, etc.

Level 2: **30 minutes or less**. A relatively nearby nuclear detonation, or an imminent chemical release.

Level 3: **1 to 3 hours**. As with a developing flood, or a wildfire suddenly changing course.

Level 4: **4 to 8 hours**. A steadily approaching wildfire, or abrupt change in the course of a hurricane.

Level 5: **A day or two**. As with a hurricane with good tracking and advance warning.

(Note: These "levels" are our own system and officials will not notify you using these terms.)

In addition to the **time factors** indicated by the "Levels" above, **extenuating circumstances** and other factors tied to the situation itself also come into play.

Extenuating circumstances that affect what you might need to take, include:

1. **The anticipated socioeconomic impact** of the disaster. Natural disasters will cause problems and cost you money, but severe terrorist attacks will cause economic ripples and maybe civil unrest for some time to come. (You'd want to take more with you in the way of assets.)
2. **The anticipated affect the incident will have on your home**. Will you have a relatively intact house to come back to, or is its destruction imminent?
3. **Your intended destination**. Different locations will impose different needs on you. You'd need less equipment if you were heading to a relative's house 50 miles away, but far more if your plan was to travel 300 miles to camp in a remote wilderness location for a few weeks.
4. **Your estimated time away**. Will you be back home in three days or will you be gone a month?
5. **Transportation**. Do you have your own vehicle? What kind is it? Is it a spacious SUV or RV, or is it a motorcycle? Does your family have more than one vehicle? Will your vehicle always be available or might you have to rely on help from others?
6. **Loading and logistical factors**. How many people are in your family? How many can help load your evacuation gear? How many cannot? How many pets do you have? How and where do you have your equipment stored?
7. **From where are you evacuating**? Where do you spend most of your time, and where are you likely to be when a disaster or attack occurs? Home? Work? School? The middle of town? The suburbs? Are home and work in the same area or in neighboring cities? How much gear would you be able to keep with you or in your vehicle?
8. **How many stops**? Will evacuation be a load-and-go thing, or will you have to get from work to home, or stops along the way to gather people, or gather gear from different locations?

We could go for page after page of cross-check tables spelling out such things as, "if this happens, take this gear," or "if you think you'll be gone this long, do this or that..." But that would only serve to limit your gear and therefore your options. So the best thing to do, in simplest terms, is **whenever you evacuate, you should carry as much of the listed goods and gear as you possibly can while exercising extreme caution in your time limits. <u>Period</u>**. There are no guarantees in any disaster and it's better to be prepared than to wish you had been. If you wind up carrying a ton of things you don't use, you can use them as barter items while on the road, or save them for next time. There will always be a next time, and if not during the remainder of your life, there will be for your children.

Let's further our overview of the Bugout Kit by looking at the main checklist covering the various component kits that will make up the full kit and its supporting elements. In the following section, "**Contents in Detail**" we'll cover the makeup of each of the main kit's components. We'll follow that section with "**Packing and Storage**" which will cover some of the strategies involved in how you package your kits, where you keep them stored at home, how to make your kits more portable, and where to keep additional gear off-property.

In keeping with the above "extenuating circumstances," we've divided our full compliment of evacuation goods and gear into various levels and support kits for **two main reasons**. **One** is **time**. In an **immediate** evacuation, you'll only have time to grab **one** kit. It needs to carry the most important items, and it has to be light enough to get out the door and into whatever transportation you may have. The **second** reason is **transportation**. If transportation is limited, or later altered (if you switch travel modes), you might be denied taking some gear with you from the start, or forced to get rid of some to lighten your load later. Having divided kits will allow you to do either more readily.

Our Bugout Kit is oriented toward self-sufficiency during a **14-day evacuation caused by a worst-case-scenario**, without relying on much in the way of outside water, food, or shelter. However, given the fact that most evacuation scenarios will involve some level of civilization, transportation, and other people, we've stopped short of making this a "survive by yourself in total wilderness" kit.

When considering all the pieces of equipment you might have to pack, keep a few helpful hints in mind to make your preparations easier and more economical:

1. The most important point of all when considering Bugout Kits and their components is that **your Primary kit MUST be packed and ready to go at all times**. In a Level 1 evacuation there will NOT be time for you to carry your checklists around the house gathering items you need to take with you. Also, prepacking helps in complete home destruction. If your whole house is destroyed in a tornado while you're gone, if you find your kit, you have everything you need. Prepacking sure beats trying to find the individual components that might now be scattered all over the neighborhood.

2. **Pack lightly but thoroughly** as if you were going to be gone a month to a place with no stores, and all you were allowed to bring was one piece of "carry-on" luggage. Carry the smallest and most light-weight version of an item possible. Going on backpacking trips will help you learn to pack like this.

3. Go only with **what you truly need**. Everyone's situation will be different. We've done the best job we could of creating a universal kit, but you still have to customize it to your needs. If there's something in a checklist that you know for an absolute fact you cannot and will not use, then don't pack it. If there's something you really, really need, you might want to pack more than one.

4. Organize and utilize the **goods you already have**. Don't acquire anything new you don't need to.

5. Out of what you need, **make what you can**. Don't pay retail price for something you can make yourself at home. You can make all sorts of backpacks, "hobo stoves," mess kits, and other items very easily. (Making some of your goods and gear could actually be a fun family project.)

6. **Shop around** for the best prices for the items you cannot make. Start at your local thrift store, military surplus, and "Dollar Stores." You'll be surprised at the deals you can find. After that, utilize all the sources at your disposal to comparison shop to find the best deals. (This is another of the many reasons we've given you so many outside sources in this book.)

7. Once you have your kit assembled and loaded, fill out the "**Last Minute List**" accordingly. You'll find a blank one in the "**Appendix**."

8. After your kits are completely assembled, keep them **strategically and securely stored**. Make them **off limits** for anything other than **emergency use**! The last thing you want to hear during a disaster is that one statement that will bring your preparedness world crashing down around you when someone in your family says, "Oh, that stuff? I borrowed it to..."

As you go through this section making detailed plans on things to pack, please keep one thing in mind: **Knowledge beats gear almost any day of the week**. It will be nice if you're able to get **to** your gear in an emergency, but situations may occur when all you have is the clothes you're wearing. The more **training** you've allowed yourself in the **earlier portions of this book**, the more alternatives you're aware of, and the more you're prepared for a "plan B," or "plan C," the better off you'll be. That being said, let's look at a general **Bugout Kit:**. This Bugout Kit Component Checklist is a list of various kits and sets of gear that will make up the larger kit. Each component kit, with its own checklist, will be discussed later.

The "Bugout Kit" Component Checklist

(Each component and kit listed below will have its own checklist and discussion on the following pages.)

Primary Kit – This is your "**Level 1**" kit; the kit you'd grab if you only had time to grab <u>ONE</u>!

<table>
<tr><td>❑ "Last Minute List"</td><td>228</td><td>❑ Personal Protection Gear</td><td>228</td><td>❑ First Aid</td><td>229</td></tr>
<tr><td>❑ Level 1 Water</td><td>233</td><td>❑ Level 1 Food</td><td>234</td><td>❑ Basic Toiletries</td><td>239</td></tr>
<tr><td>❑ Primary Clothing</td><td>241</td><td>❑ Basic Shelter</td><td>243</td><td>❑ Non-Prescription Meds</td><td>244</td></tr>
<tr><td>❑ Babies' Kits</td><td>247</td><td>❑ Communications Gear</td><td>248</td><td>❑ "Office Pack"</td><td>249</td></tr>
<tr><td>❑ Batteries</td><td>252</td><td>❑ Tools and Repairs 1</td><td>254</td><td></td><td></td></tr>
</table>

Support Kit – If you have time to grab a second set of gear (**Level 2 or 3 evac**).

<table>
<tr><td>❑ Pets' Kits</td><td>255</td><td>❑ Level 2 Water</td><td>258</td><td>❑ Level 2 Food</td><td>258</td></tr>
<tr><td>❑ Extended Clothing</td><td>259</td><td>❑ Extended Toiletries</td><td>260</td><td></td><td></td></tr>
</table>

Secondary Kit – After the safety and well-being needs of your family are met, consider secondary needs.

<table>
<tr><td>❑ More Food and Water</td><td>261</td><td>❑ Alternate Transportation</td><td>262</td><td>❑ For Helping Others</td><td>262</td></tr>
<tr><td>❑ Extended Comfort</td><td>263</td><td>❑ Extended Communication</td><td>264</td><td>❑ Extended Shelter</td><td>265</td></tr>
<tr><td>❑ Tools and Repair 2</td><td>265</td><td></td><td></td><td></td><td></td></tr>
</table>

As Able – Some evacs are a "**Level 5**" and you have a full day to evacuate. Carry these as you're able.   266

<table>
<tr><td>❑ Heirlooms and valuables</td><td>❑ Home repair & cleanup tools</td><td>❑ Entertainment</td></tr>
<tr><td>❑ Children's' Schoolwork</td><td>❑ Job related project materials</td><td>❑ Recording equipment</td></tr>
</table>

Contents in Detail

Each component of the Bugout Kit listed in the checklist above will be covered in excruciating detail here, and each will have its own checklist and discussion. **Though this section is rather wordy, we'd prefer you know all the subtle reasons an item was included, rather than you find out why the hard way in the middle of an emergency.** Though thorough and detailed, you'll still have to customize some of these checklists based on your personal situation, mode of transportation, destination, needs, and assets. In the "**Appendix**" you'll find a "**Disaster Prep Expense Journal / Planner**" that will help you create a single shopping list of the evacuation equipment items you need to buy.

The Primary Kit ——

Your Primary Kit should be the **one kit you grab on your way out the door in any Level 1 emergency**. It should **stay packed 24/7** and contain all those items necessary for your immediate survival based on the threats you're likely to face, your available mode of transportation, your primary destination, and the amount of prepping, planning, and off-site storage you've already completed.

To the experienced backpacker or seasoned traveler, these kits will seem excessive, bulky, and heavy. They are. Then again, they <u>have</u> to be since we're doing far more than making a simple, well-planned temporary excursion. We're prepping to react to an unknown emergency, relocate our lives for an indefinite period, protect ourselves along the way, and sustain ourselves in the process and beyond. We also have to keep in mind the fact that though we may have simple evacuation routes and destinations picked out, that things can change drastically without warning, unforeseen circumstances can present themselves at any time, and no emergency or disaster has ever followed a set script! You have to be ready to react to sudden and extreme changes in plans, and to not only survive, but to thrive.

We'll discuss how to organize and pack all these items in the next section, "**Packing and Storage**." We'll also show you ways to keep some goods and gear elsewhere thus lightening your initial load. For now, let's look at your Primary Bugout Kit's Component Kits and what needs to go in them.

Note: A few items will appear on more than one component kit checklist. We did this in some cases to emphasize multiple uses of some items, and because having an item in one kit doesn't mean it'll be accessible when you need the use of another kit. Though you should try to restrict size and weight of your gear, a little redundancy of some smaller items is okay. Even so, you and your family are still going to have to sit down together and decide which of the below items you truly feel you need, and the quantities involved. Let's look at the **component** kits that make up your **Primary, or Level 1, kit**:

"Last Minute List"
❏ The "**Last Minute List**" is an actual list you'd leave taped to the top of your Bugout Kit.

Overview

This list is something you'll create on your own, or by using our "**Last Minute List**" form from the "**Appendix**." It will help you list the things you have to take with you that you absolutely could not pack in advance. An example would be perishable medications. The form will also ask you to list the important things you need to **do** before heading out the door. More on that under the "Out the Door" section below.

Having this list in place is extremely important, because when the adrenaline hits in an emergency, you cannot count on your memory! That's why "911" was developed as a universal emergency number. It's quick to use and easy to remember when nerves are rattled. Make sure your list is thorough, accessible, and visible so you don't overlook it. It's <u>that</u> important.

Also, when making and storing your Last Minute List, write out all instructions in detail, since you might be incapacitated, and it might be one of your older children, a non-resident family member, or a trusted neighbor helping you collect your gear and get the family out of harm's way.

Personal Protection Equipment		
❏ Gas mask	❏ Protective suit	❏ Defensive items
❏ Safety necklaces	❏ Flashlights	

Overview

These items are your personal protective gear to protect you from hazards or dangers you might encounter in getting from a disaster area to your final destination. Your first aid kit is also considered personal protection, but it has its own checklist below.

Discussion Items

> **Gas mask and Protective suit.** These two items are elective, based on the perceived threats you face in your area. However, if your risk scores indicate you might need this gear, it should be part of your "Level 1" kit, and should be kept more accessible than most other items, as you may need this stuff to either get to your vehicle, or during your evacuation travel. At the very least, **be sure to carry a dust mask** for each person. Smoke, dust after an earthquake, and any number of scenarios could see the need for a basic breathing filter.

> **Defensive Items** will be discussed a little in this section under "Life on the Road," and more under the "Life in Isolation" portion of "**Isolation**," where we cover the recommendations and ramifications of self-defense and personal security.

> **Safety Necklace** (as described at the end of "Gas Masks and Moon Suits"). Though intended more for the kids who won't have wallets or purses with ID, this "Safety Necklace" is useful to all family members because it's essentially a tiny GYT Pack on a neck strap. Make sure each family member has an extra one packed in the Bugout Kit even though they might have one hanging in their room or school locker.

> **Flashlights.** We recommend several small, inexpensive flashlights rather than one or two big, expensive, heavy ones. Having several small ones gives you more light, greater economy, and more batteries that can be used in other things. **Hint**: The most commonly used batteries in all types of equipment are AAA, AA, C, and D (D-size is the best for your main flashlight as the batteries last longer). Decide which you use the <u>most</u> in your other equipment, and buy flashlights that use the same type(s). Also, keep one or two flashlights loaded and readily accessible in your kit. With the other flashlights, keep the batteries in **backwards** so they won't drain if the light switch is accidentally turned on. Be sure to replace all batteries when you "**refresh**" your kit. We also recommend having a flashlight or two that uses **LEDs** (Light Emitting Diodes) instead of regular light bulbs. These LEDs are brighter than regular bulbs, more durable, longer lasting, and several are used at once instead of a single light bulb.

Sources and More Information

➤ Most safety equipment information can be found earlier under "Gas Masks and Moon Suits."

➤ One generator flashlight we found is at: http://www.nightstar-flashlight.com, L.E.D. Flashlight Biz, #622-250 "H" Street, Blaine, WA 98230-5120, Phone: 604-726-4018, Fax: 604-608-3571.

➤ See more magnetic force generator flashlights at http://www.nightstar1.com.

➤ Turbo-Torch, online at http://www.brightest-flashlight.com, or contact them at #622-250 "H" St., Blaine, WA 98230, 360-474-1225, Fax: 360-656-1000.

➤ The LED Light, online at http://www.theledlight.com, or The LED Light, Inc., 3668 Silverado Drive, Carson City, NV 89705, 775-267-3170, Fax: 775-267-3108.

➤ A great deal of LED flashlight info, along with links to manufacturers, can be found online at: http://www.equipped.com/led_lights.htm.

➤ Safety Central is one of the many companies carrying the "Dynamo" hand-pump generator flashlights. You can see an example of it at http://www.safetycentral.com/dynpumflas.html.

➤ A variety of generator flashlights can be found at http://www.crankngo.com/index.htm.

First Aid		
❏ First Aid manual	❏ Storable prescription medicine	❏ Small flashlight
❏ Nitrile gloves	❏ Band-Aids ® – assortment pack	❏ Small 2"x2" gauze pads (10)
❏ Large 4" x 4" gauze pads	❏ Trauma gauze or maxi pads	❏ Butterfly band aids (20)
❏ Small roll gauze	❏ "Quick Clot"	❏ Dermabond ®
❏ Antiseptic wipes (20-24)	❏ Alcohol and Peroxide	❏ Neosporin ®
❏ Betadine solution	❏ Surgical tape & Duct tape	❏ Cotton balls and cotton swabs
❏ Petroleum jelly	❏ Ammonia inhalants (6)	❏ Mylar "space" blanket
❏ Small pair of scissors	❏ Tourniquet	❏ Triangular bandage (3)
❏ Tweezers	❏ Scalpel & extra blades	❏ Suture kit
❏ Hemostats	❏ Tongue depressors	❏ Safety pins
❏ Thermometer	❏ Syrup of Ipecac	❏ Activated Charcoal
❏ Aspirin	❏ Cold Packs and Heat Packs	❏ Burn ointment
❏ Dental floss	❏ Tooth picks	❏ Temporary filling kit
❏ Toothache medicine	❏ Wax for Braces	❏ Dental mirror
❏ Resealable plastic bags (6)	❏ Snakebite kit	❏ CPR Shield
❏ Oral Rehydration Mix	❏ Waterproof matches or lighter	❏ Stethoscope / Bloodpressure
❏ Moleskin	❏ Small roll elastic bandage	❏ Irrigation syringe & eyedropper
❏ Surgical masks	❏ Glutose ®	❏ Splint material

Overview

Even though space and weight are a concern, you'll want an extensive first aid kit as part of your Primary, or Level 1 Bugout Kit, since this is the only kit you'd have time to grab in the more dire emergencies, and it's the dire emergencies that are more likely to produce injuries. Also, if you need your first aid kit during a forced evac, you'll REALLY need it since expert assistance will probably not be readily available. A third consideration is the fact that you don't know how long you'll be gone, or what situations you'll experience, so you'll need to be prepared to handle both large and small mishaps. The checklist for "Non-Prescription Medications," below should be considered a necessary addition to this first aid kit.

Though we made a few quantity suggestions in the checklist, you'll still need to figure out how many of which items you might need based on the size and general needs of your family or group. You'll only need one of the reusable items such as the scissors, snakebite kit, bloodpressure cuff, etc. For other items, a simple rule is "at least one per person." Use your best judgement in balancing size and weight of the kit vs the safety and wellbeing of your family.

Hint: Go to the grocery store and drug store and assemble your own kits. Buying preassembled kits is always more expensive, and you don't have control over item brands. **Hint #2**: Make sure you and each of your family members has tried each and every brand of antiseptic, anesthetic, cleansing wipes, adhesive strips, etc. that will go into this kit. Off-brands, or new brands to you, might contain chemicals you're allergic to. The last thing you need during a forced evac is to create a new medical problem caused by first aid for a smaller problem.

Discussion Items

➤ **First aid manual**. Carry as large a first aid book as you can. Don't forget we have some good first aid info on our enclosed **CD**, and remember all the info you downloaded and notes you made under "Basic Training."

➤ **Storable prescription meds**. If you can get extra quantities of your regular prescription medications, by all means do, and keep them in your kit. Don't forget you may also get an **unfilled paper prescription** and keep it in your "Info Pack." This would come in handy for perishable or urgent meds such as heart or bloodpressure medication, insulin, etc. Also, do you split doses of any medications? Do you need to pack an extra **pill splitter**? **Hint**: If you're on any anti-anxiety meds or anti-depressants, make sure to get a backup prescription for these if you're unable to store backup meds. Emergency evacs would be when you need these medications the most.

➤ **Small flashlight**. Flashlights are a must item for first aid even in daylight, since you'll need to see into wounds, throats, ears, etc. Store two if you have room. Get the two-AA battery size for longer operation and brighter light, and get ones that have a smooth surface that can be easily washed after use. You may get blood or other bodily fluids on them. Keep them in their own little resealable plastic bags in case you forget to change out batteries and they leak.

➤ **Nitrile gloves**. Most people are familiar with latex gloves, however, it seems that more and more people are allergic to latex. Since you don't know who you might be rendering first aid to, nor do you know who might use your kit to help you, we suggest going with the hypoallergenic Nitrile gloves. They're cheap and plentiful, and almost as common as latex. Carry several pair.

➤ **Band-Aids** ® – assortment pack. Dump them out of their box into a resealable plastic bag and squeeze out the air before sealing. This saves space and keeps your Band-Aids dry. Be sure to include fingertip Band-Aids, knuckle Band-Aids, and some non-stick Telfa ® pads. (Or, you can use suitable generics.)

➤ **Small 2"x2" gauze pads**. Get at least a 10 pack of individually wrapped gauze pads. Take them out of their box and keep them in a plastic bag.

➤ **Large 4" x 4" gauze pads**. Same for these pads. Get a good supply of individually wrapped pads, take them out of their box and keep them sealed in a plastic bag.

➤ **Trauma gauze or maxi pads**. Carry a few of both. Carry some of the larger gauze pads, such as the 7"x8" dressing gauze and also carry some plain, unscented maxi pads. The maxi pads make great bandages and will satisfy some of your possible feminine hygiene (or barter) needs.

➤ **Butterfly Band-Aids** ® (20). Great for wound closure. You might also consider surgical tape such as Coverstrip Closure ® or Steri-Strips ®. (Though we use brand names a lot, it's to illustrate the item itself.)

➤ **Small roll gauze**. About 10 yards of a 2" width should be fine. This is too valuable an item to be without!

➤ **"Quick Clot."** "Quick Clot" a specific brand of blood coagulant. You'll find sites and sources below. **Hint**: One highly effective blood stopper is dry, out of the box, instant mashed potatoes.

➤ **Dermabond** ® is a "skin glue" meant to be used on smaller lacerations to keep them better sealed against infection. Other products will help close larger wounds, and some people have successfully used Super Glue ® to close wounds when absolutely nothing else was available. Another wound closure product is **tincture of Benzoin** which helps hold adhesive bandages in place.

➤ **Antiseptic wipes** (20-24). Alcohol wipes have more universal uses though Betadine wipes are just as effective. Not only will alcohol wipes cleanse small wounds, but you could use them to take a sponge-bath when nothing else is available, they can be used as fire tinder, and they're useful in limited decontamination.

➤ **Alcohol and Peroxide**. Keep a small phosgene bottle (about 6 oz.) full of rubbing alcohol in your kit, and keep another 6 oz bottle full of Peroxide. Both are useful for sterilization of instruments and wound cleaning, and in a hot environment, a quick sponge bath with alcohol will make you feel temporarily cooler. These are also good decontamination items.

➤ **Neosporin** is a good topical antiseptic to put on wounds as they're being bandaged, or to put on unbandaged cuts and scrapes. Plenty of other good brands of **antibiotic ointments** can be found at your grocery store, and some are **analgesic** as well as **antiseptic**.

➤ **Betadine solution**. Betadine is a good substitute for iodine and works much in the same way except that Betadine is the preferred pre-op patient cleanser over alcohol or peroxide. It's also good for washing down large scrapes, etc.

➤ **Surgical tape** & **Duct tape**. Get a small roll of each. Both will help hold bandages in place, and duct tape can provide some serious adhesion and compression for uses in splints, etc.

➤ **Cotton balls and cotton swabs**. After everything else is packed, put as many cotton balls in a plastic bag as you can to compress into the kit. You'll want as many cotton balls as you can anyway, and the padding will help protect anything breakable. Also, a cotton ball soaked in alcohol can help you start a campfire. Don't skimp on the swabs either. Put as many in your kit as you can, and keep more in your Toiletries Kit.

- ➤ **Petroleum jelly**. Whatever you do, don't forget petroleum jelly. It's useful in everything from being a skin protectant, to preventing bandages from sticking, to keeping a crown in place until you can get to a dentist. You should also include some general lubricants such as KY Jelly or its equivalent.

- ➤ **Ammonia inhalants**. With all the forms of stress in an emergency evacuation, fainting may be a problem for some. Get a small pack of these, but unlike other things, leave them in their box for better crush protection.

- ➤ **Mylar "space" blanket**. You can find these at "dollar stores" sometimes, and definitely in sporting goods or department stores. Get 2 if you have room.

- ➤ **Small pair of scissors**. Get the blunt-tipped "EMT shears." They're great for cutting almost anything, such as cutting clothing away from an injury, and are more useful than simple scissors.

- ➤ **Tourniquet**. Though almost any cord or strip of cloth could be used to make a tourniquet, you'll want a band that's specifically intended to be one. The broad, flat, surface of a true tourniquet will compress properly rather than cutting in and causing undue tissue damage like rope or cord might. Try to find ones with Velcro closures for easier use.

- ➤ **Triangular bandage**. A good triangular bandage is the most useful piece of cloth you could have. Keep at least three in your kit, and be sure to include a bandana or two with your clothing.

- ➤ **Tweezers**. Splinters aside, many emergencies will see injuries caused by glass shards and other items that will imbed in skin. Carry both regular and needle-point tweezers. **Hint**: Either the Biore or Ponds **pore cleaning strips** work well to remove splinters and shards. Simply wet and apply and remove when dry.

- ➤ **Scalpel & extra blades**. You should carry a little equipment that is well beyond your level of expertise because you never know when you'll wind up supplying someone who has the medical knowledge, or if you're in a remote location and have instructions relayed over the phone or radio. Better to have the instruments and not need them, than to need them and not have them.

- ➤ **Suture kit**. The suture kit is like the scalpel in that it's equipment above and beyond your skill level, but it's something you should carry in case you run across an expert that might need to use it on you.

- ➤ **Hemostats**. Hemostats are small scissor-like clamps used to clamp off bleeding veins and other interesting uses during surgery. They fall in with the scalpel and sutures above. You shouldn't use them unless you know what you're doing, but it's great to have them if you really need them. Some make good tweezers.

- ➤ **Tongue depressors**. These are useful as small splints, and for seizure victims to bite down on (if you can get to them in time). You can also use a clean pencil for these same purposes.

- ➤ **Safety pins**. Not only are they useful in holding triangular bandages and ace bandages, but once sterilized, they can be used to lance blisters, small boils, and to tease out ticks and splinters. Carry a small assortment, and maybe a straight sewing needle or two.

- ➤ **Thermometer**. Though the electronic digital thermometers are readily available, we recommend the old type since the old ones don't have batteries that can die at the worst possible moment. If you have room and can afford it though, carry both types, since the old style can still break.

- ➤ **Syrup of Ipecac**. See Activated Charcoal below.

- ➤ **Activated Charcoal**. The Syrup of Ipecac and the Activated Charcoal are both useful in the treatment of ingested chemical poisoning or food poisoning.

- ➤ **Aspirin**. Aspirin is the drug of choice for any medical kit. It can relieve headaches from stress, muscle aches from excess physical exertion, and can reduce fever. It's also part of the first aid for heart attacks or stroke. For aches, carry an alternative if you're allergic to aspirin, but whatever you do, make sure you have regular aspirin, and carry a fairly large bottle.

- ➤ **Cold Packs** and **Heat Packs**. The **cold packs** are available at almost any grocery or drug store. They're the kind you squeeze the bag to pop one chemical that mixes with the other to produce a cold temperature. The **heat packs** come in a variety of forms. Some are similar to cloth bags full of something that looks like iron filings, that when exposed to air, begins to heat. Another type is a bag of liquid with a small metal disk inside. Clicking the disk starts a chemical reaction that generates heat. A third type is an insulated cover that houses a burning wick. One model uses lighter fluid and another uses solid fuel sticks. You'll find all these at most camping supply stores. Both the heat and cold packs are very necessary in a Bugout Kit first aid kit because of the potential for not only strains and sprains, but for **mild exposure**. The cold packs can be used in heat exhaustion, and the heat packs can be used in potential frostbite cases. **Hint**: Some people think that sports creams actually produce heat because they feel warm. This is not the case, and these creams should not be relied on to create any actual heat (though for strains muscle aches they're okay if you have room in your kit).

- ➤ **Burn ointment**. Be sure to get one that's not only antiseptic, but anesthetic and moisturizing as well. If nothing else, carry a tube of aloe vera gel. It's soothing for regular burns and sunburns.

- ➢ **Dental floss**. Get the waxed, unflavored, unscented kind. Dental emergencies are just as debilitating as any other ailment you may encounter, and floss has dental first aid uses.

- ➢ **Tooth picks**. In addition to their obvious uses, toothpicks will work well to help apply the temporary fillings and toothache meds listed next.

- ➢ **Temporary filling kit**. Several companies make kits meant to replace lost fillings, hold loose crowns in place, etc. These are worth their weight in gold, and no kit should be without them.

- ➢ **Toothache medicine**. An example is Anbesol. You can find this stuff easily in any grocery or drug store. If nothing else you can use Clove Oil which you can also find at grocery and drug stores. In fact, you can make your own clove oil by taking whole cloves (the aromatic kind used for cooking and baking, they look like little twigs) and grinding them up in just enough alcohol (the drinkable kind) to wet them down, and letting them soak. Let the liquid sit in a warm area for about an hour. Much of the alcohol should evaporate and the remaining liquid can be used on toothaches (after it's strained through a cloth or paper towel).

- ➢ **Wax for Braces**. This is another good substance for replacing lost fillings or holding a crown in place although it's not as strong as the temporary filling mix. Braces wax can be found at any grocery or drug store. If you can't find any, you can use regular candle wax. Just warm the candle wax slightly to soften it.

- ➢ **Dental mirror**. If you're going to be replacing fillings and the like, you have to be able to see what you're doing. This is also another reason to carry a good flashlight.

- ➢ **Resealable plastic bags**. Many of the items listed in this kit should be removed from their box (but not from any protective wrapping) and placed in resealable plastic bags. This protects the dry things from getting wet, the wet things from drying out or leaking on other things, and getting rid of the original boxes makes for smaller storage space. Just be sure to keep any instructions, and don't dump medications out of their bottles. Resealable plastic bags are also useful for disposing of used bandages, etc., and in an extreme situation, they provide good transport storage for any small severed body parts such as fingers.

- ➢ **Snakebite kit**. Though you may not plan to be anywhere near a snake, the truth is you never know where you'll be or what you'll be doing. Also, in many disasters, snakes will be driven from their normal habitat and may come into yours. The Cutter ® snakebite kits are about the size of a C-cell battery and are very much worth the small expense.

- ➢ **CPR Shield**. The CPR shield protects the person administering CPR from germs and other contaminants that may be on or around a victim's mouth. **Hint**: If you don't have a CPR shield, you can make a suitable substitute out of a piece of plastic wrap or garbage bag with a hole poked through it and a thick gauze pad in place over the hole. Just make sure air is flowing through well.

- ➢ **Oral Rehydration Mix**. Stress and physical exertion, or a nasty case of diarrhea, can cause serious dehydration and throw your electrolyte balance off and affect your energy and health. While sports drinks can replenish some of the lost water and minerals, you'll need a special mix to rehydrate in the more severe cases (plain water won't be absorbed as it should be). Pedialyte ® works well in some cases, as do several over-the-counter electrolyte replacement tablets, but you might not be able to run right out to a drugstore in an emergency, and besides, this mix just can't be beat for the price. It's made as follows: Take eight level teaspoons sugar, one level teaspoon table salt, one level teaspoon baking soda, and mix it with one quart of water. If you don't carry the ingredients as part of your food supply, we suggest you make the dry mix ahead of time and keep it sealed in a small plastic bottle. You can also put individual batches in separate resealable plastic bags.

- ➢ **Waterproof matches and lighter**. These are used to help sterilize your surgical instruments. Run the instruments through the flame, put the flame out, and clean the soot off the instrument with alcohol.

- ➢ **The Stethoscope / Bloodpressure cuff** is very necessary if a member of your family suffers from hypertension. The stress of an evacuation will definitely have an affect on everyone's blood pressure! The stethoscope alone is another instrument that's better to have and not need than to need and not have.

- ➢ **Moleskin**. Evacuation may mean a lot of walking, or at least being more active than normal. This means possible blisters, and moleskin is used to pad developing blisters, or the old ones after you've treated them. You'll find this stuff at most grocery and department stores wherever they keep the foot products.

- ➢ **Small roll elastic bandage**. A couple of ACE ® bandages will prove useful in holding dressings in place and in supporting sprained joints. In cold weather you can wrap your head, hands, or feet, to keep warm.

- ➢ **Irrigation syringe & eyedropper**. This can be a regular hypodermic-type syringe or a squeeze-bulb. You'll need something that can force water through a wound, or gently across eyes and ears in order to clean or disinfect. You'll also want to be able to suction fluid out. Speaking of eyes and ears, you'll also want an **eyedropper** for gentle irrigation or to administer medicines. Speaking of irrigation and wound care, you might want to carry a foot or two of sterile, 1/8" diameter **vinyl tubing** for major wound drainage.

- ➢ **Surgical masks**. It's good to have one or two of these in your kit for the same reason you keep the surgical gloves; you never know who you'll be working on in a first aid situation. Also, if your intended destination is a public shelter, you may want one of the heavier cloth surgical masks (not the cheap dust masks) per family member to help reduce the possibility of communicable disease transmission such as with SARS.

> **Glutose ®.** Being under stress, over worked, and oddly fed, those with hypoglycemia are going to feel the results. Glutose ® helps provide glucose for not only hypoglycemiacs, but for treatment of accidental insulin overdoses. (As with the administration of any medicinal substance, make sure you're properly trained.)

> **Splint material.** Here you have a number of choices. Several companies make little fold-up "wire splints" which are like lengths of heavy metal window screen that come in a small box about the size of a deck of cards. You simply straighten it out, bandage it to support the affected limb and you're good. However, an item that will pull double duty is **newspaper**. A few dozen pages of newspaper can be folded up to provide a pretty rigid splint. If you fold it right and are careful enough, you could even make a temporary neck brace. In addition to these first aid uses, newspaper is also good for sanitation, insulation, shade, and fire. Take several sections of newspaper, wrap them in a garbage bag and put them in the bottom of your kit.

Sources and More Information

> Several dental first aid products are available through DenTemp, at http://www.dentemp.com, or reach them at Majestic Drug Co., Inc., P.O. Box 490, 4996 Main Street, South Fallsburg, NY 12779, 1-800-238-0220.

> DenTek also has temporary fillings and other supplies. You'll find their products in most grocery or drug stores, you can find them online at http://usdentek.com, or contact them at 307 Excellence Way, Maryville, TN 37801, 800-4DenTek (433-6835).

> A complete dental first aid kit is the EDK, available online at http://www.dentalkit.com, or call 800-365-2839.

> First-Aid Product.com, online at http://www.first-aid-product.com, with direct contact at American EHS, Inc., 449 Santa Fe Drive, Suite 127, Encinitas, CA 92024-5134, 760-944-1048.

> You'll find an interesting potato-based blood coagulant online at, http://www.traumadex.com, or call Phone: 800-558-6270 · Fax: 800-558-1551.

> Though we suggest buying your first aid supplies piece by piece locally, we found one fairly comprehensive first aid supply company, online at http://www.cpr-savers.com, with direct contact at Suite 108A -217300 Carlsbad Village Drive, Carlsbad, CA 92008, Toll free: 800-480-1277, Fax: 760-720-6278.

Level 1 Water		
❑ Canteens with Cups	❑ Collapsible bucket	❑ Water purification tablets
❑ Water filter pump	❑ Water filter straws	❑ Plastic sheeting
❑ Garbage bags	❑ 5-gallon collapsible container *	❑ Compressed sponges

Overview

In a rapid Level 1 evacuation, you'll have scarce little time to gather your gear. That means only what you can carry in one trip. Water is our most necessary supply, yet unfortunately, it's one of the heaviest. In addition to some water to keep you going, you'll need to carry gear to help you gather more water later. **Carry all these items regardless of your intended destination**. Water is that important.

Hint: Determine how much weight you're physically able to carry. If your Bugout Kit is lighter than that, keep adding water until you reach your max carry weight. For example, if you can lug around a 60 pound backpack, and your Bugout Kit only weighs 40 pounds, you can add another 20 pounds of water. We say again, **water is that important**.

Discussion Items

> **Canteens with Cups.** After your primary Bugout Kit is full, load it up with as many canteens of water as you can, but where you're still able to carry the pack. Military surplus canteens are inexpensive, durable, and plentiful. Get a few of these and make sure each person has at least one canteen cover and a canteen cup. Metal canteen cups are also useful as small cooking pots. Another military style product is the "Camel Back" water system. It's basically a rubber bladder with shoulder straps that you wear on your back to carry water. You can also make additional "canteens" out of plastic screw-top soda bottles wrapped in duct tape, with a carrying strap made out of ¼ inch nylon cord. Whatever you do, don't carry glass jars or bottles. (**Hint**: During a 14-day evacuation, you'll utilize **14 gallons of water per person**, and this does not include washing or bathing. Carry as much water as you can.)

> **Collapsible bucket.** Though some scenarios will see a total loss of water service, many will not. You'll probably be able to find drinkable water somewhere along your route. But... you have to have something to

carry it in. Collapsible buckets can be made of heavy plastic, or in the case of military surplus, rubberized canvas. Water carrying is another of the many reasons you want to invest in heavy resealable plastic bags (to help get water from source to storage).

➤ **Water purification tablets**. You can get iodine tablets or Halazone. Iodine has an extremely long shelf life, but carries some very minor health risks. Halazone treats the water through chlorination but has a limited shelf life. A third option is colloidal silver.

➤ **Water filter pump**. Katadyn ® and Sweetwater ® are the two most popular manufacturers. These are small, lightweight, plentiful, and highly recommended. Both are hand-pump units that utilize a silver-impregnated charcoal filter, and both are worth their weight in gold if you're caught with your buckets empty. Get an extra filter cartridge and a pre-filter, or "silt strainer" to filter out some particulate matter before water gets to the main filter. You might want to buy two sets. One to keep in the Bugout Kit, and another to keep at home if you're isolated and without water.

➤ **Water filter straws**. Several companies make filter units that you simply stick down into slightly dirty water and drink through them like a straw.

➤ **Plastic sheeting**. You should carry heavy plastic sheeting for a number of reasons, one of which is to collect rainwater if needed. A 6-feet by 6-feet piece should work nicely, as will a plastic shower curtain.

➤ **Garbage bags**. Carry about a dozen of the thickest 30-gallon size you can find. These are extremely useful in a number of water-related situations. You could line your backpack with two to create either a washtub or a rain collection barrel, you could use them to line small holes you dug in the ground to create wash and rinse sinks, etc. This is one list that's limited only by your imagination. (You can also use "contractor bags.")

➤ **5-Gallon collapsible container (optional).**   We're referring to the enclosed containers that have both a carrying handle and an off/on water valve. Having such a container will allow you to collect ample water after you've made it out of an immediate danger area. The only problem is that they're a little on the bulky side even when empty and folded up. If you can readily include it in your Level 1 Bugout Kit, then by all means get one. You could also keep one stored in your car (more about this later under "Packing and Storage").

➤ **Compressed sponges**. Normal sponges will work, but we like the compressed ones since they take up less space. You'll find compressed sponges here and there in the cleaning supply section of grocery or department stores. They're wafer-thin until the first time you unwrap and wet them. Then they expand to normal size. But as we said, normal sponges are fine. Sponges are useful to leave out in the rain to help collect and retain rain water, and you can use them to soak up water droplets and dew off plants and clean surfaces. If you run out of other water sources, you'll need these to get by. If you have a food vacuum sealer, you can put some regular sponges in a pack and shrink them down, making them easier to pack.

Sources and More Information

➤ All items listed can be found at most military surplus and camping supply stores.

➤ If you're concerned about the water you've just purified with your at-home or in-the-field methods, you can test it with a kit offered by the Water Safe company. They're online at http://www.watersafetestkits.com, or you can contact them at Silver Lake Research Corporation, P.O. Box 686, Monrovia, CA 91017, Toll Free: 1-888-438-1942, Fax 626-359-6601.

➤ We found an interesting source for colloidal silver at http://www.lightwatcher.com/preparenow/prepare.html.

Level 1 Food		
❑ MREs	❑ Freeze-dried entrees	❑ Coffee and Beverage Mixes
❑ Protein bars	❑ Hard candy	❑ "Trail Mix"
❑ Condiments	❑ Sterno or small camp stove	❑ Extra stove fuel
❑ Mess kit	❑ Aluminum foil	❑ Canteen Cups
❑ Eating utensils	❑ Food prep utensils	❑ Plastic or paper plates
❑ Plastic and paper cups	❑ Can opener	❑ Bottle opener
❑ Paper towels	❑ Plastic food service gloves	❑ Waterproof matches & lighter
❑ Tinder strips	❑ Fishing kit	❑ Hunting gear *

Overview

We've said it before and we'll say it again, as it's one of the more important points this book makes, that **in any emergency situation morale is crucial**. Food is a key component of morale. We have quite a few menu item points listed below, and more will follow later with "Level 2 Food." In planning your menu items and quantities, remember we're shooting for keeping ourselves supplied during a **14-day evacuation**. You'll want to include enough food to where each person in your family has at *least* one full meal per day. This would be a tall order for cramming all that into just one suitcase or backpack, but we'll address that obstacle later under "Packing and Storage." Also, even though MREs top our list due to ease of preparation and no need for water, you might actually pack more freeze-dried items to save space and weight in your Level 1 Kit, and put more MREs in your Support Kit.

Granted, there's a lot in this checklist. We have food items, acquisition gear, cooking gear, and utensils. That's a lot to consider given the fact that all you may do is travel 2 miles away to a local shelter during a major storm. However... as we'll constantly say... you never know where you'll be, what you'll be doing, or what's going to happen. Be prepared for every eventuality. Give your priority to the food items themselves, and if necessary, push the cooking items into your Support Kit, though you should keep your food service items like the mess kits, eating utensils, etc. in this kit. This will give you some cooking gear, and if someone else is supplying food, you'll want to be able to partake.

Note: This checklist becomes even more important if you have **specialized dietary restrictions**. Though your destination or local emergency shelter might have plenty of food, they might not have food that meets your requirements. Definitely plan on packing your own.

Discussion Items

➤ **MREs**, or "Meals Ready to Eat," are the current military food rations. They're called "ready to eat" since they require zero preparation. All you have to do is open the bag and eat. A full MRE contains entree, side items, and an accessory pack containing plastic spoon, napkins, and other highly useful items. Be sure to check the sources listed below. The only drawbacks to MREs are their size and weight. To adjust some weight and bulk, consider dividing your MRE load between full MREs and MRE entrees (without the side items or accessory packs). **Hint**: To save money and add variety to your menu, buy only the MRE entrees and accessorize them with side dishes such as Ramen Noodles ®. **Hint #2**: If kept in a cool, dry place, most MREs will last up to 10 years. There's an elaborate code on each pack and the first digit is the year it was packaged. Here's how to tell which decade:

● MREs produced in the '80s have square corners on the retort bags (the plastic pouches containing individual entrees; not the main housing bag.)

● In the '90s these retort bag corners were rounded.

● In 1996 the main outer bag housing the meal was changed from a dark brown to a "Desert Sand" color.

● Also in '95 they quit putting coffee in the accessory packs. They changed to various Kool-Aid type fruit drink powders.

● In the early to mid-'90s the cardboard case the MREs came in was changed from a "lay-flat" case to more of a "sit-upright" case.

● At http://www.sbccom.army.mil/products/food/MRE_Improvements.htm, you can find year by year MRE menu changes made between 1995 all the way through projections for the upcoming production year.

● And, in the mid '90s the military issue MREs were all stamped "Government Issue, Resale Prohibited" or words to that effect. Civilian MREs are pulled from the same production lines, but marked differently. The goal was to keep troops or supply personnel from selling off MREs.

➤ **Freeze-dried entrees** are the ones most commonly carried by backpackers due to their minimal size and weight. It's this size and weight factor that make freeze-dried items ideal for your kit. The drawback is the fact that they require water and heating. Use the MREs when you have no water and/or can't cook, and use the freeze-dried foods for meals once you're semi-settled at your destination but have access to nothing else. Also in the "freeze-dried" category are the numerous *instant* foods from the grocery store. Start there before you buy the slightly more expensive freeze-dried stuff. **Hint**: Whatever you choose to pack, make sure you know how to prepare it, that you've tried it before and find it agreeable, and that you reduce its packaging to take up less space in your kit. For example, some "soup in a cup" items come in a spacious cup meant for you to add water and eat out of the container. Dump the contents into a plastic bag and seal it, then use a marker to label the bag and list the expiration date and any prep directions. When you're ready to eat, use your canteen cup to hydrate, heat, and eat. Here are a few ideas for your instant food collection:

❑ Full "camping" entrees	❑ Powdered milk	❑ "Instant breakfast"
❑ Soups	❑ Ramen Noodles ®	❑ Oatmeal
❑ Grits	❑ Mashed potatoes	❑ Rice

For an overview of freeze-dried foods see http://home.howstuffworks.com/freeze-drying.htm.

➤ **Coffee**, **Protein bars**, and **Hard candy** are not only food items, they're comfort foods as well. During any emergency, morale becomes critical, and food is one way to keep yours high. Carry as many energy and comfort foods as you can. To the "beverage mix" list, we caution you to avoid high-sugar content drinks, and to include such things as Gatorade ®, or Gookaid ERG ® mixes. As jars are bulky, make sure all your drink mixes come in packets, or that you've transferred them from jars into resealable plastic bags. **Hint**: Hard candy is useful in some blood-sugar related first aid situations, and comfort foods are great for barter. **Hint #2**: Speaking of comfort and barter, if you smoke, be sure to carry a few packs of cigarettes. The stress of an emergency may make you want one. Also, smokes are certain to be the king of the barter items.

➤ **Trail Mix** is a simple combination of nuts, raisins, dry cereal, etc. It makes a great snack and a great morale booster. Carry a few small bags. A good companion to this are the various "cereal" bars like Granola bars.

➤ **Condiments,** spices, and other such items can make the difference between terrible and tolerable with emergency food items. You don't need much in the way of volume, but it's a good idea to have a little variety. For sources, you might collect the little packets that come with fast food orders, you can buy containers meant for camping and fill them yourself, or you can buy **small travel size packs**. Whatever you get, make sure the packs are small and well sealed. You'll also want to put all the small packets in a larger resealable plastic bag for additional protection. Here are some suggestions on what to include:

❏ Sugar	❏ Salt	❏ Pepper
❏ Ketchup	❏ Mustard	❏ Powdered butter flavoring
❏ Tabasco ® sauce	❏ Cayenne pepper powder	❏ Garlic powder
❏ Lemon zest	❏ Cinnamon	❏ Chicken bullion cubes
❏ Beef bullion cubes	❏ Artificial sweetener	❏ Honey

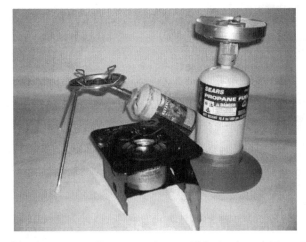 

➤ **Sterno or small camp stove**. This topic could make a book of its own. As heating and cooking is such an important part of camping, and indeed daily life, many companies have manufactured all manner of cooking gear. Cooking gear also helps when staying in a motel room that has no kitchenette. Let's look at a few:

● In a perfect world, you'd have room for a full-sized Coleman gas stove. If you do have room and you can afford the stove, go for it. They're good products. However, we don't recommend you carry one of these unless it lives in your camper or RV and is ready to go. A full-sized camp stove is too large, heavy, and bulky to put in a pack you have to carry, regardless of how short the distance. You have to save room for more important items.

● Several companies make single-burner gas stoves that use small gas canisters as the ones shown above, left. If you invest in one of these, practice with it to see how it operates, and how much gas is used cooking each meal. Though great units, these are still a little bulky.

● Even more plentiful and less expensive (not to mention smaller) are the various camping and military surplus cookers that use the tablet style, or "tab" fuel (either Trioxane or Hexamine) like the ones in the picture above and to the right. These are by far the most workable cooking option. Carry one or two of these small cooker units even if you choose another of the stove setups. Use these only in well-ventilated areas.

● You can also find any manner of small **Hibachi** grills. The only drawback is that you'll also have to carry some **charcoal** and **lighter fluid** which can be rather bulky. (Remember that bulk and weight won't really be that much of a problem if you're keeping all this gear prepacked in a camper or trailer.) However, after you run out of charcoal, you can always use wood or a couple of the "Buddy Burners" listed below.

- The least expensive alternative is to make several "**Buddy Burners**" out of tuna cans with coiled up corrugated cardboard and wax, and a "**Hobo Stove**" as shown to the right. **Hint**: Put a birthday candle in the middle of your Buddy Burner to make it easier to light.

- One thing to include too, especially if you're carrying the MREs as recommended, is the **MRE heater**. A heater set is usually included in each full case of MREs, and is simply a heavy plastic bag with a chemical pouch in the bottom. You add water to start the heating process, then you drop in your MRE entree pouch and let it heat. They're simple, lightweight, inexpensive, and can also be purchased separately. We recommend you carry a couple of these, since if nothing else, they can keep you warm in cold-weather.

- Interestingly enough, one of the sources listed below, Alpineaire Foods, supplies a built-in heating unit with some of their meals that works on the same principal as the MRE heater units.

- If you use any type of open flame, whether a stove or campfire, be sure to have a **cover**, water bucket, or some other way of instantly **smothering the flame**. Fire is a danger and you should make sure you're able to control whatever you use.

- **Hint**: Whichever stoves you choose, be sure you actually use them now and learn how to cook on them safely. Don't wait until you're on the road during an emergency to learn how to use your gear!

➢ **Extra stove fuel**. Without it, your cooking appliance is useless. Whatever fuel you use, make sure it's properly and safely stored.

➢ **Mess kit**. Fortunately, one mess kit will probably suffice for a small family, since all you really need it for is cooking. You can find these at any camping or sporting goods store, and the commercial models are better than the military issue as they contain more useable parts. Make sure yours has two lids, an extendable handle, a small cooking pot, and a cup. For other members of your family, you can make improvised mess kits from the plastic trays that many frozen, microwavable, dinners come in. Collect several, stack them together, and add a set of plastic eating utensils. Wrap the set in a resealable plastic bag. **Hint**: Only *serve* food in the microwave plates. Don't use them to heat things in a second time.

➢ **Aluminum foil**. Aluminum foil is the king of expedient camping goods, since among other things, you can make both cooking pots and serving wear for food preparation. Get a large roll of the heavy-duty kind. You can take it out of its box and slightly flatten the roll. This will take up less space in your kit and yet you can still use the foil easily enough.

➢ **Canteen Cups**. These were already mentioned under "Level 1 Water" above, but important enough to mention again. These cups can be used for drinking, eating, and cooking. Make sure you have at least one per person. They stow nicely as they fit around your canteen. See our Hobo Stove picture above. That's a canteen cup sitting on top.

➢ **Eating utensils**. Carry one set of metal eating utensils per person and at least 3 or 4 disposable sets. Use the metal ones when you have wash water, and the disposable sets when you don't. **Hint**: If you find a good deal on the alcohol prep pads (for first aid), you can also use them to clean your eating utensils. The same goes with waterless soaps or hand sanitizers.

➢ **Food prep utensils**. Most of the foods we've mentioned are "heat and eat" types. However, since you never know where you'll be or what you'll be doing, we suggest you carry some food prep and cooking utensils. This simple list includes a couple of **serrated steak knives** (also useful in a defensive role), a **large fork or two, a couple of table spoons**, and a good set of **tongs**. Your work gloves mentioned under "Clothing" will double as oven mitts. While we're on the subject of durable utensils, we'll mention that you should also carry a **"dunk" bag**. It's a simple nylon mesh bag that will hold your utensils while packed and you can use it to "dunk" them into boiling or soapy water to wash and rinse. You can find them in camping supply stores or you could make one from a pantyhose leg.

➢ **Plastic and paper plates**. Paper plates take up very little room, add very little weight, are disposable when you don't have wash water, and can be used to light fires. Plastic plates are the same in weight and space, but washable. For plastic plates, save the plastic plates or bowls that come with frozen, microwavable dinners. Save a few of the same type and size (so they're stackable), wash them, and seal them in a plastic bag. You don't want to cook in them again, but they're great for serving. **Hint**: You can also use these deep-sided, divided plastic plates as small wash bins for shaving and brushing your teeth, and you can use them in first aid situations where you may need a small sink. Mark which one you use for clean or dirty water.

➢ **Plastic and paper cups**. If you have canteen cups, you won't need the plastic cups. The best paper cups to carry are the conical "water cooler" type that come in a long sleeve. Take what you think you might need out of the sleeve and flatten the stack somewhat (sideways like you did the roll of aluminum foil) so it takes up less space. **Hint**: In addition to being disposable, these conical cups are also useful as funnels, and as eye patches in protruding-object injuries. **Hint #2**: In addition to the conical cups, carry a few **styrofoam or heavy paper cups** meant for coffee. You'll want something that will handle hot soups or beverages.

➤ **Can opener**. All you need is the little military surplus P-38 can opener. These are small and useful enough to have a couple around all the time. Put one on your key chain, keep one in the car, and certainly keep one or two in your Bugout Kit. If you're unable to use something that small, then the standard kitchen hand-held can opener is fine.

➤ **Bottle opener**. Here we're talking about the little 50-cent type, one piece metal thingy with a puncture type can opener at one end, and the bottle opener at the other end. You'll find these at any grocery store. Keep it on a cord together with your P-38 can opener.

➤ **Paper towels**. These are very necessary for food prep as well as for personal hygiene, but unfortunately very bulky. They'll flatten out somewhat, but not much. If you don't have room for the semi-flattened rolls, try this: Unroll the paper towels and fan fold them one sheet by one sheet. You'd lay out one sheet and then fold the next one on top of it and keep going accordion style. Next, take your flat stack of accordion folded paper towels and put them in a "Space Bag" so you can seal them in and force the air out of the bag, effectively vacuum sealing the pack. You'll find a source for "Space Bags" under "Packing and Storage." Doing this will drastically reduce the amount of space needed, and it will keep them dry and protected. In addition, you should keep several small individual packets of **facial tissue** packed in your kit. Tissues can double for both dinner napkins and toilet paper. In an emergency situation, **morale is a crucial factor and cleanliness is a prime component of morale**. Try to carry at least one roll per person.

➤ **Plastic food service gloves** (not the latex type). These are very necessary given the fact that water may be scarce. These gloves will keep food clean as you prep it, and your hands clean too so you'll have less hand washing to do. Also, in a first aid situation, these gloves (if worn doubled) can work when the latex gloves have been used up. You'll find these gloves in inexpensive boxes of 100 at most grocery stores. As always, take them out of the box and keep them in a resealable plastic bag for better protection and less space.

➤ **Waterproof matches and lighter**. Though matches and a lighter were mentioned under first aid, you should also pack an extra set here. Keep your matches packed separately from your lighter (get the butane kind). Put them each in doubled plastic bags, seal the bags, and then tape over them once they're sealed. **Note**: A lot of companies make various little "matchless" fire starters, but if you think about it, each of them are more labor-intensive to use, cost more, weigh more, and take up more space than matches. Go for several sets of matches and lighters.

➤ **Tinder**. Tinder is anything that will help you get a fire going. If you have paper plates and paper towels, you've already got some material you can use. However, one little useful trick is to take small strips from an old rubber tire, each about ¼" wide and about 3" long. This rubber can't stay wet, lights easily, and burns long enough to get your main fire going. You should also take along a candle or two. The size we recommend are the "standard" candles that are about an inch thick and about 6" long. A candle will not only help you light a fire, but it can provide light as well. Check your camping supply store for long-lasting "compressed wax" candles." These are less prone to melt in your pack and they burn cleaner and longer than regular candles.

➤ **Fishing kit**. Somewhere, somehow, most people will come in contact with some sort of body of fishable water, even in a short-distance evacuation. As your stored food is depleted, you'll need methods to gather more. Since fishing gear really takes up such little space (in fact you could put everything below in a matchbox), you should carry some with you. Include the following:

❑ At least fifty feet of monofilament fishing line.
❑ Ten assorted fishing hooks.
❑ Six small fishing line weights.
❑ Two or three small corks.
❑ Four or five small, assorted lures.

➤ **Hunting gear ***. **This gear is optional** depending on your intended destination, your travel route to get there, and your proclivity for hunting and prepping game to eat. If your evacuation carries you to or through wilderness or rural areas, you might consider the following:

● Non-firearm weapons such as a BB or pellet gun (we recommend a pump-up air pistol, with rifled barrel that accepts both BBs and pellets), sling shot (the type using surgical tubing for the bands), blow gun (a 3-foot long take-down model), etc. for small game. **Hint**: The BB gun or the blow gun are going to be easier to learn to use, and more accurate.

● Waxed, unscented dental floss for making a snare for small game.

● A small caliber firearm for small to medium game.

- Game prep items such as a skinning knife, filet knife, small hatchet, plastic gloves, and plastic bags.

If your routes do not take you through wilderness areas, and you're absolutely sure food will not be a problem, then you can relegate this gear to your "**Support Kit**," or "**Secondary" kit**.

Sources and More Information

- Long Life Food Depot. http://www.longlifefood.com, P.O. Box 8081, Richmond, IN 47374, 800-601-2833, Fax: 765-939-0065.
- MREs are still available through Crown Point, the civilian outlet company. They can be found at http://www.crownpt.com/MREs.htm, or you can contact them at 118 S. Cypress St., Mullins, SC 29574, Toll free: 800-CROWN-PT (800-276-9678), Ph: 843-464-8165, Fax: 843-464-8598.
- Alpineaire Foods, http://www.alpineairefoods.com, DG Management ®, P.O. Box 1799, Rocklin, CA 95677, Local: 916.624.6050, Fax: 916.624.1604, Toll Free: 800.322.6325, or 866.322.6325.
- See Campmor, online at http://www.campmor.com, and with direct contact at Campmor, 28 Parkway, Box 700, Upper Saddle River, NJ 07458, 1-800-525-4784.
- For a good selection of freeze-dried meals, see Mountain House, online at http://www.mountainhouse.com, or contact them directly at 525 25th Ave SW, Albany, Oregon 97321, Phone #: 1-800-547-0244.
- While considering what kind of stove to use, visit the Solar Cooking Archives at http://solarcooking.org.
- Also be sure to visit the "Wings, The Home-Made Stove Archives" online at http://wings.interfree.it.

Basic Toiletries		
❏ Toothbrush	❏ Toothpaste	❏ Toothpicks
❏ Dental floss	❏ Denture products	❏ Cotton swabs
❏ Cotton balls	❏ Soap	❏ Small metal mirror
❏ Washcloths and hand towels	❏ 2 Rolls toilet paper per person	❏ Comb or brush
❏ Deodorant	❏ Nail file and clippers	❏ Feminine hygiene needs
❏ Waterless hand sanitizer	❏ Moist towelettes or baby wipes	❏ Razors
❏ Items for contacts	❏ Garbage bags	❏ Latrine kit

Overview

Though seemingly a lot to carry, these items are small and can be strategically packed to be almost unnoticeable. Personal hygiene goes a long way toward both good health and good morale, so each of these items will prove doubly valuable. Additionally, if you wind up carrying some extra items, you have some powerful barter tools!

Discussion Items

- **Toothbrush**. Though the little "fold up" travel toothbrushes are fine, you should carry the kind you're used to using. You'll also want to keep them protected for sanitation and health reasons. All you really need is a plastic bag as a cover and a rubber band to hold it in place. **Hint**: Leave the electric or battery operated toothbrushes at home.
- **Toothpaste**. The travel size of your favorite brand would work well, or you can try the old **salt and baking soda** mix. Take equal parts of salt and baking soda, mix them together and keep them sealed in a waterproof container that will keep them dry. Dipping wet toothbrush bristles into the powder will cause just the right amount of powder to cling to the brush for use.
- **Toothpicks**. Toothpicks are as much a comfort item as a hygiene item. We suggest carrying as many of the flat type as you can fit inside an old chapstick tube (a good way to carry them).
- **Dental floss**. Again, as much a comfort item as anything else as being clean and comfortable are crucial to morale. Also, dental floss has numerous uses. It can be thread for sewing repairs, cord for small campsite uses, and string to make snares for hunting purposes. We suggest you carry the waxed, unscented, unflavored variety.
- **Denture products**. If someone in your family wears dentures, they'll certainly need them fully functional in order to eat. Be sure to carry **denture cleansers, adhesives**, and other products. Though we mention dentures, this item should apply to **any dental equipment**, such as braces, partials, etc.
- **Cotton swabs**. Though you have a few in your first aid kit, you'll still need a few more here.
- **Cotton balls**. Same with cotton balls. They're too light, cheap, easy to pack, and indispensable to leave behind. Consider this cheap trick. In lieu of any other kind of dust mask (or along with) how about half a cotton ball lightly stuffed in each nostril? And don't forget their use as earplugs!

➤ **Soap**. Rather than a large family-sized bar, we suggest you carry several smaller bars. Smaller, individually wrapped bars are easier to ration, and you can use extras for barter. **Hint**: There's nothing wrong with using soap as **shampoo**. Also, several companies make several types of environmentally friendly camping soap that cross purposes. We've seen several that you can use to wash you, your hair, and your dishes. If you try these brands, make sure you test them first, and that no one in the family is allergic. **Hint #2**: Make these soaps a hypoallergenic type, and devoid of moisturizers, scents, or other additives. It makes them easier to use for other cleaning purposes, easier to rinse in a limited-water situation, and easier to trade.

➤ **Small metal mirror**. Remember, don't carry glass! Your metal mirror is also useful as a signaling device.

➤ **Washcloths and hand towels**. You'll want 2 washcloths per person, one for soaping, one for rinsing. We suggest two as it takes extra water to rinse out your soap cloth in order to use it as a rinse cloth, and also, it may be a while before you can do laundry. You'll also want **1 hand towel per person**. Bath towels are a little too bulky to include in a Level 1 kit. However, you'll need some sort of towel, so make sure each person has a hand towel. They're small enough to carry and large enough to do most of what you may need them for. Remember to make sure they're brightly colored in case you need signal flags.

➤ **2 Rolls toilet paper per person**. Carry more if you have room (remember, you never know how long you'll be away). To save room, slide the cardboard tube out of each roll and flatten the roll. Keep each roll in its own plastic bag. By the way, toilet paper is high on our list of potential barter items. The vacuum pack "Space Bags," or the kitchen vacuum sealer are great ways to store toilet paper since they'll shrink the space used, and protect the paper as well.

➤ **Comb or brush**. If you get a brush, try to find the fold-up travel variety. Space in your kit is limited, and we have a lot on our lists.

➤ **Deodorant**. Though this could be considered a vanity item, it's actually a necessity. We can't say this enough that good morale is crucial, and cleanliness is important for morale.

➤ **Nail file and clippers**. In keeping with the morale factor, and with health considerations, keeping your nails clean will help with both goals. Though some might consider these vanity items, they lend themselves more to health. More germs live under your nails than on your skin.

➤ **Feminine hygiene needs**. As we'll discuss later, make sure you remove items from their main boxes in order to save space. In the case of feminine hygiene needs, you should go with what you normally use and nothing less. Remember that maxipads are also useful as compress bandages for serious wounds, and that any product in this category is high on the barter list.

➤ **Waterless hand sanitizer**. There are 2 types to choose from. One type is Purell ®, which is a quick-evaporating antibacterial rinse. The other type is Go-Jo ® which is a self-liquefying soap that can be used without water and wiped off. We suggest carrying both types. Purell ® is good for simple post-restroom and pre-meal hand washing, and Go-Jo ® is good for cleaning off serious grime such as grease and heavy dirt.

➤ **Moist towelettes or Baby wipes**. These are literally a dime a dozen, they're easy to store, and if absolutely nothing else, they're a great way to take a quick "sponge bath." You can buy these by the box full at most grocery stores, or save towelettes from your takeout dinners. If space is severely limited, go with just the towelettes. However, the baby wipes, being larger, work better, and the baby wipes generally have extra moisturizers, etc., that towelettes don't.

➤ **Razors**. Though we must plan for water restrictions, we don't have to have a fatalist attitude about it. In a waterless environment, you wouldn't bother to shave, but, you may have plenty of water, and a good shave (both ladies and gentlemen) can make you both feel better and look better. Carry a couple of the small disposable types per adult. Also, if you're in a desperate first aid situation and forced to perform some sort of supervised emergency surgery, you'll have to shave the incision site. Don't leave your razors at home!

➤ **Items for contacts**. Carry a spare set of contacts plus wetting and cleaning solutions, storage case, etc.

➤ **Garbage bags**. This is another of those multi-purpose items that are worth their weight in gold. Wherever you might have a tiny open space in your kit, you should slide in a trash bag. They can be slit open to make tarps, ground covers, in-ground sinks, porta-potty liners, laundry bags, ponchos, tube tents, and the list goes on. In fact, if one of your kits is housed in a hard-sided suitcase, you could line it with a garbage bag (or heavy "contractor's bag"), fill it with water, and you'd have a place to wash clothes or a bathtub for the baby. You'll also need them for storing dirty laundry.

➤ **Latrine kit**. You'll need a latrine kit, or some similar setup, for those situations where you really have to go, and you're in an unpopulated area miles from the nearest exit. The reason we include this is because in a mass evacuation, it's likely that traffic may move so slowly as to make this a likely scenario. You'll at least need a **small garden shovel** to dig a small latrine, and a set of **heavy resealable plastic bags or jars** since you might not be able to get out of your car to go dig a latrine! You could also go with a commercial "porta-potty" made for campers, or a 5-gallon plastic bucket with a snap-on lid, garbage bag liners, and kitty litter (more on this under the "Privy Deprived" portion of the "**Isolation**" section). This is another reason we suggest you carry a light tarp, since you might need a privacy screen.

Sources and More Information

> Most all these items are available at your grocery or drug store. Others are available at your local hardware store, and outdoors or camping supply shop.

> If you want to do more shopping online to stay away from crowds, you can try Drugstore.com, online at http://www.drugstore.com, or you can call them at 1-800-DRUGSTORE (1-800-378-4786).

> For the ladies interested in alternative feminine hygiene accessories, take a look at the "Keeper" online at http://www.keeper.com.

Primary Clothing		
❏ Durable pants	❏ 2 under shirts	❏ Short sleeve outer shirt
❏ Long sleeve outer shirt	❏ Short pants	❏ 2 pairs cotton athletic socks
❏ 2 pairs underwear	❏ Long underwear	❏ "Shell" clothes
❏ Belt	❏ Work gloves and mittens	❏ Eyeglasses with strap & case
❏ Sunglasses	❏ Bandana	❏ Durable hiking shoes
❏ 2 soft hats	❏ Light jacket	❏ Poncho
❏ Eyeshades and earplugs	❏ Detergent	❏ **"Bolt clothes"** *

Overview

Adequate clothing is such a necessity, yet it's always so bulky. If you were packing for a simple camping trip or a short excursion somewhere, this list would be a simple one since you'd know exactly where you were going, how long you were going to be gone, and you'd have a good idea what the weather will be like along the way. Emergency evacuation under unknown circumstances, to an uncertain destination, and in unknown weather conditions and time lengths, is a different story entirely.

When packing these clothing items, make sure you've gathered everything you need to cover your hands, head, ears, neck, feet, and any other spot through which you could lose warmth. In the heat, it'll be easy enough to shed clothing, but, being the climate-controlled creatures of habit we are, we need to be able to protect ourselves from not only freezing temperatures, but from a steady exposure to cool or cold temperatures that could cause a cold or make one worse. If nothing else, cold weather can sap our energy as our bodies try to maintain body heat. We need to do all we can to protect ourselves.

One general note on clothing: Make sure it's all **durable** and **washable**. No flimsy dry-clean-only garments!

Discussion Items

> **Durable pants**. Denim jeans or military fatigues are probably the best. Make sure your pants are comfortable or slightly roomy to allow for ease of movement, since you'll be doing a lot more physical work than you think, and you'll want room for layers underneath to keep you warm.

> **Under shirt**. A regular T-shirt, or a printed variety. Make sure it's made of a durable, washable material.

> **Short sleeve outer shirt**. This can be a casual pullover, or simply another T-shirt.

> **Long sleeve outer shirt**. Go for a slightly heavier, more durable, material, and preferably make this shirt a turtleneck for better warmth and protection. Your long sleeve shirt may be protecting you while you do physical things, or from bugs or weather. **Hint**: Get as much warmth and protection out of a piece of clothing as possible. For example, if you pack a sweatshirt, make sure it fits like it's supposed to (for warmth, comfort, and layering) and that it has extra features such as pockets and a hood. Similarly, make sure all pants have closable cargo pockets, etc. **Hint #2**: We mention turtlenecks not only because they're warmer in the cold, but because the neck can be pulled up to hold a breathing filter in place, or simply to protect your mouth and face during really cold weather.

> **Short pants**. These can be a pair of running shorts or gym shorts. You'll need something cool and comfy that can be worn in public.

> **2 pairs cotton athletic socks**. This will give you a pair to wear while the others are washed, or simply two pairs to rotate. Just like with your GYT Pack, a clean pair of socks can be used as a breathing filter or used as mittens in the cold.

- ➢ **2 pairs underwear**. This includes all sexes. Carry boxers, briefs, panties, bras, etc. One to wear, one to wash. One trick is to pack a **swimsuit** as one pair of underwear. They're durable, usually comfortable, they'll provide you an outfit to wear in the heat, and who says that a temporary stay at an evac destination has to be awful? You might get to go swimming!

- ➢ **Long underwear**. Some sets of long underwear are surprisingly thin and easy to store. No region is immune from cold or freezing temperatures, so make sure you and yours are prepped.

- ➢ **"Shell" clothes**. This is a class of clothes having a lot of different nicknames, brand names, and descriptions. Some examples are thin, long-sleeved T-shirts, flannel "lounging" pants, or even a set of long sleeve, long legged pajamas. The idea here is to provide a set of clothing that can be worn as clothing if you have a chance to wash your main clothing, as clothing if you need a thin set in the heat, or as an extra layer if you need the warmth. Whatever you choose, make sure it's lightweight fabric, lightly colored, loose fitting, and made from a comfortable material. For colder climates, go with flannel, and in hotter climates, go for loose-fitting, lightweight cotton, either white or royal blue in color to reflect the sun's heat.

- ➢ **Belt**. Since you might be attaching things to it, be sure to carry a durable leather or nylon ("ballistic" nylon) belt. Another reason to carry one is the fact that we've asked you to carry comfortably fitting pants which might need help staying up. Also, you might lose a little weight during a lengthy post-disaster recovery.

- ➢ **2 soft hats**. Carry something to protect your head from either winter cold or summer sun. Your hot weather hat should be light in color, have ventilation holes, and a brim all the way around, or at least have a ball cap where the brim shades the face. For cold weather, carry a wool knit cap or fur lined cap with ear covers.

- ➢ **Work gloves and mittens**. Make sure these are durable whether they're made of cloth or leather. These will protect your hands from abuse while laboring, and provide warmth in the cold. We also mention mittens as they're warmer than gloves in the extreme cold since they keep your fingers in contact with each other.

- ➢ **Eyeglasses with strap and case**. If you wear glasses, carry a spare set. This spare set should also have a neckstrap to help you hang on to them, and a carrying case to protect them while not in use. If you had to run out the door in a hurry, these might be your only pair. By the way, ALL eyewear, sunglasses included, gets a neckstrap.

- ➢ **Sunglasses**. Get the prescription kind if you need them, but be sure to keep a neck strap on these too as well as having a protective cover. Also, add a pair of inexpensive **safety goggles** (or swimmer's goggles) to cover either set of eyewear, or simply to use as protection.

- ➢ **Bandana**. Like the triangular bandages in your first aid kit, bandanas have a million and one uses. You can use them in first aid situations, cover your head in the hot sun, wear it as an eyeshade, use it as a dust mask, or filter some debris out of water you're about to purify. Carry several, and make sure they're brightly colored for use in signaling.

- ➢ **Durable hiking shoes**. These can be full boots if you have room for them, or they can simply be heavy sneakers. Whatever you choose to pack make sure they're broken in, and that they can protect your feet over all kinds of terrain and in all types of weather. You might also consider some **light walking shoes**, but only if you have room (you might want your one and only pair of light walking shoes to be the ones stored with your "bolt clothes" mentioned below). If you packed boots, you'd probably appreciate a pair of lighter tennis shoes, sandals, or the like. Also, don't forget to pack **extra shoe laces** for all footwear you pack.

- ➢ **Light jacket**. In a cold region, a full, heavy coat would be preferable. However, we're space-limited here, so go with a light windbreaker type jacket, preferably one made of Gore-Tex ® fabrics.

- ➢ **Poncho**. In other sections we referred you to the little flimsy "dollar store" ponchos. Here we're talking about a real one, preferably military surplus. Ponchos can protect you from rain, act as a coat in the winter, and can be made into an awning or small tent.

- ➢ **Eyeshades and earplugs**. Two things are certain; one, you never know where you'll wind up, and two, rest is important. These two little items can help you block out both light and noise to help you get some much needed shuteye. A bandana can make a good blindfold-type eyeshade, but for the earplugs, get an inexpensive pack of the foam-rubber type from your hardware or sporting goods store.

- ➢ **Detergent**. Carry a very small pack of detergent. Since you've only got one set of clothes, you'll want to be able to take care of them if water is available. **Hint**: If you carry something like **Dawn** dishwashing liquid as your clothing detergent, you can also use it on dishes, on you, and in **emergency decontamination efforts**. (Whether you carry powder or liquid, be sure to seal it well in a couple of layers of resealable plastic bags.)

- ➢ **"Bolt clothes"**. Since we're talking about a rapid, Level 1 evacuation, we'll finally get around to discussing "bolt clothes". These are the items of clothing you should have prepped and ready, **outside your kit**, that you can grab in case you need to "bolt" out the door. We mentioned this under "Fire Drill" since one of the worst times for an emergency is when you're asleep and wearing next to nothing (or less). When creating a set of bolt clothes, keep all the following in mind:

❏ Keep the entire set of clothes **together**, either all on one hanger, or in one bag. You'll want to be able to find them and get into them quickly, just as a Fireman would.

❏ Cover the hanging or bagged clothes in a **brightly colored plastic bag** to both protect it and identify it. You'll also want to mark the outside bag (as you should all your emergency packs) with **strips of reflector tape or glow-in-the-dark paint**. Many scenarios could see you hunting for your evac gear by flashlight. Mark the things that need to go with you!

❏ These clothes should be **kept near where you sleep** (if not near your kit) since if you were anywhere else, you'd more than likely be fully dressed. If this is the same location as your safe room, you can have your bolt clothes pull double duty by using them to hold some of your **safe room's emergency gear**. This way the gear is in your safe room if you need it, and the items are already packed if you need to take them with you. Keep this in mind as you read the items below that should be stored in your clothing's pockets.

❏ Use clothing that is **easy to get on**. The fewer pieces the better. We suggest a one-piece set of coveralls of a durable, medium-weight material that can be worn in pretty much any weather. Get a set that has several large cargo pockets to hold the other items that need to be a part of your clothing. Along with the coveralls, be sure to have: ___ a pair of durable walking shoes (to put on immediately), ___ a set of long underwear, ___ an undershirt, ___ a belt, preferably already run through the loops and ready to fasten (if your coveralls don't have belt loops, sew some on since you'll need a belt for other things).

❏ In these **cargo pockets** be sure to keep: ___ a pair of heavy work gloves, ___ a pair of heavy socks, ___ underwear, ___ duplicate wallet with credit cards, blank checks, and cash, ___ an extra set of keys, ___ small plastic poncho, ___ "Safety Necklace" as mentioned earlier, ___ small flashlight, ___ soft hat, ___ bandana, ___ pocket first aid kit, ___ extra 911 phone if you have one, ___ small toiletries kit including toilet paper. (Ideally, you'd keep a duplicate GYT Pack with these clothes.)

Sources and More Information

➤ All of these clothing items are commonly available at any clothing store. Our suggestion is that you try local thrift stores for some of the outer garments in order to save money.

Basic Shelter		
❏ Family tent	❏ Tube tent	❏ Sleeping bag with cover
❏ Inflatable mattress	❏ Inflatable pillow	❏ Mosquito netting
❏ Plastic sheeting	❏ Tarp	❏ Light blanket
❏ Roll-up mesh hammock	❏ Rope	❏ Cord
❏ Duct tape	❏ Umbrella *	

Overview

Shelter is a broad term used to discuss equipment that will help protect you from the elements. These items are fairly dependent on your mode of travel, travel routes, and intended evacuation destination. Even though your route and destination may have plenty of shelter available, nothing guarantees the shelter is going to be accessible or inhabitable. You should carry what you're able (especially the smaller items) since you never know exactly what's going to happen, and you have to be ready to adapt.

Discussion Items

➤ **Family tent**. Fortunately, tents are self-compensating in their weight and bulk. The larger a tent you need probably means you have more people that can help carry it and set it up. For smaller tents, many of the "pop-up" dome types are economical, light weight, and easy to set up.

➤ **Tube tent**. This is exactly what it says it is. It's a thin, nylon or plastic tube about 6 feet long and 3 feet in diameter, that when strung up with a cord through its length, becomes a small one-person tent. These are so small (they can fold up to about the size of a wallet) and inexpensive that you should carry one per person as an emergency option. You can also make one out of two or three garbage bags and some duct tape.

➤ **Sleeping bag**. Sleeping bags, though potentially bulky, are worth the expense and effort, especially in cold weather. In warm weather they can provide a little extra cushioning to lie on. In lieu of a full sleeping bag, you can make a light bedroll out of two blankets rolled up together. Whatever you use, make sure it has a cover to keep it clean and dry during transport. A wet (or contaminated) sleeping bag is useless, and you'll need your rest whenever you can get it.

➤ **Inflatable mattress**. This is an **optional** item though recommended if you have room. Morale is crucial, energy is key to morale, and **sleep is key to energy**. Also, the higher you sleep off the ground or floor, especially in public shelters, the better off you'll be due to all sorts of creepy-crawlies, and more importantly, fleas. A very inexpensive alternative to a camping inflatable mattress would be the type made for use in a swimming pool. You could also carry a foam pad, or a vinyl covered foam pad (like a light exercise mat).

- ➢ **Inflatable pillow**. Along with bedding comes pillows. Most inflatable mattresses or rafts have a pillow section. If yours doesn't, you can get a separate one, or you can inflate a gallon-sized, heavy duty resealable plastic bag. This is such a small item that can provide so much comfort, so be sure to pack it.

- ➢ **Mosquito netting**. This is also optional, but useful enough to seriously consider, especially if you might be dealing with a flood. Your immediate area will become a mosquito haven, not to mention the normal assortment of other pesky insects. Healthy adults may not need the netting, but small children and the elderly will. If nothing else, buy large sheets of cheesecloth which will work rather well. Put the cheesecloth in a "Spacebag" to squeeze out the air, make the package much smaller, and to protect the cloth.

- ➢ **Plastic sheeting**. Plastic sheeting can make an expedient rain cover or ground cloth, and is also highly useful in gathering rain water for drinking. Also, it's good to have around should you lose a vehicle window for whatever reason. You can get varying thicknesses of plastic sheeting and dropcloths at any hardware store, or department store with a paint or hardware department. A simple and inexpensive alternative that bridges the gap between plastic sheeting and tarps would be a plastic shower curtain. These are cheap, plentiful, you can get clear or opaque ones, and they have holes or grommets along a reinforced edge that make them easier to tie down.

- ➢ **Tarp**. If your home and evacuation destination are both metro areas with plenty of available shelter at the destination and along the way, we recommend you at least carry the tarp as emergency shelter and give second priority to tents. This is especially true if you have a fairly decent sized vehicle. The vehicle can be your shelter, and the tarp an additional awning, or a good ground cloth to sleep on. You can also use it as a privacy screen around any outdoor latrine or shower.

- ➢ **Light blanket**. Even if you do pack a sleeping bag, take along a light blanket. It's extra warmth in the winter, and it's cover if warmer weather has you using your sleeping bag only as padding.

- ➢ **Roll-up mesh hammock**. These are cheap and plentiful and fold up to about the size of a baseball. With one of these, a couple of trees, and a tube tent, you're set for a cozy off-the-ground nap. If nothing else, these can be used as fishing nets or as a cargo net.

- ➢ **Rope**. Remember under "Basic Training" how we asked you to learn all you could about knots? Rope is priceless. You'll need it if you have to set up camp, tow your car, ford a river, rescue a family member, or tie down your gear. We suggest at least a 50-foot length of ½ " nylon braided rope. This is best kept stored in your vehicle as you may need it in day to day travel mishaps, and it's rather large and heavy for a kit that you'll have to get out the door in a hurry.

- ➢ **Cord**. We suggest a 50-foot length of ¼ " nylon braided cord, preferably parachute cord. With the potential for making tube tents, temporary awnings and the like, cord is indispensable.

- ➢ **Duct tape**. With all the uses for your plastic sheeting, you'll need duct tape for shelter setup and for tent repairs, pack repairs, etc. Carry a small roll.

- ➢ **Umbrella***. We gave this an asterisk as it's really an optional item, though it would prove useful in providing comfort and protection against some elements. If you don't have room in your main kit, set your umbrella aside to be carried in your Support or Secondary kits. Besides, you should have one that lives in your car.

Sources and More Information

- ➢ All of these items are available at department stores and camping stores. We also recommend you visit thrift stores and yard sales to find good deals on serviceable equipment.

Non-Prescription Medications		
❑ Aspirin or ibuprofen*	❑ Multi-Vitamin/Mineral *	❑ Anti-Diarrhea Med*
❑ Laxative*	❑ Antacid*	❑ Alka Seltzer ®
❑ Anti-Nausea / anti-vomiting *	❑ Non-prescription sleep aid*	❑ Non-prescription stimulant*
❑ Chapstick*	❑ Cough drops, Throat lozenges*	❑ Decongestant
❑ Nyquil ® and Dayquil ®	❑ Vick's Vapo-Rub ®	❑ Eye drops or Artificial Tears ®
❑ Nasal spray	❑ Ear drops (antibiotic if available)	❑ Allergy medication
❑ Antihistamines (Benadryl ®)	❑ Topical antihistamine	❑ Motion sickness meds
❑ Hemorrhoid meds	❑ Anti-itch / topical analgesic	❑ Calamine Lotion ®
❑ Topical anti-fungal	❑ Monistat ® or similar	❑ Cayenne Pepper Capsules
❑ Hand and body lotion	❑ Sunblock	❑ Insect repellant
❑ Flea collars	❑ Asthma inhaler	❑ Talcum powder

Overview

This might seem like a lot to carry in a small kit, especially since we've already told you to pack light, but these items can all be packed into half a shoe box if you use the small travel or sample sizes.

Most items in this list are self-explanatory as to what they are and how they're used. What's not readily obvious is why we've included so many items in what should be a light-weight, grab-in-a-hurry kit. Any time you're forced to leave your normal surroundings, make a stressful trip, and take up temporary residence in an unfamiliar location, you're going to be exposed to a whole new world of stressors, irritants, ailments, and other things you weren't prepared for. The items in this kit will take care of the vast majority of the little things that WILL be affecting you during an evacuation and that COULD evolve into major problems if not handled. Also, the items that you might not need directly may definitely be needed by someone in your group. If nothing else, these items will keep your morale high by keeping you comfortable, and they make <u>great</u> barter items as will some of the toiletry items covered earlier.

We've put together a few general tips and hints about these items:

1. The items marked with an **asterisk *** are considered to be the basics and the items most likely needed. You'll want to carry a **normal or family sized bottle or pack**. Use your judgement about the quantities of other items.

2. **Reduce the bulk** of all this by buying small travel sizes, getting sample packs, or by taking the items out of their bulky boxes or bottles and put smaller amounts in either resealable plastic bags, or vacuum seal bags that we'll discuss in more detail under "Packing and Storage." Remember to label everything with name and expiration.

3. Anything **liquid** that has a topical application will have **individually wrapped towelettes or pads** produced somewhere by somebody. If available, get those instead of tubes, vials, or bottles. They're easier to pack, use, and trade.

4. Check everything, especially items you repackage, for **name, directions, dosage, and expiration dates**. For items not clearly marked, use an indelible marker to **write the info on the package**.

5. Go through this list a couple of times keeping in mind the **particular needs of each of your family members**. Is anyone prone to any particular ailment, allergy, or reaction? Do you have young children? Does this list remind you of any **prescription medications** that need to be packed? Speaking of prescription medications, do you need to pack a **pill splitter** for half-doses?

6. Make sure that you've **personally used every item** in your kit. The last place you'd want to find out you have an allergy to a new brand is while you're in the middle of an emergency evacuation!

7. Similarly, since many of these items are useful for barter, make sure **whoever you're trading these items to knows exactly what they are, and have tried them before with no ill effect**.

Let's look at some more detail regarding these items:

Discussion Items

> **Aspirin or ibuprofen**. As we mentioned under "First Aid," we suggest you carry both. You'll need something to help you with the aches and pains of all the labor involved in evacuating, something to help keep a mild fever in check if you catch cold (which is more than likely with increased stress), and aspirin has a definite use in first aid for heart attacks and strokes.

> **Multi-Vitamin/Mineral ***. Your energy needs are going to be increased, stress will be increased, and the quality and quantity of food will be down somewhat, as will your general health and resistance. You'll need all the dietary help you can get!

> **Anti-Diarrhea Med***. Stress and new foods are one thing, and various "bugs" you might catch along the way are another. Be prepared.

> **Laxative***. Stress and new foods may also have a blocking effect. Be ready for that too.

> **Antacid***. Speaking of system upset due to stress and dietary changes, we'll remind you to include items to remedy heartburn and indigestion. Consider items like Pepto Bismol ®, Tums ®, Tagamet ®, DiGel ®, etc.

> **Alka Seltzer** ®. Alka Seltzer sort of bridges the gap between acid stomach upsets and nausea. You'll also find an Alka Seltzer that treats cold or flu symptoms.

> **Anti-Nausea / anti-vomiting ***. Sometimes vomiting is good, since your body will try to expel something irritating. Sometimes it's the vomiting that's irritating and you need to keep it in check. Some anti-nausea meds, such a Phenergan ®, are by prescription only. Other home remedies include peppermint (for very mild nausea), the Chamomile tea we previously mentioned, and Ginger (the spice) which can be found in capsules or teas at most grocery stores and many vitamin stores.

> **Non-prescription sleep aid***. You definitely need your rest when you can get it, and stress and adrenaline will be your number one enemies. Whatever you carry, make sure it's something you've used before and you know how it will affect you. One of the things we recommend is a simple herbal tea made from

Chamomile. You can find it at almost any grocery store. The type we've tried is "Celestial Seasonings' Sleepy Time Tea." Chamomile tea is a very mild natural sedative that basically "takes the edge off" and lets you sleep rather than puts you to sleep.

➤ **Non-prescription stimulant***. For some people, stress will have the opposite effect, and will create sluggishness and lack of energy. We suggest something mild such as a low-dose caffeine product for those times you need to be alert and ready. Similarly, be sure to stock any **anti-smoking products** such as various patches, or Nicorette ® gum. If you recently quit smoking, the stress of an emergency will make you want a cigarette. If you smoke and run out of cigarettes, these items will take the edge off for a little while.

➤ **Chapstick***. Chapped lips notwithstanding, in an exposure emergency, chapstick (or similar balms) can be applied to all exposed skin to prevent sun or wind damage, and to even offset a couple of degrees worth of temperature exposure. A workable substitute (which should be in your kit anyway) is **petroleum jelly**.

➤ **Cough drops, Throat lozenges***. These can probably be considered a comfort item, but certain levels of comfort are important morale boosters. You should also consider a small sample size of cough suppressant if you're subject to coughs. Also, you'd probably wind up being the hero of the shelter if you supplied some cough suppressant to someone who was hacking up a storm and keeping everyone awake at night!

➤ **Decongestant**. You'll notice that this and the following few items cross purposes with colds and allergies. You'll more than likely experience a touch of both during an evacuation excursion not only because of the fact that you'll be exposed to so many new people, allergens, and irritants, but that your immune system will be stressed while doing it.

➤ **Nyquil ® and Dayquil ®**. We mention these not so much for brand name, but for the fact that they're the most popular cold symptom relief products. Carry what works for you.

➤ **Vick's Vapo-Rub**. Since we're talking colds, allergies, and being able to breathe, we thought we'd mention Vick's. We also include this as it's a "home remedy" that Vick's and other menthol aromatics pull double-duty by keeping mosquitoes and some other bugs away.

➤ **Eye drops**. Instead of the ones that mask the irritation by getting the red out, we suggest you carry eye drops that soothe and moisturize. Chief among these would be simple saline solutions that we discussed under the section covering the "Get You There" pack earlier.

➤ **Nasal spray**. If you do have allergies, you'll know how valuable this stuff can be. Also, you'll know that simple saline solution for the eyes can pull double duty. If you've never had allergies, you still might want to carry some spray just in case.

➤ **Ear drops**. Though not as high on the probable-use list as previous items, an earache can become a major annoyance if it crops up while you have so many other things happening. In addition to soothing ear drops, carry a brand that has antibiotic capabilities as well.

➤ **Allergy medication**. Many people have some type of allergy or sinus problem, and if you're going to have any sort of outbreak, it'll be during an evacuation. If you have prescription meds for this, then we hope you've paid attention and already have them packed. Otherwise, include a good supply of your favorite over-the-counter brand.

➤ **Antihistamines** (Benadryl). Though this overlaps with allergy medication above, a good antihistamine has several direct applications. Chief among them is helping stem a mild allergic reaction to foods, insect stings, or other similar ailments not cured by generic allergy meds.

➤ **Topical antihistamine**. In addition to oral antihistamines, you'll want a small tube of the topical stuff for mosquito bites, skin allergies, and anything else not covered by the above two items.

➤ **Motion sickness meds**. Dramamine ® is the most popular, but the Ginger that we mentioned previously for nausea is also good for motion sickness. You'll want some of this in your kit in case the increased stress of the situation makes some of your evacuation companions car sick, or if part of your evacuation transportation might include boats or planes.

➤ **Hemorrhoid meds**. Anytime you mix stress, dietary changes, and physical exertion, you run the risk of developing or aggravating a hemorrhoid. We also found that Tucks ® pads are recommended by some campers as being a great product for sponge baths, though we haven't tried it. If nothing else, hemorrhoid treatment products are great barter items!

➤ **Anti-itch / topical analgesic**. Many products contain hydrocortisone, which is good for general rashes and reactions to things like poison ivy, etc.

➤ **Calamine Lotion ®**. In keeping with the risk of rashes, irritations, and poison ivy and the like, we'll remind you of how valuable and useful Calamine is.

➤ **Topical anti-fungal**. During an extended evacuation trip, or stay at a shelter, or other location, you may be in damp clothes for longer than normal, showering in new locations (if you can), and generally running the risk of developing such annoying things as "jock itch," "athletes foot," and related ailments. Carry a topical anti-fungal such as Tinactin ®, Mycatin ®, Mycitracin ®, or some of the Gold Bond ® products.

➢ **Monistat ® or similar**. The conditions mentioned just above will also increase the likelihood of a yeast infection for the ladies. Again, if nothing else, this is a great barter item.

➢ **Cayenne Pepper Capsules**. For those of you versed in herbal medicine, you'll know that Cayenne pepper is touted to have certain anti-microbial properties, especially with regards to the food with which it's consumed. This means that it may protect you against a tiny bit of unwanted bacteria in food. However, THIS DOES NOT MEAN YOU CAN EAT ANYTHING AND EVERYTHING AND THEN TAKE A CAYENNE CAPSULE AND ALL WILL BE OKAY! It just means that it might make the difference, and it can't hurt to have that little extra safety margin regarding possible tainted food (though the rule stands with bad food – "when in doubt, throw it out"). Try using cayenne before packing a few capsules as part of your kit. If you haven't done so already, you should learn everything you can about herbs and alternative medicines.

➢ **Hand and body lotion**. You may find yourself exposed to wind, cold, or sun for longer than you'd like. Having a little lotion, such as baby oil or petroleum jelly, can protect your skin from dryness, excess sunburn, and can protect your exposed skin just a bit from the cold. Roman soldiers used to coat their bodies with olive oil to help protect against the cold. Another type of moisturizer, **Aloe Vera Gel,** is one of the wonder medicines of the herbal world. It's great for soothing burns, healing skin, and even washing hair and bathing.

➢ **Sunblock**. Definitely carry sunblock! One evacuation scenario would see you simply run out of your home because it was destroyed by a tornado. Naturally, you'd come back to your house later to salvage what you could, and that can mean lots of work outside. Having the right products with you already, will save you unnecessary trips to the store. Carry the highest SPF you can get, especially if you have infants or young children that might be forced into long sun exposure periods.

➢ **Insect repellant**. Flood situations increase the likelihood of mosquitoes. Plus, you won't know whether or not the emergency shelter you're going to be staying at has a lice, flea, or tick problem. You may also need the repellant when you get back home and are outside a lot during your cleanup efforts. It's best to be ready.

➢ **Flea collars**. Speaking of fleas and ticks, one trick to keep you protected is to wear flea collars around your ankles (though not on bare skin) while you're walking, and at the head and foot of your bed while sleeping. This flea collar idea is optional depending on your intended destination and whether or not you have a good insect repellant.

➢ **Asthma inhaler**. Some asthma products are by prescription only, others are not. Any of the stressors or irritants mentioned already could trigger an asthma attack. If you or anyone in your family has ever had any indication that asthma might be a problem, then one little inhaler, which doesn't take up much room at all, could be a lifesaver. Or, you could save someone else's life.

➢ **Talcum powder**. It may be a while between showers. You can also substitute corn starch and/or baking soda, which would actually have more uses than regular talcum powder.

Sources and More Information

➢ All items listed can be found at most grocery stores, or drug stores.

➢ **Hint**: Some of the above items can be found at your grocery store in the form of herbal teas. One example is the natural sedative effect of Chamomile tea found in a product called "Sleepy Time Tea" ® produced by the Celestial Seasonings Company. Another example is an instant beverage product called "Thera-Flu" ®.

Babies' Kits		
❏ Diapers	❏ Bottles, liners, and nipples	❏ Powdered formula
❏ Pacifier and/or Teething Ring	❏ Teething medicine	❏ Garbage bags
❏ Baby wipes	❏ Talcum powder	❏ Baby oil & Diaper rash ointment
❏ Pedialyte ®	❏ Baby food and utensils	❏ Bibs
❏ Distilled water	❏ Sunblock	❏ "Papoose" carrier

Overview

One might debate the necessity for adding these items into a Bugout Kit for two reasons. One, most of these items are probably in a diaper bag already, and two, babies will quickly outgrow the need for these particular items. Our suggestion is to compromise and keep a small amount of each in your kit, augment it with the diaper bag that you'd more than likely grab in response to an evacuation, and rotate items out of the kit as your baby grows up. The safety, comfort, and wellbeing of your baby is extremely important, so it's better to have extra supplies and not need them than to need them and not have them. **Note**: We're assuming that you've packed extra **baby clothes** along with clothing for other family members.

Discussion Items

- **Diapers**. Remember our "Space Bags?" You can use them to pack more diapers in a smaller space and make sure they're kept dry and protected. We suggest a mix of both disposable and washable diapers.

- **Bottles and nipples**. Don't forget **liner bags** if your bottles use them. Also, be sure to pack the plastic bottles, even if you normally use glass. Glass is heavy and breakable. Play it safe.

- **Powdered formula**. Simply pack what you normally use.

- **Pacifier and/or Teething Ring**. These certainly go without saying.

- **Teething medicine**. Same here.

- **Garbage bags** are extremely important when you consider needing a sanitary sheet to change your baby, a new home for the dirty diaper afterward, and a liner to make an expedient tub.

- **Baby wipes**. Double up on these since you might be in a water-restricted situation. These are not only needed for cleaning after diaper changes, but you might need them to bathe your baby.

- **Talcum powder**. Corn starch is also a good substitute. Whatever you pack make sure it's something you normally use and your child reacts well to it.

- **Baby Oil & Diaper rash ointment**. Same here. Make sure there are no allergies.

- **Pedialyte ®**. Babies are sensitive creatures and will feel your stress almost as much as you do. Add dietary changes to the mix and you might have some upset stomach, or diarrhea, and you'll need to replenish lost fluids and electrolytes. If you pack nothing else on this list, make sure you pack some Pedialyte ® or equivalent, such as the "Oral Rehydration Mix" mentioned in the "First Aid" checklist above.

- **Baby food and utensils**. Though baby food will more than likely be on the Last Minute List and not prepacked, we feel you should prepack some solid foods that are formulated for infants and have a good shelf-life. Don't forget to carry **feeding utensils** since the full-size forks and spoons you packed with your mess kit might be a little too large.

- **Bibs**. Don't forget you'll need to clean up in situations where water may be limited. Carry the plastic covered bibs rather than cloth, since the plastic covers can be wiped off with some sort of sanitizer.

- **Distilled water**. Though you'll be carrying water for your family, some of it might be treated with a little extra bleach for storage. Since babies shouldn't be exposed to that kind of thing, we suggest you keep a liter or two of distilled water in your kit. As it's not treated for long-term storage, you should **change out this water twice annually**.

- **Sunblock**. Be sure to carry sunblock or other skin protection. We listed sunblock earlier in the "Non-Prescription Medications" checklist, but infants are worth the additional mention.

- **"Papoose" carrier**. In modern terminology, it's a hands-free device for your infant. We've all seen these "baby backpacks" which as small cloth packs you can put your baby in, and you basically wear them like a backpack, only they ride against your chest. These packs are small, easy to pack, plentiful, inexpensive, and extremely useful in these kinds of situations.

Sources and More Information

- All items can be found at most department stores, or infant specialty stores such as "Babies R Us," online at http://www.babiesrus.com.

Communications Gear		
❏ Walkie Talkie	❏ CB radio	❏ 911 cell phone
❏ AM / FM Radio	❏ Extra cell phone battery	❏ Cell phone recharge adapter
❏ Cigarette lighter adapter	❏ Cell phone battery adapter	❏ Whistle
❏ Chemlight sticks	❏ Aerial flares	❏ Smoke signal

Overview

Good communication in any emergency situation is critical. We gave you a longer checklist under the "Communication" portion of the **"Foundation"** section. Here we'll cover items that should go in your Level 1 Bugout Kit. Later, we'll cover items recommended for your "Support Kit" or "Secondary Kit."

Discussion Items

➢ **Walkie Talkie**. Here we're referring to the hand-held, two-way radio of your choice. We recommend that you have at least one pair of **FRS or GMRS radios** that you keep in your kit.

➢ **CB radio**. We suggest that you have a 40-channel CB radio that **lives in your vehicle**. Make sure yours has plug-in power adapters as well as the ability to use **standard batteries**.

➢ **911 cell phone**. Even if you have a regular cell phone, you should still have a second cell phone that lives permanently in your kit. Try to make it the same type as your regular phone so that your accessories can be used on either.

➢ **AM / FM Radio**. These can be as small or as large as you like. You can find small "sports" models that are about the size of a small 9-volt battery. If you have room, we recommend you get a combo AM/FM radio and either cassette tape player or CD player. Though larger, we're going to suggest in a later checklist that you carry a couple of cassettes or CDs for entertainment and morale. (Don't forget, if you can afford it, get a hand-crank powered radio.)

➢ **Extra cell phone battery**. It's good to have a second battery if your main one develops a memory and won't keep a charge. Your cell phone is too valuable not to accessorize.

➢ **Cell phone recharge adapter**. This one is the type that plugs into a wall outlet. There's no guarantee you'll have access to electricity, but if you do, you should be able to take advantage of it. Don't count on being able to gather up the one you normally use at home. This adapter should live permanently in your kit.

➢ **Cigarette lighter adapter**. This is the **cell phone** recharger that plugs into your vehicle's cigarette lighter.

➢ **Cell phone battery adapter**. As we discussed earlier under "Communication," you should invest in a recharge adapter that works off regular batteries. Make sure you pack one of these in your kit and that it uses the type of battery you carry the most of.

➢ **Whistle**. Carry a small sports whistle. You'll note we recommended this earlier with the "Safety Necklace."

➢ **Chemlight sticks**. Not only are these good for signaling, you can use them as small flashlights.

➢ **Aerial flares**. The small "Skyblazer" flares, or the various "pen gun" flares are small enough to pack in your kit should you need a night time distress signal. Some people might consider these and the smoke signals below to be optional based on location or destination, but we feel these have a variety of uses in a number of situations, so they should be carried.

➢ **Smoke signals**. Most civilian smoke signals manufactured as distress signals are small, about the size of a film canister or C-cell battery. Carry a couple of these for daytime distress signals.

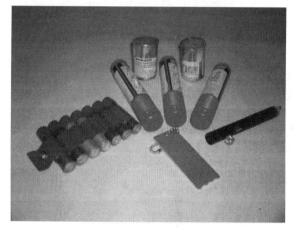

All sorts of signaling devices can be found at the average camping supply store.

Sources and More Information

➢ All communication gear accessory sources can be found in "Communication" in the "**Foundation**."

"Office Pack"		
❑ Info Pack	❑ Notepad or small notebook	❑ Assorted pens, pencils, etc.
❑ Sharpie ® markers	❑ Envelopes	❑ Stamps
❑ Change of address forms **	❑ "Important Contacts" lists**	❑ "Update Cards" **
❑ Spare checkbook	❑ Credit card duplicates	❑ Cash
❑ Keys	❑ "Evac Atlas" **	❑ Maps
❑ "Destination Identifiers"	❑ Small flashlight	❑ "Entertainment"
** These items will have forms in the "**Appendix**."		

Overview

We started this discussion earlier in the "Info Pack" section of "Planning Ahead for Phase II." If you only read that section once, we strongly suggest you go back and read it again. It's that important.

This "Office Pack" is the tool that is going to keep all your paperwork, personal information, contact information, written communication, and other important documentation protected, useful, and useable. In today's society, documentation is your life! **Hint**: As this documentation is so important, make sure all the important parts of your Office Pack are protected inside a resealable plastic bag.

Discussion Items

➤ **Info Pack**. This is your collection of important documentation as discussed in the "**Foundation**" section. Though for security's sake you wouldn't want to keep too many originals in your Bugout Kit, you do want to keep *some* originals. Most important of your original documents would be a **secondary ID** such as a non-drivers license state issued ID or a **passport**. This Info Pack will also include the contact and data forms you filled out for your family. Make sure each is divided and organized accordingly so you don't waste time flipping through pages looking for the info you need. You can probably use a paper binder/folder and tab-dividers to keep everything organized.

➤ **Notepad or small notebook**. You want to be able to leave notes, write letters, draw maps, etc., and you may need to write down **instructions, directions, or other info provided by officials**. Also, you might want to keep a journal of your experiences. Others may learn from what you do, and depending on the nature of the disaster, you may one day look back on it as an adventure.

➤ **Assorted pens, pencils, etc**. Pens and pencils are good for your notebook, and **chalk** was mentioned in the "Communication" section as being useful for leaving notes or symbols at rendezvous points or message drops. You'll need some other specialty writing implements as well. You should also carry a **grease pencil** and a **dry-erase marker**. You never know what you'll have to write on to leave a message, so be ready. These last two items are also useful for writing on your maps' **plastic sheet covers**.

➤ **Sharpie ® markers**. Emergency notes you leave for others may need to be written in **indelible ink** to withstand sun and water. Make sure you have both medium and broad-points. Some messages may need to be read from a distance.

➤ **Envelopes**. Make these a bright, noticeable color that is used only for emergency communications. Also carry a few regular envelopes as you might want to try the mail as a backup to other communication methods. **Hint**: As you may have to leave these exposed to a little weather, you might want to carry extra resealable plastic bags to leave them in.

➤ **Stamps**. Carry both regular and postcard stamps to send your written commo.

➤ **Change of address forms**. You might have to relocate indefinitely, or simply want a temporary stop to the mail being delivered to your home that was destroyed by a tornado. Having a form with you will save you a trip to the post office. Due to USPS changes, the proper hardcopy form is no longer downloadable from the postal service website. **We recommend you go to your local post office and get the proper change of address form now** (PS Form 3575) **and keep it in your kit**. If you need one on the road, you'll need it then and you won't have time to waste standing in line at the Post Office only to find they've run out of forms. (We've included a reasonable facsimile of the form in the "**Appendix**" but it's best you get the official paper form.) We've also included some "Change of Address Cards" in the "**Appendix**" and on the CD. Use these in addition to the postal forms.

➤ **"Important Contacts" lists**. You'll find these in the "**Appendix**." Be sure to make a copy of each and every one and keep it in your kit along with a photocopy of your address book. **Hint**: Though we stress making sure everything is in hardcopy format, carrying a backup disk is okay.

➤ **"Update Cards"**. These can also be found in the "**Appendix**." Send these directly to all the pertinent people from your "Important Contacts" list so they'll know where to find you and how you are. Having preprinted cards like this will save you a lot of writing time and will keep you from forgetting important detail. Also, be sure to carry a few **blank, pre-stamped postcards** for simple, less urgent, notes.

➤ **Spare checkbook**. Though our society deals primarily with cash or plastic, many places still take checks. Make sure you have some on hand. You might **also** want to carry a **few traveler's checks** (don't invest much since we don't know what the economic conditions will be or whether or not you'll be able to cash these). If you do, get the kind that provide for instant replacement if they're lost or stolen (very important given the fact that the theft of your kits is a prime security concern). **Hint**: As you may have to negotiate a loan somewhere along your evacuation travels, get in the habit of keeping your most recent **paycheck stubs** in your car or Office Pack. These will help verify employment in some financial situations.

➤ **Credit card duplicates**. Keep a duplicate of every credit card you own. In fact, you should also have a card or two in your kit that has a zero balance, that's not used at all, and that you don't even keep in your wallet. This way you're **sure to have an empty card** for use in emergencies.

➤ **Cash**. Cash is king in an emergency. Merchants might not trust checks or credit cards temporarily, the power might not be on to process credit cards, ATMs might not function, and vending machines don't take checks or plastic. Carry cash in small bills, nothing larger than a twenty, and carry plenty of coins (keep them in a plastic pill bottle). The next question is, "how much cash to carry?" **Pack as much as you can spare** in your kit and leave it be. If you can afford it, since we're talking about a 14-day evac minimum, pack $100 per person in your family, and add twice that total for your vehicular needs. **Hint**: Take your credit cards, checkbook, and paper money, put them in an envelope, put the envelope in a resealable plastic bag, wrap it tightly in a couple of layers of duct tape, and put it in the bottom of your kit. This will keep you from dipping into your emergency funds during a shopping impulse. It'll also protect your checks from water, and it might save them from theft if someone rifles through your kit and doesn't recognize the pad of tape as something valuable. **Hint # 2:** Divide your cash and other items into **two** batches, wrap them the same way and put them in different parts of your kit. This keeps you from putting all your eggs in one basket.

➤ **Keys**. Keep a copy of **every key you own** in your kit. If absolutely nothing else, pack copies of your car, house, work, and especially **safe deposit box** keys. Your safe deposit box is probably where all your original documents (in addition to some other useful items) will be stored. You also want to make sure you have **keys to any of your off-site storage units, storage trunks at your evacuation destination,** etc.

➤ **Evac Atlas**. The "Evac Atlas" will be discussed in detail below under "Planning the Destinations." This Evac Atlas contains all the maps and data necessary to the successful planning and utilization of emergency services at your evacuation destination and along the routes. **This is an important tool and your Office Pack will not be complete without it.**

➤ **Maps**. You'll be carrying your Threat Map (which will become a part of your "Evac Atlas"), and you'll also need assorted road maps of your immediate area, your evacuation destinations, and the routes along the way. You'll want to keep your maps protected with **plastic sheet covers**.

➤ **"Destination Identifiers."** This is one of those little "ounce of prevention" things that's good to have in place since we don't always know what's going to happen in an emergency. Let's say your evacuation destination is a friend or relative's house that's fairly close to your home, or rather the disaster area. Depending on the nature of the disaster, local authorities at your destination may try to **control access** to the area in order to regulate number of refugees, lessen the strain on local infrastructure, etc. You'll need to have in your possession some sort of documentation that identifies you with that destination, and proves that you do have business being there. Let's look at some suggestions:

● **Postmarked letters, both <u>to</u> you and <u>from</u> you,** at the address you're heading for.

● It could also be a **notarized letter of agreement** from the local resident saying that you are invited to come stay at their house in the event of an emergency. (You in turn should send them a similar agreement regarding your home.)

● You might also carry some kind of documentation showing you have a secondary **safe deposit box** or a rented **storage facility** unit in the town (having either or both of these at your destination is a good idea anyway), copies of **utility bills** in your name (you might be responsible for the relative you're visiting), or something else showing some sort of official or commercial connection.

● If you have **property** in the area, carry an original copy of a **mortgage bill, tax bill, utility or other bill**, mailed to you at that address, or a "certified as true" copy of your **deed**.

● If you acted early and made a hotel or **motel reservation**, be sure to get a **confirmation number, or print a hardcopy receipt** if you made the reservation online.

● Medical conditions might have you selecting a destination based on available facilities. Be sure to carry documentation supporting your medical claim so you won't be denied area access.

Though not legally binding, this type of simple documentation might mean all the difference in the world in being allowed into an area, or told to "keep moving."

➤ **Small flashlight**. Since you may need to write letters or read maps after sundown, this is a good place to keep a small flashlight (store it with the batteries in backwards) or an extra **chemlight** stick or two.

➤ **"Entertainment."** Since we just mentioned reading after dark, and since the Office Pack deals with recorded media, this seems as good a place as any to discuss the need for some sort of entertainment, diversion, or morale boosters. What are we about to say? Yep, that's right, morale is crucial. Therefore, think about packing some of the following things as space allows:

● A **book** to read. Pack one you've never read before, but make sure you know what it's about. Choose something **positive, uplifting, entertaining**, and maybe even **inspirational**. The last thing you need after a disaster is something that may depress or anger you.

● **Cassettes, CDs, or DVDs**. Carry a couple of musical or inspirational audio works. Same rules apply to depressing material. Stay with what makes you feel good. If you have a vehicle with a video player, keep a few extra tapes or DVDs in the vehicle to keep passengers occupied and their mind off of what's going on.

● **Coloring or activity books** for the children. Don't forget a small pack of crayons. Actually, some adults would like a good crossword or puzzle book, or maybe just something to doodle with.

Hint: If you happen to be a **Notary Public**, try to take your stamp and seal with you. You, more than anyone, will understand the need for all the documentation we've mentioned so far. In an emergency, families may have to put their documentation to use, or may be forced to create new documents (such as in filing official documents or claims while on the road, creating powers of attorney, wills, living wills, etc.) If you're a notary, you already know your state laws outlining what you can and cannot do, but if you have your stamp and seal, you'll be in a much better position to help those around you.

Sources and More Information

➤ All the items above are available at your local office supply store.

Batteries		
❏ Flashlights	❏ Cell phone(s)	❏ FRS radio(s)
❏ CB radio	❏ AM / FM radio	❏ Pager
❏ Hearing aids	❏ GPS unit	❏ Laptop
❏ PDA	❏ Wheelchair	

Overview

Since batteries are so useful, we felt it necessary to give you a separate checklist to help you see which of your gear needs batteries and what kind of batteries to stock. In the "**Appendix**" you'll find a "**Battery Worksheet**" that will allow you figure how many of what types you need, and make shopping easier.

We've come up with some **helpful hints** regarding batteries:

1. For logistical purposes, try to buy gear that uses the **same kind of batteries**, including your flashlights. This way you can rob one item to power another.

2. Make sure each piece of equipment (where possible) also has both a cigarette lighter adapter and wall outlet adapter. Rather than carrying different adapters for each piece of equipment you carry, go to Radio Shack or other electronics outlet and get a "**Universal Adapter**." If you still wind up with a few different adapters, make sure each is labeled as to what it goes to. You'll want everyone in your family able to use all emergency gear.

3. If at all possible, see if any of your equipment will **recharge batteries** when plugged into alternate power sources.

4. In any event, have a mix of disposable batteries and **rechargeables**. You'll find there are some pretty good rechargeable battery types available.

5. Don't forget our suggestion to install a "Quick Start" system or similar (with **power inverter**) in your car. These units and similar can be used to jump your own car battery as well as power DC and AC current equipment, essentially turning your vehicle into a gas powered electric generator.

6. Try to keep a battery rack and **battery tester** in your workshop. When it's time to "refresh" your Bugout Kit, test the batteries to see if they need to be replaced.

Discussion Items

➤ **Flashlights**. As we mentioned earlier, you'll want to get flashlights that use the most common battery shared by all your other equipment. Something else to keep in mind is the fact that there are several flashlights that don't use batteries at all. We've listed their sources in other areas. Also, keep your main flashlights ready to go, but others that are kept deeper in your kit, should have their batteries put in backwards so they can't be accidentally turned on and drain the battery. Another school of thought is to not put batteries in your secondary flashlights at all and just leave the batteries in their packs and all the packs together to make refresh easier.

➤ **Cell phone(s)**. Keep a second cell phone battery, and the types of regular dry-cell batteries that your backup cell phone adapter uses.

➤ **FRS radio(s)**. For your radios and other essential communication gear, get the "high energy" or "longer life" batteries. They're a tiny bit more expensive, but worth it.

➤ **CB radio**. Though your CB will more than likely use a cigarette lighter adapter, make sure you have some high energy batteries for it as well.

➤ **AM / FM radio**. The best AM / FM to keep in your kit is one of the smaller sports radios that's about the size of a 9-volt battery. The only drawback is that they generally use the small "button" type battery that's a little more expensive than most. If you don't have anything else using a button battery, you might carry a slightly larger radio that uses AAA or AA batteries.

➢ **Pager**. Pagers aren't on our list of emergency equipment that you *have* to have, but since they have their communication uses, and you may have one on you when you evacuate, be sure you have the batteries for it. (Standard pagers aren't high on our list, but two-way text messaging units such as the "Blackberry" are.)

➢ **Hearing aids**. These will use the small button-type batteries, and it's not likely anything else will use the same size. However, hearing aids are important, so if you have one, make sure you have extra batteries.

➢ **GPS unit**. GPS, or Global Positioning System, units are wonderful gizmos. However, we don't recommend you invest in one unless one of your evacuation destinations or routes takes you cross-country on foot. And you guessed it, if you have a GPS unit, make sure you have extra batteries to fit it. A close cousin to these are the rather expensive "**Personal Locator Beacons**" recommended primarily for back-country backpackers and as lifeboat equipment. These units activate a homing device and alarm that is monitored by a private firm specializing in rescue. When you activate your unit, this company will come pick you up. However, YOU pay the bill.

➢ **Laptop**. We recommend you carry your laptop computer as part of your "Support Kit," or Level 2 kit. Backup batteries for your laptop are very expensive, so we recommend you do the same thing you did for your cell phone, and **create an adapter** that uses standard batteries. This is in addition to having a cigarette lighter adapter and a universal plug-in wall adapter.

➢ **PDA**. As we mentioned before, we don't recommend that you go out and buy a PDA just for the purposes of having one. However, if you already do have one, they're rather useful in communication if you have the type with cellular connections and email. This is another device that will need the high-energy batteries.

➢ **Wheelchair**. For our mobility challenged friends, we remind you to carry recharge accessories and an extra battery along during an evacuation, because you just never know where you'll be or how long you'll be there. If your unit uses a "wet cell" battery, be very careful about leaks or letting the battery tip over.

Sources and More Information

➢ "Batteries.Com," online at http://www.batteries.com, or at Batteries.com, 6024 West 79th Street, Indianapolis, IN 46278-1727, 888-288-6500.

➢ Battery Mart, online at http://www.batterymart.com, or at Battery Mart, 1 Battery Drive, Winchester, VA 22601-3673, Tel: 540-665-0065, Fax: 540-665-9623, Toll-Free 800-405-2121.

➢ Battery World, online at http://www.batteryworld.com, or at Battery World, Inc., 23164 Ventura Blvd, Woodland Hills, CA 91364, Phone: 818-225-0478, Fax: 818-337-7544.

➢ The Battery Station, online at http://www.batterystation.com, or contact them at The Battery Station, 17857 Rock Bluff Rd., Ottumwa, IA 52501, 641-935-2039.

➢ Battery Country, online at http://www.batterycountry.com, or contact them at, Battery Country, 127 East Jackson Street, Thomasville, GA 31792, Phone: 1-800-BCOUNTRY (800-226-8687), Fax: 229-226-7314.

➢ E-Batteries, online at http://www.ebatts.com, or at eBatts.com, 703 Rancho Conejo Blvd., Newbury Park, CA 91320, Toll Free 800.300.1540, International 805.499.4332.

➢ Phone Batteries, online only at http://www.phonebatteries.com.

➢ Code Red Batteries are a specially manufactured D-cell size battery that are only activated when you twist the top. The manufacturer claims this gives them a shelf-life exceeding ten years. Most of the general equipment providers we've listed will carry them.

➢ Recharging alkaline batteries. A few companies market rechargers touted to recharge alkaline batteries. Two we've seen are the ReZap ® and Battery Xtender ®. We have not tried these although we've done research into the process. The overwhelming data seems to indicate that it's not a good idea to recharge alkalines as the chemical process inside the battery will produce explosive hydrogen gas. Some alkaline batteries will take a small recharge but won't maintain power and can only be recharged a couple of times.

➢ Cellboost™, online at http://www.cellboost.com, offers disposable battery packs for different makes of cell phones. See them online or call them at 1-800-833-1070.

➢ To provide a little bit more consumer information on batteries, we've located comparison charts for you:
  - http://www.powerstream.com/Compare.htm.
  - http://support.radioshack.com/support_tutorials/batteries/batgd-c01.htm.
  - http://www.zbattery.com/zbattery/batteryinfo.html.
  - http://www.greenbatteries.com/documents/Battery_FAQ.htm#.

Tools and Repairs 1		
❏ Leatherman Tool ®	❏ Swiss Army knife	❏ Duct tape
❏ ¼" rope or cord (50 ft)	❏ Dental floss	❏ Small roll of wire
❏ Sewing kit	❏ Small tube "Super Glue"	❏ Eyeglass repair kit
❏ Rubber bands	❏ Safety pins	❏ Medical equipment tools

Overview

Your gear will do you absolutely no good if it breaks and you can't fix it. Even the minor equipment items are on this list for a reason, so it's logical that they should always function. These listed repair tools are small, inexpensive, and when your important gear malfunctions, these tools are worth their weight in gold.

Discussion Items

- ➢ **Leatherman Tool ®**. This is our favorite tool. It's a combination of pliers, a knife blade, file, and assorted screwdrivers. Several companies make worthwhile "multi-tools."

- ➢ **Swiss Army knife**. This is the perfect companion to your multi-tool. We suggest a larger model that has scissors, a saw blade, and magnifying glass among its many other attachments.

- ➢ **Duct tape**. No repair kit is complete without duct tape. Carry a small roll (about 10 yards worth).

- ➢ **1/4" rope or cord** (50 ft). If you don't already have this as part of your shelter equipment, you should carry a small roll for repairs. If absolutely nothing else, you may wind up traveling with others and your kits might have to be lashed to the top of a vehicle. You don't want to waste valuable time hunting for something so small and so easily prepacked.

- ➢ **Dental floss**. Similarly, if you don't already have dental floss in your kit, repairs are yet another of the many reasons you should. Carry the waxed, unscented, unflavored variety of floss, not dental *tape*. In repairs, floss can be used as both twine and rather strong thread. In addition to all the other reasons to carry floss, here's an interesting trick to remove a ring from a swollen finger: Take about a three-foot length of floss. Tuck about two inches of one end between the ring and the finger and pointing toward the palm of the hand. The long end should hang out towards the tip of the finger. Coil the long end firmly and carefully around the finger just in front of the ring, without overlapping the coils. After it's all coiled up, start pulling the ring towards the tip of the finger as you pull the other end of the floss through the backside of the ring and around and around as it's unwound. The ring should move forward over where the long end of the floss had been wound. Repeat until you get the ring off. Though a tiny little trick, swollen hands are common in stressful situations, and you might not have the tools to cut a ring off. Now that we've mentioned this little trick, you're ready to deal with one more problem.

- ➢ **Small roll of wire**. This is useful in situations where duct tape and floss just won't cut it. Carry about 10 feet rolled up in a small coil. Another way to carry wire is to carry a few **paper clips** and a few **twist-ties** that have the little wire running through them. All of these are great for repairs, including some electrical repairs.

- ➢ **Sewing kit**. You can find tiny sewing kits in most grocery stores. They'll come in a little box, not much bigger than a matchbox, or on stiff paper cards. Usually, they'll have a few needles, a few lengths of different thread, some buttons, and maybe a thimble. The first time you experience ripped pants, a tear in your tent, or a hole in your backpack, you'll be glad you had a sewing kit. Not only that, but if your clothing supply is limited, you'll want to be able to keep them repaired.

- ➢ **Small tube "Super Glue"**. With so many of your electronics coming in plastic housings, and with battery covers being so delicate, you'll wind up breaking a few. You'll break other things as well, so it's best you carry a small tube of "super glue" (some come in multi-packs of single-use tubes) to repair the tiny things that could make a big difference. Also, as we mentioned earlier, super glue can be used as skin glue or sutures in a dire emergency when there's absolutely nothing else.

- ➢ **Eyeglass repair kit**. If you carry glasses, you'll need to be able to put them back together if a screw or a lens falls out. These little kits are about the size of a golf pencil and can be found at almost any grocery or department store. They're usually among the little odds and ends you find near the checkout register. Essentially, the kit is a small screwdriver with some tiny screws and an eyeglass necklace, all housed in a small plastic tube.

- ➢ **Rubber bands**. You just can't have a bonified repair kit without rubber bands. These are close cousins to duct tape in their millions of uses. Carry about a dozen of varying sizes (including bands for dental braces).

- ➢ **Safety pins**. If your sewing kit doesn't already have safety pins, carry about a dozen in varying sizes. Not only are these useful in mending clothing, you'll need them to hold some bandages in place in first aid situations, and you can pin small items such as chemlight sticks to clothing.

- ➢ **Medical equipment tools**. We mentioned wheelchairs above. If you depend on any sort of medical equipment, and they require specialized tools that can't be substituted, make sure you have backup tools as part of your kit. Here's a short reminder list:

❑ Wheel chairs	❑ Walkers (or a spare cane)	❑ Oxygen equipment
❑ Limb and joint braces	❑ Dental braces	❑ Denture repair kit
❑ Hearing aids	❑ Catheter / Colostomy items	❑ Prosthetic devices

Sources and More Information

- ➢ These tools can be found at your local hardware or department stores, and the medical equipment parts and accessories should be available from the equipment's local provider or manufacturing headquarters.

Support Kit.

Here's our **Support Kit** components portion of the main Bugout Kit's Component Checklist from the beginning of the section. This kit will actually be smaller than your Primary Kit as it's **simply an expansion and accessorization** of the Primary Kit's gear. (That's why most of these checklists have no "Sources and More Information" section.) If you have time to grab a second set of gear (Level 2 or 3 evac), it should be composed of these component kits and accessory groups:

Support Kit – If you have time to grab a second set of gear (**Level 2 or 3 evac**). Checklists are on pages indicated.					
❑ Pets' Kits	255	❑ Level 2 Water	258	❑ Level 2 Food	258
❑ Extended Clothing	259	❑ Extended Toiletries	260		

In a Level 2 evacuation, you have a little more time (though not much) to gather your goods. Though we're adding more equipment and supplies, you still have to remember that space and weight are at a premium. So add what you need, but don't go overboard. Let's look at these kits in more detail:

Pets' Kits		
❑ Pet Info Pack	❑ Carrier	❑ Leash & collar or harness
❑ ID and Rabies tags	❑ Dry food and wet food	❑ Water
❑ Can opener, lid, and spoon	❑ Food and water bowls	❑ Bedding
❑ "Pooper Scooper" and baggies	❑ Litter and scoop	❑ Newspaper or "Piddle Pads"
❑ Baking soda	❑ Treats and toy	❑ Stake and rope
❑ Muzzle	❑ "Doggie Downers"	❑ Pet first aid kit
❑ Prescription meds	❑ Scented dryer sheets	❑ Flea treatment and comb

Overview

As far as we're concerned, pets are as much a part of the family as any person, and you need to prep for them as such. Period.

Naturally, all the items listed here are not going to fit into the same container housing your Level 1 Primary Kit. This kit will have to be separated physically, but should be gathered in an emergency evac as if it was part of the main kit. The only reason this kit is listed as part of your Level 2 gear is because retrieving it may require a second trip into the house (unless you have a family member specifically assigned to retrieve the pets and their gear).

Though this checklist is primarily oriented towards small mammals such as dogs and cats, it's easy to transpose and adjust for horses, birds, reptiles, or other pets you might want to take with you. Make sure you plan for transportation, feeding and watering, cleaning and care, medical needs, possible boarding, and reassurance.

The safety and wellbeing of farm animals is a closely related subject. All life, including animal life, is a subject near and dear to our hearts. However, for the purposes of subject division, we're treating the evacuation of farm animals as a "business continuity" operation. Therefore it will be mentioned in our "**Cooperation**" section, as it may become a volunteer group's project. It will also be a key component of our upcoming book **<u>Disaster Prep 201</u>** which is oriented toward the homeland security operations of state, county, city, and municipal governments, as they develop their preemptive mitigation efforts, improve their post-disaster planning, develop improved business continuity measures, and more closely integrate themselves with the civilian volunteer groups we hope to assist with this book.

Discussion Items

- **Pet Info Pack**. Just as people need info packs with all their pertinent information, so do pets. You'll need this information if you get to an emergency shelter that doesn't allow pets and you have to board yours with a volunteer group's nearby facility. Your pet's info pack should contain:

  - ❏ Your name and contact Nos.
  - ❏ Pet's medical history
  - ❏ Photocopy of tags
  - ❏ Pet's name and breed
  - ❏ Behavioral problems
  - ❏ Photos of pet
  - ❏ Veterinarian's name & contact
  - ❏ Dietary restrictions
  - ❏ Handling hints

  To make it easy for you, we've included a "**Family Member Data – Pets**" form in the "**Appendix**." While we're on the subject of "info" we want to mention that you should label every piece of pet gear just as you would your personal gear. Put your pet's name and your last name, for example, "Fido Smith," on everything. ID their carrier, leashes, collars, medications, food bowls, cans of food, water bottles, bags of litter, everything. Among numerous reasons for doing this, it will help operators of a boarding facility keep the right supplies with the right animal. **Hint**: Give your pet a "Safety Necklace" noting evac destination info.

- **Carrier**. Small pets should definitely have a carrier, and the only debate should be about carriers for large dogs since space in your evacuation vehicle will probably be pushed to the limit. Carriers do several things. They protect your pet, they give your pet an enclosed refuge, they make it easier to handle a stressed pet, and if nothing else, you can pack all these kit items in the carrier until time to pull out the gear and put in the pet. Speaking of carrying gear, big dogs can carry **saddle bags** will all their gear in it if they're too big for a carrier (don't make the bags too heavy). Get a carrier that has built-in food and water bowls, and for cats, get the kind of carrier that has a snap-on litter pan on its base. Also, get a set of **wheels** for your carrier to make it easier to pull behind you when it's only being used to transport gear. (Don't pull your pets around in a wheeled carrier. They'll be stressed enough as it is.)

- **Leash and collar or harness**. If we're envisioning such extreme emergency evacs that you might not even have time to grab your wallet, you're certainly not going to be able to run after every little piece of pet gear such as leashes and collars. Carry spares in this kit. Use whatever your pet is accustomed to whether it's a simple collar or a harness. **Hint**: Though they won't like it, you can use a small dog harness and leash on your cat. It might be the only way you can allow them out of their carrier, depending on where you are.

- **ID and Rabies tags**. Your pet should have an ID tag on both their normal collar and the extra collar in this kit. As for rabies tags, make sure your pet wears theirs at all times. If it's an indoor pet that never goes out, such as an indoor cat, you might keep the tag in your kit so it's there when you need proof of vaccination. You should also **photocopy** (certified as true) your pets tags to keep with their records in their info pack.

- **Dry food and wet food**.   Even if your pet doesn't normally eat dry food, you should carry some since it's lighter in weight than wet food, and it keeps longer. If your pet normally eats dry food, some wet food will not only be a nice treat (pets need comfort foods for morale too), but will help hydrate them to a small degree.

- **Water**. Same rule as with people. Carry as much as you can.

- **Can opener, lid, and spoon**. If you're carrying wet food, you have to be able to serve it. Also keep a plastic lid that fits your pet's open can of wet food.

- **Food and water bowls**. Remember that you can use plastic trays from microwavable frozen dinners, or you can get pet carriers with bowls built in.

- **Bedding**. Remember that your pet will need to stay warm and comfortable just like you. Keep an old towel, blanket, or other bedding packed in the carrier. You might also pack an extra sweater if your pet wears one.

- **"Pooper Scooper" and baggies**. You might be in a location where you have to clean up after your dog.

- **Litter and Scoop**. To care for your cat, you'll need a supply of litter and a way to keep it clean. For litter we recommend you get one of the scoopable, bacteria and odor control brands.

- **Newspaper or "Piddle Pads."** Some situations may see you stuck indoors and unable to walk your dog. "Piddle Pads" are specially made to hold wetness and scented to let a dog know that it's okay to "go" there.

- **Baking soda**. In case your pet misses the newspaper or piddle pads, you might want to take along a small box of baking soda to sprinkle on the spot after you've cleaned up what you could.

- **Treats and toy**. Again, just as people need comfort, so will your pets. They'll sense stress as much as you, and treats or a toy will help calm them down. Pack a toy your pet has previously played with and likes.

- **Stake and rope**. You may be in a location where you can put up a stake and let your pet roam a little provided you have them on a rope so they can't run off or bother others.

- ➤ **Muzzle**. You never know how stressful situations will hit your animal. Be ready to keep them muzzled to protect other humans or other people's pets around you. You also might need the muzzle to keep your pet from chewing at any first aid dressings applied if they were injured.

- ➤ **"Doggie Downers."** Numerous brands of over-the-counter mild sedatives are available for dogs and cats. Make sure your pet has taken the brand you pack and that it agrees with them.

- ➤ **Pet first aid kit**. Most items contained in your regular first aid kit will work for animals. Bleeding is bleeding, broken bones are broken bones. So if you have pets, you might want to increase the amounts of some things in your main first aid kit. However, some items can be different:

  - ❏ Pets will need their own **thermometer**. While we're on the subject, let's mention some values: **Normal temperature** ranges for both dogs and cats are from 101 to 102.5° F. Below 100 or above 103 are abnormal. **Heart rates**: Dogs: 70 to 160 beats per minute. Cats: 160 to 240 beats per minute. **Respiratory rates**: Dogs: 10 to 30 breaths per minute. Cats: 20 to 30 breaths per minute.

  - ❏ Carry a **"pill popper"** if your pet might need medications, but doesn't like taking pills. You'll find these at almost any veterinarian's or pet supply store. Also carry an **eye dropper**.

  - ❏ Dry, out of the box, **instant mashed potatoes, or corn starch**, are good to help stop bleeding from minor wounds.

  - ❏ Carry **Vaseline** ®, Laxatone ®, or similar, to orally administer to cats to prevent hairballs.

  - ❏ Carry an extra towel or two for bedding and warmth, and some **tube socks** to keep bandages or splints on paws or legs.

  - ❏ For other useful first aid items and considerations, visit the sites below, ask your vet, check with your local County Extension office or Department of Agriculture, or check your local pet supply store for first aid and care considerations (and instruction manual such as a **first aid book**) for your particular pet(s).

- ➤ **Prescription meds**. As with people, pack any long-term use prescription meds that your pet takes. Note the expiration date and make sure it's on your kit's "refresh" list. For temporary meds, or meds with a short shelf life, make sure you list them on your "Last Minute List".

- ➤ **Scented dryer sheets**. Cats are self-grooming, but dogs aren't. After a while without a bath, Fido will probably be smelling pretty ripe. Carry some scented dryer sheets to rub your dog down with. It won't cure the whole problem, but it's a quick and easy way to make them smell better. **Hint**: you can do the same thing with your clothes if you haven't had a chance to wash or bathe. **Hint # 2**: You can make a "dry shampoo" for Fido if you have any clean odor control, clay-based, kitty litter. Grind some up, rub it into your dog's fur, and comb it all out with the flea comb. You can do the same with corn starch and/or baking soda. (Once again, you can do the same to you.) The point is, you can keep your dog clean without using up your valuable water.

- ➤ **Flea treatment and comb**. If you have indoor pets, you might not keep flea collars on them. However, flea and tick protection will be necessary once you have them out and about and possibly exposed to other pets in a boarding facility. Pack a flea collar or one of the topical flea and tick repellant treatments such as Advantix ® or Frontline ®. Automatically included in this concept would be a **flea comb**, as physical removal of the little buggers helps a great deal. Don't forget a pair of **tweezers** in case of ticks.

Sources and More Information

- ➤ Pet equipment can be found through Jeffers Pet, online at http://www.jefferspet.com, or call 800-533-3377.

- ➤ Medications can be ordered through 1-800 Pet Meds, online at http://www.1800PetMeds.com or by calling 1-800-PET-MEDS (800-738-6337).

- ➤ American Pet Association, http://www.apapets.com, HQ – PO Box 725065, Atlanta, GA 31139-9065, 800-APA-PETS (800-272-7387), Fax: 305-294-8964. They have an online vet finder.

- ➤ American Society for the Prevention of Cruelty to Animals, http://www.aspca.org, American Society for the Prevention of Cruelty to Animals (ASPCA), 424 E. 92nd St, New York, NY, 10128-6804, (212) 876-7700. Visit their preparedness page at: http://www.aspca.org/site/PageServer?pagename=emergency.

- ➤ Contact the **Humane Society** at: http://www.hsus.org/ace/352, The Humane Society of the United States, 2100 L Street, NW, Washington DC 20037, Phone: 202-452-1100, or find their local office in your phone book. Also see the Humane Society disaster page at: http://www.hsus.org/ace/18730.

- ➤ The American Veterinary Medical Foundation: http://www.avma.org, 1931 North Meacham Road, Suite 100, Schaumburg, IL, 60173, Phone: 847-925-8070, Fax: 847-925-1329.

- ➤ Clemson University's "Animals in Disasters": http://virtual.clemson.edu/groups/ep/animal.htm.

- ➤ Information on **Emergency Animal Rescue** at: "United Animal Nations" http://www.uan.org, United Animal Nations, 5892A South Land Park Drive, P.O. Box 188890, Sacramento, CA 95818, Tel: (916) 429-2457, Fax: (916) 429-2456.

- Animal Disaster Preparedness links: http://www.horsereview.com/Links_Disaster.htm.
- Visit FEMA's "Animals in Emergencies" page at http://www.fema.gov/library/anemer.shtm .
- See the info on http://www.petcaretips.net.

Level 2 Water		
☐ More water	☐ Garden hose	☐ 5-gallon container
☐ Solar shower	☐ Collapsible bucket	☐ Plastic sheeting

Overview

Your Primary, or Level 1, kit earlier had *some* water , but it focused more on water *acquisition* since weight was such an important consideration. If you have a chance to grab your Level 2 kit, that means you either have a bit more time on your hands, and/or that your transportation dictated that the bulk or weight of your kits were not big obstacles.

Discussion Items

- **More water.** This is a no-brainer. If you have time and space for a second kit, make sure it has more water. To be completely self-sufficient and not need outside water sources, you'd need to carry at <u>least</u> one gallon per person, per day that you're planning to be gone (more in hot weather). That's a lot of water. However, do what you can. Carry more water as you're able in durable containers, and include the extra water-related equipment below.

- **Garden hose.** The garden hose may prove useful in some situations where you have to siphon water, or run it from or to a source you can't readily access. Some new models of garden hose are made of fabric and roll up flat to save space. Ten to twelve feet is a decent length to pack. Also, make sure it has its couplings on each end.

- **5-gallon container.** A collapsible 5-gallon container is extremely useful for storing water you've gathered after evacuating. Pack it in your kit, or keep it permanently stored in your vehicle. If time, space, and weight limits allow, fill your container *after* you get it in the vehicle (easier to load that way).

- **Solar shower.** A solar shower is essentially a collapsible 5-gallon water container made of black plastic and with a shower head attached to its drain valve. It'll let you carry drinking water and then create a shower later on if you so desire. If your plan is to evacuate to a remote wilderness area to camp, or utilities at your destination might be a problem, this becomes an essential item.

- **Collapsible bucket.** We listed one in your Primary kit, and since they're so useful, we suggest you pack another one in your Support Kit.

- **Plastic sheeting.** We mentioned this earlier as a shelter and rain gathering tool. If you didn't pack it in your main kit, pack it here.

Level 2 Food		
☐ More food	☐ Extra mess kit	☐ "Coffee pot"
☐ Camp griddle	☐ Second "stove"	☐ Extra stove fuel
☐ Utensils	☐ Extra fishing gear	☐ Hunting gear

Overview

As we did with water above, we're going to add a little more food here. Additionally, we're going to add a few more food acquisition and prep items, though they are elective where packing more food is not.

Discussion Items

- **More food.** Yet another no-brainer. Where before we suggested you emphasize the freeze-dried foods to conserve weight, here we're going to say it's okay to go a little heavier on the MREs and other items. You should increase the amount of:
  - ☐ MREs. Carry more of the complete meal pouches.
  - ☐ Protein bars, energy bars, granola bars, hard candy, and trail mix.
  - ☐ Spices and condiments.
  - ☐ Add dehydrated and freeze-dried items as space allows.

- ➢ **Extra mess kit**. Having one or two mess kits in your Primary Bugout Kit, adding this one, and possibly adding the items below should give the average family all it needs to cook pretty much anything, anywhere.

- ➢ **"Coffee pot."** This isn't so much a coffee pot as it's just a small covered pot in which to boil water. You can use a camping coffee pot made to sit directly on hot coals, a stainless steel tea pot, or something similar. This makes it easier to boil water for purification, and heating water to prepare your dehydrated food items, heat your MREs, and make tea or instant coffee.

- ➢ **Camp griddle**. You can use a regular kitchen griddle if you'd like, just make sure it's not a Teflon type (doesn't last long over a campfire), and remove the handle for easy packing. Essentially you're looking for a flat metal surface to cook on. Though **optional**, having a griddle will allow you to alter some of your foods and their preparations for a little variety.

- ➢ **Second "stove."** Earlier you chose between carrying a large stove, a small one, or a couple of small ones. Here you can carry the one you didn't choose, or simply another of what you chose in the first place. Better yet, if you went with a full-sized gas camping stove earlier, here would be a good place to add one or two of the small tab-fuel stoves or even more "Buddy Burners" to power your "Hobo Stove." The small ones are great for heating single meals or cups of coffee, and having more than one stove will let you cook different things simultaneously. We still recommend the smaller stoves. **Hint**: You can improvise a larger Buddy Burner by using a roll of toilet paper and a number 10 can. Put the whole roll of toilet paper inside the can, pour in just enough rubbing alcohol to wet the roll (but not puddle in the bottom) and light it. Exercise good caution while you're doing this. (Don't do this if you're in a confined space or low on toilet paper.)

- ➢ **Extra stove fuel**. Same as before. Whichever stove you choose, make sure you have the fuel for it. **Note**: Before you opt for a second stove, your first consideration would be to invest the money, space, and weight on **extra fuel** and any necessary accessories for your **main stove**.

- ➢ **Utensils**. If you decide to go with the extra mess kit, coffee pot, griddle, or maybe another stove, you might want to carry a couple of extra food prep and service utensils. Consider these:

  - ❑ Large hunting knife. This serves dual function in the kitchen and in the field. We recommend the Marine Corps Combat Knife, or some of the Buck hunting and camping knives.

  - ❑ Spatula. Makes it easier to turn pancakes or pan-fried fish on your griddle, or in your mess kit pan.

  - ❑ Tongs and forks. If you only carried one large fork and no tongs in your main kit, this is a good place to make sure you have another fork, and to definitely bring tongs.

  - ❑ Slotted spoon. There's no telling what you might have to boil and strain.

  - ❑ Extra pot holder. Extra stoves might mean more people helping cook. They'll need hand protection.

- ➢ **Extra fishing gear**. We had you add a tiny collection of fishing equipment in your Primary kit. You can augment it here by simply carrying extra hooks, line, and sinkers. Having extra will let you set up a larger "Trout Line," a series of hooks along one line that lets you catch multiple fish. You just set it up, bait the hooks, lay it out in the water, and then come back and check it later. Be sure to check the military survival manual on the enclosed **CD** for more fishing info.

- ➢ **Hunting gear**. If you didn't carry any hunting gear in your Primary kit due to space or weight restrictions, you should consider including it here.

Extended Clothing		
❑ Extra socks	❑ Extra underwear	❑ Outer shirt
❑ Long pants	❑ Athletic shoes	❑ Cold weather items
❑_____	❑_____	❑_____

Overview
Again, as this is a "Support" kit, most of these checklist items will be "more of the same" for your Primary kit. In "Discussion Items" below, we'll cover the new items present on this list.

Discussion Items

- ➢ **The extra socks**, extra underwear, additional outer shirt, and long pants are all merely "more of the same" for items in your main kit. As size and weight are still restrictions, you want to keep each item down to one, meaning only one more shirt, or one more pair of pants, etc. You don't have room for a wardrobe change.

- ➢ **Athletic shoes**. In your main kit, you needed sturdy shoes for protection. Though **optional**, you might consider a pair of sneakers or tennis shoes in this kit to make life more comfortable when you get where you're going. Again, these are **optional** and the decision to include them should be based on your kit's **size and weight** restrictions. (Extra shoes for a whole family will take up a **LOT** of room.)

 ➢ **Cold weather items**. Cold temperatures can be found in every area of the country. Make sure you have adequate coats, long underwear, gloves, mittens, or other items that will help protect family members against low temperature exposure.

Extended Toiletries		
❑ Augment the Primary kit	❑ 1 bath towel per person	❑ Minor cosmetics

---

Overview

As usual, the first step is to gather more of the same for your Primary kit. For toiletries though, we've added a couple of new items that may help you feel cleaner, neater, and "more human." Being dirty, unkempt, and miserable during tough times will not help your mood.

Discussion Items

 ➢ Augment the Primary kit. Since toiletries will generally be in short supply and great demand, you should do here what you've done with food and water, and simply carry more of the important items. Some of our suggestions include more:

  ❑ Toilet paper.
  ❑ Feminine hygiene products.
  ❑ Paper towels.
  ❑ Premoistened towelettes or baby wipes.
  ❑ Garbage bags.

 ➢ **1 bath towel per person**. Though we should still conserve space, we can probably go ahead and include one bath towel per person provided we use some sort of vacuum bag to shrink it down in size a bit. This way you can pack the big fluffy towels and not worry too much about space. Remember that your towels can also provide extra warmth as a bed cover, and since you've also remembered to carry **brightly colored** towels, they can be used for **signaling**. As every good "hitchhiker" knows, you always carry a good towel.

 ➢ **Minor cosmetics**. We're sure the debate will rage over what exactly "minor" means. To some, a minor kit will be 2 full suitcases worth, while to others, merely a tube of lipstick. If you have the space, carry just enough to make you feel good about your appearance. Morale is important. (And if nothing else, cosmetics are high on the barter list.)

Secondary Kit

The Primary and Support Kits lend themselves to immediate physical survival, basic comforts, and minimal "return your life to normal" assets. The **Secondary Kit** is designed to help sustain you through longer evacuations, add a little more comfort, provide more equipment for your "return to normal," and add gear and goods for helping others. Carry your Secondary kit with you if you feel your evacuation to be a potentially long one, and/or if the destruction of your home has occurred or is highly likely.

Let's look at the Secondary Kit's section from the main Bugout Kit Component Kits checklist:

Secondary Kit – After the safety and well-being needs of your family are met, consider secondary needs.

❑ More food and water	261	❑ Alternate Transportation	262	❑ For helping others	262
❑ Extended comfort	263	❑ Extended communication	264	❑ Extended shelter	265
❑ Tools and repair 2	265				

In our discussion of these component kits, we'd like to point out that some of the items listed below are not specific pieces of equipment, but *types* of equipment or goods.   At this level of support gear, you'll customize your lists based more on level of evacuation, your intended destination, how much you intend or need to help others, and your actual assets, rather than on specific survival or safety need. Let's look at these component items:

Overview

Though in our Secondary Kit we're adding more comfort items, and discussing goods for a "more leisurely" Level 4 or 5 evacuation, let us not forget that any evacuation is still serious business and should be conducted as rapidly as possible to get you on the road as fast as you can regardless of how much time you have. Outrun the threat, beat the rush, and get to your destination while it still has supplies.

This section of "more food and water" will introduce the concepts of "**rake and run**" or the "**mad minute**" as we clear the pantry in our dash out the door.

As we mentioned before, you should keep extra plastic tubs, or clean plastic trash cans on hand. The tubs you can keep tucked up under ones that you use for storage, like one paper cup stacks up under another, and you can keep an extra trash can, like the tall ones you might use in your kitchen, the same way. In this situation, these will give you instantly accessible bins to **rake** food items off your pantry or cabinet shelves and then **run** them out to your vehicle after you've loaded your Primary and Support kits. You can also use wheelbarrows or wheeled trash cans for this.

Though this seems like "anything goes," you'll want to keep some important points and priorities in mind:

1. **Water** is still the most important item, so make sure you load all your bottled water from your fridge and freezer. Use the plastic bottles of water from your freezer to cool any refrigerated food you take with you.

2. Don't forget that **size and weight** are still considerations. Before you go grabbing just any empty tub or trashcan, make sure the kind you've kept ready all this time will actually fit in your vehicle after your vehicle's loaded with other gear and people. Also, you'll want to be able to still carry your tub after you've filled it with food and water.

3. **First food priority** should be given to items that will give you the **most useful yield** per ounce. For example, your 6-oz box of "minute rice" will give you more to eat than a heavier can of green beans.

4. **Second priority** goes to foods that offer more of a **complete meal in one package**. For example, though those green beans might be great, you'll get more benefit out of a stew-type soup, one that has a variety of vegetables and other ingredients depending on your diet. Choose **non-condensed soups**, or ones that you don't have to add water to. These soups are good foods, not only because of the variety of ingredients, but because they don't require extra water, and the water they contain will help hydrate you.

5. **Third priority** goes to **single-item energy foods** such as peanut butter, cans of nuts, canned meats, extra protein bars, etc.

6. **Fourth priority** goes to **canned side dishes** such as your can of green beans.

7. **Fifth priority** is to augment your **comfort foods**. Potato chips and the like offer little in the way of nutrition, but they can perk up someone's spirits which is always a good thing.

8. **Sixth and final priority** is the items in your **freezer or refrigerator**, since these will only last so long. But... if you have room for it, if you can finish all of it within a day and a half, or if it's all you have, then be sure to take it with you. That's why some of the items in this checklist are **coolers**.

9. Don't forget **extra pet food**.

10. Don't take anything that's packaged in **glass**. You'll be doing all this as fast as you can, and glass will probably break either while your raking everything into your bin, or somewhere along your trip. An easy way to help avoid glass is to select food items that come in plastic already. Many brands are getting away from glass anyway.

Discussion Items

> **Coolers**. Even if you don't take anything with you from your refrigerator or freezer, you should consider making at least one of your bins a plastic or styrofoam cooler. It may prove valuable down the road with ice and other items you might accumulate later.

> **Plastic tubs and bins**. These are the extra storage tubs and trash cans we mentioned earlier. They make it so much easier to rake everything into them so you can rush it all out the door.

> **Old blankets or comforter**. These are optional, but you can make a "poor man's cooler" by using a plastic tub, loading it with your cold food items and frozen bottles of water, and wrapping it with an old blanket or comforter. This won't last as long as a regular cooler, but it will keep cool longer than a plastic tub alone.

> **Duct tape**. Use the duct tape to seal the opening of your regular cooler, or your "poor man's cooler" to keep it sealed and the items colder longer.

- ➤ **Thermos**. You'd only want to do this during a Level 5 evac, but if you're about to hit the road for a long trip, and you're tired from all this loading, a thermos full of nice hot coffee might just hit the spot. Also, you never know when you might be needing help from some volunteer group's soup line, and a thermos would be great to keep food hot.

Sources and More Information
- ➤ Virtually all of this gear can be found at your local grocery, hardware, outdoors, or department store.

Alternate Transportation		
❏ Bicycle(s)	❏ Kayak or canoe	❏ Inflatable raft

Overview
Granted, these alternate transportation suggestions have very limited applications. However, one of our duties in this book is to open your eyes and mind to possibilities and alternatives, and to give you more options than you had before. We'll give you scenarios for using these alternate forms of transportation as we discuss them below.

Note: We only recommend these items be used by those experienced in their use and physically capable of operating them. We offer these scenarios for use only as an example of possible alternatives and not as a specific recommendation or specific reaction. As always, use your knowledge, but control it with reason, and temper it with your best judgement.

Discussion Items
- ➤ **Bicycles**. A bike will be more useful than these other items. You might wind up at a crowded emergency shelter and not want to give up your vehicle's parking space to go run nearby errands. Or, while at your destination, you might send family members in different directions at once to run important errands.
- ➤ **Kayak or canoe**. As we said before, the earth is 75 percent water, so you might as well be able to take advantage of it. Imagine this: Suppose you're an experienced kayaker, you're evacuating from a nuclear explosion that occurred 30 miles behind you, and you're stuck in traffic that'll be gridlocked for hours. You look in your rearview mirror only to see that winds are starting to shift and the dwindling mushroom cloud is pointing your direction. You're stuck in traffic, unable to move, and there, 100 feet in front of you, is a river that could take you out of the area in under an hour. Yet your kayak is at home. Remote scenario? Definitely! But you have to think like this. So... do you have a kayak or canoe? Do any of your "Emergency Risk" items dictate a possible level 3 to 5 evac (where you'd have time to load the Kayak)? Does your "Threat Map" indicate that any of your primary evacuation routes will take you near a useful waterway?
- ➤ **Inflatable raft**. Though not nearly as fast in the water as a kayak or canoe, a small inflatable raft takes up far less room and can still allow you to navigate a calm waterway. A small raft could also be towed behind your canoe to carry some of your gear. If you decide you **truly** need this option, get a raft that has a good pump or has a compressed air cartridge to inflate it. If you have to exercise this option, you won't have time to sit there and blow up your raft by mouth.

Sources and More Information
- ➤ Most all your bikes and boats will be available through local sporting goods stores, or through the outdoor equipment sources we've listed at the end of this "Contents in Detail" section.

For Helping Others		
❏ Volunteer Member Equipment	❏ Volunteer Group Equipment	❏ "Rescue Packs"

Overview
Life involves others, and you, being the responsible one, realize this far more than most. You know that an emergency may not involve an evacuation of you or your area directly, but that it's a rescue mission where you have to go get a friend or family member out of harm's way. You also know that you may have to help your fellow citizen at some point in an evacuation, and you know that since you're part of a volunteer group, you'll be called upon to help after your own family is situated.

Discussion Items

➤ **Volunteer Member Equipment**. This would be your personal equipment you need to function as a member of your volunteer group. This might include any uniforms, hats, IDs, radios, etc.

➤ **Volunteer Group Equipment**. This would be your share of equipment that allows the group to perform its function. For example, your group's function in a limited evacuation might be to stock and manage the local emergency shelter. Your share of the supplies might be paper cups and paper plates. Be sure anything like that is a part of your Secondary Kit. (It's not a part of your Primary Kit because you and your family come first). More on the Member and Group equipment under "**Cooperation**."

➤ **"Rescue Packs"**. Rescue packs are completely optional, but could be very important in the overall scheme of helping others. Consider this: You may have a friend or family member living not too far away that falls victim to some sort of mishap that leaves them without goods or gear. This could be a natural disaster that causes them to lose their home, or their town to run out of supplies. You might be the one that has to make a road trip to go in and get them out, or to go in and bring them supplies. Having small packs made up in advance (very small versions of your "Primary Kit") will make it easy to go "rescue" someone. Depending on the disaster, you could also ship them the kit(s) via US mail or even FedEx. (This is another likely scenario in which you'd need your documentation showing you had business in the area. Otherwise, you might not be let into a disaster area.)

Sources and More Information

➤ Since this is our only direct discussion concerning going from one area to another for the purpose of a rescue, we thought we'd pass along a site we found that will help you set up a medical rescue for an American traveler in Mexico: http://www.binationalemergency.org.

Extended Comfort		
☐ Extra toiletries	☐ Non-prescription medications	☐ Cosmetics
☐ Reading material	☐ Tapes or CDs	☐ Adult beverages
☐ Cigarettes	☐ Birth control	☐ Alarm clock

Overview

Extended comfort is the little bit of insurance you'll need in the way of items that will keep your spirits up in the event your evacuation or relocation is longer or more difficult than expected. This list will include items to make you feel better physically, emotionally, and to help prevent disruption to your lifestyle as much as possible. As with everything else in these kits, keep your size and weight restrictions in mind.

Discussion Items

➤ **Extra toiletries**. Carry more in the way of toothpaste, soap, shampoo, toilet paper, and other items to keep you clean, healthy, and comfortable.

➤ **Non-Prescription Medications**. Carry extras or refills of any non-prescription medication you regularly use.

➤ **Cosmetics**. Again, to some this might mean another entire suitcase, and to others an extra tube of lipstick. The goal is to carry just enough to make you feel good. No more, no less.

➤ **Reading material**. Add an extra book or a couple of magazines you've never read. Again, make sure the material is comforting, inspirational, or at least distracting. You don't want stories that keep you focused on your situation, or that make you depressed or emotionally distraught. Definitely include books, coloring books, puzzle books, etc. for the kids. Keep them occupied, happy, and distracted from what's going on.

➤ **Tapes or CDs**. As with books, make sure the music you pack, or the "books on tape" you bring are all uplifting. Again, don't forget the kids.

➤ **Adult beverages**. Under "Health" we recommended you quit drinking or cut down. In officially designated shelters, alcohol will be prohibited. When driving during an evacuation you should not drink a drop. However... if you finally get where you're going after a harrowing emergency situation and a glass of wine or two fingers of Scotch will take the edge off and make you feel worlds better, then by all means go for it. Tuck away a few of the small plastic "airplane" bottles (which also make great barter items) in your pack if you want. Remember, no glass, watch your kit's weight, and don't crowd out space that could be used for water, food, or other vital items.

➤ **Cigarettes**. Same with cigarettes (or other tobacco products). We suggest you quit, and none will be allowed in shelters, however... (Again, great barter items. Carry a carton.)

- ➤ **Birth control**. Not that you'll be making new friends during evacuation, but you'll want to carry all the little things that will prevent your lifestyle from being interrupted anymore than it has to be.
- ➤ **Alarm clock**. Seemingly an odd item to pack away for emergencies, an alarm clock does have its uses (provided your watch, cell phone, or PDA doesn't have an alarm feature). For example, shelters may have certain meal times set, or may have scheduled shower times, etc. Other supply providers outside a shelter may have a set start and shutdown times. You wouldn't want to miss any of your opportunities. Having a watch or clock will also make it easier to sleep in shifts.

Extended Communication

❑ Disposable camera	❑ Portable television *	❑ Laptop computer *
❑ Digital camera *	❑ PDA device *	❑ Personal Locator Beacon *
❑ Land-line phone	❑ _____	❑ _____

Overview

Knowledge is power, and getting it in a timely manner is priceless. So is being able to communicate both during and after a disaster that forced you out of your home. The * items are considered important, but having them should be based on your ability to afford them.

Discussion Items

- ➤ **Disposable camera**. We consider this an important item in your Secondary Kit as you may need to immediately document damage to your home on your return.
- ➤ **Portable television**. As discussed under the "Communication" section, a small hand-held television is not only inexpensive, but extremely useful in keeping you informed of news, current events, and more importantly, official word and reports following an EAS broadcast. If you have a hand-held TV and don't keep it packed as emergency gear, at least make sure it's on your "Last Minute List."
- ➤ **Laptop computer**. If you have a laptop, make sure it's on your "Last Minute List" since it's not an item you'll want to keep packed. Having your laptop, you can keep up with email, look up vital information on the web, and depending on the length of your temporary relocation, you could keep up with some work projects, file claims online, and the kids can keep up with some of their school work. You can also use your laptop for movies, music, or books on CD or DVD. This way you might not have to pack an extra player. **Hint**: If you use a dial-up connection for your internet service provider, go ahead and look up the **local connection access phone numbers** you'd need for internet access from your evacuation destinations and likely spots along the way from which you might want to log on. **Hint #2**: Remember what we said earlier about the advantages offered by cellular modems and web-cams.
- ➤ **Digital camera**. Don't ditch the disposable camera in favor of the digital. You might not have time to find the digital even if it's on your "Last Minute List," and it still needs batteries. However, if you do have the digital, you should make an effort to take it. It'll help document disaster damage, and you also might want to create a photographic journal of your evacuation travels.
- ➤ **PDA device**. Same rules as the laptop. Don't go buy one just for evacuation gear purposes, but if you've already got one be sure to gather it and take it with you.
- ➤ **Personal Locator Beacon**. These are <u>really</u> optional and only listed in the event your planned evacuation destination and routes are remote wilderness locations. PLBs are basically "Lo-Jack" units for people. They're GPS (Global Positioning System) based transmitters that will act as a locator and transponder unit sending your location and a distress signal back to the PLB's company office when you activate the alarm. Previously, this type of system was used by ships' life rafts as a way for the Coast Guard to locate sea disaster survivors. Now several PLBs are manufactured for use on land.
- ➤ **Land-line phone**. Though this is optional, and **not** high on our list of recommendations, a small land-line phone might prove useful in some instances. We suggest a small, "slim line" corded phone with a variety of jack attachments. The purpose is to have a phone if you have access to wiring, but no actual phone. This might be the case if you come back to a damaged or looted home, or take up refuge in a functional but unfurnished evac destination. See if you can find the kind of phone that is a complete unit, all packed within the receiver to cut down on size and weight. Add to it the following attachments:
  - ❑ Standard modular jack connector.
  - ❑ Four-pronged plug jack. Your phone store or local Radio Shack can tell you more.
  - ❑ Four alligator clips in case you have to connect your phone wire by wire.
  - ❑ Make sure the phone you choose can be set to either "tone" or "pulse" dialing.

Sources and More Information

➢ GPS and PLB units will be available at most outdoor and sporting goods stores. You'll also find a few intermingled in the items carried by the equipment providers listed at the end of this section.

Extended Shelter		
❑ Tents or tarps	❑ Bedding	❑ Fold-up hammock
❑ Umbrellas	❑ Rope	❑ Nylon cord

Overview

The first thing you should consider under "Extended Shelter" is including the items you excluded from your main kit (such as **mosquito netting**). The second thing to consider is that there might be something from the main kit that you feel you should **duplicate**. You might want another tarp, more plastic sheeting, more rope, another pop-up tent, or other items. This checklist will contain a few of the items that we do suggest you duplicate, and we've added a couple of other small items that will help further your protection from the elements.

Note: We keep pushing cooking and shelter equipment because there is **no guarantee** you'll be able to make your planned destination, in the time limits you desire, or along the route you'd hoped. Also, there's no guarantee everything will function according to your wishes once you're there. Be prepared.

Discussion Items

➢ **Tents or tarps**. Whatever you didn't pack in your Primary kit, you should include here. Go with enough to protect your whole family from the elements should you be diverted from your planned destination or route.

➢ **Bedding**. Same as above. You should consider bedding items you left out of your Primary Kit due to size or weight restrictions. Rest is vital to energy and you'll need your energy. Take another look at air mattresses, foam pads, or even fold-up camp cots if you have the room to take them. Don't forget mosquito netting or extra flea collars.

➢ **Fold-up hammock**. Now this, you really should've packed in your Primary Kit due to it's small size, light weight, and many uses. Pack it now if you didn't, or pack an extra if you're able.

➢ **Umbrellas**. Ponchos can be used to protect you from rain, but a small collapsible umbrella can protect open car or tent windows from rain, provide shade from the hot sun, and even provide a small wind break in the cold or while you're cooking on an open flame stove.

➢ **Rope**. Rope and cord are essential if you wind up forced to rig outdoor shelter.

➢ **Nylon cord**. Get the braided 1/4-inch type. Fifty feet should do nicely.

Tools and Repairs 2		
❑ Hatchet / hammer combo	❑ Fold-up wood saw	❑ Small machete
❑ Axe	❑ Crowbar	❑ Nail and screw assortment
❑ Siphon hose	❑ _____	❑ _____

Overview

Our "Tools and Repairs 1" checklist earlier focused on tools and parts needed to keep your primary equipment functional. Here we'll add a few pieces geared toward managing your immediate environment such as in an outdoor survival situation, or for beginning cleanup at your home after a disaster.

Though listed here, this is more a reminder checklist than anything else. Most of these tools should already be stored in your car or in your house. These are just a few we felt you should consider carrying with you **IF time, space, weight, and cost** allow.

Discussion Items

➢ **Hatchet / hammer combo**. You'll need both, so save space with a combo tool. It has a hatchet edge on one side and a hammer head on the other.

➢ **Fold-up wood saw**. At the bare minimum you should carry a "wire saw" which is a short length of wire with a ring on each end. The wire is specially configured to have a saw-tooth edge to it so you can basically "floss" down small trees. A better tool is the fold-up saw. You might need one of these to set up a temporary shelter, clear away fallen limbs blocking your car, cut limbs to lever your car out of a ditch, etc.

➤ **Small machete**. Great for those chores where a knife is too small and a hatchet too big.

➤ **Axe**. An axe is rather large and therefore optional as far as most evacuation plans go. However, if any of your plans take you to or near rural or wooded areas, you might need to clear a path for you car, gather firewood for a temporary camping stayover, or fashion larger levers if your vehicle is larger than the average car and needs help getting unstuck.

➤ **Crowbar**. You'll never fully appreciate the value of seemingly simple tools until you're in a situation where you needed one and didn't have it. If you don't keep a crowbar in your vehicle, keep one in your pack. Not only will it pry open your car door if it jams after a minor fender bender, it'll let you into the locked trunk inside your storage facility when you've lost the keys.

➤ **Nail and screw assortment**. Another of the "better to have it and not need it" tools, a small collection of assorted nails and screws can be a lifesaver.

➤ **Siphon hose**. You might want to share some fuel with others, or they with you. Also, you might come across a gas station where the power is out and so are the pumps. With a 15' hose and cash, the station owners might be able to sell you some gas after all. You can buy lengths of a variety of diameters of vinyl tubing at almost any hardware store. Get a ¼ to ½ inch diameter length of hose. **Hint**: If you wind up not using it to siphon gas, you can use it with water.

As Able

Some evacs are a "Level 5" and you have a full day to evacuate. Carry these items as you're able. About the only things dictating a Level 5 would be an approaching wildfire or hurricane, both of which put your **home** at risk. Therefore at this level, you might want to think about two main things; one is your **irreplaceables such as heirlooms or valuables**, and another is **tools that might help you when you have to come back to repair a damaged home**, or search through a destroyed one (such as the items listed in the checklist above).

There are other things to consider too, so let's look at simple discussion items as checklists don't apply.

➤ **Heirlooms and valuables**. If you took our advice about packing up the irreplaceables, this would make them easier to find, and you'd be less likely to forget anything. At the bare minimum, be sure to list them on your "Last Minute List" so when your doing your "Mad Minute" or your "Rake and Run" through the house, you don't forget anything. **Hint**: In addition to the spare plastic bins we suggested for this, you could also keep an **empty duffel bag** rolled up with your Bugout Kit to grab these last minute items.

➤ **Home repair and cleanup tools**. If you have the transportation to carry it, the time to load it, and destruction or damage to your home is guaranteed, you should pack your entire tool collection as listed under "**Basic Home Prep**." If you come back to a damaged home, don't expect the local hardware store to have any tools left, even if the hardware store is still there. You should include **cleaning supplies** as well.

➤ **Comfort**. You might also want to consider, based on where you intend to go, carrying extra physical comfort items. These would include better bedding (such as a cot), fold-up lawn chairs, battery-operated fans, and maybe even a **small portable generator**. You'll also need these items when you get back home if you have to camp out in your yard while rebuilding your home.

➤ **Entertainment**. Carry more in the way of tapes, CDs, books, magazines, etc. At this evac level, you might also want to bring your larger radio, tape, CD, or DVD player.

➤ **Children's' Schoolwork**. If there's any indication at all that you'll be gone for an extended period and your children are in school, be sure to bring their books. As we'll discuss in the "Cooperation" section, some schools may prepare absentee lesson plans, or they may post lesson plans on their website.

➤ **Job related project materials**. Similarly, if your job is one that allows you to do any work at home or telecommute, be sure to pack your stuff so you can maintain your job while on the road.

➤ **Recording Equipment**. (video cameras, still cameras, tape recorders, journal notebooks). Depending on the nature of the disaster, you might regard your evacuation as an adventure and want to record parts of it, or you might need some tools to record the damage done to your home when you return, for insurance purposes. Besides, some items such as your cameras and video cameras are valuable and should be taken with you if you have the room and time.

Miscellaneous Gear and Equipment Sources

These are but a few of the supply sources we've found that carry a variety of the items mentioned in this section. As always, don't buy what you won't need, try to make some of your needed items at home, and for what you do buy, shop around for the best deal.

➤ Magellan's Travel Supplies. A good source of assorted gear and gizmos. http://www.magellans.com, 800-962-4943, Fax: 800-962-4940, Magellan's, 110 W. Sola Street, Santa Barbara, CA 93101-3007.

➤ Sportsman's Guide. Military surplus, hunting, and camping. http://www.sportsmansguide.com, HQ Gov't Surplus, 411 Farwell Ave., South Saint Paul, MN 55075-0239, 800-888-3006, Fax: 800-333-6933.

➤ US Cavalry. Military surplus, hunting and camping, and police tac gear. http://www.uscav.com, 2855 Centennial Ave, Radcliff, KY 40160-9981, 800-333-5102, or 888-888-7228, Fax: 270-352-0266.

➤ Major Surplus and Survival, Inc. Good source of low cost surplus and camping gear. http://www.majorsurplusnsurvival.com, P.O. Box 3796, Gardena, CA 90247, or at 435 W. Alondra, Gardena, CA 90248, 800-441-8855, 310-324-8855, Fax: 310-324-6909.

➤ Nitro-Pak. Great source of miscellaneous self-reliance equipment. http://www.nitro-pak.com, 13309 Rosecrans Ave., Santa Fe Springs, CA 90670-4940, 800-866-4876, 310-802-0099, Fax: 310-802-2635.

➤ Cabela's. Big supplier of hunting, fishing, and camping gear. http://www.cabelas.com, 400 East Avenue A, Oshkosh, NE 69190, 800-237-4444, Fax: 800-496-6329.

➤ Visit EMERGENCY PREPAREDNESS SERVICE at http://www.emprep.com, or at, 309 – South Cloverdale # B-10, Seattle, WA 98108, Phone 206-762-0889, FAX 206-762-1040, TOLL FREE 1-888-626-0889.

➤ Check the universal gear page at http://www.equipment.net, to search for tools and supplies. Contact them directly at EQUIPMENT.NET, INC., PO Box 810461, Boca Raton, FL 33481, FAX: 305-675-5896.

➤ Four Star Military Surplus, 301 West Main Street, Hartselle, AL 35640, Toll Free: 1-877-751-0979, Phone: 1-256-751-0979, Fax: 1-256-751-0971, http://www.4starmilitarysurplus.com.

➤ For outdoor and food and water gear, visit Emergency Resources, at http://www.emergencyresources.com, or at P.O. Box 360634, Melbourne, FL 32936-0634, Phone: 888-567-1707, Fax: 321-254-4320.

➤ Phoenix Systems, Inc., carries hard-to-find equipment. They're online at http://gear4locks.com, or you can find them at 6517 S Kings Ranch RD #185, Gold Canyon, AZ 85219, 1-480-474-1226.

➤ Emergency Essentials, online at http://www.beprepared.com, or contact them at 362 S Commerce Loop, Suite B, Orem, UT 84058, 1-800-999-1863.

➤ Brigade Quartermasters, online at www.actiongear.com, or contact them at PO BOX 100001, 1025 Cobb International Dr. NW, Ste. 100, Kennesaw, Georgia 30156-9217, Telephone 770-428-1248 • 1-800-338-4327, Customer Service 1-800-228-7344, Fax 1-800-892-2999.

➤ Campmor, online at http://www.campmor.com, or contact them at 28 Parkway, Box 700, Upper Saddle River, NJ 07458, 1-800-525-4784 and 1-888-226-7667.

➤ REI, online at http://www.rei.com, REI, Sumner, WA 98352-0001, Phone: 1-800-426-4840, 253-891-2500, Fax: 253-891-2523.

➤ See "Quake Kare," online at http://www.quakekare.com, or call 1-800-2-PREPARE (800-277-3727).

➤ Visit Lifeline Security, online at http://www.lifelinesecurity.com, or call 1-800-249-2311.

➤ You might find some hard-to-find items at 911Supplies.com. They're online at http://www.911supplies.com, or contact them at 11693 San Vicente Blvd. #109, Los Angeles, CA 90049, Fax: (805) 725-4258.

➤ Numerous safety and preparedness supplies can be found online at either http://www.safetycentral.com, or http://www.preparedness.com, or by contacting Preparedness Industries, Inc., 311 E. Perkins St., Ukiah CA 95482, Phone #707-472-0288, Fax #707-472-0228.

➤ Safety and preparedness gear can be found online at http://frontiersurvival.com, or by contacting Frontier Survival, 75 South Main Street, Manti, UT 84642, Phone #1-435-835-8698, or toll-free: 866-877-4166.

➤ Visit Camping Survival, online at http://www.campingsurvival.com, or at JHL Supply, P.O. Box 720, 191 W First St North, Fulton, NY 13069, Toll Free: (800) 537-1339, Local: (315) 592-4794, Fax: 315-592-4796.

➤ Raytech first aid and emergency gear, online at http://www.raytechcatalog.com, or contact 222 Fashion Lane #111, Tustin, CA 92780, Tel: (714) 544-8864, Fax: (714) 544-8489, Toll-Free: (800) 838-5898.

➤ Survival Incorporated, online at http://www.ultimatesurvival.com or call (888) BE-READY (888-237-3239).

➤ See James' Survival Kits, online at http://www.sharplink.com/jkits, or call 888-577-2257.

➤ While shopping, you're sure to come across a dozen or so brands and types for each piece of equipment you're considering. When in doubt, check things out through either Consumer Reports, the Gear Review site at http://www.gearreview.com/default.asp, or Backpack Gear Test at http://www.backpackgeartest.org.

➤ Plans, projects, and patterns for making camping gear http://www.backpacking.net/makegear.html.

Two crucial factors that will affect the speed and efficiency of your evacuation are going to be **how** you have your gear **packed** and **where** you keep it **stored**.

For many people, the concept of packing for an emergency evacuation coupled with a possibly extensive stay at a distant destination conjures up images of war refugees travelling in long columns with what belongings they could salvage piled high on the backs of ox carts or trucks.

In this portion, we're going to try to have you a bit more organized than that with all your necessary evac gear stowed neatly in a backpack or suitcase or two, with more gear stored in strategic locations away from home. Let's start with how to pack all the things we just covered under "Contents in Detail" above.

Packing

As we stated earlier, when packing your evacuation gear, you should pack as if you were going on a two-week camping trip and could only carry what would go in your backpack. Your Secondary and Support kits can be larger and more cumbersome, but your Level 1, Primary Bugout Kit has to be lean and mean.

Let's consider a few concepts concerning the **type of pack** or bag you'll choose to house your main kit(s):

1. **It needs to be small and light enough to be manageable and portable**. Your Level 1 Bugout Kit is limited to what you can carry in **one trip**. Also, you may wind up travelling with others which may mean limited luggage space. Therefore, pack as if you were taking a flight with only carry-on luggage. Remember too, that your kit must be small enough to be loaded by lesser able members of your family.

2. **Your kit needs to be easy to carry**. Some people keep emergency gear in plastic tubs (or wheeled trashcans). This is fine for household isolation gear, but goods that are to be carried out of the house need to be in containers made for transport such as backpacks, duffel bags, or small wheeled suitcases.

3. **Durability is a key factor**. Whatever container you choose, it needs to be one that can stand some abuse. You may wind up tossing your kit out a window to get out of the house in a hurry, it may ride on the outside of your vehicle, you may have to travel some distance with it, and you may live out of it for an extended period of time.

4. **Security is also a factor**. Can your kit be locked? Can it be chained or fastened to something else to prevent its being stolen, either while in storage or while in use?

5. **Our recommendation for a Level 1 Primary Bugout Kit** is a medium to large **backpack** and/or **small wheeled suitcase** with telescoping handles. (The average person can carry a backpack while rolling the suitcase.) However, one alternative (or addition) worth mentioning, is the good ol' reliable **5-gallon plastic bucket** with a snap-down lid and handle. You can carry quite a bit in it, it can be used as a seat, sink, toilet, water bucket, wash basin, and any number of creative things.

6. **Mark these packs with strips of reflector tape and/or glow paint**. Do this to <u>all</u> your emergency kits since numerous scenarios will see you having to hunt for your gear by flashlight. It'll also help you identify your gear as being yours if it's packed with other people's gear during your evacuation travels.

7. **Pets' kits**. Keep everything you can packed together and near your Bugout Kit. For storage purposes, some things might be packed in a small duffel and the duffel packed inside the pet's carrier. Instead of a duffel, you might consider getting a set of "saddlebags" for larger animals such as large dogs or horses. Make sure you know your animal's pack-weight limits.

Now that we've discussed what you're going to put everything in, let's cover some tips on **<u>how to pack</u>**:

1. **Rule 1, no glass**. Glass is heavy, breakable, and the broken bits are dangerous. Don't pack anything in glass unless you absolutely have to.

2. **Load by access priority**. In backpacking, you load primarily by weight, with heavier items going in the bottom of the pack toward the wearer. For Bugout Kits, you'll want to pack items according to how rapidly you might need them with weight being a secondary consideration (unless your primary evacuation plan is to backpack to a remote location). Let's look at a few examples:

   ➤ Keep a **flashlight** and your **spare keys** in an outside pocket where they're accessible for instant use, and tied to the pack with a length of cord, or wire fishing leader with connectors, so they don't get lost in your hustle out the door, or while frantically searching for other items.

   ➤ Keep items you think you might **need on the road**, like first aid kits and other emergency gear, cash, snacks, or your latrine gear, more accessible than items you'd use only at your destination.

   ➤ Also keep a **short length of cord and a few trash bags** in an outside pocket. Since you don't know how your travel plans might change, and you might have to lash your gear to the outside of someone else's vehicle, you'll want to be able to do it immediately without wasting valuable time trying to find cord, and you'll want the trash bag to protect your kit from the weather or other unforeseen hazards.

3. **Create packs within packs.** You don't want to just dump everything inside your kit. For starters, disorganized gear takes up more space. Secondly, you'll want to keep like items together for when you need them, and if forced to leave some gear behind, you'll want to be able to pull what you need in a rapid and orderly fashion. Therefore, each of your kits listed above should have its own pack or bag even though it's going in a larger bag. Also, you might want to keep a small cloth backpack (with no frame) or duffel bag rolled up with your Primary Bugout Kit, in case you need to divide your goods for whatever reason. (For example, you might not have enough family members to load the gear, or you might be forced to evac in a smaller vehicle.) This way you can keep from having to walk around trying to carry half a dozen individual smaller packs or bags. **Hint**: Some of your extra items can be placed in different parts of your bag. For example, your additional rolls of toilet paper can be tucked where there's space in your bag, provided you've already included a couple in your toiletries kit.

4. **Secure everything**. You might be in situations where you're hurriedly fumbling around in your pack in the dark to find something before moving on. That's the time you'll lose something important. Having packs within packs as mentioned above will help you keep from losing a lot of the small things, but some things should have their own tether. For example, your flashlight in an outside pocket, or your Leatherman tool, compass, etc. should have a small cord keeping it attached to your pack. The more important a single item is, the more it needs to be secured.

5. **Keep everything in your pack redundantly sealed** in its own plastic bag whether it's a garbage bag or a resealable plastic bag. There are several reasons for this:

   ➢ First, you'll want everything in your kit **protected** from **water** and any number of **contaminants**. You might wind up throwing your kit out a window in the pouring rain to escape your house, you might be forced to change your transportation and your gear might have to be lashed to the outside of a vehicle, or any number of things could happen. You and your gear might also be exposed to a **biochemical** incident. Redundant sealing makes it easier to **decon** your gear. You'll also want to contain liquid items that might leak.

   ➢ Secondly, **plastic bags and jars are worth their weight in gold** in a life-in-the-field situation. All water gathering and sanitation uses aside, plastic bags and garbage bags can be used to seal contaminated items in a decontamination situation.

   ➢ Thirdly, extra sealing will **extend the shelf-life of some perishable items**.

   ➢ Fourth, you don't know what kind of **abuse** your packs will receive, and the bags will help provide a small amount of cushioning to protect some of your more valuable gear such as your radios. **Plastic jars** with screw-top lids also serve the same purpose, and are especially useful in protecting slightly fragile items such as your radios or spare cell phone. They're also useful as drinking cups and for water storage.

   ➢ Line your main pack with a **trash bag** to keep everything doubly protected. This is your lifesaving equipment, and it deserves all the protection it will give you in return. Better than one trash bag is a double layer of trash bags, or a "**contractor bag**" which is an extremely tough and durable plastic bag made for use at construction sites. Your local home improvement store or hardware store should have a few rolls in stock.

   ➢ Lastly, and least important of all, having some air pockets in your kit will allow it to be used as a **flotation device**. Why? Because you never know where you'll be, what you'll be doing, how you might be traveling, or if you might accidentally drop your kit (and/or you) in a body of water.

6. **Use vacuum bags**. Space is at a premium in your kits, so whenever you can shrink something down to make it smaller, it's a good thing. One way to do this is to take some of your "fluffier" items, like clothing, and store them in plastic vacuum bags. This gives the protection of the sealed plastic and helps conserve space. For example, if you were to pull the cardboard tubes out of your rolls of toilet paper and vacuum seal them individually, you could get three rolls into the space of one. The same holds true for your clothing, paper towels, cloth towels, etc. There are two types of vacuum bags available, one is the "squeeze type" and the other uses a pump and sealer.

   ➢ The "squeeze type" bag looks similar to a large resealable plastic bag with a zipper, but the seal is actually a one-way valve. You load your bag and then roll it up or press on it to force out the air which stays out. The best example we've seen is the "Space Bag" produced by Coleman. You'll find them at http://www.spacebag.com, or contact them at New West Products, Inc., 7510 Airway Rd., Suite 4, San Diego, CA 92154, 1-800-469-9044, or 619-671-9022. These bags are good for your larger cloth items like clothing and towels. **Hint**: Some of the larger bags made by this company are made to hook up to your **vacuum cleaner** to draw out the air. These larger bags are great for your sleeping bags and bedding.

   ➢ Several companies market **vacuum pump and seal kits**. These come with heavy duty plastic bags that you fill with whatever it is you want sealed, insert the end in the counter-top unit and the unit mechanically pumps out the air and then heat-seals the plastic bag when done. This electromechanical pump creates a better vacuum than the squeeze type and will therefore work better on things like the rolls of toilet paper or paper towels. One we found is called the FoodSaver. It's online at http://www.foodsavertv.com, or contact them at Tilia, Inc., P.O. Box 194530, San Francisco, CA 94119-4530, 1-800-777-5452, or 800-219-8855, Fax: 415-

896-6469.   To reach Tilia's headquarters: Phone us: 1-415-371-7200, Write us: Tilia, Inc., 303 Second Street, North Tower, Fifth Floor, San Francisco, CA 94107-6302.

7. **Load heavier items in the bottom**.   After deciding which might be rapid-access gear, and which might be destination gear, you should load by weight.   This is important with backpacks as it keeps heavier items closer to your center of gravity, and in wheeled suitcases, it keeps the weight near the wheels making it more stable to roll. Besides, you don't want heavy objects up top breaking fragile items underneath.

8. **Give kids their own packs**.   If kids are old enough for grade school, they can have their own pack.   This'll make it easier to ensure everyone has enough **personal clothing, toiletries**, etc.   (You can keep the packs in their rooms in the safety boxes that house their fire escape ladder if there is one.)   When the child has their own pack, they feel more included and protected.   Sending them to get their pack also keeps them occupied with a positive activity, in a known location, and out from under your feet for a second or two.   **Make sure their packs don't contain anything critical,** only extra clothes etc., since you might not be able to get to these auxiliary packs.

9. **Label everything**.   Since it might be a family member helping you access items in your pack, they need to know what everything is.   Label your smaller packs so they'll know their contents without having to rummage through everything.   Also, make sure everything is labeled with your name for purposes of ownership verification at shelters and so on.   Also label clothing as it makes it easier to tell children's clothes apart, etc.   You can mark some of your items with glow-in-the-dark paint, or with a small stitch of uniquely colored thread in a hidden spot. Labeling also makes it easier to ID and collect your gear in the event a tornado hits your house while you're gone one day and sends your emergency kit down the block somewhere.

10. **Print instructions**.   You should label each piece of equipment and its accessories so that you'll not only remember which goes to what, but so will the family member that will use the gear if you're incapacitated.   In addition to equipment and accessories, you'll want to make notes on anything not packaged in its original container, sealed sets of clothing (who it belongs to), instructions for use, expiration dates (all food and medicines should have expiration or purchase dates written on them anyway), etc.

11. **Put small locks on your kits**.   Keep one key on your key ring, and another taped to hidden spot on pack itself. Keeping your gear locked may prevent some of those "property disputes" mentioned above.   It'll also make it very slightly more difficult for gear pilfering if your pack happens to be lashed to the outside of your vehicle.

12. **Track your perishables**.   Pack all perishable medicines together and pack all batteries together (but not in the same bags).   When you refresh your kit (change out the perishables) you won't have to root through the whole pack.   We suggest you refresh your kit each spring and fall, swapping out the batteries and some of the medications.   The kids will be going back to school and might need some of the meds, and you can use your old batteries in some of the gifts you might give during holiday season.   A good way to remember the timing is to refresh your kits when daylight savings time changes.   When you rotate your clocks, rotate your stock.

13. **Kits vs cargo**.   How big is your family?   How many people in your volunteer group?   How often do you travel? These factors can make the decision on whether to pack simple backpacks and suitcases, or larger carriers. Some examples we've seen include car-top luggage carriers, small "U-Haul" type trailers, pop-up campers, etc. all of which have been kept secured but ready to go, and loaded with non-perishable emergency gear.   Don't forget bike racks if you have a bike that you could load during a Level 5 evac.

14. **Hint**: Your "Bugout Kit" as a whole doesn't have to be kept in one container.   Your Primary kit should be together, but other kits and redundant items can be kept in separate packs and sometimes at different locations.   Let's look at Storage next.

Storage

Moving from our rather short "**how** to pack" section above, let's continue in kind and look at **where** to keep your gear so that no matter what happens, you're set and ready to go.   Remember, things can happen at any time, and "Murphy's Law" dictates they will happen at the worst possible time and in the worst possible place.

You might have to evacuate your city during the middle of the day while you're at work.   You might be forced to take an evacuation route that carries you the opposite direction of your home and your main stash of gear.   You may experience a severe Level 1 evacuation that causes you to leave all your Support and Secondary gear at home.   You might experience a tornado or other destructive event that destroys your main supply of gear.   Or, you might be part of a panicked evacuation with all your gear, only to be robbed by someone who failed to plan ahead. Then there's the fact that several scenarios could occur in the middle of the night and cause you to evac with only your pajamas.

Here we'll cover how to evenly distribute your gear, and how to protect it wherever it's distributed.

In discussing **why** you should keep goods and gear **in different locations**, keep the following in mind:

- ➤ Each of us spend our days and nights in different ways. Determine where you spend a good portion of your time and realize that may be where you are when an emergency occurs. You'll want to keep in mind **duplication** and **equal distribution** of some of your emergency gear among: ❑ Home ❑ Vehicle ❑ School ❑ Work ❑ Place of worship ❑ Volunteer group (and at your ❑ Destination, but we'll cover that later).

- ➤ Naturally, you'd keep extra gear where you spend most of your time. For most of us, we'd say that's home. However, consider how much time you spend at work. Keep what you can at work, and you might even talk to the boss about the company stocking some emergency supplies of its own if it doesn't already.

- ➤ You'll also want to consider storing some goods and gear on opposite sides of any major travel obstacle. Suppose you live on one side of a major river and work on the other. You'd want some accessible supplies stored somewhere on each side of the river since any large-scale emergency would keep you from crossing the river, either because of traffic or because of damage to the bridges. Plan accordingly.

- ➤ Later, under "Planning the Destination," we're going to ask you to pick an evacuation route and destination along each of the **four compass directions**. Since you really don't know where you'll be forced to head, or how little or how much you might be able to get ahead of time, it's best to have at least some things available in a couple of locations away from your potential evacuation starting points.

- ➤ Any number of scenarios could see your home **destroyed** before you even got a chance to get there and get your gear. Other scenarios could see you **driven out of your house** so fast you had zero chance to take anything with you other than your own skin.

Let's start with gear distribution. Here's a small worksheet, filled in with examples, to help you plan where you keep your goods stored. You'll find a blank copy in the "**Appendix**." Following this worksheet is a discussion of miscellaneous considerations to keep in mind when locating some of your equipment at these various sites. The sample list below isn't a suggestion on specific locations, only an **example** of how to fill out the worksheet.

Evacuation Supply Distribution (example)

Items	Home	Car 1	Car 2	Work 1	Work 2	Dest. 1	Dest. 2	Deposit Box	Cache 1	Cache 2	Other
Personal Protection Equip.	M	✔	✔					✔			
First Aid	M	2		✔							
Level 1 Water	M			2		2			✔	✔	
Level 1 Food	M								✔		
Basic Toiletries	M	2	✔			✔					
Primary Clothing	M	✔	✔			✔			2		
Basic Shelter	M								2		
Non-Prescription Meds	M	✔				2					
Babies' Kits	M	✔			2	2	✔				
Communication Gear	M	2									
"Office Pack"	M	✔					2				
Batteries / Accessories	M	2									
Tools and Repairs 1	M	2									
Pets' Kits	M					2					
Secondary Water	M								2	✔	
Secondary Food	M								2	✔	

"M" = Main Supply, "2" = Secondary Supply, "✔" = Minor Amount Only, and "X" = None.

Dest. = Any preplanned **destination** where you might store some family supplies.

Deposit Box = Your **Safe Deposit Box at the bank**. A fireproof safe would not be considered a separate location as it would be stored at your home.

Cache = Any **secure storage location** where you have decided to store extra supplies or gear. One cache example would be a **rented storage unit**. Another example might be a **waterproof container buried in a remote location**.

This small worksheet will not only help you **evenly distribute** your evacuation supplies and assets, but will help you and your family members **remember <u>where</u> you have gear distributed**. You'll find a full-sized blank copy in the "**Appendix**." On the sheet you fill out be sure to include all of the components of your Support and Secondary kits.

This worksheet applies to **specifically allocated evacuation supplies** and not to general possessions. Also, this does not apply to **isolation** supplies it's assumed you've logically located the majority of your isolation supplies at your primary residence. Most of your "**M**" notations will be where your **main or Primary Bugout Kit** is stored (you'll note in this example, most of the items listed are component kits of your Primary kit). However, not all gear will be kept in the main kit, and there's no guarantee you'll keep your main kit at home. Where do you spend most of your time? Where are your rendezvous points in relation to your evacuation routes and your home? You may find it smarter to keep your main kit elsewhere.

Where this worksheet tells you <u>where</u> things are distributed by item, the "**Evacuation Gear Inventory**" sheet in the "**Appendix**" allows you to record exactly <u>what</u> you've stored at each location.

Now that we've looked at where you might want to store things, let's go over a few miscellaneous details concerning each of the probable locations:

Home

> You'll want to keep your kit both **protected and accessible**. You'll want to keep it in an area that has a low danger of fire, no exposure to outside elements or extreme temperatures (due to food and meds), secured from theft, and either in your safe room or near an exit.

> Keeping your kit in your **safe room** will not only keep it protected, but it will give you access to some of your emergency gear during non-evacuation emergencies. (However, we recommend completely separate gear for evacuations vs isolation or other emergencies.)

> If you keep your kit(s) away from your safe room and in a small closet nearer an exit, you might want to consider giving this closet the same **structural reinforcement** mentioned for your safe room in the "Structural Considerations" portion of "Basic Home Prep." With regards to your kits and the possibility that your home might be destroyed while you're away, you'll want your kits to be protected and retrievable as opposed to destroyed or strewn about the neighborhood.

> In this same location, you might want to consider some **inexpensive fire protection** such as the "Range Queens" or "Dryer Watchguard" mentioned earlier. Though designed for other locations, these little fire systems *might* keep your emergency gear from being consumed by fire in destructive scenarios taking place while you're gone.

> An important consideration is **protecting your kits from theft** during a random burglary. You don't want all your emergency gear, spare credit cards, and personal information printed in your "Info Pack" to go out the door with a thief. Though you keep your kits accessible, there's no reason they don't have to look like standard luggage (unattractive to thieves), and they can be slightly hidden by hanging clothes, etc.

> Another consideration in where to keep your goods is, "**what is your primary risk**?" For example, if it's anything biochemical, you don't want to keep any of your goods stored outside where you might have to expose yourself in order to retrieve them. Keep them inside. Or, if your primary threat is flood, keep your goods stored high off the floor, or on an upper floor.

Car

> Since your main vehicle is more than likely your main evacuation transportation, you should **keep as much gear loaded in it as you can**, especially during the higher alerts.

> Try not to keep very much food or medications in your car due to **temperature** extremes, and what you do store, try to keep it insulated.

> When you have extra gear loaded in your car, especially valuables or sensitive info, be sure to exercise all **security precautions** at your disposal. Sudden emergencies, or even a higher alert, can make people do irrational things such as steal transportation. Don't lose yours.

> **Hint**: If traveling long distance and you've overloaded the car, stop at a safe distance from the emergency you're fleeing and ship some of your less-crucial gear to your location via FedEx, UPS, DHL, etc.

Work

➢ Your evacuation origin may be work, and your destination may well be home. If you work in a larger metro area that's the target of a terrorist attack, you may have to evacuate there and make your way home. Problem is, you might have to go the long way around. If that's the case, you'll want to keep goods always packed in your car, or carry a spare "**Bugout Briefcase**" (something halfway between your GYT Pack and your Primary Bugout Kit) to work. Make sure you carry items you know you'll need in an emergency. For instance, high heel wearers should keep comfortable walking shoes at work. If you think you might use alternate transportation, add a small cloth backpack to your briefcase kit so you can load your gear into that if you have to take your bike or kayak. **Note**: Always take your GYT Pack and your "Bugout Briefcase" with you when heading out work's door. There's no guarantee you'll be able to get to your vehicle or normal mode of transportation, or back into your office. Have everything ready to go and keep it with you.

Destination

➢ If any of your primary evac destinations are the private residences of friends or relatives, you should consider keeping some personal supplies stored at these locations.

➢ Our suggestion is that you keep a simple trunk full of assorted items listed in the "Support" and "Secondary" kits. Be careful about foods and meds based on the temperatures your trunk might be exposed to (in case it's stored in the attic and not an interior closet), or on how long it might be before you could refresh that kit.

Deposit box

➢ Your safe deposit box at the bank is actually a pretty good little storage location. Besides the originals of your important documentation and a full copy of your "Info Pack," you can store **another GYT Pack or a small survival kit**, and maybe a personal protection item or two.

➢ The only drawback to safe deposit boxes is that they're only available during **business hours**, so don't place any crucial emergency gear there.

Cache

➢ A "cache" is an extra or hidden storage location. It could be a rented storage unit, or something you buried in a remote spot in the woods. Whatever you choose, it should be **accessible 24 hours a day, 7 days a week,** and be protected from the elements and from thieves.

➢ If you can afford it and choose to rent a small **storage unit**, make sure the door has good lock, your gear is stored in **trunks** with locks, and the trunks are locked to each other or to spots inside storage unit (keep **keys** to the gate and trunks on your main key ring, in your car, and in your Bugout Kits). Looting will be a big problem in some scenarios and you don't want to find all your gear gone. **Hint**: Remember that storage units are subject to **extreme temperatures**. Don't try to store any food items other than the freeze-dried or dehydrated varieties. Even then, keep them in **plastic bags** (in fact, everything in the trunk needs to be in plastic for protection from humidity) and in the center of a trunk surrounded by clothing or other items that will insulate it somewhat from sudden temperature changes. Also remember that any water stored may freeze so be sure it's in plastic containers made for water storage and that you don't fill them more than ¾ full to allow for ice expansion. Keep your **water** containers upright and **wrapped in plastic bags** so they don't leak and damage other gear. **Hint #2**: If you wind up locking a lot of things, bite the bullet and spend a little extra on **combination locks** that can all be set to the **same combination**. You can still protect things from thieves, yet any family member can readily access the family's emergency goods at any location.

➢ Make sure any storage facility has 24/7 access and that the gate locks can be manually opened if a power outage has shut down keypads or other electronic access. Also, list the **owner / manager's contact info** on your "Important Contacts" forms. **Hint**: We'll mention this again under "**Cooperation**" but it's a great idea for a small group to go in on a storage location together. Everyone has a key to the main door, and inside the unit are **locker units** which are individually locked with only the locker's occupant having a key.

➢ If you decide to **bury some gear** somewhere, which is a very common practice, keep the following in mind:

❏ Make sure your container is both strong and completely **waterproof**. Water is your biggest enemy. Your second biggest enemy is the **weight of the dirt** that's going to be on top after you bury the thing. We suggest using a 4-foot length of 8-inch diameter **PVC** pipe with end caps glued in place and redundantly sealed with caulking and duct tape on the outside of the seams.

❏ Don't bury your unit in **ground that stays wet** regardless of how well you think it's sealed.

❏ Make a **desiccant pack** out of **charcoal** briquettes and regular dry **white rice** packed in a section of nylon stocking. Make the pack about the size of a softball. This will absorb a certain amount of humidity and incidental moisture inside the tube. (You should also do this for trunks you have stored at a storage facility.)

❏ Any **metal items** such as tools, should be given a very thin coating of **oil or grease** to protect them from moisture the desiccant pack can't absorb.

❑ Keep all internal items wrapped in plastic bags or plastic jars, and then everything wrapped inside a heavy plastic contractor bag, just as you would your main Bugout Kit, and for the same reasons.

❑ **Don't store batteries** in your buried unit. Anything you bury is likely to stay there for a long time, much longer than the lifespan of any battery. They'll only die and corrode putting your other stored items at risk of irreparable damage.

❑ Though you might store water acquisition and purifying equipment, **don't store water**. Containers may leak, and water will only stay good for so long. Your container might stay buried for a long, long, time.

❑ Watch the **weight of the container** after it's packed since you don't know how much energy you'll have after an emergency causes you to dig it up. Bury some items to keep you going, but don't go overboard, since you'll have to drag this thing out of a hole and lug it back to your vehicle.

❑ Make sure your location is **not likely to be dug up** during construction, etc. and that the area is seldom traveled. You don't want to lose your tools to someone curious about the freshly dug dirt and deciding to dig and investigate. Consider burying things underneath a garden, flower bed, or along a fence line.

❑ Also make sure the **cache location is clearly marked** on your Threat Map and the maps in your "Evac Atlas," and that key family members know how to find it if needed.

Sources

➢ The first place to look for packs and suitcases is at your local thrift store. After that, shop around for the best features and prices at the various supply locations we listed earlier at the end of "Contents in Detail."

➢ Storage trunks and the like can also be found at thrift stores.

➢ For PVC and other items you might need if burying some supplies, check your local hardware or building supply store.

Notes:

Planning the Destinations

We like to take things in a logical progression. We started this "**Evacuation**" section with the assumption that you were at home, or whatever your "Point A" happened to be. We had you start there and gather the things you'd need to take with you in an evacuation. Now that you have your stuff packed, the next logical question is, "Where's Point B?" Let's start our work on Point B by deciding, "Where ya gonna go?"

Goals of Planning the Destination:

1. To give your evacuation efforts a **decisive direction**.
2. To cover the necessary **criteria for selecting a suitable destination**.
3. To give you **more than just a single destination** and thereby give you options.
4. To **circumvent potential hazards and speedbumps** of getting you from point A to point B.
5. To help you **research and organize the many travel safety options** you'll need at your disposal.

We've divided this section into:

1. **Where Ya Gonna Go?**
2. **Planning the Route.**
3. **The Evac Atlas.**

Where Ya Gonna Go?

Since none of us truly know how the next attack or disaster is going to play out, the three most important things we need are options, options, options.

Continuing with "threes," we're going to ask you to choose **three** *types* of evacuation destinations within **three** *distance* categories. **Note**: Later on, we're going to ask you to pick evacuation routes in each of the four compass directions. Technically this could give you a potential twelve destinations. However, our goal is simplicity, so remember that the purpose of discussing these potential destinations and routes is to open your eyes to various types of destinations and the advantages and disadvantages of each. In reality, you may only have one or two possible destinations, but a variety of routes to get you there should your main route be compromised. Our goal is a choice of destinations within limits of logic and reason.

The three **types of destinations** are:

1. **Officially designated shelters**. The most recognizable examples of local shelters would be the older civil-defense fallout shelters, and more commonly, schools or other structurally sound buildings acting as a temporary weather shelter, or a temporary relocation center for people driven out of their homes by an area disaster. One other type of "shelter" would be a "tent city" set up to house refugees evacuating a massive disaster or attack. These tent cities may be set up in towns adjacent to an attack or disaster site, or may be placed on nearby public lands. Remember all the agencies we listed earlier under "Planning Ahead for Phase II?" These are the folks you'll need to contact in order to find out the location of existing or planned emergency shelters in your immediate area or region. You'll be asked to list these locations in your "Evac Atlas" we'll be creating later.

2. **Privately operated shelters**. We'll cover this a bit more under "**Cooperation**," but some groups may operate shelters and refugee centers for their members, and maybe for the public. Chief among these groups would be Houses of Worship, or Volunteer Groups specifically organized for emergency preparedness. One trend we're trying to foster is the "business as the community" where a business would treat their entire staff as family and set up shelters for their employees. In another example, a volunteer group might have a rental option with a farm in a neighboring state so that members would have a place to pitch their tents after a group evacuation. Destination choices, like many options mentioned in this book, are limited only by your imagination.

3. **Private property**. Private property can be broken down into another group of three: Friends' or relatives' homes, hotels or motels, or camp grounds. You'll already have your friends and relatives listed on your "Important Contacts" form, and we'll give you the contact info for some hotel and motel chains, as well as some campground contact info below under "The Evac Atlas." **Note**: As you may plan on having a family member or friend as your destination, you should plan on being *their* evacuation destination. Fully stocking your home with the supplies listed under "**Isolation**" should prep you to host for a while.

The **three distance categories** are:

1. **Local**. In the event of a locally destructive event such as a tornado or earthquake that damages your home, you'll need to go only so far away as there are no major secondary considerations to avoid (like fallout), and you'll want to be close by to clean up and rebuild, and maintain your job or school activities in the process. In fact, in those instances, your local "destination" may well be a temporary campsite set up in your own backyard.

2. **Regional**. Hurricanes are a great example of an evacuation scenario that would see you forced out of an immediate area and require you to travel quite a distance to avoid the storm. You'd need to find a destination far enough away for safety, but close enough to easily return after the weather cleared.

3. **Long distance**. The more cataclysmic the event, the more damaging, the more alarming, and the greater the secondary repercussions, the farther away from ground zero you might want to be. A good example is a nuclear detonation. You'll want to be as far upwind as you can from fallout, and you'll probably want to be completely out of the region due to civil unrest and other negative eventualities that might result from such a catastrophic attack.

Hint: For each threat you studied under "**Reaction**" you should determine the "minimum safe distance" for each danger. For example, with a nuclear detonation, you may have to travel 20 miles upwind, and with a hurricane you might evac 100 miles inland or to a hardened local shelter.

In addition to types and distance, we need to consider evacuation **direction**. The simplest suggestion is to pick an evacuation destination in **each of the four main compass directions**, and/or in directions dictated by local, area, or regional **travel bottlenecks or obstacles**. Different disaster and terror attack scenarios could see you hit from different directions, and each scenario could compromise one or more of your evacuation destination(s) or route(s). It's best that you be able to head in a variety of directions in order to react to a variety of threats. For example, a hurricane could come from one direction, while fallout from a nuclear detonation 50 miles away could drift into your area from another direction. Or, you may be in different parts of your city, county, or state for that matter, during different times of the day, and your current location at the time of an event will dictate the direction you have to travel.

Now that we've defined the different types of evacuation destinations and the distances we might have to travel to reach them, let's look at some of the important criteria for making our final selections.

For **each** of the <u>local</u> destinations (other than your backyard) you're considering, answer the following questions:

How to Choose a Local Evacuation Destination
The marked questions must have a "yes" answer. For the rest, the more "yes" answers the better.

		Yes	No	
Must be Yes	1			Is this location a designated shelter, commercial lodging facility, or friend's house?
	2			Will this location be under private or official control with representatives present?
	3			Does the building have adequate room, plumbing, food storage, and kitchen areas?
	4			Is this shelter away from any danger area noted on your Threat Map?
	5			Are the routes to this shelter always going to be passable 24/7? (No bridges, tunnels, etc.)
	6			Is the building structurally sound enough to survive the most common destructive events?
	7			Is the location pre-stocked with emergency supplies for emergency visitors?
	8			Does the site have its own auxiliary electric generator?
	9			Does this location have extra water storage on site?
	10			Will this shelter accept pets, or will a boarding area be set up?
	11			Is alternate transportation (public or private) available to and from the shelter?
	12			Do you know other people planning to use this shelter?
	13			Does everyone in your family know where this location is and how to get there?
	14			Can all family members get to this location from wherever they might be at any given time?
	15			Can this location be sealed against a biochemical attack?
	16			Can you store some goods and gear at this location in advance?
	17			Will you be able to leave this location of your own free will?

Be sure to record your local shelter choices on your "**Evacuation Destinations**" sheet and Threat Map and list them in order of priority or preference. It will be important for family members to know where everyone else is likely to be.

For planned **regional** and **long-distance** destinations, **include all the questions above**, and also ask the following:

How to Choose an Out-of-Town Evacuation Destination
The marked questions must have a "yes" answer. For the rest, the more "yes" answers the better.

	Yes	No	
Must be Yes 1			Is your destination far enough away to avoid any disaster you may flee?
2			Are the routes to this destination open 24 / 7 all year long with no travel obstacles?
3			Are you familiar with this location and its immediate area?
4			Is your primary transportation capable of taking the longest route there?
5			Does this destination have adequate services such as hospitals, food, water, etc?
6			Does this town have its own Police and Fire Departments?
7			Is your destination upwind or well away, from any wind patterns of your home area?
8			Is this destination accessible by at least two distinctly different routes?
9			Do you own property in this area?
10			Do you know someone at this location?
11			Do you have documentation showing you have a connection to this location?
12			Have you traveled each of the various routes to this destination?
13			If fleeing a terror attack, is this destination fairly safe as far as terrorism goes?
14			Are there alternate forms of transportation that run regularly to and from this location?
15			Does public transportation routinely use your primary route to this destination?
16			Does your destination score low on the "Emergency Risk Score?"
17			Are you part of a group that has chosen this location as its evac destination?
18			Is this location close enough to reach within 3 days by your primary transportation?
19			If this is a likely destination for many others, is there adequate infrastructure and supplies?
20			Are there adequate auxiliary services to handle any need such as pets' needs?
21			Do the routes along the way have an adequate number of stops with services?
22			Is this destination part of an official evacuation route or official list of destinations?
23			Is your area the only one likely to evacuate to this destination?
24			Though far away from danger, is your destination close enough to easily return home later?
25			Can you store some goods and gear at this location in advance?
26			If you leave this location temporarily, will you be allowed to return?

Now that you've considered all this, go ahead and **pick your potential local, regional, and long-distance destinations and write them down** (also noting the location of friends' houses that can be used as "safe houses"). Remember that you may be doing this for each of the four compass directions, so be sure to record your destinations in **order of preference**. This will help later when deciding on **routes** and **rendezvous** points. Note your chosen destinations on your "**Evacuation Destinations**" page, and in the **"Evac Atlas"** we'll have you creating in just a bit.

Keep in mind that although we started this section wanting "options, options, options," that there's genius in simplicity, and sometimes having too many choices only causes confusion. So, when planning all your possible evacuation destinations, pay attention to your options, narrow them down with the different criteria we've listed above, and apply your own common sense. You know your location, situation, potential threats, and probable reaction obstacles much better than we do, so ultimately, you're the only one who can decide on which potential evacuation destinations are truly workable and which ones are "maybes." The reasons we included this section are to give you food for thought concerning options you hadn't considered, bases you haven't covered, and questions you might not have asked.

By the way, when planning your evacuation, you'll want to make sure you know all you can about your destination. Be sure you include it when gathering all the information later as you create your "Evac Atlas," and not just simply list it as being the destination.

While you have all your maps out, let's look at the **routes** you'll take to get from point A to point B.

Now that you have an idea of where you might be able to go in an evacuation, you'll need to plan avenues to get there. We'll talk about actual transportation later, but for now, let's discuss routes.

Keeping all of the above considerations and criteria in mind, get out your local Threat Map, your county and state maps, and the maps of adjoining states to take a look at the places you're likely to be when an evacuation order is given (home, work, etc.). Next, look at the places you chose as evacuation destinations, and the various routes in between. Go over the maps keeping in mind all the above plus all the threats and bottlenecks you avoided on your local Threat Map, and apply the same considerations to your potential destinations and the routes to get you there.

There are actually two halves to your "route." The first half is your local travel that will get you out of the immediate area of a disaster, allow you to rendezvous with your family members, and then get from there to the transportation artery or main route that will start you on your way to your destination. The second half of your route is this main artery that we'll cover below. For now, let's focus on this first half and look at some things you'll need to keep in mind when formulating your most immediate travel plans:

1. First, get your Threat Map showing target areas, reaction obstacles, and travel bottlenecks.

2. Next, consider where you would be during the four most common time periods:

   ➢ One is when the family is at home at night, or other times when the family is normally together.

   ➢ The second is during travel to and from work and/or school and the corresponding times of day.

   ➢ Third comes the bulk of the day when family members are at work or school.

   ➢ Fourth is leisure time, such as on the weekends. Where do family members go for leisure?

3. Go through all of your potential threats, that you listed under "Emergency Risk Score" and on your Threat Map, and decide how each threat would affect your reactions, travel, and rendezvous, during each of the four time periods mentioned above. While matching threat to time periods, keep the following points in mind as the decisions you make in this planning session will dictate the reactions your family take in an actual emergency.

   ➢ On one of your maps, show the actual location of every family member during each of these time periods. These would be considered your "starting points."

   ➢ First, how many ways away from your starting point are there? For example, some neighborhoods have only one entrance. Will any scenario find you in a place where you could be boxed in?

   ➢ Besides the obvious traffic bottlenecks and obstacles, where will the most likely traffic jams be if this scenario happens at this time of day? Besides regular traffic, which roads would be tied up with emergency vehicles? Which areas might see civil unrest?

   ➢ How will various family members be paired up at this time of day? In other words, for each time period, which family members are in closest proximity to each other? For example, does Mom work on one side of a river near where one child goes to school, while Dad works on the other side of the river near where an older child works? Who's closest to whom?

   ➢ Which of the logical travel routes away from probable disaster sites will take you closest to family members you might be near, and that offer you the best chance of going to get them? (This will help you determine rendezvous points later.)

   ➢ Is there any single route or roadway that always seems to be open in each scenario at each time period? (Though choices are good, too many can sometimes lead to confusion. There's an advantage to simplicity.)

   ➢ How many different roadways could get you to the main road you wanted to take in that particular situation?

   ➢ Where would you go next? Consider this question for each potential disaster or attack. Would you need to leave the area entirely? If so, which way would you head? Would you merely take a circuitous route home? Or, would you head to a friend's house nearby to wait things out?

4. Once you have the answers to these questions, sit down with your family to make sure everyone knows the basic plan for various situations during different times of the day. Mark these local routes on your Threat Map and other maps, and keep these maps handy. We'll be using them again later as we plot potential rendezvous points.

Now that we've looked at the local "first half" of your evacuation route, let's turn our attention to the second half, or the part where you get on the main roadway that will take you out of harm's way and to where you want to be.

When reviewing the following, keep in mind that you'll need to pick **direct and alternate routes** to your intended destinations. The direct routes will be better, but there's no guarantee they'll be accessible, or as useful as you hope, so you should always have options at your disposal. Depending on the threat, you might also have to take a round about route to get to your destination in order to avoid secondary hazards or obstacles resulting from the disaster.

Apply the questions and criteria below to each of the routes you're considering:

How to Choose an Evacuation Route
The marked questions must have a "yes" answer.   For the rest, the more "yes" answers the better.

		Yes	No	
Must be Yes	1			Have you collected maps of this route?
	2			Have you compared this route to your Threat Map criteria, and is it safe?
	3			Does this route have adequate services along the way? Food, gas, medical, etc?
	4			Does this route avoid major bridges, tunnels, or railroad crossings (if at all possible)?
	5			Is this route open and passable 24 hours a day, 7 days a week, 365 days a year?
	6			Can your primary transportation handle this route?
	7			Is this route part of an official evacuation route system?
	8			Have you traveled this route before?
	9			Is your primary transportation capable of making the distance?
	10			Can you get off this route or change directions if desired?
	11			Do other forms of transportation, such as bus lines, use this route?
	12			Do these alternate forms of transportation have stops available along the way?
	13			Is it possible to reach your destination using Interstate Highways as an option?
	14			Are you able to completely avoid Interstate Highways if so desired?
	15			Is your route simple enough to easily give directions to others?
	16			Is your route immune to primary or secondary effects of any disaster possible for your area?

After you've decided on your destinations and routes to get to each, mark everything on each applicable map. Next, list all of your destinations and each route in **order of preference, or at least in order of probability**. This will help you decide on your "Rendezvous Points" later.

Keep your maps handy. You'll need them to complete your "Evac Atlas" that we'll now get around to explaining.

The Evac Atlas

There's no such thing as knowing what's going to happen in any emergency, large or small. As we've said before and will continue to point out, all urgent situations are fluid and dynamic with an ever changing life of their own. We won't know what's in store for us from one instant to the next, so we need to be able to react and adapt quickly. One way to react quickly in an evacuation situation is to have as much information as possible concerning your options, alternatives, and sources of assistance along each of your evacuation routes. The "**Evac Atlas**" is a collection of that very information. Though your Bugout Kit and other preparation measures are geared toward keeping you ready when <u>nothing</u> else is available, you should be just as able to utilize the services and sources that <u>are</u> available.

The heart of the "Evac Atlas" will be the **forms** you find in the "**Appendix.**" What we'll give you here is an overview of **what the forms will entail**, and some **sources for finding the information**.

A LOT of information is going to go into this Evac Atlas. If you're taking off for a while because your house burned down or was demolished by a tornado, will you need this much detail? Maybe, for your local area. If you're temporarily relocating because of a hurricane, are you going to need this kind of information? Probably. During a panicked and rapid mass evacuation, such as after a nuclear strike, where hundreds of thousands of others are going to be hitting the roads, do you need all this data? **Absolutely!** Disasters and terror scenarios aside, what about a family emergency where you needed roadtrip assistance and had **ZERO** time to waste looking around for help?

Your Evac Atlas can be described like this: **It will be your own customized travel atlas of emergency services and supplies available along each route to each of your chosen evacuation destinations**. Where your Threat Map showed threats to your safety, the Evac Atlas will locate assets that will help *ensure* your safety. Above, we asked you to choose out of town evacuation destinations and to pick a main and alternate route to each location. We're going to put together a detailed collection of useful information on each of your chosen destinations, and each of the main stops along the way of each route.

This is a lot of information to gather. But, knowing where you can immediately find an emergency item or service can make all the difference on whether your evacuation is a tragedy of errors, or an adventurous excursion. The one asset that's definitely **not** at your disposal in an evacuation is **time**, so if you need emergency assistance or supplies, you'll need to know where to find it immediately, without having to waste time searching, and without counting on the phone systems working. You also need to know all alternative sources should a mass evacuation drain area supplies.

What if you're forced out of your home area without your gear and you need to stock up enroute? What if you break a toe in your mad dash to load the car? Where's the first doctor listed as one of your insurance company's providers along one of your routes and outside any danger area? If a major malfunction occurs with your vehicle, where are the qualified service locations along your evacuation route? What are the emergency info radio station channels of the little towns along your route? Which towns have hospitals with emergency rooms? Which have "immediate care centers?" If your vehicle absolutely breaks down, where are the alternate transportation stations or depots? Which of any of these services are found at your intended destinations?

The list of things you might need to know could go on and on. In fact it will. Let's look at a checklist and the info sources that follow, that you'll be using to fill in your "Evac Atlas" forms found in the "**Appendix**."

The "Evac Atlas" Data Items

Medical & Emergency Services (Map code "M")

❏ Hospitals w/ emergency rooms	❏ Immediate care facilities	❏ Insurance Co. Providers
❏ Dentists	❏ Drug stores / pharmacies	❏ Veterinarians
❏ County Health Dept.	❏ Motorist Aid call boxes	

Law Enforcement and First Responders (Map code "L")

❏ Emergency number, if not 911	❏ Police Dept. non-emergency No.	❏ Sheriff's Dept. non-emergency
❏ Fire Dept. non-emergency No.	❏ Emergency Management office	

Food and Water (Map code "F")

❏ Grocery stores	❏ Chain restaurants	❏ Locally owned restaurants
❏ "Mini Marts"	❏ Farmer's markets	❏ Pet supply stores
❏ Ice houses or plants	❏ Fishable lakes and streams	

Transportation (Map code "T")

❏ Gas stations	❏ Mechanics / garages	❏ Towing services
❏ "H.E.R.O." units	❏ Auto parts outlets	❏ Authorized vehicle dealers
❏ Travel club affiliates	❏ Bus stations	❏ Car rental facilities
❏ Public airports	❏ Private airports	❏ Taxi companies
❏ Train stations	❏ Navigable waterways	❏ Docks and marinas

Communication (Map code "C")

❏ TV stations	❏ Emergency radio stations	❏ Regular radio stations
❏ Cellular "touch star" services	❏ Online traffic info sites	❏ Local municipal websites
❏ Cellular service outlets	❏ Local PD & FD scanner codes	❏ White and Yellow page copies
❏ Post office locations	❏ Private mailing / fax centers	❏ Western Union locations
❏ Public libraries / Internet Cafes	❏ Payphones	

Shelter and Lodging (Map code "S")

❏ Official emergency shelters	❏ Hotels and motels	❏ Friend's or relative's house
❏ Camp grounds & RV parks	❏ Rest areas	

General Resupply (Map code "G")

❏ Branches of your bank(s)	❏ ATM locations	❏ Check cashing centers
❏ "Super center" stores	❏ Malls and shopping centers	❏ Sporting goods outlets
❏ "Dollar" stores	❏ Thrift stores	❏ Hardware stores

Relief and Support Agencies (Map code "A")

❏ Volunteer group reps	❏ Local relief agency offices	❏ Charitable organization offices
❏ Gov't relief agency offices	❏ Relief missions & public shelter	❏ Charitable "Church" groups

County and City Administrative (Map code "P")

❏ City Hall	❏ County Courthouse	❏ County Commission office
❏ Chamber of Commerce	❏ County Gov't. Complex	❏ County Co-op or Extension

Overview

Under "Discussion Items," we're going to list all the checklist items above, give a short explanation of what each means, why you might need the info, and then we'll follow each with some sources you can use to find the necessary data without having to do much actual leg work. First, let's go over a few key points concerning the Evac Atlas itself.

1. Like everything else, your "Evac Atlas" should **start at home**. Begin by marking the location of all the above services, and assets on your current **Threat Map**. This will help you plan your local rendezvous points and to quickly find local assistance to help you in a local disaster.

2. **Regardless** how you plan to evacuate an area, by plane, boat, bus, train, or automobile, you should prepare an Evac Atlas for **each anticipated route**. Suppose you evac by private aircraft. Which towns along your flight path(s) offer a hospital with emergency room? Which have a service center able to service your type aircraft? Where can you switch from air to ground transportation the easiest? If your plan is to take your pleasure boat and skirt along the coast to another city, which stops along the shore offer services you might need? Besides, if you're evacuating due to a terrorist attack, you don't know if your travel modes will have to change. The FAA may ground all private flights, or the Coast Guard may order all private craft in to port. "Where ya gonna go?"

3. Your Evac Atlas will actually be made up of **several smaller Evac Atlases, each one dedicated to one particular route to one of your intended destinations, to the destination itself, and to all the stops along that route**. For example, if you were evacuating Savannah, Georgia, you could take an Interstate highway in three different directions, or you could evacuate by boat. You could evacuate north to Columbia, South Carolina, west to Macon and eventually Atlanta, or south to Jacksonville, Florida. You'd create an Evac Atlas for each of these three land directions, and one for each boat direction. Each Evac Atlas would show all the above information for each stop along that route, and for the final destination.

4. Each smaller Evac Atlas you create should have the following:

   ➢ A **map of the single route** this Evac Atlas applies to. This map will show each of the stops along the way for which the above information has been collected, and will show the location of the more important facilities such as rendezvous points, hospitals, food sources, banks, etc. That's why each category in the checklist has a **"Map Code" letter** by it. On your map you can note which location has an important facility just by using the letter-code and drawing a small arrow. (**Note**: You won't need to mark the position of each and every service, only the most important of the ones you might need to actually visit such as a hospital ER. Locations such as radio stations don't need to be marked, just record the station channels on your forms.)

   ➢ You should also include a **county** map and a **local street** map of each location so that the listed services and assets can be physically located if no other form of communication is available.

   ➢ In addition to the stops and sources you'll want to list, you'll also want to treat these maps as **expanded Threat Maps** and show everything on them that you did on your **local** Threat Map you created earlier. Chief among these threats would be traffic obstacles that might box you in or cause you unnecessary delays in getting back on the road.

   ➢ For each stop along the route, all the above information will be collected for that location and the data recorded on the forms provided. This way, **regardless of which route you wind up taking, you'll have all the information necessary** for quickly contacting and/or finding important services and sources at any given point in your travels and you won't waste valuable time or effort searching for things you'll need immediately.

5. Though there's a lot of information needed, we do suggest a **limit**. We feel that you'll only need your *detailed* Evac Atlas to cover the first **300 miles** of any route (keeping in mind "minimum safe distance"). Beyond that, it's likely that you're past the point of needing *rapid* emergency assistance, evacuee traffic will have probably dissipated somewhat, and if you need some sort of resupply you'll probably have more time to search. **However**, you're never out of the woods until you get where you're going. So if you enjoy creating these maps and locating sources, keep going. Knowledge is power and the right kind of knowledge in an emergency situation is priceless.

6. However, we do suggest you **prepare an atlas for your final destination** as there's no telling how long you'll be there, and this information will make it easier to you to utilize services if you're not familiar with the area.

7. **Note**: It's VERY important that you pay extra **close attention the smallest of towns** along your route, especially the ones where you'd think, "Oh, nothing useful could be there..." Here's why: Everyone else on the road will be thinking, "Oh, nothing useful could be there..." and while they're all waiting for hours in line at more publicized service centers, you're taking advantage of the services offered at small "Mom & Pop" locations where there is no waiting line. Time factors aside, the lesser known locations are less likely to run out of supplies. Think about this when you're listing ATMs, restaurants, grocery stores, auto service centers, or anything from the checklist above.

So, how do you go about collecting this information? If this were a perfect world, you'd be able to take some vacation time to travel each of your routes and stop along the way to tour each town and see what they have to offer. If you can do this, great (as you'd learn more and would probably make some new friends), but most of us can't so let's look at some different sources for **finding as much of this information as possible without leaving the house**. Let's start with some sources that will help you find what you need with all of the above categories of information. Then we'll look at some specific sources as we discuss each of the checklist items below. (You can also enlist the aid of the people at your final destination since *you* may be *their* destination if the tables are turned, and they'll be using the same routes and will need the same information.)

When contacting these general sources, let them know exactly what you're doing, the specific route you're researching, and that you're needing information regarding each of the categories in the checklist above. It's doubtful that the larger state and county-level sources will have <u>all</u> the information you need, so you should also ask them for useful publications. Among these are: ❏ **Phone books** (if you can get them for free), ❏ a copy of the **local newspaper**, preferably the edition containing the most ads, ❏ a copy of any **leisure publications**, ❏ "**Welcome Wagon**," "**Mover's Guide**," or "**Relocator's Information Packs**," ❏ a **county map**, and ❏ a **city map**. Let's look at some of these general sources:

➢ **Start your search online**. The easiest way to search is on the net without leaving your desk. We've located a few comprehensive sources that will not only allow you to find the government entities we suggest you contact, but will let you find the direct business or location you're looking for. Later on, we'll give you specific websites and service locations that will help you pinpoint your information search. For now though, let's look at the websites you'll want to visit first:

● You can access most state's website with this address, http://www.state.XX.us, where you replace "**XX**" with your state's two-letter abbreviation.

● Yahoo has a great state and city info site at: http://local.yahoo.com/u_s_states. After clicking on a state, click on the "**City Guides**" link under that state. Also see http://www.hometownlocator.com.

● Global Computing has a good info site at: http://www.globalcomputing.com/states.html.

● A similar service is offered by: http://www.50states.com.

● Roadmaps.org is a good place for mapping links as well as informative links for states, counties, cities, and travel info: http://www.roadmaps.org/links/state-links.html.

● The private group "US Highways" has a links collection of travel-related services that can be found along our numbered highway system. They're online at http://www.us-highways.com. A related site can be found at http://www.roadguides.com, and a cousin of theirs shows info about the I-95 corridor, with other travel-related info at http://www.usastar.com.

● Try the "Ultimate Trip Planner" at http://www.theultimates.com/trip.

● For some of the searches you'll be doing, you'll need **zip codes**. The US Postal Service will let you search them online at http://www.usps.com, or call (800) ASK-USPS (275-8777).

● You can also use the **online phone books** located at http://www.yellowbook.com, http://www.realpages.com, http://www.whitepages.com, http://www.switchboard.com, http://www.smartpages.com, http://www.yellow.com, http://www.therealyellowpageslive.com, http://www.infobel.com/usa/default.asp, federal blue pages at http://www.usbluepages.gov, http://bp.fed.gov, or the multi-government info page at http://usofa.com/pages/government.

● The site http://usofa.com/pages/government will let you search federal, state, county, or local government websites, as will the other US Government sites we gave you under "Information Research."

● As for **newspapers**, revisit all the news links we gave you under "EAS Alerts."

● **Note:** We've provided many more **links** in the **two links collections** on our **enclosed CD**.

● **Hint**: Anytime you're looking up information online, you can usually **copy and paste** the information from a **website** into a **wordprocessor** file to **save and print** it that way rather than fill out our forms long hand. If you choose the copy and paste option, be sure to print your lists to go in your Bugout Kit. Saving to disk is fine for a backup, but, during an evacuation, there's no guarantee you'll have your computer with you, or that other computers will be accessible or have electricity. The same holds true for the Yellow Pages selections. Cut out pages and sections of the services you'll need and tuck them in your kit rather than rewrite all the information into the forms.

➢ **State Department of Travel and Tourism**. Check your state's website to see if they have a listing for such a department. Other titles are "Conventions and Visitors Bureau," Tourism and Industry," and so on. For our friends without internet access, call your state capital and ask for the information. Also, the larger towns along your route might have their own tourism office.

➢ **State Highway Department or Department of Transportation**. If the state tourism office doesn't have the information you need, your state's DOT might. After all, they're the ones that usually engineer traffic flow at interstate exits, and might have a listing of businesses and services at each. Each state is going to be different, so you have to check all sources. New Jersey's Department Of Transportation site maintains a list of links to each state's DOT at http://www.state.nj.us/njqi/Transportation/OtherTransLinks.htm, and Roadmaps.org offers the same thing at http://www.roadmaps.org/links/state-links.html.

- ➢ **Your state's Congressional delegates in DC**. Senators and Representatives love to keep their constituents happy, and many will either have quite a bit of free information, or will know where to send you. You can find congressional addresses at either http://www.house.gov, http://www.senate.gov, http://thomas.loc.gov, http://www.vote-smart.org, or http://www.firstgov.gov/Contact/Elected.shtml or you can call the congressional switchboard at: 800-839-5276, or 202-224-3121.

- ➢ **Your area's rep in your state legislature**. If you're having a hard time finding the information you need, your local representatives of your state's legislature should be able to offer some good advice. Most **state websites** will offer links to the legislature. If yours doesn't, you can try: http://www.vote-smart.org, or http://www.firstgov.gov/Contact/Elected.shtml.

- ➢ Another source of governmental contact info can be found online at http://www.statelocalgov.net/index.cfm.

- ➢ **Department of Commerce**. You should be able to find one at a state level, and in some instances it'll be combined with the tourism offices listed above. These people will be a treasure trove of good information and leads to other sources.

- ➢ **Census Bureau**. When planning your evacuation, especially your destination, you'll want to make sure you know all you can about it. One source for a good variety of data is the US Census Bureau. They're online at http://www.census.gov, or you can contact them directly at U.S. Census Bureau, 4700 Silver Hill Road, Washington DC 20233-0001, 301-763-INFO (4636).

- ➢ **Chamber of Commerce.** Most cities will have a Chamber of Commerce, and if the town is too small, the county will certainly have one. Ask them for a **phone book**, a **county** and **city map**, and the **contact info** of the various **civic organizations** in that area. Some of these organizations may have decided to offer the kinds of disaster assistance services we mention in our "**Cooperation**" section. We also found you two useful sites: http://www.2chambers.com, and http://www.uschamber.org.

- ➢ **State EMA.** Depending on how advanced they are, your state's Emergency Management Agency might have all this information listed for you. You should also go back and contact each of the **offices and agencies** listed earlier under "Planning Ahead for Phase II" (especially for NEMA). Some of these offices and agencies in your area may have already noticed this need and gathered the necessary information. Here's a link to each state's Office of Homeland Security: http://www.ready.gov/useful_state.html.

- ➢ **The County Commissioner's office**. In some cases, you'll find this to be a close cousin to the Chamber of Commerce. Utilize both. (You'll find listings in your phone book's blue pages and on the county websites.)

- ➢ **The Department of Labor**. In some states the Dept. of Labor will have relocation information packages available to people moving to new areas in the state. Relatedly, local **Economic Development Councils** might have the same thing. Look these various offices up online or in your phone books.

- ➢ **County Co-Op or Extension office**. Many assistance programs run by and/or through a county will be associated with the County Co-Op or Extension office. Call the offices in each county you may be evacuating through and tell them which information you're trying to gather, and ask about any planned victim assistance they might offer in the wake of a regional disaster. They might give you contact or procedural information for things not even on this list.

- ➢ **City Hall**. After you've contacted the Chamber of Commerce at each location you're researching, you should call City Hall. They'll be able to provide you whatever it was the C of C couldn't. The two most important things you'd want at this level are a **local phone book** and a **city map**, as well as suggestions for more info sources. You can call various City Halls, or stop by if you're passing through. You'll probably get more info by stopping by. Also, ask City Hall about local Civic Organizations and Volunteer Groups that may be of assistance in a disaster.

- ➢ **Moving companies** may also have a certain amount of relocation information and city info.

- ➢ **Your roadside assistance or travel club.** If you're a member of AAA or similar, you can ask them for travel planning and route info help. Let them know exactly what you're gathering info for and you'll probably find them more than willing to go the extra mile in helping you prepare.

- ➢ **AAA** autoclub offers a Palm-Pilot type device called the "**Auto Pilot,**" containing stored data on various gas, food, lodging, and emergency services listed and organized by route and exit. Call them at 800-322-2100, or you can see it online at http://www.aaaautopilot.com.

- ➢ A similar device, called the "**Road Whiz**," is available online at http://roadwhiz.ultimatecorner.com/index.cfm, or by calling 800-747-2605, ext. 46.

- ➢ **Your local library**. The research department of your library will probably have travel atlases and various city-info books, as well as phone books for locations in your region or immediate area. If you travel your routes and stop at City Halls, you should also visit the local library to look at their information.

- ➢ **Ride the routes**. As we said, the best way to gather this info is to travel the routes and stop and look around. If you do this, not only should you stop and gather all the above info from all the mentioned sources, but you should **photograph and videotape your trip**. You'll find things in the pictures you might not have noticed, and seeing the images again will refresh your memory as to the location of certain places and services. You might also take some of these pictures and paste them into your atlas. If you have children, take them on these excursions and tell them what you're looking for. You'd be surprised at what kids are able to notice and remember.

Discussion Items
As you gather this information you'll want to have your **Evac Atlas forms** from the "**Appendix**" handy so you can fill them out as you go. Though the directions will be repeated on the forms, don't forget you'll want to: ❏ **Mark singular locations on your main map.** For example, not every small town will have a hospital with an ER, so you'll want to note on your map which ones do (that's why we gave you a "**Map Code**" letter with each category). ❏ **Record the address and contact info** for each location you find, since you'll want to be able to call them ahead of time if phones are working, or find them quickly on your own if phones are down. ❏ **List *why* a source or location was singled out.** For example, you might list a specific restaurant because they can cater to your dietary needs. You'll want to write this down for two reasons: One, in an emergency, you can't be counted on to remember everything, and two, this information might be used by someone else who's helping you.

Medical & Emergency Services (Map code "M")

1. **Hospitals with Emergency Rooms**. That blue "H" sign might send you 20 miles north when the next exit 5 miles away has a better hospital closer. You'll need to know your options in advance.

   ➢ Since knowing where all the ERs are is one of the most important topics, it should be the first thing looked up in each phone book or local information packet you receive from the towns and stops along your route.

   ➢ Ask your health insurance provider for a listing of authorized or preferred providers in the areas you might travel. This will give you a listing of hospitals.

   ➢ For general **online hospital locating services** try:

   ● "DoctorDirectory.com, Inc.," offers an extensive hospital locator service. They're online at http://www.hospitaldirectory.com/, or you can contact them at DoctorDirectory.com, Inc., 32 Broadway, Suite 200, Asheville, NC 28801, (p) 828.255.0012, (f) 828.255.0442.

   ● You can also search the "American Medical Association," or AMA, website at http://www.ama-assn.org/ama/pub/category/3158.html, and search for providers by location.

   ● Visit HospitalWeb's listing of thousands of hospital websites and related information. You'll find them online at http://neuro-www.mgh.harvard.edu/hospitalweb.shtml. Click on "USA Hospitals."

   ● Try ER365, online only at: http://www.er365.com/find-a-hospital-emergency-room-365.htm.

   ● American Hospitals, online only at: http://www.americanhospitals.com/hospitals.

   ● Use Allhealth's online locator at: http://www.allhealthnet.com/hospitallocator.htm.

   ● Though primarily a job-search site for nursing, the "Hospital Soup" website offers a decent search page at http://www.hospitalsoup.com/hospitalsearch.asp. To find emergency rooms, select "Emergency Services" where it asks you to "Search by Facility Category."

   ● Try the Hospital Connect page at http://www.hospitalconnect.com/cgi-bin/mqinterconnect.

   ● Visit the American Hospital Directory at http://www.ahd.com.

   ● Also try Hospital Link at http://www.hospitallink.com.

   ➢ Each state will have its own **medical licensing board** and hospital accreditation board. Most of the time you can find this through the state's **Secretary of State office**. They'll have a listing of every medical facility in the state including the immediate care facilities ("Doc in the Box") and dentists. You can find a link to your state's Secretary of State office through the National Association of Secretaries of State, online at http://www.nass.org/sos/sos.html. Otherwise, look for yours in your phone book's blue government pages.

   ➢ The site, http://www.the911site.com, offers a comprehensive links and connectivity service to numerous emergency service sites and information.

   ➢ The "Wellness Concierge," online at http://www.wellnessconcierge.com, offers a great deal of health information for travelers. While most of this centers on leisure and business travel, there is still quite a bit of directly applicable evac information.

2. **Immediate care facilities**. Your minor ailment that flared up while on the road might not need a full emergency room, and it's best to avoid crowded ERs anyway. You'll want to stay out of their way, and you don't want to pick up a germ for something nastier than what you're trying to cure.

   ➢ The best way to find these care facilities is through the state boards and sources listed above or through the local phone book. Make sure you've marked these locations on your city maps.

3. **Physicians**. Though most insurance companies will let you use any health care provider, especially in an emergency, many offer much better rates for using those providers on their preferred list.

> ➢ Ask your **health insurance provider** for a list of preferred providers in the areas you might travel.

> ➢ Show your planned evac routes to your regular physicians and ask them about colleagues, affiliates, and any network members you might find along the way.

> ➢ "DoctorDirectory.com, Inc., offers an extensive locator service. They're online at http://www.doctordirectory.com, or you can contact them at DoctorDirectory.com, Inc., 32 Broadway, Suite 200, Asheville, NC 28801, (p) 828.255.0012, (f) 828.255.0442.

> ➢ Search the Doctor Page, online only at http://www.doctorpage.com.

> ➢ Visit DocFinder, online-only at http://www.docboard.org.

> ➢ You can also search the public section of the "American Medical Association," or AMA, website at http://www.ama-assn.org/ama/pub/category/3158.html, and search for providers by location.

> ➢ Also try the online locator service at http://www.unmc.edu/library/consumer/docfinder.html.

> ➢ Many **travel and roadside assistance clubs** offer a small form of medical, or emergency insurance or assistance. Ask if they use specific providers, or if they can list all providers by location.

> ➢ Once you find your out-of-town doctors, check them against the disciplinary register at http://www.questionabledoctors.org. Don't want to flee one problem only to run into another now do we?

> ➢ A similar "doctor discipline" site can be found at http://www.healthgrades.com.

4. **Dentists**. A toothache or other dental problem is just as annoying and debilitating as many other ailments. Know where your treatment sources are.

> ➢ If you have dental insurance, ask for a listing of all providers along your probable travel routes. Remember that your **routes may cross state lines** so make sure you get all the data you need.

> ➢ The American Dental Association may be able to help you find dentists along your planned route. They're online at http://www.ada.org, or you can contact them at American Dental Association, 211 East Chicago Ave., Chicago, IL 60611-2678, 312-440-2500.

> ➢ The Dental Care Net is online at http://www.dentalcarenet.com/locate.phtml, or you can reach them at PO Box 219, Hyannis, MA 02601-0219, 1-800-949-7372, FAX 508-778-2217.

> ➢ Another seemingly thorough service, complete with a mapping resource, can be found online at http://www.1800dentist.com, or by calling 1-800-DENTIST (336-8478).

> ➢ We've also found a couple of good online dentist locators for you at:

> > • Visit Dentists.com at http://www.dentists.com, or by calling 1-800-436-8977.

> > • WebDental has a good search function at http://www.webdental.com.

> > • Visit http://www.emergency-dentist.com.

5. **Drug stores / pharmacies**. In addition to prescriptions you may get from the hospitals or immediate care facilities above, you might have packed a pre-issued written prescription for an important or non-storable medication such as insulin. In any event, you'll want to be able to get any prescription filled as quickly and easily as possible so you can be on your way. You'll also want contact info for pharmacies, so that your primary care physician can call in a prescription to have it waiting on you.

> ➢ Again, your **insurance provider** may have a list of participating pharmacies or its own online locator service.

> ➢ When showing your evac routes to your physician for doctor and hospital info, ask them about pharmacies.

> ➢ Along with insurance info, check your other memberships and travel club benefits to see if they offer a prescription discount and/or pharmacy listing.

> ➢ Below, we'll ask you to compile a list of **grocery stores**. You should note which ones have a pharmacy.

> ➢ Next, check the **national offices of various drug store / pharmacy chains**, or stores that have pharmacies as part of their service, to get their individual listings:

> > • CVS Pharmacy has an online locator at http://www.cvs.com, or you can call 888-607-4287.

> > • Eckerd's online locator is at http://www.eckerd.com, or call 1-800-Eckerds (1-800-325-3737).

> > • Walgreen's Drugs, online at http://www.walgreens.com, or call 847-914-2500.

> > • Safeway, online at http://www.safeway.com/PharmacyLocater.asp, or call 1-877-SAFEWAY (723-3929).

> > • Rite Aid's locator is at http://www.riteaid.com/stores/locator, or call 1-800-RITE-AID (1-800-748-3243).

> > • Wal-Mart pharmacies can be located at http://www.walmart.com, or call 800-966-6546.

> > • K-Mart pharmacies can be located at http://www.kmart.com, or call 1-866-KMART4U (1-866-562-7848).

> ➢ We've also found a couple of **independent online pharmacy locators**:

- Blue Cross & Blue Shield at http://www.bcbsmo.com/pharmacy_finder/PharmacyFinder.asp.
- "Fine Pills" at http://www.finepills.com/related_pharmaceutical2.php.
- American Health Care pharmacy locator http://americanhealthcare.com/pharm_Locator.html.
- "Serve You" locator at http://www.serve-you-rx.com/search.php3, or call 1-888-243-6890.

6. **Veterinarians**. Pets are part of the family. Treat them as such. If one of yours becomes sick or injured during your trip, take care of them as soon as you can. **Hint**: If you have large animals such as horses, you should locate all large-animal feed and supply stores, boarding stables and facilities, and qualified veterinarians. You should also locate potential transportation for them now and record the transportation's contact info on your "Important Contacts" pages.

   ➢ Show your potential evac routes to your regular vet and ask them to help you locate colleagues and affiliates along the way, especially those offering **24-hour emergency services**.

   ➢ Veterinarian insurance is becoming much more common place than in the past two or three years. If you have such insurance, ask them for a **list of providers**.

   ➢ We've found a few pet insurance sources to let you learn more about the concept and to comparison shop:
   - Veterinary Pet Insurance, online at http://www.petinsurance.com, or call 800-USA-PETS (800-872-7387).
   - Pet Care, online at http://www.petcareinsurance.com, or call 1-866-275-PETS (866-275-7387).
   - Petshealth Care Plan, online at http://www.petshealthplan.com, or call 1-800-807-6724.
   - Pet Assure, online at http://www.petassure.com, or call 1-888-789-PETS (888-789-7387).
   - Premier Pet Insurance, online at http://www.ppins.com, or call 877-774-2273.

   ➢ There will also be **state boards** covering veterinarian licensing, and many will be searchable through the Secretary of State's office.

   ➢ Go back and check the website and source listings we gave you for pet care under the Pet's "Support Kit" portion of "Prepping the Gear" earlier. Several will have veterinarian locators.

   ➢ In addition to veterinarians themselves, you'll want to find the locations of independent **boarding and care** facilities. Though pets are a part of the family, you may be in a situation where you have no choice but to leave them at a safe place to be cared for and picked up later. If you're faced with that kind of choice, you'll need to know exactly where to take them.

   ➢ See the Veterinary Emergency and Critical Care Society, online at http://veccs.org, with direct contact at 6335 Camp Bullis Rd, Suite 14, San Antonio, TX 78257, Phone: 210-698-5575, Fax: 210-698-7138.

   ➢ Pets 911, who has an affiliation with the US Humane Society, offers a locator service for pet services, shelters, adoption centers, pet friendly lodging, and a host of other services. They're online at http://www.pets911.com, with a quicklinks page at http://www.pets911.com/partners/hsus, or you can call them at 1-888-PETS-911 (888-738-7911).

   ➢ For some various **vet locators** try:
   - Try the Vet Locator, online at http://vetplace.com.
   - Emergency Pet Clinics at http://www.petswelcome.com/milkbone/vetmap.html .
   - Emergency vets, http://www.1888PETS911.org, online, or call 1-888-PETS-911 (738-7911).
   - For emergency clinics, search http://www.healthypet.com, or call 800-252-AAHA (2242).
   - State by state listing at http://www.talktothevet.com/VETERINARIANS/veterinarians.HTM.
   - A variety of links can be found at http://vetmedicine.about.com/cs/vetfinders/index.htm.
   - **Banfield** and **Petsmart** have a relationship and each has an online store locator. Find Banfield locations at http://www.banfield.net, or at 1-800-838-6738. Find PetSmart locations at http://www.petsmart.com, or by calling 888-839-9638.
   - See the Humane Society Disaster Center page at: http://www.hsus.org/ace/18730.

7. **County Health Dept**. Though you'll probably only be eligible for treatment in the county you reside in, county health departments are a good source of information and you'll want to record the contact info for the departments of each county you'll travel through, and **especially your destination**. If nothing else, the health department will be a good way to keep tabs on current health problems and epidemics if that's why you're evacuating. They'll also be a good source of info regarding emergency rooms and immediate care facilities should things change since you created your Evac Atlas.

   ➢ Most state websites will have links to counties within that state, and the county websites will have health department listings. You can also use your phone book's blue government pages.

   ➢ The CDC maintains an online listing of Health Departments at http://www.cdc.gov/other.htm#states, and at http://www.cdc.gov/mmwr/international/relres.html.

> You can find your state's Public Health director and office online at http://www.statepublichealth.org.

8. **Motorist Aid call boxes**. Many states have a program where every mile or so there's a roadside telephone that will automatically connect you with emergency services. When making your Evac Atlas maps, be sure to note which sections of which highways will have these phones. You'll want to know where they start, where they stop, etc. (Remember, your trip may take you across state lines into areas with a different way of providing services.)

> The two best places to find this information is through the state's Department of Transportation, or Law Enforcement agencies within that state or county.

Hint: Since these medical assets are so important, you might give them a better notation on your map than the simple map code "**M**" we suggested above. For example, use "**M-ER**" for Emergency Rooms.

Law Enforcement and First Responders (Map code "L")

9. **Emergency number if not 911**. Odd as it may seem, at the time of this writing, not all areas of the country are on board with a 911 system. Many of those not on board are the small towns we're asking you to gather info on.

> When contacting the various Chambers of Commerce and City Halls, ask them about emergency numbers.

> If different departments have their own emergency number, collect the numbers for Law Enforcement, Fire Department, and Ambulance Service, etc., putting each on the appropriate Evac Atlas page.

10. **Police Dept. non-emergency number**. As we'll discuss later under "Life on the Road" there are several reasons you'd want to contact the administrative offices of a Law Enforcement agency. One evacuation example is knowing the accessibility status of a smaller town, its hospital location, or any other emergency source or service. In a massive evacuation, calling the local PD is the quickest way to find out if any service is available or if it's best to continue on to a town farther out. (Naturally, in an emergency situation you'd call the emergency number.)

> Non-emergency numbers is another of the many reasons we suggest you collect phone books.

> When contacting the local sources just mentioned, ask about direct, non-emergency contact for all area first responders, and also any public information office they might have.

11. **Sheriff's Dept. non-emergency number**. The county Sheriff is usually the top Law Enforcement authority in the county. Try them if the local PD can't give you the information you need.

> You'll find Sheriff's office info on the county websites and the phone book's government pages.

12. **Fire Dept. non-emergency number**. Fire Departments usually have a strong tie to the Emergency Management operations of a particular area. To find out the emergency service status of a location, for example if a location has set up a refugee center, contact the FD along with the Emergency Management office listed below.

> The county websites will have listings for the county Fire Department.

> The local phone books will have listings for the city and **volunteer** Fire Departments.

> Many of the websites we gave you under "Fire Fighting Training" in the "Basic Training" portion of the "**Foundation**" section will have contact links for Fire Departments in various locations.

13. **Emergency management office**. During an emergency, the EMA would be more involved in the role of coordinator rather than direct contact with the public. However, don't cross the EMA off your list.

> State and county websites should have a listing for the state EMA headquarters and for local offices.

> Be sure to check the blue government pages of the phone books you collect.

> Remember to check the sites http://www.ready.gov/useful_state.html and http://www.nemaweb.com.

Food and Water (Map code "F")

14. **Grocery stores**. Though in worst-case-scenario situations like we're planning for, you can't count on grocery stores being open, or stocked, you shouldn't cross them off your list. As we stated earlier, some of the larger, more well-known outlets may be overwhelmed, but smaller "Mom & Pop" stores may be surprisingly accessible. Know where all your food sales outlets are. As far as **water** goes, any place with a **faucet** will do.

> Each county health department will be responsible for conducting inspections of any facility selling or serving food. They'll have a list of all locations.

> Though the larger chains *might* be closed or emptied, don't *assume* they are. List them too as you'll want every option at your disposal when a need arises. (By the way, these sites might be useful for finding **coupons** and other specials to save you money in your day to day shopping.)

- Kroger, online at http://www.kroger.com, or call 866-221-4141.
- Publix, online at http://www.publix.com, or call 800-242-1227.
- Winn Dixie, online at http://www.winndixie.com, or call 1-866-WINN-DIXIE (1-866-946-6349).
- Bi-Lo, online at http://www.bi-lo.com, or call 1-800-862-9293.

- Wal-Mar, at http://www.walmart.com, or call 800-966-6546.
- K-Mart, at http://www.kmart.com, or call 1-866-KMART4U (1-866-562-7848).

➤ Don't forget local phone books, and be sure to look for websites of other grocery chains.

➤ **Hint**: Some grocery store chains have "membership discount cards." If getting one is free, get one from any store you can even if you don't shop there regularly. Having one will save you a little bit of your valuable cash if you have to stock up at another store along your evac route.

➤ **Hint #2**: Price gouging during a disaster is illegal. Call the Federal Trade Commission now, at 877-FTC-HELP (382-4357), to ask them who to call in your state (or states you may evac through) to report price gouging. We mention this here, as price gouging happens a lot with food and fuel.

15. **Chain restaurants**. The farther away you get from a disaster area, the more likely it is you'll find normal business and services operating. Though we've prepped you with all sorts of self-sufficiency supplies including food, we're sure you'd rather live as normally as possible in order to keep everyone's morale high. Eating at your favorite restaurant will help greatly.

➤ A listing of all known chain restaurants could drastically increase the size of this book. List about five of your favorites and look up their national office online.

➤ Also, the chain restaurants are the ones most likely listed in tourism books and info sources.

➤ Don't forget your local phone books and county health departments for listings as well.

16. **Locally owned restaurants**. Chain restaurants, especially those closer to highways and interstates, might be filled past capacity. Therefore, list locally owned and out-of-the-way restaurants that are less likely to be full.

➤ In addition to phone books, the local Chamber of Commerce and Health Department are the best sources for finding the locations and contact info of all the local eateries.

17. **"Mini Marts"**. Later on, when we ask you to find gas stations, you'll want to be sure and mark the ones that have a small food mart as part of their service. Combining the two will save you valuable time on the road so you don't have to make two separate stops for fuel and food.

➤ See "Gas stations" below under "Transportation."

18. **Farmer's markets**. Getting away from prepared or processed food, you might be able to find what you need at a farmer's market since, being less obvious, these would be less likely to be emptied of their supplies than grocery stores or restaurants. Also, some farmer's markets will have little stands where some of the goods are prepared on-site or home-canned for sale.

➤ In addition to the usual info sources regarding cities and counties, you might find some area farmer's market listings through a state or county department of agriculture.

➤ You might try local school clubs such as **FFA** or **4-H**.

➤ We've also located a few online sources for finding a **farmer's market**.
- http://www.localharvest.org.
- http://www.smallfarms.com.
- http://www.openair.org.
- http://www.ams.usda.gov/farmersmarkets/map.htm.
- http://www.nal.usda.gov/afsic/csa/csastate.htm.
- http://www.homestead.com/marketfarm/marketsites.html.

19. **Pet supply stores**. Pets are part of the family too. If you can't find what you need at a grocery store, you need specific medicines, or if you need to get a carrier so your pet will be allowed into a shelter, you'll need to know all available sources. Your best bet in an emergency resupply situation is to get what you can where you stop for people items. In addition to those sources, look at the following.

➤ Look at the previous contact information we gave your for the Humane Society, etc., as well as the above information for vet finders.

➤ While you're contacting vets or stores, ask if any of them offer boarding on premises, or are associated with local **boarding or kennel facilities**.

➤ Here are a couple of websites and contact info for some of the more popular pet sites.
- Pet Smart. Find PetSmart locations at http://www.petsmart.com, or by calling 888-839-9638.
- PetCo, online at http://www.petco.com, or call 1-877-738-6742. (PetCo has a newsletter.)

20. **Ice houses or plants**. If you're in a Level 5 evac, you brought perishable food with you, and all the grocery stores and minit marts are out of ice, it'll be nice to know where you can get your hands on more. This doesn't figure high on our "emergency needs" list, but we see it as being a useful item.

> Every area is bound to have an ice production facility, especially agricultural areas and locations with ranches and/or meat packing plants. Look up ice houses in the phone book or ask the local farmer's market, county extension office, county agriculture department, or local restaurants. Remember, don't carry dry ice in your car since it releases carbon dioxide as it "melts."

21. **Fishable lakes and streams**. If no other food is available...

> If it's large enough to be seen on a map, it's probably fishable. You'll want to note these locations for those instances where there is absolutely nothing else. Having this marked on your evac map will remind you this is an option if you need it.

> You can also buy fishing location maps for your region from most bait and tackle or camping supply stores.

Transportation (Map code "T")

22. **Gas stations**. There is no evacuation without transportation, and transportation doesn't go far without fuel. Since gas stations figure prominently into trade and tourism, it'll be easy to find gas station location information through sources already mentioned. In addition to the stations out by interstates and highways, you'll want to know where the ones are in town. Try these sources.

> Most gas credit cards have some sort of travel info service available on their website. Try them to find the locations of all their stations in a given town. Remember, the stations by the interstate will be depleted first, while the ones in town may have gas.

> If you require **diesel**, you'll definitely want to mark the location of stations that carry it, as not all will.

> In addition to the gas cards you already have, try these sites and sources:
>   - Shell, online at http://www.shell.com, or call 1-888-GO-SHELL (467-4355).
>   - BP, online at http://www.bp.com, or call 1-800-850-6266.
>   - Chevron, online at http://www.chevron.com, or call (925) 842-1000.
>   - Exxon, online at http://www.exxon.com, or call (281) 870-6000.
>   - Phillips 66, online at http://www.phillips66.com/brands/marketing.htm, or call 281-293-1000.
>   - Citgo, online at http://www.citgo.com, or call 1-800-356-8832.

> For those of you driving vehicles using alternative fuels, the Department of Energy has a link to an online interactive fueling station locator map at http://www.afdc.doe.gov.

> **Hint**: Price gouging during a disaster is illegal. Call the Federal Trade Commission now, at 877-FTC-HELP (382-4357), to ask them who to call in your state (or states you may evac through) to report price gouging. We mention this here, as price gouging happens a lot with food and fuel.

23. **Mechanics / garages**. After gas comes mechanical operation. For those malfunctions you didn't learn to fix under "Basic Training," you'll need to know how to find a qualified and reputable mechanic.

> If your vehicle requires a certain type of mechanic, or if the vehicle is still under warranty, etc., **ask your local dealer** to help you locate certified or qualified mechanics along your routes.

> Some **states regulate mechanics** like they would other professions. Check the Secretary of State's listings.

> Many **travel clubs** carry a listing of qualified mechanics.

> Your **auto insurance provider** may also have a list of mechanics and garages that are on their provider list. Though few mechanical problems will be covered by the average auto insurance policy, getting the list will make your research easier.

> When marking the locations of **gas stations** above, note which ones offer **full garage services**.

> Some of the sites and sources we gave you earlier in the "**Automotive Repair**" part of "Basic Training" may offer links to finding repair centers.

> Though these will probably be the **first to be overwhelmed**, you still need to know the locations of national chain services. Here's a random sampling of various chains:
>   - Sears Auto Center, online at http://www.sears.com, or call 1-800-349-4358.
>   - Penny's Auto Center, online at http://www.jcpenney.com, or call 1-800-322-1189.
>   - PEP Boys, online at http://www.pepboys.com, or call (215) 430-9000.
>   - Aamco, online at http://www.aamco.com, or call 1-800-GO-AAMCO (402-2626).
>   - Mr. Transmission, online at http://www.mrtransmission.com, or call 800.377.9247.
>   - Mr. Goodwrench, online only at http://www.mrgoodwrench.com.
>   - Midas, online at http://www.midas.com, or call (630) 438-3000.
>   - Brake-O, online at http://www.brakeo.com, or call 972-242-6200.

- Jiffy-Lube, online at http://www.jiffylube.com, or call 800-344-6933.
- **Hint**: This is only a partial listing of chains and services available, and doesn't include specific parts or equipment such as tires. As you're putting your Evac Atlas together, customize your emergency sources based on your potential needs.

  ➤ If an RV figures into your evac plans, be sure to visit http://www.gorving.com, and sign up for their newsletter, or call them at 888-GO-RVING (888-467-8464).
  ➤ For boats and planes, check the links and sources we gave you earlier under "Basic Vehicle Prep."

24. **Towing services**. How are you going to get to this garage if you're completely broken down? **Note**: Get a listing of **several** towing companies, since in a mass evacuation mechanical problems will be rather common, and it may be hard to find an available tow truck, assuming they can get through traffic to get you. However, it's better to have this info and not need it, than to need it and not have it.

  ➤ Your **auto club or travel club** will have a listing of providers.
  ➤ Again, some towing companies, as they're probably bonded and required to be licensed by some states, may have a listing on the **Secretary of State's website**.
  ➤ When gathering the info for the **gas stations and garages** above, ask which ones offer towing. "One stop shopping" is usually the quickest, easiest, and most economical way to go.
  ➤ Here's a collection of online Tow Truck Associations:
    - http://www.towman.com/tow.htm.
    - http://www.towtimes.com/associations.htm.
    - http://www.towsigns.com/links.php.
    - http://www.the911site.com/911tow.
    - http://www.uhaul.com/finder.html.

25. **"H.E.R.O." units**. Larger metro areas may offer a "HERO," or Highway Emergency Response Operator service where a truck is dispatched to the site of an accident or mechanical breakdown. Many of these units carry enough tools and equipment to handle minor malfunctions and can get you up and running without having to tow you to a garage. Some of these units are dispatched through the 911 emergency service. In other areas similar services are privately owned and operated.

  ➤ The Highway Departments or Departments of Transportation mentioned earlier should have all necessary info to contact these units (or their equivalent). You'll note on the form in the "**Appendix**," we've left a blank for "radio channels monitored" and "cell phone" so you can find out which of your CB or FRS radio channels you can use to call these folks.
  ➤ When checking with Civic Organizations and Volunteer Groups headquartered along your routes, see if any of them offer this type of service.

26. **Auto parts outlets**. Since you took our advice and learned how to perform your own car repairs, the only thing you might need is a place to buy the parts.

  ➤ You'll find auto parts chain stores listed under "Basic Vehicle Prep" earlier.

27. **Travel club affiliates**. Some of the larger auto or travel clubs will actually have offices in various locations. Though they might only have a stand-alone office in larger metro areas, they may have affiliates in some surprisingly small places.

  ➤ Ask your travel club, or look on their website, for locations of any offices, providers, or affiliates that may help you with some emergency repairs or other travel needs.

28. **Bus stations**. You might need to trade transportation modes and continue your long-distance evac by bus, or you might simply need to know how to contact the local bus line at your evacuation destination.

  ➤ Your phone books, City Hall, and Chamber of Commerce can tell you about local transportation.
  ➤ Contact the state or county Department of Transportation to ask about all public transportation available.
  ➤ For cross-country busses contact:
    - **Greyhound**, online at http://www.greyhound.com, or call 1 (800) 229-9424.
    - **Trailways**, online at http://www.trailways.com, or call 703.691.3052.

29. **Car rental facilities**. Again, keeping in mind you might have to change transportation, you'll need to know just as much about where to find a car, as where to find a bus.

  ➤ Many **travel clubs** will not only have car rental listings, but will offer a discount for being a member.
  ➤ Check your **gasoline credit cards** next. Some of them may offer an affiliated discount.
  ➤ If you need to know the location of a few rental companies, try these:
    - Alamo, online at http://www.alamo.com, or call 800-327-9633.

- Avis, online at http://www.avis.com, or call 800-831-2847.
- Budget, online at https://rent.drivebudget.com/Home.jsp, or call 800-527-0700.
- Dollar, online at http://www.dollar.com, or call 800-800-4000.
- Enterprise, online at http://www.enterprise.com/car_rental/home.do, or call 800-325-8007.
- Hertz, online at http://www.hertz.com, or call 800-654-3131.
- Payless, online at http://www.paylesscarrental.com, or call 800-237-2804.
- National, online at http://www.nationalcar.com, or call 800-328-4567.
- Try this service to look at current competitive rental prices: http://www.autorentalguide.com.

30. **Public airports.** Though many scenarios would see a temporary grounding of all non-military aircraft, other situations might see roads so clogged that some airlines may offer to help get evacuees out of an area. Though not likely, if this becomes the case, or if you just want the extra option, you'll need to know where to go.

➢ Most **road maps** will show airport locations. If you hand draw your route map onto your Evac Atlas page, be sure to mark airport locations.

➢ The best source of airport info is always the **FAA**. You'll find them online at http://www.faa.gov, and you can contact them directly at the Federal Aviation Administration, 800 Independence Ave, SW, Washington, DC 20591, (800) 255-1111 (problems and incidents desk).

➢ A massive collection of airline 1-800 numbers can be found at http://www.geocities.com/Thavery2000.

➢ A quick way to make flight (or hotel) reservations while on the road is to use the online travel agencies:

- Travelocity, at http://www.travelocity.com.
- Orbitz, at http://www.orbitz.com.
- Hotwire, at http://www.hotwire.com.
- Cheaptickets, at http://www.cheaptickets.com.
- Expedia, at http://www.expedia.com.
- Priceline, at http://www.priceline.com.

31. **Private airports.** If you're a private pilot with your own aircraft, you probably have a map and listing of all airstrips in your area. However, if you're not a pilot and your family is in the middle of a jammed evacuation and you decide you have the funds to secure a charter flight, you'll need to know where to find one. Also, some private groups may offer evacuation assistance. It's best you find out now if any groups are contemplating this type of assistance, and how to contact them when necessary.

➢ You should ask about commercial and private airports while gathering other transportation information from the Chambers of Commerce and other municipal info sources.

➢ The Aircraft Owner's and Pilot's Association, online at http://www.aopa.org, will have an Airport Directory as well as other useful info. You can also reach them at 800/USA-AOPA (872-2672).

➢ If you think air evacuation might really figure into your plans during one scenario or the other, you'll want to find chapters of the Civil Air Patrol in your area. This civilian branch of the air force will figure prominently into any local plans for air evac or air relief. You'll find them online at http://www.cap.gov, or you can call their headquarters at 800-FLY-2338 (359-2338).

➢ You'll find plenty of airport location and contact info online at http://www.airnav.com/.

32. **Taxi companies.** Another of the costly options as far as a mode of travel goes, yet you may well need any transportation you can get, or you may need intermediate transportation to get you from a garage to a bus station, etc. Or, you may simply need a taxi to get around your destination.

➢ The best way to find all local taxi companies is through searching the yellow pages and white pages, either online, or from the books you collect.

➢ If you have a PDA device, several companies offer taxi finder files that you can download on your computer and install on your PDA:

- PDA Central offers freeware at http://www.pdacentral.com/palm/preview/71308.html.
- Taxi Numbers has a download at http://www.taxinumbers.com.

➢ Online taxi locator: http://www.whereonearth.com/cgi-bin/phpdriver?MIval=int_taxi_demo_start.

➢ Taxi Numbers listed above at http://www.taxinumbers.com, has an online locator service.

33. **Train stations.** Amtrak is the most widely known passenger train. However, others exist that make limited runs, and in an "any port in a storm" evacuation, you'll certainly want to know all your options.

➢ When researching all the above transportation options through the various municipal and transportation department sources, make sure trains are on your list.

> **Amtrak** is online at http://www.amtrak.com, or you can call 1-800-USA-RAIL (1-800-872-7245).

34. **Navigable waterways.** Remember our Kayak scenario earlier where you're stuck in traffic as the winds shift and fallout clouds begin to drift your direction? Depending on the situation causing the evacuation, you may wish to consider waterways an option.

> The first thing to do is simply make a note of waterway locations on your standard road map. The mapping sources we gave you earlier under Threat Map will have several selections of navigable waterway maps.

> If you seriously consider this as a viable option, you'll want to compile another set of Evac Atlas pages for locations along this waterway (as it will then become a separate route) so you'll know what services are available where.

35. **Docks and Marinas.** Some of you may choose to evacuate entirely by water, just as some of you may choose to fly instead of drive. We always suggest going with what you have and what you feel most comfortable with, provided you've considered all the criteria and options we've given you along the way. If you're an avid boater, it's likely you've collected your waterway and nautical charts and know where docking and other boat facilities are. Here, we've tried to help you locate more sources:

> The various Departments of Transportation and municipal contacts will have a listing of facilities.

> Don't forget your phone books, especially the **yellow pages**.

> The better **road maps** will also show docking and marina facilities.

> We've also located a few random **online dock and marina finder sites** for you:

- Marina Finder, at http://www.marinafinder.com.
- Marina Life, at http://www.marinalife.com.
- Boating Resources, at http://www.cbel.com/boating_resources.
- Marinas.com, at http://www.marinas.com/Default.htm.
- Boatstation Electronics, at http://www.boatstation.com/marinas.
- Sliphunt, at http://sliphunt.com.
- Freeware Palm has a free marina locator at http://www.freewarepalm.com/internetpqa/nauticalpro.shtml.

Communication (Map code "C")

36. **TV stations.** It's easy to find TV channels on your set, but there are a few other things you'll want to know too. You'll want to know the broadcast coverage area of local stations so you'll know where to tune your portable set as you travel from one area to another. You'll also want the phone number for these stations so you'll have one more intel option if other sources aren't giving you what you want. For example, an emergency in the city you left might not be as important to stations along your route and consequently you'll hear regular programming with little news. Most stations can give you an update on regional news if you phone the news room.

> All television broadcast stations should be listed in the phone books.

> Search the FCC television station database online at http://www.fcc.gov/fcc-bin/audio/TVq.html, or call the FCC Video Division at (202)-418-1600.

> Other independent online television station databases can be found at:

- Directory USA Television, at http://television.directoryusa.biz/states.php.
- Disaster Center, at http://www.disastercenter.com/form3.htm.
- TV Radio World, at http://www.TVradioworld.com.

37. **Emergency radio stations.** Many areas have radio stations that are news-only and that are used as the "emergency info" radio station for that station's broadcast area. We're trying to get legislation enacted to make 91.1 FM and 912 AM the universal emergency radio frequencies (they're the closest numbers to "911.") Until then, you'll need to know which stations offer emergency news and instruction, and you'll want to know what their frequency is as well as their station's phone numbers.

> The FCC has two radio finder databases. One for AM and one for FM:

- Search AM stations at http://www.fcc.gov/mb/audio/amq.html, or call (202)-418-2700.
- Search FM stations at http://www.fcc.gov/mb/audio/fmq.html, or call (202)-418-2700.

> Other **independent online radio station databases** can be found at:

- America's Radio Travel Atlas, at http://www.amfm2go.com.
- Radio Locator, at http://www.radio-locator.com.
- Radiostation.com, at http://www.radiostation.com.
- Radio Enthusiast, at http://www.radioenthusiast.com/am-fm_DX_links.htm.

- Yahoo's links, at http://asia.dir.yahoo.com/news_and_media/radio/stations/web_directories.
- Radio List, at http://www.radio-list.com/.
- Find Radio News, at http://www.journalismnet.com/radio.
- Public Radio Finder, at http://Publicradiofan.com .
- International Radio Stations online, at http://www.radiotower.com/.

38. **Regular radio stations**. In addition to the news stations listed above, you'll want to know which stations in each area offer "general programming," which will have some news, and "specific format" radio which may be automated with specific programming and offering no news at all. These will be great distractions to ease stress.

➢ The sources listed above should have what you need.

39. **Cellular "touch star" services**. Some cellular service providers offer "touch star" services for rapid dialing in some areas. A couple of examples in Georgia are Cingular's *GSP for the Georgia State Patrol, and *DOT for the Department of Transportation's road information line. This feature makes it easier to remember contact info for key information services.

➢ Call your cellular service provider, or look them up online, to see which rapid dial features are automatically included as part of your service. You'll find some of their contact info earlier under "**Communication**."

➢ **Hint**: As you actually travel these routes, mark on your map the areas in which you do _not_ get a cell phone signal. Later, compare this with the coverage you're _supposed_ to be getting.

40. **Online traffic info sites**. More and more locations are starting to offer online traffic information. Some even let you view traffic cameras online. While **you shouldn't waste valuable evacuation prep time** by looking up road information before leaving, you should know the information sources in case you make a stop along the way, or if you have a laptop with a cellular modem. Also, this _may_ be an alternate way to find out if any stops along the way have **closed themselves off** to refugee traffic.

➢ We've located a few online sites that will allow you to look up traffic web cams to view:
- Take a look at Traffic Look, online-only at http://www.trafficlook.com.
- Also see its cousin site at http://webcambiglook.com.
- See the Motor Portal traffic cam links at http://www.motorportal.com/Traffic/traffic_cams.htm.
- Tommy's List of Live Cams Worldwide http://www.rt66.com/~ozone/cam.htm.
- For road conditions check out http://www.usroadconditions.com.
- See more road condition news at http://www.nwa.metronetworks.com/nwa/nwdirect.htm.
- For weather conditions see http://www.weatherlook.net.
- Weather and traffic cameras http://www.weatherimages.org/weathercams/usa.html.

➢ Accutraffic seems to have some pretty good coverage and info online at http://www.accutraffic.com.

➢ Rand McNally's road construction site at http://www.randmcnally.com/rmc/tools/roadConstructionSearch.jsp.

➢ You can get updated weather forecast maps, etc., at http://www.usroadconditions.com/

➢ See the US DOT's road closure information site at http://www.fhwa.dot.gov/trafficinfo/index.htm.

➢ In many states you can **dial 511 for traffic information**. The US DOT's website at http://www.fhwa.dot.gov/trafficinfo/511.htm, will explain more.

➢ Smart Traveler, online at http://www.smarttraveler.com, offers limited links to traffic cameras and info in various areas. Their links may increase in the future. See them online, call for current updates at 800-44-ANITA (800-442-6482) and for other services call 617-494-8100.

➢ Call your own state's Highway Commission or Department of Transportation and ask about update info, numbers to call, and webpages to try. New Jersey's DOT site maintains a list of links to each state's DOT at http://www.state.nj.us/njqi/Transportation/OtherTransLinks.htm, and Roadmaps.org offers the same thing at http://www.roadmaps.org/links/state-links.html.

41. **Local municipal websites**. Along with the traffic info sites, you might be able to find some current news concerning emergency efforts and services offered (or closed) at stops along your route.

➢ You should've already located all these sites in your efforts to find all the previous information.

42. **Cellular service outlets**. Communication is vital during any emergency. Suppose your phone was damaged somewhere along your trip. Suppose you were forced out of your home too fast to grab your gear. Suppose you want to temporarily change services due to your destination's coverage area. Suppose some cellular providers were providing free phone usage at their stores to help evacuees. The list of reasons you'd need to find a cellular service center could go on and on. The key is knowing where they are in advance so you can find a convenient location that can take care of all your needs in advance. Also, in some instances, cellular outlets might offer a few free minutes to evacuees. At least that's one of the things we're pushing for. More under "**Cooperation**."

> We gave you cell phone service sources earlier under "Communication." Go back to look at coverage maps and look up service center locations.

43. **Local PD & FD scanner codes**. Scanners, if you've packed one as part of your "Secondary Kit," are great for keeping up with emergency actions and reactions of municipalities along your evac route and at your final destination. In fact, as you'll notice on the Evac Atlas forms in the "**Appendix**," we've left blanks for you to fill in scanner frequencies.

> If you missed the portion on **scanners** in the "Communication" section earlier, go back and look at it now. We listed scanner info as well as frequency search sites.

44. **Copy of yellow pages**. Phone books are great information sources, but so cumbersome. Above we mentioned several **phonebook websites** where you can look up a lot of information online and where you can even order a phone book on CD. Some of this is free, but some can be a little expensive. That's why we suggested earlier that you get free phone books where you could, either from City Hall or from local Chambers of Commerce.

> While online, be sure to **download and print** pertinent sections. For example, instead of looking up one location offering one potentially necessary service, print and save the entire section. Instead of listing just one garage, print the whole section of garages and mechanics and put the printout in your Evac Atlas.

> Do the same with a real phone book. Instead of looking up single locations or carrying around the whole book, **photocopy or cut out useful pages or sections** and tape them into your Evac Atlas.

45. **Copy of white pages**.

> Same as above.

46. **Post office locations**. The mail will eventually get through. There's no doubt about it. You might want to know the location of Post Offices at your destination so you can put in a temporary change of address. We've even known some people to lighten their load by mailing some supplies to themselves in care of their final destination, from points along their route.

> The best place to find Post Office locations is through the Post Office. Try http://www.usps.gov, or you can contact them directly at (800) ASK-USPS (275-8777).

> By the way, we've included a Change of Address form in the "**Appendix**," though we strongly recommend you go to your local Post Office to secure an official form to go in your "Office Pack."

47. **Private mailing / fax centers**. Private mailing centers are good because they offer a variety of services. Most will offer FedEx, UPS, DHL, US Mail, and fax services. Though there's no guarantee all of those will be operational during all emergencies, they'll eventually be back up and running. Of equal importance is the **fax** service. This is a good way to send a message where you don't have to stay on the phone waiting to get a signal or waiting for lines to be restored. All you have to do is leave a fax message, pay for it, and tell them to keep trying to send while you continue your journey. You'll find some of these centers at:

> Visit Federal Express online at http://www.fedex.com, or call them at 1-800-GO-FEDEX (800-463-3339) to locate drop boxes or facilities. (This is another good way to send some of your gear ahead.)

> The UPS Store, formerly "Mailboxes Etc.," has a useful store locator feature online at http://www.theupsstore.com/about/abotheups.html, or call (888) 346-3623, or (858) 455-8800.

> DHL is online at http://www.dhl.com, or call 800-CALL-DHL (800-225-5345).

> Kinko's online at http://www.kinkos.com, or call 800-2-KINKOS (254-6567).

> Office Max online at http://www.officemax.com, or call 800-283-7674.

> Office Depot online at http://www.officedepot.com, 1-800-GO-DEPOT (463-3768).

> Staples online at http://www.staples.com, 1-800-3-STAPLE (378-2753).

48. **Western Union locations**. Western Union is another great way to have a message sent when you don't have time to wait around. It's also a great way to send or receive funds during an emergency.

> Western Union has a store locator at http://www.westernunion.com, or you can call them at 1-800-325-6000.

49. **Public libraries / Internet Cafes**. Any library is a good source of local info, as most will have city and county maps in their reference section along with phone books and a host of other publications. Many will also have internet access though you may need a library card to use it. This is okay if you've evacuated your home town and are still in your county. You should know the location of your own libraries, plus all the libraries along your route. You should also know the location of any "Internet Cafes" so you can send and check email.

> Mark the location of the last library in your county (the ones you'd have a card for) along each of your routes, and show which ones have internet access. If the library was outside the danger area and you had time to stop, you wouldn't want to miss a chance to email updates to loved ones in your YahooGroups list.

> **phone numbers** for your dial-up internet service. You'll want to be able to use local numbers to access the internet from your final destination as well as from some of the more likely rest stops along the way.

> **Libraries** can be located through the municipal contact and online phone book sources already covered.

- To find "**cyber cafes**," or "**internet cafes**," we've found a few random **locator sites** for you:
  - The Cyber Cafe search engine at http://cybercaptive.com.
  - Cybercafes.com at http://www.cybercafes.com.
  - AAA Cyber-Cafe search at http://www.cyber-cafe.com/icafesearch.asp.
  - Internet Cafes at http://www.globalcomputing.com/CafesContent.htm.
- One of the things we'll suggest under "**Cooperation**," is that your group urge local office supply and computer stores to set up some terminals to allow refugees and evacuees that might be passing through your area an opportunity to send an email message to loved ones. Keep this in mind when you're listing all the potential fax and mail centers we mentioned above, as you'll also want to list **computer stores**.
- **Hint**: If you have a laptop that you might take with you in an evacuation, you should know the **local access**

50. **Payphones**. Different phone systems operate from different line trunks. It may surprise you to find you might be able to get an outside long-distance line on a payphone when nothing else is working.

- There's really no central source that's going to tell you where every payphone is. In fact, most of the various supply locations listed above will have a payphone somewhere. Our suggestion is that when you actually travel these potential evacuation routes and you run across a gas station, minit mart, etc. that has **no** payphone, that you're sure to mark it as such on your map. This will help you avoid an unnecessary stop if it's only a payphone you're hunting.

Shelter and Lodging (Map code "S")

51. **Official emergency shelters**. Remember in the "Planning Ahead for Phase II" section where we had you list all sorts of emergency planning and relief agency contacts? This Evac Atlas is why. These are the folks that will tell you if and where an emergency shelter has been set up when existing ones are either full or uninhabitable.

- Revisit the sources listed under "Phase II" and record all the contact info in your Evac Atlas so you can call these offices while on the road to get more information if news sources and other sources are lacking data.
- Be sure to note the location of any pre-planned evacuation shelters on your maps.

52. **Hotels and motels**. In some cases, your final destination may be a motel down the road from where your house once stood before the tornado. In all situations, you'll want all the contact info you can get so you can call ahead to make reservations, or to see if there's any room left. When contacting motels, plan ahead and ask if they take pets. Do what you can within the circumstances to conserve funds. If less expensive lodgings are available, take them. The contacts we'll give you below will help you shop around a tiny bit. However, don't shop around so much that you lose shelter.

- Your travel club or roadside assistance club will be able to provide you a list of lodging facilities, and may offer you a discount.
- As with everything else, get a listing of local lodging, including **Bed & Breakfasts, rooming or boarding houses, and rent-by-the week apartments** from the municipal contacts you try.
- If the event you're fleeing is covered by insurance, ask your agent if certain lodging locations are, or are not, reimbursable. Make sure you get the list in writing.
- Most of the **online travel services** we listed above under "Transportation," have a hotel locator.
- Also take a look at some of the hotel locator services:
  - Hotels.com, online at http://www.hotels.com, or call 800-2-HOTELS (1-800-246-8357).
  - Hotel Locator, online at http://www.hotellocators.com, or call 800-423-7846.
  - Hotel Finder, online at http://www.1hotelfinder.com, or call 800-823-9653.
  - Cheap Rooms, online at http://www.cheaprooms.com, or call 800-311-5192.
  - Discount Hotels, online-only at http://www.discount-hotel-finder.com, where you'll also find a links section for the major chains.
- You might also have to search by chain. We've found some of the more popular chains for you. When contacting these chains and searching through their information, see if any offer the following FREE services: ❏ Discount memberships ❏ Trip planning ❏ Concierge or equivalent.
  - Best Western, online at online http://www.bestwestern.com, or call 800-780-7234.
  - Comfort Inn, online at http://www.comfortinn.com, or call 877-424-6423.
  - Courtyard by Marriott and Marriott, online at http://www.marriott.com, or call 800-932-2198.
  - Days Inn, online at http://www.daysinn.com, or call 800-DAYS-INN (329-7466).
  - Hilton, online at http://www.hilton.com, or call 800-774-1500.
  - Holiday Inn, online at http://www.holidayinn.com, or call 800-263-9393.

- Howard Johnson, online at http://www.howardjohnson.com, or call 800-446-4656.
- Hyatt, online at http://www.hyatt.com, or call 800-233-1234.
- La Quinta Inn, online at http://www.laquinta.com, or call 866-725-1661.
- Motel 6, online at http://www.motel6.com, or call 877-4-MOTEL6 (466-8356).
- Radisson, online at http://www.radisson.com, or call 888-201-1718.
- Ramada Inn, online at http://www.ramada.com, or call 800-272-6232.
- Red Roof Inn, online at http://www.redroof.com, or call 800-RED-ROOF (733-7663).
- Sheraton, online at http://www.starwood.com, or call 800-325-3535.

➢ Other travel info, along with a good collection of links, can be found at "Joe Sent Me – The homepage for business travelers," at http://www.zyworld.com/brancatelli/MEMBER/j.htm.

➢ "Pets Welcome" has a listing of lodging locations that accept pets. Go to http://www.petswelcome.com, or contact them at Petswelcome.com, PO Box 504, Hughsonville, NY 01253, (845) 297-5150.

➢ Also for pets, visit the Travel Pets site at http://travelpets.com.

53. **Friend's or relative's house**. Though your final destination might be a friend or relative, you still need to record the contact info for other **people along your evac route** that could offer assistance or a place to stay overnight. Be sure to write their contact info along with everything else you've recorded and mark their location on your map. In an emergency situation, all of us are going to be stressed, and none of us can be counted on to remember everything. So when in doubt, write it out.

➢ Don't forget to note if your friend or relative is associated with any necessary goods or services in this list.

➢ Also, don't forget to have "**proof of contact**" for any of your associates along the way, just as you do for a person who might be your final destination. Remember what we said earlier. Local authorities might have an area closed off and you'll want some sort of proof, whether it's a simple letter, or mail in your name, that connects you with the person you're trying to visit.

54. **Camp grounds**. Sometimes there's just no room at the inn. While many evacuees may be crowding out the local motels, most will not be as prepared as you with alternate shelter. Therefore, you might find that a local campground will have plenty of room for you to pitch a tent overnight.

➢ Your state department of travel and tourism should have a listing of all state parks offering campground services. If not, check with your state's Department of Natural Resources, or department of the interior.

➢ Your travel club or roadside assistance club should be able to help you locate campgrounds (provided they offer a trip-planner service).

➢ Woodall's offers a fairly comprehensive service for camp ground info and site searches. They also offer a good line of hardcopy camp directories. You'll find them online at http://www.woodalls.com, or you can reach them at Woodall Publications Corp., 2575 Vista Del Mar Drive, Ventura, CA 93001, (877) 680-6155.

➢ Contact the National Park Service, at http://www.nps.gov, or at Director-National Park Service, 1849 C Street NW, Washington, DC 20240, Phone: (202) 208-6843. Also try their park locator feature at www.nps.gov/parks.html. Try the other government parks site, online-only at http://www.recreation.gov, or http://www.nationalparks.com, to find other types of parks.

➢ Though a Canadian company, **Camp Search** will let you search for camping accommodations as well as various theme "camps" in the US. They're online at http://www.campsearch.com, or you can contact them directly at CampSearch.com, 219 Dufferin St., Suite 200A, Toronto, Ontario, Canada, M6K 3J1, 416-588-6375, Fax: 416-588-8753.

➢ This online service, Reserve America, at http://www.reserveamerica.com, will let you search by type of facility, browse by maps, and their contact page has state park service phone numbers for different states.

➢ Naturally, we've found a few comprehensive online sites to help you with your search:
- Camp the World, at http://www.camptheworld.com.
- Camp the USA, at http://www.camptheusa.com.
- US Camping, at http://www.uscamping.net.
- About's State Index, at http://camping.about.com/cs/cgdirectoriesindex/l/blaast.htm.
- Go to Discover the Outdoors, at http://www.dto.com/dto/map.jsp, and click on "Camping."

55. **Rest areas**. You'll definitely need to mark the locations of all the rest areas along your route. Here's why: Suppose you're on the road, worn out and looking for a place to sleep. You've called ahead and all area hotels, motels, and campgrounds are full. Is the next rest area just up ahead, fifty miles ahead, or three miles behind you? Also, some local volunteer groups may decide to "adopt a rest area" and provide some needed supplies or services to evacuees.

➢ The state's Highway Department or Department of Transportation should have information showing the location of all rest areas, and they should also be able to tell you which ones have restrooms, vending areas, payphones, etc.

➢ When traveling, be sure to mark the ones you find, noting which of the above services they offer.

➢ Many (but not all) roadmaps will show the location of some rest areas.

General Resupply (Map code "G")

56. **Branches of your bank(s)**. Though you'll have a lot of barter items in your various kits, **cash** is still king of the road. Checks might not be accepted and some places might not be set up for plastic. You'll need to know where you can cash checks, and access and manage your liquid assets from wherever you might be on the road.

➢ Check your **bank's website**. Most will have a locator service on there somewhere.

➢ When searching grocery stores and other resupply locations, ask about on-site or nearby banks, since if you do stop, you'll want to accomplish as much resupply as possible at one time.

➢ Investigate your bank's online account management options to see if **wire transfers** are offered. This may allow you to transfer funds to a different bank if you wind up in a town having no branch of your regular bank, or an uncooperative ATM system.

➢ Additionally, some banks offer online banking services that will allow you to pay bills, transfer money, and manage your accounts from wherever you may be.

➢ Remember that Western Union can wire money. We listed their contact info above.

57. **ATM locations**. If all you need is a little cash, you might not need to visit a bank branch at all, and an ATM will do fine. However, due to heavy traffic, many ATMs might be empty so you'll need to know where **alternates** are located without having to waste valuable time hunting for them.

➢ Your bank's locator site might also have ATM locations listed.

➢ Your travel or roadside assistance club will have similar information.

➢ We've also located a few online locator sources for you:

● Try the ATM Locator site, at http://www.atmlocator.info.

● Visa has an ATM locator at http://www.visa.com/pd/atm/main.html.

● MasterCard has one at http://www.mastercard.com/cardholderservices/atm/.

● "Credit Union 24" locator, at http://www.cu24.com/atm/default.html.

● The SUM ATM locator, at http://www.sum-atm.com/ATM_Locator.htm.

● Fastbank, online at http://www.fastbankatm.com/locator/search.jsp.

58. **Check cashing centers**. If you need cash, there's no branch of your bank and all the ATMs are empty, you'll need to know where you can cash a check. We recommend this as a last choice because most of these places charge a fee for check cashing, and it's a larger fee than your ATM would charge. Also, in many scenarios, there is no guarantee that a check will be taken anyway. In addition to the phone books you've collected, try your travel and assistance clubs, and ask for this info from the municipal contacts and Chambers of Commerce.

59. **"Super center" stores**. Suppose you were forced out of your area and you were unable to get to your kits or your stored supplies. You'll want to know where you can stock up on food and supplies with one stop shopping, and you'll need to know the location and contact info for these stores so you can find out if they're open before you pull off the road. Remember, we're doing this so you can stay prepared, yet be on your way and on the road for less time. Speaking of time, you'll notice on our forms in the back we've left room for "**hours of operation**" for most all entries. You'll need to know who's closed so you don't waste time trying to find them, and if you're needing goods in the middle of the night you'll need to know who's **open 24 hours**.

➢ Wal-Mart can be located at http://www.walmart.com, or call 800-966-6546.

➢ K-Mart can be located at http://www.kmart.com, or call 1-866-KMART4U (1-866-562-7848).

60. **Malls and shopping centers**. As not every little town will have its own supercenter, you'll need to know where is the most target-rich environment for stores.   When gathering info on these locations, make sure you get the contact for the management office as well as the main store or two at the location. This should give you plenty of contact info so if you're able to call from the road, you'll know for certain if the location has remained open.

➢ The best places to find out about malls and shopping centers is through the phone books you've collected or found online. The Chamber of Commerce and City Hall will also provide good info.

➢ You can also ask for this info from the grocery stores you contacted earlier as they may be located nearby.

61. **Sporting goods outlets**. Suppose you couldn't get your gear, all the motels are full, and you'll have to stay at a campground overnight. You'll need to know where to find the appropriate supplies right? Also suppose all the

local grocery stores are empty, yet there's a nice lake just up the road. Need fishing gear? Need freeze-dried food that other evacuees might not think about until later?

> Check the phone books and Chamber of Commerce.

> Check with the state wildlife commission, or its equivalent, to find out which stores at a given location sell hunting and fishing permits. These locations will usually have gear as well.

62. **"Dollar" stores.** Every time we mention "dollar" stores in this book, we're using this as a general term regarding stores where everything is literally a dollar. If you have to resupply on the road, you'll want to be as frugal as possible, and you'd be surprised at the stuff you can find in a dollar store. If you need a few things and one of these stores is convenient, shop there before hitting the more crowded and more expensive mall and chain stores, especially when resupplying **toiletries**.

> Local phone books and municipal offices are going to be the best way to find locations of local outlets.

> You can also try the sites http://www.dollarstore.com or http://www.dollartree.com.

63. **Thrift stores.** The best place for clothing, and possibly some camping items are thrift stores. They won't have the current fashions, but they'll clothe you at a price you can't beat. Also, they'll have the usual array of pots and pans for cooking over a campfire or in a motel kitchenette, and they may have tents, tarps, or other shelter items.

> Look up "thrift stores," "consignment shops," and "second-hand stores" in your phone books.

> Goodwill Industries has a locator feature on their website at http://www.goodwill.org, or you can contact them at Goodwill Industries International, Inc., 9200 Rockville Pike, Bethesda, MD 20814, (240) 333-5200.

> The Salvation Army is every bit as large and useful as Goodwill. However, at the time of this writing, their website, http://www.salvationarmy.com, has no central locator, and their office has no central information line. However, it's worth the time to find locations along your evac routes. The site http://www.salvationarmy-usaeast.org/find/locations/thrift_window.htm, will let you search for stores in the eastern US, and http://www.qso.com/tsa/stores/index.html, will let you search the southwest. Otherwise, use your collected phone books and municipal offices for info.

64. **Hardware stores.** Just as you should know the location of auto parts stores to help you fix those things you know how to fix, you should know the location of hardware stores for the same reason. Who knows what kind of tools, parts, etc. that you might need to help you along your route, at your destination, or in preparation for repairs on returning home. **Hint**: Knowing the out-of-town hardware store locations will allow you to stock up **before returning home to "ground zero"** where local stores will be depleted.

> Again, the first place to check is the phone book and with the municipal information contacts.

> You might also try these national chains' locator services or store listings:

   ● Home Depot, at http://www.homedepot.com, or call 1-800-553-3199.

   ● Lowe's, at http://www.lowes.com, or call 1-800-44LOWES (1-800-445-6937).

   ● ACE Hardware, at http://www.acehardware.com.

   ● True Value Hardware, at http://www.truevalue.com, or call 773-695-5000.

Relief and Support Agencies (Map code "RA")

65. **Volunteer group reps.** You really should be part of a volunteer group. There's no higher civic duty than neighbor helping neighbor in a time of crisis. One of the things your volunteer group should do is **network with other groups** that might have members or chapters along the likely evacuation routes for your area. This way, if you're evacuating and need help, you'll have someone to call.

> If you liaison with a larger outside group coordinator, such as Americorps or Citizencorps, they will help you link up with other organizations in other cities in your region.

> Also try the National Voluntary Organizations Active in Disaster: http://www.nvoad.org, along with some of the others you'll find listed in the **"Cooperation"** section.

66. **Local relief agency offices.** Most of the information from post-disaster relief agencies will come over the news networks. However, as we suggested under "Planning Ahead for Phase II," you should contact these folks to learn of the various locations they might set up relief centers, and the procedures involved in requesting assistance or filing a claim. The way this fits in to an evacuation, is that some of these agencies may have off-site contingency plans in place, to include aid and shelter, should something happen to your area.

67. **Charitable organization offices.** As an example, the Red Cross may immediately set up an aid station in an area outside a ground zero. It's best to know where these aid sites might be rather than pass one up if you really need help. Remember this: With any official program, a large percentage of the population will never know about it until it's over and gone. These agencies specialize in aid, not in marketing. So while they're there to help, it's not likely they'll do the best job in the world of informing everyone where they are. You have to be proactive and know these things ahead of time.

68. **Gov't relief agency offices**. Same as the previous two, except in this case it's government agencies such as FEMA, or your state's Emergency Management Agency, as opposed to private sector.

> Everything you need should be listed under "Planning Ahead for Phase II."

69. **Relief missions and public shelters**. Though low on our list of recommendations, relief missions still offer potential aid should there be absolutely nothing else. Though most missions focus on the homeless, many will try their best to aid victims and refugees of a disaster.

70. **Charitable "Church" groups**. Blending relief groups with volunteer groups, keep in mind potential assistance from religious organizations. Some will cater only to their own followers, while others will help anyone in need.

> Check with your place of worship to find out more about any existing evacuation plans and/or the relationship with other locations in the region.

71. **County Co-op or Extension**. Many county programs are handled through, or coordinated by the Co-Op or Extension office. We listed this earlier as a possible source for gathering some of this Evac Atlas information. While on the road, if communication systems are working, this office(s) may be able to answer some of your emergency need questions that couldn't be answered any where else. This option is certainly low on the list, but an option nevertheless.

> The sites mentioned above, along with the government blue pages from the phone books, will help you locate each county extension office.

> Remember, as you evacuate, you may cross several county and/or state lines. Keep this in mind when gathering extension office info, and when gathering any of the above data.

Give a copy of your Evac Atlas to your contact / rendezvous person, so they'll know where you might be, in much the same way you'd file a flight plan. Or, they can use it themselves if they evac towards you.

Notes:

_____
_____
_____
_____
_____
_____
_____
_____
_____
_____
_____
_____
_____
_____
_____
_____
_____
_____
_____
_____
_____
_____
_____
_____

Getting There

Well now, we've covered gathering, packing, storing, and distributing your gear; we've discussed what's involved in deciding where you might go, and we've had you collect all manner of intel about the routes and destinations you've chosen. You're ready and set, so the next logical topic is, "Go!" Let's talk about some of the details involved in actually evacuating and making it to your final destination.

We'll discuss your immediate reactions on learning you have to evacuate, coordinating your efforts in getting everyone together, your various transportation considerations, and we'll talk about your travels.

Goals of Getting There:
1. To allow you to be able to **react and evacuate quickly, efficiently, and safely**.
2. To make sure you're able to **gather your family together** in order to evacuate.
3. To educate you as to **what you need and don't need regarding transportation**.
4. To bring you up to speed on **what to expect during your evacuation travels**.

We've divided this section into:
1. **Out the Door.**
2. **Rendezvous!**
3. **Transportation.**
4. **Life on the Road.**

Let's start with those final-moment things you need to do when actually executing an evacuation.

Out the Door

The biggest questions affecting your overall evacuation are, "**Where are you going to be when an evacuation is necessary?**" and, "**What will cause the evacuation?**" Unfortunately, the answer to these is, "**Who knows?**"

We just don't know what's going to happen when, where, or how, and that's the reason we had you gather so much diversified gear, keep it packed in divided kits and locations, and then had you choose multiple evacuation destinations and routes. You have to be ready for anything.

Since we don't know how things are going to begin, we'll cover some considerations that will apply to all evacuation situations, and others to keep in mind for specific instances or scenarios. Remember earlier under "Planning the Route" where we had you look at where you and your family might be at any given point in the day? We'll reiterate that a bit here and drop in a few related considerations, since those factors play such a critical role in your reactions.

Let's cover the remaining and varied things you can do in the way of **advance preparation**.
  - ➤ **Divide the duties.** Your unique situation will change during the day, and from day to day, regarding how you would or could react to an emergency. A disaster occurring during the day when the family was at separate work and school locations would see one reaction, and a nighttime emergency when the family was together at home would see another. Each different situation and its associated reaction and needs should be considered and included in your planning. Chief among these considerations is the division of duties, or "Who does what?" Who goes after the kids? Who gets the Bugout Kit? What about the pets? Keep the following in mind when discussing your plan:
    - Focus on the worst possible situation, and that would be when the family was **separated**.
    - **Proximity** is the first factor you should look at. Who's near what or whom? If one head of household or able adult is near the kids, and the other is nearer the home and supplies, then dividing the duties is easy.
    - Revisit your Threat Map to look at your **travel bottlenecks** and geographic obstacles. What might look like close proximity on the surface might change drastically if it turns out you would be completely cut off from certain areas or people because of these obstacles. For example, what if you live on one side of a sizeable river, work on the other, and a major earthquake damages the area's bridges? What if a railway HazMat incident caused your evacuation and it's this railway that cuts you off from your family or supplies? You'd eventually get where you needed to be, but such a situation would certainly throw a kink in your immediate plans. Heavy traffic resulting from an emergency would have the same result in this example.

- In this planning, you should also list your available **assets**. Where might you acquire other goods or supplies? Where could you store some of your own gear to make access easier? Where do your close friends and relatives live? Do you have any mutual emergency assistance understandings with your neighbors or volunteer group members? (This would essentially be the local-info portion of your Evac Atlas.)

- Once these duties are divided and specific plans set in place, **write them down,** carry a copy with you, store a copy in your Bugout Kit along with your Last Minute List, and regularly review them with the family.

➢ **Practice.** Experience is the best teacher, and practice is the best thing short of actually experiencing a disaster requiring an evacuation. It's also the best "advance prep" you can make.

- As mentioned earlier, whenever you have a **fire drill**, carry it to the next logical level, and make it an **evacuation drill** by having the family practice grabbing their gear, **loading the car,** etc. Actually loading the car will open your eyes to several potential **stumbling blocks** that you'll need to iron out now. What if your gear won't fit or what if it has to be loaded a certain way? What if you have an unexpected passenger?

- Whenever you go on **family vacations**, if you take your own vehicle and travel by way of any of your planned evacuation routes, be sure to make little stops along the way to pick up intel for your Evac Atlas. Also, clue the kids in to what's going on and why. Most kids would love pretending to "run away" and you'd be surprised at how much detail kids will remember regarding what's where with possible evacuation resources. Besides, loading the car with all the vacation luggage is a form of practice in and of itself.

- Speaking of vacation, depending on your mode of travel, you should try to take as much of your Primary Bugout Kit or pertinent duplicated items with you since you don't know what's going to happen where you're going or at home while you're gone. What if you're on a trip somewhere and your home area goes under quarantine? It's doubtful you'll be let back in, so where are you going to go? What are you going to do? We recommend you at least carry your "Office Pack," a first aid kit (one should always be in your vehicle anyway), and a little food and water. It's assumed you'll have clothing and toiletries since you're on vacation.

- As we'll cover in the next section, "Who Else is Going?" you'll want to periodically refresh everyone's memory on where the **rendezvous points** are, and what **procedures** they should automatically follow if a disaster occurs and no communication is available.

➢ **Special needs.** Everyone has unique challenges in life. Others may help us with ours and we should help others with theirs. An emergency evacuation is certainly an area of opportunity to help, but we'll need to know in advance what kind of help is required. You'll also have to tell others what you need.

- **Mobility challenges** top the list. If you, a neighbor, or coworker, have mobility problems such as being confined to a wheelchair, be sure the proper groups know about it and are prepared to assist during an evac. Chief among the groups to notify are your Neighborhood Watch group, the Fire Department, Police Department, and your area's Emergency Management office. In addition, make sure the mobility challenged individual has **printed a list** of medical assistance instructions, equipment operating instructions, spare parts needs, and any other necessary instruction as a part of their evacuation preparations (use a copy of the appropriate "Family Member" data form from the "**Appendix**"). You should also look into getting an "escape chair" for the mobility challenged. It's a light version of a wheelchair.

- **Sight or hearing impaired** individuals may need the same consideration shown them. Also, in the case of some sight impaired individuals, you will need to take the needs of their **seeing eye dog** into account.

- **Generally challenged.** Let's never forget to offer help to our friends and their families who are mentally or generally challenged in other ways.

- **Language barriers** are next. If any neighbors or coworkers have language difficulties, make it a point for you, your Neighborhood Watch group, and your disaster-aware friends at work to **learn appropriate warning phrases in the language they speak**. You'll only have to learn a few things like, "evacuate," "emergency," "go to your house," etc. Being the responsible one, you might also want to know who can act as interpreter (record their contact info on your "Important Contacts" sheets).

- **Transportation issues**. Not everyone owns a vehicle. If you have a vehicle, can you carry someone who doesn't? Can you help them, find someone else who can help them evacuate if you can't? This is another subject for your Watch group, first responders, and Emergency Management office. Notify them as you did above with the mobility challenged. Also, one of the many, many useful projects a Volunteer Group could pursue is to provide limited emergency transportation.

➢ **Advance warning** of a pending disaster (Level 4 or 5 evac) provides enormous prep opportunities. That's why we gave you a whole section on "Emergency Alert" earlier and listed ideas on how to make sure you got the word. **Even if a warning doesn't become an actual event,** it's still great to have a "heads up." A hurricane could be heading your way, or maybe a massive wildfire. We also might see another situation similar to the Cuban Missile Crisis where we could be facing a nuclear attack. In any event, warnings will allow you to make some **last minute preparations** so you can be that much more ready should something actually happen. Let's look at a few things that we didn't cover under "Orange Alert" or "Red Alert."

- If you're at **work**, see if you can go **home**. If your at home and you *have* to go to work, carry some evac gear with you and make sure the rest of the family is as together and prepped as possible. The more people that are together and ready, the better.

- If you think you might have time, go top off the car's gas tank, check the oil, fluids, tire pressure, and all that. When you get home, back the car in, leaving it facing outwards and ready to bolt.

- Give a "heads up" to your out-of-town emergency contact people. Let them know your situation and for them to stay by the phones.

- Check road condition info sources for the current status of your primary and secondary evac routes.

- Email important files to yourself and make a quick backup disk from your computer.

- Double-check your "Last Minute" list and make sure everything's ready to go.

- If your family's at home, you should start activities that will keep everyone busy, and get you ready to go if you have to. You can start with a Fire or Evac Drill telling the family, "Let's practice what to do in case this thing gets too close." This way, you can be ready, but there's less panic involved since to the kids it's just another drill. However, on this drill, have everyone load the Bugout Kits, their personal packs, and the pets' gear. You'll also want to extend the "drill" to the point that you hook up trailers, bike racks, rooftop cargo carriers or the like, "just in case," or "just for practice." Leave all the "Mad Minute" and "Rake and Run" stuff for an actual evac. After the "drill" is over, tell the kids, "We'll just leave all this stuff in the car. I'm hungry. Let's eat" (and then **actually eat**). The kids won't mind postponing some work. This way your car's loaded, the family will be fed should you have to leave soon, and a meal will keep everyone occupied that much longer. Later on, if the threat diminishes, unloading the car will be easy and it was still a great exercise.

- Otherwise, after a drill, **stay** as **ready to go** as you comfortably can. Keep your wallet, keys, cell phone, and all other pocket or purse goods handy and ready, maybe in a basket on your dresser. If you sleep, sleep partially dressed or with your "bolt clothes" at hand. You might also consider keeping your car keys in your pockets or pinned to your clothing while you sleep, depending on the nature of the pending disaster. Relatedly, just as you packed your heirlooms during your area's severe weather season, you might consider making a dresser, bureau, or cedar trunk, your temporary "**evac prepack**" area to store some clothes and other supplies that might be part of your not-yet-packed "rake and run" items. This way, things aren't necessarily loaded in the car yet, but they're all in **one area** if you need to gather them quickly.

- You can also use this time to talk with your family about the situation and to calm any fears that might be developing. Sit down with everyone and go over the plans you've worked so hard to create and reaffirm to everyone that you're set and ready to go, that you're fully prepared, and that everything will be alright.

- Another advantage to the above drill excuse is that half your evac work will be done should a pending disaster such as an approaching hurricane or wildfire really become a threat. Now that the car's loaded and ready to go, the only thing you have left are some of the household shutdown and prep items that we covered earlier under "**Reaction**." Remember, the goal of any evacuation prep is to have you on the road as soon as possible, not only to outrun whatever's driving you out, but so you can be on the road and maybe at your destination before the unprepared masses clog the streets. (Consequently, if you evac early you'll also be out of everyone else's way when they finally hit the road.)

- **Note**: The points listed above apply only to situations that <u>threaten</u> a Level 4 or 5 evac, but that have not developed to the point that an actual evacuation would be considered. These are the things <u>you</u> would be doing while the lesser-prepared people in your community were just becoming aware of the news.

Other advance preparations you could make were covered under "**Reaction**" when we discussed specific threats. The bulk of everything else you truly need was provided earlier when we prepared the kits, destinations, and routes. The only real thing left to do with the actual "action" of beginning an evacuation is to look at those items you'll need to perform when the order is given, or the final decision made to leave. To help you remember what to do when you need to **bolt** out the door, we have another acronym to add to your list. This one goes to the word **B.O.L.T.**:

<u>B</u>ugout Kits   **B**ugout Kits should be loaded in your car, or at least ready to be loaded.

<u>O</u>thers   **O**thers should be notified that an evac is in progress and to gather or rendezvous.

<u>L</u>ock up   **L**ock up the house after turning off utilities or making preparations listed on your "Last Minute List."

<u>T</u>ake off   **T**ime to hit the road. You have to go gather your family and beat the crowds.

Since this is such an important subject, we're going to break these down a bit more for you and add a few helpful hints. **Note**: All these extra considerations are based on **time available**. A Level 1 evac is a Level 1 evac, and you'll have to **bolt** immediately. Naturally, the more people you have helping you, the more of these functions can be accomplished **simultaneously**. If your numbers are limited, or those around you unable to help, then perform the *detail* of these functions <u>**only**</u> **as time allows**.

Bugout Kits. This not only applies to the kits you packed earlier, but to the other items you might need as well. Also, since our goal is to prep for the worst case scenarios, these items below don't necessarily apply to Level 3, 4, or 5 evacs, only to the more rapid Level 1 or Level 2 emergencies.

➢ Remember, you might have to "bolt" from work, school, leisure activities, or anywhere in between. It would be nice if all emergencies occurred while you were safely at home, but that's a rarity. Therefore, remember to keep some gear at various locations, especially in your car, and keep gear even more accessible during the higher alerts and during any severe weather season. Don't forget your GYT Pack. (You might have to evacuate the area in which you work, and home may well be in a neighboring city and your final destination.)

➢ Remember to actually practice loading your gear into the car. This will show you some of your potential obstacles. **Hint**: Though we recommend roof-top carriers, be sure to load your more valuable and/or necessary gear inside the vehicle where it can be both accessible and better protected from theft.

Others. There's more to this "others" list than simply calling your immediate family or travel companions.

➢ Alert the family members that aren't with you that they need to either come home or meet you at a **rendezvous** point. Remember the plan you created earlier, when we had you divide the duties?

➢ Round up the people you are with, including any special-needs family members or neighbors, grab your pets, and load everyone in the car. **Note**: If you wind up carrying a special-needs neighbor with you, and this neighbor was registered as a special-needs person with local first responders, be sure to **call first responders** and tell them the person is with you. If you can't get through by phone, then **leave a note** on this person's residence, office door, etc.

➢ Consider the fact that just as you may help a **neighbor**, you may also need their help with loading your gear or other things. Make arrangements now on how you would help each other.

➢ Have a message drop on your property in case you <u>have</u> to evac and there's no other form of communication available. You can **leave a secure note** as to where you're going. Have similar drops in other frequented and familiar areas (like rendezvous points). If you missed it, go back to the section on "Communication."

➢ Under "Tornado Drill," we suggested you train your pets to head to your safe room on voice command. This is also a good way to round them up in an evac. Put a baby gate across your safe room's door and you can round up pets and toddlers quickly. The only time you should ever, EVER, leave a **pet** behind is if it gets away from you in a Level 1 evac and you need to leave the area *immediately* to save human life. If the pet has a way back in, leave food and water. At your first opportunity, hand off a copy of your "Pet ID Kit" to local animal control or the Humane Society. You may have **to fax this in** once you're at a safe location. Though these offices and their personnel will probably evacuate too, they'll be back in service as soon as possible, so they may find your pet before you're able to return. Having your pet's info will help them.

➢ If time allows after your preparations are made (assuming this wasn't done earlier in situations with more warning) you should make your share of your volunteer group's "phone tree" calls.

➢ Once you have everyone with you, if commo lines are open you should call someone at your evacuation destination to let them know you're on your way and give them **the route you'll be traveling and an estimated time of arrival** and the known status and location of all family members. (Depending on the nature of the evac, you may choose to do this from the road.)

Lock Up. You'll want to do everything you can, within the time limits you may have, to protect your home.

➢ **Practice** your household shutdowns on occasion. You should be able to lock all your doors, close and lock all windows, turn off your HVAC, gas (only if you know how to restart it), water, and electricity, all within **five minutes** or less! The "**Household Shutdown**" chart below will tell you which things to turn off based on the pending disaster. Other advance prep items for each disaster listed were covered in the "**Reaction**" section. (We'll repeat the "Household Shutdown" chart on your "Last Minute List" in the "**Appendix**.")

➢ The "Household Shutdown" list will cover different disasters, but in EVERY Level 2 or slower evacuation, you'll want to do the following:

❑ Turn off all appliances. You don't want to temporarily evac only to find your home burned down because of the toaster oven. **Hint**: Label your breaker box as to which breakers to turn off in an evacuation.

❑ Turn off your HVAC system. If you're evacuating from a fire, or any type of biochemical concern, turning off your heat or AC will prevent fueling the flames of a fire should your house catch, and it will help prevent drawing in outside contaminants in a HazMat scenario. However, if you're evacuating in **freezing weather**, make sure your water is turned off and pipes drained. Naturally, this is not a step you'd take in a Level 1 evac, but a precaution you'll want to consider if performing a Level 2 evac or slower.

❑ Lock all doors and windows. Not everyone is going to evacuate as planned and depending on the disaster there may be a short period of civil unrest, looting, or desperate people looking for supplies. Better secure than sorry.

❑ Close and lock garage doors. Not only is this a security issue, but closing your garage doors will also help save your house in a high wind situation. If the garage doors are open, wind can get up under your house and lift it off its foundation.

➢ Don't shut off your power unless you have to. You don't want to lose the food in the fridge. However, if you have to, in addition to the main breaker, you can shut off your **power** by yanking your **meter** (the glass part) out of its housing. Either take it with you or lock it up inside your house so it won't be damaged or stolen.

➢ Water should be shut down from **both** the **outside** meter and any **inside** valve that controls the whole house. If you only have time to do one, make it the **outside** valve. The main goals of shutting off the water are to prevent contaminated water from getting into the house's plumbing, and to retain water in your pipes should you need those extra gallons if you return home and water service is not yet restored. One extra consideration you should give your water is if you're evacuating in sub-freezing temperatures and you've shut down your HVAC or lost your heat. Think ahead as to how you'd drain your pipes to protect them from freezing and bursting if you couldn't leave faucets dripping. See our "Household Shutdown" checklist below.

Take Off. There's nothing left to do but be on your way to pick up others and hit the road.

➢ If you didn't have time to gas up your car, use your stored **lawnmower** gasoline. That's why we suggested that if you have gas-powered lawn equipment or generators, that you keep as much gas as you can safely store on-property. (DON'T fuel your car with any gas that's mixed with motor oil. Mark the containers.)

➢ Don't forget to check your vehicle's **fluids**, especially **oil and water, and tire pressure** (though you should have always kept an eye on them all along). You don't want a minor problem becoming a major obstacle in the middle of an emergency evacuation.

➢ During the process of alerting everyone and loading the vehicle, you should have your radio or TV on to keep up with **current information** regarding the threat you're under.

➢ If you have the sources and a way to check them quickly, try to see if there is any news concerning the road and traffic conditions of your primary evacuation route. We gave you info sources above under "Evac Atlas." If there is no information at all, proceed with your primary plan.

HOUSEHOLD SHUTDOWN					
Evacuating From:	**Gas**	**Water**	**Power**	**HVAC**	**Other Considerations**
Nuclear Strike	On	On	On	**Off**	If your home wasn't damaged, do minimal shutdown and get YOU to where you need to be. (**Level 1.**)
"Dirty Bomb"	On	**Off**	On	**Off**	Keep your water off until local authorities have determined if municipal supplies were contaminated.
Biochemical Event	On	**Off**	On	**Off**	Keep your water off until local authorities have determined if municipal supplies were contaminated.
Wildfire	**Off**	On	On	**Off**	You'll need your water and power on for your sprinkler and other systems.
Hurricane	**Off**	**Off**	On	On	In addition to boarding your windows, don't forget to protect your gas meter, gas connections, and electric meter against debris. Turning off the water helps prevent contaminated water from getting in to your plumbing.
Flood	**Off**	**Off**	**Off**	**Off**	Though the bottom half of your house is soaked, turning off gas and power may prevent a fire that consumes the top half. Turning off water helps prevent contaminated water from getting in to your home's plumbing.
Dam Burst	On	On	On	On	Forget utilities. Get YOU out of the way. (**Level 1.**)
Earthquake	**Off**	**Off**	**Off**	**Off**	This will be an after-the-fact evacuation, but make sure utilities are off to prevent secondary damage.
Tsunami	On	On	On	On	Get YOU out of the way if you can. (**Level 1.**)
Note: At the bare minimum, you should remember to turn off all appliances.					

We'll sum up this section with a very important point: You should practice everything as much as you possibly can. Though we started this section with a discussion of a more "leisurely" Level 4 or 5 evac warning that *might* become an evacuation, always remember that the goal of a **Level 1** evac is to have you, your family, your transportation, your supplies, your house, and any passengers all prepped, loaded, or secured, and **on the road in <u>under</u> 30 minutes**.

Now that we've covered the basics of bolting out the door, let's look at the most important evacuation consideration of all, and that is that there's absolutely no guarantee that the family is going to be at home and together when the next disaster hits. Let's discuss how to coordinate everyone's whereabouts.

Rendezvous!

The most stressful and difficult aspect of any emergency situation, whether it involves evacuation or not, is making sure everyone in your family or group is safe and accounted for, and is able to communicate and unite with the rest of you. We hinted at this a little under "Communication," we discussed the "buddy system" under "Fire Drill" earlier, and we covered choosing some evac routes with potential rendezvous sites. This section will put the pieces together.

We'll help you prep for keeping the family together by covering the following in this portion:
1. **Knowing the existing emergency plans of work and schools.**
2. **Discussing automatic reaction procedures for everyone if no communication is available.**
3. **Setting rendezvous points and contact persons.**

Emergency Plans of Work and School
Is someone else's emergency plan going to take precedence over, or influence your family's plan?

For example, many schools have emergency reaction plans of their own where children are either evacuated to a predetermined location or the students shelter in place on school grounds. Many scenarios could prevent parents from picking up their child at school. Do you know the reaction plans of your child's school? Does the school have a plan? Do they have a specific evacuation destination? Has the school planned at the level you're researching now? Do you have children of different ages attending different schools? Will they have different evacuation destinations? What are bus drivers instructed to do if a disaster occurs while the students are enroute to or from school?

What about your workplace or your spouse's? Though not mandatory like a school's emergency plan would be for students, some businesses have thought ahead and planned to take care of their employees in the event of disaster. Does your workplace have such a plan? Is it one you could see being a part of? If circumstances dictate that you evacuate with your coworkers, do you have communication and rendezvous plans with your family that take your workplace's plan into account? You'll need to know these things so you can adjust your advance plans or possibly your actual evacuation as it's happening.

Right now we're asking more questions than providing answers. Interestingly enough, most of the answers to the questions we just asked will be answered as you fill out some of the various **data forms** in the "**Appendix.**" As we mentioned earlier, filling out the data forms and emergency contact info in the back of the book will be almost as much of an education as reading what we've written so far.

Discussing the Automatic Reactions
One night during your three-month goal period, while the family is together, discuss what everyone should do if a major disaster occurred while the family or group was separated and no one was able to get in touch with anyone else. This is also a great opportunity to reassure children who may be rather concerned about what might happen during the next terrorist attack. Essentially, this is a review of the things we mentioned while telling you to "divide the duties," so let's consider a few things important enough to reiterate:

➢ Get out your Emergency Risk Score Checklist, your Threat Map, and Evac Atlas. Review the response plans you made earlier to each type of probable threat for your area at each of your marked locations or potential target sites and for the various times of day. While reviewing your existing plans and responses, keep the following in mind:

- Revisit where family members might be, if businesses will be open or closed, how rush hour traffic might affect your plans, how traffic might be affected by different scenarios, etc.

- How will the scenario affect each family member's travel? Repercussions will dictate routes, routes will dictate which rendezvous points (or destinations) may or may not be accessible.

- Decide for each scenario where everyone should gather. Will it be at home or will it be somewhere else? (We'll cover the finer aspects of choosing a rendezvous point below.)

- Decide for each scenario which duties will automatically fall on which family member. Who's closest to home? Who should go pick up the kids? If you're a single parent, is it more logical to go home to pick up the gear and then go get the kids or vice versa? Have you coordinated any of your planning with other parents that have children in the same school?

- How capable are family members? For example, teenagers in high school are more able to make their way home or to a rendezvous point than a child in elementary school.

- Focus on the importance of good communication, and of letting other family members know if your normal daily plans change. For example, if Junior usually walks home from school, make sure he lets you know if he has stopped off at a friend's house for a little while on the way.

➤ Have agreements with neighbors. Do you have a very trusted neighbor or family member living near your home? You might consider adding them to your automatic reaction plans. For example, your neighbor may work in an area close enough to get home in certain emergencies whereas you might not. You might want to clue your neighbor in to where your evac supplies are and include them in your rendezvous and evacuation plans, letting them know what to do in case certain scenarios occur. In other scenarios, the tables might turn and you'll need to grab your neighbor's supplies for them.

➤ Some scenarios could see family or group members separated, evacuated in different directions, and unable to reunite until later. This is why we suggest having redundant contact persons and why we suggest every family or group member have a copy of your "Important Contacts" info.

➤ Lastly, what about your special needs people from above? You need a plan B to make sure they're taken care of in the event you can't get to them yourself. Either have your neighbors agree in advance to help if you can't or try to get through to first responders to let them know you can't get there.

➤ To tie these considerations together and get to the heart of your automatic reaction plan, let's cover the things that each family member must keep in mind if you're separated when an emergency occurs and they can't communicate right away. It'll be slightly different from our "**BOLT**" procedure above. Since you're not together and not ready to go just yet, you can't quite "bolt," so our acronym here is "**S.C.O.O.T.**"

<u>S</u>afety	**Safety** comes first in any endeavor. Protect yourself from the current emergency.
<u>C</u>all	**Call** <u>if</u> you can to coordinate what's going on.
<u>O</u>ptions	**Options** are your best friend in an emergency. Know yours in advance and react accordingly.
<u>O</u>rder	**Order** should dictate options chosen. Set a priority for certain actions or locations.
<u>T</u>ime	**Time** is crucial in an emergency evacuation. Set limits for each step above.

Let's explain these in a little more detail:

❏ **Safety** comes first. Teach everyone that wherever they are, make sure they're protected from whatever's going on. Revisit the "**Reaction**" section if you need to. Also, make sure all family members are as well trained as possible in each of the subjects listed under "Basic Training."

❏ **Call**. After you're safe, you'll want to see if standard lines of communication are open. If you can communicate, then coordinate your family's reaction then and there. If local calls won't go through, try a long distance call from your cell phone, a regular land line phone, and a payphone. You might be able to reach your out-of-town emergency contact person. **If not, this is when the automatic reactions come into play**.

❏ Know your **options**. Do you have any sort of transportation? What's the nature of the emergency you're in the middle of? Can you make it home? Can you make it to the nearest rendezvous point? Should you try to rendezvous and evacuate or does the situation dictate you find one of your nearest "safe houses" (a friend or relative's house) to stay put overnight and ride the situation out? Or, should you evacuate completely?

❏ Set a specific **order** of steps to take or options to choose. Should everyone head home, or should they head toward a specific rendezvous point? Or, should they go to one of those and then after a certain time period head towards the next? Rather than locations, should your "order" pertain to the options listed above? For example, depending on the scenario, you might give priority to group evacuation plans of the location you're at, with a second priority of seeking safety and transportation with friends. We wish we had more answers for you here, but order can only be set by you based on your unique situation and the threats you're most likely to face. **Hint**: Keep your options few in number and simple in procedure.

❏ **Time** is crucial, so set a time limit on some of your options and order of events above. For example, decide how long someone should wait at a rendezvous point before accepting alternate transportation. How long should a family member wait at home before loading the gear and going out to meet the others? Besides that, what is it that dictates the start time for determining how long you should wait?

The downside to this section is that there are no concrete answers we can give you, other than these points and considerations, regarding a specific *automatic* reaction plan that will cover all the ways your unique situation will be affected by a multitude of scenarios. The best thing we can tell you is to talk through the above points as much as possible and don't make your plans so complicated that people are confused, or so rigid that you can't adapt to a changing situation. The best plans are simple and straightforward, and can be applied to the most emergency situations.

Though we've been rather complicated in our descriptions of plans and our lists of equipment, the end result is simplicity. You'll have your gear, you'll know what to do if something happens, and you'll know where to go. Simple. It's been a rather wordy journey heading that direction, but the goal is to make your evacuation a simpler process.

Again, once you've reviewed and/or decided on the reaction to each scenario, **write it down**. **Make sure everyone has a copy** of your "**if this happens, do this**" plan so everyone knows what to do, when, where, and with whom.

Rendezvous Points and Contact Persons

We had you choose evacuation routes and create your Evac Atlas first since those are more important, and they, along with your Threat Map, will determine where some of your rendezvous points are located. While you have the maps out discussing reaction plans, let's look at them for rendezvous points.

Let's start with some general discussion and considerations about these points and your contact persons.

> Rendezvous are more successful when you know who's where and when. For that reason, you'll want to have friends or family members that you can call and relay messages if your immediate family or group members can't be reached. Pick at least two in-town people and two out-of-town people and list them on your "Important Contacts" pages in the order they should be called (though you wouldn't call all of them, you'd just call until you got someone).

> A key to choosing who your contact people will be is to choose people who are not only very reliable, but **consistently easy to find**. Also, the out-of-town contacts should be instructed to remain available if they hear news of any event affecting your area.

> To add to your contact options, encourage schools and workplaces to setup a YahooGroups email group or similar, to keep students, employees, and their respective families informed of any pertinent info during an emergency. This would include any evacuation option that had been exercised by the school or workplace.

> As for your specific geographic rendezvous points, you'll want to choose a **minimum of three** and a **maximum of nine**. One will always be just outside your home for purposes of evacuating a fire. Of the others, you'll want to choose at least one in-town location and at least one out-of-town location in case your scenario prevents you from staying too close to home. This way your minimum of three gives you one near the home, one in town and one out of town, and your maximum of nine gives you the one at home and four in town and four out of town (one local and one distant in each main compass direction).

> Though we say a maximum of nine locations, the truth is, the fewer the better. There's genius in simplicity regarding plans, and even though having options is good, having too many only causes confusion.

> It's also good to have an out-of-town rendezvous point because of the option mentioned earlier of a family member evacuating with a group other than the family. Your family's rendezvous point may well be that particular group's final destination. You could pick up family there and then head on to your own destination if your plans differed. For example, a school may evacuate to a safe point not too far out of town. That could be your rendezvous point. You'd gather the rest of the family, go to the school's evac destination, pick up your child and maybe be on your way to your own destination.

> Use the buddy system wherever possible. For example, in a family of four, two members might meet at one point, while the other two family members meet at another point, and then everyone meets in the middle.

> Speaking of the buddy system, we'll remind you again that it's a good plan to decide on specific duties during an emergency. Decide who goes after whom or what, and then when and where.

> Pick something memorable, accessible, and public, yet unique for your rendezvous points. If there's a big mall in your area, guess what? EVERYONE's going to choose the mall. Pick something similar but different, or an area of the mall property that no one else will use and that has easier traffic access.

> You might also base your rendezvous points on fairly common chain stores. Pick something common, but not so common that there's one on almost every street corner. This way, regardless of the direction in which you're forced to evacuate, you're already halfway to defining your rendezvous point and it will be easier for the family to remember.

> In some instances, you may be able to "check in" at your rendezvous point. For example, one rendezvous location might be the lobby of a motel on the edge of town. Whoever arrives there first should tell the desk clerks their name and that they are waiting on a party who might call there to find them. This will be the exception rather than the rule, but you have to think about all possibilities.

> Ever organize your errands by traffic flow and the "no left turns" rule? That's when you plot the order of stops you want to make during the day where you can get to all of them without having to make left turns

through heavy traffic. It keeps you from getting blocked in someplace. Keep the same principle in mind when scouting rendezvous points. You'll want to get to them and back without risking getting blocked in by traffic which may be heavy in a disaster.

➢ Know your alternate parking spots and how to get to the point on foot. For example, one of your children may be able to make it to the rendezvous point easily, whereas traffic or destruction prevents you from getting the car any closer than a couple of blocks away. Know where you can park your car in this scenario and safely make it through on foot to gather your child.

➢ Relatedly, keep in mind that some scenarios will see mass panic. This may mean heavy traffic, near riot conditions, etc. Make sure the rendezvous points you're considering are not only easy to get in and out of, but are away from danger areas, and offer some physical protection.

➢ Remembering some lessons from the "Communication" section, be sure each rendezvous point has both a very specific "meet point" at the location and that this same meet point can be used as a message drop. For example, if your rendezvous point is a gas station, and the station has three payphones outside, choose one specific phone as your meet point and message drop.

➢ Speaking of the function of your rendezvous point, you should always try to kill two or more birds with one stone. Make your rendezvous point useful to your cause. For example, you might want to choose a grocery store. You can pick up whoever it is you're meeting, and load up on a few supplies, all in one stop.

➢ In addition to supply, your rendezvous point could double as temporary protection. For example, you might pick a friend or relative's house close to your work or school location as a potential rendezvous point. This might also be a place to stay the night if a local emergency, such as a severe storm or brief civil unrest, prevents you from going straight home. We mentioned this earlier and called it a temporary "safe house."

➢ If your child rides a bike or walks home from school, know the route they take and require that they stick to it, especially during higher alerts. You may have to search their route to find them in case something happens while they're on their way home and they didn't know to hurry home or go to a rendezvous point. Also be sure to set a little rendezvous point somewhere along their route so they know where to go to make it easier for you to find them if they *do* know something's going on. Make it a safe public place or a friend's house.

➢ Speaking of children and evacuation options, teach children who NOT to ride with, and also how to ID officials, such as knowing the difference between Police, Sheriff's Deputies, Fire officials, etc.

➢ If your rendezvous point is unavoidably a location that throngs of others will be at, such as the school evacuation destination, carry something that will get the attention of the person you're there to pick up. It could be a sports whistle or an air horn, but better yet, make it something **quiet** like a small colored flag during the day, or colored chemlite sticks at night. Do whatever it takes, within reason, to make you stand out from the crowd. Make sure the person you're picking up knows what to look for.

➢ As an additional option to a fixed rendezvous point, figure out some simple rules for a **floating** rendezvous point that can be used in other situations. For example, if you're evacuating in a small convoy and the group gets separated, you might pick a common chain store where everyone will stop to meet if the group is separated. The rule could be as simple as saying, "Okay, if we're separated, the next *such-and-such store* you see, stop there and the others will catch up." This is a low-priority, but you have to cover all bases.

For each rendezvous point you're considering, give it the final litmus test and ask these questions:

Choosing a Rendezvous Point
The marked questions must have a "yes" answer. For the rest, the more "yes" answers the better.

		Yes	No	
Must be Yes	1			Are you familiar with this location and its immediate area?
	2			Is this point reachable while avoiding your listed Threat Map hazards or "bottlenecks?"
	3			Do you travel to or near this point frequently?
	4			Is this location easy to get in and out of without getting blocked in by traffic?
	5			Are there two or more routes to this spot?
	6			Is this spot conveniently located among home, work, school, etc?
	7			Is this location on or near one of your evacuation routes?
	8			Have you identified an unmistakable, precise spot at this location as a "meet point?"
	9			Is this location within walking distance (via a safe route) of work, school, etc?
	10			Are there alternate forms of transportation available to and from this location?
	11			Does the location offer an additional benefit, such as gas, food, supplies, or temporary shelter?

12		Are schools or other family member functions along the route to this location?
13		Is it easy to give directions to this location to people who've never been there?
14		Is this location near (or at) a potential local emergency shelter?
15		Is this location well within law enforcement patrol zones?
16		Are there safe, direct, routes from this rendezvous point to the next?
17		Is the location some sort of public or government facility such as a post office, etc.?
18		Does the rendezvous location offer protection from weather or other dangers?
19		Is there some sort communication nearby, such as payphones, internet cafe, etc.?
20		Have you devised a way to leave a note or message if one person misses the other?
21		Is the location a publicly accessible commercial facility?
22		Is your rendezvous point a friend or relative's house?

When your rendezvous plan is set, write it down, mark the locations on your Threat Maps and pertinent Evac Atlas maps, fill in the Important Contacts and other forms and make sure everyone has a copy.

Now that we've helped coordinate everyone's whereabouts, let's talk about how you'll pick everyone up and get them all to safety.

Transportation

Though one of the most important concepts regarding an evacuation, this section on transportation will probably be one of the shortest. We've covered a lot of what you needed to know under "Basic Vehicle Prep," and we'll try to answer some other questions here. Let's look at some basic points:

➢ We hear a lot of questions and comments regarding types of evacuation transportation. Some people like RVs since they're a house on wheels. Some like Hummers or SUVs since they have good cargo room and are just as maneuverable as a car. Others like motorcycles since they can zip in between cars in gridlocked traffic and aren't necessarily restricted to the road. Who's right? Basically, they all are. Each type of vehicle has its own strengths and weaknesses. But, as long as each vehicle fits well with its owner's evacuation plan, then all is well. More important than *type* of vehicle are the *qualities* of your vehicle:

❏ **Dependability** tops the list. No vehicle will be worth a dime if it won't get you where you need to go.

❏ **Accessibility** runs a close second. Your fully stocked, superbly maintained RV isn't going to carry you or shelter you if it's in a lot across town and you can't get it when you need it most.

❏ **Capacity** comes next. Your transportation has to be capable of carrying you and your family, and all your necessary supplies.

❏ **Maneuverability** is also important. Your vehicle has to be able to negotiate any and all of your evacuation routes and the routes to and from your rendezvous points, or each of these needs to be designed around your vehicle. For example, if you're driving an extended cab open bed pickup and towing a pop-up camper, you wouldn't pick rendezvous points that were only accessible down small alleys.

❏ **Protection** is an often overlooked quality. Your vehicle should be able to not only protect you from normal traffic hazards and dangers, but from weather, and from any "fringe" effect of your most likely threats. For example, if you live downwind from a chemical plant, then you probably wouldn't want a convertible as your primary evacuation vehicle. You'd want something you could seal easily and quickly. Or, if you were evacuating a hurricane, you might experience some advance foul weather and would want something more substantial than a motorcycle.

❏ **Crime protection** is another factor that should be seriously considered. In a mundane evac, such as from a hurricane, there is little panic. People simply pack their stuff and go on a road trip. More severe scenarios will see instances of theft and carjacking where unprepared criminal types will try to steal what they need. Riots are another possibility. The better you can lock yourself in, the safer you'll be.

➢ As we discussed earlier, carry alternate transportation with you, <u>IF</u> you have time to load it. Strap on your kayak or canoe, toss an inflatable raft in the trunk, and put the bikes on the racks. If you run out of gas, would you rather walk or ride your bike to the gas station? If you're stuck in permanently gridlocked traffic and a mushroom cloud is drifting your way, wouldn't you like to drop your kayak in that river just up ahead? We even heard of one girl in New York who kept roller blades in her desk. During the blackout while so many were walking, she simply skated home. These are remote scenarios, but when researching options for numerous life-threatening scenarios, you have to look at every concept and put creative ingenuity to work. Some ideas won't work at all for some people, but may work for you.

➤ Have **alternate transportation** options in general. With your GYT Pack we suggested you carry mass transit tokens or passes and that you also carry car rental discount cards if you have them. We also had you select quite a few transportation alternatives under Evac Atlas. You need to keep in mind that your evacuation may well be a **multi-stage operation**. You'll have to get out of where you are, pick up others along the way, get to your gear (hopefully) and then get on the road. You might need public or alternate transportation for any or all of these steps. Therefore, you'll want to maintain information on each of the following and keep it with you at all times:

● Work with others. What if your workplace or neighborhood are blocked in? Do you know ways to walk out of those areas to safe locations where someone could pick you up? Or, could you go pick up others?

● Keep **schedules and contact info** for each of your area's **mass transit** systems. This would include city and local bus lines, train and subway systems, and taxi services. Know their routes, stops, stations, terminal locations, and main terminal locations. Contact each now, and find out what their official reaction policy will be in an emergency. Some will simply shut down while others will do all they can to help evacuees. Our one warning for this and the alternatives that follow, is that you should not expect any system to be running as scheduled in an emergency. Always have options for your options.

● Gather the same kind of information for **county and intra-state transit systems**. These might be bus, van, or train lines. The sources we gave you under Evac Atlas will help you find these systems.

● You'll also want good info about **cross-country bus and train lines** such as Greyhound and Amtrak, and should have already gathered that as you created your Evac Atlas.

● Know the location of auto rental outlets in the area of your work, home, and each rendezvous point.

● We saved **air transportation** for last since the more dire scenarios may see a temporary grounding of passenger flights, and private aircraft may be used to evacuate the owner's family. However, gather the contact info since air travel certainly is an option and in an emergency, options are our friends. In addition to the normal commercial passenger lines, you should look through your city and regional yellow pages to see what kind of **private flights** are offered, and ask what their policies are for emergency situations. Some civil air patrol groups and their close cousins will actually have some evacuation assistance plans in place. Find out what's available as an option. Again, a rare situation, but options should be explored.

● Let's finish this little portion with a quick example of how the above might tie together in a multi-stage evac. Let's say you work in an office building in a downtown metro area, and something similar to 9-11 happens, yet the news says there seems to be a chemical weapon involved and the order to evacuate downtown is given. Phone systems are useless because of everyone calling everyone else. For the purposes of this scenario, let's say that you're upwind and far enough away from the site to be safe. Nevertheless, you decide to put on your hood and breathing filter from your GYT Pack, and cover all your exposed skin as you head out the door to see that traffic is already gridlocked around your office building. You go to the parking deck to get your small supplies backpack from your car and your bike off its rack. You bike your way out of the immediate gridlocked area to the nearest bus stop where you hop an outbound bus. At the terminal you bike half a block to a car rental facility and rent a car. You circumvent the gridlocked area and go to your family's main rendezvous point to pick up a family member and head home.

➤ In addition to multi-staging alternate transportation, you may have to multi-stage your own transportation. For example, let's say something similar to the above happens, but you're already at home. However, your RV is in a lot across town, or maybe your evac is a boat at the marina and a straight-line travel to either is out of the question. You might send your family and gear one way in one vehicle while you bike and bus, or motorcycle over to where your main evac vehicle is located, get it, and then meet the family at your main out-of-town rendezvous point. We heard from one individual who even had a jet ski rental agreement with an out-of-the-way shop so that in an emergency he could get a jet ski to ride into the marina from the water to get his boat if the roads to the marina weren't negotiable. Rare scenarios or options? Maybe, but the point is to <u>know</u> your options and make the <u>most</u> out of what you have.

➤ When possible, travel with others. If you're part of a group, it's likely you've chosen a common rendezvous point and destination, and that you'll travel in convoy fashion during an evacuation. Though a little harder to coordinate as emergencies increase in severity, it's the safest and most efficient way to travel. Also, if you're in a group that's heading to the same destination you can divide group gear responsibilities. One person might be in charge of all the toilet paper for the group, another would bring paper plates and so on. Lots more on this under "**Cooperation**."

➤ Try your best to keep your family all in one vehicle. Though you may own another vehicle and wish to take it with you, consider the fact that everyone is safer if together. Besides, fewer vehicles means lighter traffic.

➤ **Hint:** Though they may not be of legal age, **any level-headed child big enough to reach the pedals should be taught how to drive**. We don't know all the situations we may be involved in and we need all alternatives available to us. For example, the adult in the family may be incapacitated. This able child might be home alone when something happens and is forced to evacuate. It may well be this "child" that loads the gear and goes to get the adults. Teach your child well, give them the training, include them in what you do, give them the responsibility, and it's likely they'll reward you with sterling performance.

Life on the Road

So, you've had your gear prepped and packed for a while now, something happened, you reunited with your family, got your gear, loaded your vehicle, and are now starting on your way along one of the evac routes you mapped for your atlas. Time to sit back and relax? Nope. Sorry. Okay, you can relax a bit since some of the more inherently dangerous steps are out of the way. However, evacuation travel carries its own unique risks.

The good news is that most of these risks are associated more closely with the panicked mass evacuations. With hurricanes and wildfires, most evacuations will be a ho-hum chore. Load your stuff and go and you might even make new friends on the road. There's little to no panic involved. However, detonate a nuclear warhead or release a biochemical weapon and things change... drastically! Let's look at a few things regarding safety and logistics to keep in mind while on the road:

➢ The first thing you have to do is get out of the area and to the open road. Your biggest risk at this juncture in a worst-case-scenario is in having your vehicle and gear taken from you. So, keep your doors locked, don't stop for strangers, don't stop at all if you don't have to, and try not to advertise the fact that you're carrying survival equipment. Load as much of your gear as possible inside the vehicle and keep it covered. Anything that has to travel outside the vehicle should be locked if possible, or at least well tied.

➢ One technique criminals may employ to get you to stop and get out of your car is the minor fender-bender trick. They'll bump you with their car just enough to call it an accident in hopes you'll stop and get out. **Don't do it**. If you didn't cause the accident and both your vehicle and theirs are able to move, don't stop. Call 911 and ask for advice. Many states have laws that allow you to continue on in certain circumstances. If you want to research this now, call several sources for an accurate picture of the laws along your evac route. Call your Police Department, County Sheriff's Office, State Patrol, DMV, and your State Bar Association.

➢ While you're asking about these laws, also ask about "self-filed accident reports," and see if you can **get a copy** to carry with you in your Office Pack. Many states have these because of the numerous incidents of minor scrapes and scratches that can't be attended to, especially when something larger is going on. To add to anything they send you, be sure to check out our **Accident Report** form in the "**Appendix.**"

➢ Idiots will come out of their shell. While we're talking about staged accidents, let's mention real ones. In a mass evac, you'll see every manner of stupid and aggressive driving you've ever seen, all rolled into a few hours. People will be in the median, on the shoulder of the road, and even in creep-along traffic, you'll still have people trying to cut you off and weave ahead in front of you. Rule one is to be a good defensive driver and pay attention to traffic safety. This is another of many reasons we tell you to be on the road as soon as possible. You'll want to be long gone by the time these people hit the road.

➢ "Road Rage" will be a considerable problem. The stress factor among normally passive drivers will be incredible. You, being responsible and forewarned, should make allowances for this. If you're the victim of an aggressive driving act and "gestures" start flying at you from the another driver (even if they're at fault) do what you can to defuse the situation rather than let it escalate. Just give a smile and shrug your shoulders as if to say, "Sorry... not sure what happened." And let it go. Your goal is to get you and yours safely to your destination. Forget the idiots. Stay focused on your goal.

➢ Whoever is driving is actually going to have it easy for the first few hours since they have something to do. For your passengers, it may be a different story since they now have time to think about what's going on. This is where stress will start to show itself, and it's this stress that will probably be your biggest enemy in an urgent scenario. The trick is to keep everyone occupied. Have someone read the map and navigate. Have the kids play a game, listen to their radio for news, or just listen to a tape or CD. If you have pets, put the children in charge of reassuring and calming the pets. Sometimes the best therapy is to *be* the care giver.

➢ Another way to pass the time and keep everyone occupied is to create a journal of what you're doing. Not only will this pass the time, but it will document your actions should insurance become a factor in anything you do, and your experiences may help others by offering hints and pointing out pitfalls to avoid should a similar situation happen to them. Have everyone write notes on what they did, their thoughts on what's going on, suggestions on what could have gone better, etc. Also document your trip with photos and journal entries about where you are, where you're going, and what you did and saw along the way.

➢ In addition to simple distractions to relieve tension and stress, you might want to spend a little time now learning various "panic mitigation" techniques. If you'll look ahead to our section "Coming Home II – The Emotional Aftermath," you'll find numerous mental health sites and sources listed. Review their information and see what the pros in that arena have to offer. What we'll do here and now is give you another acronym to help you remember to stay calm. Appropriately, it goes to the word **C.A.L.M.**:

Count	**Count** your breaths. Breathe slowly, deeply, and count. Now **count** your blessings.
Affirmations	If you have any positive-thinking or religious **affirmations**, repeat them to yourself.
Language	Watch your "bedside manner." Those you're with will take their emotional cues from <u>you</u>.
Mission	Focus on your **mission**, what you have to do, and the things others can do to help.

➤ Security awareness has to continue when you make stops for refueling or restroom breaks. Use the buddy system anytime anyone leaves the car, and make sure at least one person stays with the car. Granted, 99 percent of the people around you will be good people in the same boat you're in and will be zero threat. However, in mass evacuations, that other 1 percent will definitely be a valid concern, and that's all it would take to ruin your trip. This is another reason we suggest being a part of a group. There's safety in numbers.

➤ You'll also want to set a time limit and on-site rendezvous point if you have to separate for any reason to get supplies, etc. An example might be that you pull off the road for gas and there's a grocery store nearby. You might send two people together to the store with a requirement that they meet you at the car at a certain point in the parking lot within a specified time period. Remember, you're in an emergency evac, you won't have time to play around, and you won't have room for lax security.

➤ In addition to the buddy system, you should pretty much conduct yourself like you would if visiting a third world country or high crime area. Walk with a purpose, be aware of your surroundings but keep your eyes dead ahead, don't dress as if you have money, don't dress for sexual attraction, don't carry an exposed weapon, don't strike up unnecessary conversations, but don't be rude. If someone tries to make contact and you don't want to talk just give a friendly smile, a wave, and keep walking.

➤ If you do pull off the road to stop, make the most of it. Stop where you can take a restroom break, top off the gas tank, pick up more supplies if need be, and attempt communication with your emergency contact people if you were unable to earlier. Make as few stops as possible. **Hint**: Since you might be tied up in traffic for indefinite periods or in situations where it would be best not to stop, you might consider carrying some plastic jars with screwtops in order to preempt some of those restroom stops. Not a pleasant thought, but it can head off some less pleasant alternatives. Safety and security come first.

➤ **Hint**: When you make these stops, pay with checks or credit cards if you can, and hang on to your cash at every opportunity. You don't want to use up your cash when you don't have to, in anticipation of situations down the road where you can't use anything but. Also, save ALL your receipts. You'll need them later.

➤ We're going to cover more safety and security, and especially the considerations of using force, later under "**Isolation**." However, we do need to cover a couple of things here since in prepping for your personal safety, you might consider carrying a weapon. As we said, we'll cover the bulk of that topic later, but for now let's leave you with a few things to keep in mind:

❑ In any situation, call 911 first, regardless of whether you think they can help or not.

❑ Stay within the spirit of the law regarding weapons. To find out about your area's laws, contact your local Police Department, County Sheriff's Office, State Patrol, your county Probate Judge or Justice of the Peace, and your State Bar Association. You can find some other laws online at http://www.packing.org.

❑ Even if you don't intend to carry a firearm, it might be to your advantage to become licensed to do so. In some areas it *may* make it easier for you to justify carrying a less lethal weapon.

❑ Remember that the vast majority of the people around you will not be a threat.

❑ If you have a weapon, it needs to be accessible. Locked in the trunk won't help you, and a panicked evac is one area you may need it. Don't keep it visible, brandished, or in reach of children - just accessible.

❑ As with everything else, you need to have options when considering self defense. The more ways you can protect yourself without unduly hurting another, the better. Remember, some of the people that might try and take your gear may be acting out of panic and normally wouldn't hurt a fly. This doesn't mean they have a right to take your things or hurt you, it means that it might not take much to dissuade their attempt and that they won't deserve any lasting damage.

➤ Other thefts might come to you in the form of price gouging. Gas stations might try to quadruple their prices, motels might do the same, and food...? However, disaster price gouging is against the law. If you run into this, the first thing to try is calling local Law Enforcement either through 911 or by way of one of the non-emergency numbers you have in your Evac Atlas. They might not be able to send an officer down, but they can certainly call the establishment in question to explain that gouging is against the law.

➤ The last consideration in your long road trip is getting off the road, and the only question here is whether or not traffic flow is being controlled by authorities. Remember earlier where we suggested you pack documentation that ties you to your destination? This is where you'd want to have it ready.

Life at the Destination

Okay, you prepped your gear, faced the emergency, got the family together, protected them during the evacuation, and now you've arrived at your destination. Can you rest yet? Well... okay. Get some rest. You're going to need it as we'll explain in this section.

Your journey's not really complete until you've returned home and life has returned to normal, or until you start a new home. This is another step along the way, and as usual, there are still things to work on and lots of ideas to consider. Here we'll cover everything from getting your bloodpressure back to normal after an evac, through cleaning up after whatever it was you escaped.

Goals of Life at the Destination:

1. To give you a glimpse of what a **temporary relocation** might be like.
2. To prepare you for what to expect if you have to stay at a **public shelter**.
3. To help you with **resupply, communication, and getting ready to head home**.
4. To set you up for success when putting your "**Planning Ahead for Phase II**" stuff to work.
5. To take the edge off some of the **psychological effects of dealing with a major disaster**.

We've divided this section into:

1. **Shelter Dynamics.**
2. **Coming Home.**
3. **Coming Home II – The Emotional Aftermath.**

Yep, even though we've called this your "destination," it's really only a halfway point; another step in a long process of "shifting gears," of starting, stopping, and stressing, that will have its own effect on you. Understanding what's about to happen *to* you, *with* you, and *within* you is just as important as putting together the kits that will ensure your physical survival during this time period.

Hopefully, you were able to make your destination a friend or relative's house. That will probably be the easiest of all on you as you'll be with people you care about and vice versa, and you'll have a low-pressure environment for some much needed unwinding and destressing before starting the second half of your evacuation - coming home.

If you chose a remote campsite to while away a little time, that's just as good provided you have plenty of supplies and everyone's in good health. A little peace and quiet out in nature will do a body good.

Peaceful, but a little less desirable, is to stay in a hotel or motel. We say less desirable because of the potentially high costs. Unnecessary costs will only drain your supplies and add to your stress.

Next down the list would be staying at a public shelter. We don't call this the bottom of the list because having shelter is still a good thing. The "bad" things would be not making it to the shelter, or finding yourself in situations where you had to put your outdoor survival and emergency medical skills to use. But as far as "destinations" go, public shelters carry with them the greatest need for things to remember and points to consider. If you're prepped for not making it to your destination at all, and ready to deal with life at a public shelter, then most other alternatives are a piece of cake. So, let's assume your only option was a public shelter as we continue to discuss this leg of your journey.

Shelter Dynamics

There are lots of things to consider here, not only with what to expect at your location, but with yourself, with others around you, and with the things you need to be doing during this time period to get ready to put everything back together. Many of the things listed can apply regardless of where you're staying. Let's start with the most important part and talk about you and those you're with.

You and Yours

Remember the schoolday prank of trying to burst the pipes by flushing all the toilets at once? In theory, all the flushing would generate a heavy flow of water, and when you stopped flushing, the rushing water would have nowhere to go and the overpressure would burst the pipes. Well, that's almost how it's going to be with you if you don't handle the stress that has been building up inside since you first realized an emergency had occurred. You might have been too busy to notice the stress, but now that you've stopped... Let's look at some things to keep you and yours on an even keel.

➤ Worst things first. If you or anyone in your family is hypertensive, we hope you packed bloodpressure medication in your Bugout Kit, had it in your GYT Pack, or at least included it on your "Last Minute List." This is when you'll need it most. The same goes for any anti-anxiety, or anti-depressant drugs.

➤ For the rest of you, you'll still need something to take the edge off. That's why we recommended you pack a mild over-the-counter sedative in your kits; not something to knock you out, just something to let you rest. Along with that, to help you deal with the hustle, bustle, and noise of a busy shelter, we suggested earlier that you pack eyeshades and earplugs.

➤ Keep in mind though, that even though everyone should rest, you're still in unfamiliar territory around people you don't know, so not everyone in your group can go to sleep just yet. The strangers around you will also be stressed and many will be far less supplied than you. It's better that at least one person in your group be awake to watch over the others and your supplies. Those that really need to sleep should, and those more keyed up and awake can stand the first watch. Rotate the watch as you're able, and utilize the buddy system when going for food and so on. One person should always stay with the gear.

➤ For those that do stay awake, we recommend playing a quiet board or paper game, or reading a calming book. Just rest, recharge, keep out of everyone's way and give them room to stay out of your way. Starting calmly, quietly, and slowly will allow you to make friends with your sheltermates later on since you won't be starting off by venting mutual stress at each other.

➤ Another suggestion is to make sure everyone drinks plenty of water, eats, and takes their vitamins along with any other necessary medication. You'll need to keep up your energy levels and you don't want the excitement of the day making you forget to take crucial meds.

➤ Speaking of energy levels and avoiding stressful situations with others, another reason to rest and keep to yourself at shelters is the potential for common communicable diseases. If someone there has the flu, guess what? Most will get it sooner or later. Rest, recharge, lay low, and avoid new problems.

➤ Don't forget your pets. They'll need reassurance too, especially since it's likely they're in a nearby boarding facility (since most shelters won't take pets).

Working With Others

Though we cautioned you to watch your gear and keep to yourself, we'll remind you that 99% of the people around you are in the same boat as you and are far more like you than different. You'll probably wind up making some new friends that you would never have met had you not shared adversity. Let's cover a few cautions, and several ways that everyone can work together for mutual benefit.

➤ The main caution is that since a lot of people under a lot of stress are going to be suddenly thrown in close quarters together, little things that might not normally be a problem are going to cause conflict. People are going to have frayed nerves, short fuses, and the resulting confrontations are going to do nothing but start a vicious cycle. The best thing you can do is set an example and keep mostly to yourself, use earphones to listen to music, read, smoke only outside, and stay out of people's way. Hopefully they'll return the favor.

➤ Do some things to help divert attention. If the officials running your shelter have not done so already, set up some bulletin boards.

❏ Make one for **communications**. Let people who were separated from loved ones post messages asking if anyone has news or if anyone with functional cell phones or radios can help them find out where their family may be. This may well be a need you have too.

❏ Another could be for **news updates**. People will want to keep up with current events regarding what caused the evacuation. People in the shelter with radios, TVs, and phones should post updates if there are too many people to spread the news by word of mouth.

❏ A third great bulletin board would be a "**Barter Board**." Try as they might, most people are not going to be as well supplied as you, and you might find yourself wanting a few things and needing to hang on to your cash. Post "items wanted" and "items will trade" on the board and let people work out their own swaps.

➤ While we're on the subject, let's look at a list of some goods, services, and comfort (or "vice") items (in no particular order) that might prove useful for barter. By the way, the more dire the evacuation and the less stocked the shelter, the more valuable and important barter becomes.

❏ Food and water	❏ Toilet paper	❏ Feminine hygiene products
❏ Batteries	❏ Gasoline	❏ Non-prescription medications
❏ Cosmetics	❏ Liquor *	❏ Candy and/or chocolate
❏ Insulin needles	❏ Tools	❏ Cigarettes
❏ Soap	❏ Transportation	❏ Auto parts

| ❏ Auto repair | ❏ Medical skills | ❏ Haircuts |
| ❏ Miscellaneous toiletries | ❏ Evac equipment repair | ❏ Clothing |

*Alcoholic beverages will not be allowed in official shelters. However, if you have items that stay out in vehicles, and never come inside...

➢ As for you, while you're offering things to trade, be careful about being too public about the kinds of assets you have with you. Though 99 percent are good people, there's still that 1 percent that will steal your gear. Always treat your gear and goods as if it were luggage at the airport. Never let it out of your sight and don't let others know what you're carrying. Also, you don't want to be mobbed and robbed by a few dozen people wanting your one little barter item.

➢ Speaking of the shelter not being stocked well, and possibly understaffed by officials, one thing you might consider is suggesting that the evacuees elect a constable or quartermaster amongst themselves to both ration what supplies are available and to help keep a semblance of order, and security over supplies.

➢ Under "**Cooperation**," we'll discuss the concept of your volunteer group being the ones that help stock the shelter you're planning on staying in, and we'll give you suggestions on how to divide the supplies amongst yourselves so that when everyone shows up, the shelter is fully stocked.

Prepping to Come Home

Along with rest and resupply, you'll want to set things in motion to make your coming home to repair and rebuild as easy as possible. This is where you'll want to put to use all those things you learned earlier under "Planning Ahead for Phase II." Coming home after a mass emergency evac is certainly Phase II.

➢ You'll have completed a rather necessary step for yourself and everyone else if you set up the news update bulletin board since the first thing you'll want to know is the current situation regarding your home town.

➢ Once you know how catastrophic the event was, you can gauge how much effort will be required to go back and get things back to normal. Was anything destroyed? Was the event a terrorist attack that will cause a long term area-denial, meaning you won't be able to return? **Hint**: One little trick is to call your home's answering machine (not voice mail). If it picks up it means the power is on and your house is more than likely intact, otherwise you'd get continuous ringing, a busy signal, or an error message.

➢ To begin your return home, start by calling the offices and agencies you researched under the "Phase II" section. Call your insurance agent, local relief groups, and the emergency management agency that has jurisdiction over the event. Get their advice on how you should proceed based on the situations generated by that event.

➢ Other information can be obtained from the various relief and recovery groups such as the Red Cross. (See how valuable your "Info Pack" with its "Important Contact" pages is?)

➢ Next, call the non-emergency number of your home area's Law Enforcement agency (city Police, county Police, Sheriff, etc.) and ask them the questions on your **"Coming Home Worksheet"** in the **"Appendix."**

➢ If given the all-clear and you can return home, the first thing you should do before heading home is call the various repair and cleanup services you researched under "Planning Ahead for Phase II." Go ahead and get your name on their lists early so they can get to work on your property as soon as possible.

➢ Keep in mind that you may also see a rather lengthy relocation depending on the disaster. If that's the case, you have all your personal and family business information in your Info Pack, and you have your address change forms and cards. Provided you're healthy, your loved ones are accounted for, and you have a good source of food and supplies, you should be set. Try to stress over the situation as little as possible.

➢ You can also enter a temporary or permanent change of address online via the US Postal Service at http://www.usps.com, or through other services such as One Stop Move at http://1stopmove.com/main.jsp.

➢ The possibility of a long-term relocation is also one of several reasons we told you to include your destination city in your Evac Atlas. You'll want to know as much as you can about your new location and the assets and services available to you.

However, getting back to your calls home, now that you've set wheels in motion to go home, let's go.

There's more to coming home than simply returning to the house, just as evacuation was much more than simply leaving. We'll start with things you'll need to do when returning to a damaged or destroyed home.

General Considerations

It's hard for some people to stay away from a completely devastated area, even if they know for a fact there's absolutely nothing they can do. People develop an emotional attachment to home and just can't leave. It's likely that's the case with you and it's also likely that authorities have given the all-clear to return, otherwise you won't need the following pointers. These are general considerations to keep in mind when returning to a damaged property and we'll take into account a variety of scenarios.

➢ Sometimes you have to do the hard thing and stay away, especially if there's a possibility of secondary dangers for your home town such as excessive destruction, biohazards, chemical hazards, fallout, etc., even though your immediate area has been approved for return.

➢ Keep listening to your radio or TV for important follow-up news.

➢ Start taking photos and videos of damage as soon as you're able. You'll need these anyway, but you'll certainly need them if your insurance adjuster is backlogged and can't get to your property right away. Also be sure to save any and all receipts for any equipment, service, or supply you buy to help you recover.

➢ When taking photos and video of the damage, include your whole neighborhood in some of the shots. Not only will this help verify the magnitude of your situation, you may be helping some of your neighbors who were not able to document the damage to their property.

➢ Watch for safety hazards caused by damage: Damaged roads or bridges, standing water, wet and slippery surfaces, gas leaks, broken glass, exposed wiring (may be live), and standing water that may be electrically charged due to exposed live wires.

➢ If you ever have to check to see if something might shock you, touch it with the back of your hand. Number one, just as in a fire evac you don't want to burn your hand, and two, if the device does carry an electric charge you don't want your hand positioned where an electric shock would cause your hand to close and grab hold of the live wire or device.

➢ Scan the perimeter before starting to go into the house. Check for obvious structural damage such as leaning or missing wall sections, roof damage, cracks in masonry or chimneys, the smell or hissing sound of a gas leak, the smell of electrical burning or visible sparks, etc.

➢ Call local authorities for wild animals, dead animals, etc. Don't handle them yourself. If live animals are trapped in a building after a flood, leave them a way out and they'll usually take it. (Maybe not snakes.)

➢ Turn off what utilities you can from outside without putting yourself in danger. Don't touch anything electrical if there's any moisture present on breaker boxes, switches, or the ground or floor where you'd need to stand.

➢ Unplug any appliances that may still be wet.

➢ Don't use any flame-based lighting, and turn on flashlights and all other lights outside before walking inside with them, as sparks will set off gas.

➢ Once inside, watch for loose floors, shifted furniture that may fall, cracked ceilings or hanging overhead fixtures that may fall.

➢ Pump any flooded basements gradually to avoid wall collapse.

➢ Even if the power is on, it may have been off for a while so check food in the fridge or freezer for spoilage.

➢ If there was a fire, discard smoke and fire damaged consumables. Discard canned goods if the can has expanded or is otherwise damaged. Medicines need to go too since they're not formulated to be "cooked" in a fire. The same goes for consumables and medications exposed to floodwaters.

➢ If any of your countertop water filters were submerged by floodwaters or were exposed to contaminated tap water, they'll have to be discarded.

➢ Though you're getting damaged items out of the house, don't throw anything completely away until documented by your insurance adjuster. Photo and/or video what's remove from the house.

➢ Take it easy, pace yourself, drink plenty of clean water and eat plenty of food while working to clean up. Don't forget to take your vitamins or other regular medications. Also wear plenty of insect repellant.

➢ If you can't stay at your residence while you're working, ask local volunteer or watch groups, or local Law Enforcement to help watch your property. If you can, your best option is to camp out on your property while you're working, as looting may become a problem.

Cleaning up the Home

These steps will focus more on cleaning a **contaminated** home as opposed to the steps you'd take for a damaged home as discussed above.

The best way to protect your home was to seal it as the incident unfolded. However, things happen. You might have been away when the incident occurred, or you might need to clean up after a flood has washed pollutants or hazardous materials into your home. The very first thing you'll want to do before starting any kind of cleanup on your own is to check with local officials about the severity of the incident, the type of substance(s) involved, and the official recommendation of how to treat your home and vehicles. For the purposes of this section we'll assume you were away or forced to evacuate and that this cleanup is happening after your return. Here we'll focus on the basic steps necessary to the cleanup of potentially toxic substances regardless how they arrived. **Note**: Our other assumption here is that authorities have determined toxicity levels to be within acceptable limits. However, we also caution you to use as much personal protection, through clothing and respirators, as possible. Better safe than sorry.

➢ Some scenarios will see your home or property so badly contaminated that you will never be allowed to inhabit them again. (This is why we've listed some private companies that may be able to help retrieve and decontaminate some of your more valuable possessions. It's also why we suggested that during alerts you seal up some of your irreplaceable items and store them in a secure area.)

➢ Other situations will have their own level of severity and will require the appropriate response. For instance, you may need to wear a full respirator and protective clothing when entering your house or you may simply need good durable work clothing. Though an area of town might be deemed safe, individual homes might have concentrations of toxins in dangerous levels. We recommend you prep for the worst when returning home and that you wear full protective gear, at least at first. Better safe than sorry.

➢ The first thing you'll want to do is air out your house. If the weather allows, open doors and windows and turn on any window fans or your central HVAC system's fan to circulate air through the house. Depending on the situation, you may want to leave the house open all day and come back later to start cleaning.

➢ Call animal control to dispose of any dead animals on your property. Don't handle them yourself.

➢ Start from the top down, and then from outside to inside. Spray-clean your roof with soap and water and then do your outside walls, decking, stairs, walkways, and driveway. If you don't do the outside first, you'll likely track contaminants back into the house during the cleaning process. Depending on the original substance, you may choose to add a neutralizing agent to the wash process. Ask local officials for more information, and be sure the water you're using to clean with is clean itself.

➢ Once inside, start from the upper floors and work your way down. Remember, most chemicals are heavier than air and will settle, so you wouldn't want to clean lower floor first only to have it soiled by residue kicked up while cleaning upper floors. (Though chemical levels were low enough to let you return, it doesn't mean they're gone.)

➢ Depending on how much carpeting you have, and the nature of the contaminant your home may have been exposed to, you might want to put down a good absorbing agent on your carpet to draw out any remaining agent. Cornstarch is a good economical and easy to use absorbent. Sprinkle it liberally into all your carpeting, let it stay the entire time you're cleaning the rest of the house, and then vacuum and steam clean the carpets. Ultimately, you may want to replace your carpet entirely, just to be on the safe side.

➢ Set up ventilation fans in your attic (wear protective clothing due to the insulation.). Set up one as an intake fan to bring air into the attic and another as an exhaust to blow the air out. This will only be necessary if the event involved an airborne chemical. You wouldn't do this in a flood cleanup if the attic was not affected.

➢ Keep ventilation fans running in your house the entire time you're working.

➢ Discard any open food, open toiletries, or cosmetics that may have been contaminated. Also discard open and exposed contact lenses and solutions, paper towels, napkins, facial tissue, and toilet paper. Anything you throw away should be sealed in garbage bags. Be sure to photo and video everything you throw away.

➢ Dispose of spilled or contaminated household chemicals properly. Be aware of any chemical containers that have ruptured or that may have mixed their contents, as they may cause a HazMat incident of their own.

➢ Thoroughly clean any human contact items like hair brushes, etc. Depending on the substances involved, you may want to discard them. Definitely discard "intimate" human contact items such as toothbrushes. Wash all dishes and food contact utensils. Discard wooden or other porous utensils.

➢ Packaged goods should still be okay. Be sure to thoroughly clean the packages. For cans with paper labels, use a Sharpie to relable the can, then remove the label, and wash the can. Make sure the can opener is clean as well.

➢ Wash all washable fabrics. You might want to include a simple neutralizing agent in with your detergent providing it won't affect fabrics. Remember to check with officials to see if your neutralizing agent should be a base such as baking soda, or an acid such as vinegar.

➢ Special attention should be given to human-contact fabrics, especially pillow cases. The pillows themselves should be discarded.

➢ Dry-clean-only fabrics should be aired out as long as possible and then sealed in plastic bags to be taken to the dry cleaner's. The dry cleaning company should be made fully aware of the type of chemical(s) that may be contaminating the fabric. You may wind up having to discard the items.

➢ Vacuum and then steam clean any carpets or rugs, mop tiled floors, and thoroughly clean wood floors. Be very careful when disposing of your vacuum cleaner bags. **Note**: You may have to replace all your carpeting and rugs depending on the contaminant or the event.

➢ After everything is done, replace your HVAC filter sealing the old one in a garbage bag. Do the same with any vacuum cleaner bag you may have used, and protect yourself while changing both. Call for instructions on how to dispose of discarded items. You should also have your HVAC ducts professionally cleaned as soon as possible.

➢ Have a qualified service come inspect your septic tank or equivalent after a flood.

➢ You may want to have your home professionally cleaned by a HazMat cleanup specialist. You should already have the contact info for such a company listed on your Important Contacts form in your Info Pack.

This is a lot of information to remember, and certainly too much to put in a simple checklist. Therefore, **pack this book in your Bugout Kit** to have with you on the road, and when you return to start over.

These steps are very simplistic and will work for only mild to moderate instances of contamination with less potent hazardous materials or pollutants. For more severe situations, you're more than likely going to have to call in a professional HazMat cleanup service as it's doubtful you'd be allowed to return.

Earlier, under "Planning Ahead for Phase II" we asked you to compile a list of services you may need after a disaster. If you're evacuating due to a severe terrorist attack involving biological or chemical weapons, your residence could be seriously compromised and return access restricted for some time to come. One type of industry has the potential for being able to go to your home and retrieve and decontaminate certain goods or items and return them to you. Two important things we need to point out about this: One, this service will not be cheap by any standards, and two, some of the companies we're about to list have not considered contaminated goods retrieval even though they have most of the necessary gear. These companies are Crime Scene Cleanup companies. They specialize in going into post-investigation violent crime scenes to do extensive cleanup of biological materials and make the location safe for human inhabitation. These companies have the protective gear and experience that would allow them to perform the service of retrieving and decontaminating your belongings in some scenarios. Contact them now if you feel that your area might be at risk for such an attack or incident. Here are a few examples:

➢ California's Crime Scene Cleanup techs can be located through phone books or the state's listing at 916-327-6904, or at http://www.dhs.ca.gov/ps/ddwem/environmental/med_waste/practitioners.pdf.

➢ Crime Scene Cleaners, Inc., online at http://www.crimescenecleaners.com, or at Crime Scene Cleaners, Inc., 23 Altarinda Road, Suite 103, Orinda, California 94563, TEL: 1-800-357-6731, FAX: 925-254-0557.

➢ Seattle area's "All-Covered Bio Recovery," online at http://www.allcoveredbiorecovery.com, or by phone at: Seattle: 206-568-0789, Eastside: 425-576-1471, Fax: 425-825-1938.

➢ Northern California's Asepsis Technology, online at http://www.asepsistechnology.com, with direct contact at 323 Willow Glen Court, Healdsburg, CA 95448, 1-800-593-2737 · Cellular: (707) 483-2737.

➢ The Washington and Oregon areas have "Bio Clean LLC," at http://www.biocleanllc.com, and 888-412-6300.

➢ "Bio Recovery Services of America" services areas in Ohio, Michigan, and Indiana. They can be reached at: http://www.biorecovery.net, 800-699-6522.

➢ For Ohio, Pennsylvania, and West Virginia, call "Bio-Scene Recovery, Inc." at 877-380-5500, or log on to http://www.bioscene.com.

➢ Call "Crime Scene Clean-Up" of Maryland at 800-295-5460, or log on to: http://www.crimeclean-up.com.

➢ "Mayhem and Mishaps" serves Florida and the Carolinas. Reach them at 803-419-9118 or online at: http://mayhemandmishaps.com.

➢ Southeast Texas is home to "Red Alert," online at http://www.wecleanall.com or by phone at 800-570-1833.

➢ ServPro Industries, Inc. online at http://www.servpro.com, with direct contact at 575 Airport Rd., Gallatin, TN 37066, 615-451-0200, Fax: 615-451-0291, Toll Free: 1-800-SERVPRO (800-737-8776). (Though not a crime scene service, this company advertises that they specialize in fire and water damage cleanup.)

➢ Similarly, take a look at the disaster cleanup services offered by **Service Master**. Your local phone book should have a listing if there's an outlet in your area.

> ➤ **Hint**: When dealing with any outside company or contractor in the cleanup or repair of disaster-related damage, keep the following in mind:
>   - Start with a written estimate. Don't start a service on a mere verbal estimate.
>   - Require that anything that might push the contract over budget be approved by you first.
>   - Get a contract in writing, and make sure any verbal changes or additions are put in writing.
>   - Pay no more than 20 percent down on the estimated total.
>   - If possible, go with this service as they purchase your repair supplies. You can verify the cost of goods, or you can offer to buy the materials yourself and save money.
>   - After completion of the work, make the contractor provide you with a written release of any claim or lien against your property.
>   - Do not make final payment until you are fully satisfied the work is completed.

After everything is clean and the household is back in order, REPACK YOUR GEAR. There may be a next time.

And... now that the physical household is back in order, let's take a look at the **emotional household**.

Coming Home II – The Emotional Aftermath

Since evacuation is usually the more stressful eventuality, let's talk a little about "Post-Traumatic Stress Disorder" (commonly called PTSD), how to soften the blow a bit, and maybe prevent some of it to begin with.

Any emergency situation, especially evacuations, are going to carry a "one-two punch." The first punch lands as you realize you have to evacuate. The second comes as you return home and have to deal with any damage or cleanup and then wind up with time to reflect on what's happened.

Let's start with a few lists regarding what to expect, what to look for, and how to help.

Do These in A̲d̲v̲a̲n̲c̲e̲ of an Emergency to Keep any Later Stress at a Minimum:

1. Include all able children in all planning and drills. This is another benefit of drills in that everyone feels included and prepared. In fact, all the planning you're doing now will aleviate later stress through confidence.

2. Any child large enough and smart enough to drive should be taught how as they may be called upon if you and your spouse are unable to drive. Being included in this kind of planning and being able to fill the role as needed would be the epitome of inclusion. Inclusion breeds solidarity, and participation in helping the family eases a great deal of fear. And, as it is with all family members, preparedness equals peace of mind.

3. Write down inspirational quotes, religious verses, and learn panic-mitigation techniques (the simplest of which is taking deep breaths to control your emotions) and learn them so you can maintain your calm demeanor and good "bedside manner" during an emergency. This kind of calm will set an example and will also help set the mood of others around you. Remember our acronym **C.A.L.M.** from earlier.

4. Your psychological readiness should be planned, developed, exercised, and attended to like any other asset or skill devoted to your overall preparedness. You should do everything in your power now to make sure your emotional strength has the greatest foundation possible so that you can better weather the storm and handle the stress and possible trauma associated with a disaster or attack. Numerous self-help books, tapes, and programs are available to help you along this path, and the list is too numerous and varied to give you one or two particular titles here. Check your library and bookstore's psychology and self-help sections to see what is available. You might also check the links under "Sources" below and also check with your county mental health center.

Be Prepared for Coping with Emotional Distress After a Disaster:

1. A disaster will have an emotional effect at some level with everybody.

2. Being concerned about the wellbeing of family, friends, and neighbors is normal. Anxiety can be quite high especially in situations where there's nothing you can do to help.

3. If your community offers any sort of counseling, you should attend (even if it's under the guise of taking someone else to help them cope.)

4. Focus on what you're able to do and fix, and not on what you can't. Worry is like rocking in a chair. It may give you something to do but it won't get you anywhere.

5. Don't take things out on anyone around you. They're in the same boat you are.

Consider These Points for Helping Children During an Emergency:

➤ Children will feel the stress just as much if not more than adults. Their "post traumatic stress disorder" symptoms will strongly mirror those of adults, but may be expressed through play and outbursts more than verbalization. On the other hand, the most resilient members of your family will probably be the children as most tend to bounce back more quickly.

➤ Your demeanor during the emergency is what children will respond to the most. Just as doctors practice good bedside manner, so should you in an emergency. If you freak, the kids will too.

➤ Keep them close by, explain the situation to them in a calm, matter-of-fact manner. Try to answer all their questions and don't brush them off with "let's talk about that later," or some other diversion.

➤ Keep their activities as close to normal as possible. Even if stranded at a shelter a few days, make them keep up with school studies, favorite books or games.

➤ Treat the situation as a temporary trip and not much more than a small annoyance.

➤ Give them jobs to do based on their age and abilities that will help with rebuilding. Exclusion breeds fear. Inclusion generates security.

➤ Compliment and reward efforts made by children to help. Don't let your own stress prevent that.

➤ Limit their exposure to radio, TV, or adult strangers that may be overly emotional.

➤ Maintain your child's health. Encourage proper diet, rest, and exercise.

Post-Traumatic Stress Disorder Symptoms Include:

➤ Insomnia, difficulty sleeping, or fitful sleep.
➤ Short attention span or difficulty concentrating.
➤ Obvious depression, sadness, mood swings, or crying.
➤ Low physical energy to include cold and flu symptoms.
➤ Excessive allergies or recurrent asthma.
➤ Feelings of anxiety, frustration, or hopelessness.
➤ Anger or rage.
➤ Numbness or no emotion at all.
➤ Loss of appetite.
➤ Thoughts of suicide.

➤ Bowel or bladder problems.
➤ Nausea.
➤ Headaches.
➤ Exaggeration or distortion of the event.
➤ Repetitive talking about the event.
➤ Fear of going to school.
➤ Unusual fear of darkness, separation, weather, etc.
➤ Feelings of guilt.
➤ Disorientation and difficulty communicating.
➤ Unusual feelings of fear.

To help you remember some of the **PTSD** symptoms that your family members may exhibit after experiencing a disaster, we have an acronym for you to the word **D.I.S.A.S.T.E.R.**:

Depression.

Insomnia.

Short attention span.

Allergies or Asthma.

Suicidal thoughts.

Total loss of emotion.

Energy loss.

Rage or anger.

Some Ways to Deal With This Stress Include:

1. Talking with others sharing the same experience.
2. Accept help from experienced PTSD counselors.
3. Maintain your physical health as health is directly tied to emotion. (Yet another reason we started this book with a section on Health.)
4. Realize what Will Rogers meant when he said "We can't all be heroes, some of us have to sit on the curb and clap as they go by." Sometimes there's really nothing you can <u>do</u>.
5. Stick to your daily routine as much as possible.
6. Maintain communication with family and friends.
7. Relax, knowing you followed all the instructions in this book and that, in general, you did the very best you could.
8. Attend memorial services with the community or your place of worship.

After an Emergency do These for Kids (and most will help the adults in the process):

1. Get back to routine as quickly as possible at home, school, and recreational programs.

2. If toys are damaged or lost, replace them with exact duplicates as quickly as possible.

3. Limit exposure to news coverage of the event's aftermath.

4. Talk with your children and pay them adequate attention.

5. Console only as needed. Though fragile, children are resilient and can snap back to normal rather quickly. Being overly consoling may make them think there's more wrong than they originally thought and that can reintroduce unnecessary stress.

6. Maintain your good "bedside manner" after the disaster and all the way through rebuilding.

7. If you have a pet, ask your children to focus on caring for the pet and making it feel better. Being the care giver sometimes helps far more than being the receiver. If you don't have a pet, consider getting one. It's also a good idea to occupy siblings by asking one to help comfort the other.

8. Adolescents are in an unusual position. Too old to ignore and too young for autonomy. As they're "apprentice adults" treat them as such. Give them honest but calm answers to questions, ask for their input, and include them as much as possible in the repair process and show gratitude for their efforts.

Sources and More Information

Just as we asked you to locate construction, repair, and cleanup services to help repair your home, we'll also ask you to look up the information and contact numbers for mental and emotional health services so you can put your emotional health back in order as well. Record the info on one of the "Professional Contact" pages of your "Important Contacts" forms, and keep it in your Bugout Kit's "Office Pack."

➢ The National Institute of Mental Health: http://www.nimh.nih.gov, National Institute of Mental Health, Office of Communications, 6001 Executive Boulevard, Room 8184, MSC 9663, Bethesda, MD 20892-9663, Phone: 301-443-4513 or 1-866-615-NIMH (6464), toll-free TTY: 301-443-8431; FAX: 301-443-4279, FAX 4U: 301-443-5158. The page to help children cope with disaster is http://www.nimh.nih.gov/publicat/violence.cfm.

➢ US Department of Health and Human Services', "Mental Health and Tragic Events." http://www.hhs.gov/mentalhealth, U.S. Department of Health and Human Services, 200 Independence Avenue, S.W., Washington, D.C. 20201, Telephone: 202-619-0257, Toll Free: 1-877-696-6775.

➢ Disaster-related mental health links can be found on the American Psychiatric Association site at http://www.psych.org/disasterpsych/links/weblinks.cfm, and on http://www.psych.org/disaster.

➢ MentalHealth.org: "Disaster Mental Health," http://www.mentalhealth.org/cmhs/EmergencyServices/after.asp, The Substance Abuse and Mental Health Services Administration, P.O. Box 42557, Washington, DC, 20015, 800-789-2647. Also see the root site, http://www.mentalhealth.org/cmhs/EmergencyServices.

➢ Visit http://www.nasponline.org/NEAT/unsettlingtimes.html for the National Association of School Psychologists' "Coping in Unsettling Times," National Association of School Psychologists, 4340 East West Highway, Suite 402, Bethesda, MD 20814, 301-657-0270.

➢ The National Center for Children Exposed to Violence, http://www.nccev.org, National Center for Children Exposed to Violence, Yale University Child Study Center, 230 South Frontage Road, New Haven, CT, 06520-7900, 1-877-49-NCCEV (877-496-2238).

➢ National Center for Post Traumatic Stress Disorder's "Dealing with the Aftereffects of Terrorism," at http://www.ncptsd.org/disaster.html, National Center for PTSD, VA Medical Center (116D), 215 North Main St, White River Junction, VT, 05009, Executive office (802) 296-5132, PTSD Information Line (802) 296-6300, fax (802) 296-5135. While you're there, be sure to visit the home page for sources to more answers.

➢ National Mental Health Association's "Dealing With Stress After a Natural Disaster," at http://www.nmha.org/reassurance/naturalDisaster.cfm, National Mental Health Association, 2001 N. Beauregard Street, 12th Floor, Alexandria, VA, 22311, Phone 703/684-7722, Fax 703/684-5968, Toll free 800/969-NMHA (6652), TTY Line 800/433-5959.

➢ Tips for Emergency and Disaster Workers http://www.mentalhealth.org/publications/allpubs/KEN-01-0098.

➢ A good links collection of emotional stress related information can be found online on the National Library of Medicine's website at http://www.nlm.nih.gov/medlineplus/disastersandemergencypreparedness.html. Scroll down to the section on "**Coping.**"

➢ A variety of stress and trauma information and sources can be found online at "David Baldwin's Trauma Information Pages" at http://www.trauma-pages.com.

➢ You'll find some general family-oriented resources, as well as some family disaster-coping information online at http://familymanagement.com. For more information on their many programs and resources, write them at P.O. Box 5338, Novato, CA 94948-5338, or call 415-209-0502.

- ➢ Marine Corps Combat Stress manual online: http://www.tpub.com/content/USMC/mcrp611c.
- ➢ You'll find the military manual on **combat stress**, FM 8-51, on our **CD**.

Now that you're safely back at home, let's talk about what might happen if you couldn't leave. Let's go to our next section, "**Isolation**."

Notes:

V. ISOLATION

"You must do the thing you think you cannot do." –Eleanor Roosevelt*

✓	Activity	Date Completed	Pg.
Goals of the Isolation Section			
❑	Per our "**Emergency Risk Score,**" we know why we might become isolated.	___/___/___	**54**
	Advanced Home Prep:		**325**
❑	We understand all the important details of "**sealing from the outside in.**"	___/___/___	
❑	All parts and materials for **sealing our home** have been gathered.	___/___/___	
❑	Have reviewed the pros and cons of **storm cellars and fallout shelters**.	___/___/___	
❑	At least one source of **alternate energy** has been established.	___/___/___	
❑	We've **stocked up on enough batteries** to run all necessary devices.	___/___/___	
	The Isolation Pantry:		**340**
❑	Enough **water** to last the **family** and **pets** for **21 days** has been stored.	___/___/___	
❑	**Three weeks** worth of **food** and **dry goods** have been stored.	___/___/___	
❑	We have looked into buying a separate **freezer**.	___/___/___	
❑	Our home **medicine chest** has been adequately filled.	___/___/___	
	Life Without Utilities:		**354**
❑	**Alternative heating methods** and considerations have been studied.	___/___/___	
❑	A "**space air conditioner**" has been purchased or fabricated.	___/___/___	
❑	Alternative **cooking equipment** has been made or purchased.	___/___/___	
❑	All listed **hygiene and sanitation items** have been gathered and stored.	___/___/___	
	Life in Isolation:		**366**
❑	The family has discussed and understands all **security needs and issues**.	___/___/___	
❑	We've reread all our **fire fighting training** info and created a "**Fire Trunk.**"	___/___/___	
❑	We've trained together with **neighbors regarding fire reactions**.	___/___/___	
❑	Items to provide **entertainment** and fight **boredom** are gathered and stored.	___/___/___	
	Goal Setting:		**A-1**
❑	Entered **start dates and goal dates** onto **calendar** for all the above.	___/___/___	

In the "**Evacuation**" section, we talked about "getting out of Dodge" in a hurry. Here we'll cover the exact opposite as we discuss "heading for the bunker and battening down the hatches."

In general, isolation can mean several things. You could be forced to shelter-in-place to ride out an emergency, or face a forced quarantine. You could be stranded by something that cut off all access such as a flood. Or, something may cause a temporary disruption in local supplies or services. In any event, this "**Isolation**" section covers your preparation to keep you self-reliant and self-sustained in your home.

Many people regard an isolation incident as the easier of the disaster reactions. After all, home is where everybody's stuff is. However, as we'll see below, being isolated is no cake walk when you consider what you might be isolated by, isolated from, and/or isolated without. Let's take a look at some considerations that apply to an isolation situation in general before we get into specific aspects of your planning and preparation.

> ➤ Under "**Evacuation**" we cautioned you to prepare for both evacuation and isolation dividing your money and attention 55/45 between the two, favoring which is most likely given your unique situation. Here you'll want to keep in mind the fact that evacuation and isolation might go hand in hand. An emergency might dictate a short period of isolation followed by an evacuation.

> ➤ On returning home from a forced evacuation, you still have to deal with the aftermath of what drove you out in the first place. This may mean an indefinite disruption in local services, supplies, etc. Therefore, having good stores set up for an isolation will help keep you supplied as you rebuild or recoup.

> ➤ Evacuation and Isolation might be a combination such as in camping in your yard after your home and community were destroyed by a tornado.

> ➤ Since evacuation and isolation may be coupled, you should have separate goods and gear for each.

> ➤ People always underestimate what they'd need in the way of supplies. As Murphy's Law is a very real thing, remember that an event causing an isolation may occur when you're hosting company. For this and many other reasons, always store more than you think you'll need.

For the purposes of this section, we'll focus on your home. Though quite a number of scenarios could see a temporary sheltering in place while you're at work or school, it's the forced quarantines, periods of intense civil unrest, and natural disaster damage that would see the longest isolations, and many of those will either find you at home or allow you to get there. If you were away from home, your situation would be more akin to evacuation, and the things you learned in the previous section would be more applicable.

Speaking of evacuation, we started that section with a short discussion of various evacuation "levels" and the things that might dictate them. Let's look at **levels of isolation**, which here, are levels of **difficulty**:

Level 1: A **complete restriction** to home with **no** venturing out and a **complete loss of all utilities**.
Level 2: A **complete restriction** to home with **no** venturing out and a loss of *some* **utilities**.
Level 3: A **complete restriction** to home with no venturing out and **full utility function**.
Level 4: **Partial area restriction** with a complete loss of **all utilities**.
Level 5: Closing of **public venues** with **limited public travel**, and a **temporary loss of utilities**.
Level 6: Closing of **public venues** with **limited public travel**, and **no loss** of utilities.

(Note: These "levels" are our own system and officials will not notify you using these terms.)

As with any urgent situation, numerous extenuating circumstances come into play with isolation:

1. **The nature of the beast**. What was it that forced you to shelter-in-place? Both a blizzard and a coordinated terrorist attack could dictate a Level 1 isolation, though each are drastically different.

2. **Time**. How long are you going to be isolated? A day? A week? A month?

3. **Secondary effect**. What will you have to face after the isolation portion of the event is over? Again, what if it's a blizzard? What if it's a terrorist attack?

4. **Other people**. A sheltering-in-place at work or school will see you around others, and in those situations, you're more likely to receive help from officials. What about isolation at home? Will you be by yourself? Will you have family? How many neighbors are you close to? What if they can't help you?

5. **Weather**. Are you locked inside on a nice spring or autumn day, or are you freezing in the winter or broiling in the summer?

6. **Utilities**. Despite movement restriction, which utilities will not work? How long will they be out?

7. **Civil unrest**. Is this just a mundane, annoying situation in which everyone's in the same boat and just waiting things out, or will there be rioting in the streets, looting, and an increase in crime? How prepared is local Law Enforcement? Were they more affected by the event than you or your neighborhood?

8. **Personal or family needs**. How will you handle an emergency if you can't get out or call out?

9. **Practice**. Have you turned off all the utilities in the house to spend a weekend fending for yourself? Do you ever go camping where you're living off limited supplies? Has everyone in the family been trained as to what to do in an isolation so that the household and those there can be taken care of regardless of who's in the house and who was forced away? Try one weekend for a family "dry run." The best way to truly evaluate your needs and capabilities during a worst-case-scenario is to spend a weekend with everyone at home and use NO utilities. Okay, skip the buckets and kitty litter if you want, but keep that one in mind. Also keep the following in mind.

   ➤ Turn off all the breakers except for the refrigerator. (Don't want to lose your food.)
   ➤ Turn off your water.
   ➤ Don't use your heater, AC, or appliances. (Don't actually turn off your gas, just don't use it.)
   ➤ Don't use your telephone, and don't use entertainment equipment unless it's battery operated.
   ➤ Conduct a seal-the-room-put-on-your-gas masks drill.
   ➤ Spend some time brainstorming the "Murphy's Law" factor of "what else could happen?" and see how you would or could react.

As with everything else, we're going to help you prep for the worst case scenario so that you're ready to face it if that's what happens, and so that anything less will be a breeze by comparison. In this section, we're going to get you ready to face a **Level 1** isolation caused by a situation that **knocked out all utilities** and will keep your family locked in the house for **21 days**.

The first thing we need to do to get you ready for *that,* is to get your home ready for *you.*

Advanced Home Prep

Since the focus of the "Isolation" section is to prepare you for an extended period of time being stuck in your home, let's look at your home itself to make sure it's ready for you.

This portion picks up where "Basic Home Prep" left off and will make sure your home is as physically ready and functional as you can possibly make it.

Goals of Advanced Home Prep:
1. To make sure your home can **protect you from as many eventualities as possible**.
2. To give you the finer points behind "**duct tape and plastic sheeting**."
3. To answer your questions about **storm cellars and fallout shelters**.
4. To equip your home to be **self-sufficient and self-sustaining**.

We've divided this section into:
1. **Sealing the Home.**
2. **Fortified Shelters.**
3. **Alternative Energy.**

Sealing the Home

Finally, we get to the "duct tape and plastic sheeting" section. Sealing the home is what you'd do as a reaction to a known biochemical terrorist attack or local industrial accident. Though relatively simple in procedure, there is quite a bit of important detail to be taken into consideration, and as we said in the beginning, the difference is in the detail and we'll give you the detail that makes all the difference.

Though the debate could rage back and forth about whether or not a sealed room would be a reaction you're likely to perform, we stand by our universal theory that it's better to have it and not need it, than to need it and not have it.

Let's look at picking a room or rooms to be sealed, stocking the area in advance, presealing and sealing, and final considerations and additional hints.

Picking a Room

Under "Basic Home Prep" we discussed structurally augmenting a "safe room" to protect you from criminals and/or destructive events. Let's look at some general points for choosing the room or rooms you'd select to seal to protect you from biochemical hazards, as the considerations are different.

➤ Where your structural safe room might have been low in your home, your sealed safe room to protect you from hazardous materials should be on an upper floor (if you have one) since most chemicals and weaponized biological materials are usually heavier than air and tend to settle.

➤ Choose a moderately sized room in your house that has the fewest natural openings (doors and windows, fireplaces, etc.) and no more than two walls exposed to the outside. Some folks may choose their hallway and include a hall bathroom and maybe the kitchen. Choose your room(s) by how protected it is naturally, how much sealing material you intend to keep on hand, and how much help you have to do the sealing.

➤ If at all possible, include your structurally-augmented and emergency-equipment-stocked safe room as a part of your sealed area. This will give you full and protected access to your gear, and will prevent you from having to stock a separate safe area. As we mentioned before, if you intend for your safe room to be part of your sealed room, you should make sure all seams are well-sealed during construction.

Stocking the Area

If you were able to annex your structural safe room as part of your sealed area, then you're almost done with supplies and stocking. However, if the two areas are separate and distinct you'll need to have the same goods and gear in your sealed area as you do in your safe room (unless you have adequate help to get all your gear from one location to the other quickly). To revisit this **supplies list**, see the portion on "Safe Room" in the "Security" part of the "Basic Home Prep" section. Let's look at some **additional items** to consider:

➤ Create a "**seal the room kit**" that lives complete and undisturbed in the area you intend to seal. This kit should contain:

❑ Heavy **plastic sheeting** (30 mils thick or thicker) **pre-cut and labeled** as to what it's supposed to seal. It's better to cut and label now so you know for sure you'll have enough plastic, and you won't have to worry about wasting time cutting and fitting during an emergency. When precutting, remember that each sheet needs to overlap the opening it's covering by about 4 inches all the way around. You want to seal against the wall or floor to effectively cover any open seams.

❑ **Scissors** if you decide not to precut.

❑ Large roll of both **duct tape and painter's tape**. You'll want about a 3-inch width on both. Duct tape can do heavy sealing of window seams, etc., and the painter's tape will do well at holding plastic sheeting in place over windows from the inside and will release without ripping paint and sheetrock.

❑ **Modeling clay**. This is useful to seal unusual cracks or gaps that may have gone unnoticed, or that are difficult to seal with the other materials.

❑ Enough **towels** to cover seams at the base of doors and some windows.

❑ Spare **water** to wet the sealing towels (in addition to water for drinking and sanitation).

❑ Heavy duty **stapler or staple gun**. Speed trick for room sealing: Use a stapler to give a couple of quick staples along the top of a plastic sheet to hold it in place so one person can finish taping. Hand the stapler off to another to do the same. This way you don't tie up two people sealing one opening at a time.

❑ Also, somewhere in your garage or storage area, you should keep a large **plastic sheet** or **tarp** to seal any gaping holes. Though unlikely scenarios, you might be sealing rooms against the elements after a tornado knocked a tree through a wall, or you may be sealing against potential fallout after a nuclear blast shattered windows. Just because everything is intact today, doesn't mean it will be when disaster strikes.

❑ **Hint**: Create one of these kits for each area you frequent such as work, school dorm room, car, etc. All biochemical threats aside, such a kit would help **protect you from the elements** after high winds caused something to break your windows, or knock a hole in your roof.

❑ **Hint #2**: In each kit, keep a copy of the sealing instructions you'll find below.

➤ In addition to the materials needed to seal the room, you might also want to store the gear used to seal *you*, namely your **gas masks and protective suits** (if you chose to invest in them). Cleanup or forced evacuation after a HazMat incident or biochemical attack are a couple of the scenarios we can see such suits and masks being useful. You'll want to have yours accessible from inside your safe area. Either store them there or remember to retrieve them while you're sealing the room.

➤ Have available both an **electric space heater** (no oxygen-consuming flame.) and a free-standing room **airconditioner** (not a window-unit). Since you'll be shutting off your central heat and air, or other climate control, you'll definitely want something in the room with you. Below, in the "Climate Control" section under "Life Without Utilities," we'll tell you how to make an **improvised air conditioner**, and we'll list items to help keep you warm as well.

➤ As carbon dioxide buildup will be a problem in any sealed area, you'll want to keep some oxygen-producing **houseplants** in this area. Houseplants alone won't keep the air supply oxygen-rich at all times, but they could make a difference. Any green-leafed houseplant will work, as all use CO2 and produce oxygen as part of photosynthesis. "Spider Plants" are believed to be the biggest oxygen producers.

➤ Also concerning carbon dioxide, **no one** should be allowed to sleep on the floor (including pets) since that's where **CO2** will accumulate and tracked-in contaminants will settle. Your sealed room should have a bed, cot, or inflatable mattress (thick one) for each person. You can store the inflatable or collapsible beds in the area itself or drag in mattresses from other rooms as you're sealing the room. Also, never let anyone sleep while wearing a respirator.

➤ To determine how long you can stay in an enclosed room, the following conservative "rule of thumb" formula for breathable air time can be used: (length x width x height of room in feet) x .03, divided by (number of people in the room) = breathable air time in hours.

➤ Under "Making an Air Filter" below, we'll discuss an **improvised air filter**. If you intend to make one, you'll need to be set up to install it. For that, we suggest you install a small pet door in the interior door between your sealed area and the rest of the house. Don't plan on using an exterior opening.

➤ You'll also need to make sure you have your **vacuum cleaner** in the room as that's what powers the air filter.

➤ Speaking of air filters, we recommended earlier that you buy one or more commercial **room air filters** for health reasons. The hidden benefit to doing this is that some of them may also reduce hazardous material contaminants that may have found their way into your safe area or the rest of your home. Though not specifically designed or rated for use against biochemical weapons, we recommend the "Ionic Breeze" air filter from Sharper Image, or other ionizing air filter. It stands to reason that anything that filters the air down to one micron will also filter out particulate matter and maybe some vapor droplets. Be sure to have one of these filters inside your sealed room. It may not only make the air a little safer, but it may also remove common odors and freshen stale air (though it won't provide oxygen).

➤ Considering the fact that all manner of **utilities** might be out during any sort of intentional attack, be sure to read ahead to the sections concerning "Life Without Utilities," and have all necessary supplies ready to be brought into your safe area as you seal yourself in. Chief among these are any climate control devices, and personal hygiene equipment.

➤ Speaking of gear, we'll conclude this little section by reminding you that if your sealed room is separate from your safe room and supplies, that you keep your safe room supplies packed and ready to be moved from one room to the other. Speed counts.

Before we jump from prep to execution, or actually starting to seal or take measures to preseal, we'll give you an acronym to help remember the priorities of reacting to a shelter-in-place event that comes with advance warning. Since you'll be heading for cover, we'll use the word **C.O.V.E.R.**:

Children	Get the **children** inside. Remember your "tornado drill" from earlier? (Get your pets too.)
Others	Call **others** and pass the alert.
Vehicle	Move your **vehicle** inside the garage and/or cover it. Seal it to use if an evacuation follows isolation.
Environment	Tighten your **environment** as you start protective measures from the outside in.
Refuge	Move inside to your safe area or **refuge** and prepare to seal it up.

Presealing

To really and truly seal a home against hazardous materials is a laborious and time-intensive effort. However, there are quite a number of little things you can do to help preseal in order to make a final sealing thorough, efficient, and rapid if it becomes necessary. Some of these you may choose to do **only during heightened terror alerts**.

➤ Under both "Frugality" and "Basic Home Prep," we mentioned the need for good weather sealing and general insulation. The hidden benefit to a good seal around doors and windows, and other gaps in your structure, is that it also helps seal the house against hazardous materials. **Note**: This insulation won't **do** the sealing, but it will certainly help.

➤ Get into your attic and seal around ceiling fixtures such as overhead lights or fans. Use expanding foam or insulating / sealing materials that carry a UL flame retardant rating.

➤ Also seal around roof openings such as plumbing vent stacks, exhaust fan ducts, etc., and then tape around the access door leading up to your attic from inside the house.

➤ During a higher terror alert, you might want to seal the gables and larger vents to your attic.

➤ Sleep with doors and windows closed.

➤ If possible, create an indoor space for your outdoor pets and keep them protected.

➤ Seal your outside clothes dryer vent and use an indoor lint-trap setup.

➤ If the weather allows, cover any roof vents or turbines.

➤ If it's cool weather already, seal over window-unit air conditioners from both inside and out.

➤ Keep your car covered and/or in an enclosed garage (preferably both).

➤ A final consideration is that presealing might save your life since most HazMat incidents, whether accidental or intentional, may be over or past the peak threat before the situation can be made public. In other words, your home may be exposed and the threat over before you ever know about it. Being presealed may keep you protected by limiting the amound of hazardous materials that get into your house.

Sealing the Room

Sealing a room is as much sealing the house and the associated activities that go along with it, as it is covering windows with duct tape and plastic sheeting. To effectively seal your house and then a room within a house, consider the following:

1. If you receive the order to shelter-in-place, the first thing to do is **notify all away-from-home family members** that something is up. Scenarios will rarely unfold with everyone conveniently at home. Also, if you feel threatened by something imminent such as a fire at a nearby chemical plant that you feel may cause a HazMat release at any moment, don't wait for official word. React on your own and either seal the house (at least heavily preseal) and/or evacuate. Better safe than sorry.

2. The more people in the family that can help with this sealing process the better since you'll want as many steps completed **simultaneously** as possible. Speed is crucial.

3. **Start from the outside in**, rapidly shrinking your perimeter and sealing yourself in as you go. It's almost like intentionally painting yourself into a corner. Start by gathering all outdoor pets and bring them and their food inside (provided it's safe for you to go out).

4. If a power outage is probable and you have an **electric generator** outside, fuel it and start it now. You might not be able to go back out later.

5. Make sure your **car is covered and/or inside your garage**. If you have a garage with doors, make sure the doors are closed. You'll want your car to be contaminant free if you have to evacuate later.

6. **Close and lock all exterior doors and windows** and close the fireplace damper if you have one. Locking doors and windows usually holds their seal tighter, and in situations where public panic is possible, you'll want all the security you can get.

7. Close all **interior doors** as you work your way back to your safe area.

8. **Turn off** your central heat and air system, any window-unit air conditioners, all exhaust fans (kitchen and bathroom), and all electrical appliances except for those in your safe area or the room air filters that you'll leave running in other parts of the house. Tape over any appliance that vents to the outside. Be doubly sure your bathroom exhaust fan vents and the vent in the hood over the stove are covered and sealed. **Do this even if the power is out** since you don't want the power coming back on and any appliance that happened to be on, sucking in contaminants from outside.

9. Seal over any **in-floor heat register or vent** that opens to the outside or into the basement.

10. Bring any last-minute **supplies** into your safe area. For a list of items to have on hand, revisit the safe room section of "Basic Home Prep."

11. Start sealing your safe area by sealing **window frames** and any **window-unit heating/air-conditioning**. Use duct tape or its equivalent to seal seams where windows open and along the outer edges of the window frame where it joins the wall. For the AC unit, make sure the unit is completely covered with plastic and the edges of the plastic sealed all the way around.

12. Now that seams are sealed, use your pre-cut heavy **plastic sheeting** to seal **over** each window, applying painter's tape along all edges of the plastic and sealing the plastic against the surrounding **wall** and not along the edges of the window. Overlap the window a good 4 inches.

13. Do the same for any **heat/AC vent** coming into the room from a central heat and air system. Also seal **electrical outlets** with tape, overlapping to the wall and not just covering the sockets. For outlets you intend to use, plug in an extension cord and then seal around the plug and socket. You'll also want to thoroughly seal any **door knobs or locks** since all of them are openings.

14. **Seal all doors to the room** except for the main entrance, which should enter the room from an interior portion of the house and not from outside. For other doors, use painter's tape to seal the gaps between door and frame, and then use plastic sheeting and painter's tape to seal the door(s) as you did the window(s) with the sheeting being tape-sealed to the surrounding wall well beyond the door or frame to give a good overlap and good coverage. If the bottom of the plastic sheet is taped against carpet, augment the seal with wet towels.

15. Once in the room, **seal the entrance door behind you** as you did the other openings with two applications of tape and heavy sheeting sealed to the wall. Double the seal along the bottom of the door with wet towels.

16. Your gas masks and protective suits may provide good redundant protection against a chemical agent attack or accident. Don't put them on unless you know for sure a release has occurred and officials say it's a bad one. Follow the instructions for gas mask usage and do not let anyone sleep in their masks. Sleepers tend to draw shallow breaths which may not pull fresh air through mask filters and the sleeper may suffocate. Also remember to not let anyone sleep on the floor where both carbon dioxide and any tracked in contaminants will settle.

17. Though thoroughly sealed, you should have enough air in the average sized room (10x10) for a family of four (two adults and two children) to comfortably stay for about 6 hours. If you're in your room longer than that, you'll need your air filter that we'll discuss below.

18. Even if you do not utilize your main hallway as part of your sealed area, seal the hallway that leads to your sealed room in the same manner anyway if you have time. Close all doors and seal them. This will create an antechamber to your safe room to keep your family safe if someone has to venture out and back for whatever reason. (Venturing out is one of those few scenarios where a gas mask and protective suit would be useful.)

Final Considerations and Additional Hints

➤ If you happen to be a scuba diver, bring a dive tank or two of compressed air inside the room with you and open the valve only a <u>tiny</u> bit (you'll hear just a very slight hiss) after several hours of being cooped up in the sealed room. This will freshen the air without violently increasing the pressure in the room. In fact, there is an entire school of thought that says a way to help "seal" a room (in addition to the tape and plastic sheeting) is to subtly increase the internal pressure through the use of compressed air so that no air can enter the room from outside. This is only a hint. Don't worry about going out and getting compressed air. It's just one of those little "icing on the cake" things if you happen to scuba dive.

➤ One caveat to the use of compressed air: Don't use an air compressor to drive air into your room as it will be drawing on contaminated air from outside.

➤ Even if sealed in only one room, keep your room air filters running in the rest of house to keep rest of house clean as possible. Only run them in rooms not open to the outside lest air currents pull things in.

➤ Create your own **"seal the home" drill** where everyone takes turns leading the family on a walk-through and describing in great detail each step that would taken in sealing the home.

➤ If you decide to actually seal a room as part of a drill (which we recommend), you should test your efforts by having someone burn several **incense** sticks inside the house, and douse the outside window ledges of your sealed room with cheap **perfume**. If you can smell anything inside your sealed room, you have a leak.

➤ Depending on the nature of the threat, you may have to spend a couple of days in your sealed room. At this point fresh air will become critical. You'll need a special air filter (that you should make <u>now</u> and not later).

Making an Air Filter

The purpose of this air filter is to purify the air you bring in from *outside* your safe room, but from <u>*inside*</u> your house. Here's how to make your air filter. The following homemade filter is not rated by NIOSH although it should still serve you rather well as it's constructed on the same principles as most filtering respirators. An alternative to this, if you can afford it, is to use the **filtering canister** from either an Israeli gas mask or a current US military issue MOPP system in the same manner described below, with vacuum cleaner hoses coming off each of the canister's openings, and the seams sealed. You'll get a better idea of what we're talking about as we proceed.

➤ First, you'll need the following **equipment, parts, and materials**:

❏ A vacuum cleaner that accepts hoses (preferably with a sealed canister and HEPA filter).

❏ Spare vacuum cleaner hoses.

❏ Four HEPA air filters (any size will do so use the same size your HVAC system uses).

❏ A cardboard box wide and long enough to hold your HEPA filters and deep enough to hold eight of them instead of just four. We'll be using the extra space.

❏ Spare cardboard strips or pieces of wood to make spacers and risers.

❏ Scissors and/or box cutters to cut the box, tape, and T-shirts.

❏ A couple of old T-shirts.

❏ Large roll of Duct tape.

❏ Elmer's glue or a hot glue gun.

❏ A plastic spatula.

❏ Baking soda.

❏ Activated charcoal. (You can get this at most drug or vitamin stores.)

➤ Here's how to **assemble your parts**.

❏ Cut four squares of material from your T-shirts, the same size as your HEPA filters.

❏ Mix separate pastes, moderately thick, about the consistency of thin Elmer's Glue ®, out of your baking soda and your activated charcoal. Use water for each, but don't mix the two pastes with each other.

❏ Use the spatula and spread the baking soda paste thoroughly and evenly across two of the T-shirt pieces. Saturate the material, but don't let the paste cake into lumps, since air has to pass through.

❏ Do the same thing using the other two pieces of material and your activated charcoal paste.

❏ After the pieces of material have dried, glue them across the intake sides of your HEPA filters, pasted side up, and seal them completely around each edge. Use one piece of material on each filter.

❏ We're about to stack the filters horizontally in your cardboard box, but first we need to put spacers in the bottom of the box so the bottom filter will be about 2 inches off the bottom of the box to give you room to attach one of your vacuum cleaner hoses to the bottom of the box as shown below. This is the hose that will lead from the filter to your vacuum cleaner.

❏ After you've put in the spacers, cut a small hole in the side of the box and as close to the bottom as you can get. Insert one of your vacuum cleaner hoses a few inches and glue and tape around the seam both inside and out where the hose meets the edge of the box.

❏ Now take one of the HEPA filters that has a baking soda material panel on it and use that as your bottom filter. Put it in place using lots of glue to seal around the edges where it meets the box. If the glue isn't sealing well, seal the seam with duct tape. In fact, use both if you can. You want air being forced *through* the filters and not *around* them.

❏ Put your other baking soda treated filter in next repeating the same seam sealing process.

❏ Now install your two activated charcoal sets, sealing each in the same manner, and doubly sealing the seam around the top filter. Hopefully, this has left you with a 2-inch space at the top of your box. This will put both your charcoal filters on top, your baking soda filters on bottom, and the marked air-flow direction on the HEPA filters pointing downward.

❏ Cut another hole in the side of the box near the top for your other vacuum hose. This hole should also be on the opposite side of the box from the bottom hose. Seal it in place just as you did the bottom hose. Test your hoses to make sure the proper hose has the correct connector to attach to your vacuum cleaner. Also be sure to change the hose if you get a new vacuum.

❏ Close the lid of your box and wrap the entire thing in two layers of duct tape. You don't want any air coming into your filter except that which has been drawn through the hoses and forced through the filters.

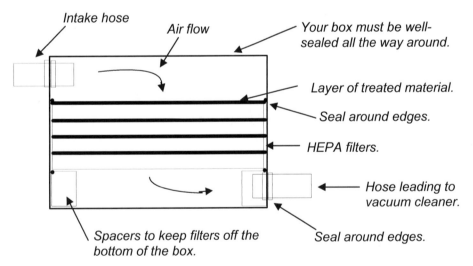

Intake hose

Air flow

Your box must be well-sealed all the way around.

Layer of treated material.

Seal around edges.

HEPA filters.

Hose leading to vacuum cleaner.

Spacers to keep filters off the bottom of the box.

Seal around edges.

Though easy to make from inexpensive materials, your air filter must be constructed so that no outside air can get in and all air coming through the filters flows through the filters, and not around them, as shown by the air-flow arrows in this side-view drawing.

Remember to make this filter now so you'll have it on hand when you need it. You won't have time to fabricate one in the middle of an isolation.

➤ **Here's how to use your filter**.

❏ First, you'll have to draw in some air from outside the room. Recirculating and filtering the air already in the room will not remove carbon dioxide. Our caution is that you do **NOT** draw in air from **outside** the house. You'll want to draw air in from an opening that leads to the interior of your house. Use the size of your house as a partial filter, since what little chemical that may have gotten in to your house may settle before it gets drawn into your filter.

❏ The hose coming out of the top of your filter box is your **intake** hose. Poke it through some sort of opening that leads from your sealed room into the house, and yet one that can be sealed around with duct tape after the hose has been passed through. We suggest having a small pet door in the door of the room you intend to use as your sealed room. The only drawback to this is that it places your intake hose near the floor where carbon dioxide and chemicals may lurk. If you have an opening higher off the floor, use it. You can also remove the door knob and use that hole, provided you duct tape around the hose after it's in place.

❏ Attach the filter's bottom hose to your vacuum cleaner and turn the vacuum on.

❏ Keep your filter box positioned so the filters inside are lying horizontally with the air-flow direction going top to bottom. The main purpose for this design is to let gravity help hold your charcoal and baking soda sheets in place as air is forced down through them. The charcoal will help absorb some contaminants and that's why it's the first layers the air will pass through. The baking soda will help neutralize some types of remaining contaminants if any.

Instead of making your own filter box, you could attach your intake and exhaust vacuum cleaner hoses to a spare respirator filter. Just make sure that your connections are well-sealed.

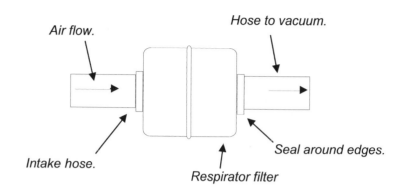

Air flow.

Hose to vacuum.

Intake hose.

Respirator filter

Seal around edges.

Sources and More Information

➢ Air Purifiers, Cleaners, and Smoke Eaters at http://www.breathepureair.com, or Breathe Pure Air, Dr. Craig Donnelly, 49 Calhoun Street, Suite B, Charleston, SC, 29401, Voice: 1-877-688-2703, Fax: 843.579.9454.

➢ "Ionic Breeze" air filter from Sharper Image, http://www.sharperimage.com, 800-309-2759.

➢ Oreck has an ionizing air filter as well. They're at http://www.oreck.com, 800-289-5888, or 800-805-4208.

Moving right along, what if you're faced with a situation that duct tape, plastic sheeting, and air filters just can't handle? We touched on tornado shelters under "Reaction," but what about serious tornado shelters or nuclear blast or fallout shelters? Do you really need one? Let's see.

Fortified Shelters

As promised, we're going to elaborate just a bit on a subject touched on in the "**Reaction**" section.

The most expensive and labor-intensive change you could make on your property, is the installation of either a storm cellar, or a fallout shelter. Most of us will not be able to afford either, but in the event you feel this concept to be vitally important to your unique situation, let's discuss some of the detail you'll need to know. The first question is should you even consider building an underground shelter?

Should you *consider* building an underground or reinforced shelter?	
Yes	**No**
➢ If you live in a tornado-prone area.	➢ If flood is your primary threat.
➢ If you live near a very likely nuclear target or immediately downwind from one.	➢ If your area has a high water table or bedrock.
➢ If post-nuclear evacuation is not an option.	➢ If other shelter is readily available for tornadoes.
➢ If you have a house with a back yard or sizeable basement.	➢ If your immediate area has maintained its civil defense shelters.
➢ If your home provides no physical protection, such as with mobile homes.	➢ If your home has a well-stocked, solidly reinforced basement.
➢ If you can easily afford one and having one will give tremendous peace of mind.	➢ If you really cannot afford one, and will focus on other options.
➢ If you're a care giver and your evacuation will place others in grave danger.	➢ If you feel evacuation should be your primary reaction.

Other considerations lie in the advantages and disadvantages inherent in cellars and shelters. Let's look at a few:

Advantages

➢ A storm cellar would give you much greater protection against a tornado than a reinforced frame house.

➢ A fully equipped and stocked shelter would greatly increase your reaction options to a nuclear event, many types of civil unrest, and depending on its construction, equipment, and location, it would help protect against certain chemical attacks or accidents.

➢ You'd also have an on-site "second residence" if you needed a place to stay while rebuilding or repairing damage to your house after a natural disaster.

➢ If nothing else, you'd have a secure location for storing supplies to protect them from looters after a disaster.

➢ Under certain circumstances, this is a cost that could be shared by neighbors or family members living close by, and who would also share in the shelter's use.

Disadvantages

➢ Cost. Even the fabricated "half and half" units that see a lot of your own elbow grease invested will still be rather expensive. These costs would include not only parts and materials, but excavation, air filtering systems, water sump pumps, etc.

➢ Also, to be considered fully functional, your shelter will have to be heavily stocked which means you may have to buy double the supplies you would normally have on hand.

➢ Space. Though you don't need a huge yard to install a cellar or shelter, you do have to have some room that's accessible by heavy digging equipment. And by the way, your real septic tank and its leach field, or public utility lines for gas, electricity, cable, etc., will also be a factor to consider.

➢ Most are useless in any sort of situation in which flooding is a possibility, and if they're high enough out of the ground to survive flooding, they may be slightly less effective against storms or radioactive fallout.

➢ Maintenance. No fully functional shelter is going to be an "install it and forget it" thing. You'll have to keep it up almost like you would a separate residence.

➢ Having a shelter might make you want to stay during a post-nuclear fallout period when in some scenarios evacuation may be your best bet.

Though you might not be able to build an underground shelter, you may still be able to build a reinforced above-ground shelter, or a "half and half." A half and half is a shelter that is half above ground and half underground with a solidly reinforced roof and a few feet of earth or concrete on top. Though less attractive and less secretive, these shelters are easier to build and therefore more economical than a completely-underground shelter.

Cellars and shelters can be constructed in a variety of ways out of a number of parts and materials. The only thing limiting these structures would be your pocketbook, imagination, and space to put one. Some common types of storm cellars or fallout shelters include:

1. A new, clean septic tank installed for storm cellar purposes (most are too small for anything but a storm cellar for a small family).
2. A structurally reinforced underground basement (the most common type of individual fallout shelter during the peak of the cold war and probably the easiest to make).
3. Large diameter corrugated steel or concrete pipe, or "culvert" pipe, buried in the yard, in a hillside, or made into a "half and half."
4. Fiberglass or steel units specifically designed and engineered to be used as underground fallout shelters (the most expensive).

Should you decide you really want to invest in either purchasing or constructing an underground shelter, let's look at a few general considerations you should keep in mind. As you'll see in our discussion below, we lean away from making a fallout shelter a "do it yourself" project unless you're well versed in the finer points of constructing, equipping and maintaining one.

➢ Your shelter's entrance will need to be close enough to the house to get to in a hurry yet far enough away so the house can't collapse on it and trap you. Therefore, you might need more yard space than you think.

➢ With that in mind, you should make sure nothing can collapse on your entrance, and you should also consider having more than one entrance or emergency exit from your underground shelter.

➢ Air filters and bilge pumps (to pump out any water that may eventually seep into your shelter) require specialized equipment. See the providers listed in "Sources and More Information" below.

➢ For adequate protection against radioactive fallout, you'll need these thicknesses of materials, any of which will supposedly stop 99 percent of radiation:

- 3 feet of water.
- 2 feet of packed earth, or 3 feet of loose earth.
- 16 inches of solid brick.
- 16 inches of hollow concrete blocks filled with mortar or sand.
- 5 inches of steel.
- 3 inches of lead.

➢ As we mentioned above, the easiest of the underground shelters is the reinforced basement. If you choose this option, make sure you have adequate shielding as mentioned above with "X" number feet of earth, etc. Also make sure that you have sufficient structural enhancements to keep your house from collapsing on top of you and that you equip and stock your basement to be as autonomous as a second residence. Keep in mind too that the needs for a second entrance, air filters, and water pumps apply to a basement just as they would a separate shelter.

➢ By the way, if you bury a shelter in your yard, call before you dig. Make sure you're not about to destroy some utility for your neighborhood.

Sources and More Information

To help you better understand the complexities involved in building a shelter and to help you find some sources of parts and shelters if you decide you need one, we've helped by locating the following:

➢ Disaster Shelters offers a good online overview and collection of pertinent information on the construction of your own underground shelter. They're online at http://www.disastershelters.net.

➢ American Saferooms, online at http://www.americansaferooms.com, or contact them at PO Box 292, Lynchburg, TN 37352, U.S. Call Toll Free 1-866-271-0400, International 1-931-759-5333, Fax 931-759-5444.

➢ Radius Engineering, online at http://www.nuclearshelters.com, with direct contact at 222 Blakes Hill Rd, Northwood, NH 03261, 603-942-5040, Fax: 603-942-5070.

➢ Tips for constructing an underground shelter are online at http://www.plansfordummies.com/shelter.htm.

➢ The British have a rather useful and informative Civil Defense / Nuclear Shelter info brochure still available online at http://www.cybertrn.demon.co.uk/atomic/shelters/shelter.htm. Be sure to visit their Civil Defense subject page at http://www.cybertrn.demon.co.uk/atomic/index.htm.

➢ More detail on culvert-pipe shelters can be seen online at http://www.waltonfeed.com/old/cellar3.html.

➢ Shelters Direct, 10756 Rhode Island Ave., Beltsville, MD 20705, 800-865-5555, www.sheltersdirect.com.

➢ Nuclear Fallout and Shelters: http://people.smu.edu/rmonagha/shelter.html.

➢ For a list of shelter manufacturers, see: http://www.surviveanuclearattack.com/ShelterManufacturerlinks.html.

➢ See **Radius Engineering** at 222 Blakes Hill Rd, Northwoods, NH 03261, http://www.radiusdefense.com, 603.942.5040 (voice), 603.942.5070 (fax).

➢ Visit the Fallout Shelters & Nuclear Civil Defense FAQ page online at www.radshelters4u.com, to download plans and books, and for more details on ready-made shelters.

➢ A good links collection on Underground Shelter Resources: http://www.earthmountainview.com/shelters.html.

➢ Visit http://www.tornadoproject.com for their links to other shelter information sources.

➢ Some of the sources we listed earlier under "NBC Data" will have fallout and shelter information.

➢ We've added several how-to shelter publications on our **CD**.

Alternative Energy

The subject of alternative energy could fill enough books to dwarf the Encyclopedia Britannica, and well it should. Too many of us rely on overpriced single sources of gas or electricity. When you think about how much we rely on energy, all disaster prep needs aside, it seems rather foolish to place all our stock in one single source of each anyway.

Rather than cover the entire spectrum of alternative energy information in this one little section, we're going to simplify things and give you an overview of the subject, a few suggestions of our own, and enough information to help you choose your own direction. Then we'll provide you a number of sources to not only continue your education but to help you find economical short term, intermediate, and long term solutions to your energy and self-sufficiency needs.

In this portion we'll cover alternate sources and energy production. We'll get into applications of this alternate energy, as well as other methods of heating, cooling, and cooking under "Climate Control" and "The Manual Kitchen" below.

Alternative energy ties directly to our earlier section, "Frugality." You'll need money to buy some of the equipment you may require, or to make some of the adjustments necessary for energy efficiency. All this also ties in to frugality as many of the changes you make will see a reduction in your current energy costs. However, as we mentioned before, make sure the investment you make in any equipment or changes designed to save money on your utility bills can be recouped in a fairly short period of time.

As we go through this section, we'll look at some general concepts to keep in mind, various power sources, and then sources of more information or equipment. We'll also mention here that many of the suggestions below will be difficult to put to use by apartment dwellers. Still, there will be a few things here for you so don't skip this section.

General Considerations

Let's look at being able to more readily utilize the energy sources available to you. Keep these in mind and make any desired changes as you can afford to.

> The first thing you'll want to look at is how diverse your household's energy source needs are. Are most of your larger appliances electric or gas? What do your stove, oven, water heater, central heating system, or dryer use? Are they all electric, are they all gas, or is there a mix?

> Based on what you find, your first consideration might not be alternate energy *sources*, but alternate *equipment* so that you'll have appliances that use different energy sources should one source not work for a while. Here are some examples:

● If you have a **gas** stove and oven, you should also have an **electric** microwave, hotplate, and toaster oven. Each are inexpensive and will allow you to cook with electricity if the gas is out.

● Similarly, if you have an electric stove and oven, your might want to get either a **gas grill**, or cooking utensils that will allow you to prepare food in your **fireplace** if you have one.

● Alternate your home's heat as well. If you have electric or gas heat, you should get the opposite in space or room heaters. We'll touch on this again under "Climate Control" below. **Note**: Even though your main central heat may be gas, it will still take **electricity** to run, whereas gas room or space heaters will not.

● Regardless of what type hot water heater you have, you should look into getting a solar heater to run in tandem. It's simply an insulated box that sits on your roof as black-painted water pipes snake back and forth through it to be warmed by the sun. This is not only an economical method of reducing your bills, it will also provide you with rather warm water in the event all energy sources are down.

> If your household is all electric, consider getting a few small appliances that run on **propane**. You can use the small propane tanks like a gas grill would use, or you can have a large propane tank installed and serviced by a local gas company. In some cases, you can have a gas line run to your house by your local natural gas company. Though this is a more expensive option, it's also one that could be shared by neighbors if you set this up to be a shared gas source.

> While we're on the subject of gas, let's cover alternating gas sources just as we've discussed alternating between gas and electricity. Even if your home is set for natural gas, there may be times when that source will be down either because of problems at the main facility, or with pipes along the way. You might consider having a large propane tank installed on your property and buying **propane conversion kits** for all your gas appliances. (You would not use the propane or install the kits unless the main gas lines were down and expected to be out for a while.) Most propane conversion kits are inexpensive, easy to install, and easy to switch back when your main source of gas comes back on line. Though potentially a pricey endeavor, it will certainly give you options, and in disaster planning options are our friends.

> If your main source of gas *is* a large propane tank, look into storing a few smaller propane tanks to use as a backup if your main tank runs out and no one is able to come service you for a while.

> **Hint**: If your piped-in natural gas is supply is lost, **turn your gas off** at the meter or at the first main valve inside the house. Pilot lights will go out when the gas does and you don't want the gas coming back on when you don't know it's been restored, and filling your house with gas from the unlit pilot lights.

> With regard to electricity, before we get into alternate production sources, let's consider battery backups for some smaller items, namely communication gear (such as a cordless phone or small TV), and any necessary medical equipment (such as an oxygenator). You can either get a battery backup for the individual piece of equipment, just as your laptop has, or you can look into getting a battery backup system for the house that powers certain outlets in the event of a power loss. A small example of battery backups can be seen in emergency exit lights in public buildings that come on when power is lost.

> Another battery backup example is the "Jumpstart" system we described under "Basic Vehicle Prep." A third example is a simple setup where you'd keep a spare car battery hooked to a trickle charger (standard car battery recharger) all the time and then use the battery for your radio or TV in a power outage.

➢ While we're on the subject of batteries, get your "Battery Worksheet" from the "**Appendix**," the one you used to help plan your battery needs for your evacuation gear. Go through the house and make sure you have enough batteries on hand to operate all your flashlights, radios, fans, portable TVs, and any other piece of emergency equipment that will take a battery.

➢ You can make small lanterns by putting Votive or "Tea" candles in a metal can with holes punched all around the sides. At home you can use mason jars, jelly jars, etc., but we recommend against carrying glass in any kits if you're making these lanterns for evac gear.

➢ **Hint**: Lighting will be a major concern during a power outage. Though a million places will sell you all kinds of lanterns and candles, our suggestion is that you go with battery-powered or "wind-up" light sources whenever possible, using your flame-based light sources sparingly and carefully. Flame carries its own fire-safety considerations as well as oxygen consumption concerns.

Alternate Sources and Energy Production

When most people hear the subject "alternate energy" they think of backup generators, solar power, and water and wind power. That's exactly what we're going to talk about here. Though our immediate goal is to have temporary energy to help you survive an extended isolation that may involve the loss of utilities, we're also going to throw in some info to help you develop regular energy sources that may reduce your dependence on commercial sources.

Backup generators:

➢ IF YOU MAKE ANY ONE MAJOR PURCHASE, MAKE IT AN ELECTRIC GENERATOR. Get a generator that runs on what's easiest to get. For example if you regularly have lots of gasoline for your lawnmower, etc., get a gasoline powered generator. If you have a big propane tank like the one mentioned above, and very few appliances that use the propane, then get a generator that runs on propane. Better yet, get one that uses gasoline and comes with a propane conversion kit. In the sources below, you'll find contact info for suppliers and manufacturers of quite a number of different types and sizes of backup home generators.

➢ **Hint**: Many generators are small enough for even our apartment dwelling friends to keep one out on a balcony or porch. Regardless of which model you get, make sure you have it securely chained and locked against theft, and that the fuel used to run it is safely stored and protected against both fire and theft.

➢ Another interesting concept many people don't think about is that if you have any sort of internal combustion motor available, you have the heart of a generator. This could be anything from a gas powered chainsaw or lawnmower, all the way up through your family car. Though you have the "heart" of the generator, you'll still need some more assets in the way of parts and your own mechanical ability.

❑ First, you'll need the gizmo that produces the electricity. This can be something as simple as a car's alternator that you purchased from a salvage yard. An alternator acts as a car's generator. All you have to do is figure some way, through drive belts or old bicycle chains, to connect your alternator to your engine so that it turns. Several of the sources below will have plans for homemade generators that work using this principle and you can see some examples of setups and connections on their sites.

❑ Riding lawnmowers, especially those that use drive belts, may be the easiest to convert since all you have to do is slip the drive belt off the mower blades and loop it around your alternator's pulley.

❑ After the alternator, you'll need a device called a "power inverter." This will not only change the current from DC to AC, but many are also transformers that will step up the voltage to the same level as house current. We've listed some power inverter sources below too.

❑ Moving up to the more sophisticated engines, your car is probably the easiest to use of all. Many systems such as the "Jumpstart" system we described under "Basic Vehicle Prep" will simply plug into your cigarette lighter to charge, and can run a couple of small household appliances at once. This effectively makes your vehicle a ready generator. In addition to a jumpstart system, you can also use a simple "power inverter" (available at most department, automotive, and hardware stores), which will plug into your cigarette lighter outlet and allow you to plug in regular electrical appliances.

❑ As an innovative alternative, you can jack up the drive wheels of your car and run a pulley belt off of one of the wheels. Run this belt to a generator or other piece of machinery you need to operate (such as a well pump) and simply put the car in gear and let it run.

❑ The main caution with your car is to make sure it doesn't overheat from running for long periods of time while stationary. To help with this, especially if you're running your car as a generator in the summer, open the hood, open the windows, turn on your car's heater, set the climate controls to "fresh air" or "outside air" and turn the fan on high. Your car's heater will act as a secondary radiator and help keep the engine cool.

❑ Another caution with your car is to make sure no one can steal it while it's running. Chain it to your concrete anchors, block it with another vehicle, park in a fenced-in area, etc.

➢ We also have a couple of **suggestions on how to use your backup generator**:

❑ First, look into getting a battery backup system that the generator can power. Having a battery system coupled with a voltage regulator and power inverter greatly increases the number and size of appliances that can be powered with your one generator.

❑ Remember that you can't power everything in your house, and out of what you can power, you shouldn't try to power them all at once or use your generator 24 hours a day. Therefore, you'll need to prioritize your energy needs and learn when and how to alternate appliance usage.

❑ Mini-freezers, small space heaters, and small AC units are cheap and won't drain a generator like a central heating and air unit or a window-unit air conditioner would. Use your **light timers** we asked you to buy under the "Security" section of "Basic Home Prep," to automatically alternate generator output usage between your fridge, freezer, and these other necessary appliances.

❑ For most appliances, you'll simply run an extension cord to your generator or to the junction box that's run from your generator inside the house. If there was any one room in the house that you would want "straight wired," meaning a permanent wire and outlet has been installed between this room and your generator, it would be your safe room and/or the room you intend to seal. The reason for this is that if you have scarce little time to fuel and start your generator, you certainly won't have time to locate and run extension cords.

Solar Power:

➢ Solar power can be utilized in more ways than the big solar panels most people visualize when they hear the term. However, since that's the most popular concept, let's cover a few points:

● Small solar panels are actually plentiful and cheap. Several of the sources below will tell you where to find these smaller panels that will let you power your cell phone, laptop, television, or even trickle-charge your car's battery. Since these are so available and inexpensive, and since even apartment dwellers can put them to good use, we suggest you definitely look into these.

● The medium to large panels that can power appliances, or power your whole house are rather expensive and carry with them hidden costs. For starters, the cost of such large panels is rather high, and so is the equipment necessary to fully utilize this power source. You'll need a battery system that is charged by the panels, a voltage regulator to make sure that the power running from the batteries to your home's outlets flows evenly and without spikes. You'll also need a power inverter to convert the battery system's direct current (DC) to alternating current (AC) which is what your appliances use. You may also need a tilt and rotate base for your solar panels that allow them to track the sun as it moves across the sky.

➢ You don't need photovoltaic cells to harness the sun's power. Any time you simply need heat for the sake of heat, or heat energy to run a device, you can use the sun.

● Numerous companies make "solar ovens" constructed of reflective metal, and designed to use this reflected sunlight for cooking and baking. See "Sources and More Information" below.

● We talked to one individual who had a very small greenhouse next to his home. He set up a small system of inexpensive duct work, and during winter months he would circulate air from his home through the greenhouse which helped warm his house.

● Another family had a quick solar solution to heated showers during a water or power outage. They kept an old waterbed mattress that happened to be dark navy blue. Their plan is to lay the mattress out on their elevated deck (reinforced to handle the weight), fill it with water and let it heat in the sun. From the valve they'll run a hose down to a shower head and set up a privacy-screened shower area under the deck. It's an idea with limited applications, but you have to think about these things.

Wind and Water Power:

➢ We'll cover water power first since it's the most limited. There are really only two ways to use water as a source of energy production. One is if you have a running stream on your property and can utilize it to turn a generator. The other is to fabricate a steam powered generator.

➢ Wind power is a different story entirely and it's in this arena that many new developments in the area of home power are being made. Ironically, wind has the strongest potential yet will have the shortest section in this energy portion. We've already covered considerations related to power such as the battery system, voltage regulators, etc. These apply to wind-generated power just as they would to gas-engine generated power. As far as unique considerations applying to wind energy itself, we only have a few items to consider:

● The two biggest obstacles that wind power researchers have had to overcome are the difficulties involved with friction and other forces making generators hard to turn, and the efficiency of the generator involved since wind speed is so variable and inconsistent. Technology has made great strides in both areas over the past decade and wind generators have increased in efficiency and dropped in price. We feel wind to be a source to consider.

- Along with an increase in efficiency and a decrease in price, wind units have also decreased in size. Therefore, anyone with a roof top or even a balcony railing (such as those of you in apartments) could put a wind-powered generator to work for them. The sources listed below will not only fill you in on the rest of the story, they'll tell you exactly how to put wind energy to work for you in other ways.

➢ By the way, check with your state's version of a "Public Utilities Commission." In many areas, power companies are required to pay you if your power production winds up overflowing back into the main grid. Your commission or power company can better explain this to you.

Sources and More Information

➢ **Edmund Scientifics** is a good source of parts and tools. See http://www.scientificsonline.com, or contact them at 60 Pearce Ave. • Tonawanda, NY 14150-6711, TEL: 800-728-6999 • FAX: 800-828-3299.

➢ Find the Mother Earth News book, the **Handbook of Homemade Power**. Contact Mother Earth News online at http://www.motherearthnews.com/index_js.html, or at 1503 SW 42d Street, Topeka, KS 66609-1265, 800-234-3368.

➢ A collection of generator links can be found online at http://www.equipment.net/list/generatorequipment.htm.

➢ Visit The Energy Alternative, online at http://www.theenergyalternative.com.

➢ Northern Tool and Equipment has a selection of generators. They're online at http://www.northerntool.com, or contact them at P.O. Box 1499, Burnsville, MN 55337-0499, 800-533-5545, Fax 952-894-0083.

➢ The Epicenter carries a page on how to convert a lawnmower engine into an electric generator. It's online at http://theepicenter.com/tow082099.html, or you can contact The Epicenter at 384 Wallis #2, Eugene, OR 97402, 541-684-0717, Fax: 541-338-9050. While there, be sure to visit their power page at http://theepicenter.com/power.html, and you'll certainly want to visit their homepage.

➢ The Solar4Power, Advanced Energy Group, online at http://www.solar4power.com, offers quite a bit of educational information on solar power as well as a free catalog of solar and wind power equipment. While there, be sure to visit their two educational pages:

- http://www.solar4power.com/solar-power-glossary.html.
- http://www.solar4power.com/solar-power-education-links.html.

➢ Oasis Montana offers a large collection of links to renewable energy info, as well as a newsletter. Go to http://www.oasismontana.com/PV_index.html, or contact them at Oasis Montana, Inc., Renewable Energy Supply and Design, 436 Red Fox Lane, Stevensville, MT 59870, 1-877-627-4768, Fax: 406-777-0830.

➢ Defend Yourself.com offers a large links collection relating to generators at http://www.dfendyourself.com/link_directory/emergency_generator.html.

➢ We also found you one source of electric generators with Electric Generators Direct, online at http://www.electricgeneratorsdirect.com. You can also contact them at 1-800-860-1387.

➢ A good bit of educational information about generators can be found on the How Stuff Works page at http://www.howstuffworks.com/emergency-power.htm.

➢ A rather complex, and yet seemingly powerful home-made generator system is described in detail, complete with downloadable information and files at: http://www.schmitzhouse.com/Johns_Electronics_02.htm.

➢ SolarVerter offers small, individual solar panels to power your battery-operated devices. They're online at http://www.solarverter.com, or contact PATRICK TECHNOLOGIES, 4970 Varsity Dr., Lisle, IL 60532, Toll-Free (U.S. Only): (888) 858-2801, Phone (Intl. & Domestic): (630) 719-9020.

➢ Visit Backwoods Solar at http://www.backwoodssolar.com, or contact them at Backwoods Solar Electric Systems, 1395 Rolling Thunder Ridge, Sandpoint, Idaho 83864, Phone: 208-263-4290. While you're there, look at their selection of propane-powered and other alternate power freezers and refrigerators at http://www.backwoodssolar.com/Catalogpages2/refriger2.htm#SERVEL PROPANE.

➢ For each of your natural gas appliances, call the manufacturer to ask about propane conversion kits.

➢ Be sure to visit the government's National Renewable Energy website at http://www.nrel.gov. While there, be sure to visit one of the solar pages at http://www.nrel.gov/clean_energy/teach_solarpower.html, for a listing of solar ovens and other gear, complete with downloadable plans and instructions.

➢ More info on solar cooking is on Solar Cooking.org's documents page at http://solarcooking.org/docs.htm. Visit their homepage at http://solarcooking.org, and click on their solar ovens plans page.

➢ Visit PicoTurbine, online at http://www.picoturbine.com, for a discussion of wind and solar power including plans for wind-generators and solar ovens at http://www.picoturbine.com/picoturbine-plans.htm.

- See Solar World at http://www.solar-world.com, for large and small photovoltaic panels and kits.

- Also see BP Solar, online at http://www.bpsolar.com, for larger solar panel applications.

- Visit the Solar Online Learning Center at http://www.solenergy.org, for educational information about solar power and its applications.

- You'll find even more of the same at http://www.solarenergy.org.

- Also see http://www.infinitepower.org.

- See the HomePower downloads page at http://www.homepower.com/magazine/downloads.cfm.

- Propane conversion kits are available through US Carburetion, Inc., at http://www.propane-generators.com, or you can contact them at 281 Coral St., Building B-1, Canvas, WV 26662, or call 1-800-553-5608.

- For an extensive selection tutorial and comparison of different types of home electric generators, visit both http://agitator.dynip.com and http://www.nerstrand.net/genfaq2.html.

- Visit "Mr. Solar," at http://www.mrsolar.com, and with direct contact at Online Solar, Inc., Mr. Solar.com, P.O. Box 1506, Cockeysville, MD 21030, (410) 308-1599, Toll Free: 877-226-5073, Fax: 410-561-7813.

- If you're serious about turning your home over to alternate energy sources all the way, there are five magazines that should become part of your library:

  • Popular Mechanics, http://www.popularmechanics.com, or call 1-800-333-4948, 212-649-2853, Fax: 212-586-5562. Popular Mechanics discusses installing a permanent generator into your home's wiring at: http://popularmechanics.com/home_improvement/home_improvement/1998/3/install_backup_generator/print .phtml. (Remember that our "DP1 Book Links" file on the **CD** will help you log on to these sites.)

  • Nuts & Volts, online at http://www.nutsvolts.com, or contact them at 430 Princeland Court, Corona, CA 92879, 909-371-8497, Fax: 909-371-3052.

  • Popular Electronics, now "Poptronics," online at http://www.poptronics.com, or contact them at 7065 W Ann Rd., #130-999, Las Vegas, NV 89130, Fax: 1-877-275-1829.

  • Backwoodshome at: http://www.backwoodshome.com, or Backwoods Home Magazine, P.O. Box 712, Gold Beach, OR 97444, Toll free: 800-835-2418, Phone: 1-541-247-8900 Fax: 1-541-247-8600. See their two online generator articles at http://www.backwoodshome.com/articles/thomsen28.html and http://www.backwoodshome.com/articles/thomsen43.html for a discussion of gasoline vs diesel generators.

  • Mother Earth News at: http://www.motherearthnews.com/index_js.html, 1503 SW 42d Street, Topeka, KS 66609-1265, 800-234-3368.

- For smaller, self-generating products, such as radios, flashlights, etc., visit Crank'n'Go, online at http://www.crankngo.com/index.htm, or contact them at 1-877-9-ARDENT (877-927-3368).

- For renewable energy equipment and high-efficiency appliances, visit Solardyne, online at http://www.solardyne.com, or you can reach them at Solardyne Corporation, 1001 SE Water Ave, Suite 120, Portland, OR 97214, Phone: (503) 830-8739, Fax:(503) 595-0145.

- Freeplay, a South African company, offers a range of self-generator powered radios, cell phone chargers, etc. You'll find them online at http://www.freeplay.net.

- Even more of the same, including the hand-crank cell phone charger, can be found through Innovative Technologies, online at http://windupradio.com, or call them at 1-250-386-2556.

- Another tutorial for selecting a home generator can be found at : http://www.nerstrand.net/genfaq2.html.

- For a list of solar info links and equipment sources go to http://www.survivalcenter.com/solarequipment.html.

- Check the universal gear page at http://www.equipment.net, to search for related tools and supplies. You can contact them at EQUIPMENT.NET, INC., PO Box 810461, Boca Raton, FL 33481, FAX: 305-675-5896.

- One source of info and equipment for a variety of alternative power sources can be found at http://www.utilityfree.com, or you can contact Utility Free directly at 800-799-1122.

- A variety of solar equipment calculators and worksheets are available through Big Frog Mountain, online at http://www.bigfrogmountain.com, or reach them at 100 Cherokee Boulevard, Suite 321, Chattanooga, TN 37405, (423) 265-0307, (423) 265-9030 FAX.

- Northern Arizona Wind and Sun carries solar and wind-powered generators and related information. They're online at http://www.solar-electric.com, or contact them at Northern Arizona Wind & Sun, 2725 E Lakin Drive, Flagstaff, AZ 86004, Flagstaff Office: 800 383-0195, 928 526-8017, or 928 527-0729 (FAX).

- Info on wind energy can be found through the American Wind Energy Association, at http://www.awea.org, or you can contact them at American Wind Energy Association, 122 C Street, NW, Suite 380, Washington, DC 20001, (202) 383-2500 | Fax: (202) 383-2505. While you're at the website, visit their FAQ page at http://www.awea.org/faq, and scroll down to the section on "Home and Do-It-Yourself Systems."

- Visit Horizon Solar at http://www.horizonsolar.com, and the links page http://www.horizonsolar.com/links.php.

➤ Also visit Oasis Montana, Inc. online at http://www.lpappliances.com/FrosTek.html, to see their propane freezer. You can also contact them at Oasis Montana Inc., Renewable Energy Supply and Design, 436 Red Fox Lane, Stevensville, MT 59870, 406-777-43221 or 4309, Toll Free Order Lines: 1(877-OASISMT) 1(877-627-4768), or 1(877-OASISPV), 1(877-627-4778), Fax: 406-777-0830.

➤ Among the alternate energy-production items, you can find propane-powered refrigerators and freezers at the Alternative Energy Store, online at http://www.altenergystore.com, or you can contact them at the Alternative Energy Store, 65 Water Street, Worcester, MA 01604, 1(877) 878 – 4060 or 1(508) 421-8201.

➤ Visit Propane Refrigerators online at http://www.propanerefrigerators.com, or contact them at Sparling's Propane Co. Ltd., 82948 London Road, Blyth, Ontario, Canada, N0M 1H0, Telephone: 519-523-4256.

➤ For a continuation on our energy sources discussion, along with a great collection of links to more info and sources, log on to http://www.greaterthings.com/News/FreeEnergy/Directory.

➤ We've put several publications dealing with solar cooking and the construction of solar ovens on our **CD**.

Notes:

The Isolation Pantry

You may have noticed here and there in the book that we've done a few things in "threes." The human body also relies on "threes." Three minutes without air, three days without water, or three weeks without food and you've had it. Hopefully we covered air well enough under "Sealing the Room." Here we'll make sure your needs for water and food are met so that no isolation will leave you at risk, or even wanting. Here's something else to consider; all physical needs aside, having plenty of water and food on hand in a shelter-in-place emergency can make all the **psychological** difference in the world. If you're well supplied, you're more likely to be content and at peace with the situation than you would be if sustenance were a major concern. This contentment can make the difference between your isolation being a major nuisance, or a life-threatening ordeal.

Goals of The Isolation Pantry:

1. To provide everything you need for your **immediate physical survival**.
2. To reiterate the fact that **the old "72-hour supply of food and water" rule just isn't enough**.
3. To secure enough supplies so that you **won't have to venture out** and expose yourself to danger.
4. To **keep your morale high** in an isolation situation by **making sure you're not left wanting**.
5. To make sure you have basic **medical supplies** to **stay healthy and to prevent other problems**.

We've divided this section into:

1. **Water**.
2. **Food**.
3. **Other Pantry Items**.
4. **The Medicine Chest**.

Water

As with any instruction, the biggest hurdle to clear is in convincing readers that the need for learning the material or performing the exercise is real. To some of you, the goal of supplying yourself during a 21-day isolation might seem like an unrealistic length of time. After all, most "experts" say to stock enough for just 72 hours. Why 21 days?

Consider this: Any number of natural disasters; earthquakes, floods, hurricanes, tornadoes, blizzards, and the like can knock out roads, bridges, supply routes, utility lines, pipes, or production facilities, and there's no guaranteed time limit for having any of them repaired and functional. As Georgia residents who weathered the "Flood of '94," North Carolina residents who rode out Hurricane Floyd, and Florida victims of Hurricane Andrew will tell you, they were without water and numerous other supplies for longer than just three weeks. Add to those the fact that any number of terror-related incidents can carry much more dire effects and necessary reactions, such as the fact that the incubation period for Smallpox is 14 days, and you can see that the list of reasons we should prep for a three-week stay at home could be a rather lengthy and scary one. (This doesn't even include public panic and disorder.) As we never print detailed terror attack scenarios, we'll just hope you're convinced of the need and we'll leave it at that.

Let's see now, 21 days worth of water at a minimum of **one gallon per person per day**? That can be an awful lot of water. Remember now, that **gallon per day** is <u>only</u> for **drinking and minor hygiene**, so the more you can store for cooking, washing, and any necessary cleaning, the better. And... DON'T forget your **pets**.

There are two basic ways of ensuring you have enough water. One is through adequate **storage**, and the other is "**production**" or water collection and purification. Since storage deals with what you've already got, and it's the only way you'll have water if you're confined to the house, we'll start there. But first, let's look at a short checklist of water gear you should have on hand that's redundant and separate from your evacuation gear. Make sure this gear is packed away with other isolation gear so that you'll be sure it's always on hand when you need it most.

"WATER WORKS"

❑ Gallon of unscented bleach	❑ Eye dropper	❑ Two bottles iodine tablets
❑ 5-gallon water storage jugs	❑ 50-gallon tank (optional)	❑ Extra water heater (optional)
❑ Two-liter plastic soda bottles	❑ Canteens	❑ Hand-pump water filter
❑ Two large metal stock pots	❑ Plastic bins and buckets	❑ Plastic sheeting or tarp

---

Water Storage:

Storage is actually easy, even if storage space around the house is at a premium. You can store some water in large containers, or even small bottles tucked away in little nooks and crannies around the house, as well as in any empty spot you have in the fridge. Most important of all, if you're in a situation where you cannot venture out, storage is the only way you'll have water. Let's cover some general concepts:

➢ Before we get into containers and places to keep them, let's discuss treating the water. Safety before quantity. Some of you might be asking, "And how could water spoil?" Well, it doesn't, but bacteria and other things can grow in it. Here are some simple ways to prevent that problem:

● When you get containers for water storage, whether you buy the container, or save containers such as two-liter plastic soda bottles, make sure they're clean. Wash them, rinse them well with a mild bleach solution, and then rinse them out with clean water before use.

● For water that is to be stored at room temperature or warmer, make sure all the air has been removed from the container before sealing it. Flexible containers can be squeezed, and rigid containers can be filled to overflowing. Use a piece of tape and a magic marker to label the container as to when it was filled.

● If your tap water is chlorinated, then use it as is. For immediate *drinking* we suggested filtered water to tap, but the chlorine in the tap water will help keep bacteria from growing.

● To treat your already-clean water for storage add **one drop of household bleach** (non-soap variety) per **half-gallon** of water. Instead of bleach, you may add **one iodine** water purification tablet per **gallon** of stored drinking water provided you have no dietary restrictions against iodine.

● "**Rinse residue**" is another quick way to treat water with chlorine. This is quicker and easier than figuring out drops per gallon and it helps clean the container in the process. Mix a 30% solution of unscented, non-soap household bleach. We suggest one cup bleach to two cups water. Dump the solution into your water storage container and put the cap back on. Slosh the solution around in the container to coat all inside surfaces, and then pour the bleach mix back into the cup. The amount of bleach remaining as "residue" on the container walls should be about right for treating the volume of water that container will hold. The larger the container, the more bleach that will remain clinging to its walls. If you go to use your water later and notice a bleach smell, simply pour the water into a pot or sauce pan and let it sit for about two hours. Any remaining bleach will evaporate. To speed up evaporation, heat the water until it starts to steam.

● Store your water out of direct sunlight. Sunlight helps some types of bacteria, molds, and algaes grow.

➢ The simplest way to store water is in medium to small water jugs or bottles kept around the house wherever space permits. Many department stores and camping supply outlets carry plastic canteens and other containers made for water storage. Many of the larger ones come equipped with a water spout to make water dispensing easier. Though the five-gallon size is the most common don't discount other sizes.

➢ Another great water storage container is the empty two-liter plastic soda bottle. Just make sure you clean it out well. Fill these up, label them as to when they were filled, and store them anywhere they'll fit. You can put them in the pantry, cabinets, closets, out in the garage, etc.

➢ Your freezer and refrigerator are also great places for storing these plastic soda bottles full of water, and keeping water here has hidden benefits. Let's explore this idea some more.

● Fill your plastic bottles about 3/4 full. As you swap bottles in and out of the freezer this allows room for the water to expand as it freezes.

● A full fridge or freezer is more efficient, so be sure you have water bottles tucked wherever you can place them. Once everything is cold or frozen, the more items you have in the fridge or freezer, the longer it takes for the temperature to rise to the point the unit has to cycle and cool again. The fewer cycles, the less energy used. This will also make it easier for your emergency generator to keep the refrigerator running.

● As you fill your freezer with food items and move frozen bottles of water down in the fridge, the ice helps cool the fridge.

● If your power fails, your frozen and cold water bottles will keep food fresh almost twice as long as without them. In this instance you'll want to move some of the frozen bottles down into the fridge area to help keep it cool. Speaking of your refrigerator and freezer, go ahead and turn them to their coldest setting at the first warning of any emergency so food gets colder and lasts longer if you lose power.

● Being frozen, the water in the freezer will keep longer than water stored elsewhere. So basically, you won't need to treat the water you store in your refrigerator or freezer.

● Emergencies aside, you can use the frozen bottles in place of ice in your cooler on trips and picnics. When the ice melts, it stays in the bottle rather than soaking everything, and you can drink the water. The same principle holds true for your refrigerator after a power failure.

● You can also use your frozen bottles to help make an "improvised air conditioner" that we'll discuss later under "Climate Control."

➤ Don't forget to keep canteens of water in your vehicle along with any other emergency gear.

➤ Moving on to larger containers, let's discuss a few ways your plumber can help you out:

● The average hot water heater holds 30 to 40 gallons of water. If you have room, you can have an old hot water heater added somewhere along your inside water line. It doesn't need to be right next to your regular water heater, and it doesn't even need to be a working water heater. It just has to be clean and not leak. Doing this could give you up to an extra 80 gallons of water stored between the two tanks.

● If you ever replace your existing hot water tank, see if you can get a higher-capacity model. Not only will it give you longer hot showers, you'd have more drinking water in emergencies.

● A slightly more expensive option are various water storage tanks made to install in the same manner.

● Some of these in-line water storage tanks can be mounted in your attic. This not only helps with water storage, but depending on how the tank is plumbed, it can act as a mini water tower and help increase your home's water pressure. Your plumber can tell you if this is a workable idea for your home and water system.

➤ Other commercial water tanks are stand-alones meaning they don't have to be hooked to any plumbing. You just fill them, treat the water, and leave them until it's time to rotate your stores. These tanks can be found in sizes anywhere from five gallons up to a 200-gallon size (and more from some companies).

➤ Speaking of rotating your stock, change out all your water stores, even the treated batches, at least once a year. We suggest a staggered rotation so you can use the stored water along and along, rather than have to dump any of it. Use your water for drinking, watering the pets, the plants, washing the car, etc. Do anything but waste it.

➤ Let's also discuss "stored" water that you can use if all your bottles and jugs run out.

● As we mentioned above, your hot water heater will hold 30 to 40 gallons of water. Turn the unit's power or gas off, let the water cool a while, and drain the water into buckets. You'll find a faucet or spout near the bottom of the unit. You want to turn it off even if the power is out so you don't burn up the internal heating element when the power comes back on and there's no water in the unit. In fact, we suggest you go write this instruction on your water heater now.

● The average toilet tank (no, not the bowl) holds about 2 gallons of water. Earlier, under "Orange Alert," we told you to remove any automatic cleaner or freshener from your tank so that the water would be drinkable, or at least useful for washing, etc.

● There will still be a few gallons of water in your plumbing (provided you turned it off at the meter and nothing is contaminated). Turn on the uppermost faucet in the house, which will allow air into the pipes, and then drain the pipes by turning on the lowermost faucet. Be sure to have buckets ready. **Hint**: This difference in upper and lower faucets could be as simple as sink vs tub. If your pipes don't drain readily, find a way to blow air into your uppermost faucet to help force the water through the pipes.

➤ Also under "Orange Alert," we asked you to keep your tub scrubbed and to collect extra plastic bins and trash cans that were kept empty and clean. If you know an emergency is coming, like an approaching hurricane, you'll want to give your tub a final cleaning and fill it with water. Do the same with your extra bins. This will give you gallons and gallons of extra last-minute water in case you lose your source of clean water.

➤ **Important note**: As soon as all your containers are full, use your best judgement about turning off the valve at your water meter. If you're about to face any sort of threat that has the potential of contaminating the water supply, then turning off your valve seals your plumbing system against water that could come in and contaminate your larger storage tanks.

➤ **Hint**: Diversify your water storage areas and keep containers all over the house. In a destructive event, you don't want "all your eggs in one basket." Having separate stores helps protect your supply.

Water Production

Simply put, water "production" is being able to find and filter more water should your stored water run out. This carries with it two major considerations, and those are **sources** and **purification**. Let's start with sources. However, keep in mind that very few of these sources will help you during a flood, or in a situation where you can't venture out of the house. Storage is still your best bet.

➤ Add an alternate source of water. If you're on municipal water that's piped in from a purification plant, you should also see if you can have a well drilled and pump installed on your property. Or, if you're only on a well system, you should check and see if municipal water is available. Having two sources of drinking water is always a good thing, all emergency scenarios aside. Just be sure you've set up your alternate energy sources to run your pump if normal power is out.

➤ If you purchased a couple of garden sprayers to fill with water and keep on hand as fire extinguishers, remember that you can drink the water from those (provided you never used them for anything else) and replace it with either raw water or waste water (such as soapy wash water). Just be sure that once you've swapped the clean water for dirty that you mark the unit somehow so you don't go back for another drink.

➤ You can drink water from your swimming pool provided it's uncontaminated by your current threat, and that you treat it just as you would raw water from a lake or river. You'll want to lose all the excess chlorine and any other chemicals present.

➤ If there's a local lake, creek, or river near you it may be a source of water, but there are some things you'll need to consider first.

- Will this water be accessible?

- If you don't already know if the water is already hazardous, ask your County Extension Office or local conservation service if the water in that lake or stream is too polluted to try to purify.

- Be sure to check the stream's route or the lake's area against your Threat Map. If you're tucked away at home due to anything involving hazardous materials, you'll want to know if this water is in any danger of contamination through runoff. (You'll want to know this anyway since this nearby body of water could bring contaminants your way that could affect you without your even trying to drink the water.)

- Even if the water is not drinkable, you can use it to flush toilets.

➤ While we're on the subject of natural water, let's look at **rain water**.

- Though rainwater seems pure, you still have to filter and **purify** it. Air pollution has taken its toll in general and if you live downwind from an industrial area you'll have to be extra careful about your rainwater.

- How you **collect** your rainwater will also have an effect on its purity. Many people have rainwater collection barrels at the base of their gutter downspouts. If you have a metal or tile roof, your water will be cleaner than if you have asphalt shingles since chemicals from the shingles can leach into the water. We have some rain barrels located under "sources" below.

- As far as **roof runoff** goes, you'll also want to pay attention to **first rain vs second rain**. If it hasn't rained in your area for a week or more, dust and possibly some pollutants will have settled on your roof. The first rain will be heavy with these contaminants as it washes off your roof and collects in your rain barrel. A second rain following closely behind the first will be much cleaner. (During a heavy first rain, dump any rain barrels after the first two hours to dump the dirty water and collect the cleaner water that will follow.)

- In addition to barrels, you can collect rain water by setting up a tarp or plastic sheet in your yard. Tie the corners of the tarp off in such a fashion that the tarp forms a natural trough that will direct rainwater down into a collection bucket. Make sure you use a clean tarp. You can also use clean, empty "kiddie" pools, or even just leave clean buckets out in the rain.

➤ Since we keep mentioning the need for water purification, let's talk about a few ways to do it. Let's assume you have a big bucket of nice brown river water. You'll run into several risks with this type of water with things that range all the way from germs and parasites to chemicals. Let's run through some steps that will take care of as many of these problems as possible.

❑ First, as you're collecting the water, you should pre-filter. Dip with one bucket and pour the water through a double layer of cloth, such as a bandana, into a second clean bucket. This will filter out some of the silt or floating debris, twigs, leaves, etc.

❑ Boil the water for at least ten minutes at a full, rolling boil. Depending on your altitude, you might have to do this in a pressure cooker. If you normally have to cook with a pressure cooker, then use it to boil your water. This will drive off some of the more volatile chemicals, and kill most of the bacteria and some parasites. Notice we said *some* parasites. **Hint**: Boil your water in an area where you have a little **ventilation** since you don't want to be inhaling any of the volatile chemicals you're trying to boil off.

❑ At this point, you can use one of your camping water filters such as Sweetwater ® or Katadyn ®, or you can chemically treat the water. We recommend treating first and then filtering. (**Note**: Most of your camping filters are rated to filter raw water straight from the source. It's our feeling that the more purification you can do before filtering, the better off you are and the longer your filter will last. Better safe than sorry.)

❑ To treat: Use 6 drops, or 1/8 teaspoon unscented, non-soap bleach per gallon of clean water (twice that if the water is cloudy). Let sit 30 minutes. Repeat this treatment if no bleach is smelled from the water after the first treatment. Let it sit 30 minutes to kill organisms, then an hour to allow the bleach to evaporate.

❑ Instead of bleach, you can use water purification tablets such as iodine or Halazone ®. Just follow the directions on the bottle. Regarding iodine tablets, they're superior to other types of water purification tablets, as iodine tablets have an incredibly long shelf life.

❑ Another treatment option low on our list is "Colloidal Silver." Colloidal silver supposedly works in water to bind all the available oxygen effectively depriving the bacteria of oxygen and killing it. See "sources" below.

❑ At this point, your water is *probably* safe for drinking, and definitely suitable for washing, bathing, and so on, but go ahead and filter it now just to be safe. (**Note**: Countertop filters will <u>not</u> handle raw water.)

❑ Once filtered, you might want to pour it back and forth from one clean pot into another a few times. This will aerate it somewhat and keep it from having that "flat" taste.

❑ Drink up. Just as with any other survival situation, make sure you're properly hydrated. Health problems can develop or become worsened by diminished water consumption. **Hint**: Add a powdered drink mix.

➢ One miscellaneous thought regarding water is that you can let your pets drink water leftover from cooking (such as from canned vegetables) provided your pet is in good health and doesn't have a problem with things like sodium. This will save a little drinking water for you.

➢ Now that we've gone over a few sources of water and ways to make it drinkable, let's look at a few sources of water you should **NOT** utilize regardless of how much purification gear you have.

❑ Do not try to purify flood waters. It's likely that the concentration of contaminants in flood water will be too great to filter out or treat. Remember, flood water will be washing out sewage plants, chemical plants, cemeteries, and so on, and not only do you have to worry about local sources of contaminants, you'll also be dealing with everything from upstream.

❑ Do not use your well water during a flood unless your well shaft is sealed (and most aren't). The water in the aquifer should be okay, but flood water may seep down your well shaft and be drawn in as you pump.

❑ Though some people say you can use condensation from your air conditioning unit, we suggest you steer clear of it. Condensation collection units are notorious for harboring molds, mildew, and fungi, all of which can cause numerous health problems. "Legionnaire's Disease" is an example. True, some of these might be treatable or filterable, but you've still got to collect the water to treat it. Best to leave it alone.

❑ Again, do not collect rainwater if your situation involves an airborne hazardous material incident.

❑ If you have a couple of choices of outdoor water to purify, opt for the flowing stream over the stagnant pond. Still water has had more of a chance to breed bacteria, collect local runoff of pollutants, etc.

❑ Other unsafe water sources include toilet tank water that still has a heavy concentration of "freshener" in it, steam radiator water, or water from your waterbed (fungicides and chemicals from the plastic) though it could be used for washing and bathing.

Sources and More Information:

➢ Though pricey, the non-electric water distiller featured on http://www.naturalsolutions1.com/waterwise1.htm looks like a useful item. You can contact them at Natural Solutions Environmental, Inc., #113, 4238 N. Arlington Heights Rd., Arlington Heights, IL 60004-1372, 847-577-7000, Fax: 847-577-7045.

➢ Some rain water collection barrels can be found at http://www.watersavers.com.

➢ A great selection can be found at http://www.rainbarrelguide.com.

➢ A discussion of rain barrels and water cisterns can be found in the online discussion of storm water collection at: http://www.lid-stormwater.net/raincist/raincist_construct.htm.

➢ Even more information on rainwater collection can be found in the Mother Earth News article at http://www.motherearthnews.com/199/rain_harvest.html.

➢ And more can be learned at the Abundant Earth site http://www.abundantearth.com/store/rainbarrelkits.html.

➢ A discussion on colloidal silver: http://www.lightwatcher.com/preparenow/prepare.html.

➢ Remember that we listed a few water tank sources under "Basic Home Prep" as their installation is more of a structural consideration.

Food

Water may be more important for survival, but food is the king of energy and morale. Here we'll cover the strategies for what kinds of food to have on hand, short term and long term considerations, food storage, and food acquisition.

Many advisors will have you believing that if you're not making your own beef jerky, or drying fruit or fish, you're not prepared. We agree that these are some skills that you should definitely learn and we even gave you some info sources earlier under "Outdoor Survival." However, it's our feeling that your biggest challenges are going to be in the short run while you're adjusting to the initial hit of whatever has happened. That's the most crucial period and the one in which you need to stay the calmest and healthiest, and therefore a good supply of your normal food will probably be better than anything else. We'll also cover some long-term food supply considerations, but our heaviest focus is going to be on the short-term. Besides, the vast majority of us are modern urban dwellers with limited budgets, limited space, and we see more asphalt than wooded fields full of natural foods.

However, as we've advised before, you should do all you can to ensure your current and future safety and wellbeing, and food is no different from any other asset that should be gathered and set aside for that "rainy day." In fact, given the very real threat of agroterrorism, we're sure to see instances where cash is useless and food is priceless.

Let's hit the ground running and talk about the types of food to keep on hand. This will naturally lead us to our discussion of storage and also to long term considerations. Let's begin with your short term food considerations. **Hint**: Start making a large shopping list while you're reading through this section, and keep in mind that if you have dependent relatives living nearby, you may need to supply them too.

Short Term Food Supplies

In simplest terms, your short-term food supplies are going to be more of what you eat on a regular basis. This is important for a number of reasons. One, you'll actually save money on groceries if you buy in bulk. Two, dietary changes in an emergency might compound the stress you'll be experiencing, and lead to things like diarrhea, which will only make matters worse. Three, and most important of all, **the psychological value of normalcy during a disaster cannot be overestimated**. The more you have in the way of your regular diet and foods, the better off you'll be. Let's look at some considerations:

➢ First, simply **increase the stores of goods you normally consume**. Buy in bulk and learn to shop when your supplies drop **down** to about **two week's** worth. In other words, go out on one huge shopping trip and stock the house for a **month's** worth of food, and then shop every two weeks. You always want to have at least two week's worth of your regular food in the house at all times and never dip below that level. That way if an emergency hits, you're set. This will prevent you from having to run out and "panic buy." You'll have more important things to do with your time and energy at that point, and that's assuming any food will be available at all. **Hint**: Don't forget **pet food** and other pet supplies.

➢ **Hint**: Stocking up on food will also allow you time to plan subsequent shopping trips so you can look for sales, volume deals, and collect coupons in order to save money.

➢ When considering the amount you use in a two week period, always add in an additional family member. If you have a family of four, plan for a family of five. This way, you're certain to have enough, and if an emergency occurs while you're hosting company, you'll need the extra food. Plan to provide, at the absolute minimum, one full balanced meal per day per person. Naturally, three meals a day is preferable.

➢ **Hint**: If your "normal" diet consists mainly of frozen dinners, you'll certainly want to add a variety of canned and packaged dry goods, not only because of basic dietary concerns, but because your main supply of frozen food could be lost if the power goes out and your generator quits.

➢ Secondly, in this increased supply of your normal foods, place a **heavier** emphasis on your normal foods that require **little to no water** as there may be long periods you're without water (and make them non-spicy and non-salty items so you won't get thirsty). For example, canned vegetables, and especially non-condensed, stew-type soups are great. With the soups, you have a fairly balanced meal that is simple to prepare (heat and eat), and not only doesn't require water, but will actually help hydrate you.

➢ You'll also want to make sure you add some **energy foods** such as simple or complex carbs that also carry good nutritional value. **Peanut butter** is a good example. So are various "**protein bars**" that offer balanced nutrition in a compact form. Make sure you get brands your family has tried and likes.

➢ Be sure to include all **condiments** in your two-week supply. Salt, pepper, mayo, ketchup, salad dressing... make sure you've considered it all.

➢ Also increase your selection of **beverages**. You'll want both complete beverages and mixes. You might consider a few bottles of sports drinks or meal replacement drinks (both of which will last for a long, long time in the fridge), tea bags, regular and instant coffee, or any number of powdered drink mixes. You'd be surprised just how palatable purified river water becomes after you mix in a packet of flavored drink mix.

➢ Place a lighter emphasis on extremely **perishable** items such as fresh fruits and vegetables. We're not saying cut down on your consumption of these healthy items, just don't count them in your two-week inventory. Consider them as extras that you might shop for weekly.

➢ Get a chest or upright **freezer** if you're able. This will help greatly with our next concern, which is storage. Fill your freezer with normal freezer items; meats, vegetables, frozen dinners, etc. and don't forget to include several plastic bottles of water as we discussed earlier. Bread and milk will also keep rather nicely in the freezer. Get milk in the plastic jugs, and all you have to do is put the jug in the freezer (plastic jugs will allow for expansion). When you thaw it out, make sure it's all the way thawed and then shake it up to remix it as the water content tends to freeze first. **Hint**: Keep your freezer locked 24/7 regardless where it is. This will help keep it closed in a destructive event and also help prevent food theft in a post-disaster panic-looting situation. Keep the key nearby where family members can get in, but otherwise, keep it locked and secured.

➢ **Morale boosters**. Somewhere, hidden in your freezer and wrapped to protect against freezer burn, you should include a pack of your family's favorite **junk food**. It might be chocolate, cookie dough, ice cream, or some other freezer-friendly food. Other things not meant to be cooled or frozen, such as popcorn, or hard candy, can be packed in a box and put in the bottom of an indoor closet. Remember, our food stores are just as much there for psychological support as physical sustenance. And in some instances, these foods are what? Yep. Barter items.

➢ **Hidden supply**. Along with these morale boosters, you should consider having a hidden supply of normal staples that you absolutely don't want to run out of. For example, for many people, this might be coffee. Whatever you want to have a hidden supply of, make sure it's labeled and dated and you don't forget to eventually rotate it out. Similarly, your "hidden supply" could bridge the gap between evacuation and isolation. Keep a box of food that will be rotated in with your normal stock eventually, but that's kept set aside in case of an isolation, or an evacuation where you were gathering extra food for your "Secondary Kit." Having a balanced package of foodstuffs already set aside will speed up your evacuation.

Food Storage

"Food storage hot, food storage cold, never leave it in the pot, one hour old." Spoiled or outdated food is not only a waste of money, it's a health danger. The last thing you need in a disaster is to create additional problems for yourself by either losing food or eating bad food. Since we're asking you to increase the amount of food you have on hand, let's cover ways to make sure it all stays good and useful, so it will keep you good and useful.

➢ Our first suggestion is "rotation, rotation, rotation." It's important to point out that rotation is preferable over storing a separate "stash" of food. By rotating, you'll consume everything, and you won't lose money by having to dispose of foods that have reached their expiration date. Consume the old stuff before the new, and set up your system of knowing which is which now, and always follow it. One way to do this is in the way you physically arrange your food stores. Put newer items on the shelves or in the fridge, in the back or on the bottom, and move older items to the top and/or front to be consumed next.

➢ Another rotation consideration is to make sure nothing stays on your shelves that's over two years old. In fact, for most canned goods and grocery store items, about one year is all you'd want to go even though most items may easily keep up to three years. Make sure all your stock is rotated, and at the end of the year if anything is approaching a year old, you can donate it to your local food bank (as it will still be good) and you can restock with newer goods. **Hint**: You can change out your supplies when you change time for daylight savings time. Simply put, "When you rotate your clocks, rotate your stock."

➢ Next, consider marking the purchase date on items that don't have an obvious expiration date. This is useful with canned goods as most companies use their own unique date coding system that none of us understand. Use a marker to write the date of purchase on top of the can, and then shelve it in the appropriate spot. Do the same thing on boxed goods such as mixes, frozen dinners, etc. Label anything that doesn't have an expiration date visible. You should also label and date any leftovers or restaurant "doggy bags," and eat them within a week unless frozen.

➢ Learn the date codes. The vast majority of companies that produced packaged consumables will have a toll-free information phone number somewhere on their package, and probably a website. Call them and ask how to decipher the date codes on their products if it's not obvious to you, and ask what the average shelf-life for their product is. Write this info down so the whole family will know. (See "Sources" below.)

➢ Another thought is to buy smaller portion sizes on items that require refrigeration only after opening. One example is mayonnaise. Once you open a jar of mayo it has to go in the fridge or else it will spoil. If you had several small jars instead of one big one, you wouldn't have to worry about your whole supply, just the little jar you opened. This way, if the fridge dies, you'll only lose your one small portion while others in the pantry are still good.

➢ Learn and look for signs of spoilage. With some things it might be an odor such as with bad meat or sour milk. It might be mold, such as you'd find with bread, or it could be a swollen package such as you'd see with canned goods that go bad.

➢ To keep things from spoiling in the first place, make sure they're properly stored. For freezer items, make sure they're double-wrapped to protect from freezer burn. Refrigerated items should be kept properly sealed either in the original package or in another clean and resealable container (preferably labeled and dated). All other items should be kept in a cool, dry place.

➢ Get a good set of resealable plastic wear made for use in either the freezer or fridge and from there, the microwave. You might also want to look at a kitchen vacuum sealer (see "Sources" below). Both storage methods will be useful if you buy in bulk and have to repackage some of your perishables. This vacuum sealer might also prove useful when you're packing your Bugout Kits as you can shrink and seal all sorts of "fluffy" items like toilet paper, towels, or clothing items.

➤ Your cool, dry place doesn't have to be your pantry or even your kitchen. Some times you might want to box a certain amount of canned goods or your "morale boosters" we mentioned, and keep them in the bottom of a closet. Don't forget they're there as you'll eventually want to rotate them in with your regular stock to make sure they're consumed and don't go to waste.

➤ As a few final miscellaneous considerations for your cabinets or pantry, consider these points:

● Stock your shelves keeping an earthquake or other destructive event in mind. Stock your lighter items, like boxes of cereal and so on, on the higher shelves. Put your smaller canned goods on the middle shelves, and put glass jars and larger cans on the very bottom.

● While you're in the midst of all this shelf arranging and box storage, gather your family and look over all the items, deciding which would be included in your "rake and run" during a Level 4 or 5 evac.

● If you have a separate pantry, you might reinforce it like you did your safe room so that if something like a tornado hits while you're away, your house will have that much more structural reinforcement, you'd have another safe area, plus your food stores might be saved from being strewn about the neighborhood.

The "Intermediate" Foods

Now that we have a good handle on "stocking the larder" and a few storage hints, let's go over some "intermediate" food supplies. These are food items that either fall outside the list of things you'd normally consume, items that have a much longer shelf-life, or that can be used in meal preparation (or barter). As we're planning on a three-week isolation, the normal foods you stocked earlier should get you through the first two weeks. We didn't want to plan longer than that with the regular foods since there's no way to gauge spoilage, etc. These intermediate foods will take you through your third week and hopefully a couple of weeks past that if the need arises.

➤ First, we'd like to point out that if any of these items are already part of your normal two-week's consumption, that you consider **adding a little extra** supply and keeping it all tucked away in a cool, dry place like we mentioned with "hidden supply" above.

➤ If you didn't include **beverage mixes** earlier, you should now. Get them in dry form and keep them sealed and protected so they'll last. Some mixes can prove beneficial, such as powdered sports drinks, and the others will prove useful for morale. Consider these:

❑ Baby formula	❑ Instant and regular coffee	❑ Teas (bags or instant)
❑ Powdered milk	❑ Protein drink mixes	❑ Powdered sports drinks
❑ Fruit drink mixes	❑ Tang ®	❑ Chocolate milk mix powder
❑ Hot chocolate drink powder	❑ "Instant Breakfast"	❑ Electrolyte drink mixes

➤ In keeping with "flavorings," be sure you've topped off your **dry** spices and flavorings. Not only will these allow you to improve on the same old bland food if you're cooped up for a while, some have medicinal uses, and others will be useful for barter with the neighbors if you can venture out. Therefore, you might stock a little heavier on the first three items listed here:

❑ **Sugar** (white and brown)	❑ **Salt** (table and "rock")	❑ **Pepper** (red and black)
❑ Beef bullion cubes	❑ Chicken bullion cubes	❑ Garlic powder
❑ Onion powder	❑ Powdered Ginger	❑ Cinnamon
❑ Artificial sweeteners	❑ Powdered coffee creamer	❑ Powdered butter flavoring
❑ Powdered cheese flavoring	❑ Be sure to include other spices you customarily use.	

➤ While we're on powders, be sure to stock the following even if you don't normally bake or make bread: **Flour, baking soda, corn starch, baking powder, and packaged yeast powder.** As you already know, things like flour and baking soda have alternate uses besides cooking. Flour can be used to make glue and baking soda can be used in everything from biochemical decontamination to deodorizing a room. Also, anything good for baking is good for barter.

➤ Other powdered mixes can make a meal or a side dish. Include a selection of **gravy or sauce mixes, instant puddings, powdered gelatins, pancake mix,** etc.

➤ Some instant foods to stock would be **oatmeal, grits, various soup mixes, cream of wheat, powdered eggs, instant mashed potatoes** (also useful to stop bleeding, remember?), and so on. Most of these come with individual serving pouches inside a larger box. These foods will keep for a long, long time if kept in a cool, dry place.

➤ Dried staple items would include **rice, beans, peas, numerous pastas, and the various Hamburger Helper ® and Chicken Helper ® kits,** etc. These foods will probably keep the longest of all in this little section. Make sure your long-term items are redundantly sealed. One suggestion is to put a lot of these items together in clear plastic tubs. They'll be protected and you can still see what you have.

> To wrap up this little portion, remember how we suggested you take "home ec" classes under "Basic Training?" Knowing how to cook can not only save you a lot of money, but it'll help you be creative with your meals when you're stuck with a limited selection of ingredients. We've listed a couple of good, basic, cooking instruction books under "Sources" below.

Long-Term and Ultra-Long-Term Foods

If you've stocked everything we've listed so far, you're probably good to go for a month or longer. However, as we have no idea what repercussions will follow a disaster, we need to make sure you're as self-sufficient as possible for as long as possible. Therefore, let's look at some varieties of long-term-storage foods and at food "production."

> What we consider "long-term" foods can be covered quickly and easily. In fact we've already done it. Go back to "**Evacuation**" to look at "Level 1 Food" under "Prepping the Gear," and revisit our discussion of **MREs and freeze-dried "camping" entrees**. Remember though, you want to keep your evacuation gear separate from your isolation supplies since your reaction to an emergency may be a combination of the two and you'll need redundant supplies for each. In your isolation supplies, store enough in the way of MREs or freeze-dried foods to provide each family member at least t**wo full meals per day for two weeks**. Also remember to keep either of these meal sources in a cool, dry place, and don't freeze your MREs.

> **Hint**: Your camping trips will help you rotate your stock of MREs and freeze-dried meals.

> For ultra-long-term storage foods, nothing beats the freeze-dried meals packed in **nitrogen** and specially sealed to last a **decade**. Take a look at "Nitro-Pak" and other sources listed below.

> While looking through the nitrogen-packed meals, take a look at the different packages of **seeds** for growing your own fruits and vegetables. You can either get a good supply of seeds from those sources, or you can use your vacuum sealer to seal a variety pack of seeds that you put together yourself.

> **Gardening**, as mentioned before under "Frugality," is not only a good way to save a little money on food, it may be your only source of food under certain circumstances. Go ahead and grow a small garden now to have a little fun and to get the experience for when you may really need it. Later on, if you have to, you can garden in a room (you might have to, to protect your food from theft). All you need is a sunny room or some "grow lights," pots or planting troughs, and soil. Your local garden store can help you with the rest.

> While you're learning to garden, don't forget we also suggested earlier under "Frugality" that you learn how to **can and preserve** your food. And yes, this will include the drying and jerky making we mentioned above.

> Speaking of cans and long-term storage, watch for dented, swollen, or corroded cans when looking through your pantry for something to eat. You're much better off eating no food than eating bad food.

> If you live near a lake, river, or stream, you should own **fishing gear** and know how to fish even if you don't do it on a regular basis.

> Similarly, if you live in a rural area or near woods, you should **consider hunting as a source of food** even if you don't normally hunt. However, be wary of fish and game after a biochemical or nuclear incident.

> To include foraging in the mix, and that is going out and **gathering wild foods** such as nuts, berries, and various plants, we remind you to go back and visit all the sources we gave you earlier under the "Outdoor Survival" portion of "Basic Training."

Sources and More Information

> We've found a few sites to help explain the various processing and packaging date codes found on packaged foods, along with acceptable storage info:
> - http://www.a1usa.net/gary/expire.html.
> - http://www.mealtime.org/about/about_coding.aspx.
> - http://www.amarogue.com/foodcode.html.
> - http://www.fsis.usda.gov/oa/pubs/dating.htm.

> Visit the National Center for Home Food Preservation online at http://www.uga.edu/nchfp, or contact them at The University of Georgia, 208 Hoke Smith Annex, Athens, GA 30602-4356, FAX: (706) 542-1979.

> A very useful food storage and shelf-life information page is also available from UGA online at http://www.ces.uga.edu/pubcd/b914-w.html.

> We've also found several links to good recipe sites specializing in full meals with limited supplies:
> - http://www.melborponsti.com/index.htm.
> - http://www.macscouter.com/Cooking.

- http://www.allrecipes.com.
- http://www.cooking.com.
- http://www.my-meals.com.
- http://www.mealsforyou.com.
- http://www.three-peaks.net/cooking.htm.

➢ Another source of long-term-storage dehydrated foods is Walton Feed, online at http://waltonfeed.com, and with direct contact at 135 North 10th, P.O. Box 307, Montpelier, ID 83254, Voice 800-847-0465, Fax: 208-847-0467. While you're there, be sure to sign up for their free newsletter. Be sure to visit their food storage subpage at http://waltonfeed.net/grain/faqs.

➢ Visit the government's food safety site at http://www.foodsafety.gov. They'll have everything from the latest on the fight against agroterrorism, at http://www.foodsafety.gov/%7Efsg/bioterr.html, to food safety and handling hints for the home, at http://www.foodsafety.gov/~fsg/fsgadvic.html, and especially at http://www.foodsafety.gov/~fsg/september.html.

➢ The USDA has a food safety and storage page at http://www.fsis.usda.gov/oa/pubs/dating.htm, giving an overview of product expiration dates, and storage times for various types of food. You can also call them for more information at 1-888-MPHotline (888-674-6854). – Toll free Nationwide 1-800-256-7072 (TDD/TTY).

➢ Though more for commercial food and agricultural operations, the US government's food safety site offers some useful information for household-level food safety. Visit them online at http://www.foodsafety.gov.

➢ Visit the National Food Safety Database, online at http://foodsafety.ifas.ufl.edu/HTML/consumer.htm.

➢ You'll find several food safety training manuals online at http://www.fstea.org/resources/manuals.html.

➢ Also see the food expiration dates chart at http://www.a1usa.net/gary/expire.html.

➢ You can find good info on home canning at http://www.homecanning.com, or by calling 1-800-240-3340.

➢ In the 1800s, people made more with less. A collection of old cookbooks can be found online at http://digital.lib.msu.edu/cookbooks/index.cfm.

➢ Homestead Harvest has a canning guide, online at http://www.homesteadharvest.com/canning101a.html, along with a good collection of gardening links and info. You can read their material free online, or contact them at 15941 S. Harlem Ave. No. 212, Tinley Park, IL 60477, (708) 633-1840. While online, be sure to visit the home page at http://www.homesteadharvest.com.

➢ We suggested earlier that you take a good course or courses in the category of "Home Ec." This kind of education will let you do so much more with so much less when it comes to food. To this education, we suggest you add these two books to your library:

- **Cooking for Dummies** by Bryan Miller and Marie Rama.
- **The Joy of Cooking** by Irma S. Rombauer and Marion Rombauer Becker.

Other Pantry Items

Regardless of the types of food you stored above, where you stored it, how you stored it, or even how you plan to secure food later on, there are certain things you'll need to have on hand to prep and serve your food and to keep your household in running order while you're isolated. Here we'll continue our "pantry" discussion by listing other household items you might need to stock up on.

You should have a three-week supply of these items, and all of them should be packed away and kept separate from your normal daily use goods and from your evacuation gear to be used only when the need arises. When the need does arise, it will be a certainty that going shopping will not be an option. None of these items are so expensive that you can't set aside a separate supply. In fact, in most cases, a single package should give you your three-week supply. For example, a large roll of heavy-duty aluminum foil should easily last a month. **Hint:** If your family happens to run into a financial hardship one day, you'll find that's as much an "emergency" as a natural disaster. If push comes to shove, having these supplies on hand and already paid for will help.

Let's start with a basic checklist of extra household items to keep on hand and we'll follow it with a discussion of the items and the reasons why some were included. Some of these will be familiar to you from the **"Evacuation"** section.

HOUSEHOLD ITEMS FOR ISOLATION

Food Prep and Service

❏ Paper plates	❏ Plastic eating utensils	❏ Paper cups
❏ Paper napkins	❏ Plastic food service gloves	❏ Plastic wrap
❏ Aluminum foil	❏ Wax paper	❏ Resealable plastic containers
❏ Resealable plastic bags	❏ Manual can opener	❏ Kitchen thermometer

Dual-Use Food Items

❏ Baking Soda	❏ Salt	❏ Sugar
❏ Unprocessed dry white rice	❏ Cornstarch	❏ Instant mashed potatoes
❏ Regular, non-instant oatmeal	❏ Vinegar	❏ Plain flour
❏ Honey	❏ Ground Cayenne pepper	❏ Pie Crust Mix

Sanitation

❏ Toilet paper	❏ Paper towels	❏ Facial tissues
❏ Soap	❏ Dish washing liquid	❏ Clothes detergent powder
❏ Waterless hand sanitizer	❏ Moist towelettes	❏ "Baby wipes"
❏ Non-soap, chlorine bleach	❏ Liquid surface cleaner	❏ Rubber cleaning gloves
❏ Trash bags	❏ Old newspapers	❏ Scoopable kitty litter
❏ Moth crystals	❏ "Fuller's Earth" or equivalent	❏ Spare towels and washcloths
❏ 5-Gallon plastic buckets	❏ Mouse traps and moth balls	❏ (Also see "Medicine Chest")

Let's discuss these checklist items by category:

Food Prep and Service

These are some of the little detail items that can make all the difference in the world on how well you and yours handle being cut off from the outside world. As with all the other goods listed in the checklist above, pack these and keep them safely stored where you won't dip into them unless it's an emergency. As most of the things in this list are non-perishable, you can box them up and keep them in the attic, the garage, in a closet, or in an outbuilding. Make sure to label your box and seal it well.

> **Paper plates.** You might not have water to wash dishes. The need for fresh dishes aside, you'll want to get rid of the remaining food residue to keep pests at a minimum. With paper plates you can burn them in the fireplace or on your grill. This gets rid of the residue, reduces your trash output at a time when trash pickup services won't be coming, and in the winter it'll help generate a little heat. Since we're planning on a three-week stay, make sure you have a bare minimum 63 plates per family member. That'll cover a person for 3 meals a day for 21 days. Add more for food prep. In addition to paper plates, be sure to save the plastic trays and bowls that come with frozen dinners. You'll find them as useful (if not more so) as the plates.

> **Plastic eating utensils.** Same as above. They're disposable if you don't have water for washing.

> **Paper cups.** Same as paper plates, except you might want to have a few more cups than plates. Try to get a sleeve of the waxed paper cups. Additionally, get some paper cups made for **hot beverages**.

> **Paper napkins.** Again, same as above. **Hint:** Since you may wind up burning these, get the plain white variety. They'll be less expensive and won't have inks that may give off harmful fumes when burned.

> **Plastic food service gloves.** It's important to note that these aren't the latex or Nitrile surgical gloves you'd put in your first aid kit. These are the thin plastic variety that you see restaurant workers use. 100-packs of these gloves are extremely inexpensive and can be found at grocery stores. Get a 100-pack per person. Since you might not have water, use these gloves when doing anything that might normally require you to wash your hands before or after doing.

> **Plastic wrap.** Plastic wrap not only helps reseal food to keep it fresh, but it has quite a number of first aid and sanitation uses. Be sure to store a box even if you don't normally use it.

> **Aluminum foil.** As we said earlier, the uses of aluminum foil could make a book of their own. In an isolation situation, you may need it to reseal foods, cook with, block sunlight in the summer, or even make a fireproof base under your Sterno stove. You can also line cooking pots with it and "clean" the pot by removing the foil.

> **Wax paper.** Stash a box of wax paper even if you normally don't use it. It's great when working with dough, and it's another one of these things that it's better to have it and not need it than to need it and not have it.

- ➤ **Resealable plastic containers**. Though your power *may* be out, there's no guarantee it *will* be. Therefore, when you have leftovers, it'll be nice to have something to put them in to keep them in the fridge or freezer. Also, some of these containers make nice soup bowls. Get the inexpensive type like Gladware ®.

- ➤ **Resealable plastic bags**. You can never have too many resealable plastic bags. Store a box each of different sizes. These can be used for anything from food storage, through first aid and sanitation needs.

- ➤ **Manual can opener**. Though your power may be on, you can't always rely on that electric can opener. You should get two types of can opener. One should be the type that removes the whole lid, and the other should be the little combo tool that punctures cans with one end and opens bottles with the other. Speaking of bottles, if you pack away a bottle of wine, be sure to have a **corkscrew**.

- ➤ **Kitchen thermometer**. In a survival situation you don't want to take any chances with food. Use your thermometer to make sure everything you cook is thoroughly cooked.

Dual-Use Food Items

We thought we'd separate some food items for a couple of reasons. One, these items have more uses than just food or food prep. Two, we wanted to single them out in case you would not normally buy them. Another good thing about all these substances is that they have an incredibly long shelf life and can be stored away from your main pantry area so they won't get used up and not be there when you need them.

- ➤ **Baking Soda**. Baking soda is one of those wonder chemicals and there's probably no such thing as having too much of it around the house. You can use it to clean, freshen, cook, and if you remember from previous sections, it's used in an "Oral Rehydration" mix, and in some instances, biochemical decontamination.

- ➤ **Salt**. In addition to being used in an "Oral Rehydration" mix, salt is used in food preservation, toothpaste, and can become an extremely valuable barter item. Be sure to keep extra on hand.

- ➤ **Sugar**. The main ingredient of the "Oral Rehydration" mix, sugar is also used in food preservation and is another <u>great</u> barter item.

- ➤ **Unprocessed dry white rice**. In addition to being a great food item, rice makes a great desiccant for keeping some of your stored gear moisture free.

- ➤ **Cornstarch**. Useful in cooking, yes, but cornstarch also has a place with first aid for your pet, biochemical decontamination, and makes a great substitute for baby powder.

- ➤ **Instant mashed potatoes**. Again, a great food, but also useful in first aid. A topical application of dry, uncooked, instant mashed potatoes can help stop severe bleeding by aiding in blood's coagulation.

- ➤ **Regular, non-instant oatmeal**. Oatmeal is another food that will last forever if stored properly. It's also useful in first aid in situations where you might need a Calamine Lotion ® substitute, and it has its uses in decontamination when dealing with an unknown liquid substance.

- ➤ **Vinegar**. Actually, there are two types of vinegar to consider. One is the standard "white" vinegar and the other is "apple cider" vinegar. White vinegar is useful for cooking, cleaning, and first aid, and apple cider and wine vinegars are considered health foods.

- ➤ **Plain flour**. Like cornstarch, flour has absorbent properties of its own. Not only can it help clean up liquids, it can be used to make a paste almost like Plaster of Paris.

- ➤ **Honey**. Not only is honey a great condiment, it has medicinal properties as well. It can help cure sore throats, and in fact, it can be used as a topical antiseptic.

- ➤ **Ground Cayenne pepper**. Cayenne is good for cooking and herbal remedies. In an isolation, it's also useful for sprinkling around stored garbage and refuse to keep rodents and stray animals away.

- ➤ **Pie Crust Mix**: In the "olden days" people would bake food into a thick crust (almost like a burrito with no holes) as an early form of canning. Food would keep a full day at room temperature. If your fridge quit and food were about to go bad (but was still good) you might be able to save some of it by baking it into a crust.

Sanitation

Disease control is a serious concern. Illness, dietary changes, and emotional stress from whatever situation you find yourself in can cause vomiting and intestinal disorders on their own, so we don't need household germs adding to the problem. Another concern is injury. Blood and other bodily fluids must be cleaned immediately. Similarly, food residue will encourage rodent and insect infestation. And last but not least? Cleanliness will help improve morale.

- ➤ **Toilet paper**. We recommend a minimum of twelve rolls per family member. That'll give 4 rolls a week for three weeks. We also suggest you stock extra just to be on the safe side, and to have some to barter with.

- ➤ **Paper towels**. Just as with your napkins, get the plain white version of your favorite brand as you don't want to risk any ink fumes if you burn your garbage. We suggest a minimum of four rolls per person. This will give plenty for average cleaning needs. You might also want to stock a few extra rolls for trade.

- ➤ **Facial tissues**. This is optional depending on your stores of toilet paper.

- **Soap**. We suggest storing a type of plain bar soap that your family has used before and likes. We suggest a plain type as they tend to be more hypoallergenic which is good for a family under stress, and it easier to barter with. Also, try to use a brand that is low in scents or moisturizers, as you'll want something that rinses easily if you're in a water-deprived situation.

- **Dish washing liquid**. We suggest a type that has good solvent or degreaser properties, such as Dawn ®. Not only do you need to make sure you have plenty of sanitation supplies on hand, you can use dishwashing liquid in decontamination if necessary.

- **Clothes detergent powder**. If you have water and power, you'll want to wash clothes to keep everyone happy, clean, and healthy. Keep a hidden stash of detergent so you don't run out at a critical time. We suggest powder as it tends to keep better than some of the liquids that might separate or settle. Another reason to keep a large stash is cleanup after a destructive event. Suppose you have to clean up after a flood. Local stores may be closed, or out of supplies, and even if they're open and stocked, the last thing you need is to waste time or money shopping. This is another of the many reasons we want you to stock extra supplies for any disaster. Time and money will be as valuable as any other emergency commodity.

- **Waterless hand sanitizer**. If you don't have water, you'll still need to wash your hands. With this, as with any of these supplies that have liquid or moisture, be sure to keep it individually sealed in a resealable plastic bag. You may be storing these supplies in an out-of-the-way place for a long time until needed, and the plastic bags will help prevent loss of moisture.

- **Moist towelettes**. In addition to the waterless hand sanitizer, moist towelettes will help you keep your face and hands clean. As these have a moisture content to them, be sure to store them in a resealable plastic bag even though they're already in individual packets.

- **"Baby wipes."** Besides cleaning hands, faces, and babies, these wipes are great for quick sponge baths when there's no water. Keep them sealed in a plastic bag to help them retain their moisture during storage.

- **Non-soap, unscented, chlorine bleach**. We've already covered the zillion uses for bleach. Make sure you have a secret stash set so you never run out. Rotate this stock annually. Try to find a powdered variety of a chlorine-based bleach. We'll be putting it to use in our "Privy Deprived" section below.

- **Liquid surface cleaner**. This can be regular household cleaners such as 409 ®, Windex ®, etc., or something like Lysol ® or Pinesol ®. Make sure you always have something to clean up food residue, and fight germs while everyone's cooped up in close quarters. Since you are cooped up, get the unscented type.

- **Rubber household cleaning gloves**. You'll need them for the reasons they were invented.

- **Trash bags**. We've covered the seemingly unlimited uses for trash bags in other areas of this book. Make sure you always have a good supply on hand. As with everything else on these lists, trash bags will help supply you as you clean up after whatever it was that caused your isolation in the first place. Make sure you get the heaviest plastic (in mil thickness) you can. (Don't forget heavy-duty "contractor bags.")

- **Old newspapers**. Under "Climate Control" below, we'll tell you how to make logs for your fireplace. In sanitation, old newspaper can be used as a drop cloth, table cloth, or other disposable (and burnable) protective cover. Newspaper is also useful for cleaning utensils, as you'll want to wipe them down first if you're on limited water rations. Did you know you could also use layers of it as a blanket? It's also especially important if you have dogs and are in a situation where you can't venture out of the house.

- **Scoopable kitty litter**. Cats will have to go to the bathroom too, and even if you don't have cats, the section "Privy Deprived" below will explain kitty litter's use. In addition to those, kitty litter is useful for soaking up liquid spills of various types including blood, vomit, and unknown liquid substances.

- **"Fuller's Earth" or equivalent**. We covered Fuller's Earth, or Zeolyte, earlier. It's a chalk-like substance except much more absorbent. You can find it at many hardware or janitorial supply stores, and it's much better at soaking up liquids than the kitty litter mentioned above. It's also touted to be an environmentally friendly insecticide though this is an untested claim.

- **Spare towels and washcloths**. There's no guarantee that a disaster is only going to hit when you have all your laundry done. Having an extra set of towels and washcloths will make sure you have everything you need to keep everyone clean and happy. Also, don't forget what we said about the possibility of you hosting company when a disaster hits. Always plan for more than you think you'll need.

- **5-Gallon plastic buckets**. You should probably allocate about two per person. As the section below, "Privy Deprived," will explain, they're useful for everything from washing clothes to making expedient toilets.

- **Mouse traps and moth balls**. Well, actually we recommend moth crystals. You can sprinkle them around your house, and especially your garbage, to keep some vermin away. We covered this concept earlier with the mention of Cayenne pepper. Mouse traps are meant to keep mice in check, as mice and rats may become a problem if sanitation and garbage services are down for a while. Keep a few stored with the rest of your non-consumables. By the way, peanut butter is better mousetrap bait than cheese.

> **Hint**: When storing all these items, pack them like the bag boys do, "like with like." Perishable go together with each other, non-perishables the same way. No consumables mixed with non-consumables, and pack dry with dry, wet with wet, etc. Label all tubs as to packing date and contents. Any perishable contents should have purchase date and expiration date recorded. Anything not in its original container should be labeled as to contents, instructions for use, and expiration date.

Sources and More Information

> In addition to these, be sure to look at all the goods and gear sources we listed earlier under "**Evacuation**."
> Survival Unlimited carries a wide variety of isolation-related goods and gear. They're online at http://www.survivalunlimited.com, or contact them at Picou Builders Supply Co., Inc., 235 N. Airline Hwy, Gonzales, LA 70737, 1-800-455-2201, 225-647-2171, Fax: 225-647-7899.
> Here's a link to another educational site concerning what to put in your second pantry http://outreach.missouri.edu/webster/webster/y2k/Survival.html.
> Also see "Grandma's Pantry" online at http://www.grandmaspantry.com/millennium.html.
> Though out of print, try to get your hands on a copy of **Henly's Book of Formulas**. It's a compendium of formulas for homemade "chemicals" and other products made from commonly available components.

The Medicine Chest

Just as you need to be self-sufficient with energy, water, food, and household supplies, you need to be able to handle small to moderate ailments so that they not only don't get out of control, but to keep everyone in your household as well and happy as possible in the given circumstances. That's one of the many reasons we gave you so many outside sources and information regarding health and first aid.

Though extremely important, this section is easy to write as we've already covered what you need. Go back to the "**Evacuation**" section and look under "Prepping the Gear" at the checklists for the following:

1. **First Aid Kit.** Though evacuation and isolation may be linked or overlap, they'll have their own unique characteristics, timing, and location. Therefore, each eventuality must have its own equipment. Duplicate your Bugout Kit's first aid kit and keep it where you keep your family's first aid supplies. We recommend keeping this kit sealed in a plastic bin in your safe room or area. This will keep it handy and protected from the elements should you experience a destructive event.

2. **Basic Toiletries.** Though we've already mentioned things like toilet paper and trash bags above, you should go back and reproduce your evacuation toiletries kit to have on hand in an isolation. Redundantly seal everything just like you would for evac, so as to protect it while in long-term storage. Anything with a moisture content goes in a resealable plastic bag. Towels and washcloths go in a vacuum seal bag for both protection and reduction of space. As very little of these toiletries are perishable, and as you'll have a fairly good supply kept in the house anyway, these items can be stored in a plastic bin and kept in the attic along with some of your non-perishable pantry items listed above. Doing this will ensure that you have critical items available regardless of the situation.

3. **Non-Prescription Medications.** Unlike some of your toiletries and dry-goods that can be stored in the attic and left there as emergency resupply, your non-prescription meds will have to be kept in the house in a cool, dry place just like some of your foods and other perishables. These items should, however, be kept together and away from your normal household supply so that you don't accidentally run out, and then find yourself unable to get to the drug store. Be sure to mark the purchase date on each package if it doesn't have an expiration date (call the manufacturer to see what the shelf-life is) and pack everything together in a plastic bin to be kept in your safe room or interior closet. Be sure to mark the bin as to what its contents are, and the date you packed it away. Also be sure to notate the location of this bin and its contents on your "**Refresh List**" so you can rotate everything in with your normal supplies to use all of it rather than let it expire and be thrown out.

Note: When gathering this gear, keep in mind the special medical needs of family members, and pets.

Being such an important subject, this is as good a place as any to reiterate the need for redundant and separate gear for evacuation and isolation. You never know what's going to happen, where, when, or in what combination. Therefore, you'll never know if an isolation will be followed by an evacuation, or if after returning from an evacuation, how long you'll be on your own and cut off from regular supplies and services while you clean up after whatever it was that caused you to evacuate. Again, what are two of our favorite sayings? "Redundancy is our friend, and it's far better to have it and not need it than to need it and not have it."

Sources and More Information

> See our sources listed under "**Evacuation**" for items you can't find at your local drugstore or grocery store.

Life Without Utilities

Modern man has had modern utilities for far less than two centuries, and these utilities are merely comforts that we have grown to call necessities. It's not that we've changed biologically and are now physically dependent on these conveniences, it's just that we've not experienced what the generation or two before us took for an everyday way of life. Let's look at some of the things we'll need to know and do should we have to live a few weeks like our forefathers did. It's really not hard... if you're ready.

Goals of Life Without Utilities:

1. To show you that **living without modern utilities is not only possible, but really not that rough**.
2. To help **protect you from the heat of summer or chill of winter**.
3. To **keep you fed with hot meals**.
4. To give you helpful information for **maintaining your "indoor outhouse."**
5. To **keep you safe, healthy, and happy** in conditions you might have once called "**adverse**."

We've divided this section into:

1. **Climate Control.**
2. **The Manual Kitchen.**
3. **Privy Deprived.**

Climate Control

As we asked earlier, "When an emergency hits, are you locked inside on a nice spring or autumn day, or are you freezing in the winter or broiling in the summer?" Though spending a few cold or hot days or nights might seem merely an inconvenience, temperature extremes could turn minor health problems into major ones, they could cause damage to your home, and they carry with them a variety of concerns. It's always the extremes that are the problem, so let's look at how to set yourself up for protection during a severe Level 1 isolation, while exercising safety in the process. For the following, we'll assume you were unable to secure an alternate source of energy.

Warm in the Winter

Anyone can handle the chilly nights where all you need to do is throw an extra blanket on the bed. In this portion, we'll look at the detail you need to know to keep yourself warm in continual sub-freezing temperatures when all the utilities are out. If you're prepped for this, anything less is a walk in the park.

Weatherproofing

Frugality aside, one of the reasons we suggested earlier that you weatherproof your home as much as possible, was to make it easier for you to control the inside temperature in situations such as this. Let's look at a few more sealing and weatherproofing considerations.

➢ In a scenario where there is a complete loss of heat, one of your weatherproofing concerns is going to be your plumbing and keeping it protected. If your water is still on, make sure you've collected as much as you can in various containers in case it's lost due to the same problem causing your other utility outages. After that, leave your faucets dripping steadily to prevent the pipes from freezing. If your water is off, do what you can to drain your pipes so there's no water in them to freeze. Do what we suggested earlier under "water sources" in cutting off the water at the meter or first main valve inside the house. Next, open your uppermost faucet and then drain the pipes at the lowermost faucet which may be simply the tub's faucet, or the drain faucet on your hot water heater. Save all the water you drain for consumption purposes. **Hint::** Be sure to **turn off your hot water heater** so that when the power comes back on you're not heating an empty tank which can cause damage to the tank and create a fire risk.

➢ Next, seal all openings that aren't being used. **All biochemical threats aside, your "seal the room" kit can help seal your house against weather.**

❑ Tape over the seams of all exterior doors that will not be used and cover the doorway with plastic. If you have enough spare blankets, you might hang one over the door like a curtain.

❑ If you have a tile or hardwood floor built directly on a slab foundation, you might want to lay down a few blankets or throw-rugs to keep from losing heat through the floor.

❑ For the exterior door that is being used yet has no outer storm door, tape a pair of plastic sheets over the door and have them slightly overlap in the center of the door. This will create a flap that you can push aside to use the door, and then close once you're back inside.

❑ Close storm shutters over windows that receive no direct sunlight during the day.

❑ Tape over the seams of all windows and then cover the window openings with plastic.

❑ Draw the curtains or hang blankets over these sealed windows that get no incoming sunlight.

❑ Close off any unused rooms on the shady sides of the house. As they provide no warmth from incoming sunlight, and since they will not be used, you shouldn't waste what heat you're able to create by heating these rooms.

❑ For sunny rooms, you'll still want to seal the windows with tape and plastic, but leave the rooms open to allow daytime warmth to help heat the house. Close these rooms at night when trying to heat living and sleeping areas by other means.

❑ Since we're assuming the furnace isn't working, don't let vents or ducts transmit cold air from other areas. Close and seal duct vents and cover over any open floor registers.

❑ Tape around the entrance to your attic if it happens to be inside the house.

People Packaging

Now that we've insulated the house, let's insulate you. While we're on the subject, we'll also talk about body heat and personal heating items. When talking about people and cold, it's important to point out that every portion of the country is capable of experiencing extended periods of lower than normal temperatures. Though not always sub-freezing, it can seem freezing to you and will carry some of the same risks. Therefore, people who live down south need to be just as prepared as those living up north.

➢ For starters, make sure you have all the following items stored somewhere in your house:

❑ At least two extra blankets per person.

❑ Cots or inflatable mattresses. All of you might have to pile into one room to share heat, and no one should sleep on the floor where the cold will settle.

❑ Insulated sleeping bags. (This is one of a very few pieces of equipment that can be shared between your evacuation and isolation gear.)

❑ Wool caps or insulated hats, gloves or mittens, thermal long underwear, and heavy socks for each family member.

❑ At least one "pocket warmer" per person. These can be the chemical or flame type. If you can find any, you might even consider a pair of "battery operated socks" that warm up with the aid of a small battery. Most outdoors or hunting supply outlets will carry them.

➢ For every two blankets on top of you, there should be one under you. Most mattresses are not as well insulated as people think, and if you're sleeping on a cot or waterbed, you're losing even more body heat.

➢ If you don't have extra blankets, layers of newspaper will work surprisingly well.

➢ If your electric generator can't run your space heater, it might be able to run heating pads which can help warm up bedding before you retire for the night.

➢ Dress warmly, but not so warmly you sweat. Damp clothing will only make you colder at night.

➢ Try to stay active during the day, but not so active you break into a sweat.

➢ The more people and pets sleeping in a room together at night the better. Shared body heat will help warm the room. Close the door, but don't seal it. Though this bedroom might not be sealed against a biochemical threat, remember our cautions from "Sealing the Room" earlier. DON'T let anyone sleep on the floor as that's where the cold and carbon dioxide will settle.

➢ If you have a choice of floors and each are equally sealed and stocked, make your sleeping area on an upper floor since heat rises. **Note**: You'd chose an upper floor anyway if you happen to be sealing yourself against a hazardous materials incident as most chemicals tend to settle.

➢ Eat warm food and drink warm liquids, such as soup broths, when able. Be sure to eat well since maintaining body temperature in cold weather uses up a lot of energy.

➢ Don't drink alcoholic beverages as alcohol is a depressant and will slow down your metabolism, reducing your body heat when you need it most. It's only an old folklore tale that alcohol warms you up. Remember, in extremely cold weather, you don't want to take any kind of depressant.

Fireplaces and Grills

Fire is the oldest source of warmth, and fortunately a tool readily at our disposal, even for apartment dwellers. Here we'll give you a few "left field" ideas for putting fire to use, and some safety concerns as well so you don't wind up being a statistic rather than a survivor.

➤ If you have a fireplace, make sure you always have a good supply of firewood and/or synthetic logs during the winter months. Do this even though you may have had gas logs installed. If the gas is out, you can remove the logs (be sure to save the instruction manual) and still burn wood.

➤ The problem with most modern fireplaces is that they're built for decoration and not for efficient heat production. However, most are spacious enough inside to let you use them for cooking. Therefore, if you have a fireplace, you might want to invest in a cheap set of pots and pans (all metal, no wooden handles, no Teflon ® coating) to use for open-flame cooking. You learned about cooking over a fire when you went on your campouts right?

➤ One way to make your fireplace a more efficient heater is with the use of "U" shaped pipe. This can be a single piece of pipe that's laid in the fire and has air circulated through it, or a specially designed log stand that has pipes running from underneath the logs over their tops to catch the heat, and also are set to have air circulated through them. You'll find the log grate at hardware or home-supply companies, and you can have a "U" shaped piece of 3" pipe custom made. Whatever you do, make sure the metal is made for this purpose and that it will not produce toxic fumes. You'll also want to use it a few times while the room is well-ventilated so that you can burn off any oil or grease that may be present on the metal.

➤ If you choose to use your fireplace for warmth and have everyone sleep in the room the fireplace is in, be sure to use all fire safety precautions. Have all family members sleep up off the floor, keep all flammable materials at least six feet away, and have the fireplace's protective screen(s) in place. If this room is a small one, don't seal it up completely as the fire will consume needed oxygen and will give off a certain amount of carbon monoxide.

➤ Don't use a gas range to heat your kitchen unless you have ventilation. Since the ventilation would negate the heat, you might want to put your gas stove lower on your list of heat sources.

➤ Also, regardless of whether or not you sleep near the fireplace, remember two very important air rules. **Number one**, never use any type of flame for heat unless you have good **ventilation** (this includes candles, Sterno ® stoves, etc.). **Number two**, **never heat with charcoal** inside the house as it produces a higher amount of carbon monoxide than other fuels. Regular, untreated wood should be your first choice for heating and cooking use. Synthetic logs may give off fumes unsuitable for cooking, and treated lumber may give off fumes unsuitable for humans.

➤ To make some of your own synthetic logs, try this. You'll need some newspapers, charcoal lighter fluid, a small wooden dowel about two feet long and a half-inch thick, and some string or wire. Take a couple of sections of the newspaper, leaving it folded, and wrap it around the dowel as if you were rolling the paper for delivery. Wrap the paper about an inch and a half thick around the dowel and tie it off with string. This will be your "inner log." Soak your inner log in just enough charcoal lighter to get it wet. Then set it aside (outside) to dry. Once dry, wrap another inch and a half to two inches of newspaper (that was soaked in water and allowed to dry) on top and then tie it off with your string or wire. This should burn well to provide heat, but we don't recommend using it for cooking.

➤ If your fireplace will not be put to use, be sure to close the damper and any solid glass cover or screen to keep cold air out. You'll also want to cover it with a plastic sheet taped in place.

➤ If you're able to venture out of the house, you can use your outdoor gas or charcoal grill, or even a medium-sized hibachi to help keep you warm. Build a medium fire (do all this outside) and heat fire-glazed bricks (never concrete or other bricks as they may explode when heated) and/or large rocks (that you've tested under heat before) until they're good and hot. Pick them up with tongs, put them in a metal bucket or old metal frying pan, and carry them into your sleeping area to help heat the room. Put the bucket on a fireproof stand that won't tip over and is not near flammable materials. You can make a suitable stand out of a small metal cookie sheet with a few layers of aluminum foil on top and bricks to set the bucket on.

➤ If you use this "hot rocks" idea, remember to only use dry materials, and don't wet them down trying to create steam. You don't want steam in your room as that will only make everything damp and that much colder once the heat wears off.

➤ Speaking of venturing out and having outdoor fires, there is one idea that some of you may be able to use depending on how much yard space, tools, parts, and time you'd like to invest in preparing for this possible cold-weather isolation. This idea involves digging an outdoor fire pit, close enough to the house for this purpose, but far enough away for safety. Across the top of this fireplace is a metal pipe making a "U" shape across the pit just as we mentioned above with the fireplace, with a small battery-operated fan circulating the air. In fact the ideas are similar except where the pipe in the fireplace circulated hot air in the room, this one is circulates air into your basement or crawlspace in order to heat the underside of your house. To put this to work, you'd need to maintain a wood or charcoal fire in the pit at night when the temperatures were lowest. If you choose this option we suggest the start and end-points of your duct work be under the room you intend to gather the family in order to share warmth. Again, this is not an idea that everyone can put to use, but you have to use this kind of creative ingenuity.

➢ In keeping with this warm air circulation theory, we saw another idea we liked. We saw a "miniature greenhouse" that looked like a shallow black refrigerator with a glass front. This box was kept on the back porch of a house where it faced the sun for the better part of the day. Air was circulated through this box, which warmed the air with sunlight, and it worked well enough to keep an outdoor workshop warm.

Appliances

So far, we've focused on things you can do to stay warm during a complete lack of electricity, gas, or alternate energy. Now we'll touch on things you can do utilizing certain appliances, or other equipment.

➢ Under "Alternative Energy" above, we told you to diversify and have similar appliances that use the *opposite* type of energy from the main appliance. For instance, if you had a gas stove, we suggested you have an electric hotplate and toaster oven. Here we'll tell you to *augment,* meaning that you should **get additional appliances using the *same* type power** in case it's the *appliance* that malfunctions and not the *energy* source. Just as earlier you wanted the same function regardless of power, here you want to be able to utilize the power source regardless of the particular appliance. If you have an electric stove, still keep an electric hotplate and toaster oven. Even if you have electric heat, still keep electric space heaters. If you have gas heat, maintain gas space heaters, etc.

➢ Keeping the above example in mind, learn to use alternate appliances. For example, if your gas furnace goes out, you might want to sleep in the kitchen and let the electric oven heat the room.

➢ If absolutely nothing else, if you have some sort of electric power, keep as many electric lights burning in your sleep room as possible to help heat it. You can also use blow dryers and toaster ovens to help heat a room for a bit before retiring for the night. (Don't leave such items running.)

➢ Speaking of ovens, consider getting a wood-burning stove. Not only are these stoves decorative, they're functional, and can be used for both heating and cooking. **Hint:** If you have room to be creative with the vent stack coming off your wood stove, route it under upper-floor rooms or through the attic where some residual heat from the exhaust can help heat the house rather than go to waste.

➢ If you use these stoves, gas heaters, or kerosene heaters, be sure to follow the manufacturer's recommendations for fire safety and proper ventilation. Better yet, hire a professional installer or inspector.

➢ Speaking of appliances and obscure considerations, keep this in mind. Your power may be out, you might not have an alternate source of electricity, yet you've managed to heat a good portion of your home through other means. Ironically, the food in your refrigerator may be at risk. If that's the case and it really is freezing outside, you might keep some of your foods outside to stay fresh and frozen.

Miscellaneous

As freezing weather and alternate heat sources both carry their own risks, you need to know the symptoms of some of the physical problems that could be created.

➢ Know the symptoms of both **carbon monoxide** and **carbon dioxide poisoning** (remember, you should have battery-operated carbon monoxide alarms).

❑ **Carbon monoxide** will cause headaches, dizziness, nausea, and will result in passing out.

❑ **Carbon dioxide** will cause shortness of breath, difficulty in breathing, rapid pulse rate, headaches, hearing loss, hyperventilation, sweating, and fatigue.

➢ Know the symptoms of both **frostbite** and **hypothermia.**

❑ **Frostbite** symptoms include numb and/or very pale fingertips, toes, earlobes, or the tip of the nose. These points are usually the first to be effected.

❑ **Hypothermia** symptoms include drowsiness, disorientation, uncontrollable shivering, slurred speech, memory loss, disorientation, and severe fatigue. Hypothermia is when your core body temperature gets down to 92 degrees F. If your core temperature gets down to 82 degrees, your heart and lungs will shut down. Stay warm.

Sources and More Information

➢ More information on cold weather survival can be found in the Army's Training Circular 21-3, online only at: http://www.adtdl.army.mil/cgi-bin/atdl.dll/tc/21-3/toc.htm.

➢ Wood burning stoves and all manner of fireplace equipment can be found at your local hardware or home improvement store.

Cool in the Summer

When staying warm in the winter, if you only did three things; close the doors and windows, dress warmly, and build a fire, you'll have increased your warmth by a drastic percentage. Staying cool in the heat of summer, especially if you're in a room sealed against a hazardous materials threat, is a great deal more difficult and carries with it its own peculiar details and considerations.

Weatherproofing

Again, the extra insulation you performed under "Basic Home Prep" will help you maintain a comfortable temperature inside your home while the outside temp is unbearable. However, hot weather will carry quite a number of considerations different from winter weather.

- ➤ In winter you wanted to use the sunlight, in summer you want to block it out. On the sunny side of the house, close any storm shutters, close the windows, draw the blinds and curtains, and close off any unused rooms that have an exterior wall on the sunny side of the house.

- ➤ Close off duct vents that might carry hot air in from other parts of the house, and cover over any open floor registers that might allow outside heat to flow upwards into your house.

- ➤ Though our goal is to prep you for a worst-case-scenario, which here happens to be a biochemical isolation in a sealed house during a complete utilities shutdown in the hottest period of the year, we do realize there will be other types of isolation emergencies. Some will see you with more mobility. Therefore, let's cover a few hints and tricks to keep your home cool if you *are* able to get out a bit and you know your power outage will be a long one.

  - ● If direct sunlight on your outer walls create a problem, hang lightly colored bed sheets or tarps as extended outdoor awnings to shade these walls. You can also use aluminum foil, shiny side facing out.

  - ● If you have a dark roof, you can also lay these sheets or tarps on your roof. Lighter colors of material reflect sunlight's heat, and darker colors absorb the heat. Doing this to your roof can measurably reduce the accumulated daytime heat you feel inside the house.

  - ● If you don't have enough sheets or tarps, and water is available, you might want to run your roof's ridge-line sprinkler that we asked you to install under the "Fire" portion of "Basic Home Prep" earlier. Run this sprinkler for a limited time during the peak sunlight hours of the day to cool your roof. You can also do the same with a yard sprinkler and a hose.

- ➤ Though you're closing off rooms and ducts, we don't necessarily want you to seal everything like you did in winter unless you're sealed in against a biochemical threat. You'll want to be able to open some windows on the shady side of the house to take advantage of a breeze.

People Peeling

In winter, we want you to "people package" and bundle up to stay warm. Here we'll do the opposite and strip you down to stay cool. Let's look at a few other options to keep you more comfortable.

- ➤ Many works covering ways to stay comfortable in hot weather say to "wear loose fitting clothing that's light in color and breathes well." That's all well and good if you're out in the sun, but in this case you might be stuck in a sealed room. When stuck indoors, wear as little as possible.

- ➤ If you have long hair, tie it up to keep it off your neck and out of your face. If you have a beard, you might want to shave.

- ➤ Drink plenty of water even if you don't feel thirsty or feel like you're sweating much. You'll definitely be losing moisture through your skin and respiration, and it needs to be replaced.

- ➤ Regarding water, drink cool water if possible and don't eat much in the way of hot foods (temperature or spicy). In winter we wanted you to eat well and regularly, but in the heat of summer we want you to eat lightly. However, you should still take your regular vitamin/mineral combo as you'll want to keep your electrolytes in balance.

- ➤ One of the things we suggested earlier was a sedative herbal tea made from Chamomile. This doesn't put you to sleep, but it "takes the edge off" and lets you relax. We suggest that you relax as much as possible and avoid any unnecessary physical or emotional exertion.

- ➤ Avoid caffeine or other stimulants, alcohol, or anything else that has diuretic properties.

- ➤ Remember all the normal little heat tricks. Make a fan out of paper, keep a small spray bottle full of water to occasionally "mist" your face, and remember that most "dollar" stores carry little battery-operated personal fans that will get a couple hour's run time off a single battery.

- ➤ Another trick that doesn't necessarily reduce the temperature, but will make you feel better is the sponge bath. This is one of the many reasons we keep suggesting you stock moist towelettes and baby wipes. Every now and then take a quick sponge bath with rubbing alcohol. You can use alcohol prep pads that you may have stocked, or you can use it from the bottle on a washcloth. Don't do it too often as it will dry out your skin, and certainly don't do it if you're in a sealed room, because of alcohol vapors.

➤ One thing that will reduce your temperature are the first aid "cold packs" that you squeeze and mix the chemicals to produce cold. You'll need to have these on hand anyway for first aid, but having a few extras for comfort can't hurt. The best places on the body to put one are across the forehead, behind the neck, on the wrists, or on your ankles as these are the locations your veins are the closest to the surface and will exchange body heat more readily.

➤ If you're not sealed in against a biochemical threat, and you're not isolated by a flood, you might want to consider staying on a lower floor of your house where it may be cooler.

➤ You might even consider sleeping on a thin pad on the floor (if not in a sealed room) since that might be cooler than an insulated mattress higher off the floor (since heat rises).

Appliances

Our biggest suggestion of all regarding appliances is that you take our earlier advice about getting a generator and a room floor air conditioner.

➤ The average home generator can't run everything. You should either manually unplug and plug in whichever appliances you want to run at the time, or you should put your "light timers" that you bought for security purposes to good use by having them automatically alternate appliances that draw generator power.

➤ Alternate between your freezer, refrigerator, and your floor or "space" air conditioner and small room fans.

➤ If you didn't buy a space air conditioner (which are too inexpensive to not have one) you can make a fairly effective one out of a cooler. Better yet, make two or three of these units.

❑ Get a large and inexpensive styrofoam cooler with a lid. You'll want it to be around three feet long, two feet wide and two feet deep. (Or at least 2'x1.5'x1.5')

❑ Cut a two-inch hole in each end. On one end, make the hole high near the top edge. On the other end, make the hole low, about 2 inches from the bottom.

❑ Take some of your frozen bottles of water out of the freezer and set them inside the cooler spacing them so that air can still flow between them.

❑ Arrange one of your battery-operated personal fans near the upper hole on one end of the cooler. This will draw air into the cooler, across the bottles and force the air out the bottom hole on the other side. If you have two of these battery fans, use one as an exhaust. Also put them on opposite sides as shown below.

❑ Having the exhaust hole near the bottom of the cooler and the first one near the top will force air across and down between the bottles and vent the colder air that settles. Also, having your exit hole two inches up from the bottom will allow a lip at the bottom of the cooler to collect condensation rather than let it dribble on the furniture in your room. You may collect this condensation to wet washcloths for cold compresses.

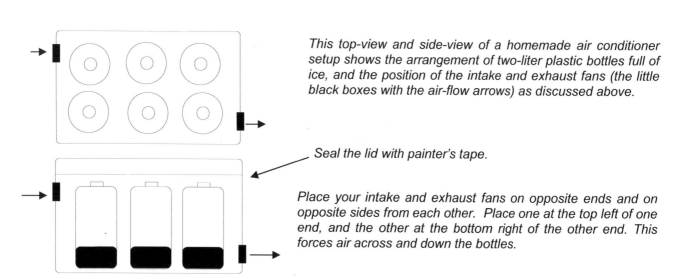

This top-view and side-view of a homemade air conditioner setup shows the arrangement of two-liter plastic bottles full of ice, and the position of the intake and exhaust fans (the little black boxes with the air-flow arrows) as discussed above.

Seal the lid with painter's tape.

Place your intake and exhaust fans on opposite ends and on opposite sides from each other. Place one at the top left of one end, and the other at the bottom right of the other end. This forces air across and down the bottles.

● **Note**: This little unit won't cool your room down like a real air conditioner would, but in the average sized room (about 10x10), a unit using 6 frozen 2-liter bottles will drop the room's temperature only a few degrees. However, in high heat situations, this could really help. Having 2 or 3 of these units is even better.

➤ As your bottles thaw out, you can swap them out for other frozen ones from the freezer, or you can drink the cool water. If you drink the water, you can replace it with soapy water, or other waste water you may have on hand and refreeze it for use in your little homemade AC unit.

Miscellaneous

As any heat situation carries with it several risks, you should know what they are.

➤ Know the symptoms of heat exhaustion and sunstroke.

❏ **Heat Exhaustion**: Heavy sweating, and skin may be pale, cool, or flushed. Weak pulse, with fainting, dizziness, nausea, or vomiting.

❏ **Sun stroke** (or heat stroke): High body temperature, hot, dry, red, skin (usually with no sweating), rapid shallow breathing, weak pulse. Sun stroke is the more dangerous of the two.

➤ Children, the elderly, or those with poor health will be affected by heat the most.

➤ Exertion in extreme heat also carries an increased risk of heart attack or stroke.

➤ Under "Outdoor Survival" we suggested you go camping as much as possible. Here we'll suggest that you lock yourself in the house over a weekend and shut off **ALL** utilities to see how prepared you are or aren't for such a situation.

➤ **Hint**: If you're able to venture out of the house, you can sit in your vehicle and run the AC for short periods as long as you're aware of carbon monoxide problems and take care not to overheat your engine.

Sources and More Information

➤ Most information on cooling systems can be found at your air conditioning supply store.

➤ The camping and outdoor survival information sources we gave you earlier (including the manuals located on our **CD**) will give you additional information on heat, heat exhaustion, and sun stroke.

➤ The Department of Aging has established a heat wave hotline. The telephone number is 1 (800) 339-6993.

➤ We've found an economical product similar to our "homemade air conditioner" above. It's called the "KoolerAire," and can be found at http://www.kooleraire.com, or contact them at KoolerAire Inc., 600 Chestnut Ridge Rd., Chestnut Ridge, NY 10977, 845-425-8800, Fax: 845-425-8868.

The Manual Kitchen

Now that we have you set with a more prepared home, an alternate source of energy, have you sealed and protected if need be, and have you set to face some uncomfortable temperature extremes, it's probably time we fed you, right?

You'll need to eat and you'll want to eat, but you'll want to do it easily, and you'll want to do it all without burning the house down and without giving anyone food poisoning. Let's look at some things that will help.

If your alternate power source or generator is capable of running your refrigerator and freezer, and your hot plate, toaster oven, or microwave, you're set. In this section we'll cover food and cooking without these benefits. Let's start by looking at a few general considerations.

Food Safety

What's the point in having food if you can't eat it, and what's the point of protecting yourself from all manner of external threat if you're only going to fall ill from your own emergency goods?

➤ As your situation might see you without utilities and only *somewhat* isolated, know all your **local sources of ice, including ice production facilities** (list them on an "Important Contacts" page), and be ready to get to them early at your first indication that you'll be without utilities. **Hint**: Ice production facilities cool their ice well below freezing, usually taking it down to zero degrees. This way it lasts longer on its trip to the stores. Your freezer is probably set down in the high twenties, not as cold as this factory ice. Therefore, getting the factory ice can chill the items in your fridge and freezer down to a little below what they're normally stored at and this will make your food items last much, much longer than they would have with just your ice bottles.

➤ Similarly, locate local sources of **dry ice** as that will serve you just as well as far as keeping fridge items colder longer. Your only caution here is to **not** use any dry ice in any sealed room as it's nothing but frozen carbon dioxide, and you don't want it crowding out your breathable air.

➤ Speaking of fresh items in your refrigerator or freezer, what do you do during lack of power or if the appliances just break down at the wrong time? Let's look at a few ideas:

❏ First, keep the fridge or freezer door closed. The average refrigerator, without the benefit of ice bottles, will keep food cold for about 4 hours once the power is out.

❑ In a complete power outage, cover your refrigerator front and back with blankets or a comforter. If you have a generator, cover just the front and door seams leaving the coils in the back unobstructed. This will increase the refrigerator's insulation making it easier on the generator.

❑ Second, we hope you took our advice about freezing plastic bottles of water. This will keep your units cool much longer than without them.

❑ Next, in order to keep from wasting food, you'll want to change your consumption priority. Normally you'd want to eat whatever you desired at the moment. Now you'll want to eat what might go bad first.

❑ Eat the items in your fridge as they'll be the first to warm up. Prioritize these items by which of them might go bad first. Raw meats would be first, milk and eggs a close second, previously cooked leftovers third, followed by packaged and processed meats, cheeses, etc. However, don't eat anything that's been room temperature or higher for over an hour.

❑ The items in your freezer will last longer and should be consumed next, and in the same order as the fridge items. Raw meats first, and so on. Frozen items offer one testing advantage though. If the food item still has ice crystals in it or on it, it's probably still good.

❑ **Hint**: If you've only got a small generator and have both a freezer and a refrigerator / freezer, operate the full freezer. Use it to freeze your water bottles, and use them to cool the fridge.

➤ When handling anything having to do with food, remember your plastic gloves. These gloves will protect the food from your dirty hands and protect your hands from getting dirty with food. You'll want to do this since you might not have any spare water for washing.

➤ Speaking of wash water, you should get in the habit of washing any food that needs it, before putting it in the fridge as you might not have water later.

Sanitation and Food Safety

Speaking of washing food before putting it away, how in the world are you going to keep a safe and sanitary kitchen when you have no running water, and possibly no way to refrigerate food? It's not as complicated as you might think. If you took our advice earlier and added camping as a hobby, you're certain to have gathered some "cooking in the field" experience that will help you in this situation. To that experience, let's add a few important considerations:

➤ The first rule of thumb is to **cook everything thoroughly**, even though you might normally like food at a certain temperature or cooked a different way. Safety takes precedence over simple pleasures, and thorough cooking will help ensure food safety. Remember to use your kitchen thermometer.

➤ Some of the items we've mentioned earlier for other uses have direct application in kitchen sanitation. Be sure to keep plenty of the following in stock:

❑ Plastic food service gloves   ❑ Newspaper   ❑ Kitty litter
❑ Waterless hand sanitizer   ❑ Alcohol prep pads   ❑ Plastic garbage bags

➤ If you're in a water-deprived environment, remember to use a fresh set of **plastic gloves** for anything you do that normally required washing your hands. This includes brushing your teeth, going to the restroom, preparing or eating food, and cleaning up afterward. The less you have to wash your hands, the more drinking water you save.

➤ When cleaning up after cooking, wipe out any pots or pans and clean food off reusable utensils with **newspaper**. Burn the paper in your fireplace or grill, if you have one, or roll it up and place it in the trash.

➤ An unscented **kitty litter**, either a clay-based or wheat-based brand can be used as a dry scouring powder. The litter will absorb excess oils and food particles still clinging to containers and utensils and will also provide a mild abrasive action to help clean. Remember to wear your plastic gloves when cleaning.

➤ Next, you might apply a coating of **waterless hand sanitizer** such as GoJo ® or Purell ® and let your utensils sit a minute in order to kill bacteria.

➤ Afterwards, wipe them down with newspaper again and then wipe each utensil thoroughly with an **alcohol prep pad**. The alcohol should remove any remaining hand sanitizer and kill off any remaining bacteria. Also, the alcohol will evaporate off your utensils leaving them dry.

➤ We don't recommend bleach unless you have plenty of water, and you have good ventilation.

➤ During cleanup, burn what trash you can in your fireplace or grill (not while cooking other food) and seal any remaining trash as well as possible to reduce the threat of disease and/or an infestation of insects or rodents.

Appliances

By now, you know what kind of equipment you'll have at your disposal and what kind of energy sources you can put to work. As we said before, if you can utilize any of your modern conveniences, you're set. In this section of "Appliances" we'll look at the things you need to keep in mind when using your more primitive or improvised equipment. You'll want to be able to use each to their full potential, but exercise safety in doing so.

- Camping is going to teach you the most of what you need to know. The more accustomed you are to preparing food in the most primitive settings, the easier cooking at home in less than ideal conditions will be.

- If you'll go back to the "**Evacuation**" section at the discussion of camp stoves under "Level 1 Food," you'll find all the equipment you need to have on hand for cooking at home during an extended isolation. Remember, get redundant and separate gear.

- The biggest concern when using any source of flame for cooking is safety. Therefore, when using any of your cooking gear, exercise the following cautions:

  ❑ Do not use any sort of flame in a sealed room as it will consume useful oxygen and may produce smoke and other toxins. Only use flame in well-ventilated areas.

  ❑ Never cook with charcoal indoors. It is known to produce more carbon monoxide than other fuels. However, for your outdoor grill, charcoal tops our list of recommended fuels as it's easy and safe to use, and the heat it produces is uniform and easy to cook with.

  ❑ Similarly, do not use synthetic fireplace logs or newspaper logs to cook with unless your food is in an enclosed container. Some toxic chemicals may be transmitted to exposed food.

  ❑ If using any sort of flame, make sure the platform it's on is stable and protected. Always use either a metal cookie or baking sheet with lipped edges or make a place mat with raised edges out of aluminum foil.

  ❑ Never use a flame so high that it comes up around the edges of the container you're heating or cooking in. You'll burn your food, and greases and oils in the food may catch fire. Use a medium to low flame.

  ❑ Always have a fire extinguisher nearby, as well as the pan's lid and some baking soda.

- One cooking method, not fully explored yet, is the charcoal grill. We've mentioned it a couple of times but didn't really cover it. On one end of the emergency cooking gear spectrum you have the large appliances like wood stoves and the modern camping gear like compressed gas stoves, and your more expensive outdoor gas grills. On the other end of the spectrum you have your little canteen cup heaters, "Tommy Cookers," "Buddy Burners," and "Hobo stoves." In the middle are the average charcoal grills. Here are a couple of things to keep in mind:

  • Backyard charcoal grills and Hibachis are plentiful and inexpensive. The Hibachis are small and many models of grills can fold up to take up minimal storage space. Regardless of your other cooking appliances, you should have a charcoal grill of some sort.

  • Charcoal briquettes and charcoal lighter fluid both have incredible shelf lives and can be safely stored for long periods of time. **Hint**: Charcoal briquettes that are not pretreated with lighter fluid can also be used in your homemade desiccant packs.

  • The biggest consideration of all? Don't forget **matches**.

  • **Hint**: During a situation where food is scarce, you might not want to grill outdoors. The aroma will announce to the whole neighborhood that you have food, and that might become a security issue.

- Regarding your other cooking appliances, since you're at home where things can be safely stored for longer periods, you should carry a good supply of Sterno ® or gas canisters (preferably both) in order to run your improvised stove for indefinite periods. Remember, though we're planning for a 21-day isolation, we really don't know how long it might last. Be ready.

- **Hint**: If you're cooking over **charcoal** or **campfire embers**, it can be tough to cook certain items because you won't know what the **temperature** is. Try this: hold your hand about an inch or two over the coals and count the seconds you're able to hold it there. Less than three seconds is around 400 degrees, or **high** heat. Between three and four seconds is around 350 degrees, or **medium** heat. Four to five seconds is about 300 degrees, or **low** heat.

- Another consideration that we've mentioned regarding cooking over open flame is your hardware. You might not want to use your good pots and pans, especially those with Teflon ® over flame as flame is so hard to regulate. Consider the following:

  • If you have a good selection of camping gear that's separate from your evacuation gear, then you probably already have another mess kit and assorted cooking gear.

  • If not, then go to local thrift stores and yard sales to get the basics. You'll want a cast iron frying pan or "skillet" (with lid to make a "Dutch Oven"), a large metal stock pot, a tea kettle, one or two assorted sauce pans and soup pots, and maybe a Fondue pot. These should give you plenty for cooking and for purifying water, and should work well whether you're cooking over charcoal, Sterno ®, gas, or in your fireplace. It should also give you extra cookware to use different heat sources simultaneously. You might have one dish cooking in your fireplace, another on the grill, and so on.

  • You should also have a separate set of utensils such as knives, forks, and tongs set aside for use with open flame. For this, all you really need is a set of "barbeque" utensils.

> While we're discussing appliances, don't forget we're prepping for a complete loss of utilities which means no power to run electric can openers or mixers. Make sure you have good ol' hand-operated can openers, bottle openers, and mixers (egg beaters) as part of your gear.

Sources and More Information

> Revisit the various sources we provided under "**Evacuation**" at the end of the "Contents in Detail" section.
> We also mentioned that the uses for aluminum foil could make a book of their own. There is a book. Go visit http://www.reynoldskitchen.com or call 1-800-745-4000.

Privy Deprived

What would you do without your bathroom? Since the main things in your bathroom are your toilet, your sink, and your tub, let's see what it will take to temporarily replace their function in an extended isolation.

Your bathroom depends on incoming water and a functional sewage hookup in order to work. A loss of either would cause problems. You might lose incoming water or your sewage system or septic tank may be useless to you (as they would be in a flood). Here we're going to plan for both being unusable, but since we like to cover everything, we'll also give you a few hints on how to compensate if either one is available. Let's start our discussion with the more pressing of the three bathroom functions and once again ask the question, "Where ya gonna go?"

Toilets

For most civilized people, this will be the biggest concern of all regarding a 21-day total isolation, provided survival needs are met. Many people think all you need is a 5-gallon bucket and some bleach. To this **bleach** option we give a resounding, "**NO!**" Let's see why, and discuss your options while we're at it.

> Regarding bleach, try this: Pour a small amount onto a small saucer or plate, place it in your bathroom and close the door. Come back in an hour or so and see how long you can stay in the room. Add to that various other "fumes" and the fumes from resulting chemical reactions, and you can see why we say no to the use of bleach in what may well be a sealed environment. (Besides, bleach "keeps" or stores for less than a year.)

> Also, dip a finger or two in the bleach, rub it on your fingers a bit and then immediately wash it off with soap and water. See how long the bleach smell will actually stay with you even after washing. **In a sealed, or water-deprived isolation situation, you do <u>not</u> want to be handling anything that might require "flushing with copious amounts of water" as its first aid treatment**, and you certainly don't want bleach (or anything else)... well, ummmm... "splashing back up on you."

> Instead of bleach, we recommend a **dry** system using **kitty litter**. Let's look at the whole setup:

❑ We recommend a clay-based, scoopable, odor and bacteria control kitty litter. Not that you'll be scooping, but the scoopable kind absorbs liquids better than the other types. We suggest having a 20 lb. supply per family member (which will store for years) in addition to litter you may keep on hand for your cat.

❑ In addition to kitty litter, you'll need: Heavy-duty trash bags, at least one five-gallon bucket with snap-on lid per family member, sturdy twist-ties and/or duct tape, baking soda, and some powdered bleach.

❑ To set up a "toilet" take the seat off your regular toilet to use on your bucket (or you can cut the seat out of an old chair and place the bucket underneath). In the bucket place three trash bags, one inside the other. One bag will always stay, and the other two will be removed to empty the unit. Place about a half inch of litter in the bottom of the lined bucket. Be sure to label each bucket, as each person should have their own.

❑ After using the toilet, sprinkle just enough litter to cover what you just "deposited." Be conservative so you don't run out of litter. Sprinkling in a little **baking soda** might help as well. Take off the toilet seat and gather up the edges of the trash bags, twisting them lightly together and tuck them lightly down into the bucket where you can pull them back out for the next use. Rest the bucket lid over the top, holding it in place with a heavy object. Don't snap the lid back in place as you don't want to slosh the bucket getting the lid off.

❑ When the bucket becomes **half-full** it's time to empty. Cover your final deposit just as you would any other, and while you still have your plastic gloves on, pour about a quarter-cup of bleach on top (the only time you'd want to use bleach in an indoor situation). Leave the outer trash bag in place and twist the tops of the other two liners together and gently pull the bag out of the bucket. Twist the bags firmly at the top and secure the top with a twist-tie, or better yet, duct tape. Set the bag with other refuse that's to be disposed of later. Better than liquid bleach in this case would be powdered bleach. Make sure you get a powdered bleach that has some type chlorine or chlorine compound, such as sodium hypochlorite, as its main ingredient. Many non-chlorine powdered "bleaches" are simply whiteners that have no anti-bacterial properties. **Hint:** Some powdered automatic dishwasher detergents have bleach in them.

❑   Insert two new trash bags into the bucket and sprinkle a half-inch layer of litter in the bottom. This is yet another of the many, many reasons we tell you to keep plenty of heavy-duty trash bags on hand at all times.

➢   If you're caught in a situation without bleach or kitty litter, you can use the same bucket and bag setup and use shredded newspaper along with flour, cornstarch, baking soda (or other absorbent powders), and powdered dishwashing detergent or other powdered bleach source. Again, the goal is to have a dry system.

➢   Some companies make a "travel" or "camper's" chemical toilet complete with frame, seat, bags, and a non-bleach chemical. These are good, but might be rather pricey when considering supplying each family member for 21 days. We also do not know how well your refuse will store with these chemicals, or the shelf-life these chemicals have during long-term storage. Kitty litter will keep for years.

➢   Whatever type toilet setup you're able to use when restricted indoors, keep your "bathroom," as well as your refuse storage area, as far away from your living area as possible.   This is the main reason we suggested you stock some solid air fresheners in your isolation kit somewhere. Solid air fresheners will store longer than liquids so you'll be sure they'll work when needed. We also recommend solid air fresheners because in a sealed-in situation you **don't** want to use aerosols.

➢   If you're able to venture outside and you have a yard with any opportunity for privacy at all, even if it's a screen or tarp you set up, then you might want to consider a simple latrine.

❑   Dig a trench about a foot wide, three feet long, three feet deep, and at least fifty feet away from any water source. This is called a "straddle trench" because that's exactly how you use it. No toilet seat, just straddle the trench. Of course if you can build a bench or something to hold a seat, even better.

❑   After using the latrine, sprinkle dirt back into it just as you would the kitty litter in the bucket setup. Also, being outdoors and being your yard, you can use a little bleach after each use, though you still want to be careful about getting any on your skin.

❑   When you've refilled your latrine within a foot from the top, pour in more bleach, cover it back all the way and then dig a new trench at least three feet from the first one. By the way, do you know where in your yard is safe to dig? **Call your utility companies now** so you'll know.

➢   Also, if you're able to venture out, you have a source of outside water, and your sewer system is still functional, you can use outside water to fill your toilet tank to keep it flushable. Just because you might lack incoming water doesn't mean you have to use the bucket or latrine system.

➢   Don't forget your pets in all this. Cats will have their own litter boxes, but for dogs, you'll need to make sure you have plenty of old newspapers on hand and even more trash bags.

Sinks
Fortunately it's really easy to work around a lack of a sink.

➢   Sinks can be replaced by buckets, tubs, bins or other waterproof containers. Simple as that.

➢   You'll need at least three types of containers for your improvised sink system. One will be for source water, the other for your "ready" water or the clean water you're about to use, and the next for waste water.

➢   For clothes or dish washing you'll need four containers; your source water, wash water, rinse water, and waste water.

➢   Let's define each just a bit more since the way you manage your water supply is important.

●   Your source water will be a container holding your main water supply. It should be kept clean and covered, and nothing should be dipped into it but a ladle or cup to dip out the water you're about to use (unless you have a valve or faucet attached). Keep the ladle or cup on the outside so dirty hands don't have to fish around in the clean water to retrieve it.

●   Your "ready" water is that you'll actually be using and it's what you've just dipped out of your source container. Use this to wet your toothbrush, your clean washcloth, or whatever it is you're using. You might want to have one "ready" container per person so each can have their own. This container can be something as simple as a small plastic bowl you saved from a food item. The smaller it is, the more effective it will be at helping to ration your water usage.

●   Your waste water goes in a dumping container where you'd spit toothpaste, wring out your washcloth, etc. Everyone can use the same dump bucket so get something fairly large like a 5-gallon plastic bucket with a lid. If your sewer is still functional you can dump your waste water into your toilet's tank and use it for flushing. (Don't waste water by pouring it down the drain.)

➢   For clothes washing and dishwashing, use 5-gallon buckets for both your wash and rinse water, but use only as much water as you need. Remember though, that we suggested paper plates and burnable or disposable utensils so you don't waste valuable water for unnecessary washing. The same holds true for clothing. Don't wash anything you don't have to.

➢   As far as hand washing, teeth brushing and things sinks are commonly used for, there are substitutes.

- Your ready-water and waste-water bowls from above are, in fact, little sinks.
- Using your plastic gloves will help reduce the number of times you have to wash your hands.
- Store a good waterless hand sanitizer like Purell ®, and a waterless soap such as Go-Jo ®.
- You can also use alcohol prep pads, moist towelettes, and baby wipes for hand washing.
- If your sewage system is usable, you can use your sink even though you may have to use another source of water.
- **Hint**: If you're on limited water and about to go do manual labor, dig your fingernails into a dry bar of soap. This will get soap under your fingernails which will help seal out dirt as you work. Later on, it's easier to get your hands clean using less water.

Tubs and Showers

Though primarily used for bathing, your tub may have other uses. Since we've covered cleanliness so many times in other sections, there's really little to be added here.

- Remember, your tub should be full of clean water you collected in anticipation of this situation.
- If your drainage works okay, you can use your sinks or emptied tub for washing you, your clothes, and your dishes, depending on how much water you can acquire or spare.
- For general bathing, use your towelettes and baby wipes for sponge baths.
- Babies can be washed and rinsed in small plastic tubs or bins in the same manner we suggested for sinks.
- If you're able to venture out and have a fairly good water supply, you can set up a privacy screen and set up a "solar shower" which you'll find at any camping supply outlet. It's just a 5-gallon black plastic water container with a valve and a shower head, and leaving it out in the sun during the day heats the water.
- Though you'd purify it before drinking, river or lake water should be fine for bathing. If you have easy access, privacy, and environmentally friendly soap, you can bathe *in* the river or lake.
- Speaking of bathing and things like that, and of babies and bath water, you should have a supply of both disposable and cloth diapers on hand in your isolation supplies if you have an infant. If you have no water and can get outside, use the disposables. If you're sealed inside and have water and sewage, use the cloth diapers as it's better to wash them than let the disposables pile up inside.

Sources

- See the Army field manual, FM 21-10 Field Hygiene and Sanitation, and its re-release, FM 4-25.12, on our enclosed **CD**.
- For commercial toilet options visit "Restop," online at http://www.whennaturecalls.com, with direct contact at 888-924-6665, or Fax: 760-741-6622.

Now that you're set with the goods and gear you'll need for an extended isolation, let's look at how to deal with some of the more dire problems that might crop up while you're isolated, cut off from others, or cut off from outside services.

Notes: _____

_____

_____

_____

_____

_____

_____

_____

_____

_____

_____

_____

_____

_____

Life in Isolation

We would be sorely remiss in our duties if we discussed isolation supplies and left you without covering what life in isolation might be like, and some of the associated hidden risks. In an isolation you're cut off. You're cut off from not only supplies, but from other people, and from services. What's it going to be like? We wish we could give you a specific and certain list of things to expect, but we can't. What we can do is cover some of the dangers you may face as well as provide you some pointers to make your little "vacation" just that; a little vacation that finds you safe, and none the worse for wear.

Goals of Life in Isolation:

1. To help you be **self-reliant without the benefit of public safety services**.
2. To help you protect your family and your property from **looters, other criminals, and civil unrest**.
3. To help protect you from one of isolation's biggest physical dangers; **fire**.
4. To help you better understand and prepare for the **emotional aspects of being cut off**.

We've divided this section into:
1. **Security and Defense.**
2. **Limited Firefighting.**
3. **"Cabin Fever."**

Security and Defense

We've said it before, and we'll stand by it here, although we'll add a little to it, that we believe 99 percent of everyone around you are basically good people and that in an emergency situation most will be in the same boat as you and will not be a threat to you. In fact, the stories of friendships developed through adversity are uncountable. However, there is that 1 percent you always have to watch out for.

Now let's let the other shoe drop. Though this 99 percent is composed of basically good people, you never know what's going to happen inside the minds of some people when an emergency or disaster strikes. Crime itself is a big enough concern, but add to that the riots and looting that commonly occur during periods of even minor civil unrest and you can see a threat you'll have to be concerned with. Now take these same people and put them not only in a period of <u>great</u> civil unrest, but in a situation of survival when they're unprepared. Where will they go? What will they do? To whom will they do it? Our concern is that they may decide to take what they need from those who were wise enough to prepare in advance.

Some people may think that all isolations might be automatically safe since no one can get to you in physical disasters, and curfews and quarantines will be obeyed. We wish that were so. The truth is, the more severe the situation, the greater will be the civil unrest and the threat of violent crime. Some problems might be caused by panicked individuals looking for supplies to protect themselves. Others, such as those caught up in a biological attack quarantine, might feel that they're dead anyway and nothing they do will matter so why not go on a crime spree just for fun? These people won't obey curfews or adhere to quarantines, and there just aren't enough Cops in the world to make sure they do. You may be on your own. Let's look at how to protect you and yours.

When discussing personal and family protection from crime and violence, there are really only two schools of thought; **prevention** and **reaction**. Since we covered "prevention" when we discussed security considerations under "Basic Home Prep," here we'll focus on "reaction." We'll talk about what to do should crime come knocking on your door, or worse, crawling through a window. We'll give you options and we'll cover some basic legal, moral, and emotional considerations of self-defense.

There are two main reasons we included this section in this book. First and foremost, you'll need the information. Secondly, though we fully stand by everyone's right to own a firearm and defend their families, we also feel that there are quite a number of options you can exercise before resorting to a firearm. There's a quote by Abraham Maslow that says, "To the man who only has a hammer in the toolkit, every problem looks like a nail." So it is with weapons. If you only have one option, that's the one you'll go for in a dangerous situation. We want to give you more in your toolbox than just a hammer.

If you want to know the latest on firearms, there are plenty of books and magazines on the subject available at any newsstand, so we're not going to clog book space here. However, if you want to explore some creative options you may have at your disposal, read on. Let's look at a few major considerations that can answer a lot of questions in and of themselves regarding defense.

1. **Use more than an ounce of prevention.** At your house we're sure you have smoke detectors, a sprinkler system, a few dry-chemical fire extinguishers, and the hose that stays connected. If you have a fire we're sure you're ready to fight it tooth and nail. But... all things considered, we're also sure you'd rather not have a fire at all. That's the way we feel about personal protection – we'd rather avoid the situation altogether if possible. If you want to keep the vermin out of your garden, you put up a fence. So it is with social parasites that think they have the right to harm you and yours or use you as their personal supply depot. You do what you can to lock them out to prevent an incident from occurring in the first place. So buy good locks, have an alarm system, be aware of your surroundings, work together with your neighbors, and if trouble starts, **the first thing you should do is call 911** (whether you think they're able to come or not). Also, the things we're discussing here also apply to **protecting yourself while on the road during an emergency evacuation**.

2. **Have options.** Like the quote says, "To the man who only has a hammer in the toolkit, every problem looks like a nail." With <u>some</u> people, if all they have is a firearm, then that's what they'll react with in most given situations. Granted, some situations are instantly grave and some people are physically incapable of many alternative forms of self-defense. When you can though, always have options at your disposal. Anything less than lethal force and certainly anything that <u>prevents</u> you from getting to the level of lethal force should be considered.

3. **Rules of engagement.** Every potential criminal situation will encompass any number of unforeseeable variables, and will start at different levels and escalate from there. Whatever the variables are and however the situation escalates, you need to be able to react accordingly and appropriately. Before we look at specific options regarding force, let's look at general pointers regarding these situations and the use of force or weapons.

    ➢ The deterrence offered by locks and general security comes first and if you get any sort of warning that a situation is about to occur you **call 911**. After that, you may be forced to deal with a situation that requires you to make judgement calls with little to no information beforehand. The rest of these "rules" will hopefully help you make the right decision as to how to react.

    ➢ When dealing with force or violence, the only force you're allowed to use yourself is just enough to stop whatever it is you're facing.   Bluff, threaten, and scare if you can to prevent a situation from getting to the direct confrontation stage. After that, we're assuming you're in a situation where you might be hurt or worse, so cause your would-be attacker pain rather than injure them, injure them rather than seriously damage them, damage them rather than kill them, but resort to lethal force **ONLY** if they are about to kill you or yours. But never, <u>ever</u> get to that level without using everything in your power to prevent it, for their sake and yours.

    ➢ If weapons are included in your list of options, be sure of two things. **One**, that the weapon is legal in your area, and **two**, it's an appropriate weapon for the level of force you're allowed to use. For example, you shouldn't use a knife against someone when pepper spray would do the trick.

    ➢ Remember that whatever you do, you'll eventually be asked to justify it by Law Enforcement, and by your own conscience.

4. **Let's look at some options.** We'll give you a few suggestions based on our little escalation statement above about "threaten rather than hurt..." **Not all of these will be applicable.** However, if you learn them, this knowledge will become part of your "arsenal" and will give you options to choose from when you need them.

    ➢ If you have the opportunity to safely bluff, you should. Notice we said <u>safely</u>. Don't run out your front door to face down an armed gang and tell them to "get off your property or else." That's where statistics come from. Instead, we suggest that from the safety of locked doors you bark a **verbal command** like, "**STOP!!** Cops are on the way and I'm armed!!" or words to that effect.

    ➢ Scare comes next, and **noise** tops the list. If you let loose with a long blast on an air horn, which might well be the alert signal for your neighbors, some less-committed felons will turn and run. **Lights**, if you have them, are next. This could be floodlights outside your house, or spotlights aimed out a window at your potential intruders. Roaches usually run when the lights come on.

    ➢ So far, we've given you a couple of minor options for people **in your yard** while you're enjoying the safety of braced doors. What if they get in? What if they're already in before you know they're there? Confrontation is something best avoided, but unfortunately, avoidance can't be guaranteed. As we said above, your response should be dictated by the threat you face. Let's discuss some "less than lethal" options.

5. **Weapons.** Presented in no particular order are some "less than lethal weapons" you should consider becoming familiar with. These are called "less than lethal' rather than "non-lethal" because they still hold the potential to cause serious damage including death. Though you won't have these with you all the time, becoming familiar with them gives you more "tools in your toolbox."

    ➢ **Unarmed self-defense**. We've recommended a few times that you take a course in unarmed self defense. It's great for keeping you healthy, it will help give you confidence and help you develop that "readiness mindset," and in this case you may need it for its true purpose. If nothing else, the more adept you are at unarmed combat, the calmer you'll be in a confrontational situation, and it's that calmness that will allow you to think through your options.

> **Pepper spray**. Anything that will allow some distance between you and the bad guy is a good thing. You can buy pepper spray almost anywhere, and it's legal almost everywhere. If you can't find pepper spray try keeping a toy squirt gun full of some sort of hot pepper sauce from your spice cabinet. **Note**: If you have to use your pepper spray in an enclosed environment, you'll feel some of the effects too.

> **Electronic stun gun**. Closing the distance gap a little would be the electronic stun guns that temporarily incapacitate an attacker with a safe but high voltage shock. Contact stun guns, the type you actually have to touch to a person, are inexpensive and easy to come by. A much more expensive model called a "TASER" fires darts into an attacker and transmits an electric current through wires trailed by the darts.

> **Baseball bat**. The list of things you could find around the house to use as a bludgeon is a rather long list. You can use heavy bottles, hammers, golf clubs, metal flashlights, figurines, etc. and any one of them can pack a nice wallop. Another impact weapon we found to be rather novel was a crossbow used to fire a "bird bolt." A bird bolt is a short crossbow arrow with a solid rubber ball on the end rather than a point. **Note**: If you have to resort to a bludgeon weapon, try not to hit anyone in the head unless it's a life or death situation. Despite what you see in the movies, a blow to the head is a very serious injury.

> **Kitchen knife**. Knives definitely walk the fence between "less than lethal" and lethal weaponry. The most important thing to learn about using a knife in self defense is that you still have options. You don't have to stab someone in a vital area to end a confrontation. Many times a nasty nick to an extremity is all it takes.

> **Air guns**. BB guns and paintball guns have a popularity all their own, and both are capable of serving in a defensive capacity. Both will let you confront from a distance rather than face to face, both can startle and inflict pain without causing serious damage, and both are relatively inexpensive and plentiful. BB guns are a little higher on the potential injury scale as they will penetrate skin. Paintball guns give you some pretty good options as you can buy balls filled with a dry pepper spray, and others with a dye that will mark your attacker for later ID by authorities. (Both are good against four-legged vermin as well.)

6. **Rules for using a weapon**. Now that we've covered a few weapons as a topic separate from defense and use-of-force issues, we'll give you a few rules that apply specifically to weapons use.

> First and foremost, know when **NOT** to use a weapon.

> Know your local laws regarding use of force.

> Be sure you're able to exercise safety concerning yourself, your family, and bystanders when using your weapon, and know how to use your weapon in its most minimal capacity. For example, with the knife, you should know all the ways to nick rather than stab a vital area.

> Anything you are considering using or keeping as a weapon, become as familiar with it as possible. This is important not only for having options, but for your safety. If a class is offered that teaches you how to better handle any of the types of weapons we've listed above, take it.

> If you use a weapon, you've committed yourself to the defense, so keep up your defenses until the attack has been neutralized. Don't make one small retaliatory move and then drop your defenses expecting your attacker to turn and run.

> Make sure anything you choose as a weapon is legal in your area and legal for you to posses.

> Again, if you use a particular weapon, be ready to justify not only your use of force, but your choice of weapon and the *way* in which you used it.

7. **Your firearms options**. Now that we've touched on the less-than-lethal options, we'd be remiss in our duties if we failed to cover lethal weaponry, in this case firearms. The worst case scenario is one that would see you forced to use a firearm in self-defense. However, you've still got a couple of non-lethal options at your disposal. For example, one of the primary home defense firearms we recommend is a 12-gauge pump shotgun. If someone decides to break into your house and ignores the verbal commands, then the sound of a pump shotgun action being cycled carries its own little deterrent factor. After that, guess what? You've <u>still</u> got less-than-lethal options, even with a shotgun. We suggest making the first two rounds in your shotgun less-lethal rubber buckshot or beanbag rounds. Less-lethal ammo aside, your other "option" is marksmanship. It's much better to shoot to "wound and stop" than to kill. Forget what you've heard about shooting to kill in order to not be sued. It's easier to defend an injury suit filed by a violent social parasite, than to fight a wrongful death suit filed by the deceased parasite's "grieving family." In today's society, you're going to be sued regardless, so go for the lesser of two evil suits. More important is you having to live with having killed someone. Your physical safety and wellbeing comes first, and if that can be achieved while protecting your psychological wellbeing, then so much the better.

8. **Firearms suggestions**. As mentioned, one of the primary home-defense firearms we recommend is a 12-gauge pump shotgun. It meets numerous requirements for non-lethal versatility, user safety, reliability, and economy, and it fills equal roles between home defense and hunting. Our second choice, or our choice for a companion

firearm, would be a rifle or handgun chambered for the .22 "long-rifle rimfire" cartridge. Firearms ownership is your right, but it is also a personal choice, that only you can make, as to whether or not you want to include any in your emergency equipment. Therefore, we'll end our firearms recommendations here. Some of the sources listed below will allow you to continue your own firearms ownership research should you wish to continue.

9. **General firearms considerations.** Firearms ownership is a right as well as a responsibility, and it's also a personal decision. We would never insist that someone who's never owned a gun before go out and buy one just to "have." There are a lot of things to be taken into consideration in such decision, and here are but a few:

➢ The most important thing to learn with firearms is when **NOT** to use one!

➢ Make sure you know the firearms laws in your area and that you are in compliance.

➢ If you ever question whether you are even-tempered enough to keep a firearm, don't get one.

➢ If a safety class is available to you take it, even if you're an experienced owner. We can all use refreshers, especially where safety is concerned.

➢ If local Law Enforcement puts on a class take that too. They'll have good material regarding legal aspects of firearms usage as well as good info on how to properly deal with officers arriving on scene where you've used a firearm in self-defense. (**Basic rule**: If possible, never have a gun in your hand when Officers arrive, even if you're holding someone at gunpoint. Though you're the good guy, Cops won't know who's who.)

➢ If you have children in the house, their safety is paramount. Numerous gun safes, locks, and other devices are available that will keep your firearm inaccessible to children, but available if you need it.

➢ Next is bystander safety, and choice of ammo comes into play here. Buy appropriate ammo so if you ever have to shoot, your rounds don't go meet the neighbors in their house. We suggest the following for safety:

   ● For shotguns, our first choice is the less-than-lethal rubber buckshot and beanbag rounds. Following that would be number 8 birdshot as it is rather effective for hunting and defense at close range, but does not penetrate walls well. This means if you miss, your shot is less likely to wind up in the kids' rooms.

   ● For defensive handguns we recommend frangible ammo such as "Glaser Safety Slugs." These bullets disintegrate on contact with plaster walls which reduces their penetration into the wrong rooms if you miss.

➢ Though many people place their main emphasis on marksmanship, we give it a slightly lower rating. We feel you should train in safety first and then marksmanship. Marksmanship is important though for three reasons. **One**, the more you practice marksmanship, the more you handle the weapon and become familiar with it. This will increase your respect for the weapon's power and thereby the safety you exercise. **Two**, good marksmanship will give you the option of wounding to stop, as we mentioned earlier, rather than resorting to potentially fatal shots. **Three**, fewer misses means fewer innocent bystanders at risk. **Hint**: Fewer shots mean the same thing, so never shoot unless you absolutely have to.

➢ If you're going to a public shelter, or utilizing any sort of public transportation in an evacuation, then you do not want to take along any long guns such as a rifle or shotgun. Though you're one of the good guys, the only thing you'll do is make Cops and fellow evacuees rather nervous.

➢ **Again, firearms are a LAST resort!**

Sources and More Information

➢ For a wide variety of gun safety info and publications, be sure to visit the Americans for Gun Safety sites at http://ww2.americansforgunsafety.com, and at http://www.agsfoundation.com. You can also contact them at Americans for Gun Safety, Washington, DC, Phone (202) 775-0300, Fax (202) 775-0430.

➢ One of the best sources for firearm safety information is the National Rifle Association. They're online at http://www.nra.org, or you can call 1-877-NRA-2000 (877-672-2000).

➢ A variety of information on confrontational defensive situations can be found at http://www.stormpages.com/handtohand22/page5.htm. As with any compendium site, take everything with a grain of salt and use only what applies to your personal situation.

➢ A variety of useful information and assorted products can be found online at the Women's Self Defense Center at http://www.womensdefensecenter.com.

➢ For rubber buckshot and the like, see FireQuest Ammunition Catalog, at http://www.firequest.com, or contact FireQuest International, Inc. P.O. Box 315, El Dorado, AR 71730. 870-881-8688 or 8488.

➢ For the "Wing Chun Today" Kung-Fu video training series by the International Martial Arts Hall of Fame inductee Master Jason Lau, log on to http://www.jasonlau-wingchun.com.

Though fire fighting is a serious subject, this is actually going to be a rather short section. Why? Because it's mostly a list of what NOT to do, coupled with the few things that you CAN do in order to fight a fire safely. We've already covered the subject a great deal under "Basic Training" and when we discussed equipment in the "Fire" portion of "Basic Home Prep." We added to the mix with the "Fire Drill" and general "Fire" portion of the "Reaction" section where we covered what to do if you were caught up in a fire that you could not fight.

Now that you're in an isolation situation, where it's likely that help cannot come for whatever reason, let's discuss the options you might have at your disposal for fighting a fire, or at least keeping one in check while you wait for whatever help is available. As we'll constantly remind you, ALWAYS call 911 even if you think they can't come.

Just as we did with crime above, we'll start with a few general rules about what to do and we'll give you some information that allows you to react appropriately as a fire escalates.

➤ Remember our **F.I.R.E.** acronym from earlier. **Find** everyone in the house, **Invite** the Fire Department (even if you think they can't come), **Restrict** the flames by closing doors, and **Evaluate** quickly, as to whether you can **Extinguish** the flame or if you should **Evacuate** entirely.

➤ Examining that a bit more, you'll note we started with people first. People take precedence over property, and insurance will take up where you left off as far as your house and belongings go. Another thing to point out with the above is inviting the Fire Department. **Regardless of whether or not you think they can come, you should call 911**. Any time any fire looks like it might be more than a simple pan or trashcan fire, or if you think it has any kind of potential for getting away from you, then you call in the pros. Don't let a fire get out of hand because you *thought* you could handle it by yourself. (Reading this section won't make you a professional Firefighter any more than our section on First Aid will make you a doctor.)

➤ If we could have you do any one thing in your self-paced firefighting training, it would be for you to **develop a healthy respect for fire**; its unpredictable nature, the damage it can cause, the difficulties involved in controlling one, and the danger to the one fighting it. If you ever have the opportunity to attend a live demo hosted by your local Fire Department, go. Or, for a smaller lesson, provided you can do this **safely**, build a small bonfire or campfire, let it get going good, and try to put it out with just your garden hose (but **only** do this with assistants present, while obeying local burning and fire laws, and certainly while using common sense). See just how difficult it is to extinguish the fire and to *keep* it out.

➤ The next consideration of any fire situation is **speed**. The only way you're going to win against fire is with early detection and rapid reaction. Your fire drills will help with people reactions, but what about your reaction speed with your equipment? How quickly can you get to your fire extinguishers? How about your hoses? What about the gear you have outside in your fire trunk? **Practice** is the key to all of these. We suggest that when you have family fire drills that you extend them on occasion to include equipment drills. Grab an extinguisher, drag out the hose, or pull out the fire trunk and see how quickly you can get into it. Remember, **a fire will double in size every thirty seconds**.

➤ Fight simple fires with simple methods. For example, if a pan of grease on the stove flames up, you don't need to grab the extinguisher and dust the whole kitchen. You should simply put the lid on the pan or dump in some baking soda that you should keep handy for these occasions. Keeping in mind our earlier discussion of types of fires and what to use and not use on them, let's look at some simple reactions so they can become part of your arsenal.

● Again, for grease fires, keep the **pan's lid** handy along with some **baking soda**. This should be your standard operating procedure when doing any open pan cooking. Remember to never use water on a grease fire as you'll only spread the flaming grease as it floats on water.

● For simple class A or paper or fabric fires such as a wastebasket, your kitchen trashcan, etc., start with the first available **liquid**. Are you drinking anything? Toss it on the flame. How close are you to the sprayer on your kitchen sink? You can safely drag a flaming waste basket (provided the flames are really low) **backwards** toward the sink and just spray the flames out.

● **Hint**: It's never a good idea to pick up a flaming pan, or a flaming *anything* for that matter. However, if unusual circumstances dictate that you absolutely <u>have</u> to pick up a flaming item and move it, make sure you have good hand protection, and that you hold the item **behind you** as well as possible (or **walk backwards** with someone's help) and move it slowly but steadily. Holding the thing in front of you and walking forward will only fan the flames towards you.

● Other simple fires are easy to **smother**. Let's say an overturned candle has set your fabric tablecloth on fire. You can turn up the edges of that same cloth and smother the burning part, or you can lay a book, a jacket, floor mat, place mat, or other **less flammable** object over the burning cloth and gently **press** out the flame. Notice the two things we said; *less flammable object*, meaning don't use anything that will only make the fire bigger, and *gently press the flame out*, meaning **don't swat** at the flame as you'll only fan it and make it bigger. Lay your item on top and let it smother the flames, just as earlier we told you to roll slowly if your clothes catch on fire and you have to do the "stop, drop, and roll" thing.

- Another good tool for smothering small flames is a **wet mop** or a broom with a wet towel wrapped around the end. Remember to press and smother, not swat and fan.

➢ As fires get a little bigger and you want to keep your distance, your fire extinguishers will come into play. After choosing the right one for the job, dry chemical for flammable liquids and electrical fires, and water for flammable solids, you'll want to know the **P.A.S.S.** method. This is the acronym (not ours) for remembering how to use an extinguisher. You **P**ull the **P**in, **A**im the nozzle at the base of the flames, **S**queeze the lever, and **S**weep back and forth across the base of the flames to put out the fire.

➢ Let's discuss a very important point concerning the "**base of the flames**." Typically, the base of the flames refers to what's actually burning. If you were burning a pile of leaves in your yard and the flames were about six feet high, the base of the flames would be the burning leaves and not the top of the flame. This is important because we've seen quite a few people simply point an extinguisher towards a fire and discharge the unit hoping the fire will magically go out. Doesn't work that way. You have to aim your extinguisher at what's burning so the maximum amount of chemical or water will hit directly on the source of the fire.

➢ Let's change our example. What if your drapes catch on fire. Would the base of the flames be the bottom of the drapes near the floor? Nope, it would be whatever portion of the drapes was actually burning. You'd want to sweep your extinguisher up and down the drapes aiming directly at them so as to put the maximum amount of chemical on the fire source, or as Firefighters are apt to say, "Put the wet stuff on the red stuff."

➢ This drapes example leads us to another concept and that's "**containment**." Suppose your drapes are on fire, the flames are licking the ceiling, and also starting to catch the carpet on fire. You already know that the base of the flame is the drapes themselves, but what of the items that are *about* to catch on fire? You'll want to contain the fire, or keep it from spreading, by including all the **flame contact points** in your extinguisher or garden hose sweeping. In this example, you'd sweep up and down the drapes, sweep a little on the ceiling that might catch, and also on the carpet. Keep going back and forth, working like you're trying to sweep all the fire back to its point of origin. In our burning leaves example above, if the flames were about to set an overhanging tree branch on fire, you'd spray your hose on the pile of burning leaves on the ground, then a little on the tree that might catch, and then back on the burning leaf pile.

➢ Speaking of containment, don't let the fire contain you. Whenever you fight any fire inside the house, whether with an extinguisher or the hose behind your washing machine, remember this rule: ALWAYS POSITION YOURSELF WITH YOUR ESCAPE ROUTE IMMEDIATELY ACCESSIBLE. In other words, fight the fire while backing out the door. Never allow any risk of being trapped by fire. If fire was paint, we'd be reminding you to not "paint yourself into a corner." Always fight a fire from the vantage of your escape route.

➢ The concept of containment is a close cousin to the concept of "**restriction**." In this case you may be letting a fire burn, but keeping it from spreading, in effect restricting it. You might also restrict a target such as when you prep your house for a wildfire. You're not going to try to put out the forest fire, but you'll isolate your house with water sprinklers and the like so it will be isolated from the fire and not burn. Another example of the restriction concept can be seen in the "bucket brigades" of old. We've all see westerns and other movies where an old wooden building catches fire and everyone forms a line passing buckets of water up to the front where they could be tossed on the fire. Actually, much of that work was intended to keep the fire from spreading. Little could be done back then with just buckets, regarding an established fire in a wooden building, so folks did what they could to keep the neighboring structures from catching fire.

➢ At this point, you may well be using your garden hose (the one you attached behind your washing machine maybe) to fight the fire. If so, be sure you are staying low to the floor, fighting the fire from as far away as possible, and have a quick exit just behind you. Actually, the best thing you could do here is lock your hose nozzle on a medium spray, aim it toward the fire, and hold it in place on a chair or piece of furniture somehow, so you can leave while letting the hose continue to do its job.

➢ If our interior house fire examples have become large enough to cover *containment* and *restriction*, it's time to tell you to get out of the house. This is one of our more important points: **ANY TIME A FIRE CANNOT BE READILY EXTINGUISHED BY THE METHODS YOU HAVE AT YOUR DISPOSAL, IT'S TIME TO GET OUT OF THE HOUSE AND CONTINUE ANY FIRE FIGHTING EFFORTS FROM OUTSIDE.** In other words, the best thing you can do is run away. It's okay though, we're not retreating, we're about to attack from a different direction. We're assuming of course, that you've already moved everyone else out of the house well before this point. (Remember, we covered exterior fires under "**Reaction**.")

➢ Before running out the door though, let's make sure you've done a couple of things or that other family members did them on *their* way out the door <u>earlier</u>.

❑ All exterior windows should be closed. (But never go back in the house to close them.)

❑ All interior and exterior doors should be closed but <u>not</u> locked.

❑ The heating and air system (HVAC) should be turned off (or the electric meter yanked from outside if you don't need electricity to pump your fire-fighting water).

❑ The gas should be turned off at the tank or at the meter (provided you can reach it and return safely).

❑ All family members and pets should be accounted for via the buddy system, and their safety double-checked by the heads of household.

➢ Now we get down to the nitty-gritty, because this is where you're officially fighting a large fire. At this point there are two important points to make.

❑ First, and most importantly, at this point you should be working with a partner, either a capable family member or a neighbor. This is one place the buddy system rules.

❑ Secondly, at this point, your dry chemical extinguishers and garden sprayers full of water will be totally ineffective. From here on, it's the garden hose as a bare minimum.

➢ Now that we're outside and possibly about to tackle a sizeable fire, let's go over a few general rules that will prove helpful at this point.

❑ If the fire was big enough to force you out of the house, **stay out** unless human life is in danger. Everything else can be done from outside. After all, you have your "fire trunk" full of extra gear, right?

❑ Fight the fire from as **far away** as you can possibly get. Even in the normal house there's a danger of explosion from heated air, flammable gasses from heated wood framing, temperature interchanges, backdrafts and backflashes, and a host of other potential dangers.

❑ Wear as much **protection** as you can. Your fire trunk should have a hardhat, coveralls, goggles or protective mask, gloves, and a respirator. With all the heat (which has to be felt to be believed) and dangers from smoke and potential explosion, you'll need more protection than this, but these are the minimum.

❑ Add any sort of **shield or barricade** you can to protect yourself from heat and possible shrapnel should something explode. Stand behind a tree, a vehicle (but not if it catches fire), a wheel barrow, a wet piece of carpet, a wet mattress, or if nothing else, use a metal trash can lid as a shield. Put as much as you can between you and the fire. Protect you first!

❑ Pay attention to secondary threats. Which way is the wind blowing? Toward you? Is the fire about to reach a room containing potentially hazardous or explosive materials? Has the fire spread outside the house to reach vegetation that might start burning around you? Are you doing anything that might block the arrival of Fire Department vehicles or personnel?

❑ We mentioned working with a **partner** above, because they'll be helping you in a couple of ways. First, if you're the one close to the fire using your hose, they might stand further back and have a hose trained on you to keep you cool if needed. Next, they should keep an eye out for additional hot spots breaking out by watching for new smoke and flame locations. Additionally, they're a more objective participant than you and may well protect you from doing anything more dangerous than what you're doing already.

❑ A partner, especially a second partner, can also be helpful if they've got a **second hose** coming from another water source. This gives you two water streams and twice the volume of water. You can spray from different angles to cover more area of a burning room, or you can focus on two hot spots at once.

➢ The only remaining question under this "things you can do" portion is to outline some actual fire fighting techniques, which given the situation of being an almost-solo civilian operation, is rather limited. However, there are some things you can do:

● While reading everything below, remember that the three components of fire are oxygen, fuel, and heat, and that removing one of those elements (or better yet, two or all three) will put out the fire.

● You've closed all the windows, you're outside the house, and you want to fight an interior fire. Remember in your "fire trunk" how we suggested you keep a brick or rock with which you could break a window? This is what it's for. You want to be able to breach your window in some way and you won't have time to waste looking around the yard for a suitable projectile. So, put a hole in your window with your brick. **Note**: Try to break only one pane of glass. You don't want too much air coming in and feeding the fire. You only need a hole big enough to allow a stream of water from your hose. Nothing more, nothing less.

● Throw your rock from as far back as you can. When the window breaks, **duck**. Fresh air may cause a backflash which could result in a small explosion throwing flame and your broken window right back at you.

● Open up a steady stream of water into the room and aim it as best you can in the direction of the flames. Remember to contain the flames by aiming water at the base of the flame and the areas that might catch fire next. Move you from side to side in order to change the angle of the incoming water. Don't get any closer to the fire than you have to. **Note**: Always remember that your garden hose is far superior to a portable dry-chemical fire extinguisher, but it's nowhere near as effective as even the smallest hose on a fire truck.

● One way to keep your distance is to attach a sprinkler head to your hose, turn on the water , and then lob the sprinkler into your burning room. You might have to twist the hose to turn the sprinkler in the right direction, but you can walk away and let the sprinkler hose the room. This is even more important for upper-floor windows, or windows with drawn curtains or blinds. You don't want to be up against the window or wall trying to hold your hose, especially while standing on a ladder. Any number of things could happen including the wall or roof collapsing on you. Either use a sprinkler and toss it in the room, or tape your squeeze-adjustable hose nozzle to the end of a broom handle, hook it to the window opening, and walk away.

- In fact, any time you can train a hose on a fire through an open window and then walk away from the hose it's a good thing. You can tape your nozzle to a stick and set it in the ground, you can hook it to a tree, you can hook the nozzle to the top of a ladder and lean the ladder against the window where the hose is spraying into the room, or you can do any number of things limited only by safety and by your imagination. The key concept here is to get water on the fire while keeping you at a safe distance.

- If your fire is in a room with no windows, you may need to breach the wall. That's where your fire trunk's axe and sledge hammer (for brick or block) come into play. The average home's walls are thinner and more fragile than you might think, even with brick. In fact, if you had some sort of spike, like the pointy end of a pick-axe, you could poke a hole right through most walls and aim your hose through that. Make only a small hole, set your hose up to spray through that, and walk away if you can. Let the hose do what it can while you get away from a wall that could collapse on you.

- Fortunately, this is even easier to do in mobile homes due to their thin wall construction. We say fortunately because mobile homes tend to have fewer windows and also tend to burn more rapidly in a fire. You'll have to "breach and drench" as fast as possible.

- If the room you're spraying all this water into has a lot of electronics, you might want to have someone turn off the power either at the meter if there's a switch, or by yanking the meter if you don't need electricity to pump your water. You don't want to be hosing live wires that could send a current back up your water stream and electrocute you.

- As you're doing all this, **stay low and keep your distance**. Remember the danger of explosion, from collapsing walls, the danger from heat, and even though you're outside, there's still a danger of smoke inhalation as there will be LOTS of it. If smoke starts to get thick, tie a rope around yourself and have a partner stand farther away holding the other end. They can pull you out of the way, or you can follow the rope if you can't see as you're leaving the area.

- Leave yourself an **escape route**. Professional Firefighters are taught to follow their hose to get back out of a building if they get caught up in a dangerous situation. Following your hose won't help you if it's connected to the house. This is another reason we mentioned the buddy with a rope tied to you. You might also want to tie a rope to a tree to lead you back in a safe direction. In any event, make sure you have a way to get out of danger should a situation take a turn for the worse.

- Watch for any flame spreading to outside vegetation. This could start a small grass or tree fire that could spread and endanger you or your neighbors. This is yet another good reason to have a buddy with you.

- Keep pouring in as much water as you're able and keep an eye on the fire's progress. If it looks like you're winning, don't stop. Keep pouring on water since you need to make sure the fire is out and you'll want to cool the items that were burning so they don't reignite.

- If the fire is starting to spread, your next decision is whether to turn on your "Poor Man's Sprinkler System" valve over that room.

➤ If the fire is one that is burning "outside in," such as a brush fire approaching your house, or your neighbor's house is on fire and threatening yours, revisit the things we told you in the "Fire" portion of "Reaction," on how to prep your house and yard for protection. You won't have time to do it all, but the advice about rapid landscaping and roof and yard sprinklers should help.

➤ Also in this situation, remember the concepts of containment and restriction. Your neighbor's house fire might be too far along to fight, but you have to do what you can to keep it from spreading to your house. Safety first though. If the neighbor's house is fully involved, if the wind is blowing toward your house, or if you're really feeling the heat, then the best thing you can do is get out of the way.

➤ To help you remember some of the above concepts, we have yet another acronym for you. Some people say you fight fire with fire, but we say you fight fire with **W.A.T.E.R.**:

Water	For safety's sake, make sure you always have a source of **water** for fighting fires.
Assistance	The buddy system rules. Never do anything alone if you can find an able **assistant**.
Threats	Be aware of all **threats** such as heat, smoke, propane tanks, collapsing walls, etc.
Escape	Always leave yourself an **escape** route. Don't get trapped in a dangerous situation.
Reason	Above all, exercise **reason**. People come before property. Stay safe and don't take risks.

Guess what? That's about all you can safely do with the limited personnel and equipment at your disposal. From this point on, let's assume you won and doused the flames. We'll continue with a few important points on what to do next.

➤ Just because the flames are out, don't think it's all over. Some objects can stay so hot they bake off the water that had smothered the flame and they reignite. Flame can also "burrow" into small areas that you can't see and smolder until big enough to start a new fire. Keep a steady watch for at least twelve hours, looking for new smoke or flame. (We're still assuming something has prevented the arrival of the professional Firefighters you called when the fire started.)

➤ After that, if there seems to be minimal structural damage, you might open exterior doors, or a window or two, and set up exhaust fans to draw out the remaining smoke. Perform these steps from outside if possible, and don't forget that remaining smoke and fumes can hurt your lungs. Don't expose yourself for too long.

➤ The best thing to do is stay out of the house overnight, or for at least twelve hours. After that, you still don't want to go in if there's any structural damage to the roof system, walls, or floors. Wait until your isolation emergency is over and professional firefighters and building inspectors can inspect your property.

➤ If damage was minimal enough that you're sure there is no structural damage, you might make limited trips into the house. Remember to maintain a "reignition watch" for at least twelve hours.

➤ Once you're able to venture in and out freely, you'll want to follow our guidelines for cleaning up a damaged house, as outlined under the "Evacuation" section's "Coming Home" portion.

➤ As you're cleaning up, be sure to document all the damage and all your losses. Though things are in turmoil in this isolation situation, insurance companies and other services will be back online eventually. Make sure you're ready for them and "prepared for phase two."

Sources and More Information

➤ Take a look at the Cold Fire ® fire extinguishing compound. Visit FireFreeze Worldwide, Inc., online at http://www.firefreeze.com, or contact them at 272 Route 46 East, Rockaway, NJ 07866, Phone: 973-627-0722, Fax: 973-627-2982. More info is at http://www.greaterthings.com/News/ColdFire/pr031122.html.

➤ Most of the other outside sources you might need for this section were listed either under "Basic Training" or "Basic Home Prep's" "Fire" section.

Cabin Fever

Well, you're tucked away in the house, you're fed, watered, fairly well protected against the elements, and we just helped protect you from criminals and fire. Ready for your biggest enemy? Well then, go look in the mirror. You and your own emotions are going to be one of the biggest hurdles you'll have to clear in an extended isolation. Though many of us are home bodies, we still hate being confined. In fact, the only segment of the population that will be happy with an extended quarantine are agoraphobics.

Granted, having your physical survival needs, your personal safety, and your creature comforts needs met play a huge part in being able to handle confinement. However, this subject is just like any other and there are large advantages in the smallest detail. Let's look at some of the detail involved in helping you handle "cabin fever."

➤ Just as with an evacuation, stress is going to take its toll on you. Your best bet is rest and being able to recharge your batteries. Remember what we said under "**Evacuation**" concerning mild sedatives, eye shades, and ear plugs.

➤ If nothing prevents it, you should continue with light exercise (except in severe heat). It will give you something to pass the time, and exercise will work off some of your stress.

➤ Aside from stress, boredom is going to be major player in how you weather your situation. It's not easy staying inside for three weeks straight. We suggest you stock up on the following things now and store them out of the way with the rest of your emergency isolation goods:

❏ Video tapes of movies or TV shows that you have not seen. Tape them but don't watch them. Save them for entertainment during your isolation.

❏ Store books and magazines that you have not read. Remember to make them positive, uplifting, and inspiring titles and subjects.

❏ Store coloring books, crayons, puzzle books, etc. for the kids. Store lots of them. The kids will be the hardest to pacify, and even the adults would like a good crossword puzzle or two.

❏ Pack away some board games the family has never played, but make sure they're not the type to cause arguments. Some games are entertaining, some draw out competitive natures.

❑ Get an extra walkman, or portable CD player and new tapes or CDs to go with it. You never know if the old players are going to be broken at the wrong time. Make sure you have plenty of batteries to run them.

❑ Get a few new game cartridges for any game system you may have.

❑ Keep a blank notebook and pens and pencils. You'll need them for everything from Tic-Tac-Toe to keeping a journal of your isolation experiences. Who knows? We might publish what you have to say.

❑ If you have a favorite hobby that can be enjoyed quietly at home, remember to stock extra supplies for it. Keep these supplies tucked away so they'll be there if you need them to fight cabin fever. Some examples are knitting, needlepoint, painting or drawing, scrapbooking, etc.

❑ When gathering all this entertainment, keep a fair balance between group and individual activities. You'll all want to be together at some times and off in your own little world at others.

❑ Also, balance all these items between those that need electricity (or batteries) and those that don't. You don't know what's going to be working.

➤ Plan for your vices. Cravings are powerful forces and the denial imposed by captivity will only make them grow stronger. Plan now to satisfy any urges or desires that you know will present themselves in isolation. For example, if you knew Grandpa would go bonkers without his favorite Scotch, then by all means, pack away a bottle of Scotch. We touched on this earlier when we suggested you store some of your family's favorite junk food in the freezer. In addition to the freezer items, consider stocking the following in various hiding spots around the house:

❑ Tobacco products   ❑ Adult beverages   ❑ Nuts or other snacks
❑ Chocolate and hard candy   ❑ Birth control   ❑ Coffee

➤ Continue your normal life as well as you can. Consider these:

● Earlier we asked you to establish a plan for telecommuting with your job if that was an available option. This is where you'd need to put that option to use.

● We also urged you to see if your children's schools had any sort of lesson-continuity plan in place that would allow them to keep up with their studies during any disruptions. This is the reason we suggested that during the higher alerts and severe weather seasons that children bring home all school books every day.

➤ As a final note on keeping your emotions on an even keel, think how rough things would be for you had you not made all these preparations.

Wrapping it Up

We started this book by telling you there's genius in simplicity but that simplicity was borne of information, pared down by education, and customized by practice.

Now that we've run you through the "**Foundation**" and its Basic Training, the "**Reaction**" section with its numerous drills and responses, all the "**Evacuation**" steps, and finally "**Isolation**," you should now have all the education to put your family's plans together. Most plans are certainly completed by now, but we wanted to give you a final list of things to do in order to consolidate your plans before we left the individual and family portion of the book and started work on group activities.

1. Write down all the plans you create including drills, emergency reactions, safety and protection locations, rendezvous points and evac procedures, and all steps necessary for beginning an isolation.

2. When formulating your various plans know all your available assets and options, but keep the number of reactions within reason focusing on the most useful or probable options or reactions. Options are good, but too many may cause confusion.

3. Find the best ways to teach all of your plans' aspects to your family. Use drills, repetition, text, acronyms, poems, short stories, pictures, and so on. Use different methods as different people learn things in different ways.

4. Discuss your plans with family members now and then, especially in severe weather seasons, terror alerts, or when news stories remind you.

5. Follow these discussions with a drill or practice session.

6. Discuss the drill afterward, searching for improvement or necessary changes to your plans.

7. Record your notes and changes to your written family plan.

It's this constant planning and practice that will keep your family as prepared and as safe as it possibly can be. Now that your family is ready, what about the others around you?

In this final "**Isolation**" section we've mentioned neighbors and the buddy system. We also reiterated the need for advance planning with jobs and schools. Throughout the book up to this point, we've pointed out the numerous benefits of working with others and reminding you that 99 percent of the people around you are just like you and will be in the same predicament when the next disaster or attack comes down the pike.

Since we've spent so much time *mentioning* and *hinting*, let's take things to the next level and explore "**Cooperation**" where we push networking, teamwork, and mutual protection to the max.

Notes:

VI. COOPERATION

"We cannot live only for ourselves. A thousand fibers connect us with our fellow men." -
Herman Melville

✓	Goals of the Cooperation Section		
	Activity	Date Completed	Pg.
	Forming or Joining a Group:		378
❑	The **basis, type, and purpose** of the group has been established.	___/___/___	
❑	**Internal organization** has been agreed upon and set.	___/___/___	
❑	**Incorporating** as a **non-profit organization** has been completed.	___/___/___	
❑	**PR and fundraising** methods have been discussed and implemented.	___/___/___	
	The Group in Action:		394
❑	The **specific functions** the group will perform have been identified.	___/___/___	
❑	A group **website** and a **Yahoo Group** have been created.	___/___/___	
❑	A group phone tree has been organized.	___/___/___	
❑	"**Skill and Asset**" lists have been completed for all members.	___/___/___	
❑	**Sources for project-related material and equipment** have been located.	___/___/___	
	Training and Equipment:		414
❑	All members' **home emergency preparations** have been completed.	___/___/___	
❑	**Basic training** needs and instruction sources have been identified.	___/___/___	
❑	**Individual member equipment** needs have been identified and acquired.	___/___/___	
❑	**Advanced group equipment** has been identified and acquired.	___/___/___	
	External Liaison:		418
❑	**Larger volunteer groups** with **similar purpose** have been contacted.	___/___/___	
❑	Regular **communication channels** with **local authorities** are open.	___/___/___	
❑	All **aspects of governmental liaison** have been reviewed and discussed.	___/___/___	
	Goal Setting:		A-1
❑	Entered **start dates and goal dates** onto **calendar** for all the above.	___/___/___	

Groups; to Form or Join

As we've stated throughout this book, the primary responsibility for your safety and wellbeing rests squarely on your own shoulders. Various agencies will be there to help with disasters, but even for the best of them, help can only go so far or be there so quickly. Everything else will be up to you and to those around you.

By "those around you" we mean the people you associate with every day whether it's family, neighbors, coworkers, fellow students, whomever. You may need to rely on them, and they in turn rely on you. You're always around somebody to some degree or another, and one of our highest civic duties is to make sure we help each other. This is in no way a slight to our very brave and dedicated Fire Departments, Law Enforcement, EMTs, or other emergency workers, but sometimes in an emergency, our true <u>first</u> responders are neighbors and bystanders.

The desire to help others is an innate quality in the vast majority of us. In fact, most of us would like to do far more than just helping the neighbor on either side of us, or the occasional stranger in distress. This is evidenced by the fact that after a major disaster gaining national attention, one of the most common sources of stress for those not directly involved is the fact that they can't actually <u>DO</u> something to help.

This "**Cooperation**" section will help you get set up to help others.

Goals of Groups; to Form or Join:
1. To stress the **extreme importance of working with others** during an emergency.
2. To help you **form either a small and informal group, or a large networked organization**.
3. To assist your group in **picking a focus and direction**.
4. To ease your group through its difficult first stages of **forming, filling, and funding**.

We've divide this section into the following categories:
1. **Base, Function, and Scope.**
2. **Form and Organization.**
3. **PR and Fundraising.**

First, let's look at a few reasons why you'd want to either form or join a group:
1. **Patriotic duty**: There's no higher civic duty than neighbor helping neighbor.
2. **Mutual protection**: There's safety in numbers, even if it's only the "buddy system."
3. **Support first responders**: In any kind of large-scale emergency, official first responders will be temporarily overwhelmed. Any time a person or a group can look after themselves, that frees up emergency workers to help others. Also, as more first responders are called up to serve in other areas or positions, volunteer groups will be needed more than ever to help fill local gaps.
4. **Economy**: A pooling of resources means less of a burden on the individual or on first responders.
5. **Prevention**: The best way to deal with an emergency is to stop it before it starts. If your group stops a terrorist, or even a criminal, before he can strike, all the better.
6. **Speed**: Neighbors are usually on the scene before officials. If these neighbors are organized and trained to give actual assistance, that can make all the difference in the world.
7. **Genuine effort**: Many volunteer groups exist, but some exist more on paper than in actual members. Having a local group, even though liaisoned with others, means actual people in your area ready to assist after a disaster.

If you'd rather join a larger preexisting group or form a loose relationship with neighbors, you don't have to worry about the following organizational and business portions. Skip down to the section on "**Preexisting Organizations**" under "**External Liaison**" for a listing of groups and more information.

For those of you wanting to form your own group, let's move on to the actual steps you need to take in order to become an established entity.

Base

Is your group going to be a loose collection of friends and neighbors committed to helping each other? Or is it going to be a nationally networked volunteer organization? Maybe something in between?

Deciding who you're going to help, and how, is your first decision. This will overlap with other steps, but for now, let's discuss the concept of "base," or the people that will compose your group.

The kind of group you become a part of will probably be dictated by your associations and with whom you spend more time, or with whom you're closer, either geographically or emotionally. Let's go over some types of organizations to give you a better picture of the various ways you can group together and help each other. You may want to consider being a part of more than one group. Why? Because of the two universal rules: One, "redundancy is our friend," and two, "you never know where you're going to be or what you're going to be doing."

Here are some examples of the more common **bases for volunteer groups**, and a couple of examples of the functions they may provide. A more detailed discussion of function and goals will be found later under "**The Group in Action**." Read through these to decide what kind of group you might like to be a part of, or help form.

> ➢ **Neighbor helping neighbor.** This is the simplest and most common type of group as most people's attention is focused on their home and home area. For a neighborhood group very little needs to be formally organized unless you were forming an official watch group and wanted a relationship with your local Police Department. A neighborhood group could be anything from a Neighborhood Watch group (which we'll cover in more detail under "The Group in Action"), to a "disaster-only" covenant where neighbors relayed emergency alerts (phone tree), helped each other in evacuations, or helped clean up debris or shared utilities after a destructive event. A "neighborhood" could be defined as a residential area, a block or section of a street, an apartment complex, a condo owner's association, etc.

> ➢ **Houses of Worship.** Most religious organizations tend to be close-knit groups that look after each other anyway. Adding a few disaster-related functions to existing projects and programs should be an easy task. These programs or projects could be for members only, or could be expanded into a neighborhood, community, or city-wide project. More on that under "**Scope.**"

> ➢ **Civic Organizations.** Every town has several civic organizations. These groups are formed around other ideals besides emergency reaction, but they do offer the potential to provide assistance to their members or the surrounding community. We may also see a grouping based on non-local memberships such as AARP members forming various Advisory Boards or Think Tanks to help their local municipalities.

> ➢ **Purpose or function.** Many groups are organized purely by what it is they intend to do, and not by geographic boundaries or personal or business relationships. A national example is the Red Cross whose purpose is aid, comfort, and relief following any disaster in any location. Under "**The Group in Action**" we'll give you quite a number of different function and project ideas.

> ➢ **Work.** The company you work for could also be the basis for an emergency-mitigating organization. This "group" could be the whole company, or just a small group of employees banded together. The goals of a business group might be employee protection, business continuity, or public service to the community.

Leaving this intro list, and before we get into "Function," let's take a more detailed look at how work or a business could be the basis of a volunteer group, since this will be an isolated topic in this volume of Disaster Prep.

The Business or Workplace as the Volunteer Group Base

We've long held that the family was the basic unit as regards emergency planning. We've also felt that neighborhoods should band together in times of need. However, there's something else we need to consider. Most of us are as close or closer to our coworkers than to our neighbors. Therefore, in times of disaster, it might be more natural to band together with our fellow employees. Also, depending on the timing or location of the disaster, you may well be forced into a survival situation with your coworkers.

What if the corporation or business was the "community," and was the main focal point for emergency planning by and for employees and their immediate families?

The company, corporation, or business may well want to become the center for disaster preparedness planning, and set the wheels in motion to protect its extended family of employees should disaster strike.

Let's provide some food for thought on how a business could operate as, or with, a volunteer group:

1. **Volunteer Involvement:**
   - The company can be its own volunteer group with either internal or community-oriented goals.
   - The company can encourage its employees to be part of independently organized volunteer groups.
   - The company can organize its own volunteer group utilizing both employees and outside volunteers.
   - In lieu of people, the company can contribute goods or services to volunteer groups, or can pledge to provide goods or services to government first responders in the event of an emergency.
   - The company can offer its building(s) (portions thereof, an unoccupied warehouse, etc.) as an emergency shelter to its own employees or as a public shelter depending on the timing and affected area of a disaster.
   - Through the use of phone trees, emergency generators, radio equipment, banners, sirens, pennants, and red spotlights, the company can utilize its office building(s) as relays for local Emergency Alert Systems.
   - The company can sponsor first responder type training for its employees to aid them in their own disaster prep needs or emergencies.

2. **Evacuation:**
   - In a limited evacuation, assuming the company's buildings are not in the area being evacuated, the company could offer its building as the rendezvous point and emergency shelter for its employees.
   - In a widespread evacuation, where the company would have to vacate the entire area, the company could have a series of preselected rendezvous points and shelters for employees and their families. This can be company property such as a warehouse or company-owned recreation area, or it may be a distant motel or rental property where the company has arranged to subsidize the costs of its employees staying a few days.
   - Relatedly, the company may provide some mass transit for its employees and their families to take them to a preselected shelter.

3. **Isolation:**
   - Several scenarios may see employees restricted to the buildings. Anticipating such a scenario (this also ties in to using the office building as a shelter), the company should stock emergency supplies such as first aid and medical supplies, food, water, bedding, and general hygiene supplies, to ensure adequate safety and comfort for its employees and guests. Also, it's a good idea for any office building to have a backup electrical generator
   - The opposite may occur and employees are forced to isolate or quarantine in their homes. It's a good idea to go ahead and plan for that and set up programs and policies for mass telecommuting where possible, and for temporarily shunting or outsourcing other business aspects, for the purpose of business continuity.
   - The company might provide locker space in the building for employees to bring personal emergency supplies. (For general security reasons, these lockers should be kept secured except during emergencies.)

Function

Group function is not only what your group is going to *do*, but *how* your group intends to do it. Most functions can be divided into **service, supply, or consultation**, and any of these can be provided as **before, during, or after-disaster assistance**. The later section, "**The Group in Action**," is going to be a detailed study of numerous things a group could do (and in any combination). To get started and help you pick a direction, let's look at some examples.

Service. "Service" would be all the direct, hands-on, activity-related, or labor-oriented functions. This is where you actually go out and DO something such as gather food for the local food bank, assist at a blood drive, or go help clean up the rubble after a destructive event. Let's look at some simple examples:
   - **Security**. Set up an extended Neighborhood Watch program.
   - **Transportation**. Help with either an evacuation, or running supplies to a temporary shelter.
   - **Shelter**. Displaced locals may need temporary shelter, or your town might play host to refugees.
   - **Support services**. Help the Red Cross operate a blood drive, or provide goods and services to first responders or victims.
   - **Education** and **training**. Organize free classes in first aid, CPR, and related topics for your community.
   - **Communications**. Relay emergency alerts, augment first responder systems, or help disaster victims get messages to family.
   - **Cleanup**. Post-disaster debris removal, or even a small-scale household chemical HazMat cleanup.
   - **Rebuilding**. Help repair members' houses or those of general disaster victims.

➤ **Supply Distribution**. Collect or store various post-disaster supplies and aid in their distribution.

Supply. This is more a focus on goods and materials rather than any sort of labor or activity, even though the definitions could overlap. You could collect and store these goods and supplies and either give them directly to officials or established relief agencies, or you could distribute them yourself. (Under "**External Liaison**" we'll discuss establishing a working relationship with other groups.)

➤ **Donations**. Host drives for donated clothing, tools, or other items needed after a disaster, or help with fund raising to support local services or established relief groups.

➤ **Essential goods**. Provide food, water, and clothing to disaster victims.

➤ **Tools** and **materials**. Supply the cleanup and rebuilding efforts of others.

➤ **Utilities**. Provide fuel oil, natural gas, water, or electric generators to provide utilities to a stricken area.

Consultation. If there's one thing that would be of benefit to most any community, it's a good source of expert advice and educated opinion. Instead of, or in addition to, services or supplies you could offer the benefits of your own group's brainstorming sessions or professional experience.

➤ **Citizen Advisory Boards**. Many municipalities have established boards to provide input on any number of specific subjects. Become part of these, offer input to these, or find a new need to fill.

➤ **Volunteer Consultants**. Provide expert input based on your area of expertise. For example, a security consultant could volunteer time to infrastructure sites to help them improve security.

➤ **Think Tanks**. As we'll see under "**The Group in Action**," there is a world of subjects and areas that require attention. There is plenty of need and opportunity to provide brainpower.

Scope

Simply put, scope is the width, breadth, and depth of your group's intended involvement. How big a geographical area will you try to support? How many people will be in your group? Are you going to focus all your efforts in-house or to helping others? Are you going to have group chapters in other areas or simply provide the services of your single group within a certain distance or area? Any and all of the "functions" listed above could be provided within any scope, large or small. Let's go over a few examples.

➤ **Members only**. Many groups are set up for the mutual protection and benefit of their members and do not plan to extend their efforts outside the group. There's nothing wrong with limiting your involvement, as any form of one person helping another in a crisis is a good thing.

➤ **Community**. A community-wide scope would see goods or services provided to the community as a whole, whether the community residents were functional members of the core group or not. A "community" can be defined in a similar fashion as "neighborhood" earlier. It could be a residential area, apartment complex, condo owner's association, section of a city, etc.

➤ **City**. A simple expansion of community, though the amount of labor, goods, materials, or money involved will certainly be greater and will require more members to carry out group goals.

➤ **County or region**. Some groups may choose to expand beyond city limits to include the whole county or the surrounding counties or region, either within the state, or regardless of state lines. (Victims don't care about borders.) This county or regional scope may mean that you have additional chapters of your group in other areas, or that your one group would react to an emergency within this given area.

➤ **National**. It's rumored that the Citizencorps division of FEMA has listed over 500 individual groups that are trying to organize and expand to become national volunteer groups with chapters in every state. This is not to discourage you from trying to do the same if that's a meaningful goal, but rather to let you know you're in good company, and that there is plenty of opportunity to combine your efforts with others. (You'll find contact info for Citizencorps in the "**External Liaison**" section.)

➤ **Current single-site disaster**. Again, your scope could be national whereas your chapter and its members are local. Your group's function might be to provide services at any disaster within a certain geographical area or distance, and/or to provide goods or other donations to any disaster site in the country. Another good example of a local group with national scope lies within advisory boards or think tanks. Ideas and input are not limited to any single location. Your group's brainstorming product can be passed along to your local, state, and national government, or to any corporation. We live in the information age, and public safety is one area where we certainly need more information.

➤ **Need-specific**. Getting away from geographical boundaries for a moment, your group's scope could be based on type of recipient. Your function might be focused on helping only one type of victim (such as fire victims) or in supplying one type of goods or service. Your intended recipient will dictate the kinds and amounts of services, supplies, or other assistance you provide.

Now that we've covered your group's plans, goals, and desires, let's look at the nuts and bolts of putting a group together. Let's start with the business aspects of such an entity, cover the form of business your group should take, and some suggestions on organizing the personnel involved.

Form and Organization

Our first *order* of business is to find out if you need to decide on a *form* of business. Naturally, this will only apply to those wishing to establish a specific group for the purposes of providing goods or services within a specific capacity.

If you're already formed as a preexisting group, such as a civic organization or place of worship, or if your group is an informal mutual-aid agreement amongst associates such as in a neighbor helping neighbor situation, you probably won't need to set up any sort of separate business entity.

For those setting up an independent group with a specific purpose, let's look at a few of the business and organizational steps you'll need to take.

Business Form

The most common form of business taken by organizations specifically created to recruit and manage volunteer members for the purposes of public wellbeing is the 501(c)3, or "nonprofit" organization. It's a tough process, but this nonprofit status will give you certain tax exempt benefits, some liability limitations, and will open the door for several types of grants and funding we'll discuss under "PR and Fundraising."

Let's go over the steps needed to become an official nonprofit organization :

1. **Create a Business Plan**. Anyone and everyone you deal with during the steps following this one will want to know who you are, how you plan to raise money, and what you intend to do with it.
   - The best people on the planet to help you through this part is the SBA, or **Small Business Administration**. Contact them at: http://www.sba.gov, or, Small Business Administration, SBA Answer Desk, 6302 Fairview Road, Suite 300, Charlotte, North Carolina 28210, 1-800-U-ASK-SBA (1-800-827-5722).
   - You should also find your closest SBDC or **Small Business Development Center**. The SBA (at http://www.sba.gov/sbdc) or your local Chamber of Commerce should have their contact information. An SBDC will be instrumental in helping you with parts of your business plan, with the remaining steps of business setup, and with grants and funding.
   - The Small Business Development Center has an online set of procedural links to help you with various steps along the path of setting up the business portion of your volunteer group. It's online at: http://sbdcnet.utsa.edu/SBIC/gettingstarted.htm. The cousin site to this one can be found at http://www.business.swt.edu/sbdc/files/formingplan.htm. Check your phone book for your closest office.
   - Your local Chamber of Commerce or County Extension Office should be able to give you contact information for other offices and agencies that can offer local assistance in setting up a business.
   - In your business plan, you should set in stone how money will be acquired, handled, safeguarded, accounted for, and distributed. Financial problems are the number one source of dissent within nonprofit groups, and the lack of financial planning the number one cause of loan or grant refusals.

2. **Incorporate**. The SBA and SBDC can also help you decide if you need to be an "S" Corp, a "C" Corp, or an "LLC," and they can explain the benefits and drawbacks of each.
   - For the specific process of incorporating, you'll need to contact your state's **Secretary of State**. You can find your Secretary of State's office online at http://soswy.state.wy.us/sos/corps.htm, or http://www.nass.org/, or you can find the information in the government section of your phone book. The Secretary of State's office can provide you all the information and forms you'll need for incorporating in your state. **Note**: Don't waste your money on middle-men or "incorporation services." The process is usually not that complicated and in most states is very inexpensive.
   - If you plan to operate as a national group, your headquarters chapter will need to be incorporated in its home state, and it may be more advantageous for each individual chapter to be a separately incorporated nonprofit entity. The SBA and SBDC can help you make that decision as well.

> You'll also want to get "Publication 557" from the IRS. It's the information you'll need to file for nonprofit tax exempt status, and it will have some of the data you need in order to make sure you've dotted all your "i's" and crossed all your "t's" during the business plan and incorporation stages. Visit your local IRS office for this publication or log on to http://www.irs.gov to download it. You can also call the national office at 202-622-8100 or write them at Internal Revenue Service, US Department of Treasury, 1111 Constitution Ave., NW, Washington, DC 20224.

3. **Apply for an "Employer Identification Number."** After incorporating, you'll need your Employer Identification Number, or EIN, even though you may not technically "employ" anyone. This is essentially your corporation's "Social Security Number" and it's the number used to track your corporation through the tax system.

> The one group that handles EINs is the IRS. They have a separate office for this. Contact them at: Internal Revenue Service, Attn: Entity Control, Mail Stop 6271, P.O. Box 9941, Ogden, UT 84201, (801)-620-7645. You can download the forms in Adobe Acrobat's .pdf format from: http://ftp.fedworld.gov/pub/irs-pdf/fss4.pdf.

4. **Apply for Tax-Exempt status**. This is what sets your corporation apart as a nonprofit organization, as opposed to a regular corporation, and it's what opens the doors for grants and funding.

> Again, the only group that handles this procedure is the IRS, and they have separate phone numbers to help you along the way.

• Information can be found online at http://www.irs.ustreas.gov/plain/bus_info/eo/ , or call 877-829-5500.

• You can also download the "Tax-Exempt Organization Tax Kit" by logging on to http://www.irs.gov and looking for the "Tax Exempt" section. You can also get them by calling 800-TAX-FORM (800-829-3676).

> The SBA and your SBDC can also help you with this process.

5. **Register with your state**. Once you've applied for tax exempt status with the feds, apply to your state as well.

> With most states you'll be asked to complete a "**Unified Registration Statement**" or URS.

> Your SBA, SBDC, Secretary of State's office, or Chamber of Commerce can give you the contact information for the appropriate state office, and what the registration is called if different from the URS. In some states you'll contact the Attorney General's office, and in others the state tax office.

6. **Red tape details.** Now that the big hurdles have been jumped, you'll want to get all the little things out of the way such as local business license (if any), etc. The SBA, SBDC, and your local Chamber of Commerce, along with your local business licensing office can tell you which details you need to complete for your state, county, or city.

> One source to help you keep up with news items affecting non-profits can be found at: http://news.gilbert.org.

> For a collection of links to necessary government services and other information, visit http://www.firstgov.gov/Business/Business_Gateway.shtml.

> For more information on volunteer group legal issues, visit the "Nonprofit Law" page at: http://www.nonprofitlaw.com, or reach the office directly at Pfau Englund Nonprofit Law, P.C., 3213 Duke Street, #622, Alexandria, VA 22314, Voice: 703-304-1204. While on the home page, be sure to sign up for their **free email newsletter**. For more legal info and risk management education, visit these two sources:

• The Nonprofit Risk Management Center, online at, http://www.nonprofitrisk.org, or contact them at 1130 Seventeenth Street, NW, Suite 210, Washington, DC 20036, Phone: (202) 785-3891, Fax: (202) 296-0349.

• PERI, the Public Entity Risk Institute, online at http://www.riskinstitute.org, with direct contact at 11350 Random Hills Road, #210, Fairfax, VA 22030, Phone: (703) 352-1846, Fax: (703) 352-6339.

> Though a New York centered group, you can learn some things about non-profits from the Nonprofit Coordinating Committee of New York. See them online at http://www.npccny.org, or contact them at: 1350 Broadway, No. 1801, New York, NY 10018, phone 212-502-4191, fax 212-502-4189. While you're there, be sure to read about the "Volunteer Protection Act of 1997" at http://www.npccny.org/info/gti2.htm.

• More "Protection Act" info is at http://www.senate.gov/~rpc/releases/1997/VOLUNTEE.LO.htm, and at http://www.races.net/voluntr.html.

• An alternative perspective of the "Volunteer Protection Act of 1997" can be found online at http://www.asae-aon.com/vpa.php, and at http://www.hotairballooning.org/info_volact.htm.

• A State-by-state listing of "Good Samaritan" laws can be found through Citizencorps, online at http://www.citizencorps.gov/councils/liability.shtm.

By the way, most of the above agencies and companies, including the IRS, will help walk you through every stage necessary to becoming a corporation and then a nonprofit organization. So, you can probably complete the process without the aid of an expensive middleman service. However, if you have a good **accountant**, they should be included in the process, and if your group has any intention of dealing with the general public, a good **attorney** will be helpful. An attorney can help you set up your operations to meet any local legal requirements or liability guidelines.

Now that we have the bulk of the red tape out of the way, let's move on to other organizational needs.

Officers and Committees

During the process of incorporation, you'll be asked to provide a list of officers. In most cases you'll be asked for at least a president and secretary, and in some cases a "registered agent" whose function is being the corporation's "legal contact."

For the purposes of actual group operations (whether you're incorporated as a nonprofit or not), let's go over the basic offices, but more importantly, the necessary qualities of volunteer group officers, and the suggested committees that will be the backbone of your operations.

Hint: As you'll want to be involved with various government agencies during both the setup and operational aspects of your organization, you may want to pay close attention to who is part of your core officer group. Doing a little advance research into grant and funding programs, and the needs and business relationship requirements of various government agencies should provide you with the data you need to make informed choices regarding your officers.

➤ **President**. Usually, the person who comes up with idea of creating an organization becomes the president. In other cases, the president is chosen on their value as a spokesperson, group image representative, or for their leadership or administrative abilities. As your organization's focus is the safety and wellbeing of people, and the direction of activities associated with emergencies and disasters, you'll want to chose a president based on leadership, charisma, experience, organizational skills, and motivational abilities. Administrative functions can be handled by other officers.

➤ **Secretary**. Normal secretarial duties include taking minutes at meetings, and making sure all appropriate and necessary corporate filings and internal and external communications are handled. Choose your secretary based on their ability to handle the bulk of your administrative and communications needs, and on their writing skills. Computer literacy is a plus.

➤ **Treasurer**. You'll find that an experienced accountant who can set up a secure system of money-control checks and balances, knows their way around business administrative needs, corporate filing requirements, and any of the steps mentioned above for incorporating as a nonprofit organization, is worth their weight in gold as a treasurer.

➤ **Sergeant at Arms**. An older, and almost out of style position, the Sergeant at Arms was traditionally charged with helping keep order at meetings and controlling entry of latecomers, etc. While deviating somewhat from this traditional role, your Sergeant at Arms may fulfil more of a security role given the nature and needs of a group organized to provide goods or services at a disaster. If your group plans on active participation at a disaster site, your Sergeant at Arms may well become your security director in charge of establishing security procedures both at group meetings open to the public, and at supply sites at disaster scenes where unruly crowds could become a problem. This is just one small example, but if your group will have any direct interaction with the general public, you may want a security director. Try to find a retired Police Officer with good people skills and a level head.

Good committee chairpersons will be the backbone of your group's operations. Here are a few suggested committees and associated duties and functions. The committees you actually create will be based on the function your group serves. **Note**: Even if you choose not to form individual committees (as you might be a rather small group) you'll still need to consider the tasks and purposes being mentioned since the same functions should be performed within your group.

➤ **Public Relations**. If you're forming a group that will assist more people than just the immediate group membership, it'll help if the general public knows you exist and what you're about. You'll want to publicize the fact that you've formed a group, and what your purpose, scope, and function are. Also, you'll want the public to know about the goals you've reached, and the details involved each time your group performs its duties. The PR chairperson should also act as an "Historian" by keeping documentation and records (to include newspaper clippings and video of television news, etc.) of all group events whether simple meetings, training exercises, or actual group projects. A simple example would be an extensive scrapbook with photos and newsclippings. Your PR chairman should be someone who's good with people, who knows the ins and outs of marketing, and maybe has contacts with local media.

➤ **Membership**. A group is not a group without members. The membership committee should work closely with the PR committee in publicizing the group's existence and purpose, and in organizing membership drives. The Membership Committee chair would be in charge of advertising membership recruitment meetings, organizing and hosting the meetings, registering new members, and maintaining membership rosters. Depending on the intended size and scope of your group, your membership committee may be rather complex. For example, your membership committee might have one person in charge of public service announcements, another in charge of arranging meeting locations, another in charge of refreshments for the meetings, and still another in charge of recording who shows and who joins.

➤ **Fund Raising**. Fund raising is a "maybe" committee because many groups operate without money, and others operate solely on grants. If your group intends to raise money as one of its public benefit functions, or if your group's function requires equipment or capital, you'll need a chairperson to organize your fundraising operations. Your fund raising chair should naturally be someone trustworthy, and should be a separate person from your treasurer simply for the safety of division of duties and of checks and balances.

➤ **Communications**. This is also a "maybe" committee since it's main function is arbitrary and could well fall under the responsibilities of others. The Communications Committee could be in charge of either the actual communication between the group and its members, or of communications equipment or media. For example, the committee could oversee newsletters sent out to members, it could be in charge of setting up a group website, or it could be in charge of communication equipment such as phones or walkie talkies used during group functions. If your Communications function is set up as a separate committee, it should also be in charge of creating the phone tree by which members contact each other to warn of a disaster.

➤ **Quartermaster**. Your Quartermaster would be in charge of all group equipment. This would include storing or acquiring any tools or equipment used in group functions. For example, if your group's function was to operate a temporary shelter in a catastrophe, the Quartermaster would be in charge of acquisition and distribution of beds, restroom supplies, food, utensils, etc. The Quartermaster would also be in charge of finding economical and reliable sources for any of your non-donated equipment and supply needs. Too, if your group's function were to collect durable goods to give to charities or victims of disasters, your Quartermaster would be in charge of cataloging and storing the goods collected.

➤ **Training Director**. The bare minimum function of each and every volunteer group should include the safety and wellbeing of its members. This might mean that the main duty of the Training Director is in making sure all group members are trained in all areas of preparedness mentioned in this book. Members will not be able to come to group functions in the event of a local emergency if their families are not safe and taken care of. The other duty of the Training Director would be to make sure all group members were trained to perform any special function within the group. For example, your group's function might be to help dig through the rubble of a disaster scene. Members will need to be trained on dealing with hazardous materials, safety measures in debris removal, and first aid if they find a live victim. Your Training Director should also be in charge of constantly researching new and updated information regarding technical aspects of your group's function, or preparedness in general, and should communicate these updates to the membership.

➤ **Rotating Watch**. If a group is to act, they should know when. Someone in the group should be watching the news, monitoring a scanner, able to receive calls from local officials, and keeping track of current events every day. If this is to be a shared duty, the Rotating Watch chairperson should know the schedule of all watch volunteers and coordinate their efforts so that no one person is tasked with the responsibility, and yet the watch is covered at all times with no gaps. This is especially important during the higher terror alerts and severe weather seasons.

Naturally, the entire concept and focus of various committees will depend on the size of your group and the function you intend to perform.

Group Dynamics

Once any group is created, it goes through a certain number of predictable growth steps. These are commonly called "**form, storm, norm, and perform**." They go like this: First the group is **formed**. Then it goes through a period of organizational upheavals, maybe some minor "**stormy**" power struggles, some focus issues, and what are generally referred to as "growing pains." "**Norm**" refers to a settling to normal after all the initial issues are handled, and everyone is ready to "get down to business." This "getting down to business" is when the group starts to **perform**.

To the above, we'd like to add two observations of our own and they are "**swarm**" and "**warm**." When groups first start performing, all members are usually extremely enthusiastic about their group and its goals. They "**swarm**" into the meetings and to everything the group does. As time goes on, the enthusiasm usually cools down a bit from the "red hot" swarming the group knew at its beginning. This is the **warm** phase. This is where you want to keep the group. You never want to let it get cold. Let's cover a few considerations that will help you through all the growing pains and will hopefully help keep the group alive and active for some time to come.

- **Organize first**. Before you commit to recruiting members, publicizing your group and its intentions, or trying to liaison with outside entities, you should complete all your business and management organization. Have the machine itself in place before taking the next step.

- **Serve a specific purpose**. Few people are going to devote time, effort, or money to a group whose goal is, "If something happens we'll try to help." Generic reaction is hard to identify with, and most people join functional groups because they believe in the goal or purpose.

- **Set appropriate and realistic starting goals**. No one will stay with a group without a purpose, and a group that sets goals that are either pointless or unreachable, essentially serves no purpose. For example if you do an outstanding job of recruitment and wind up with 1500 members, you're going to lose them if all you plan on doing is to help feed one volunteer Fire Department when they go out on calls. The flipside of the coin is also true. If you only have very few members but plan on providing evacuation transportation for a tri-county area, your goals will be unreachable and you'll lose your members.

- **Match your membership to your goals, purpose, and needs**. People only stay where they feel needed and useful. If your members can't perform the group's function, they'll feel useless and leave. If your group cannot perform its function due to lack of qualified membership, it serves no purpose. For example, if your intended function is to help first responders dig through rubble after a destructive event, you probably wouldn't recruit members from the "retired ladies' sewing club." However, if you were forming a think tank, these same ladies might be rather valuable.

- **Pace yourself**. Burnout can result from too much too soon, and lack of interest will be the result of too little too late. Though we suggest three months as the time limit for having all personal protection in place, group formation and operations are different. You'll want to have meetings, training sessions, etc. just often enough to keep everyone involved and up to speed, but not so often that it wears everyone out, or so seldom that people lose interest. Our suggestion is to have a quick series of organization and training meetings, and then one function per quarter after that. Bridge the gaps with a mailed newsletter or email updates. Naturally, the pace you set will depend on your group, and its function and members.

- **Divide the workload**. Some people will try to do all the work, and some people will try to do none of it. Divide the workload and involvement as best you can. The committees we listed above are a great way to make sure everyone is feeling involved and useful, but keeps individuals from being overloaded.

- **Reward the members**. Besides feeling useful and involved, people also want to feel appreciated. Though volunteer groups are not formed with the intention of paying the members, there should be some sort of tiny reward be it recognition, simple thank you letters, or some minor benefit bestowed on the group as a whole.

- **Trust**: Don't ever do anything to lose the trust of the membership. Always have obvious and unquestionable safeguards in place for any money handling, always follow through with scheduled events or other commitments, and never breach confidence by doing such things as mishandling funds, selling membership rosters, or divulging member information. Also, officers should remember they represent the group 24/7.

- **Consider personalities**. Some groups will need every member they can recruit, and will choose to have an "open membership" meaning anyone who wants to join can. Other groups will want to choose their members based on geographic location or specific skill. Other groups still, may be less dependent on external requirements, but may desire to have a group that will be a close-knit team. These groups might want to recruit members on an invitational basis and carefully consider the personalities of each prospective member and determine if the candidate will be committed to the group's goals and get along well with the other members. Many civic groups and most employment situations use this approach. An example would be that if you wanted to form a think tank where members synergistically fed off of each other's camaraderie and enthusiasm to come up with bigger and better ideas, the last person you'd want to include is someone who was unintelligent, negative, obnoxious, or argumentative.

- **Mix the viewpoints**. Though you want to choose people with a good nature and even temperament, you should consider alternate points of view to get the best results from any effort. There are three types of learned people associated with any given subject. There's the one who went to school to receive formal training on the matter, there's the person who found themselves thrust into the experience and gained "on the job training," and there's the thinker / philosopher who with great aptitude and intelligence, thinks problems through in incredible detail. You need all three types in your group. Just make sure everyone understands the concept of the goal being the most important thing, and that disagreements and discussions can be held in a positive atmosphere without argument and anger.

Membership

Now that we've covered how to organize your group, discussed officers and committees, and talked about the dynamics of the group as a whole, let's talk about how to recruit and organize your actual members, and how to manage them.

Recruiting: Recruiting may or may not be a function of your group. You may have automatic members such as in the case of congregation members looking after each other, or a business operating as a defacto group to look after employees or provide a community service. How and where you look for members will depend on your group's form, scope, and focus, so that's why we put those first. Also, you might want to take a look at the "**Group in Action**" section below to further cement your group's function before you commit to recruiting members. Let's look at a few things to consider:

> The first decision, unless it's already made for you is the **type** of membership. Will it be **open** and anyone can be part of your group, or will it be by invitation only? Will it be based on **geographical** limitations, or will it be based on particular **skills**?

> Advertising your group and announcing your **open** membership drive can be handled through standard media. Being a non-profit volunteer group, local TV stations, radio stations, and newspapers may give you free advertising in the form of PSAs or, "Public Service Announcements." Another good avenue for open membership recruitment would be a short presentation at local Parent Teacher Association meetings at area schools, or at local civic group meetings.

> For **geographically** targeted membership, you'll have to use geographically targeted publicity. This usually includes direct mail, or fliers distributed door to door or at area civic group meetings. An example of a geographically targeted group would be a Neighborhood Watch where members live in the neighborhood.

> If your group is searching for members with a particular **skill**, you may use the same PSAs mentioned above, but you should also investigate **trade journals**, professional **associations**, or make your announcement through applicable area companies or clubs. For example, your group may want to locate all qualified heavy equipment operators in your area. You could contact local construction companies, labor unions, etc. to ask if they'd let you distribute membership info. **Hint**: A good source list for trade journals is the various "**Writer's Market**" or "**Writer's Guide**" books and magazines. You can also find hundreds listed at http://www.freetrademagazinesource.com.

> You'll find a "**Services, Skills, and Assets**" form in the "**Appendix.**" You'll also find a "**Volunteer Application**" form and a "**Volunteer Group Meeting**" planning checklist.

> Depending on the size of your group, you might want to create a simple database that would allow you to call up members based on their skill, geographic location, or other factors. You'll find an Excel file entitled "WatchGroup" on our enclosed **CD**.

Communication: People tend to support that which they help create and/or that which makes an effort to include them. Good communication is a two-way street and is necessary to maintaining a solid membership in your group.

> **Establish a group website** with a members' section that's accessible only with a password. Use this section for sensitive info, commo to members, and a bulletin board section so members may contact each other. (The main page of your site is useful in your Public Relations activities.)

> **Set up a "Yahoo Group."** Go to http://www.yahoogroups.com to set up a group for communications to, with, and from your members. Setup is simple and instructions are easy to follow. Yahoo Groups is also a great source of instructional discussion groups on a variety of subjects. Log on and see what they have to offer. **Hint**: You'll want to set your group up as an *unlisted* and *invitation only* group so non-members can't peek at what you're doing.

> **Set up a phone tree**. The most important reason to set up a phone tree is to relay emergency messages to the group. However, it's also a good way to pass the word about meetings, training, and non-emergency functions. In fact, passing along the non-urgent messages is a great way to test the phone tree itself.

> **Create a mailed newsletter**. Though the internet is a wonderful tool, the truth of the matter is that only about 25% of the country is computerized. This newsletter only needs to go out once a quarter or so, or more often if you feel it necessary, to keep members updated on the group's efforts, the subject of your focus, and dates of upcoming meetings and training sessions. Therefore, you might consider making it primarily a hardcopy newsletter which has the hidden benefit of delivering a tangible item directly to members rather than an internet posting or email that they have to remember to log on to find.

Screening: The question of open membership vs invitation-only membership, and the possibly delicate nature of group purpose sometimes leads to the question of screening and background checks of prospective members. Given the litigious nature of our society, the current status of privacy legislation, and the double-edged sword of any involvement with others, we felt it best to give you some general information and guidelines on screening.

> First and foremost, if you <u>absolutely</u> feel that some type of screening needs to be performed, YOU SHOULD **NOT** DO YOUR OWN SCREENING. Screening and background checks should only be performed by licensed and bonded companies who specialize in background checks and who are up to date on laws, procedures, and other regulations, and who also carry the proper liability and E&O, or "Errors and

Omissions" insurance. Your local SBA and SBDC can tell you more. (Some groups screen only those who handle money.)

➢ Law Enforcement will sometimes do their own screening of volunteer groups that provide assistance. A case in point is in searches for missing children. As volunteers show up to an area to conduct a physical search for a missing child, there is usually a volunteer registration table. The sign-up information is often used to do a quiet screening to make sure the search group does not contain any pedophiles or sexual predators.

➢ Smaller volunteer groups whose only purpose is mutual protection in a major disaster, will not need any sort of screening any more than you'd do background checks on sinking-ship survivors who're clamoring aboard a lifeboat. In dire emergencies, any assistance is welcome and it doesn't matter who offers it.

➢ However, should you do occasional support work for various community functions, you might want a better idea about who composes your group. The last thing you want is some social parasite infiltrating your group and its functions only to fake an injury for the sole purpose of suing the group and the function you were supporting. Not only would it be a financial burden defending yourself, it would probably be the last time your group was allowed to participate in local events. You also wouldn't want a career criminal being part of your Neighborhood Watch group. Don't want to let the fox into the hen house now do we?

➢ The best solution is to allow local government, if your group function calls for a government entity liaison, to set the requirement that background checks be conducted, and that they do the screening themselves. This takes the liability away from your group and yet still does the job.

➢ If you truly feel you need to have some sort of screening conducted by a qualified service, or if the governmental agency with which you have a support agreement wishes screening, we've included a couple of useful forms in the "**Appendix**." One is the "**Volunteer Application**," and the other is the "**Services, Skills, and Assets**" form. If screening is required, you'll probably be asked to fill out a **background check release form**, and each professional service will have their own.

➢ A good example of volunteer registration forms can be seen on the Washington Military Department, Emergency Management Division website at http://emd.wa.gov/6-rr/e-ops/sar/sarforms/sarforms-idx.htm.

Meeting Organization: Now that we've covered setting up the group and getting your members, let's cover a few random considerations on how to keep your meetings organized and productive.

➢ To help, we've included a "**Volunteer Group Meeting**" planning checklist in the "**Appendix**."

➢ Plan the topics and agenda of your meetings ahead of time to keep them organized, efficient, interesting, productive, and mercifully short. If you don't have an agenda and you try to have large meetings in an open round-table discussion format, they'll drag on forever with everyone wanting to speak their piece.

➢ Distribute the meeting agenda to members ahead of time so they'll know what to expect at the meetings and can contribute accordingly.

➢ Offer refreshments at the start of the meeting and/or during a break in the middle.

➢ Allow a short period of time at the beginning and end of the meeting for members to network, and socialize.

➢ The larger your meeting, the greater the need for security. For example, if you advertised a city-wide open-membership meeting, you may draw a large crowd and there's no guarantee that everyone's agenda will be the same. It would be a good idea to have an off-duty Police Officer or two to help keep order and provide an official air to the proceedings. Later meetings with established members will probably not need such heavy security.

➢ Most meetings follow a simple order:

❑ A speaker or presentation if there is one. (Put your speaker first so they can speak and leave and not have to sit through the internal business portion of your group's meeting.)

❑ Minutes from any previous meetings are usually read by the Secretary.

❑ "Officer's Reports" such as a Treasurer's report usually comes next.

❑ Committee Reports generally follow officer's reports.

❑ Next in the common agenda is a discussion of any "unfinished business."

❑ Following unfinished business would be a call for "new business" which is usually the time general suggestions, new projects, upcoming training sessions, etc. are announced.

➢ For a more in depth look at meeting organization find a copy of **Robert's Rules of Order**. It's an extremely common book and can be found at almost any library or bookstore. You can view an older version of the rules on our enclosed **CD**.

➢ A great book on creating and organizing boards and panels from the general public is **The Public Involvement Manual** by James L. Creighton, published by Abt Books.

Benefits: A true volunteer group is exactly that; a collection of people volunteering their time without thought of pay or benefits. Depending on the size of the group, the scope of its function, and time and effort requirements of the officer corps, you'll need to consider a certain amount of income or other benefits for some of your key players. Whatever you do, make sure the entire membership knows the details, that everything is above board, and that everyone is comfortable with the decision to pay officers and the amounts paid. Let's look at a few considerations.

➢ Any kind of payroll is going to add a headache factor to any group. Make sure you have a competent accountant on board before even thinking about pay.

➢ For non-profit organizations that intend to have chapters in different areas, it might be more advantageous for bookkeeping purposes, for each chapter to be incorporated as a separate non-profit entity in the area they're based. This will alleviate a few headaches as bookkeeping and payroll would be the responsibility of each individual chapter. This is only a suggestion and it's a decision you'll have to make on your own with the help of a competent accountant and attorney; two professionals you should have as part of your team before ever making the decision to be a nationwide group.

➢ Examine non-cash ways to "pay" some of your officer corps. As an example, a local restaurant may pledge a certain amount of food each month as their way of contributing to your efforts. Some of these meals could go to officers as part of or instead of a direct income. Other businesses could provide other services.

➢ Any time a company donates any goods or services in support of your group's staff, distribute as much as you can to the general membership. The better you care for your members the more devoted they'll be.

➢ The fundraising section below will provide several grant sources. Some grants may include funds for payroll.

➢ Examine economical ways to reward your general membership. These can be simple things like letters of recognition, buttons or plaques given out at small awards ceremonies, etc. Also, when your PR director sends a news release to the local media about a project the group just completed, make sure as many members as possible are included in photos, or mentioned by name. People tend to support that which makes an effort to include them.

➢ One group called S.H.A.R.E., or Self-Help and Resource Exchange, provides food at a 50% discount to people who volunteer their time with civic projects. To get the details and see if your group might be eligible, contact them at SHARE, 6950 Friars Rd., San Diego, CA 92108, 888-742-7372, http://www.worldshare.org.

➢ The **Corporation for National Service** offers different forms of monetary benefit to members of volunteer groups that dedicate a certain amount of time to community functions. The CNS operates both independently and through AmeriCorps and SeniorCorps. Contact them at http://www.nationalservice.org, or at 1201 New York Avenue, NW, Washington, D.C. 20525, phone: 202-606-5000, TTY: (202) 565-2799.

Hopefully, this section has given you a broad picture of what setting up and organizing a group is all about. Once you have your group in place, let's look at a few concepts to keep it up and running.

PR and Fundraising

If your group is a smaller group, or a mutual-aid agreement group or similar, you can breathe a sigh of relief because you won't have to read this section either. This portion of the "**Cooperation**" section covers the necessary evils of keeping a larger organization afloat; marketing and finance. Or, in our case, Public Relations and Fundraising.

Public Relations

If you want a decent-sized membership, if you want to maintain good relations with government entities and first responders, and if you want the general public to recognize your group when it comes time for actual operations before, during, or after a disaster, you're going to have to exercise good public relations.

➢ Always remember that public relations is far more than mere marketing. It's the image of your group that's held by the general public, and this image will need constant nurturing and attention.

➢ The first segment of the "public" with whom you'll want to ingratiate yourself is local government officials. Always start with the Mayor and top Law Enforcement officials in your area. This initial courtesy will certainly be a feather in your cap later on. Always invite them to any of your public meetings.

➢ Remind all members that your group carries no real authority, and that the group is simply there to help. Therefore, when the group is in action, they will only generate resentment if they try to act as an authority and order citizens around. Always present the attitude of being there only to provide aid and assistance.

➢ Publicize all your meetings, projects, and accomplishments. You can do this by sending press releases to radio and TV stations, and to newspapers. It's also advisable to invite local media representatives to any exercises you conduct, training sessions, or any other PR or fundraising activities you perform. Try to get photos and film footage any time your members are out and about in any sort of group uniform.

➤ Become neutrally involved in local politics and government events to a certain degree. For example, don't support a particular candidate or party, but do send the Mayor or other officials a card on their birthday, etc.

➤ Your group should also show support for positive events or accomplishments of other individuals or groups. For example, if a particular Police Officer or Fireman does something rather heroic, be sure to send them a personal letter as well as make a small announcement through radio, TV, or newspaper that your group congratulates this individual. Also, under "Think Tank," we'll point out certain brainstorm topics and examples that can only be fully realized through legislative action. Though your neutral group should steer clear of political affiliation, this does not mean you should stay away from providing valuable input to legislators and officials. In fact, sending in good ideas is great PR.

➤ Explore the tremendous opportunities offered by local cable access channels. **Disaster Prep 101** will soon have a **video** counterpart. This video will touch on the basics of disaster preparedness and will be designed as a sponsored presentation, meaning your group could air the video on cable access as a service offered by your group. Our contact info is listed in the "**Appendix**." Keep watching for news on the video's release.

➤ We found one group that helps you with management, public relations, and funding issues. Log on to http://www.idealist.org, to visit the "**Action Without Borders**" site. You can also contact them at Action Without Borders, Inc., 79 Fifth Avenue, 17th floor, New York, NY 10003, Tel: 212-843-3973, Fax: 212-564-3377, to ask about membership and various programs.

➤ A similar organization is "Guidestar" which carries an extensive database of nonprofits and charitable organizations, along with information of benefit to each. You'll find them at http://www.guidestar.org, or at Guidestar, 427 Scotland Street, Williamsburg, VA 23185, (757) 229-4631. (This group is also useful if you want to verify information about a nonprofit organization or charity before joining or donating money.)

➤ A close cousin to both of these, in that it's an information clearinghouse for nonprofit organizations is the "Internet Nonprofit Center," at http://www.nonprofits.org. You can also reach them through The Internet Nonprofit Center, c/o The Evergreen State Society, P.O. box 20682, Seattle, WA 98102, 206-329-5640.

➤ Lastly, make friends with local media since you'll need to work with them on activating phone trees, EAS expansion, situational awareness, and PSAs to educate the public and local government about who you are and what you do.

Fund Raising

For most, the prospect of fund raising will bring back those old school memories of selling candy bars for this, lightbulbs for that, and late afternoon car washes. It's the kind of thing a lot of us grew up with. Interestingly enough, some of the cash supply efforts we'll mention here will overlap with those olden days, but we'll try to add a new twist or two. We'll break this little section down into **products, services, grants**, and **creative bartering**.

Products: For our purposes here, we'll define "products" as tangible goods. The trick is not only to find products that will be popular in sales and provide a good profit margin, but items that identify with your group and its purpose, and will not conflict. For example, a Volunteer Fire Department might sell fire extinguishers and smoke alarms to raise money, but you won't see them selling matches. One other thing to consider is that any fund raising activity is also a PR opportunity. Make the most of it.

The short list below is by no means a comprehensive cash generating primer, but hopefully you'll find a few useful ideas and maybe some things that will spark your imagination.

➤ Don't discount the old "standard" fund raisers like the ones we just mentioned, and the following:

❏ Bake Sales	❏ Candy Sales	❏ Rummage Sales
❏ Raffles (if legal)	❏ Car Washes	❏ Yard Service Teams

➤ Any disaster preparedness equipment item mentioned in this book could make good sales items. You could sell a simple product, or sell and install any number of things we've talked about in the various sections. The true list is a lot more extensive than just these few items:

❏ Fire Extinguishers	❏ Smoke Alarms	❏ Fire Escape Ladders
❏ First Aid Kits	❏ Road Safety Kits	❏ Walkie Talkies
❏ Intercom Systems	❏ Deadbolts and Chains	❏ Peepholes
❏ Water Cisterns or Tanks	❏ Electric Generators	❏ Local Evac Atlas
❏ Backpacks	❏ Map Atlases	❏ Bugout Kits for Pets

❏ **Disaster Prep 101** (We've included a reseller's coupon in the back for volunteer groups.)

Services: The Car Wash we listed above should actually go here. You don't always need an item to sell, since you could sell your time and expertise to raise cash. However, as you did with products above, make the services you provide tie in with disaster prep, with your group's function, and make it something that provides a public benefit. Let's cover a few examples:

> **Teach basic computer use and intro to internet**. As we stated earlier, the internet is too valuable a tool to not learn to use. However, only about 25% of the country is computer literate. All you have to do is utilize publicly accessible computers and meet your "students" there.

> **Scan photos and documents**. As we mentioned under the "Info Pack" portion of "Planning Ahead for Phase II," it's a good idea to scan in all your old photos and documents so you'll always have a softcopy backup of irreplaceable items. You can provide this service for a small fee as both a small fundraiser and a good public education and PR opportunity. If you can also provide a Notary service, so much the better.

> **Theft safeguard**. While you're helping people scan in their important documents, you could also help them engrave, photo, and video their valuables while creating a log of all property owned.

> **House numbering**. As we'll see under "The Group in Action," house numbering is actually a disaster prep need. After Hurricane Andrew, everyone from first responders through insurance adjusters were having a terrible time locating properties because all the houses and their numbers were destroyed along with all street signs. Your group could charge $1 each to stencil the address onto the curb in front of each house (stencil street names at intersections for free as a group function and public relations activity). **Note**: Be sure you have permission from the city before you do this. If they don't like the paint option you could make small drive-in-the-ground plaques with the address numbers on them and charge accordingly.

> **Product installation**. Some of the "products" mentioned above don't necessarily have to be sold. You could **install** them for people who buy them on their own from independent sources. The water tanks and electric generators listed are a good example. You don't sell them, you just set them up.

> **Service Projects**. The projects mentioned under "**The Group in Action**" are intended to be given freely by the group to those who need the help during a disaster. Period. However, some projects, like house numbering above, could be turned into a before-the-fact service project to raise money. Read through the "Group in Action" section and see if it gives you any new ideas.

> **Training**. Any of the training we mentioned earlier under "Basic Training" would make a great service function (not to mention good PR). Among these would be:

❑ First Aid Classes   ❑ CPR Training   ❑ Basic Fire Safety
❑ Limited Fire Fighting   ❑ Canning and Food Prep   ❑ Home Safety Courses
❑ Automotive Repair   ❑ Boating Safety   ❑ Basic Home Prep

Note: If you choose to offer any sort of class to the general public, make sure any **instructors** have the proper **certification and/or licensing**, and that you fulfil local requirements for offering instruction for a fee.

Grants. No discussion of funding would be complete without covering the topic of grants. We've found some general grant info sources:

> The first and best place to look for federal grant information is in the "Catalog of Federal and Domestic Assistance." You can log on to http://www.cfda.gov, or to order a copy of the catalog, call or write to the **US Government Printing Office** online at http://www.gpoaccess.gov, or write to Superintendent of Documents, 732 North Capitol Street NW., Washington, DC 20401, 202-512-0000, and 888-293-6498, or PO Box 371954, Pittsburgh, PA 15250-7954, 1-866-512-1800. Your local library should also have a copy.

> Since you're now a nonprofit organization, you should check out the "**Non-Profit Gateway**" at http://www.nonprofit.gov. This is a collection of direct links to federal grant programs, giving you contact info and sometimes necessary forms from each office. This link will direct you to the FirstGov.com collection of resources for nonprofits, and is only available online.

> Additionally, you should contact "**The Foundation Center**" at http://www.fdncenter.org or at The Foundation Center, 79 Fifth Ave., New York, NY 10003, Phone: 212-620-4230, Fax: 212-691-1828. They're set up to help you find foundations that offer grants to nonprofit organizations. In addition to the central office, the center has regional offices and libraries.

> For information on how to write a grant along with some additional grant source information, contact the "**Grantsmanship Center**" at http://www.tgci.com, or at The Grantsmanship Center, P.O. Box 17220, Los Angeles, CA 90017, phone (213) 482-9860; fax (213) 482-9863.

➢ When working with the **Small Business Administration**, and the Small Business Development Center, ask them about any loan or grant programs they're familiar with. They'll have a better handle on the individual programs offered through state and local sources.

➢ Ask your local **Chamber of Commerce** about grants and loans. Also ask them if your area has an "**Economic Development Council.**" Though an emergency-reaction group such as yours is not high on the list as far as direct economic stimulation through regular business, your actions will save lives, enhance relief and rebuilding efforts, and therefore get a community back on its "economic feet" faster than it would on its own. This is why these agencies should help you.

➢ While not a direct grants source per se, you should contact the **National Association of Community Action Agencies** at 1100 17th St., NW, Suite 500, Washington, DC 20036, phone: 202-265-7546, fax: 202-265-8850, or http://www.nacaa.org. This group helps coordinate the activities of various local groups and agencies that are recipients of federal Community Services Block Grants, of which there's a strong overlap between what the grants are intended for and the services your group may offer.

➢ Consider joining the **Association of Fund Raising Professionals**. They're online at http://www.nsfre.org, or find them at 1101 King St., Suite 700, Alexandria, VA 22314, Phone: 703-684-0410, Fax: 703-684-0540.

➢ For a good listing of grants and loans through state and private organizations, pick up a copy of Mathew Lesko's "**How to Write and Get a Grant.**" You can find this book through your library, local bookstore, Amazon.com, or directly from Mathew Lesko at http://www.lesko.com, Information USA, 12079 Nebel St., Rockville, MD 20852, 1-800-955-POWER (7693).

➢ This is a Homeless Shelter grant, but its wording lists it as a grant for "emergency housing." If your group is planning on adopting or creating a shelter during a disaster, this grant may apply. Contact the **Office of Special Needs Assistance Program**, Department of Housing and Urban Development, Room 7266, Washington, DC, 20410, 202-708-4300.

➢ Here's a public transportation grant that may bleed over into **development of an evacuation transportation system**. Contact the Federal Transit Administration, Office of Grants Management, Office of Capital and Formula Assistance, 400 Seventh St., SW, Washington, DC, 20590, 202-366-2053.

➢ Some grants may be available to your group, depending on its intended function, from the **National Community Development Association**. Contact them at http://www.ncdaonline.org, or directly at 522 21st St., NW, #120, Washington, DC 20006, 202-293-7587.

➢ Visit the **Council on Foundations** grant information. You'll find most of their information online at: http://www.cof.org/index.cfm?containerID=76&menuContainerID=0&crumb=2&. As the council was established to help nonprofit organizations and other foundations, contact them to look at the various programs they offer. Their homepage is http://www.cof.org, and you can contact them directly at Council on Foundations · 1828 L Street, NW · Washington, DC 20036 · 202/466-6512 · FAX: 202/785-3926.

➢ Similarly, take a look at the "**Independent Sector**" site at http://www.independentsector.org, and contact them at Independent Sector, 1200 Eighteenth Street, NW, Suite 200, Washington, DC 20036, 202-467-6100 phone, 202-467-6101 fax, and 888-860-8118 for publication orders.

➢ Other developmental programs and funding may be found through the **National Association of Counties.** Contact them online at http://www.naco.org or directly at 440 First St. NW, Suite 800, Washington, DC 20001, 202-393-6226.

➢ If absolutely nothing else, and you can't find a directly applicable grant or loan program, contact your **Congressional representatives in Washington DC**, and your representatives in your state legislature. Tell them what your group's purpose and goals are and ask them to help you find funding. If you don't know your reps' contact info you can find federal info by calling the congressional switchboard at: 800-839-5276, or 202-224-3121. These sites will also help:

❏ http://www.congress.org
❏ http://www.senate.gov
❏ http://www.vote-smart.org

❏ http://www.house.gov
❏ http://thomas.loc.gov
❏ http://www.firstgov.gov/Contact/Elected.shtml

➢ Also realize that some of your "grants" might come to you from local government in the way of **matching funds** contributed along with what you generate yourself, or a pledge or promise of **reimbursement** should your group ever be needed to assist local first responders. Remember to get everything in writing since municipal administrations will change after each election.

Creative Bartering: Who says all your income has to be in the form of cash? As we mentioned before, some companies who wish to contribute to your cause may do so by donating goods or services your group or its members utilize, or that you'll transfer to victims later.

➢ This barter list is limited only by your group's **imagination**.

➢ A business may donate **meeting room space** for your public functions.

➢ Print shops could donate a certain amount of **printing** to help you with your fliers or newsletters.

➢ Radio and TV stations, and newspapers could donate **PSAs** or regular slots to help you with your public relations efforts.

➢ A hardware or tool rental store may donate a **certain amount of goods** or provide a certain **discount** to your organization.

➢ Your highway department might give you **free reflective road paint** to use in your house numbering project.

➢ In "The Group in Action" below, we'll discuss CCAPs and VCAPs, which are ways a business or individuals could provide goods or services through your group to help the public or victims of disaster. Since we'll mention them frequently, we'll go ahead and define them for you:

● **CCAPs**, pronounced "see caps," are **C**ivilian / **C**orporate **A**id **P**ledges such as you'd find in situations we've mentioned where businesses or individuals might pledge goods or services in the event of a disaster.

● **VCAPs**, pronounced "vee caps," are **V**olunteer **C**onsultants for **A**sset **P**rotection, such as a group of professionals that might act as a volunteer consulting group to save a hospital from going out of business.

Notes:

_____

_____

_____

_____

_____

_____

_____

_____

_____

_____

_____

_____

_____

_____

_____

_____

_____

_____

_____

_____

_____

_____

_____

_____

_____

The Group in Action

The way we've divided the activities, functions, and purposes your group could serve is done solely to organize the information in this book, and is not intended to limit or dictate what you do. Read through all these sections since our definition of a concept may not be the same as yours. If anything here sparks your imagination and makes you come up with a completely new idea, then this section will have served its purpose since no one can come up with every conceivable public service project.

Goals of The Group in Action:

1. To brainstorm the various things **group members could do to help each other**.
2. To point out the numerous ways **your volunteer group could be of service to others**.
3. To stimulate your imagination and **help you think of things not even on this list**.
4. To give you **ready-made projects and goals** so you can start helping others immediately.
5. To provide one more little way in which **this book might help make our country a better place**.

We've divided this section into:
1. **Neighborhood Watch.**
2. **The Service Functions.**
3. **Supply and Donation.**
4. **The Think Tank.**

Neighborhood Watch

We're starting this "Group in Action" portion with one concept that we felt needed to be discussed in depth and presented as a rather complete program. It's one way a **Neighborhood Watch** could be set up and operated. The ideas present in this section are taken from a weekly civil-defense / homeland security issues column we write called "Paul's Corner." It can be found on **Col. David H. Hackworth's** [1] **"Soldiers For The Truth"** site at http://www.sftt.org. The direct link to the archives is http://www.sftt.org/paulscorner.html. Following this Neighborhood Watch idea, you'll find a list of assorted service project ideas and discussion.

Taking Neighborhood Watch to a Whole New Level

There are *always* any number of angry, dispossessed, vengeance seeking groups strewn about the planet at any one time, and the only way they know to express themselves is through violence. On September 11th it was Afghani al Qaeda. Next week it might be twisted individuals from a nation we've always been friendly with. Who knows? Point is, there is certainly more coming and we have to do the little things that make a big difference in our own protection. Chief among these is increasing the awareness and preparedness level of the average citizen, and taking "Neighborhood Watch" to an entirely new level. Want to know how? We've made a list. Each of the sections below apply to civilian groups, and cover efforts that can take place before Law Enforcement or Government Officials are involved or can respond, whether the situation is a disaster, terror attack, or crime in general.

Basic Reasons for Helping to Fight Crime

➤ Our own protection. Each of us truly is our "brother's keeper." We'd rather call 911 and report suspicious gang activity in our own area than read in the paper that one of our neighbors was shot in a robbery attempt. None of us need to experience the personal loss, possible injury, financial loss, the downtime, or the potential trauma of a criminal act. In cases like this, an ounce of prevention is worth a *ton* of cure.

➤ Freeing up official resources. Question; which is easier on our first-responder resources, sending out the Cops to check on potential criminals, or sending out numerous Cops, and possibly Fire and Rescue units to the scene of a crime? Everything we can do of a preventative nature not only prevents damage to our own safety and security, but frees up resources that are needed elsewhere and for other things.

➤ "Lock up the tools." Who knows? The vehicle theft thwarted today might prevent the getaway car or car-bomb of tomorrow. Terrorists steal their tools, and they can't do it if we're watching.

---

[1] Col. David H. Hackworth, USA Retired, is one of the most highly decorated US soldiers. Books he has authored include **About Face**, **Steel My Soldiers' Hearts**, **Hazardous Duty**, and more. Information on Col. Hackworth and SFTT can be found respectively at http://www.hackworth.com and at http://www.sftt.org.

➢ Cut the funding. Have you seen the anti-drug commercials on TV? They talk about how if you buy drugs, you may be helping to finance terror activities. This is above and beyond the damage that drugs do anyway. It's tough for dealers to deal if we're all watching and reporting.

Networking and Training

➢ Contact your local Sheriff's Office or Police Department for information on how to start and organize a Neighborhood Watch group. They'll set you up with a liaison officer (or should).

➢ If things are handled correctly in your area, you'll be involved in training programs alongside organizers of other groups in your city. You'll have a chance to network and learn from the experience of others.

➢ You should receive some sort of training from your liaison officer. Much of this will involve things to watch for and the protocol of what to do if you witness suspicious behavior.

➢ Add as much additional training as possible for your group members on numerous related topics. See if there is an EMT, Red Cross rep, etc. in your area who is willing to provide extensive first aid training. Ask your local Fire Department to give a community talk on fire prevention and some instruction on limited fire fighting. People are going to try to help anyway, so they should at least receive some instruction on what to do, or more importantly, what NOT to do.

➢ Consider writing a free quarterly newsletter for all residents to keep them informed, interested, and involved.

Equipment

➢ Since the biggest function of a Watch group can be defined as "observe and report," the first item on your equipment list should be communications gear. This should include cell phones and either CB or FRS radios. You'll need the radios should the phone system be down. Also consider a backup, battery operated, intercom system between homes should your watch area be an apartment building or small complex.

➢ Some Law Enforcement liaisons will be willing to give your group one of their radios locked on a certain frequency. If they don't mention this to you, mention it to them.

➢ Observation equipment. If you're going to report suspicious activity, you have to be able to see it, right? Encourage your group members to own binoculars, cameras with telephoto lenses, and / or video cameras (Sony makes a great night-vision video camera called "Zero Lux"). Also encourage members to own microcassette recorders so they can verbally note important details such as suspects' descriptions, vehicle tag numbers, etc. without having to waste valuable observation time trying to write.

➢ Illumination. Flashlights are a must for every household for a multitude of reasons. Encourage group members to also own hand-held spotlights. We recommend the 1,000,000 candlepower models. Not only are these good for signaling, but you'd be surprised at the intimidation effect they have when shined on would-be burglars trying to break into a car down the street.

➢ Post-disaster supply. In addition to knowing who in your group can provide certain tools or gear, you might consider creating a community stockpile of tools, food and water, or clothing and other recoup supplies.

Regular Activities

➢ Start with member preparedness by making sure everyone's evacuation and isolation needs are met in advance and that everyone helps everyone else with their "Basic Home Prep" suggestions. Just like "barn raisings" of yesteryear, the group as a whole can go from member house to member house, making the structural safety changes recommended in "Basic Home Prep." To make for good relations with all members, do officers' houses last.

➢ Patrol your own neighborhoods. That's a given. You can organize foot patrols, assign members to certain scheduled "watches", or simply promote general awareness coupled with good communication. One asset you should consider is shut-ins or homebodies. You'd be surprised at how well neighborhoods are watched by people who spend all their time at home. You should also know who in your community is away, whether on vacation, in the hospital, etc., and **how to contact them** wherever they may be.

➢ Set up a communication system with redundant overlaps that will alert your members to neighborhood emergencies, all the way through national emergencies. The earlier "Communication" portion of the "**Foundation**" section will cover numerous ways to do this. Chief among these for small local groups is a simple phone tree. Since we've mentioned this a few dozen times already, let's cover some hints to make yours as effective as possible:

❑ Since it's likely this commo method will be used in an emergency, keep in mind that members will be frantically trying to look out for their own family and won't have much free time for the first few hours. Therefore, no one should be responsible for calling more than **two** other people. Officers and committee chairs might need to call four or five, but regular members should call only two.

❑ In addition to receiving a list of the people they're supposed to call, each member should have the contact info for the person who is supposed to call them. This way, the **phone tree can work upwards** or downwards and can even be started from the bottom and work its way back to the top.

❑ Just as families have **out-of-town contacts**, so should your group. Find a distant friend or relative of a group member, a person not likely to be affected by the same emergency you're in, and give them a few random member's numbers to call. Being random, this will help ensure that all the people on your phone tree's commo chain are reached, even if some of your members are unable to complete the calls they're supposed to make. The more out-of-town callers you can find, and the more people they can each call, the better. However, make sure that no member *receives* more than two calls. They won't have any more time to answer calls than they do to make them. Also, the less you tie up local phone lines the better.

❑ Speaking of phone time, you should decide in advance the **exact message** to be relayed for any scenario. This does two things: One, it saves time since a short, specific message is easier to relay. Two, it prevents the panic created when messages are blown out of proportion as they're relayed from one person to the next. Keep your messages short and clear. Some examples might be, "_____ alert! Turn on your TV," or, "_____ has occurred, we'll all meet at _____," and so on. Keep it all short, and to the point.

❑ You should form the tree so that one person at every few levels makes a call back to the top of the tree. This way, the officers know that pretty much everyone has gotten the word. These calls can go to officers or to your out-of-town phone tree assistants.

❑ An overlapping concept between the phone tree and neighborhood safety is to make sure your group is on the phone tree list of any neighboring industry or other facility that might pose a potential threat. An example might be a factory that may have a fire or HazMat incident, or a prison that may have a escape.

❑ R.E.D. Alert, online at http://www.redalertsystem.com, with direct contact at 4800 Curtin Dr., McFarland, WI 53558, 800-356-9148, or 608-838-4194. R.E.D. stands for Response to Emergency Deployment. This company essentially offers a commercial counterpart to your group's phone tree.

➤ Organize community safety programs. Organize community training programs on everything from fire safety to first aid. Also organize crime prevention seminars even for those that won't be active "patrol" members of a group. One related suggestion is to encourage neighbors to get car and home alarms.

➤ Encourage local hospitals to put on first aid classes, Fire Departments to teach fire safety, etc.

➤ **"Needs and Intel"** books. The group leader should compile a data book and safely store this sensitive private information for emergency reference. This book should list information about community members (you'll find forms in the **"Appendix,"** and a **spreadsheet** on the **CD**). Include:

❑ List shut-ins or handicapped neighbors so emergency first responders will know which houses might have someone who can't come to the door and who may need special assistance. This should include special-needs children.

❑ Medical oxygen. Relatedly, firefighters may need to know who might have something like this on-site that could be a potential HazMat problem during an emergency.

❑ Hazardous materials. While not industrial toxins, private citizens may own certain things that might present a hazard during an emergency. Examples would be medical oxygen, oxy/acetylene welding units, work related materials like painting supplies, insecticides, etc.

❑ Medical information. It would help first responders if they had limited medical info on victims, such as family doctor contact info, medical conditions, allergies, blood type, etc.

❑ Emergency contact. Your group should know how to reach each family's "Emergency Contact."

❑ Skills. Know who in your area might be a doctor, nurse, EMT, electrician, plumber, etc.

❑ Sex offenders. You need to know who in your neighborhood might be a potential threat.

❑ Pets. Pets need care too. Residents should have door or window stickers that tell first responders how many pets might need rescuing. Your data book should show the same info. You'll also find a "Dear First Responder" info template in the **"Appendix."**

❑ Create a graphic map showing the location of each house in your watch zone, the street number of the house, the name and phone number of the residents, and if you have any intel forms completed on them. **Hint**: You might be able to find a good graphical map showing houses and other useful features through your local Fire Department, Public Works department, or Building and Zoning office. This would make a much more accurate and easier to use map. **Hint #2**: After a disaster, you'll want to use this same map to mark the location of structural damage, non-ambulatory victims, and other problems so you can hand the map off to first responders as they arrive on scene.

❑ **Note**: Be very careful with all this info! It's private and secure and should be treated as such. Make sure all the info is stored with a trusted individual, kept updated, and only shown to rescue personnel, and only during an emergency. The "Important Contacts" and "Family Data" forms will help you gather info.

➢ Gameplan by scenario. Come up with your own reactions to potential scenarios and situations. These can include general crime like burglaries, drug deals, etc. Include all items from the "**Reaction**" section, and extend the concept to cover unexpected things like a plane crash. (This is a good subject for brainstorming.)

➢ Overlap with other groups. You should communicate with other Watch groups. Something might be happening a few subdivisions away that will affect your area, or vice versa. Also communicate with security people at nearby locations. Examples would be local schools, malls, hospitals, etc. They may warn you of potential problems or vice versa. Two heads are better than one.

➢ Community stockpile. Consider encouraging group members to create a small community "stockpile" that is kept securely stored at a local storage facility; something away from the neighborhood but nearby. Each family could store backup emergency supplies, the group could create its own "goodwill" collection to give to others or keep in reserve for itself, etc. Stored items can include clothing, water, food, tents, tools, etc.

➢ Set up a **neighborhood rendezvous point** or marshalling area. You might use this spot to hold meetings, as a gathering point and treatment area for the "walking wounded" after a destructive event, as a potential command post location for first responders, or as an evacuation pickup point for those neighbors that will travel together. Choose an area free from potential hazards, accessible by all members, easy to give directions to, and easily reachable by emergency vehicles. Make sure your the Law Enforcement liaison officer for your Watch Group knows this is where you'll gather after a disaster or community emergency.

➢ Establish a block "safehouse" for kids after school, or during an emergency. See if any residents are home all the time, preferably with kids of their own, and would volunteer to be the place that the neighborhood kids could go in times of trouble if their own parents weren't home.

➢ Set up a sort of "escort" service for the elderly or other special-needs people, or potential crime victims. Protect them while they're out and about, offer to help with some of their transportation needs (you might even car-pool a group of elderly people once a week so they can do their shopping), offer to call them at a particular time each day to check in, and generally be there to help keep an eye on their needs and safety.

➢ Look for potential sources of outside danger. The biggest threat to a community might not come from inside the neighborhood. An example of this is a chemical plant or other industry located nearby. If something happens there, your group should be aware of it so they can notify all members. Make sure you're on the "notify in case of emergency" list of any local factory, etc. that might pose a HazMat threat. For example if the chemical plant 3 miles upwind from your neighborhood experienced a large fire with the potential for explosion, you'd want to know, right? An alert from this site would allow you to alert your neighbors for a possible evacuation or a shelter-in-place. (This is another reason you'd want to overlap phone trees.)

➢ Relatedly, regarding a potential attack or disaster, your rotating watch commander should maintain copy of a community events calendar so you'll know where your neighbors might be at any given time. For example, if the rotating watch commander is alerted to a tornado approaching a downtown area, and they know that many residents are at the theater downtown, they should start making calls to warn those people.

Post-Disaster Activities

➢ Plan potential evacuations or isolations as a group. Determine what each neighbor will need, and what each can offer. **Hint**: If your neighborhood is a large one, or if everyone in your neighborhood is an active participant in group functions, you might consider sub-dividing your group into sections of 10 or 12 homes each to help organize your efforts. Put a "division chairperson" in charge of each.

➢ Security patrol after a disaster or destructive event. Looters are one of the biggest threats after the primary threat of a destructive event has passed. Though your watch members are certainly going to have a lot of work to do with their own families and homes, you'll need to band together for security reasons.

➢ Create a "Needs List" after a disaster. First responders and city officials will want a damage report to include people and property. Help them by gathering this info early.

➢ Also create a "we know to be okay" list just as you'd create a "casualty list" in order to turn over to Law Enforcement. For example, if a neighborhood is hit hard by a tornado, land lines will be down, cell towers may be down, debris will prevent vehicle use, and relatives of residents will be calling to find out if loved ones are safe. Volunteers should help first responders compile a list.

➢ Directing traffic. Odd as this may seem, consider this: Many neighborhoods and residential areas are built as secluded circles with only one entrance / exit to cut down on through-traffic. What if there's a massive evacuation and no one can even get out of the neighborhood for all the traffic on the main road? What if there's a small accident, downed tree or telephone pole right there? How's everyone going to get out? Your Watch Group needs to step in and help until officials can arrive (and especially if officials can't arrive).

➢ If a disaster or destructive event is localized to your area and the rest of your city is functional, you might designate a "shopper" or someone to go after supplies so that all the others can stay and help with rescue and/or rebuilding activities.

Possible Extended Missions

"Extended missions" concerns helping to patrol areas outside your immediate neighborhood. These might be sites that pose a potential threat to your neighborhood, or they could be sites that have a high probability of attack and need all the security help they can get.

> There might be a nearby mall where many of your group's children gather. Consider volunteering some time as an "undercover observer". If you decide to do this, you might want to approach mall security with "here's what I'm going to do, how do I communicate with you guys?" rather than "do you need some help with..." Better yet, ask your Police Department's Neighborhood Watch liaison officer do it for you.

> Patrol infrastructure sites adjacent to your neighborhood. The Police will be doing this anyway, but as we said earlier, two heads are better than one. Keep an eye open for anyone attempting entry, conducting surveillance, or "acting suspicious", around power substations, telephone relay stations, fire hydrants, water towers, water pumping stations, fuel storage facilities, private airports, etc. Network with any on-site security.

> Speaking of patrols, though many groups prefer to patrol in groups and on foot, there's nothing wrong with using individual or even multiple vehicles. Just make sure you have a way to communicate with each other.

> The same applies to nearby industrial sites. Some sites might pose a secondary danger to your neighborhood, or they might provide a source of tools or materials to terrorists.

> If your neighborhood is near an airport, consider this additional "extended mission." You'll want to watch for anyone trying to use a shoulder-fired anti-aircraft missile. Though a remote possibility, it's still a distinct possibility and the only way to stop it from happening is by early detection and rapid reaction. Remember these 2 things. One, the people "casing" your neighborhood might actually be looking for a good launch point for a Stinger or other missile. Two, if they succeed, the aircraft has to come down somewhere, and if you live near an airport, that might mean it'll come down on your house. Stay alert.

> Finally, you might want to use your core Watch Group as a civic volunteer group unto itself, and perform any number of other functions we'll list throughout this section. Chief among these functions would be disaster prep training for your community, or a Think-Tank, as we'll discuss later.

Sources and More Information

> One of the best info sources is the National Sheriff's Association's Neighborhood Watch Program. They're online at http://www.usaonwatch.org, or you can reach them at the National Sheriff's Association, 1450 Duke Street, Alexandria, VA 22314-3490, 703-836-7827, Fax: 703-683-6541.

> Also contact 1-800-WE-PREVENT (800-937-7383) or log on to http://www.weprevent.org/.

> Visit the National Crime Prevention Council, online at http://www.ncpc.org/, or find them at 1000 Connecticut Ave. NW, Thirteenth Floor, Washington, DC 20036-5235, 202-466-6272.

> The state of Michigan has a good explanation of the basics of setting up a watch program at http://www.preventcrime.net/NeighborhoodWatch.htm.

> As an example of how to start a neighborhood group on a smaller scale, visit http://www.3steps.org.

> If you decide to extend your watch to cover neighboring infrastructure facilities, you might want to consider a liaison with InfraGard. You can find out more at http://www.infragard.net, or through the National Infrastructure Protection Center page at http://www.nipc.gov/infragard/infragard.htm.

> National Association of Search and Rescue: http://www.nasar.org, 4500 Southgate Place, Suite 100, Chantilly, VA, 20151-1714, Telephone: (703) 222-6277, FAX: (703) 222-6283.

> One group that offers a reward for many types of felonies and serious misdemeanors is "We Tip." They can be reached at 800-78-CRIME, or online at http://www.wetip.com.

> To report someone who may be purchasing ammonium nitrate fertilizer for criminal activity, call the Bureau of Alcohol Tobacco and Firearms (ATF) at either 800-800-3855, or at 8880-ATF-BOMB (800-283-2662).

> Visit the National Association of Citizens on Patrol. You can find them online at http://www.nacop.org, or contact them at P.O. Box 727, Corona, CA 92878-0727, 909-898-8551, Fax: 909-279-1915.

> The US State Department has a specific counter-terrorism reward established. It's called "Heroes" and you can find more information by writing P.O. Box 96781, Washington, DC 20090-6781, or by calling 1-800-HEROES-1 (800-437-6371), or email: Heroes@Heroes.Net

> For some downloadable information on how to put your "Suspicious Activity" forms (found in the "**Appendix**") to good use, visit http://www.huachuca.org/huachuca/index.php?lcat=SR*.

> For ways your group could help crime victims, visit the National Center for Victims of Crime at http://ncvc.org, or at 2000 M Street NW, Suite 480, Washington, DC 20036, 202-467-8700, Fax: 202-467-8701.

- Also contact the federal Office for Victims of Crime, online at: http://www.ojp.usdoj.gov/ovc, or contact them at Office for Victims of Crime Resource Center, National Criminal Justice Reference Service, P.O. Box 6000, Rockville, MD 20849-6000, Phone: 1-800-851-3420, (TTY 1-877-712-9279).

- To visit the UK Watch site, go to http://www.neighbourhoodwatch.net.

- We've included a couple of really good Neighborhood Watch brochures on our **CD**.

- To **report specific information you feel may be indicative of a pending terrorist strike**, call these numbers: 1) Call your local police department, 2) Call your local FBI branch office (www.fbi.gov) 3) Call or write: 1-800-USREWARDS (800-877-3927) or www.rewardsforjustice.net.

- An example of a novel approach to enhanced security (or a model of something you could set up locally) can be seen through the US Homeguard's approach to internet monitoring of security cameras by volunteers. Visit their site at http://www.ushomeguard.org/index.html for more detail.

The Service Functions

Now that we've covered Neighborhood Watch in depth, let's move on to some general concepts regarding different types of services you and your group could offer to your members or to others. Here we'll focus more on contributions your group can make that center on a function or activity. Though the "Supply and Donation" section that follows will also require activity, the focus will be more on the items supplied than the act of collection and distribution. Let's cover a few service categories and ideas. Remember, your group can either recruit the properly trained members and provide a service directly, or you can simply locate all qualified personnel or equipment in a given area and create a database for use during or after a disaster. Our recommendation is to do a little of both and offer the service directly, and also have a relationship with any other outside sources you may ask to volunteer later. Here are some examples:

- **Labor**: Simple manual labor. This would be the group of outside help that might assist first responders or disaster victims. Some forms of direct labor could include:
  - The group as a whole helps prep each member's car and home.
  - Cleaning up debris after a disaster, stacking sandbags in a flood, etc.
  - Search and rescue, as with a missing child.
  - Help in serving food to disaster victims or first responders.
  - Setting up shelters as a solo project, or assisting at established shelters.
  - Replacing street name signs after they've been blown away by a destructive event.

- **Skilled labor**: This would be people who could help when infrastructure sites were damaged, or when public shelters needed to be repaired, etc.. Examples would be:

❑ Heavy Equipment Operators	❑ Private Pilots (fixed wing or helicopter)
❑ Electricians	❑ Welders
❑ Carpenters	❑ Plumbers
❑ Brick Masons / Concrete Workers	❑ Truck Drivers and CDL Holders
❑ Ham Radio Operators	❑ Chefs and Cooks
❑ Computer Technicians	❑ Security Guards to help Police

  ❑ Professional disaster cleanup and reconstruction services to help citizens rebuild.

- **Volunteer professionals**: This would include the "white collar" help in non-labor areas. This category is much bigger than can be adequately covered here. For that matter, so is the "skilled labor" category above.
  - Medical personnel could donate time and expertise at aid stations if not already utilized elsewhere.
  - Psychologists can host "post traumatic stress" radio talk shows or group seminars after a local disaster.
  - Veterinarians can volunteer to staff mobile kennels to care for animals at a refugee shelter.
  - Attorneys, accountants, or insurance agents may help by filing aid or insurance claims.
  - Interpreters can make sure all victims or refugees can communicate.

- **First Responder relief**: Anything you can do to directly assist or to free up first responder personnel is a good thing. A few examples are:
  - Watch groups can patrol their own neighborhoods to free up Police Officers for other duties.
  - Watch groups may expand outside their immediate areas, as explained earlier, to help patrol adjacent infrastructure, or even direct traffic if need be.
  - Licensed Security Guards can volunteer to take over lower level Police functions.

- Staff minor first aid stations at a disaster scene (depending on your training) to assist EMTs. (Your group can at least handle Band-Aids, splinters, etc. leaving EMTs free to handle more severe cases.)
- Help animal control round up strays, or care for abandoned pets.
- One of the best ways your group could help first responders is to help them control all the "convergent" or "walk-up" volunteers that may show up at a disaster scene. By coordinating and controlling the walk-up volunteers, you would be taking a major headache away from the scene commander. Your group could register, quickly interview, and organize the walk-ups for duty, and keep them in a safe area until the scene commander asks your group to send people in. Organize these folks by the same criteria you'll find on your "Service, Skills, and Assets" form. Give each walk-up a color-coded name badge or vest based on their skill, ability, or desired involvement. That way you can call them up by groups, and all of your members will know from the name badge or vest, which function the walk-up will be involved with.

➢ **Database or direct involvement**: Any of the general labor or skilled labor functions listed above could be provided in either or both of two ways. The simplest and most common is direct involvement where your group members would provide the labor or service listed. Another service would be to create a database of area or regional people offering a particular skill. For example, after a severe disaster, a particular area may need all the heavy equipment operators they can get their hands on. Similarly, first responders may need additional helicopters. If your group already had the contact information and some loose relationships established through CCAPs or VCAPs, all local authorities would have to do is ask you to call in the help.

➢ **Communication**: In any kind of emergency, communication is critical. Officials will need all manner of communications options. Families will need to contact loved ones for the purposes of checking their safety or arranging a rendezvous, and people forced to evacuate will need a way to contact those they left behind.

- At the bare minimum your group should have an internal phone tree so that in the event of an emergency, every member will know about it in the shortest time possible.
- Utilize part of your group's public **website** as a bulletin board for citizens needing to contact other family members after a disaster. This will not necessarily tie up phone lines needed by officials since the website could be accessed from outside a disaster area after an evacuation. As an added service, some of your group members could monitor this bulletin board webpage and take an active role in trying to reunite separated families, either through phone and internet attempts, or through Ham Radio operators or groups.
- If you adopt a rest area, as we'll discuss below, one of the things you could do is set up a **communications kiosk** where you'd accept written messages from evacuees and try to make emergency phone calls or find other ways to pass messages for them. You can use the "Update Cards" found in the "**Appendix**" as a form for evacuees to use. Have the sender use the blank side of the card for all the "from" and "to" contact info. (Naturally, you'd want to limit the number of messages a family could send.)
- Area **Ham radio** operators can offer a communications option to local civilians and officials if phone lines are down or other means of communication overloaded. See the "Communication" section for sources.
- **Yahoo Groups** is a great way to set up a mass member email group. You should already have one set up for your group's internal communications, but you should also set one up for your community, region, or volunteer group's coverage area. Encourage local schools and businesses to become members of the group to post quick and informative messages to students, employees, and their respective families.
- Your group could provide runners or **couriers** to physically carry messages from one location to another in lieu of any other form of commo. This could be done from shelter to shelter as family members relay messages to find each other, or it could be for first responder command posts as a way to offset any loss of electronic communication. One way to help local officials communicate in a situation where regular commo is down, is to form a bicycle courier team and relay written messages to and from command posts, etc.
- In addition to runners, **signaling** is another form of service you could provide if you had the people trained to do it. Morse code via spotlight, and semaphore are two examples. (See "Communication.")
- In keeping with signaling, your group could do your community a tremendous service by helping to upgrade your local EAS system, or by relaying alerts to your community. See the "Communication" section and also see our EAS suggestions in the file named "Studies" on our enclosed **CD**.
- Set up physical pen and paper communication **bulletin boards** at local emergency shelters, and collect and transmit notes left by victims and refugees. Do everything you can to help families communicate with each other and find missing members. In fact, you can have evacuees fill out the "Update Cards" you'll find in the "**Appendix**," and use those as the notes to be passed along to loved ones.
- After the main part of a disaster has been handled and it's okay to use the phones, local **cellular phone** stores can offer to let refugees come to their store to use a phone for free for a certain amount of time. Local payphones should offer the same free service for a limited time.
- Local libraries, computer stores, or internet cafes could allot a certain amount of free **email** time so people could locate missing friends and relatives.
- Have a **cell phone drive** to provide "911 phones" for group members and special-needs people.

➤ **Education**: Though we mentioned some of these as fundraising projects earlier, training in various areas and subjects is sorely needed by the general civilian population in most areas and should be done for free where possible (except for cost of any educational materials). You could offer these as community courses:

❑ First Aid Classes ❑ CPR Training ❑ Disaster Prep Training
❑ Basic Fire Safety ❑ Limited Fire Fighting ❑ Home Safety Courses
❑ Automotive Repair ❑ Boating Safety ❑ Canning and Food Prep

● Teach local businesses how to shelter and care for employees.

● Teach businesses how to help their customers in disaster prep. For example, pet stores should post reminder checklists of emergency supplies for pets. Hardware stores in hurricane areas should post how-to materials and preprinted shopping lists for covering windows.

● Teach your community about various government and private programs that are available to them for free. The government is great about creating programs, but falls short in the marketing department. Most people don't know about the majority of programs that exist. Most important would be post-disaster relief, whether grants, general assistance, or mental health assistance. List all programs on the public awareness section of your website or newsletter.

➤ **Transportation**: All manner of transportation during an emergency should be considered and added to your list if your group decides it wants to focus on transportation as its service. Let's look at a few examples:

● The Civil Air Patrol is a great example of a group focused on transportation. Relatedly, your group could arrange CCAPs with private aircraft owners, to help in mass evacuations.

● You could also make similar arrangements with boat owners if your city lies on a waterway.

● Help your neighborhood evacuate in an emergency by arranging for transportation to pick up residents who gather at your neighborhood's meet point as discussed earlier.

● The sole purpose of your group might be for an evacuation scenario only. In this case your focus would be on transportation and maybe on establishing a set and secure destination.

● You could specialize in helping special-needs people evacuate. This would include the mobility, sight, or hearing impaired, or single parents with more than one child. (Your group should at least help these people register their needs with your local EMA, or Fire Department.)

● Help first responders evacuate another neighborhood in a smaller disaster situation.

● Set up a rendezvous point to pick up group members in order to help each other evacuate.

● If your area is flood-prone, store boats for use during a flood.

● Help evacuate livestock, or help farmers line up CCAPs with trucking companies or local truckers who would be in a position to help if needed.

● Using a full-size school bus, you could set up a transportation-only ambulance. You could carry a large number of stabilized patients in one trip, freeing up regular ambulances for the more traumatic cases and allowing EMTs to stay on-scene longer at a disaster.

● Instead of waiting for a disaster, you could help the elderly get out and about, help carry cancer patients into their treatment appointments, etc. The need for available transportation is an extremely large one.

● Transportation assistance doesn't have to involve vehicles. You could help local officials during an emergency by directing traffic at various intersections freeing up Police Officers for more important duties.

● Additionally, you could perform service projects to make transportation more efficient. One idea was discussed earlier, and that was using reflective road paint to paint house numbers on curbs and street names on curbs at intersections, making it easier for first responders to find locations, and to make it possible to locate individual properties and locations after a disaster or attack has leveled all house numbers and street signs.

● You could also offer to paint numbers on the roof of each local school bus. Many jurisdictions do not require this, but it's a highly useful feature either in the search for a hijacked bus, or in airborne coordination a mass municipal evacuation.

● Relatedly, since so many entities now use GPS devices, your group could use a GPS device and go from intersection to intersection recording the coordinates on a map to be copied and given to all local first responders. If all street signs are blown down and your street name painting project not yet complete, this map could be a life saver as it would allow first responders to navigate via GPS, and without signs. In addition to the paper map, you can list your GPS points in softcopy format in a simple database such as Microsoft's "Access." After intersections were plotted, you might offer to plot individual locations.

● Your group could install or rig certain flood level indicators around town (after checking with local emergency management) to let motorists know if roads are still passable. One way to do this is paint water-level marks at the bottom of sign posts or telephone poles. Mark these increments using green, yellow, and red paint. Tell people, "If you can see green, go. If you can only see yellow, go slow. If only red? Whoa!"

- We've seen other groups take the small glued-on-the-road highway divider reflectors and put one in the middle of a road lane in front of each and every fire hydrant. This makes it easy for Firefighters to quickly locate needed hydrants in low-light situations.

➢ **The "School Bus" projects:** (By the way, if you live near water, any of these school bus ideas could be completed in, with, or for a **boat**.)

- **Evacuation:** A full-size school bus could be used in its originally intended purpose and that's transportation. You might provide this as a service your members, or you could help your city's first responders by using your bus to evacuate an immediate disaster area.

- **Ambulance:** Though you wouldn't necessarily provide the medical assistance, most disasters will see a large number of "walking wounded." Rather than tie up ambulances or increase traffic by sending one person per car to the hospital, your bus could take quite a number of people to the nearest ER in one trip.

- **Command Post:** Not all towns are able to afford their own mobile command post. We've seen examples where volunteer groups took vehicles and outfitted them with all manner of communications equipment, computers, cameras, and basic creature comforts, and offered the vehicles to their local first responders as their very own command post. Being private ventures that were assembled by economy-minded people, these units were built at a fraction of what most command posts would cost.

- **Communications Post:** Take a large or small bus, install a generator, and fill it with Ham Radio setups; CB, FRS, GMRS, and FM transceivers, relays, and repeaters; laptops with cellular modems, and with cell phones from different providers and you have a rolling communications center that can either augment and assist your Command Post, or you could help refugees at your local shelter communicate with loved ones.

- **MASH Unit:** You wouldn't necessarily supply the medical expertise, but you might help supply equipment to those who do. First responders are realizing how inefficient it is to ferry massive numbers of victims to ERs at one or two per ambulance. Therefore, many have begun making plans to set up the best on-site aid stations possible and to reserve the ambulances for the most critical cases. Your group might outfit a bus with supplies for these aid stations. You might provide anything from tents and cots, to scissors and bandages. As you liaison with your various first responder groups, ask them what they need most.

- **Flood Prep:** In this case, your bus would be a mini-warehouse on wheels for flood related supplies. You could have water pumps, empty sandbags, shovels, plastic sheeting, plywood and 2x4s, and could tow a trailer with a few canoes or John boats. If a flood threatens, your group and supplies would be already packed and ready to be on-scene.

- **Mobile Kennel:** Due to health concerns, publicly run refugee shelters will not allow pets. A bus or RV outfitted as a kennel would mean the world to pet owners forced to evacuate. Work with your local vets.

- **Generator:** A bus outfitted with either numerous home generators, or one large one, could either be the main power supply at a disaster scene, or it could help supply power for a stricken neighborhood. You could also have another vehicle outfitted with a propane or natural gas tank.

- **Water:** Water will be vital after any disaster that disrupts the supply. A tanker vehicle could literally mean a life or death difference to some people, and delivery to a neighborhood would mean fewer and smaller crowds gathering elsewhere for water. Having fewer and smaller crowds is critically important if your emergency involves disease.

- **Cafeteria:** Your group's sole purpose may be to supply the needs of first responders. A great way to do this would be with food, coffee, water, etc., distributed out of a "cafeteria truck." A variation on the theme is to deliver food supplies to local disaster victims at shelters, or those stuck in their own neighborhoods.

- **Motorist Aid:** Some communities have H.E.R.O., or Highway Emergency Response Operator, trucks that patrol local sections of highway helping stranded motorists and assisting Law Enforcement at accident scenes. These units carry basic tools and supplies to help motorists get on their way, and basic safety equipment for the accidents. Even if your area already has something like this, it could probably use more. Also, you might help evacuees of a distant disaster if your area is near any sort of official evacuation route.

➢ **General shelter:** Any number of places within a city could offer to be emergency shelters. As official facilities become overloaded, private gyms or spas could turn basketball courts into hospitals, storage facilities could turn empty units into emergency housing, individuals could pledge a spare room at their house, apartment complexes or motels could pledge empty units, and so on. Your group could coordinate gathering pledges for this alternate shelter.

➢ **Adopt a shelter:** Your group could actually adopt a shelter to be used as an emergency shelter. Many places across the country have structurally sound buildings that at one time were part of the civil defense network. Though the supplies at most locations have long ago been redistributed for other uses, the buildings may still be used. Your group might want to make an arrangement with the building's owner to allow its use as an emergency shelter. Your group's job would be to advertise the shelter's availability, supply the shelter when needed, and supervise its use. Or, instead of adopting an entire shelter, you could provide partial support, or services at a government-controlled shelter. Some suggested services are:

- Providing a mobile kennel since most shelters won't allow pets due to health concerns.
- Serve food and/or water, or supply eating utensils such as paper plates, plastic utensils, etc.
- Provide sanitation supplies such as "Porta Johns," toilet paper, paper towels, soap, etc.
- Offer your resupply shuttle service to handle the needs of this particular shelter's guests. An example would be emergency prescription refills.
- Operate a "Commo Bulletin Board" at the shelter where guests place notes for people they're trying to find or contact and your group takes the information and tries to help get the message delivered via Ham radio or other communications networks you've developed.
- Disaster news or other current events, could be the focus of another bulletin board. Shelter guests will want to know what's going on and when they can return home.
- You could also operate a "Barter Bulletin Board" where shelter residents in need of certain things, or who have items or services to trade, could post notices.
- If your shelter is playing host to refugees from other cities, you could provide "local info" packets, similar to your Evac Atlas that offered information about crucial local services such as hospitals, free phone calls or internet access locations, or other emergency services or supplies.
- Provide an "Entertainment Director" to help keep shelter guests occupied, and their minds off the situation. Boredom and "too much time to think" are going to be enemies to victims.
- Provide a "Constable" or have your Sgt. at Arms act in a security capacity to keep a semblance of order among your shelter guests. **Note**: If you decide to operate your own public shelter, make sure you get the proper insurance, and any necessary certificates or credentials from your local government.
- If you have members with any medical experience, and they aren't required at the disaster site, ask them to be on hand at your shelter where minor injuries and stress-related illnesses may need attention. What if a pregnant woman experiences stress-induced labor?
- Depending on the expertise of your members, you could offer any number of "recuperate" services designed to help get victims back on their feet and their life back in order. This could range from any sort of professional counseling, all the way through legal help, or claim and document filing. Is anyone in your group an attorney, insurance adjuster, or Notary Public?

➢ **Fallout Shelters**. If your group truly feels your area is at risk of a nuclear strike or resulting fallout, and you want to take on the project, there is no reason you could not adopt a shelter and fully outfit it to be a fallout shelter, or you could have your own constructed. The only decision is where you'd locate this project and if it would be a public facility, or only for your members.

➢ **Victim Assistance**: Operate an independent H.E.R.O.-type unit (Highway Emergency Response Operator) to help stranded motorists. We've seen quite a few groups outfit an old school bus or panel van with all the necessary highway safety gear, first aid kits, fire extinguishers, and basic automotive tools to provide service on the roadways.

- In addition to the motorist aid unit, your group could "adopt" a section of highway in your area. This could prove highly useful if this highway were part of a planned evacuation route from another area. For example a coastal city near you might be evacuated during a hurricane. Your unit could aid motorists experiencing problems as they passed through your area. If you do this, you might want to **monitor channel 9 on your CB**. Officially it's the emergency channel, but these days very few Law Enforcement agencies monitor the channel for calls.
- You could also "**adopt a rest area**" if your location lies along an official evacuation route. You could set up a small station offering free coffee and snacks, communication assistance, maps and local information, limited auto repair services, or access to emergency supplies people might have forgotten like pet food, etc. You could also augment the rest area's facilities by bringing in "Porta-Johns," and supplies.
- If your group's plan is to do either of the above two items, you might also want to consider creating an **Evac Atlas** for the particular route you're on and the most likely destination(s) along that route. Evacuees are going to need all the information they can get about services, sources, communication, etc. and many will not have planned as they should. **Note**: If you adopt a rest area, streamline your services so that evacuees will be back on the road as quickly as possible. If they stay around the rest area it will fill up and others in need will not be able to get in.
- Operate a supply shuttle to provide goods to refugees stranded at shelters in your area. If you have a state-wide network, you could have out-of-area members comb their stores for supplies to bring to shelters in your area. **Note**: If you intend to do this, it's crucial that you have a relationship with, at the very least, city officials and that you receive some sort of credentials that recognize your role in this service. After a destructive event, movement to and from an area may be restricted by authorities.
- Similarly, you may live along a designated evacuation route between a disaster site and a larger metro area that's a designated evacuation destination. Your local authorities may decide to block access to your town in order to preserve local supplies and infrastructure and keep evacuees heading to the proper

destination. If this is the case, various supply facilities such as **gas stations at highway exits** may still be accessible to evacuees, and will more than likely be overwhelmed and drained of all supplies in short order. Your group might offer to set up some sort of **resupply shuttle service** for these outlets in the event such a plan is executed. The goal is to provide an ample supply of goods and gear evacuees might find necessary. You could also "adopt an exit or gas station" the same way you'd "adopt a rest area" as mentioned above.

- One way a volunteer group can help its community after a terror strike is by helping the community, and victims in general, get back on their financial feet. The Office for Victims of Crime, online at http://www.ojp.usdoj.gov/ovc, has numerous services, publications, and grant information. You can reach them online or contact them at the Office for Victims of Crime Resource Center, National Criminal Justice Reference Service, P.O. Box 6000, Rockville MD, 20849-6000, 301-519-5500, Toll-Free: 1-800-627-6872, TTY: 1-877-712-9279, or the main numbers 1-800-851-3420, TTY: 1-877-712-9279. They also have a Terrorism Victim Hotline: 1-800-331-0075, TTY: 1-800-833-6885.

➢ **Utility Pooling**: The only drawback to being technologically advanced, is our dependence on utilities. Many will be unavailable after a disaster.

- Water supply tops the list of necessary utilities. Small tanker trucks with pumps and hoses are not difficult to make. You'll need pumps and hoses because you might not be able to get in to some immediate areas that will need water. You may need to run a hose to pump the water in to stricken neighborhoods.

- In another take on the water supply, small neighborhoods could install a central cistern, or your group's service could be to install water tanks in people's homes. (Yet another useful fundraising and PR concept.)

- Neighbors could also pitch in and get a "community" electric generator. This would work well in a smaller, closed community such as with a condo association.

- After a disaster that left some homes damaged and others not, some utilities could be run by neighbors from functional houses to damaged ones. Water and electricity would be easy, though natural gas might be tough unless you took our advice earlier about always having the tools and expertise necessary to deal with the problems you may face.

- Your group could also supply natural gas, fuel oil, or even fabricate a much larger mobile generator to aid stricken neighborhoods or first responder command locations that need extra power.

➢ **Supply storage**: Not everyone can afford the monthly rental fee of a storage unit. A storage unit would allow one to keep a certain amount of family supplies safely stored in an away-from-home location. One volunteer group we've seen has provided this as a service for its members. A secure storage facility in a safe area of town (high ground, away from hazards, safe spot on the Threat Map, etc.) was chosen, and a couple of units were rented. A local school donated a few sets of lockers and these were placed inside the units. Each member family had their own locker in which they could keep an additional Bugout Kit, or a backup of non-perishable supplies. Each family locked their own locker and group officers kept the keys to the storage units. **Note**: You'd only want to do this for local group members or others you trust. You wouldn't want just anybody storing who-knows-what on property you've rented in the group's name.

➢ **Miscellaneous**: One of the best and yet most generic ways to help is to **find your local municipality's or first responder's biggest need and help fill it**.

- Again, we'll mention the mobile command post idea. If your community can't afford one, make them one.

- Set up a barter service to help disaster victims. Cash will be in short supply for most, but people may still have items or skills to barter. Your group could either help coordinate swaps by matching the supply to the need, or set up a center where items were brought in to trade for others, and your group safeguarded the goods and set fair trade values.

- Set up a hazardous materials reclamation center. After a destructive event, the amount of general household chemicals that are accidentally dumped into the environment are staggering. Get with the EPA and/or your local Fire Department on the best way to go about helping

- Depending on your group members' HazMat expertise, you could offer a service that goes in to permanently contaminated and off-limits residences to reclaim and decontaminate personal property for the home owner. Some scenarios could see a permanent area-denial situation where this type service would eventually prove rather useful.

➢ **Spread the word.** As we've said before, since you're reading this book and taking these steps, you're probably the more responsible one in the neighborhood. Please mention this book to folks that need it and make sure they know we're trying to help. For that very reason, you'll find a coupon for this book in the "**Appendix**." It's there to help you help your friends.

Remember, none of the services mentioned will ever be used unless citizens and officials know they're available. Good PR is going to be as necessary as the service you offer.

As we mentioned above, some services provided could either come directly from you, or you may simply provide a collection of available sources, such as knowing who all the licensed heavy equipment operators are in your area. Supplies are the same way. Your group can actively collect, store, and distribute supplies directly to victims, to first responders only, to larger volunteer groups, or you can arrange for their acquisition in the event of an emergency.

Arranging for supply acquisition and emergency pledges is a common practice, and in some areas is called CCAPs (pronounced "see caps") and stands for **Corporate / Civilian Aid Pledges**. Sometimes, people and businesses don't want to contribute anything unless there's a dire need. However, once the need arises, they'll be there as promised and without hesitation. Don't dismiss anyone simply because they can't give you time or goods at the moment. See if they're interested in agreeing to a **CCAP**.

In a severe disaster, local supply will be more important than service, as resupply from outside the local ground zero will be a while in coming. For example, after Hurricane Andrew, absolutely no one was allowed into the area from outside regardless of their cargo or intent. It's extremely important that every group include some sort of supply as part of its function, even if the supply is a secondary function intended only for group members. You can help ensure this supply either through efforts after a disaster, or through helping people in their disaster preparedness.

➤ **Water**: Water will be the number one need.

- Start with group members and make sure each has a sufficient supply at their home for their family.

- Educate the general public through your PR and educational efforts on how to store water at home and how much to store. The more self-sufficient the general public is, the better off they'll be. Our "**Isolation**" section earlier has all the info you'll need.

- As we stated before, you can take on the project of installing additional water tanks in your members' homes, in various closed communities, etc. Or, in your public education you can urge that this be done. Individual citizens aren't the only ones who could make an effort at storing additional water. Any and all businesses could do the same.

- Bridging the gap between water and food is ice. Ice can keep food fresh, and after the ice melts, you can drink it. We saw one Church group that had an interesting perc for its congregation members. The Church had several chest freezers in its basement hooked to backup generators in the event of an emergency. In the freezers were numerous 2-liter plastic bottles almost full of water and frozen. In the event of a mass power outage, congregation members could get 2 or 3 of these bottles to keep their food fresh longer and to provide some drinking water later.

➤ **Food**: Eventually, food will become scarce. However, disasters are not the only time food supply would be useful or appreciated. Let's look at a few considerations.

- Just as your members stored water, they should store enough food for their family.

- You could help your local food bank in order to feed the poor and to help stockpile in case of a disaster.

- Your group could also secure pledges from area restaurants for a certain amount of food or coffee, etc. to feed first responders in certain situations. Naturally, restaurants would not want to see you come collecting food during each and every little fire, but during a larger emergency, most would be glad to help.

- Similarly, if a restaurant or grocery store were in a stricken area, their freezers and coolers disabled, or they were at risk of being looted and stripped of their goods, most would rather their goods be loaded up and used to feed local victims or emergency workers than let it spoil or go to looters. Contact restaurant managers and grocery store owners to ask them what their contingency plans are in such a situation.

- If restaurants or grocery stores are reluctant to pledge to the general public or an unknown group, you could be the go-between for these local food supplies and the Red Cross who could certainly use the food in their local response actions.

- Just as we mentioned ice as water above, ice is important for food storage. Your group could send trucks to neighboring cities to stock up on ice (regular and dry ice) to bring back for water and food storage.

➤ **Sanitation**: In addition to the immediate survival items of food and water, disaster victims may need sanitation assistance before services can be repaired. There are several ways your group could help:

- Find all area suppliers of "Porta Johns" or other portable chemical toilets. You can arrange for these suppliers to provide your area a discount, you can simply keep the database on hand for officials, or your group could foot the bill for some of these units to be provided to your area. (Bring in supplies of plastic buckets, garbage bags, and kitty litter if nothing else is available.)

- You might need to put "Porta Johns" at any public shelter you may adopt or assist, in key locations in some residential areas, or near a central disaster site to help rescue workers. You also might "adopt a rest area" along an evacuation route and set up sanitation and supply services to help refugees or evacuees who may have been forced out of their area without adequate supplies. Having extra facilities on-site helps people get in and out quicker.

- Sanitation also includes other simple items such as soap, facial tissue, toilet paper, toothbrushes and toothpaste, paper towels, waterless hand sanitizer, baby wipes, moist towelettes, feminine needs, etc. In fact, one of the supply projects your group could perform is to buy (or have donated) large amounts of the travel-size variety of each item mentioned and assemble them into little kits to be handed out at a shelter. (The gear listed in the "**Evacuation**" section, especially the "Basic Toiletries," and "Non-Prescription Medication" items are a good example of things to offer in your supply service.)

➢ **Clothing**: Though not a direct survival item in most cases, clothing will be needed to protect victims against the elements, and a change of clothing will certainly go a long way toward improving the morale of someone who's lost everything.

- Just as many areas have food banks, you should start a clothing bank in your area. Not only can this be used to help the poor get back on their feet, it can be used as a resupply for families who've lost everything in a destructive event. In addition to standard clothing, make sure you have some environmental protection as well. Include ponchos, sleeping bags, blankets, etc.

- You could get a CCAP from local stores, especially thrift stores, who could offer a certain amount of donated goods or a set discount for victims identified through voucher by your group or by local officials.

➢ **Shelter**: Though mentioned above under "Service," there are additional points to make regarding CCAPs.
- Hotels near hospitals might pledge rooms for non-critical patients after an area attack or disaster.
- As mentioned before, group members with spare rooms could offer them to victims and refugees.
- Hotels, motels, or apartment complexes might temporarily offer empty units to visiting relief workers.

➢ **Tools and Equipment**: Need will rapidly outstrip the available supply of tools and equipment.

- Your group could arrange CCAPs with local hardware stores and tool and equipment rental facilities to provide equipment (or a discount) to your group in the event of a disaster. Some useful tools and equipment might be chainsaws, debris trailers, power tools, generators, heavy equipment, and all manner of vehicles.

- You could also create a database of area or regional tool and equipment sources just as you did with various skilled labor. You should also see what you can do about arranging a CCAP with these non-local sources, or at least maintaining the contact database for local officials to use.

- Stockpile sandbags, plastic sheeting, tarps, small boats, and other gear to fight flooding. In fact, you could host a drive to collect material scraps and old pants (for the legs) and sit around like they did with "Quilting Bees" of old, making sandbags. If you wanted to take it one step further, you could sell the resulting short pants (the ones you'd have after cutting off the pants legs) as a fund raiser and PR move.

➢ **Communication**: This is such an important concept that it deserves mention again.

- Go talk to the drug squad of your local Law Enforcement agency. Cell phones are usually confiscated at any drug bust and are rarely claimed by their former owners. Rather than being auctioned as confiscated property, these phones should be donated to worthy causes needing 911 phones for their members. Among the likely candidates would be battered women's shelters, any neighborhood watch group, school bus drivers, teachers, and any group such as yours who will be performing emergency assistance before, during, or after a disaster. (You could also hold various drives to get people to donate old cell phones.)

- Broker the purchase of a large number of inexpensive FRS radios (or help pay for a commercial radio system) for bus drivers, various neighborhood groups, or fellow disaster-prep groups such as yourselves.

➢ **Miscellaneous**:

- In addition to creating stores of supplies for the population of potential victims, you should create a smaller store of goods set aside for your members if their families fall victim to the disaster.

- You might consider maintaining a good working relationship with a local gas station or two and set up some sort of deal so that your group members will have access to fuel after a disaster.

- When we were mentioning skilled labor and equipment earlier, and discussing actually providing the skill, or simply finding those who could, there's one type of service we left out. Dogs. Dogs can benefit everyone from the Police through the physically impaired. The service you could perform is to locate all regional sources of seeing eye, guide, search and rescue, cadaver search, or other civilian dog trainers. After a massive disaster, all types of dogs will be needed, and your group would provide a valuable service by having all this info gathered in advance.

- Lastly, if nothing else, your group's sole disaster prep function could be to raise cash for larger, preexisting organizations. This would be a good function for locally established civic groups that have been around for a while and are trusted, and have also established a known relationship with the larger group. (It might not work so well to start a new group and all you do is ask for money.)

Hint: At any Service function or Supply and Donation location, especially after a disaster, make sure you have **adequate security** to keep your group from being mobbed and robbed by a desperate crowd. It WILL happen.

Remember... You can either fill needs directly, or you can arrange for others to fill them. In any event, your efforts, whether direct or indirect, can make all the difference in the world before, during, or after a disaster or terrorist attack. Don't sit back and do nothing thinking someone's already doing it all. They're not, and there will always be a need for conscientious and dedicated civilian volunteer groups. Do what you can, and do it now. Don't wait for the next disaster because then it's too late. Don't wait for local officials to ask you to come in and help because they won't. They've got bigger problems to deal with. Why do they have bigger problems to deal with? Because there aren't enough conscientious and dedicated volunteer groups.

The Think Tank

Essentially, the "think tank" is a group of people brainstorming ideas or discussing concepts. In this case, the need centers on public safety, disaster planning and management, and counter-terror operations. People often say, "I wish I was one of those guys that got paid to sit around and think these things up!" Well, you might not get paid, but the opportunity is there, and the need for intelligent input is great indeed.

It is our belief, that as America is the most highly educated country in the world, we have at our disposal the largest and most effective "think-tank-at-large" history has ever seen. No one in government can afford to ignore the potential of this source of knowledge and input. Contingency planning is only as good as the threats and scenarios perceived, and reaction is only as good as the thought and creative ingenuity that's been applied.

We have one of the best systems of government and public safety in the world, but there is still vast room for improvement, and applied brainpower is one of the best ways to help, whether you're a large group of experienced professionals, or just a bright individual with high aptitude and a desire to help. You just have to be confident in the fact that you can and will make a difference, and as the old saying goes, "you must enter to win." You have to communicate your ideas and input in a organized and timely fashion. This is by far, our greatest need.

Don't sit back and do nothing.

Setting up your own think tank as a volunteer group project is easy. You gather a few intelligent people together, start a discussion, brainstorm some ideas, record what you come up with, and then refine these ideas and pass them along to those who need them. Simple. In this section we'll cover two things.

1. We'll tell you how to **compose and organize your think tank,** and give you a list of **various ideas as examples**.
2. We'll provide a few minor thoughts on and how best to **communicate the concepts you generate**.

Organizing Your Group

As the only real interaction you'll have is with each other, if all you're going to do is form a think tank, there is no need to go through all the business form and setup steps mentioned earlier, unless you intend to seek grant funding.

The one consideration mentioned earlier that will be important is the role personalities play within your group. It's important you construct your group from people with an even temperament who are able to present differing opinions in a calm, constructive, positive manner, rather than complain about or argue someone else's input.

In your group, you'll want not only the most career experience possible on a broad number of fields, but you'll want the most life experience. Our suggestion is to seek as many retired individuals as possible. Most retired people in this country have a tremendous amount to offer. They've spent a lifetime accumulating expertise in various fields and in life in general, and it's ludicrous to push them out of the loop simply because they hit a certain age, or reached a certain time limit within a company. Expertise and intelligence do not "expire." In fact, AARP would be a great place to find potential think tank members in your area (unless you choose to make a national group that telecommutes). To contact the AARP (American Association of Retired Persons), look online at http://www.aarp.org, or contact the AARP at 601 E. Street NW, Washington, DC 20049, 1-888-OUR-AARP (1-888-687-2277).

To repeat a very important point we made earlier, you'll certainly want to mix your viewpoints in a think tank, otherwise you'll have an entire group of people only coming up with one idea. Just make sure everyone understands that disagreements and discussions can be held in a positive atmosphere without argument and anger.

Let's look at a couple of think tanks assembled with a specific purpose in mind.

➢ Since we've mentioned it several times already, let's define the concept of **VCAPs**, or **Volunteer Consultants for Asset Protection**. In this use, a VCAP would differ from a regular think tank as it would be comprised of people with experience in complimentary fields and empanelled to reach a specific goal or discuss one specific topic. For example, last year saw quite a number of hospital closings in our area. As hospitals are a highly valuable public safety asset, and as the state cannot afford to keep each and every one afloat, the solution would be to create a panel of volunteer consultants tasked with finding ways to keep ailing hospitals alive. Such a panel would consist of attorneys, accountants, retired hospital administrators, insurance specialists, etc. Let's look at some other considerations:

- Assemble panels on an as-needed basis, and "customize" each for the task or asset at hand.
- "Assets" could be any private concern deemed to have emergency support value to a community.
- An example may be in helping businesses create post-attack business continuity plans.
- One universal need can be filled by creating VCAP panels formed to secure **grants and other government funding** for local or regional first responders. The focus of such a panel should be on the constant grant research, and the filling and filing procedures necessary to secure these grants for your area.

➢ A close cousin to the VCAPs are the "**Citizens Advisory Boards**." These groups can be created at the request of government or other entities, and can cover as many different areas as VCAPs. You can also put one together on your own. Citizens Advisory Boards differ from general think tanks in that they usually (though not always) focus on a single topic. For example:

- **Legislative watch**: Monitor legislation in progress to keep an eye out on conflicting laws. For example, while part of our government is tirelessly tracking possible terrorist sleeper cells, another group is wanting to give all illegal aliens blanket amnesty and provide drivers licenses.
- **Waste monitors**: Monitor government spending, either nationally, state, or locally, and provide sound advice on simpler solutions. **Hint**: If you're going to expose a problem, you should also offer a **solution**.
- **Exploring simple solutions**: A professor at Georgia Tech assembled a spectrographic analyzer out of common and available parts for under $200. This device will immediately identify various biological samples in field tests. What other simple tools are out there?
- **Conflicting laws or regulations, or illogical governmental procedure**: A target-rich environment for sure, but we need to communicate to the powers that be, any area that can be improved upon with a simple clarification of legislation. For example, one state law we've run across suggests that <u>during</u> a biochemical attack, Police set up roadblocks to search private automobiles for alcoholic beverages.
- **Perform needed research**: Take on the job of performing all the time intensive research necessary to **see what other communities are doing** to elevate their level of preparedness. Your civic leaders and first responders may be too busy with their normal workload to take on detailed research.

➢ For a superb example of how a small, independent group of committed individuals can make a positive difference regarding important issues in our country, visit the website for the "**Soldiers For The Truth**" foundation. **SFTT** is a group of military analysts dedicated to the improvement of our US military, and the conditions under which our military personnel serve. Their work can be seen online at http://www.sftt.org.

The Brain-Trust Primer

What follows is a random collection of concepts we've been working on ourselves (some will have the full study on the enclosed **CD**), ideas we've heard from other groups, and general input from readers of our column. Some of these ideas apply to security, some to public safety, and others to legislative actions and government. As creative thought is what's needed most, and creativity really can't be categorized, we decided to present these suggested "brainstorm primer" ideas in no particular order. Read them all, see if there are some you want to expound upon, and see if some spark your imagination and make you think of something completely new. If they generate new ideas, then this section of the book has served its purpose. Let's dive right in.

➢ As school busses may be used to help in a mass evacuation or other emergency, each should be numbered on its roof to be trackable from the air, and should each should contain more extensive safety gear.

➢ Schools, during Orange or Red Alerts, should institute a "**take home all books**" program where students take home all their school books each day. If the school suddenly closes for any sort of extended period, students will be able to study at home. To go along with this, schools should create a "**study at home**" curriculum so students can keep up while the school is closed.

➢ How can your area upgrade its EAS alert system? (See our "**Studies**" file on the **CD**.)

➢ Automobiles should have their emergency flashers connected to the airbags. If a considerable accident occurs, the flashers automatically go on.

➢ Semi truck trailers should have brake lights across the tops of the trailers in addition to the normal tail-light positions so more people than the ones immediately behind the truck can see if the trucker has hit the brakes. These trucks, being taller, usually react more quickly to an accident or other situation up the road, and several trucks hitting their brakes at once would mean earlier warning to other motorists on the road. In some cases, this may mean fewer pileups in accident situations.

➤ If we ever have a case of suicide bombing in this country, the remains of the bomber should be automatically screened for infectious disease. A contagious disease may be the true weapon, and the bomb merely a means of dispersal.

➤ As fires will force people to crawl along the floor when evacuating a building, shouldn't some Braille and/or lighted exit signs be duplicated along the floor?

➤ Simple ordinances can stop some terror attacks. For example, to prevent some bombings, make it illegal to park a vehicle near certain sensitive infrastructure sites or high traffic areas, such as near power substations, telephone relay centers, high tension wire poles, interstate interchanges, etc. Any vehicle left unattended in these areas should be towed immediately.

➤ During a Red Alert, or during or after any sort of officially declared disaster, any centrally controlled communication utility should be made available to all. The thrust here is to reinstate phone and cable service to the small percentage of the population that might have had theirs temporarily cut off for late payments. In an emergency, people need to be able to call 911 if needed, and they need to be able to keep up with news. Also, in the case of a quarantine, they'll need commo with the outside more than ever. This should also apply to electricity and gas. No one should be cut off during a higher alert or declared disaster.

➤ How can airlines be made even safer? How can citizen patrols help prevent the use of Stinger missiles against civilian aircraft? (See our "**Studies**" file on the **CD**.)

➤ During any area emergency, or national Red Alert status, all toll booths should raise their barriers, suspend collection, and allow the roads to clear.

➤ What kind of tax breaks or incentives should be provided farmers who allow federal or state agencies, or private volunteer groups, use of their land for a temporary tent city after a disaster?

➤ Disaster Prep purchases and expenditures made by private citizens should be tax deductible.

➤ The costs of providing free training classes put on by private industry for the purpose of public safety education should be tax deductible.

➤ Should the phone company's repair number, "611" be changed to "711" to prevent accidental dialing of "911" which wastes Police time when checking out 911 hangup calls? (On the phone pad the 6 is just above the 9 and slips are common.)

➤ Insurance companies are quietly rewriting all policies to define terrorism as an "act of war" and therefore not a coverable event. How can homeowners and others find insurance protection?

➤ Strengthen the "Good Samaritan" protection laws as more and more disasters will see more and more civilians helping each other.

➤ Make 91.1 FM, or 910 or 912 AM, the universal "emergency news radio" station frequencies across the country. (They're the closest numbers to **911** and therefore easy to remember.)

➤ Study ways to make it harder to counterfeit ID cards. (See our "**Studies**" file on the **CD**.)

➤ Each state should allow citizens to have a "non-driving" ID card (with the same ID counterfeiting safeguards) if they desire, even if it's redundant with their driver's license. This will provide an extra form of official ID if something happens to the driver's license.

➤ It's difficult to get a cell phone signal in many larger office and/or public buildings due to all the heavy construction. Given the potential for any number of emergencies, shouldn't buildings over a certain size or capacity be required to install simple relay devices that will help cell phone users pick up a signal?

➤ Push for more "instant-on" emergency warning devices above and beyond Weather Alert radio. Televisions, stereos, car stereos, and the like should all have this feature. In addition, each should have a warning light that blinks in the event of an EAS alert if the device is playing a tape or CD, or otherwise on but not tuned in to normal networks.

➤ Temporary amnesty for unlicensed and/or uninsured drivers should be granted during and shortly after any mass evacuation. It would not be fair for someone to be *forced* into driving only to be charged for it later.

➤ After a mass evacuation, all publicly owned payphones in a designated refugee area should operate free of charge for a limited time. Public libraries with computers and internet access should offer free rationed time for email communication.

➤ All personal mail coming out of a declared disaster area or designated refugee center should be postage-free for 14 days.

➤ Persuade cellular service providers to offer free limited cell phone usage, from their store locations, for disaster victims or refugees.

> Given the rate at which mobile homes burn, shouldn't their window frames have escape release latches?

> All forms of public transportation should adopt a "Mushroom Policy." This stems from the old joke, "Mushroom cloud in the distance? All rides the other direction are FREE!" All forms of public transport (to include taxis) should pick up as many people fleeing an area as possible (without going back to a disaster area for a second run), and help get them to safety.

> After a large-scale quarantine or other isolation (especially one in which communications are down), authorities will want to know who's safe and who needs help. Brainstorm a list of simple signals that could be used to communicate to patrols. Examples would be: porch light on means everything's okay, an "X" made of tape placed in the window might mean some assistance is needed, and so on. No signal at all means authorities might want to investigate.

> All mass transit operators (to include bus and taxi drivers) should be trained in emergency reactions to include first aid and CPR, sealing the vehicle or its ventilation system in a HazMat incident, behavioral profiling of potential terrorists, etc.

> Each state's Department of Motor Vehicles should add a volunteer option to their CDLs, or Commercial Drivers License. It could be a simple checkbox just like the one you check to become an organ donor. In this case the goal would be to create a database of truck drivers willing to volunteer in the event of a disaster. The same should be offered to licensed heavy equipment operators.

> Each state that requires a license for Private Investigators or Security Guards should have the same "would you like to be a volunteer?" option on their licensing. Investigators and guards would be useful in a VIPS, or Volunteers In Police Service group.

> All cell phone services, pager services, and internet service providers (ISPs), should offer free EAS alert relays as part of their service.

> The FAA should consider the same volunteer option be placed on Private Pilot's License applications.

> Require that all dams have automatic breach alarms with community sirens.

> During higher alerts, all DOT and Dept. of Agriculture trucking weigh stations should be operated around the clock, seven days a week.

> In addition to the warning beeps, smoke detectors should have bright LED lights come on to not only warn the hearing impaired, but to provide immediate light in a possible fire situation when power may be out.

> Truckers and on-the-road service technicians or repairmen should be trained to spot potential terror activity.

> As an example of the kind of concepts that can be generated by a private individual, and a project that your group can actually help us further, visit http://www.pushback.com/terror/mines/HelicopterMinesweeper.html.

> "Disaster Prep 101" should be taught in high schools just as health and driver's ed are, and for the same reasons. Today's high school graduates will be tomorrow's homemakers. It should also be taught in colleges and community colleges.

> List ways businesses can help their customers prepare for emergencies. For example, pet stores should post reminder checklists of emergency supplies for pets. Hardware stores in hurricane areas should post how-to materials and premade shopping lists for reinforcing windows.

> What simple things can be done to limit human contact and exposure during an epidemic? Telecommuting? Home schooling? Phone-in grocery orders and delivery service? Free cable?

> All non-profit public safety groups should have first option at all potential safety equipment seized by Police, and for all surplus equipment sold off by any government agency. Topping this list (regardless of source) are cell phones, vehicles, old school busses, vans, computers, etc.

> Civilians can provide input on counter-terrorism. Take a look at our "**Terrorist Operations Development**" planning matrix in the "**Appendix**." Let's take a moment to discuss this as this matrix gives you **182 opportunities for input.**

  ● The purpose of this matrix is to provide you a "brainstorm organizer" that reminds you of all the steps a terrorist group must take in order to launch an attack and all the different groups that could participate in an attack prevention or response.

  ● Down the lefthand column are the eight categories, or planning steps, that a terrorist group must take in order to plan and execute an attack. Each step is subdivided into different considerations or steps that we'll discuss in just a second.

  ● Steps one through five are planning steps, and six through eight are the post-attack effects.

  ● The columns to the right headed "Federal," "State," "County and Local," "Private Sector," and "General Public," are the various groups and entities with a vested interest and possible involvement in preventing an attack from occurring and/or participating in a response after one has been executed. The fact that there is a number in each and every grid square is to remind you that there is something that each group could do to affect that step (and it's done to help you organize your conclusions and input). It's your job to brainstorm what this involvement could be and to communicate your ideas to the proper people.

- Let's go over each of the steps and give a brief description about what each means, and what each of the sub-steps refer to:

1. **Personnel:**   Terrorists will have to get their operatives from somewhere.
    A.   Sleepers – Terrorists already over here and waiting to strike. What can be done to discover them?
    B.   Illegal Entry – Smuggling people across borders to carry out an attack. How can this be prevented?
    C.   Local Recruitment – Terrorists might try to team up with domestic agitators.
    D.   Legislative Gaps – These would be immigration or other loopholes benefiting terrorists. What are some of the loopholes and how can they be closed?
    E.   Domestic Terror – Terror can be homegrown as shown by Timothy McVeigh.

2. **Target Selection:**   Terrorists have to pick their target. How do you give them fewer to choose from?
    A.   Primary Goal – This would be the target itself. How can we improve security at likely targets?
    B.   Secondary Goal – A secondary goal could be something like economic instability. What are some of the other effects and how might they be mitigated?
    C.   Selection Criteria – The harder a target is, the less desirable it is as a target.

3. **Tools / Weapons:**   Terrorists will need tools and weapons to carry out an attack. How can they be denied these weapons?
    A.   Hijacking – Some weapons will be taken as were the airliners on 9-11.
    B.   Theft – Other tools like vehicles will simply be stolen. How can this be prevented?
    C.   Smuggling – Most weapons will be smuggled into the country in the same manner as drugs. How can smuggling be curtailed?
    D.   Purchase – Many weapons will simply be purchased such as with Timothy McVeigh and the fertilizer used to make his bomb.
    E.   Construction – Other weapons can be fabricated. What are some ways to deny terrorists access to all these weapons sources?

4. **Gathering Intel:**   Information has to be gathered about the target, its security, etc. This is the best opportunity to spot a terrorist.
    A.   Observation - This includes taking pictures, pulling surveillance, etc.
    B.   Infiltration – Sometimes operatives will secure employment at a target site to learn more.
    C.   Direct Instruction – An example is when the 9-11 hijackers took flight lessons.
    D.   Internet – Online research is common.
    E.   Info Purchase – Sometimes terrorists will simply pay someone for the info they need.
    F.   Interrogation – Terrorists may also kidnap key personnel from a target site to get info.

5. **Planning:**   Once the tools and intel have been gathered, a plan has to be formulated.
    A.   Communication – Individual terrorists have to communicate with each other.
    B.   Meeting – Eventually they have to come out of the woodwork to discuss their plans as a group.
    C.   "Dry Run" – This would be an "almost" attack just as when the 9-11 hijackers took several flights in order to get used to airline routine.

6. **Initial Execution:**   This is when an attack is actually launched. The considerations for the following steps concern themselves with ways to improve our responses and mitigate, or lessen the impact of, the attack in the following areas.
    A.   Conventional – The 9-11 attacks could be considered a "conventional" attack in that it was force without the use of weapons of mass destruction.
    B.   Nuclear – What are some of the things we can do to respond to a nuclear detonation and further prepare our citizens?
    C.   Biological – Same for germ warfare. How can we fight the epidemic and soften the societal impact?
    D.   Chemical – Same with chemical warfare. How do we prepare in advance, and what do we do afterwards to recoup faster and more thoroughly?

7. **Secondary Effects:**   No attack will be a solitary event. It will have after-effects.
    A.   Economic – How can we plan ahead to prevent severe economic repercussions? What are some tings that could be done by each listed entity?

411

B. Public Health – What can we do to keep public health as strong as possible?

C. Panic / Public Order – How can we help maintain calm and prevent riots and looting?

D. Restoration – What can each entity do to help put life back in order as quickly as possible?

8. **Non-Local Aftermath:** This is when something happens elsewhere, but has an effect on your area.

A. Refugees – What can be done to prepare for handling refugees that might come to your area?

B. Aid – What can be done to provide aid to a Ground Zero?

C. Public Health – How can your area provide health assistance to a target area?

D. Panic / Public Order – How will refugees affect your area's law and order?

➤ This list, like all the items preceding this matrix, are here to spark your imagination so you can come up with your own input, solutions, and suggestions to help make our world a safer place. Always remember; no good thought is ever a waste as long as it's heard. With that in mind, let's talk about getting the word out.

Communicating What You Do

Think tanks are useful only if the ideas generated are communicated to those who can use them. This is actually easier than you might think, but it involves a little persistence and the development of a few relationships through constant effort and good PR. The important point to make here is that you should, under absolutely **no** circumstances, think your ideas and input will **not** make a difference. As the saying goes, "you can't put the Genie back in the bottle." Once a thought is out and shared by others, your efforts are already a partial success. Getting someone to act on your input may take a little longer, but no one can ignore an idea or answer desperately needed.

Another very important point to keep in mind is that logical solutions presented in a positive, cordial, manner are far more likely to be well received than any form of criticism. In other words, don't gripe, fix.

Let's look at some general guidelines to keep in mind when presenting your finalized ideas and concepts.

➤ First you should focus on solutions to problems and not on the problems themselves or on criticism.

➤ Even though you may generate several concepts in each meeting, **ideas should be presented separately**. Putting too many in one report can be confusing at times, plus you run into the problem that different ideas will be sent to different offices and agencies, and within some of these offices and agencies, input is routed to different groups or people.

➤ Record and articulate your ideas **clearly and succinctly**. Use the **NESC** method; **N**ame it, **E**xplain it, **S**upport it, **C**onclude it. Be clear, direct, logical, and organized. Describe the nature of your concept in one clear sentence or paragraph. Give a brief but thorough summary of the full idea. Provide examples of similar ideas, or research materials to support your findings. Wrap up your report with a brief summary.

➤ **Present ideas from the bottom up.** No one likes to be left out of the loop. For example, concepts you generate that may be beneficial to your city will more than likely be of benefit to many cities, and you'd want to share your input with as many people as possible. However, your first contact should be with *your* city before you go "up the ladder." Start with your Mayor and City Council, maybe your local paper, etc., before presenting your ideas elsewhere such as to your reps in DC.

➤ Always **utilize more than one channel**. For example, if you did have something of use to any city in America and you first presented your ideas to your Mayor and then your reps in state and federal legislature, you're still not done. You want others outside of government to know what you've submitted. You should consider adding letters to the editor of your local paper, press releases and PSA requests (depending on the nature and urgency of your ideas), and possibly the submission of printed magazine articles as a way to broadcast your think tank's "product."

➤ As a think tank may be but a small portion of your group, some of your ideas can be **executed by your group** (or by a group you're affiliated with), effectively providing a **working example** of your ideas in action.

➤ **Redundancy** is king. Don't send one letter to one person and expect things to get done. Write to the lowest level involved and to everyone higher up. For example, if you have ideas on how to upgrade your town's Emergency Alert System, write to the mayor, your state reps, your DC reps, and to the federal Office of Homeland Security. **Hint**: Given the current state of mail security and its associated delays, in addition to mailing letters, you should send faxes, emails, and make phone calls.

➤ Remember the principle of **erosion**. One letter won't do it. Send more than one communication on any given topic. Send a letter yes, but also send a fax, email, and a phone call to make sure your input has gotten through to someone, somewhere.

- Give the **carrot** as well as the **stick**. We all tend to gripe, complain, and point out shortcomings, but we sometimes fail to say thank you for the things that were done right. Positive reinforcement goes a long way. If your group observes that something was done right by someone, send the praise along.

Now that we've covered how to say it, let's show you where to send it. In addition to the local people we've suggested you include in the loop, consider the following offices, agencies, and sources for more info on needs and submissions.

- Some individual ideas, or groups of concepts, could be submitted locally via Public Service Announcements.
- You can submit comments on Federal documents that are open for comment and published in the Federal Register, the Government's legal newspaper. You'll find them at http://www.regulations.gov.
- You can keep up with the status of other regulations and agency info at http://www.stateside.com.
- You can find **Congressional addresses** at either http:www.house.gov, http://www.senate.gov, or http://www.congress.org. Another source for addresses (and on pending legislation) can be found at www.vote-smart.org. For more legislative contact info visit: http://www.firstgov.gov/Contact/Elected.shtml.
- For congressional information, use the Thomas Legislative Information service of the Library of Congress. It's online at http://thomas.loc.gov, and you can contact the Library of Congress at 101 Independence Ave., SE, Washington, DC 20540, 202-707-5000.
- Each State's Office of Homeland Security http://www.ready.gov/useful_state.html.
- Coalition to help developing countries develop http://www.proventionconsortium.org.
- To report government waste visit **Citizens Against Government Waste** at http://www.cagw.org.
- You can also find more sources of government info on the Center for Democracy and Technology's "Access to Government Information" site http://www.cdt.org/righttoknow.
- For general input regarding improving our "American Think-Tank," contact **Tom Ridge** directly, care of The White House, 1600 Pennsylvania Ave., Washington, DC 20502. His direct fax number is **202-456-6337**. Send copies of your letters your Congressional reps.
- Office of Congressional and Public Affairs http://www.access.gpo.gov/public-affairs/index.html.
- The Senior Corps Of Retired Executives, or SCORE, at http://www.score.org, or at SCORE Association, 409 3rd Street, SW, 6th Floor, Washington, DC 20024, 1-800/634-0245. Primarily a business consulting service, this is the kind of group that could help your think tank or citizens' advisory board save a failing local asset.
- In addition to your state's government, you can find links to other states and to their key administrators via the National Governor's Association page at http://www.nga.org.
- A sampling of one group's caveats to passed and pending counter-terror legislation can be found online at http://www.epic.org/privacy/terrorism. We present this as an example of the many points of view you should keep in mind while brainstorming your own concepts.

Now that we've gone into extreme detail on all the little things your group needs to do in order to get started, and can do after you get going, let's start talking about the specific things you need to get done in order to turn all this theory into reality. Let's go next to "Training and Equipment."

Notes:

_____

_____

_____

_____

_____

_____

_____

_____

_____

_____

_____

_____

_____

Training and Equipment

So far, most everything we've done regarding groups and working with others has focused on paperwork. We've helped you set up an organization, gave you hints and suggestions on internal organization, and filled page after page with examples of all the many ways you can help your fellow citizen. Now it's time to put the finishing touches on you and your members and get them ready to put all these ideas to work.

Goals of Training and Equipment:

1. To make sure your group members are **as prepared at home as they need to be**.
2. To make sure your group as a whole is **as ready to react as each of the members**.
3. To help ensure success by **preparing your members to perform the group's specific function**.
4. To get your group as **well equipped for their job** as you are for an evacuation or quarantine.
5. To **help your group with its professional image** which will help with official liaison later.
6. To set your group up for a **long life, success, security, safety**, and to be the **best at what they do**.

We've divided this section into:
1. **Training.**
2. **Equipment.**

Training

Training is actually going to be a very short section, as the rest of the book has pretty much covered the subject. We wanted to give training separate mention though as it's the *act* of training that will be so important to your group. Let's see why and then we'll cover some general training suggestions.

Why train? Simple questions get simple and straightforward answers.

➢ Though most of your group might be rather well trained in many of the subjects you might offer training on, refreshers will never hurt anyone, and most people learn a few new things the second time they take a class. Besides, no two classes will be identical, even on the same subject.

➢ Training together, even on simple subjects, builds teamwork, camaraderie, and helps group members get used to working with each other.

➢ The more training you offer the greater will be the perceived value that membership in your group will have to the members.

➢ Periodic training keeps members' interest up and prevents boredom with the group or a feeling of lack of purpose. (However, remember that too much training can become a burden.)

➢ Good training is also a good PR opportunity. You'll want the public to know about your group, and to know that quality training is being provided, so you can generate new members and develop a sense of trust and recognition if your group is ever called upon to help in a disaster and/or deal with the public in other capacities.

➢ Another benefit of training and preparation is the fact that it will be much easier for volunteer group members to go off and perform the group's duties if their families at home are protected.

What kind of training should you offer? Again, this is an easy one.

➢ At the very least, each of your members should be trained in all concepts mentioned in this book. Though groups may have larger goals, the first goal should always be the safety and wellbeing of the members and their families. Anything beyond that is gravy. Everyone should receive instruction in all subjects mentioned under "Basic Training" and they should also be fully prepared to react to emergencies and disasters including evacuations or periods of isolation.

➢ As we're continually mentioning outside books, periodicals, and videos, one thing you should do is create a training library for your members so they can check out videos, books, and so on. Also, your communications director, or training director, could keep track of current training info and cut and paste applicable information (within copyright law guidelines) to publish in your group's newsletter.

➢ Beyond that, your group should be trained in the function your group has pledged to perform. For example, if you're supplying food, everyone should know food handling safety, or how to cook.

Sources and More Information

As far as training sources, we've covered quite a number in this book. Providing outside sources of instruction for volunteer groups is yet another of many reasons we gave you so many websites and contacts. Here are a few more places to go for volunteer group training.

➤ Some jurisdictions offer limited Police training to civilians for help in setting up Neighborhood Watch groups. See **CERT** (Civilian Emergency Response Teams) at http://www.training.fema.gov/emiweb/CERT/index.asp.

➤ You can download the FEMA CERT materials at http://training.fema.gov/EMIWeb/cert/mtrls.asp.

➤ Contact **Citizencorps** through their website, through your state's Emergency Management Agency, or through FEMA at: 500 C Street SW, Washington, D.C. 20472, Phone: (202) 566-1600, 202-646-2500, or at http://www.citizencorps.gov. They'll have other course offerings as well.

➤ Also see **VIPs** (Volunteers In Police Service) at http://www.policevolunteers.org, or call: 1-800-THE-IACP.

➤ Visit FEMA's training site at: http://training.fema.gov/. You'll find numerous online courses and links to more information focused on volunteer group training.

➤ The US Army's online Chemical Casualty Care Division has a good collection of downloadable and online information. We put this link in the groups section because this particular website requires registration. To see if your group qualifies, log on to https://ccc.apgea.army.mil/newRegistration/LoginPage.asp.

➤ The **Agency for Toxic Substances and Disease Registry** has put together a collection of downloadable MMGs, or Medical Management Guidelines at http://www.atsdr.cdc.gov/mmg.html#bookmark03. This would be a good source of info for your group to introduce you to some of the "command and control" procedures that you might be on the fringes of in the event of a biochemical attack or incident.

➤ Though **not free**, The National Technical Information Service offers training films, videos, CDs and more produced by and for various government agencies. They're at http://www.ntis.gov, or you can contact them at National Technical Information Service, 5285 Port Royal Road, Springfield, VA 22161, (703) 605-6585.

➤ Another training video source is the Safety Council's online video library site at http://www.safetycouncil.com/workplaceSolutions/video_library.asp.

➤ The Rescue Training Resource and Guide is a comprehensive source for discussion and general information on first responder subjects and issues. They're online only at http://www.techrescue.org.

➤ For some other safety and training considerations for volunteer groups, visit Responder Safety, online at http://www.respondersafety.com.

➤ Though a South American publication, this website provides an English translation of some pretty good disaster info at http://www.paho.org/english/dd/ped/newsletter.htm. Visit the newsletter and peruse the linked pages for good disaster training material.

➤ For group training on disasters, hazardous materials, and other threats, visit http://www.all-hazards.com.

➤ Some of the different groups, agencies, and companies we've listed under "External Liaison" may also provide a certain level of instruction or training, or be able to provide you with other sources.

➤ Volunteer groups are not Law Enforcement. However, there will be some overlaps at some minor levels. Therefore, the more your group understands the equipment and function of Law Enforcement entities, the stronger your relationship will be with your liaison officers. Therefore, be sure to visit http://www.officer.com.

➤ As the safety of your members should be a topic of every training session, consider some of the articles and input offered by Responder Safety. They're online at http://www.respondersafety.com.

➤ As a volunteer group, you'll want to keep up with a little more information on governmental threat and hazard info than will the average individual. One really good source of information can be found online at http://www.chem-bio.com/links/periodicals.html. You'll find links to various agencies, and to their newsletter.

➤ Go to the North American Emergency Management site, http://www.naem.com/connection.html, and sign up for their newsletter, "The Connection – America's Bridge to Preparedness."

➤ ServiceLeader.org offers a clearinghouse of useful information on volunteers, volunteering, and volunteer groups, online at http://www.serviceleader.org.

➤ You'll find very useful information from the Association for Volunteer Administration, online at http://www.avaintl.org, or at P.O. Box 32092, Richmond, VA 23294, phone: 804-672-3353, fax: 804- 672-3368. Be sure to check their resources page at http://www.avaintl.org/resources/index.html.

➤ For educational, organizational, and leadership information, log on to The Corporation for National & Community Service site at, http://www.nationalserviceresources.org, or contact them at National Service Resource Center (NSRC), ETR Associates, 4 Carbonero Way, Scotts Valley, CA 95066, phone: 1-800-860-2684 or (831) 438-4060, TTY: (831) 461-0205, fax: (831) 430-9471.

➤ Our enclosed **CD** contains numerous **manuals** on a variety of disaster-related subjects.

Let's move on to another simple subject, and a close cousin of training; **equipment**.

Equipment

Equipment is another simple section because the bulk of your equipment needs will be automatically dictated by the type of function your group has pledged to perform. What follows here is a general discussion of personal safety equipment, and basic group equipment, that may be common to all.

We'll start with the individual and move our way up from there. Let's talk a second about personal equipment, or rather the things that each member should provide for themselves, and should always have with them during group functions. Just because we're suggesting the member supply these items, doesn't mean that the group could not help out if it were able.

➢ Your personal safety comes first. For this reason, always have your **GYT Pack** with you and make sure it's customized to the area you'll be in and the job you'll be performing.

➢ Always carry your own **cell phone** in addition to any other communication equipment the group may provide.

➢ You may wish to bring a **small backpack** (which for volunteers, we like to call a "tactical pack," or "**Tac Pac**") containing the following:

❑ Personal first aid kit to include two day's worth of any medications you may be taking.

❑ Flashlight and extra batteries.

❑ Two day's worth of food and water.

❑ A change of clothing, socks, and underwear.

❑ A minor toiletries kit. At least carry toilet paper, moist towelettes, soap, hand sanitizer, etc.

❑ Climate gear such as a light jacket, small poncho, etc.

❑ A respirator, and protective clothing if your function puts you at risk.

❑ Other protective clothing suited to the group's function, that is not provided by the group. An example might be durable boots and work gloves.

➢ When stocking your Tac Pac keep in mind that you may be in an extreme situation and need to share a few things with those around you. This may include first responders who were overwhelmed by the initial event.

➢ If you are asked to **supply your own tools**, make sure they're kept separate from your other tools and that they're always ready to go. Always remember that you are the "Minute Men" of the new millennia.

➢ In addition to tools for your own use, you might offer personal equipment for use by the group or under the group's direction. For example, if you lived in a flood prone area, and owned a small boat or two, these could be used by the group in the event of a flood.

➢ The same concept may apply to general supplies and other goods. For example, if your group's function is to supply and organize a local emergency shelter, your individual task may be to provide a specific item, such as toilet paper, to help supply the shelter. Other members may be assigned different items to bring.

➢ **Hint**: One of the many reasons we urge you to keep durable clothing in your home, vehicle, and workplace, is that you may become part of an impromptu search and rescue group digging through rubble after a disaster. The more gear you have with you and immediately accessible, the more able you are to help.

➢ You may also be asked to **supply your own uniform** per the group's standard.

Now that we've mentioned the group uniform, let's list some suggested components of it and continue with equipment for the group in general.

➢ **Uniforms** are important for a number of reasons. One, it helps the public identify your group, as well as identify your members as being part of that group. Two, it makes it easier for Law Enforcement or other groups or agencies you're working with, to differentiate you from the crowd. Three, it gives group members a greater sense of belonging.

➢ Uniforms can be complete sets of clothing, though they don't have to be. Consider the following:

❑ Standard clothing could be the basis of a good uniform. Have members wear the same kind of comfortable, durable, pants – probably blue jeans – and wear a specific color of shirt. The shirt should be suitable for the climate and the job being performed.

❑ Shoes should be comfortable but durable work shoes, suited to the climate and job to be performed.

❑ Specific types, designs, or colors of **vests** should be supplied by the group. On these you might have a customized **patch**, and/or the group's **name** across the back. We suggest your vests be a different **color** than any vest used by local first responders. Make yours unique.

❑ Your group should also issue **ball caps** of either a specific color, or with the group's **patch**.

❑ Each member should also have an **ID card**, in a plastic sleeve, either on a break-away necklace, or attached to the vest. **Hint:** You can designate the function of the member based on the **vest color**, ID card, ID card **holder**, **decals** placed on the ID card holder, the **color** or **patch** on the ball cap, etc. You can do this to show who has been trained for what function, who has been cleared to do heavy labor, who handles food, who your officers are, etc.

❑ You might also want to have **magnetic signs** made for members' vehicles, or a special decal inside the windshield or on the bumper.

➢ **Note:** The stronger the relationship between your group and first responders, and your group with the public, the more you should **guard your uniforms**. You don't want any being stolen and used for illicit purposes.

➢ Other equipment issued by the group will depend entirely on the group's function and how much money is raised to purchase equipment, or how much equipment is donated.

➢ Some things, like Police radios, might be issued by Police depending on your group's function. Other entities with which you liaison might issue other equipment.

➢ Use the "Skills and Assets" form from the "**Appendix**" to list and catalog the various types of equipment individual members can offer to the group.

Sources and More Information

The vast majority of the equipment source your group or its members will need has already been listed, or can be found in the links collection on our **CD**. However, we have found a few interesting sites specializing in group gear.

➢ Safety.com offers equipment as well as rating and selection criteria. See them at: http://www.safety.com, or contact them at SafetyHQ, Inc. 670 North Commercial Street, Manchester, NH 03101, Tel: 1-603-226-7233, Fax: 1-603-223-5003.

➢ Quartermaster, Inc. http://www.qmuniforms.com, 800-444-8643, Fax 562-304-7335, P.O. Box 4147, Cerritos, CA 90703-4147.

➢ One good supplier of fire fighting and rescue equipment we found is The Fire Store. They're online at http://www.thefirestore.com, or call them at 800-852-6088.

➢ Another is http://www.safetyequipment.org.

➢ You can find Search and Rescue planning forms online at http://www.basarc.org/forms.

Now that you're capable, organized, trained, and equipped, who will be the ones to call you into action? How do you let the "powers that be" know who you are, what you're about, and that you're here to help? Well, in addition to all the PR pointers we gave you above, we'll give you a fairly decent list of people to stay in contact with in order to allow your volunteer group to serve its community.

Notes:

_____
_____
_____
_____
_____
_____
_____
_____
_____
_____
_____
_____
_____
_____
_____
_____

External Liaison

The whole point of the "**Cooperation**" section is that life involves others. This continues to hold true even after you've banded together with a few of them. In this portion, we'll give you leads and pointers that will allow you to take your networking to a whole new level if you so desire.

Goals of External Liaison:

1. To help you **become part of a larger organization** if that is your goal.
2. To **keep you from wasting your time** if you're about to duplicate the projects of larger groups.
3. To **amplify the good you can do** by networking with like-minded groups.
4. To **provide you additional education and training** through contact with more experienced people.
5. To give you a larger **list of people who could directly benefit from your experience**.
6. To **foster good relationships** between government agencies and citizen volunteer groups.
7. To **increase communication** between all the people that truly want to help their communities.
8. To **make your volunteer group the most effective it can be**.

We've divided this section into:

1. **Federal Offices and National Groups.**
2. **Preexisting Organizations.**

Federal Offices and National Groups

We wish to point out here, that **nothing** in this entire "External Liaison" section will suggest that there is any office, agency, individual, or any other entity from whom you must get permission to be your own volunteer group (provided your group's function is positive, legal, and moral). The purpose of this section is to show you doors you can open in order to become part of a bigger picture. Remember too, that you don't have to become a chapter of any larger group. You can if you want to or your group can simply become an affiliate.

The first rule in establishing any sort of outside networking or involvement is that, "there's no place like home." Regardless of what you do, and how you intend to do it, you should network with all local officials. Every time your group performs any function, brainstorms any ideas, or wishes to help with any event, be it a disaster or not, you should contact the following local offices listed here in no particular order:

> Your Mayor's Office and City Council.
> The local Chamber of Commerce.
> Your County Commission.
> Your city's Chief of Police and any Watch Group Liaison Officer you're connected to.
> The county Sheriff.
> Both your city and county Fire Departments, and area volunteer Fire Departments.
> Your area's Emergency Management Agency representative.
> Your local hospitals and/or primary care facilities.
> Your district's representatives to the state legislature.

In fact, constantly keeping in touch with all the entities listed above, and the ones we're about to mention below should be a function of your officer corps, especially your PR Chairperson.

Moving higher, the bigger picture will always involve Federal offices. The function and purpose of the various offices listed below will vary, but all lend themselves to any type of involvement, input, awareness, communication, and purpose your group may need or want. Some may deal directly with the development of independent volunteer groups, some may cover only information and communication with other offices, and some are listed as they may actually benefit from hearing what you have to say.

In any event, one of the things we hope to do with this section, and in part of our upcoming book, **Disaster Prep 201** (for government entities), is to help create an atmosphere of greater cooperation between government and volunteers.

Let's look at a short list of federal agencies, organizations, and entities with which you'll want to stay in continual contact. Remember, we gave you numerous sources earlier under "Planning Ahead for Phase II," "Evac Atlas," and we told you how to find all sorts of contact information under the "Information Research" portion of "Basic Training."

> Always be sure to stay in contact with your representatives in Washington DC.

> The US Department of Homeland Security has a collection of links to each state's emergency management, and related offices at http://www.dhs.gov/dhspublic/interapp/editorial/editorial_0306.xml.

> Each State's Homeland Security Office http://www.ready.gov/useful_state.html.

> To find contact info for your state's Homeland Security office, visit the Center for State Homeland Security at http://www.cshs-us.org. You can also contact them at the Center for State Homeland Security, 3150 Fairview Park Drive South, Falls Church, VA 22042-4519, Phone 703-610-1623, Fax 703-610-1821.

> You can also find a national EMA listing at http://www.nemaweb.media3.net/index.cfm. Be sure to visit the National Emergency Management Association homepage at http://www.nemaweb.org.

> If the focus of your volunteer group goes beyond emergency preparedness and includes year-round community wellbeing, then you might want to contact the Community Action Partnership. They're online at http://www.communityactionpartnership.com, or you can contact them at Community Action Partnership, 1100 17th Street NW, Suite 500, Washington DC 20036, Call:(202) 265-7546, FAX: (202) 265-8850. Regardless of your group's function, you'll find some useful networking info on the CAP's links page at http://www.communityactionpartnership.com/about/links/default.asp.

> A world of **Government contact information** can be found at: http://www.firstgov.gov.

> **Citizen Corps** at http://www.citizencorps.gov.

> Visit the **Medical Reserve Corps**, online at http://www.medicalreservecorps.gov.

> **Freedom Corps** at http://www.usafreedomcorps.gov/.

> **Americorps** at http://www.americorps.org, or contact them at 1-800-942-2677 or TTY 1-800-833-3722.

> See the **America Responds** info page at http://www.nationalservice.org/news/homeland.html.

> VIPS or **Volunteers in Police Service** at http://www.policevolunteers.org.

> For additional information on the above, log on to Access to Government Information: http://www.cdt.org/righttoknow, and to http://govstar.com.

> Also try the National Association of Attorneys General at: http://www.naag.org.

Preexisting Organizations

Getting back to the possibility that you're not part of a local group and that you may want to join a larger group already in existence, or that your group might like to be a local chapter of a larger organization, let's take a look at some of the many groups that are already out there. We covered a few under "Planning Ahead for Phase II," and "Neighborhood Watch" earlier, and we'll list some new ones here.

> The Network for Good seems to act as a central locator and info service for various charities and volunteer groups. You can learn more online at http://www.networkforgood.org.

> More info on the US Security Network can be obtained online at http://www.ussn.org, or through the GA Security Council at 250 Williams Street, Suite 1001, Atlanta, GA 30303, 404-525-9991, Fax: 404-525-8977.

> For community-oriented opportunities, visit Freedom Corps' info site at http://www.volunteer.gov/gov.

> You can also contact http://www.1800volunteer.org online or call 1-800-VOLUNTEER (800-865-8683), to link up with others to coordinate your efforts.

> A similar group, Volunteer Match, will help match your desired contribution or involvement with existing organizations with the same goals. They're online at http://www.volunteermatch.org.

> Link through the Department of Health and Human Services' National Disaster Medical System site to a disaster reaction team in your state. Log on to http://ndms.dhhs.gov/team_sites.html.

> You can find groups and coordinators online at http://www.firstgov.gov/Citizen/Topics/PublicService.shtml, or you can try http://www.firstgov.gov/Contact/Directories.shtml, 1-800-FED-INFO.

> The Senior Corps Of Retired Executives, or **SCORE**, is online at http://www.score.org, or contact them directly at SCORE Association, 409 3rd Street, SW, 6th Floor, Washington, DC 20024, 1-800/634-0245. Primarily a consulting service, this group could help your think tank save a failing local asset.

> To contact the AARP (American Association of Retired Persons), look online at http://www.aarp.org, or contact the AARP at 601 E. Street NW, Washington, DC 20049, 1-888-OUR-AARP (1-888-687-2277).

> **NVOAD**, National Voluntary Organizations Active in Disaster, online at http://www.nvoad.org, or contact them at 14253 Ballinger Terrace, Burtonsville, MD 20866, P: 301.890.2119, F: 253.541.4915.

> Th site http://www.uact.4t.com/index.html, has an extensive **listing of some existing groups**.

> Each State's Homeland Security Office http://www.ready.gov/useful_state.html.

> NOVA – The National Organization of Victim Assistance. 1730 Park Rd., NW, Washington, DC 20010, 800-TRY-NOVA (879-6682), 202-232-6682, Fax: 202-462-2255, http://www.try-nova.org.

When contacting a larger organization, or gathering additional information, find answers to the following questions.

1. Where is the nearest office, and/or who is the nearest official representative?
2. Does this group offer any education or training?
3. Would this organization offer your group any equipment?
4. Does this office or group offer grants, financial investment, or assistance?
5. Does this group accept affiliate groups?
6. If yes, what is the procedure for offering your group as an affiliate to the larger organization?
7. Can you join this group as an individual?
8. Does this group have any official federal, state, or local governmental relationships?
9. If yes, with whom?
10. Are there any reaction contingency plans in place involving your area?

If you're thinking of joining or donating to a preexisting organization, be sure to check them out first.

> The Better Business Bureau runs a program called Wise Alliance Giving. They're at http://www.give.org, or write to BBB Wise Giving Alliance, 4200 Wilson Blvd., Suite 800, Arlington, VA 22203, 1-703-276-0100.

> Similar information can be obtained from the American Institute of Philanthropy. They're online at http://www.charitywatch.org, or at American Institute of Philanthropy, 3450 Lake Shore Drive, Suite 2802E, P.O. Box 578460, Chicago, Illinois 60657, Phone: (773) 529-2300, Fax: (773) 529-0024.

> To find more contact and other information on various charities, visit http://www.justgive.org.

Notes:

VII. EPILOGUE

"The mind once expanded to the dimensions of a larger idea never returns to its original size."

-Oliver Wendell Holmes

" 'Tis nothing either good nor bad, but that thinking makes it so." - William Shakespeare

Congratulations! You've made it to the end. We know there's a lot of ground this book covers. There are a lot of detailed and wordy sections, large checklists, and many requests for changes in the average lifestyle. Then again, there's no other way around it. As we said before, our goal for you is simplicity in your reactions. However simplicity requires understanding, understanding requires knowledge, and knowledge requires all the educational material you've gone through in this book. This is the accumulation of information you should have had all along.

The changes in our world to which we need to adapt have already been dumped into our lives. The risk is real, the threat is here, and it's the detailed content of this book that will help ensure your safety. We're glad you've done all you have.

We chose to start our Epilogue with two quotes because there are two major things you have accomplished.

First, now that you've learned these things, made these changes, and helped ready your family for any kind of emergency that may come your way, you're more able to handle life's other unforeseen problems. You've made preparedness, awareness, and readiness, a part of your life and a part of your mindset, and you can't forget these any more than you can forget how to ride a bike or how to drive. They're a part of you now, and if this book had only that one effect, then we will have served our purpose. These things are now a positive part of your children's lives too.

Secondly, you've realized that a great deal of the impact a disaster or attack carries, is its emotional impact on you and your family through fear of the unknown, and the fear of loss. You've also learned that this fear is exaggerated by being unprepared, and not knowing what to expect or how to react. However, through your preparations and the resulting confidence you have knowing you've done everything you can, you can put any negative eventuality into its proper perspective and better weather the storm.

Now that you're done with the basics, you can do as you would with your other safeguards and put it away, relaxing in the fact you're prepared, rather than dwelling on reasons your preparations *might* be needed. Just as with buying an insurance policy or fire extinguisher, you feel better because you were proactive, you're relaxed because you're protected, and you're confident in the fact you're capable of using either. So it is with Disaster Prep. Get it all done, and then put it away and relax, knowing you're protected. Everything you've done will be there when you need it.

All we have left to say is thank you. Thank you for staying the course, thank you for making our world a safer place for the good guys, and thank you for being the kind of person this book was written for. Thank you too, for helping ensure your safety and thereby ours. By preparing your family for disaster, you've taken yourselves out of the public safety victim equation, and freed up first responders to go help others. You've also helped to reduce the threat of terror attack in your own way by making one less target. The fewer "targets" there are, and the less "terrified" a population is, the less effective any sort of attack will be and therefore it may be less likely that one will be launched.

Thank you for ensuring peace. In closing, we'll leave you with one last acronym, **P.E.A.C.E.**

Prepare like the next attack is coming tomorrow.

Educate the ones you love.

Accept the fact that all you can do is all you can do.

Celebrate the completion of all your hard work.

Easy chair. You've earned a rest... take a load off!

Take care,

Paul

This page intentionally left blank.

VIII. Appendix ———————

"A journey of a thousand miles begins with a single step." -Confucius

Welcome to our Appendix. Below is the table of contents showing the name and short description of each set of forms and worksheets mentioned throughout the book. The files for all of these forms and worksheets are located on CD-2 in the "Appendix" folder. In the divider bars between each portion below, you'll see the filenames. You'll also find the CD's full Table of Contents near the end of this Appendix after page A-63.

On CD2: Folder: Appendix File: Calendars

90-Day Goal Calendar   Use this page with all your Disaster Prep 101 activities and goals.   Start Date: ___/___/___   End Date: ___/___/___

SUNDAY	MONDAY	TUESDAY	WEDNESDAY	THURSDAY	FRIDAY	SATURDAY
◄─ DATE						

IMPORTANT CONTACTS – QUICK LIST

Post copies of this page: ☐ By your telephone(s) ☐ In your safe room ☐ In your wallet ☐ As a cover sheet to your other Important Contacts pages.

Who	Name	Phone 1	Phone 2	E-Mail or Other Contact
Fire				
Police				
Ambulance				
Emergency Room				
Doctor 1				
Doctor 2				
Poison Control				
Veterinarian				
Spouse 1				
OC Spouse 1				
Spouse 2				
OC Spouse 2				
School 1				
School 2				
Neighbor 1				
Neighbor 2				
Family Contact 1				
Family Contact 2				
Electric Co.				
Gas Co.				
Water / Sewage				
Volunteer Group				
Phone Tree Upline				
Phone Tree 1				
Phone Tree 2				

Local **emergency news radio** station channels: _____ AM / FM, _____ AM / FM. **Emergency TV** station channels: _____, _____

("**OC**" note for spouses is "**Other Contact**" such as a coworker or friend. The "**Family Contact**" numbers are your **emergency rendezvous** contact people.)

☐ **All phone numbers programmed into house and cell phones.**

IMPORTANT CONTACTS - ONE - EMERGENCY CONTACTS

This is your list of people to notify in an emergency.

Emergency Contact – The person **not living with you** listed on your **"Notify in Case of Emergency"** card.

Name:	Address:
City:	County: State: Zip:
Home Phone 1:	Home Phone 2:
Cell Phone 1:	Cell Phone 2:
Email 1:	Email 2 (or website):
Workplace:	Address:
City:	County: State: Zip:
Work Phone 1:	Work Phone 2:
Pager:	Fax:
Note:	❑ "Find Me" sheet attached

Rendezvous Contact – This would be your other **local** person that helps relay messages for your family.

Name:	Address:
City:	County: State: Zip:
Home Phone 1:	Home Phone 2:
Cell Phone 1:	Cell Phone 2:
Email 1:	Email 2 (or website):
Workplace:	Address:
City:	County: State: Zip:
Work Phone 1:	Work Phone 2:
Pager:	Fax:
Note:	❑ "Find Me" sheet attached

Rendezvous Contact – Your primary **out-of-town** message relay person.

Name:	Address:
City:	County: State: Zip:
Home Phone 1:	Home Phone 2:
Cell Phone 1:	Cell Phone 2:
Email 1:	Email 2 (or website):
Work Phone 1:	Work Phone 2:
Pager:	Fax:
Note:	❑ "Find Me" sheet attached

Rendezvous Contact – Your second out-of-town message relay person.

Name:	Address:
City:	County: State: Zip:
Home Phone 1:	Home Phone 2:
Cell Phone 1:	Cell Phone 2:
Email 1:	Email 2 (or website):
Work Phone 1:	Work Phone 2:
Pager:	Fax:
Note:	❑ "Find Me" sheet attached

❑ **Each family member has a copy of this page.** ❑ **All numbers programmed into house and cell phones.**

IMPORTANT CONTACTS - ONE - EMERGENCY CONTACTS

Record important information about the locations where you are to meet in an emergency.

Rendezvous Point One – (main rendezvous point) Location name: _____

Contact person:	This person is: ❑ Mgr. ❑ Security ❑ _____
Address:	Suite / Apt:
City:	County:	State: Zip:
Phone 1:	Phone 2:
Email 1:	Email 2 (or website):
Pager:	Fax:

City Map Atlas page #: ____ , Grid #: ____ Location directions:

Meet point at this location:

Rendezvous Point Two. Location name: _____

Contact person:	This person is: ❑ Mgr. ❑ Security ❑ _____
Address:	Suite / Apt:
City:	County:	State: Zip:
Phone 1:	Phone 2:
Email 1:	Email 2 (or website):
Pager:	Fax:

City Map Atlas page #: ____ , Grid #: ____ Location directions:

Meet point at this location:

Rendezvous Point Three. Location name: _____

Contact person:	This person is: ❑ Mgr. ❑ Security ❑ _____
Address:	Suite / Apt:
City:	County:	State: Zip:
Phone 1:	Phone 2:
Email 1:	Email 2 (or website):
Pager:	Fax:

City Map Atlas page #: ____ , Grid #: ____ Location directions:

Meet point at this location:

Rendezvous Point Four. Location name: _____

Contact person:	This person is: ❑ Mgr. ❑ Security ❑ _____
Address:	Suite / Apt:
City:	County:	State: Zip:
Phone 1:	Phone 2:
Email 1:	Email 2 (or website):
Pager:	Fax:

City Map Atlas page #: ____ , Grid #: ____ Location directions:

Meet point at this location:

❑ **Each family member has a copy of this page.** ❑ **All numbers programmed into house and cell phones.**
❑ Copies of applicable maps or site diagrams attached. ❑ Include cache locations, if any, on the maps.

IMPORTANT CONTACTS – TWO – RENDEZVOUS!

IMPORTANT CONTACTS – THREE - MEDICAL
Note: This page for <u>local</u> providers only. Your "Evac Atlas" will list your out-of-town sources.

Doctor:	Specialty:		
Family member that sees this doctor:			
Name of practice:	Hospital or Complex:		
Address:	City:	State:	Zip:
Phone 1:	Phone 2:		
Email 1:	Email 2 (or website):		
Pager:	Fax:		
Business hours:	After hours / emergency number:		
Insurance that covers this doctor:	Agency or Agent name:		
Policy #:	Insurance claim phone #:		

Doctor:	Specialty:		
Family member that sees this doctor:			
Name of practice:	Hospital or Complex:		
Address:	City:	State:	Zip:
Phone 1:	Phone 2:		
Email 1:	Email 2 (or website):		
Pager:	Fax:		
Business hours:	After hours / emergency number:		
Insurance that covers this doctor:	Agency or Agent name:		
Policy #:	Insurance claim phone #:		

Doctor:	Specialty:		
Family member that sees this doctor:			
Name of practice:	Hospital or Complex:		
Address:	City:	State:	Zip:
Phone 1:	Phone 2:		
Email 1:	Email 2 (or website):		
Pager:	Fax:		
Business hours:	After hours / emergency number:		
Insurance that covers this doctor:	Agency or Agent name:		
Policy #:	Insurance claim phone #:		

Doctor:	Specialty:		
Family member that sees this doctor:			
Name of practice:	Hospital or Complex:		
Address:	City:	State:	Zip:
Phone 1:	Phone 2:		
Email 1:	Email 2 (or website):		
Pager:	Fax:		
Business hours:	After hours / emergency number:		
Insurance that covers this doctor:	Agency or Agent name:		
Policy #:	Insurance claim phone #:		

Include: ❏ Doctors ❏ Dentists ❏ Psychiatrists ❏ Veterinarians ❏ Medical equipment supply ❏ Pharmacy, etc.
❏ All phone numbers programmed into house phones and cell phones.

IMPORTANT CONTACTS – THREE - MEDICAL

Gather this for contact if your area doesn't use 911 and for later updates on emergency situations. This page is also important since you may have evacuated and need to call in for home-area updates.

Emergency News Radio station channels:

_____ AM / FM, Phone 1:	Phone 2:
_____ AM / FM, Phone 1:	Phone 2:
_____ AM / FM, Phone 1:	Phone 2:

Emergency News TV station channels:

Channel: _____ , Phone 1:	Phone 2:
Channel: _____ , Phone 1:	Phone 2:
Channel: _____ , Phone 1:	Phone 2:

City or County info websites:

Fire Department - Address:

Phone 1:	Phone 2:
Email 1:	Email 2 (or website):
Fax:	Scanner Frequencies:
Contact Person:	Hours of Operation:

Police Department - Address:

Phone 1:	Phone 2:
Email 1:	Email 2 (or website):
Fax:	Scanner Frequencies:
Other contact:	Hours of Operation:

Sheriff's Department - Address:

Phone 1:	Phone 2:
Email 1:	Email 2 (or website):
Fax:	Scanner Frequencies:
Contact Person:	Hours of Operation:

City Hall - Address:

Phone 1:	Phone 2:
Email 1:	Email 2 (or website):
Pager:	Fax:
Other contact:	Hours of Operation:

County Commission - Address:

Phone 1:	Phone 2:
Email 1:	Email 2 (or website):
Pager:	Fax:
Contact Person:	Hours of Operation:

❑ _EMA, ❑ FEMA, or ❑ **National Guard** Address:

Phone 1:	Phone 2:
Email 1:	Email 2 (or website):
Pager:	Fax:
Contact Person:	Hours of Operation:

❑ **All phone numbers programmed into house phones and cell phones.**

IMPORTANT CONTACTS – FIVE – RELIEF SERVICES
List the relief agencies responsible for you home area.

Association name: _____

Contact person 1:	Contact person 2:

Address:	City:	State:	Zip:

Phone 1:	Phone 2:
Email 1:	Email 2 (or website):
Pager:	Fax:

County: _____ Map Atlas page #: ___ , Grid #: _____ Location directions:

Note:

Association name: _____

Contact person 1:	Contact person 2:

Address:	City:	State:	Zip:

Phone 1:	Phone 2:
Email 1:	Email 2 (or website):
Pager:	Fax:

County: _____ Map Atlas page #: ___ , Grid #: _____ Location directions:

Note:

Association name: _____

Contact person 1:	Contact person 2:

Address:	City:	State:	Zip:

Phone 1:	Phone 2:
Email 1:	Email 2 (or website):
Pager:	Fax:

County: _____ Map Atlas page #: ___ , Grid #: _____ Location directions:

Note:

Association name: _____

Contact person 1:	Contact person 2:

Address:	City:	State:	Zip:

Phone 1:	Phone 2:
Email 1:	Email 2 (or website):
Pager:	Fax:

County: _____ Map Atlas page #: ___ , Grid #: _____ Location directions:

Note:

❑ **The "Planning Ahead for Phase II" section of the book checked for all key services to be listed here.**

Include all: ❑ Government relief services ❑ Charitable organizations ❑ Volunteer groups ❑ Local shelters

❑ **All phone numbers programmed into house and cell phones.**

IMPORTANT CONTACTS – FIVE – RELIEF SERVICES

Use this page for key players in your emergency reactions. For general contacts, be sure to photocopy your address book or day planner and keep a copy along with these pages in your "Info Pack."

Name:	Address:
City:	County: State: Zip:
Home Phone 1:	Home Phone 2:
Cell Phone 1:	Cell Phone 2:
Email 1:	Email 2 (or website):
Workplace:	Address:
City:	County: State: Zip:
Work Phone 1:	Work Phone 2:
Pager:	Fax:
Note:	❏ "Find Me" sheet attached

Name:	Address:
City:	County: State: Zip:
Home Phone 1:	Home Phone 2:
Cell Phone 1:	Cell Phone 2:
Email 1:	Email 2 (or website):
Workplace:	Address:
City:	County: State: Zip:
Work Phone 1:	Work Phone 2:
Pager:	Fax:
Note:	❏ "Find Me" sheet attached

Name:	Address:
City:	County: State: Zip:
Home Phone 1:	Home Phone 2:
Cell Phone 1:	Cell Phone 2:
Email 1:	Email 2 (or website):
Workplace:	Address:
City:	County: State: Zip:
Work Phone 1:	Work Phone 2:
Pager:	Fax:
Note:	❏ "Find Me" sheet attached

Name:	Address:
City:	County: State: Zip:
Home Phone 1:	Home Phone 2:
Cell Phone 1:	Cell Phone 2:
Email 1:	Email 2 (or website):
Workplace:	Address:
City:	County: State: Zip:
Work Phone 1:	Work Phone 2:
Pager:	Fax:
Note:	❏ "Find Me" sheet attached

❏ Address book photocopies attached ❏ Phone / Address contact manager software printout attached.

❏ **All phone numbers programmed into house and cell phones.**

IMPORTANT CONTACTS – SEVEN – REPAIR AND REBUILDING
These are the repair and rebuilding services you were asked to locate in "Planning Ahead for Phase II."

Type of service:	Company:		
Contact 1:	Contact 2:		
Address:	City:	State:	Zip:
Phone 1:	Phone 2:		
Cell Phone 1:	Cell Phone 2:		
Email 1:	Email 2 (or website):		
Pager:	Fax:		
Other contact:			
Notes:			

❑ Other info attached

Type of service:	Company:		
Contact 1:	Contact 2:		
Address:	City:	State:	Zip:
Home Phone 1:	Home Phone 2:		
Cell Phone 1:	Cell Phone 2:		
Email 1:	Email 2 (or website):		
Pager:	Fax:		
Other contact:			
Notes:			

❑ Other info attached

Type of service:	Company:		
Contact 1:	Contact 2:		
Address:	City:	State:	Zip:
Home Phone 1:	Home Phone 2:		
Cell Phone 1:	Cell Phone 2:		
Email 1:	Email 2 (or website):		
Pager:	Fax:		
Other contact:			
Notes:			

❑ Other info attached

Type of service:	Company:		
Contact 1:	Contact 2:		
Address:	City:	State:	Zip:
Home Phone 1:	Home Phone 2:		
Cell Phone 1:	Cell Phone 2:		
Email 1:	Email 2 (or website):		
Pager:	Fax:		
Other contact:			
Notes:			

❑ Other info attached

Gather this info now rather than waiting until after a disaster. You may have to call during an evac.

IMPORTANT CONTACTS – SEVEN – REPAIR AND REBUILDING

Use this page for other general contacts you wished to record separately from your address book pages.

Name:	Address:
City:	County: State: Zip:
Home Phone 1:	Home Phone 2:
Cell Phone 1:	Cell Phone 2:
Email 1:	Email 2 (or website):
Workplace:	Address:
City:	County: State: Zip:
Work Phone 1:	Work Phone 2:
Pager:	Fax:
Note:	❏ "Find Me" sheet attached

Name:	Address:
City:	County: State: Zip:
Home Phone 1:	Home Phone 2:
Cell Phone 1:	Cell Phone 2:
Email 1:	Email 2 (or website):
Workplace:	Address:
City:	County: State: Zip:
Work Phone 1:	Work Phone 2:
Pager:	Fax:
Note:	❏ "Find Me" sheet attached

Name:	Address:
City:	County: State: Zip:
Home Phone 1:	Home Phone 2:
Cell Phone 1:	Cell Phone 2:
Email 1:	Email 2 (or website):
Workplace:	Address:
City:	County: State: Zip:
Work Phone 1:	Work Phone 2:
Pager:	Fax:
Note:	❏ "Find Me" sheet attached

Name:	Address:
City:	County: State: Zip:
Home Phone 1:	Home Phone 2:
Cell Phone 1:	Cell Phone 2:
Email 1:	Email 2 (or website):
Workplace:	Address:
City:	County: State: Zip:
Work Phone 1:	Work Phone 2:
Pager:	Fax:
Note:	❏ "Find Me" sheet attached

(General Contacts are those people with no emergency involvement, but you still want to keep in touch.)

IMPORTANT CONTACTS – PROFESSIONAL CONTACT

Use this as your professional contacts list for your "Family Member" and "Household Data" pages.

Company name:	Contact person:
Address:	Suite:
City:	County:	State:	Zip:

Phone 1: | Phone 2:
Cell Phone 1: | Cell Phone 2:
Email: | Fax:
Website: | Web access ID & Password:
Account #: | PIN# or ID#:
Note:

Company name: | Contact person:
Address: | Suite:
City: | County: | State: | Zip:
Phone 1: | Phone 2:
Cell Phone 1: | Cell Phone 2:
Email: | Fax:
Website: | Web access ID & Password:
Account #: | PIN# or ID#:
Note:

Company name: | Contact person:
Address: | Suite:
City: | County: | State: | Zip:
Phone 1: | Phone 2:
Cell Phone 1: | Cell Phone 2:
Email: | Fax:
Website: | Web access ID & Password:
Account #: | PIN# or ID#:
Note:

Company name: | Contact person:
Address: | Suite:
City: | County: | State: | Zip:
Phone 1: | Phone 2:
Cell Phone 1: | Cell Phone 2:
Email: | Fax:
Website: | Web access ID & Password:
Account #: | PIN# or ID#:
Note:

Also use this as attachment for: ❏ Financial Page ❏ Family Member Medical ❏ Other pages needing more info

❏ **All phone numbers programmed into house and cell phones.**

IMPORTANT CONTACTS – PROFESSIONAL CONTACT

"FIND ME"

Give this information sheet to certain people who may need to find you in an emergency. Examples are: Each of your children, babysitters, certain relatives, children's teachers, volunteer group members, etc.

Our names:	←(Persons 1 & 2)

Home Address:	Apt.#:

City:	County:	State:	Zip:

Home Phone 1:	Home Phone 2:
Cell Phone 1:	Cell Phone 2 or Other:
Email 1:	Email 2 (or website):

Neighbors to this address:	Phone 1	Phone 2 or Other

Person 1: Name:	Place of Employment:

Title/unit/rank/section, etc.:	

Address:	Suite.#:

City:	County:	State:	Zip:

Work Phone 1:	Work Phone 2:
W. Cell Phone 1:	W. Cell Phone 2 or Other:
Email 1:	Email 2 (or website):
Pager:	Fax:

Coworkers at this location:	Ext	Cell phone	Home Phone or Other

Person 2: Name:	Place of Employment:

Title/unit/rank/section, etc.:	

Address:	Suite.#:

City:	County:	State:	Zip:

Work Phone 1:	Work Phone 2:
W. Cell Phone 1:	W. Cell Phone 2 or Other:
Email 1:	Email 2 (or website):
Pager:	Fax:

❑ Other contact info attached (if not living at the main address at top of this page, or if used for other person).

Coworkers at this location:	Ext	Cell phone	Home Phone or Other

Other General Contacts for Either Person

Name	Phone 1	Phone 2 or Other

"Ext" = Phone Extension at the work phone number. **Hint**: Use a copy of this form to give vacation info to others.

Attachments:	❑ **"Important Contacts – Emergency Contact"** ❑ **"Important Contacts – General Contact"** ❑ Other **Find Me** pages for or from other family members

"FIND ME"

Stored Documents List

Use this worksheet to track the storage location and valid dates of important documents such as titles, deeds, wills, etc., from the "**Info Pack**" checklist. **Keep this list with your Info Pack** in your Bugout Kit and periodically review (during kit refresh) it to make sure all documents are up-to-date and secure.

Periodic Documents are those that have an **expiration date** and will need **renewal**.

Document Type	Who?	Identifying #	Issued By	Issue Date	Expires On	Original Stored	Copies Stored
(example) **Passport**				__/__/__	__/__/__		
				__/__/__	__/__/__		
				__/__/__	__/__/__		
				__/__/__	__/__/__		
				__/__/__	__/__/__		
				__/__/__	__/__/__		
				__/__/__	__/__/__		

Periodic Documents

Fixed Date Documents are **permanent** in nature with no expiration date.

Document Type	Who?	Identifying #	Issued By	Issue Date	Original Stored	Copies Stored
				__/__/__		
				__/__/__		
				__/__/__		
				__/__/__		
				__/__/__		
				__/__/__		

Fixed Date Documents

Document Type = Will, Contract, Deed, Passport, etc. **Who?** = Whose name is the document in? **Original Stored & Copy Stored** = Storage location.

House Occupant Rescue Stickers

Dear First Responder, the following live here:

☐ ___ Mobility challenged individuals.
☐ ___ Medically fragile individuals.
☐ ___ Sight or Hearing impaired individuals.
☐ Language barrier. Language is: _____
☐ ___ Dogs ___ Cats ___ Birds ___ Other: ___
☐ Other consideration: _____
☐ Our Neighborhood Watch group has more info. Call:
_____ at: _____

☐ Our contact info posted by our _____ phone .

Dear First Responder, the following live here:

☐ ___ Mobility challenged individuals.
☐ ___ Medically fragile individuals.
☐ ___ Sight or Hearing impaired individuals.
☐ Language barrier. Language is: _____
☐ ___ Dogs ___ Cats ___ Birds ___ Other: ___
☐ Other consideration: _____
☐ Our Neighborhood Watch group has more info. Call:
_____ at: _____

☐ Our contact info posted by our _____ phone .

In the blanks next to the checkboxes, show the number of people, pets, etc. that fall into that category at this address.

Dear First Responder, the following live here:

☐ ___ Mobility challenged individuals.
☐ ___ Medically fragile individuals.
☐ ___ Sight or Hearing impaired individuals.
☐ Language barrier. Language is: _____
☐ ___ Dogs ___ Cats ___ Birds ___ Other: ___
☐ Other consideration: _____
☐ Our Neighborhood Watch group has more info. Call:
_____ at: _____

☐ Our contact info posted by our _____ phone .

Dear First Responder, the following live here:

☐ ___ Mobility challenged individuals.
☐ ___ Medically fragile individuals.
☐ ___ Sight or Hearing impaired individuals.
☐ Language barrier. Language is: _____
☐ ___ Dogs ___ Cats ___ Birds ___ Other: ___
☐ Other consideration: _____
☐ Our Neighborhood Watch group has more info. Call:
_____ at: _____

☐ Our contact info posted by our _____ phone .

(Where it says "Our contact info posted by our ___ phone," write down the place you posted your "Important Contacts – Quick List.")
Though printed on regular paper, you can tape these inside windows, or use self-adhesive laminating sheets to attach them to exterior surfaces.
Keep at least one visible at the front door, one at any garage doors, and one at each back or side entrance to your home.

OPEN ONLY IN CASE OF EMERGENCY

If you are reading this letter, it must mean we are in the middle of an emergency situation and need your help.

The information contained here should provide you enough data for you to help us in our absence.

As we have no idea what kind of emergency has prompted you to open this envelope, please follow these instructions in the order listed. Above all, thank you for your help!

1. If you are trying to reach us and have no other way to find us, start by trying the following:

Name	Phone 1	Phone 2	Email

In addition to the above we have:

❏ A phone list posted by our phone. ❏ Our address book is located _____.

❏ We have an address book program on our computer. The software name is_____.

❏ Our computer password is: _____. (← Note: Sensitive info.)

2. If you are securing care for us or our dependents for whatever reason, you'll need to know the following:

Medical   ("Who For?" = Who is the family member that sees this doctor?)

Type	Name	Who For?	Phone 1	Phone 2	Email
Doc 1					
Doc 2					
Vet					

Professional

Type	Name	Phone 1	Phone 2	Email
Insurance				
Attorney				

3. If you are gathering our goods and gear to help is in an emergency evacuation, please bring the following:

❏ Our "**Bugout Kit**" is in a: (number & type of packs or bags)_____
and is located in:_____.

❏ Our "Bugout Kit" has a "**Last Minute List**" located in its: _____. This list will give more info.

❏ Our main **Rendezvous Point** is: _____.

4. This copy of this Emergency Letter is being left with: (name:)_____. We may need the following <u>specific</u> help from <u>you</u>:_____

_____

5. Attachments and Enclosures: (Attach or enclose <u>only</u> what is needed by this letter's recipient.)

❏ House key is enclosed, or is hidden: _____   ❏ Alarm code: _____

❏ Safe deposit box key is enclosed or is located: _____   ❏ Other key(s) enclosed:_____

❏ Custody Release for minor child ❏ Medical Release for minor child ❏ Special instructions for pet care

❏ "Find Me" sheet(s) ❏ "Important Contacts" page(s) ❏ "Family Member Data" sheet(s)

❏ Computer disk enclosed ❏ Computer password:_____ ❏ Other written instruction attached

❏ Legal documents attached:_____

❏ Please contact the people listed on the attached sheet and notify them of our emergency. ("Important Contacts" page.)

Note: Ask where this recipient stores this letter. Note the location on your "Notify in Case of Emergency" card.

This page intentionally left blank.

My name is: _____

Address: _____

City: _____

County: _____ State: __ Zip: _____

Phone: _____

Cell phone: _____

Work phone: _____

Email: _____

In case of emergency, please notify:

Name: _____

Address: _____

City: _____ St: ___ Zip: _____

Phones: _____

Cell phones: _____

Work phones: _____

Email: _____

☐ Remind this person their instructions are located:

_____

My name is: _____

Address: _____

City: _____

County: _____ State: __ Zip: _____

Phone: _____

Cell phone: _____

Work phone: _____

Email: _____

In case of emergency, please notify:

Name: _____

Address: _____

City: _____ St: ___ Zip: _____

Phones: _____

Cell phones: _____

Work phones: _____

Email: _____

☐ Remind this person their instructions are located:

_____

This is the **front side** your **small** "Notify in Case of Emergency" card. **Cut** out along the *solid* lines and **fold** along the *dotted* lines. **The next page is the backside of the card.** Have everything photocopied onto card stock if you wish. You can fill this in by hand or call this file up on your computer to fill in on screen.

My name is: _____

Address: _____

City: _____

County: _____ State: __ Zip: _____

Phone: _____

Cell phone: _____

Work phone: _____

Email: _____

In case of emergency, please notify:

Name: _____

Address: _____

City: _____ St: ___ Zip: _____

Phones: _____

Cell phones: _____

Work phones: _____

Email: _____

☐ Remind this person their instructions are located:

_____

My name is: _____

Address: _____

City: _____

County: _____ State: __ Zip: _____

Phone: _____

Cell phone: _____

Work phone: _____

Email: _____

In case of emergency, please notify:

Name: _____

Address: _____

City: _____ St: ___ Zip: _____

Phones: _____

Cell phones: _____

Work phones: _____

Email: _____

☐ Remind this person their instructions are located:

_____

Where it says "**Remind this person their instructions are located**..." write a short note that will remind your contact person where they put their "**Open Only in Case of Emergency**" letter. They should tell you where it's kept.

Keep one copy of this card in your wallet, and another copy on your "**Safety Necklace**."

Blood type: _____ Organ donor? ___ Age: ____
Allergies: _____
Medic Alert?:_____
On medication for:_____
_____
Doctor:_____
Address: _____
City:_____ St:___ Zip:_____
Phones:_____
- -
Also contact:_____
Address: _____
City:_____ St:___ Zip:_____
Phones:_____
Email:_____
❑ Remind this person their instructions are located:
_____
❑ I live alone and have pets or _____at home.
❑ _____

Blood type: _____ Organ donor? ___ Age: ____
Allergies: _____
Medic Alert?:_____
On medication for:_____
_____
Doctor:_____
Address: _____
City:_____ St:___ Zip:_____
Phones:_____
- -
Also contact:_____
Address: _____
City:_____ St:___ Zip:_____
Phones:_____
Email:_____
❑ Remind this person their instructions are located:
_____
❑ I live alone and have pets or _____at home.
❑ _____

This is the **backside** of your "**Notify in Case of Emergency**" card. **Cut** out along the *solid* lines and **fold** along the *dotted* lines. Be sure to fill these out with a ballpoint pen, Sharpie ® or other **indelible** ink pen. The **bottom blank line** is for any other special note or instruction you might want to record such as the immediate location of additional information or instruction. You might note, "Look in glove compartment," etc.

Blood type: _____ Organ donor? ___ Age: ____
Allergies: _____
Medic Alert?:_____
On medication for:_____
_____
Doctor:_____
Address: _____
City:_____ St:___ Zip:_____
Phones:_____
- -
Also contact:_____
Address: _____
City:_____ St:___ Zip:_____
Phones:_____
Email:_____
❑ Remind this person their instructions are located:
_____
❑ I live alone and have pets or _____at home.
❑ _____

Blood type: _____ Organ donor? ___ Age: ____
Allergies: _____
Medic Alert?:_____
On medication for:_____
_____
Doctor:_____
Address: _____
City:_____ St:___ Zip:_____
Phones:_____
- -
Also contact:_____
Address: _____
City:_____ St:___ Zip:_____
Phones:_____
Email:_____
❑ Remind this person their instructions are located:
_____
❑ I live alone and have pets or _____at home.
❑ _____

Where it says, "**I have more contact info in**:," list a location such as your purse, glove compartment, Dayplanner, etc., where you might have a copy of your "**Find Me**" page or pages of your "**Important Contacts**."

In Case of Emergency

My name is: _____
Address: _____
City: _____
County: _____ State: ___ Zip:_____
Phone:_____
Cell phone: _____
Work phone: _____
Email: _____
Other:_____

Please Contact: Name:_____
Address: _____
City:_____ St:___ Zip:_____
Phones:_____
Cell phones: _____
Work phones: _____
Email: _____
☐ Remind this person their instructions are located:
_____

Please Contact: Name:_____
Address: _____
City:_____ St:___ Zip:_____
Phones:_____
Cell phones: _____
Work phones: _____
Email: _____
☐ Remind this person their instructions are located:
_____

This is the **front side** a **larger** "**Notify in Case of Emergency**" card. **The next page is the backside of the card.** It'll **fold along the dotted lines** and wind up being about the size of a business card.

In Case of Emergency

My name is: _____
Address: _____
City: _____
County: _____ State: ___ Zip:_____
Phone:_____
Cell phone: _____
Work phone: _____
Email: _____
Other:_____

Please Contact: Name:_____
Address: _____
City:_____ St:___ Zip:_____
Phones:_____
Cell phones: _____
Work phones: _____
Email: _____
☐ Remind this person their instructions are located:
_____

Please Contact: Name:_____
Address: _____
City:_____ St:___ Zip:_____
Phones:_____
Cell phones: _____
Work phones: _____
Email: _____
☐ Remind this person their instructions are located:
_____

Fold so that the "In Case of Emergency" panel winds up on the outside.

Also contact:_____
Address: _____
City:_____ St:____ Zip:_____
Phones:_____
Email:_____
❏ Remind this person their instructions are located:

_____
❏ I live alone and have pets or _____at home.
❏ _____

Other instruction:_____
_____
_____
_____
_____
_____
_____
_____
_____

Blood type: _____ Organ donor? ____ Age: ____
Allergies: _____
Medic Alert?:_____
On medication for:_____
_____
Doctor:_____
Address: _____
City:_____ St:____ Zip:_____
Phones:_____

Other medical data:_____
_____
_____
_____
_____
_____
_____
_____
_____

This is the **backside** of your larger "**Notify in Case of Emergency**" card. Though it folds up to about the size of a business card, there is more room for extra instruction if so needed. You may need this room for extra people to contact, additional instruction for personal needs or those of dependants, etc.

Also contact:_____
Address: _____
City:_____ St:____ Zip:_____
Phones:_____
Email:_____
❏ Remind this person their instructions are located:

_____
❏ I live alone and have pets or _____at home.
❏ _____

Other instruction:_____
_____
_____
_____
_____
_____
_____
_____
_____

Blood type: _____ Organ donor? ____ Age: ____
Allergies: _____
Medic Alert?:_____
On medication for:_____
_____
Doctor:_____
Address: _____
City:_____ St:____ Zip:_____
Phones:_____

Other medical data:_____
_____
_____
_____
_____
_____
_____
_____
_____

In case you need to find me.
My name:_____
Address: _____
City:_____ St:___ Zip:_____
Home phone 1:_____
Home phone 2:_____
Cell phone 1: _____
Cell phone 2: _____
Email 1: _____

I work at:
Co.:_____
Address: _____
City:_____ St:___ Zip:_____
Work phone 1:_____
Work phone 2:_____
Cell phone 1: _____
Cell phone 2: _____
Email 1: _____

In case you need to find me.
My name:_____
Address: _____
City:_____ St:___ Zip:_____
Home phone 1:_____
Home phone 2:_____
Cell phone 1: _____
Cell phone 2: _____
Email 1: _____

I work at:
Co.:_____
Address: _____
City:_____ St:___ Zip:_____
Work phone 1:_____
Work phone 2:_____
Cell phone 1: _____
Cell phone 2: _____
Email 1: _____

Your "**Find Me**" cards.   This is a smaller version of your "Find Me" *page*. Fill these out now and give them to people you're associated with that might need to keep this kind of information in their wallet.

Cut out along the *solid* lines and **fold** along the *dotted* lines. **This page is the front side of the card.**

In case you need to find me.
My name:_____
Address: _____
City:_____ St:___ Zip:_____
Home phone 1:_____
Home phone 2:_____
Cell phone 1: _____
Cell phone 2: _____
Email 1: _____

I work at:
Co.:_____
Address: _____
City:_____ St:___ Zip:_____
Work phone 1:_____
Work phone 2:_____
Cell phone 1: _____
Cell phone 2: _____
Email 1: _____

In case you need to find me.
My name:_____
Address: _____
City:_____ St:___ Zip:_____
Home phone 1:_____
Home phone 2:_____
Cell phone 1: _____
Cell phone 2: _____
Email 1: _____

I work at:
Co.:_____
Address: _____
City:_____ St:___ Zip:_____
Work phone 1:_____
Work phone 2:_____
Cell phone 1: _____
Cell phone 2: _____
Email 1: _____

Hint: Keep a set of these blank cards in your Bugout Kit's Office Pack. During long-distance or serious evacuations, fill them out with your evacuation destination info and put one in each person's and pet's "Safety Necklace."

Other contact:_____

Address: _____

City:_____ St:___ Zip:_____

Home phone 1:_____

Home phone 2:_____

Cell phone 1: _____

Cell phone 2: _____

Email 1: _____

_____

- - - - - - - - - - - - - - - - - - - -

Other contact: _____

Home phone 1:_____

Home phone 2:_____

Cell phone 1: _____

Cell phone 2: _____

Email 1: _____

Other:_____

Other:_____

_____

Other contact:_____

Address: _____

City:_____ St:___ Zip:_____

Home phone 1:_____

Home phone 2:_____

Cell phone 1: _____

Cell phone 2: _____

Email 1: _____

_____

- - - - - - - - - - - - - - - - - - - -

Other contact: _____

Home phone 1:_____

Home phone 2:_____

Cell phone 1: _____

Cell phone 2: _____

Email 1: _____

Other:_____

Other:_____

_____

This is the **backside** of your ""**Find Me" Cards**. **Cut** out along the *solid* lines and **fold** along the *dotted* lines.
Examples of who might need a card: ❑ Child's teachers ❑ Babysitter ❑ Select Volunteer Group members
❑ Tape one to your pet's leash. ❑ Also carry one in your **wallet** in case you lose it and it's found by someone.

Other contact:_____

Address: _____

City:_____ St:___ Zip:_____

Home phone 1:_____

Home phone 2:_____

Cell phone 1: _____

Cell phone 2: _____

Email 1: _____

_____

- - - - - - - - - - - - - - - - - - - -

Other contact: _____

Home phone 1:_____

Home phone 2:_____

Cell phone 1: _____

Cell phone 2: _____

Email 1: _____

Other:_____

Other:_____

_____

Other contact:_____

Address: _____

City:_____ St:___ Zip:_____

Home phone 1:_____

Home phone 2:_____

Cell phone 1: _____

Cell phone 2: _____

Email 1: _____

_____

- - - - - - - - - - - - - - - - - - - -

Other contact: _____

Home phone 1:_____

Home phone 2:_____

Cell phone 1: _____

Cell phone 2: _____

Email 1: _____

Other:_____

Other:_____

_____

FAMILY MEMBER DATA SHEETS – ONE -- DESCRIPTION

This page is designed to provide a complete description for use in a missing persons situation, and does not divulge sensitive information. Gather this material <u>now</u> and keep it along with **attached photos** in your "Info Pack." Make a page for each family member and update this info, as necessary, when you "refresh" your packs.

Name: First: Middle: Last:

Sex: Today's Date: ___/___/___ Age as of today: Height: Weight: Eye color:

Race: Color / complexion: Nationality: Language spoken:

Hair color: Hair style: Hair length: Wig or toupee? ❏ Y ❏ N   Shoe size:

Facial hair? ❏ Y ❏ N   If yes, describe: ❏ Left handed ❏ Right handed

Teeth: ❏ Normal ❏ Dentures ❏ Braces ❏ Retainer ❏ Other description:

❏ Glasses, Describe: ❏ Contact Lenses

Jewelry usually worn: ← ❏ Photos attached.

Scars: ← ❏ Photos attached.

Marks or birthmarks: ← ❏ Photos attached.

Tattoos: ← ❏ Photos attached.

Piercings: ← ❏ Photos attached.

Name engraved on jewelry, such as ID bracelet? ❏ Y ❏ N Item:_____   Name written inside clothing? ❏ Y ❏ N

In addition to photos of scars, marks, and tattoos, include **at least**: ❏ **3 facial photos** ❏ **One full length body photo**

Additional physical description, i.e. walks with a limp, missing finger, etc.:

❏ Pages attached

In the spaces below, include just enough medical information to help further identify this person.

Implants: ❏ Breast ❏ Joint ❏ Pin in bone ❏ Metal plate ❏ Glass eye: __ Left __ Right, Color:

Equipment: ❏ Hearing aid ❏ Oxygen ❏ Pacemaker ❏ Cane or walker ❏ Wheelchair ❏ Catheter ❏ Colostomy

❏ Prosthesis: _____ ❏ _____ ❏ _____

Prior surgeries for:

On medication for:

Allergies: ❏ Pages attached

Fill this out now with your regular or permanent base information:

Our regular Home Address is:

City: County: State: Zip:

Phone 1:   Cell Phone 1:   Email 1:

Phone 2:   Cell Phone 2:   Email 2:

Other:

Fill out the below when you submit this info in a missing person event. You may have temporarily relocated.

My name is: (←person submitting report)

I am next of kin. My relationship to this person is:

Today's date is: ___/___/___ Current Address is:

City: County: State: Zip:

Phone 1:   Cell Phone 1:   Email 1:

Phone 2:   Cell Phone 2:   Email 2:

Other:

This current address and contact info is: ❏ Permanent ❏ Temporary until ___/___/___, or _____

FAMILY MEMBER DATA SHEETS – ONE -- DESCRIPTION

Note: If using this page for a child, do not fill out variable info such as height, weight, hair, etc. until you hand this form off in a missing person situation. With everyone, collect and store photos now, and **put your contact info on the back of all photos** along with the name of the person in the photos, and the date taken. Update child photos regularly.

FAMILY MEMBER DATA SHEETS – TWO -- IDENTIFIERS

Collect this material and keep it safely stored. The previous page on identifiers, can be readily handed out to anyone looking for your missing family member. This page should _only_ be given to the Law Enforcement office coordinating the search, and _only_ when this material is asked for, as it's the only one you'll have.

Name of person being identified: ☐ **Description page** attached

Dental Records

Dental records available through: Copies attached? ☐ Y ☐ N

Address:

City: State: Zip:

Phone: Fax: Email:

☐ "Dental Fingerprint" impression was made on ___/___/___ , last updated on ___/___/___ , and stored at:

☐ Saved baby teeth or removed adult teeth available for DNA purposes. (Do not store teeth in plastic. Use paper bags.)

Hair Sample:	Blood Sample
Hair Sample: <u>Pluck</u> 5 hairs and tape them here. Hair follicles Hold the hairs in place with scotch tape, but **don't** cover the ends with the follicular material.	**Blood Sample** Place droplets here Allow drops to air-dry. Do not cover with tape or plastic. **Do not collect a blood sample from a person known to carry an infectious disease**.

☐ DNA profile previously run? ☐ Y ☐ N If yes, by:

☐ Doctor has been asked to maintain DNA material sample. ☐ Medical records or "Family Data" sheet attached.

☐ Dentist has been asked to retain "Dental Fingerprint" or other material or records.

As an alternative to blood (or in addition) attach a **cotton swab** that has been rubbed in the mouth, inside the cheek. Allow these to air-dry and keep them loosely wrapped in a small paper bag. (Never store wet samples in plastic.)

Fingerprints

Previously fingerprinted by: For:

Left Thumb	**Fingerprints may be made using a simple stamp ink pad. Practice on a separate sheet of paper before making prints on this sheet.**	Right Thumb

Left pinky	Left ring	Left middle	Left index	Right index	Right middle	Right ring	Right pinky

Use back of sheet for full handprint or baby footprints.

This material submitted by: Relationship:

☐ My contact info has been attached and can be found on the "Description" page with the photos of this person.

This page along with the "Family Member – Description" page make an excellent "Child ID Kit."

FAMILY MEMBER DATA SHEETS – TWO -- IDENTIFIERS

FAMILY MEMBER DATA SHEETS – THREE -- SENSITIVE INFO

Refer to "Planning Ahead for Phase II," to see why these forms are important. You'll also want to know why these forms have to be in hardcopy format and how to keep some of the data on them coded and secure.

Name (Give first name only):

Alternate mailing address (if any):

City: County: State: Zip:

Date of birth: ___/___/___   Place of birth:

Mother's Maiden Name:

Social Security Number: Personal Tax ID # (if any):

Drivers License #:   Issue date: ___/___/___   Expiration date: ___/___/___

Passport #:   Issue date: ___/___/___   Expiration date: ___/___/___

INS or "Green Card" #:   Issue date: ___/___/___   Expiration date: ___/___/___

Internet Service Provider: Computer password:

Screen Names:

Answering machine and/or voice mail access codes:

Safe Deposit Box #:   Bank name & phone #:

Do you have combination locks on anything? If so, record their numbers:

Lock is on: Combo: Lock is on: Combo:

Vehicle Type: Make: Model: Year:

Tag: State: County: Alarm ?: ❏ Y ❏ N "Lo-Jack?": ❏ Y ❏ N

Color(s): Body: Roof:

Distinguishing marks/features:

VIN: Gas/Diesel

Vehicle owner if not this person:

Insurance Co.:   Policy#:

Agent: Original start date of policy:

Address: Apt./Ste.:

City: County: St: Zip:

Phone 1:   Phone 2:

Email 1:   Email 2:

Cell Phone: Website:

Pager: Fax:

This page is for family reference and is not to be given out to anyone.

Attachments:	❏ Description ❏ Identifiers ❏ Medical Info ❏ Financial Page ❏ Student Contacts
	❏ Important Contacts page for additional employment, other vehicles individually owned, etc.

Code key: **(This is a note to remind you how you coded pertinent info.)**

FAMILY MEMBER DATA SHEETS – THREE -- SENSITIVE INFO

This is sensitive info. Give this page only to medical professionals to ID or render aid to this person.

Name: Sex: DOB: ___/___/___   SSN:

Organ Donor? ❏ Y ❏ N   Medic Alert? ❏ Y ❏ N For: _____   Blood Type:

List necessary meds and prescription numbers:

Dietary restrictions:

Allergies:

Handicaps/Impairments:

❏ Pins ❏ Plates ❏ Screws ❏ Replacement Joints ❏ Pacemaker** ❏ Colostomy** ❏ Catheter** ❏ Oxygen**

❏ Joint braces ❏ Crutches ❏ Walker ❏ Wheelchair** ❏ Prosthesis** ❏ Other:

** Provide details including **serial numbers of equipment**:

History of problems with: ❏ Heart ❏ Hypertension ❏ Asthma ❏ Lung ❏ Hypoglycemia ❏ Diabetes ❏ Hepatitis ❏ HIV/AIDS ❏ Cancer ❏ Eye problems ❏ Arthritis ❏ Broken bones ❏ Thyroid ❏ Endocrine ❏ Tuberculosis ❏ Liver ❏ Kidney ❏ Digestive ❏ Blood disorder ❏ Ulcers ❏ Pregnancy/Childbirth complications ❏ Stroke ❏ Other
❏ **If any of the above are checked, attach copies of records or a description on a separate sheet of paper explaining dates, doctors, details, and any lasting effects or current medications and precautions.**

Surgery: ❏ Tonsils ❏ Appendix ❏ Gallbladder ❏ Eye ❏ Heart ❏ Joint ❏ Gastro-Intestinal ❏ Cosmetic
❏ Skeletal reconstructive ❏ Cancer/Tumor ❏ Hysterectomy ❏ "C-Section" ❏ Vasectomy ❏ Hair plugs/transplants
❏ Other_____   ❏ Other_____
❏ **If any of the above are checked, attach copies of records or a description on a separate sheet of paper explaining dates, doctors, details, and any lasting effects or current medications and precautions.**

Doctor:		Specialty	
Address:		Hospital:	
City:	County	St:	Zip:
Phone:	Pager:	Cell Phone:	
Fax:	E-Mail:	Website:	
Dentist:		Specialty:	
Address:		Suite:	
City:	County	St:	Zip:
Phone:	Pager:	Cell Phone:	
Fax:	E-Mail:	Website:	

❏ Retainer ❏ Braces ❏ Crowns ❏ Root Canals ❏ Partials ❏ Bridges ❏ Dentures ❏ Wisdom Teeth ❏ X-Rays
❏ **If any of the above are checked, attach copies of records or a description on a separate sheet of paper.**

Pharmacist:		Store Name:	
Address:	Suite:	Customer ID#:	
City:	County	St:	Zip:
Phone:	Pager:	Cell Phone:	
Fax:	E-Mail:	Website:	

Attachments:	❏ "Important Contacts - Professionals" pages for additional doctors, etc. ❏ Medical records copies
	❏ Other written information ❏ Unfilled emergency prescriptions ❏ Dated list of **immunizations**
	❏ Written first aid or treatment instruction regarding any existing illness, injury, or condition.
	❏ All individual medical insurance providers listed on "Family Member Data Sheet - Financial."

FAMILY MEMBER DATA SHEETS – FOUR -- MEDICAL

FAMILY MEMBER DATA SHEETS -- FIVE – FINANCIAL (PART 1)

This is for individual accounts held by this person and not shared by the household. All accounts pertaining to the home or shared with other household members will be listed on the "Household Data" page.

Name: (← Use first name only or just initials as this page contains sensitive info.)

Individual Accounts not shared with the rest of the family. (Joint accounts go on "Household Data" page.)

Type	Bank Name	Account # (code these)	Cust. Svc. Phone	PIN #	Exp Date
Checking					
Savings					
Credit Card					
Credit Card					
Credit Card					
Gas Card					
Gas Card					
Gas Card					
Loan					
ATM Card					
Phone Card					

Place of Employment: Supervisor:

Address: Title/rank/unit:

City: St: Zip: --   Website:

Phone: -   x   Fax: E-Mail

Professional licensing? Give type and license #:

Detailed Contact for Select Accounts or Services - (Continued on Part 2)

Company name: Contact person:

Address: Suite:

Address 2:   City: State: Zip:

Phone 1:   Phone 2:

Cell Phone 1:   Cell Phone 2:

Email: Fax:

Website:

Web access ID & Password:

Account #:   PIN# or ID#:

Note:

This is page # ____ of ____ of financial data for this person. ☐ "Important Contacts – Professional" page attached.

Code key: **(This is a note to remind you how you coded the above info.)**

FAMILY MEMBER DATA SHEETS -- FIVE – FINANCIAL (PART 1)

FAMILY MEMBER DATA SHEETS -- FIVE – FINANCIAL (PART 2)

Use this to list other accounts and services. Include: ❑ Insurance Policies ❑ Individual Loans ❑ Attorney
❑ Accountant ❑ Bank Contact Info ❑ Furniture Storage Units, etc.

Company name:	Contact person:
Address:	Suite:
City:	County:	State:	Zip:
Phone 1:	Phone 2:
Cell Phone 1:	Cell Phone 2:
Email:	Fax:
Website:	Web access ID & Password:
Account #:	PIN# or ID#:
Note:	

Company name:	Contact person:
Address:	Suite:
City:	County:	State:	Zip:
Phone 1:	Phone 2:
Cell Phone 1:	Cell Phone 2:
Email:	Fax:
Website:	Web access ID & Password:
Account #:	PIN# or ID#:
Note:	

Company name:	Contact person:
Address:	Suite:
City:	County:	State:	Zip:
Phone 1:	Phone 2:
Cell Phone 1:	Cell Phone 2:
Email:	Fax:
Website:	Web access ID & Password:
Account #:	PIN# or ID#:
Note:	

Company name:	Contact person:
Address:	Suite:
City:	County:	State:	Zip:
Phone 1:	Phone 2:
Cell Phone 1:	Cell Phone 2:
Email:	Fax:
Website:	Web access ID & Password:
Account #:	PIN# or ID#:
Note:	

This attachment is page # _____ of _____ (# of total pages).

FAMILY MEMBER DATA SHEETS -- FIVE – FINANCIAL (PART 2)

FAMILY MEMBER DATA SHEETS -- SIX – STUDENT
NOTE: CREATE A NEW FORM EACH SCHOOL YEAR, OR WHENEVER SCHEDULE CHANGES.

Child's name:	Child's locker #:	Locker combination:

This data for school year: _____ **to** _____ **in the** _____ **quarter or semester. Today's date:** ___ / ___ / ___

School name:

Address:		Suite #:
City:	County:	State: Zip:
Phone 1:	Phone 2:	Fax:
Email:	Website:	

School Principal:

Phone 1:	Phone 2:	Email1:
Cell phone:	Pager:	

Vice Principal:

Phone 1:	Phone 2:	Email1:
Cell phone:	Pager:	

Teacher Contact

Period	Teacher	Room#	Subject	Cell Phone	Other Contact
Homeroom					
1					
2					
3					
4					
5					
6					
	Campus Police				
	School Custodian				

Bus #: Bus Driver's Name:	Driver's Cell Phone:
Radio Scanner frequency for school's bus system:	Other radio info:

Contact Data for School's Primary Evacuation Destination

Name of location:

Address:

City:	County:	State: Zip:
Phone 1:	Phone 2:	
Email 1:	Email 2 (or website):	

Contact person:	This person is: ❑ Mgr. ❑ Security ❑ _____

City Map Atlas page #: ___ , or Map Grid #: _____   ❑ Location directions and map attached.

Meet-point at this location:

Radio stations for school-related news: _____ AM/FM, _____ AM/FM, _____ AM/FM ❑ Phone numbers on back

Attachments: ❑ Photocopy of child's schedule ❑ Data for alternate evac destinations ❑ Ride info if not bus.
❑ Copy of student's school ID (if any) ❑ Certified-as-true copy of custody or guardianship papers
❑ Law Enforcement contact info at evac destination(s) ❑ Contact info for **BOE** and **PTA** officials.

❑ Each teacher and school administrator given a copy of our "**Find Me**" page in order for them to reach us.

FAMILY MEMBER DATA SHEETS -- SIX – STUDENT

We own _____ # of pets. Therefore, this page is #_____ of _____ total.

Fill this out now with your regular or permanent base information:

My name is:

Our regular Home Address is:

City:	County:	State:	Zip:
Phone 1:	Cell Phone 1:	Email 1:	
Phone 2:	Cell Phone 2:	Email 2:	

Fill out this section when you submit this info. You may have temporarily relocated.

Today's date is: ___/___/___ Current Address is:

City:	County:	State:	Zip:
Phone 1:	Cell Phone 1:	Email 1:	
Phone 2:	Cell Phone 2:	Email 2:	

Other:

This current address and contact info is: ❑ Permanent ❑ Temporary until ___/___/___ or _____

PET INFO (Fill this out now, and fill out a new sheet for each pet)

Pet's Name:

Species (Dog, Cat, etc.):   Breed (Type):

Sex: Height: Length: Weight: Date of birth: ___/___/___

Colorings / Markings:

Distinctive identifiers (scars, etc.):

Does this animal wear a collar? ❑ Y ❑ N Describe collar:

Is there an "owner's info tag" on this animal? ❑ Y ❑ N   Implanted locator / ID chip? ❑ Y ❑ N   Tattoo? ❑ Y ❑ N

Does this animal bite? ❑ Y ❑ N Behavioral problems or peculiar habits:

Answers to voice commands of:

Spayed or Neutered? ❑ Y ❑ N Current rabies tag #: _____ Vaccination month: _____

Regular flea treatment? ❑ Y ❑ N   If yes, brand: _____

Allergies:

History of medical problems/conditions/care:

❑ More info attached

Medications:

Does this pet have a medic alert tag? ❑ Y ❑ N

Dietary, care, and feeding instructions:

Regular Veterinarian:	Name of Clinic:		
Address:			
City:	County:	State:	Zip:
Phone 1:	Cell Phone 1:	Email 1:	
Phone 2:	Cell Phone 2:	**Emergency #:**	
Website:	Pager:	Fax:	

Attachments:	❑ **Proof of vaccinations** ❑ Copies of **medical records** ❑ **Three photos**: Face, side, angle. ❑ Photocopies of **tags** ❑ Copy of veterinary **insurance** or memberships ❑ Copy of pedigree papers ❑ Contact info for local animal control office ❑ Contact info for local Humane Society office

Store copies of this form: ❑ Bugout Kit's **Info Pack** ❑ Taped to pet's **leash or collar** ❑ Taped to pet's **carrier**.

HOUSEHOLD DATA

The Household Data section's purpose is to record all the important information to keep your home and its related accounts, services, and assets coordinated. See "Planning Ahead for Phase II" for the full story.

Family last name:

Address:　　　　　　　　　　　　　　　　　Apt.#:

City:　　　　　　County:　　　　　　State:　Zip:

Phone 1:　　　　Cell Phone 1:　　　　Email 1:

Phone 2:　　　　Cell Phone 2:　　　　Email 2:

Pager:　　　　Fax:　　　　Website:

Internet Service Provider:　　　　Computer password:

Screen Names:

Answering machine and/or voice mail access codes:

Alarm keypad code:　　　　Alarm company codeword:

FINANCIAL

Landlord or **Mortgage** Holder:　　　　Acct.#:

Name(s) on lease/mortgage.:　　　　$$_____ per _____

Contact/Rep:

Address:　　　　　　Suite.:

City:　　　　County:　　　　St:　Zip:

Phone 1:　　　　Cell Phone 1:　　　　Email 1:

Phone 2:　　　　Cell Phone 2:　　　　Email 2:

Fax:　　　　Website:

Website Logon ID and Password:

Note:

Bank:　　　　Acct. Type:　　　　Acct.#:

Name(s) on acct.:　　　　$$Balance:

ATM Card #:　　　　PIN #:　　　　Safe-Deposit Box#:

Note:

Contact/Rep:

Address:　　　　　　Apt./Ste.:　　　County:

City:　　　　County:　　　　St:　Zip:

Phone 1:　　　　Cell Phone 1:　　　　Email 1:

Phone 2:　　　　Cell Phone 2:　　　　Email 2:

Fax:　　　　Website:

Website Logon ID and Password:

Note:

Bank:　　　　Acct. Type:　　　　Acct.#:

Name(s) on acct.:　　　　$$Balance:

ATM Card #:　　　　PIN #:　　　　Safe-Deposit Box#:

Note:

Contact/Rep:

Address:　　　　　　Apt./Ste.:　　　County:

City:　　　　County:　　　　St:　Zip:

Phone 1:　　　　Cell Phone 1:　　　　Email 1:

Phone 2:　　　　Cell Phone 2:　　　　Email 2:

Fax:		Website:		

Website Logon ID and Password:

Note:

Bank: Acct. Type: Acct.#:

Name(s) on acct.: $$Balance:

ATM Card #: PIN #: Safe-Deposit Box#:

Note:

Contact/Rep:

Address: Apt./Ste.: County

City: County: St: Zip:

Phone 1: Cell Phone 1: Email 1:

Phone 2: Cell Phone 2: Email 2:

Fax: Website:

Website Logon ID and Password:

Note:

Credit Card: Acct. #: Exp. Date:___/___/___

Name(s) on acct.: $$Balance:

Date Opened:___/___/___ Note:

Bank Address: Contact/Rep:

City: St: Zip: Payment notes: $_____/_____

Phone 1: Phone 2: Cell Phone:

Fax: E-Mail: Website:

Website Logon ID and Password:

Credit Card: Acct. #: Exp. Date:___/___/___

Name(s) on acct.: $$Balance:

Date Opened:___/___/___ Note:

Bank Address: Contact/Rep:

City: St: Zip: Payment notes: $_____/_____

Phone 1: Phone 2: Cell Phone:

Fax: E-Mail: Website:

Website Logon ID and Password:

Credit Card: Acct. #: Exp. Date:___/___/___

Name(s) on acct.: $$Balance:

Date Opened:___/___/___ Note:

Bank Address: Contact/Rep:

City: St: Zip: Payment notes: $_____/_____

Phone 1: Phone 2: Cell Phone:

Fax: E-Mail: Website:

Website Logon ID and Password:

Credit Card: Acct. #: Exp. Date:___/___/___

Name(s) on acct.: $$Balance:

Date Opened:___/___/___ Note:

Bank Address: Contact/Rep:

City: St: Zip: Payment notes: $_____/_____

Phone 1: Phone 2: Cell Phone:

Fax: E-Mail: Website:

Website Logon ID and Password:

Credit Card: Acct. #:   Exp. Date:___/___/___

Name(s) on acct.: $$Balance:

Date Opened:___/___/___ Note:

Bank Address: Contact/Rep:

City: St: Zip: Payment notes: $_____/_____

Phone 1:   Phone 2:   Cell Phone:

Fax: E-Mail: Website:

Website Logon ID and Password:

Credit Card: Acct. #:   Exp. Date:___/___/___

Name(s) on acct.: $$Balance:

Date Opened:___/___/___ Note:

Bank Address: Contact/Rep:

City: St: Zip: Payment notes: $_____/_____

Phone 1:   Phone 2:   Cell Phone:

Fax: E-Mail: Website:

Website Logon ID and Password:

Credit Card: Acct. #:   Exp. Date:___/___/___

Name(s) on acct.: $$Balance:

Date Opened:___/___/___ Note:

Bank Address: Contact/Rep:

City: St: Zip: Payment notes: $_____/_____

Phone 1:   Phone 2:   Cell Phone:

Fax: E-Mail: Website:

Website Logon ID and Password:

Credit Card: Acct. #:   Exp. Date:___/___/___

Name(s) on acct.: $$Balance:

Date Opened:___/___/___ Note:

Bank Address: Contact/Rep:

City: St: Zip: Payment notes: $_____/_____

Phone 1:   Phone 2:   Cell Phone:

Fax: E-Mail: Website:

Website Logon ID and Password:

Other Financial Instruments: ☐ **Notes Attached**

	Institution:	Acct./ Serial #:	Value:	Maturity Date:
Money Market:				
Stocks:				
Bonds:				
"T Bills":				
CDs:				

Private Debts: Owed by or to whom:

$$ Amount: Re:

Address: Apt./Ste.:   Own/Rnt/Rsd   Yrs.:

City: St: Zip: -   County:

Phone: Pager: Cell Phone:

Fax: E-Mail: Website:

Employer: Frm: To: FT/PT

Title: Lic.#:   Salary:

Supervisor:	Other Contact:
Address:	Apt./Ste.:
City:	St: Zip: -
Phone: x	Pager: Website:
Cell Phone:	Fax: E-Mail:

Alimony: To or from whom: SSN:

$$ Amount:_____/_____   Re:

Case #:   Date of Decree: ____/____/____

Address:	Apt./Ste.: Own/Rnt/Rsd Yrs.:
City:	St: Zip: - County:
Phone:	Pager: Cell Phone:
Fax:	E-Mail: Website:
Employer:	Frm: To: FT/PT
Title:	Lic.#: Salary:
Supervisor:	Other Contact:
Address:	Apt./Ste.:
City:	St: Zip: -
Phone: x	Pager: Website:
Cell Phone:	Fax: E-Mail:

Child Support: To or from whom: SSN:

$$ Amount:_____/_____   Note:

Address:	Apt./Ste.: Own/Rnt/Rsd Yrs.:
City:	St: Zip: - County:
Phone:	Pager: Cell Phone:
Fax:	E-Mail: Website:
Employer:	Frm: To: FT/PT
Title:	Lic.#: Salary:
Supervisor:	Other Contact:
Address:	Apt./Ste.:
City:	St: Zip: -
Phone: x	Pager: Website:
Cell Phone:	Fax: E-Mail:

MISC. ACCOUNTS

Insurance:	Policy#:
Type of policy:	Policy Value:
Agency:	Premiums: $$_____/_____
Agent:	Original start date of policy: ___/___/___
Address:	Ste.:
City:	County: St: Zip:
Phone 1:	Cell Phone 1: Email 1:
Phone 2:	Cell Phone 2: Email 2:
Pager:	Fax:
Website:	Website Logon ID and Password:
Insurance:	Policy#:
Type of policy:	Policy Value:
Agency:	Premiums: $$_____/_____
Agent:	Original start date of policy: ___/___/___

Address:		Ste.:
City:	County:	St: Zip:
Phone 1:	Cell Phone 1:	Email 1:
Phone 2:	Cell Phone 2:	Email 2:
Pager:	Fax:	
Website:	Website Logon ID and Password:	

Insurance: Policy#:

Type of policy:	Policy Value:
Agency:	Premiums: $$_____/_____
Agent:	Original start date of policy: ___/___/___
Address:	Ste.:
City: County:	St: Zip:
Phone 1: Cell Phone 1:	Email 1:
Phone 2: Cell Phone 2:	Email 2:
Pager: Fax:	
Website: Website Logon ID and Password:	

Automotive Financing: Payments: $$_____/_____

Dealership:	Contact:	Value: $_____
Address:	Apt./Ste.: Acct#:	
City:	St: Zip:	Finance Date: ___/___/___
Phone:	Pager:	Cell Phone:
Fax:	E-Mail:	Website:
Finance co. if diff. from dealer:	Contact:	
Address:	Apt./Ste.: Acct#:	
City:	County:	St: Zip:
Phone:	Pager:	Cell Phone:
Fax:	E-Mail:	Website:

Major Purchase Financing: Re: Payments: $$_____/_____

Note:		
Dealership:	Contact:	Value: $_____
Address:	Apt./Ste.: Acct#:	
City:	St: Zip:	Finance Date: ___/___/___
Phone:	Pager:	Cell Phone:
Fax:	E-Mail:	Website:
Finance co. if diff. from dealer:	Contact:	
Address:	Apt./Ste.: Acct#:	
City:	St: Zip: -	
Phone:	Pager:	Cell Phone:
Fax:	E-Mail:	Website:

Security / Alarm Co: Acct #:

Date Acct. Opened: ___/___/___ Note:		
Address:	Apt./Ste.: Rep:	
City:	County:	St: Zip:
Phone:	Pager:	Cell Phone:
Fax:	E-Mail:	Website:
Website Logon ID and Password:		

Phone Co: Acct #:

Date Acct. Opened: ___/___/___ Note:

Address: Apt./Ste.:   Rep:

City: County: St: Zip:

Phone: Pager: Cell Phone:

Fax: E-Mail: Website:

Website Logon ID and Password:

Cell Phone: Acct #:

Date Acct. Opened: ___/___/___ Note:

Address: Apt./Ste.:   Rep:

City: County: St: Zip:

Phone: Pager: Cell Phone:

Fax: E-Mail: Website:

Website Logon ID and Password:

Power Co: Acct #:   Usage: $ ____/____

Date Acct. Opened: ___/___/___ Note:

Address: Apt./Ste.:   Rep:

City: County: St: Zip:

Phone: Pager: Cell Phone:

Fax: E-Mail: Website:

Website Logon ID and Password:

Gas Co: Acct #:   Usage: $____/____ Gas type:

Date Acct. Opened: ___/___/___ Note: Refill days:

Address: Apt./Ste.:   Rep:

City: County: St: Zip:

Phone: Pager: Cell Phone:

Fax: E-Mail: Website:

Website Logon ID and Password:

Cable/TV Co: Acct #:

Date Acct. Opened: ___/___/___ Note:

Address: Apt./Ste.:   Rep:

City: County: St: Zip:

Phone: Pager: Cell Phone:

Fax: E-Mail: Website:

Website Logon ID and Password:

Trash Co: Acct #:   Pickup days: Time:

Date Acct. Opened: ___/___/___ Note:

Address: Apt./Ste.:   Rep:

City: County: St: Zip:

Phone: Pager: Cell Phone:

Fax: E-Mail: Website:

Website Logon ID and Password:

Water/Sewer: Acct #:   Usage: $____/_____

Date Acct. Opened: ___/___/___ Note: Well: Y / N   Septic Tank: Y / N

Address: Apt./Ste.:   Rep:

City: County: St: Zip:

Phone: Pager: Cell Phone:

Fax:	E-Mail:	Website:

Website Logon ID and Password:

Other:	Acct #:	$ _____ / _____

Date Acct. Opened: ___/___/___ Note:

Address:	Apt./Ste.:	Rep:
City:	County:	St: Zip:
Phone:	Pager:	Cell Phone:
Fax:	E-Mail:	Website:

Website Logon ID and Password:

Other:	Acct #:	$ _____ / _____

Date Acct. Opened: ___/___/___ Note:

Address:	Apt./Ste.:	Rep:
City:	County:	St: Zip:
Phone:	Pager:	Cell Phone:
Fax:	E-Mail:	Website:

Website Logon ID and Password:

Other:	Acct #:	$ _____ / _____

Date Acct. Opened: ___/___/___ Note:

Address:	Apt./Ste.:	Rep:
City:	County:	St: Zip:
Phone:	Pager:	Cell Phone:
Fax:	E-Mail:	Website:

Website Logon ID and Password:

PROFESSIONAL SERVICES

Attorney:	Firm:

Note:

Address:	County:
City:	County:	St: Zip:
Phone:	Pager:	Cell Phone:
Fax:	E-Mail:	Website:
Home Address:	County:
City:	St: Zip: -
Phone:	Pager:	Cell Phone:
Fax:	E-Mail:	Website:

Accountant:	Firm:

Note:

Address:	County:
City:	County:	St: Zip:
Phone:	Pager:	Cell Phone:
Fax:	E-Mail:	Website:
Home Address:	County:
City:	St: Zip: -
Phone:	Pager:	Cell Phone:
Fax:	E-Mail:	Website:

Website Logon ID and Password:

Other:	Acct #:	$ _____ / _____

Date Acct. Opened: ___/___/___ Note:

Address: _____ Apt./Ste.: _____ Rep: _____

City: _____ St: ___ Zip: _____ -

Phone: _____ Pager: _____ Cell Phone: _____

Fax: _____ E-Mail: _____ Website: _____

Website Logon ID and Password: _____

Other: _____ Acct #: _____ $ ____ /____

Date Acct. Opened: ___/___/___ Note: _____

Address: _____ Apt./Ste.: _____ Rep: _____

City: _____ County: _____ St: ___ Zip: _____

Phone: _____ Pager: _____ Cell Phone: _____

Fax: _____ E-Mail: _____ Website: _____

Website Logon ID and Password: _____

Other: _____ Acct #: _____ $ ____ /____

Date Acct. Opened: ___/___/___ Note: _____

Address: _____ Apt./Ste.: _____ Rep: _____

City: _____ County: _____ St: ___ Zip: _____

Phone: _____ Pager: _____ Cell Phone: _____

Fax: _____ E-Mail: _____ Website: _____

Website Logon ID and Password: _____

Misc. Svcs: Paperboy, Babysitter, Grocery Store, Yard Service, Maid Service, Handyman, Barber, Taxi,

Security/Alarm Co., Food delivery, Exterminator, Accountant, Furniture Storage unit, etc.

PROPERTY, LAND, AND REAL ESTATE

Property/Real Estate: _____ Value: $ _____ Date of Purchase: ___/___/___ Zoned: _____

GPS Coordinates of property if known: _____

Residential: ❏ Multi-Plex; ____units:❏ Condo ❏ House ❏ Trailer ❏ Land: (Location is: ❏ Residence ❏ Rental Prop)

Commercial: ❏ Land only ❏ Developed; # of buildings:_____ ❏ Nature of commercial use: _____

Agricultural: ❏ Farm with buildings ❏ Cultivated Land ❏ Uncultivated Land ❏ Other: _____

Note: _____

Plat/Book: _____ Map: _____

Deed #s: _____

Address: _____ Apt./Ste.: _____ County: _____

City: _____ St: ___ Zip: _____ -

Property/Real Estate: _____ Value: $ _____ Date of Purchase: ___/___/___ Zoned: _____

GPS Coordinates of property if known: _____

Residential: ❏ Multi-Plex; ____units:❏ Condo ❏ House ❏ Trailer ❏ Land: (Location is: ❏ Residence ❏ Rental Prop)

Commercial: ❏ Land only ❏ Developed; # of buildings:_____ ❏ Nature of commercial use: _____

Agricultural: ❏ Farm with buildings ❏ Cultivated Land ❏ Uncultivated Land ❏ Other: _____

Note: _____

Plat/Book: _____ Map: _____

Deed #s: _____

Address: _____ Apt./Ste.: _____ County: _____

City: _____ St: ___ Zip: _____ -

DMV ITEMS

Use this section for cars and vehicles, boats, and any other transportation equipment.

Vehicle Type: _____ Make: _____ Model: _____ Year: _____

Tag: _____ State: _____ County: _____ Alarm: _____ "Lo-Jack": _____

Color(s): Body: _____ Roof: _____

Distinguishing marks/features:			
VIN:		Gas/Diesel	
Owner (if not family):			
Address:		Apt./Ste.:	
City:	St:	Zip: -	ICQ/UIN:
Phone:	Pager:		Website:
Cell Phone:	Fax:		E-Mail:
Vehicle Type:	Make:	Model:	Year:
Tag:	State: County:		Alarm: "Lo-Jack":
Color(s): Body:		Roof:	
Distinguishing marks/features:			
VIN:		Gas/Diesel	
Owner (if not family):			
Address:		Apt./Ste.:	
City:	St:	Zip: -	ICQ/UIN:
Phone:	Pager:		Website:
Cell Phone:	Fax:		E-Mail:
Vehicle Type:	Make:	Model:	Year:
Tag:	State: County:		Alarm: "Lo-Jack":
Color(s): Body:		Roof:	
Distinguishing marks/features:			
VIN:		Gas/Diesel	
Owner (if not family):			
Address:		Apt./Ste.:	
City:	St:	Zip: -	ICQ/UIN:
Phone:	Pager:		Website:
Cell Phone:	Fax:		E-Mail:

Miscellaneous

Be sure to cross reference any important forms associated with this document, in your "Stored Documents" list.

"Household Inventory" is separated as you'll want to hand off a copy of that form to an insurance adjuster. They do not need to see the sensitive information recorded here.

Fill in this information on computer to both print hardcopy and save electronically. <u>Code</u> the sensitive info.

Code key: (This is a note to remind you how you coded the above info.)

HOUSEHOLD DATA

This page intentionally left blank.

Temporary Custody Release for Minor Child

To Whom It May Concern:

1. I / we : _____ and _____ (print), being
   ❑ Parent(s) ❑ Custodial Parent ❑ Legal Guardian of: _____(print child's
   name), hereby authorize: _____ (print) to take temporary custody of our child
   in the event of an emergency or other situation preventing our personal presence.

2. This document does not constitute a permanent or irrevocable custody. It's sole purpose is to ensure the temporary
   safety of our child until such time as we are able to take personal custody. This document expires on: ___/___/___.

3. As additional verification, please perform the following:

   ❑ If I/we and/or my/our child have a regular relationship with you, you will probably have a copy of a "Find Me"
   sheet on file. Please call the listed emergency contacts to verify there is an emergency and that the bearer of this
   letter is indeed supposed to take custody of our child.

   ❑ Ask the bearer of this letter for a photo ID. A photocopy of it should be attached.

   ❑ Compare a current signature of this person to the one at the bottom of this letter.

   ❑ Compare my/our signature(s) below to any you may already have on file.

   ❑ Compare this letter to one that you may already have on file.

   ❑ Ensure that our child recognizes the bearer of this letter as a trusted family friend or relative.

4. When you release my/our child to this person, please **retain a copy of this letter** along with the following information:

   ❑ Your name:_____ (print).

   ❑ Keep a photocopy of this individual's photo ID as described above.

   ❑ Time and date the bearer of this letter took custody of our child:

   ❑ Day of week:_____ Date: ___/___/___ Time of day: _____ am/pm.

   ❑ Vehicular info and tag # of car:_____.

5. Signatures:

	Printed name	Signature	Date
Parent or Guardian 1	_____ ,	_____ ,	___/___/___
Parent or Guardian 2	_____ ,	_____ ,	___/___/___
Temporary Custodian	_____ ,	_____ ,	___/___/___
Attorney	_____ ,	_____ ,	___/___/___
Witness	_____ ,	_____ ,	___/___/___

6. Notary Seal:

Attachments: ❑ Photocopy of a photo ID of the bearer of this letter. ❑ Contact information to reach this same person.

(**Note**: You should fill out a separate sheet for each child, and/or each of your Temporary Custodians.)

(**Note #2**: Check with your attorney to make sure this form is legal in your state and covers all your needs.)

Medical Release for Minor Child

To Whom It May Concern:

1. I / we : _____ and _____ (print), being
   ❑ Parent(s) ❑ Custodial Parent ❑ Legal Guardian of: _____(print child's
   name), hereby authorize emergency medical treatment for our child in the event of injury or illness and we are unavailable
   to be present and/or to sign consent.

2. I / we agree to hold EMS personnel, physicians, associated hospitals, and all hospital personnel harmless against
   initiation of medical treatment, provided said treatment is performed within safe and acceptable medical guidelines.

3. This child's regular physician is:_____ and can be contacted at:

   Hospital, Clinic, or Office: _____
   Address:_____ Suite / Apt: _____
   City: _____County: _____State: _____Zip: _____
   Phone 1: _____ Phone 2: _____
   Email 1: _____ Email 2 (or website): _____
   Pager: _____ Fax: _____

4. This child is covered by the following medical insurance:

   Insurance company: _____ Agent: _____
   Policy #:_____ Group #: _____
   Other: _____
   Address:_____ Suite / Apt: _____
   City: _____County: _____State: _____Zip: _____
   Phone 1: _____ Phone 2: _____
   Email 1: _____ Email 2: _____
   Website: _____ Fax: _____

5. Though we are currently absent, please note the following, so that better care and/or treatment may be given.

   ❑ Attached, please find a form entitled "Family Member Data Sheets – Four – Medical," which will outline current
   medical conditions, treatments, allergies, etc. of the child listed above.

   ❑ Also attached, please locate a "Find Me" sheet which contains extensive contact info to find me / us.

6. Signatures:

	Printed name	Signature	Date
Parent or Guardian 1	_____,	_____,	___/___/___
Parent or Guardian 2	_____,	_____,	___/___/___
Temporary Custodian	_____,	_____,	___/___/___
Attorney	_____,	_____,	___/___/___
Witness	_____,	_____,	___/___/___

7. Notary Seal:

   (**Note**: Check with your attorney to make sure this form is legal in your state and covers all your needs.)

On CD2: Folder: *Appendix* File: *DP1Budget*

Disaster Prep Expense Journal / Planner **This copy is** ☐ **Budget Planner** ☐ **Purchase Record**

Use **one copy** of this sheet to plan your Disaster Prep expenditure **budget.** Use **another copy** to record actual **purchases**, as some may be tax deductible.

Item	Source	Qty	$ Each	Subtotal	Tax	Total	Date	R
							/ /	☐
							/ /	☐
							/ /	☐
							/ /	☐
							/ /	☐
							/ /	☐
							/ /	☐
							/ /	☐
							/ /	☐
							/ /	☐
							/ /	☐
							/ /	☐
							/ /	☐
							/ /	☐
							/ /	☐
							/ /	☐
							/ /	☐
							/ /	☐
							/ /	☐
							/ /	☐
							/ /	☐
							/ /	☐
Grand Total this page:								

Source = Where you did or will buy the item, **Qty** = Quantity, **Date** = Date purchased. **R.** = a checkmark column to show you saved the receipt.

Be sure to record new gear on your "Household Inventory" page. You'll also find an Excel spreadsheet version of this form on our CD. Look for **DP1Budget.xls.**

www.disasterprep101.com

HOUSEHOLD INVENTORY

Name:

Phone 1:

Address:

City:

St: Zip:

Phone 2:

Cell phone:

Email:

Page # _____ of _____ . Date this inventory taken: ___/___/___ . Copies filed where: _____

Use this page to list all your personal and household possessions for insurance purposes in the case of loss or destruction.

Item	Serial Number	Description or Other Identifying Mark or Number	Date Purch.	Value $$	Room	R
						☐
						☐
						☐
						☐
						☐
						☐
						☐
						☐
						☐
						☐
						☐
						☐
						☐
						☐
						☐
						☐
						☐
						☐
						☐
						☐

TOTAL FOR THIS PAGE

"Date Purch." = The date item purchased (if known.) **"Value $$"** = Purchase price or appraised value "**Room**" = Room where item kept. "**R**" = Receipt on file.

Insurance company contact info and policy information should be listed on "Family Member Data" or "Household Data" sheets.

Attachments: ☐ Photos ☐ Videotape ☐ Copies of major receipts ☐ Other proof of ownership ☐ Copies of warranties ☐ Appraisals ☐ "Find Me" card / page

(You'll find an Excel spreadsheet version of this page on our CD. Look for the "DP1Budget.xls" file.)

AUTO MAINTENANCE RECORD

For full details see the "Basic Vehicle Prep" section. Fill out a separate page for each vehicle owned.

Vehicle: Make: _____ Model: _____ Year: _____

Tag #: _____ State: _____ County: _____ VIN #: _____

Maintenance: Check the current condition of each of the following:

❏ Battery and Alternator	❏ Electrical System & Starter	❏ Bulbs and Fuses
❏ Engine Tune-up & Timing Belt	❏ Transmission	❏ Radiator and Cooling System
❏ Brake System	❏ Differential	❏ Joints and Bearings
❏ Steering System	❏ Belts and Pumps	❏ AC and Heating
❏ Oil & Fluid Change	❏ Tires	❏ Jack and Spare Tire

General Log of Services / Repairs Performed or Items and Equipment Purchased

Date	Mileage	"Dealer"	Service or Item	Cost	R
___/___/___					❏
___/___/___					❏
___/___/___					❏
___/___/___					❏
___/___/___					❏
___/___/___					❏
___/___/___					❏
___/___/___					❏
___/___/___					❏
___/___/___					❏
___/___/___					❏
___/___/___					❏
___/___/___					❏
___/___/___					❏
___/___/___					❏
___/___/___					❏
___/___/___					❏
___/___/___					❏
___/___/___					❏
___/___/___					❏
___/___/___					❏
___/___/___					❏
___/___/___					❏
___/___/___					❏
___/___/___					❏
___/___/___					❏
___/___/___					❏
___/___/___					❏
___/___/___					❏
___/___/___					❏
			Total expenditures this sheet:		

Keep a log with each vehicle you own.   **The "R" column shows that you saved the Receipt.**

AUTO MAINTENANCE RECORD

FAMILY BUDGET

Total monthly income related to this budget, including all income sources: $_____ per month.

Expense Categories			Income Sources and General Budgetary Considerations			
Living Expenses	**R**	**Monthly Payment**	**Income Sources**	**$$ / Mo.**	**I**	**General Considerations:**
Rent or Mortgage						❏ Use this form while reading the "**Frugality**" section to cut expenses & increase income.
Groceries						
Water & Sewer						❏ Use this form along with your "**Disaster Prep Expense Journal / Planner**."
Power						
Nat. Gas or Fuel Oil						❏ You'll find a **spreadsheet** version of this budget form on the **CD**. It's written in Microsoft Excel.
Phone / Cell Phone						
Trash						
Gasoline						❏ For extra income sources, see our coupon for a copy of "**The Complete Guide to Homemade Income**."

Insurance Costs						
Life						❏ **Fixed Payment Accts.** are loans or other long term consistent payments.
Home Owners or Renters						
Auto						❏ **Revolving Charge Accts.** are regular credit cards and gas cards.
Medical / Dental						
Roadside Assistance club						
Other						❏ The **remaining monthly cash** figured at the bottom of the page (the = $$_____) is the amount you can use to reduce debt, use for your preparedness needs, or increase savings.
Lifestyle Expenses						
Cable TV						
Internet Access						
Entertainment						
Hair care						
Gifts				**TOTAL INCOME:**		

		Total Fixed and Revolving account balances over 6 month period					
Fixed Payment Accts.		**Month 1**	**Month 2**	**Month 3**	**Month 4**	**Month 5**	**Month 6**
Revolving Charge Accts.							
TOTAL EACH COLUMN:							

Total monthly income of $_____ minus total monthly expenses of $ _____ = $$_____

"**R**" column = **Reducible** expense or next payoff target. The "**I**" column is to mark income sources to be **increased**.

FAMILY BUDGET

(You'll find an Excel spreadsheet version of this page on the CD. Look for the "DP1Budget.xls" file.)

Battery Shopping Worksheet

Batteries Needed: Use this worksheet to make your shopping easier.				
Battery Type	**Device**	**# of Devices**	**# Batteries Each**	**Total # Batteries**
(example) AAA	2-battery flashlight	3	2	6
AAA				
	Total # of AAA batteries needed:			
AA				
	Total # of AA batteries needed:			
C				
	Total # of C batteries needed:			
D				
	Total # of D batteries needed:			
9-Volt				
	Total # of 9-Volt batteries needed:			
Lantern				
	Total # of Lantern batteries needed:			
Other				
	Total # of "Other" batteries needed:			
See the spreadsheet version of this form on our CD in the Excel file "DP1Budget.xls."				

How to Use this Worksheet: **1.** Under "**Device**," list each piece of equipment that uses the battery listed. **2.** In the column, "**# of Devices**," write down how many of that device you have. For example, if you listed "radio," you'd then list how many of that identical radio you had. **3.** In the "**# Batteries Each**" column, show how many batteries that device uses. **4.** In, "**Total # Batteries**," you'd total how many batteries you'd need for all those identical items. In the **example**, we show that we have 3 flashlights that each use two "AAA" size batteries, and that we'd therefore need a total of six AAA batteries.

Evacuation Supply Distribution

Items	Home	Car 1	Car 2	Work 1	Work 2	Dest. 1	Dest. 2	Dest. 3	Deposit Box	Cache 1	Cache 2	Other
Example:												
Spare Clothing	M	2	X	✔	X	2	✔	X	X	✔	X	X

"M" = Main Supply, "2" = Secondary Supply, "✔" = Minor Amount Only, and "X" = None.

Dest. = Any preplanned **destination** where you might have already stored some family supplies.

Deposit Box = Your **Safe Deposit Box at the bank**.

Cache = Any **secure storage location** where you o store extra supplies.

This worksheet will help you **evenly distribute** your evacuation supplies and help you **remember <u>where</u> they were distributed**.

Where this worksheet distribution of <u>where</u> things are stored, the **Evacuation Gear Inventory** on the next page focuses on <u>what</u> is stored. You can use both of these, or whichever one suits you better.

Evacuation Gear Inventory

Record the Specific Evacuation Gear Stored in Each Location	
Home	
Vehicle 1 _____	
Vehicle 2 _____	
Work 1 _____	
Work 2 _____	
Destination 1 _____	
Destination 2 _____	
Destination 3 _____	
Safe Deposit Box #_____	
Cache 1 _____	
Cache 2 _____	
Other _____	

"Safety Blueprint" Page _____ of _____

Draw a floorplan of each floor, garage, or outbuilding of your house showing the location of each of the following items.

☐ Safe Room /Area	☐ Fire Extinguisher	☐ Smoke Detector	☐ First Aid Kit	☐ Bugout Kit	☐ Fire Exits	☐ Fire Ladder
☐ Fire Trunk	☐ Garden Hose	☐ Garden Sprayers	☐ Alarm Panel	☐ Isolation Goods	☐ Corded Phones	☐ Phone Jacks
☐ Fireproof Safe	☐ Defensive Items	☐ Flashlights	☐ Extra Cell Phone	☐ Hide-A-Key	☐ Fire "Valve Box"	☐ Alarm Devices
☐ CO Detectors	☐ "Safety Boxes"	☐ Elect. Generator	☐ Electrical Panel	☐ Water Shutoff	☐ Gas Shutoff	☐ Safety necklaces

Also show Isolation supply locations for: ☐ Water Sources ☐ Stored Food ☐ Dry Goods ☐ Medical Supplies ☐ Imp. Documents

This diagram is of: ☐ Main floor of house ☐ Second floor ☐ Basement ☐ Attic ☐ Other floor: _____ ☐ Outbuilding ☐ Yard ☐ Other _____

The "Last Minute List"

Keep this list on top of your Bugout Kit as a final checklist for items that could not be pre-packed.

Last Minute Items for a Level 1 or 2 Evacuation:

Remember the personal incidentals first: (Best to keep these together as a matter of habit.)

☐ Wallet	☐ Purse	☐ Car and house keys
☐ Cell phone & extra battery	☐ "Dayplanner"	☐ Secondary ID (i.e. Passport)

Next, collect the items you absolutely could not pack, but will need:

☐ Perishable medicines	☐ Personal medical kits	☐ Special-needs people gear.
☐ Misc. medical equipment	☐ Perishable special-diet food	☐ Pets' medications (all)
☐ **Support** Bugout Kit	☐ "Bolt" clothing	☐ _____
☐ _____	☐ _____	☐ _____

List other Level 1 Last Minute items (<u>and their current location</u>):

Item	Location	Item	Location
		(example) **Insulin**	**Refrigerator**

Last Minute Items for a Level 3 Evac or Higher:

☐ **Secondary** Bugout Kit	☐ Extra Food & Water	☐ Prepacked cargo carrier
☐ Bikes or boats	☐ Laptop computer	☐ "Rake and Run" list
☐ Heirlooms	☐ _____	☐ _____

List your Level 3, 4, or 5 items (<u>and their current location</u>):

Items	Location	Items	Location

Perform the following <u>ONLY</u> if time allows. We've listed these activities in relative order of importance, but left blanks for you to change the priority. This form is also on the CD so you can not only change it and reprint, but you can change the form entirely to fit your needs.

P	Activity
	Call family or group members for rendezvous.**
	Turn off all dangerous appliances such as stove, oven, dryer, etc.
	Turn off applicable utilities or HVAC based on the checklist on the next page.
	Close, lock, secure: ☐ Windows ☐ Interior doors ☐ Exterior doors ☐ Garage doors ☐ Vehicles
	Top off your car's tank with gasoline stored for your lawnmower.
	Leave a note on your home message area (i.e. the fridge) of where you're heading (rendezvous?).
	Forward calls from your home phone to your cell phone.
	Call your primary Contact Person:**_____ at:_____

"P" = Priority (assign it now).   ** These items can be performed while on the road if necessary.
See the back of this page for more last minute considerations.

List remaining items to gather, or activities to complete before heading out the door:

_____

_____

_____

_____

_____

_____

_____

_____

_____

_____

_____

_____

_____

_____

_____

_____

_____

_____

_____

Hint: Used wheeled trash cans or other easy-to-carry containers to help with grabbing last minute stuff.

Utility Shutoffs for Various Evacuations:

	Gas	Water	Power	HVAC	
Nuclear Strike	On	On	On	**Off**	**Level 1 Evac, GET OUT! (If not sheltering**)
"Dirty Bomb"	On	**Off**	On	**Off**	Leave water **off** until officials can determine
Biochemical Event	On	**Off**	On	**Off**	whether or not supplies were **contaminated**.
Wildfire	**Off**	On	On	**Off**	You'll need water and power to fight the fire.
Hurricane	**Off**	**Off**	On	On	Water and gas can cause secondary damage.
Flood	**Off**	**Off**	**Off**	**Off**	Power and gas can cause secondary damage.
Dam Burst	On	On	On	On	**Level 1 Evac, GET OUT NOW!**
Earthquake	**Off**	**Off**	**Off**	**Off**	An after-the-fact evac, shutdown anyway.
Tsunami	On	On	On	On	**Level 1 Evac, GET OUT NOW!**

You should regularly refresh everything in your kit that can expire. Use this below as a reminder.

☐ Mark due dates of intended refresh on your regular annual calendar.

Refresh Dates: Show the dates that you inspected and/or changed your kit's supply of:

Medications	__/__/__	__/__/__	__/__/__	__/__/__	__/__/__
Food	__/__/__	__/__/__	__/__/__	__/__/__	__/__/__
Water	__/__/__	__/__/__	__/__/__	__/__/__	__/__/__
Batteries	__/__/__	__/__/__	__/__/__	__/__/__	__/__/__
Child ID Photos	__/__/__	__/__/__	__/__/__	__/__/__	__/__/__
Computer disks	__/__/__	__/__/__	__/__/__	__/__/__	__/__/__
Stored documents	__/__/__	__/__/__	__/__/__	__/__/__	__/__/__
Other	__/__/__	__/__/__	__/__/__	__/__/__	__/__/__
	__/__/__	__/__/__	__/__/__	__/__/__	__/__/__

You should Refresh about every six months. Doing it during daylight savings changes is a good way to remember. So, every time the time changes, so should your stocks of perishables.

EVAUCATION DESTINATIONS

These are your local, regional, and long-distance destinations, listed in order of preference.

LOCAL

Local – 1:(location name)	Contact Person:		
Address:			
Phone 1:	Phone 2:		
Email 1:	Email 2 (or website):		
Local – 2:(location name)	Contact Person:		
Address:			
Phone 1:	Phone 2:		
Email 1:	Email 2 (or website):		
Local – 3:(location name)	Contact Person:		
Address:			
Phone 1:	Phone 2:		
Email 1:	Email 2 (or website):		

REGIONAL

Regional – 1:(location name)	Contact Person:		
Address:			
City:	County:	State:	Zip:
Phone 1:	Phone 2:		
Email 1:	Email 2 (or website):		
Regional – 2:(location name)	Contact Person:		
Address:			
City:	County:	State:	Zip:
Phone 1:	Phone 2:		
Email 1:	Email 2 (or website):		
Regional – 3:(location name)	Contact Person:		
Address:			
City:	County:	State:	Zip:
Phone 1:	Phone 2:		
Email 1:	Email 2 (or website):		

LONG DISATANCE

Distant – 1:(location name)	Contact Person:		
Address:			
City:	County:	State:	Zip:
Phone 1:	Phone 2:		
Email 1:	Email 2 (or website):		
Distant – 2:(location name)	Contact Person:		
Address:			
City:	County:	State:	Zip:
Phone 1:	Phone 2:		
Email 1:	Email 2 (or website):		
Distant – 3:(location name)	Contact Person:		
Address:			
City:	County:	State:	Zip:
Phone 1:	Phone 2:		
Email 1:	Email 2 (or website):		

❑ **Highlight locations where you may have stored supplies** ❑ **Contacts listed on "Important Contacts"**
❑ **All locations marked on maps** ❑ **Routes chosen and Evac Atlases created for each destination.**

Refer to the "Evac Atlas" portion of the "Evacuation" section for instructions.

Note: Use a separate cover sheet for other routes, even to the same location. Different routes have different towns along the way and will need their own Atlas for the info gathered.

This is route # ____ of ____ to destination: _____

This Atlas created on: ___/___/___ Last updated on: ___/___/___ (← use pencil)

Make a hand-drawn map of the first 300 miles of this one evacuation route in the space below.

Compass

(If you work better from verbal directions, write down your directions along with drawing a map.)

Use the letter-codes below to show which towns along the way have the following:

☐ **M** – Medical Services ☐ **L** – Law Enforcement Offices ☐ **F** – Food and Water
☐ **T** – Transportation & Services ☐ **C** – Communication ☐ **S** – Shelter and Lodging
☐ **G** – General resupply centers ☐ **A** – Relief Agency sites ☐ **R** – Rest Areas
☐ **X** – "X marks the spot" for all off-property cache locations, and places you have goods or gear stored.

Also show: ☐ Mile distances between stops ☐ County and State lines ☐ Potential alternate routes

Attached are: ☐ State maps ☐ County maps ☐ City maps for each town shown ☐ Other map pages
☐ White pages ☐ Yellow pages ☐ Downloaded information ☐ Other Evac Atlas pages

EVAC ATLAS – ONE – GENERAL TOWN INFO

Use a different page for each town along the route shown in the cover sheet / map for this atlas set.

Name of town:	County:	State:	Zip:

Map grid on state map: on county map: Other:

Main roads accessible from this town:

This town accessible by / at: Exit #s _____ to _____   Mile markers: _____ to _____

Other:

Emergency News Radio station channels:

_____AM / FM, Phone 1:   Phone 2:

_____AM / FM, Phone 1:   Phone 2:

_____AM / FM, Phone 1:   Phone 2:

Emergency News TV station channels:

Channel: _____ ,   Phone 1:   Phone 2:

Channel: _____ ,   Phone 1:   Phone 2:

Channel: _____ ,   Phone 1:   Phone 2:

City or County info websites:

Local phone numbers for internet dial-up access:

❏ Numbers programmed into modem's dialer

Fire Department - Address:

Phone 1:	Phone 2:
Email 1:	Email 2 (or website):
Fax:	Scanner Frequencies:
Other contact # :	Hours of Operation:
Notes:	❏ Location marked on city map

Police Department - Address:

Phone 1:	Phone 2:
Email 1:	Email 2 (or website):
Fax:	Scanner Frequencies:
Other contact # :	Hours of Operation:
Notes:	❏ Location marked on city map

Sheriff's Department - Address:

Phone 1:	Phone 2:
Email 1:	Email 2 (or website):
Fax:	Scanner Frequencies:
Other contact # :	Hours of Operation:
Notes:	❏ Location marked on city map

City Hall - Address:

Phone 1:	Phone 2:
Email 1:	Email 2 (or website):
Fax:	Scanner Frequencies:
Other contact # :	Hours of Operation:
Notes:	❏ Location marked on city map

❏ If research tells you this location has a **FEMA or State EMA** office in this town, list details on "**Evac Two.**"

Hospitals and emergency services listed on Evac Atlas page Two. Non-emergency services on Three.

EVAC ATLAS ONE – TOWN INFO

Use this page to list Hospitals, Emergency Care Centers, Emergency Mgt. Offices, etc.

Name of town:	County:	State:	Zip:

Other location identifier:

Type of service:	Company:	
Contact Person :	Hours of Operation:	
Address:	City:	State: Zip:
Map Directions:		
Phone 1:	Phone 2:	
Email 1:	Email 2 (or website):	
Fax:	Scanner Frequencies:	
Other contact # :	Hours of Operation:	
Notes:		❑ Location marked on city map

Type of service:	Company:	
Contact Person :	Hours of Operation:	
Address:	City:	State: Zip:
Map Directions:		
Phone 1:	Phone 2:	
Email 1:	Email 2 (or website):	
Fax:	Scanner Frequencies:	
Other contact # :	Hours of Operation:	
Notes:		❑ Location marked on city map

Type of service:	Company:	
Contact Person :	Hours of Operation:	
Address:	City:	State: Zip:
Map Directions:		
Phone 1:	Phone 2:	
Email 1:	Email 2 (or website):	
Fax:	Scanner Frequencies:	
Other contact # :	Hours of Operation:	
Notes:		❑ Location marked on city map

Type of service:	Company:	
Contact Person :	Hours of Operation:	
Address:	City:	State: Zip:
Map Directions:		
Phone 1:	Phone 2:	
Email 1:	Email 2 (or website):	
Fax:	Scanner Frequencies:	
Other contact # :	Hours of Operation:	
Notes:		❑ Location marked on city map

Include: ❑ Hospitals with ERs ❑ Instant Care offices ❑ Medical Providers listed by Insurance ❑ Dentists ❑ Veterinarians ❑ Emergency Mgt. Offices ❑ Planned emergency evacuation shelter locations

This is page # _____ of _____ for this route's Atlas, and page # _____ of _____ for Emergency Services

EVAC ATLAS – THREE – NON-EMERGENCY SERVICES			
Use this page for urgent but non-emergency services such as towing companies, auto repair, etc.			

Name of town:	County:		State: Zip:
Other location identifier:			

Type of service:	Company:		
Contact Person :	Hours of Operation:		
Address:	City:	State:	Zip:
Map Directions:			
Phone 1:	Phone 2:		
Email 1:	Email 2 (or website):		
Fax:	Scanner Frequencies:		
Other contact # :	Hours of Operation:		
Notes:	❏ Location marked on city map		

Type of service:	Company:		
Contact Person :	Hours of Operation:		
Address:	City:	State:	Zip:
Map Directions:			
Phone 1:	Phone 2:		
Email 1:	Email 2 (or website):		
Fax:	Scanner Frequencies:		
Other contact # :	Hours of Operation:		
Notes:	❏ Location marked on city map		

Type of service:	Company:		
Contact Person :	Hours of Operation:		
Address:	City:	State:	Zip:
Map Directions:			
Phone 1:	Phone 2:		
Email 1:	Email 2 (or website):		
Fax:	Scanner Frequencies:		
Other contact # :	Hours of Operation:		
Notes:	❏ Location marked on city map		

Type of service:	Company:		
Contact Person :	Hours of Operation:		
Address:	City:	State:	Zip:
Map Directions:			
Phone 1:	Phone 2:		
Email 1:	Email 2 (or website):		
Fax:	Scanner Frequencies:		
Other contact # :	Hours of Operation:		
Notes:	❏ Location marked on city map		

Include: ❏ Grocery stores ❏ Misc. Resupply stores ❏ Banks & ATMs ❏ Automotive ❏ Misc. Services, etc.

This is page # _____ of _____ for this route's Atlas, and page # _____ of _____ for Non-Emergency Services

EVAC ATLAS THREE – NON-EMERGENCY SERVICES

Use this page for general info about a particular stop or exit along the way. Include a short list of gas stations, chain restaurants, hotels, motels, and so on, that can be found at this one particular location.

Organize this list by exit numbers and list facilities grouped at that exit, intersection, etc.

Name of town: _____ County: _____ State: _____ Zip: _____

Map grid on state map: _____ on county map: _____ Other: _____

This town accessible by / at: Exit #s _____ to _____ Mile markers: _____ to _____

Other: _____

_____

_____

_____

_____

_____

_____

This is page # _____ of _____ for this route's Atlas, and page # _____ of _____ for Miscellaneous Services

You may wish to attach a small, hand-drawn map of this particular exit showing service locations.

EVAC ATLAS FOUR – MISCELLANEOUS

EVAC ATLAS – FIVE – DESTINATION

Make sure you fill one of these out for each planned final destination and include in each Evac Atlas.

Your intended final destination:	
Name of location, or name of residents:	
Address:	

City:	County:	State:	Zip:

Phone 1:	Phone 2:
Email 1:	Email 2 (or website):

Emergency News Radio station channels:

_____ AM / FM, Phone 1:	Phone 2:
_____ AM / FM, Phone 1:	Phone 2:

Emergency News TV station channels:

Channel: _____ , Phone 1:	Phone 2:
Channel: _____ , Phone 1:	Phone 2:

City or County info websites:

Local phone numbers for internet dial-up access:

❑ Numbers programmed into modem's dialer

Fire Department - Address:

Phone 1:	Phone 2:
Email 1:	Email 2 (or website):
Fax:	Scanner Frequencies:
Other contact # :	Hours of Operation:
Notes:	❑ Location marked on city map

Police Department - Address:

Phone 1:	Phone 2:
Email 1:	Email 2 (or website):
Fax:	Scanner Frequencies:
Other contact # :	Hours of Operation:
Notes:	❑ Location marked on city map

Sheriff's Department - Address:

Phone 1:	Phone 2:
Email 1:	Email 2 (or website):
Fax:	Scanner Frequencies:
Other contact # :	Hours of Operation:
Notes:	❑ Location marked on city map

City Hall - Address:

Phone 1:	Phone 2:
Email 1:	Email 2 (or website):
Fax:	Scanner Frequencies:
Other contact # :	Hours of Operation:
Notes:	❑ Location marked on city map

Attached are:	❑ City maps ❑ Phone book or pages ❑ Additional contact info for final destination. ❑ Downloaded info. ❑ Documentation to show connection with destination.

EVAC ATLAS – FIVE – DESTINATION

List the relief agencies responsible for your evacuation destination or major stops along the way.

Association name: _____

Contact person 1:	Contact person 2:

Address:	City:	State:	Zip:

Phone 1:	Phone 2:

Email 1:	Email 2 (or website):

Pager:	Fax:

County: _____ Map Atlas page #: ___ , Grid #: ____ Location directions:

Note:

Association name: _____

Contact person 1:	Contact person 2:

Address:	City:	State:	Zip:

Phone 1:	Phone 2:

Email 1:	Email 2 (or website):

Pager:	Fax:

County: _____ Map Atlas page #: ___ , Grid #: ____ Location directions:

Note:

Association name: _____

Contact person 1:	Contact person 2:

Address:	City:	State:	Zip:

Phone 1:	Phone 2:

Email 1:	Email 2 (or website):

Pager:	Fax:

County: _____ Map Atlas page #: ___ , Grid #: ____ Location directions:

Note:

Association name: _____

Contact person 1:	Contact person 2:

Address:	City:	State:	Zip:

Phone 1:	Phone 2:

Email 1:	Email 2 (or website):

Pager:	Fax:

County: _____ Map Atlas page #: ___ , Grid #: ____ Location directions:

Note:

This is page # ____ of ____ for this route's Atlas, and page # ____ of ____ for Relief Services

Include all: ❑ Government relief services ❑ Charitable organizations ❑ Volunteer groups ❑ Local shelters

❑ **"Planning Ahead for Phase II"** checked for services to list.   ❑ **Numbers programmed into cell phones.**

EVAC ATLAS – SIX – RELIEF SERVICES

"Update Postcards".

Print or photocopy these on card stock to use as postcards for either mailed communication, or use as notes to have passed along by hand.

Day: _____ Date: __/__/__ Time: _____ AM / PM
From: _____ To: _____ (names)
☐ All safe and sound. ☐
☐ _____ (name) is unaccounted for.
☐ _____ is in hospital, condition is
☐ Experienced property damage including:
☐ Relocating to: _____ Expected to Arrive: __/__/__
☐ Contact us: ☐ ASAP ☐ As you're able. ☐ No need to contact us.
We are: ☐ at home ☐ relocated __ in __ out of town (address on other side)
☐ Staying here for time period: _____ (# of days, weeks, etc.)
☐ Contact us by way of:
☐ New phone number is:
☐ We have info about:
☐ We need to know about:
☐ Note:
Signed:

Day: _____ Date: __/__/__ Time: _____ AM / PM
From: _____ To: _____ (names)
☐ All safe and sound. ☐
☐ _____ (name) is unaccounted for.
☐ _____ is in hospital, condition is
☐ Experienced property damage including:
☐ Relocating to: _____ Expected to Arrive: __/__/__
☐ Contact us: ☐ ASAP ☐ As you're able. ☐ No need to contact us.
We are: ☐ at home ☐ relocated __ in __ out of town (address on other side)
☐ Staying here for time period: _____ (# of days, weeks, etc.)
☐ Contact us by way of:
☐ New phone number is:
☐ We have info about:
☐ We need to know about:
☐ Note:
Signed:

Day: _____ Date: __/__/__ Time: _____ AM / PM
From: _____ To: _____ (names)
☐ All safe and sound. ☐
☐ _____ (name) is unaccounted for.
☐ _____ is in hospital, condition is
☐ Experienced property damage including:
☐ Relocating to: _____ Expected to Arrive: __/__/__
☐ Contact us: ☐ ASAP ☐ As you're able. ☐ No need to contact us.
We are: ☐ at home ☐ relocated __ in __ out of town (address on other side)
☐ Staying here for time period: _____ (# of days, weeks, etc.)
☐ Contact us by way of:
☐ New phone number is:
☐ We have info about:
☐ We need to know about:
☐ Note:
Signed:

Day: _____ Date: __/__/__ Time: _____ AM / PM
From: _____ To: _____ (names)
☐ All safe and sound. ☐
☐ _____ (name) is unaccounted for.
☐ _____ is in hospital, condition is
☐ Experienced property damage including:
☐ Relocating to: _____ Expected to Arrive: __/__/__
☐ Contact us: ☐ ASAP ☐ As you're able. ☐ No need to contact us.
We are: ☐ at home ☐ relocated __ in __ out of town (address on other side)
☐ Staying here for time period: _____ (# of days, weeks, etc.)
☐ Contact us by way of:
☐ New phone number is:
☐ We have info about:
☐ We need to know about:
☐ Note:
Signed:

Having cards preprinted saves you a lot of writing time if having to update a lot of people, and with cards, you won't forget what to tell them. Print several sets of these and keep them in your "Office Pack." Though other forms of commo may be down for a while, the mail will eventually get through.

Change of Address Cards

Send these to friends and family, or account or service providers, as redundant update showing temporary or permanent address changes.

Card 1

Day: _____ Date: __/__/__ Time: _____ AM / PM

Name: _____

Old Address: _____

City: _____ State: _____ Zip: _____

Phone: _____ Email: _____

Account number: _____

New Address: _____

City: _____ State: _____ Zip: _____

Phone: _____ Email: _____

Phone 2: _____ Other: _____

This change is ☐ Permanent ☐ Temporary until: _____

Card 2

Day: _____ Date: __/__/__ Time: _____ AM / PM

Name: _____

Old Address: _____

City: _____ State: _____ Zip: _____

Phone: _____ Email: _____

Account number: _____

New Address: _____

City: _____ State: _____ Zip: _____

Phone: _____ Email: _____

Phone 2: _____ Other: _____

This change is ☐ Permanent ☐ Temporary until: _____

Card 3

Day: _____ Date: __/__/__ Time: _____ AM / PM

Name: _____

Old Address: _____

City: _____ State: _____ Zip: _____

Phone: _____ Email: _____

Account number: _____

New Address: _____

City: _____ State: _____ Zip: _____

Phone: _____ Email: _____

Phone 2: _____ Other: _____

This change is ☐ Permanent ☐ Temporary until: _____

Card 4

Day: _____ Date: __/__/__ Time: _____ AM / PM

Name: _____

Old Address: _____

City: _____ State: _____ Zip: _____

Phone: _____ Email: _____

Account number: _____

New Address: _____

City: _____ State: _____ Zip: _____

Phone: _____ Email: _____

Phone 2: _____ Other: _____

This change is ☐ Permanent ☐ Temporary until: _____

Note: If including sensitive info such as new unlisted phone numbers, or sensitive account info, mail these in an envelope instead of as a post card. Print several sets of these and keep them in your "Office Pack." Though other commo may be down for a while, the mail will eventually get through.

BEFORE YOU FILL OUT THE CHANGE OF ADDRESS FORM (PS FORM 3575), print the City, State and ZIP Code of your old address in the proper spaces on the other side of the form. Then complete items 1 through 10. Remember to sign the form in item 9.

1. WHO'S MOVING?
- If it's just you, check the INDIVIDUAL box.
- If it's some members of your family with the same last name and others are staying, fill out a separate form for each mover and check the INDIVIDUAL box.
- If it's some members of your family with different last names, fill out a separate form for each mover and check the INDIVIDUAL box.
- If it's everyone in your family with the same last name, just fill out one card and check the ENTIRE FAMILY box.
- If it's your business, check the BUSINESS box.

2. IS THIS A TEMPORARY MOVE?
- Check YES if you plan to return to your old address within 12 months.
- Otherwise, check NO.

3. WHEN SHOULD WE BEGIN FORWARDING MAIL?
- Fill in the date you want us to begin forwarding your mail to your new address in the START DATE field.

4. RETURN DATE
- For a temporary move, indicate the date when you want to stop forwarding mail to the TEMPORARY address. If this date should change, notify the post office that serves your OLD ADDRESS when to stop forwarding your mail.

5a. LAST NAME OF MOVER
- Fill in only one LAST NAME.
- If anyone with the same last name is moving to a different address, use a separate form for each person.

5b. FIRST NAME OF MOVER
- If you checked INDIVIDUAL, give us your FIRST NAME.
- If you checked ENTIRE FAMILY, print the first name of the head of household and any commonly used middle names or initials.

6. BUSINESS NAME
- For a BUSINESS move, print the name of the business. Each business must file a separate form.

7. OLD ADDRESS
- Print your complete OLD ADDRESS, including an APARTMENT NUMBER, SUITE or PO BOX NUMBER, if appropriate. Include City, State, ZIP.

8. NEW ADDRESS
- Print your complete NEW ADDRESS, including an APARTMENT NUMBER, SUITE or PO BOX NUMBER, if appropriate. Include City, State, ZIP.

9. SIGNATURE
- To make this change of address valid, print and sign your name.

10. DATE
- Fill in the data you signed this form. Be sure to read the "Note" and "Privacy Notice" statements on the reverse side of the Change of Address Form.

 **UNITED STATES**
POSTAL SERVICE®

NO POSTAGE
NECESSARY
IF MAILED
IN THE
UNITED STATES

BUSINESS REPLY MAIL
FIRST-CLASS MAIL PERMIT NO. 73026 WASHINGTON, DC

POSTAGE WILL BE PAID BY ADDRESSEE

TO: POSTMASTER (OF OLD ADDRESS)
UNITED STATES POSTAL SERVICE

CITY OR POST OFFICE   STATE   ZIP CODE

This is only a facsimile of US Postal Form 3575 for change of address. This is provided here in case you have absolutely nothing else to use.

It's our suggestion that you go to your local Post Office now to get an official change of address card to pack in your Bugout Kit in case you have to evacuate your home area for an extended period of time.

If you're in an evacuation situation, you won't have time to go to a new Post Office to stand in line for forms. Also, there's no guarantee they'll have any in stock due to the demand in that kind of scenario.

Minor Accident Report Form

Personal Info **(fill this section out now)**

Insurance Co:_____Policy #:_____ Phone:_____Agent:_____

Auto Club/Road Service:_____Phone:_____ ID#:_____

Preferred Auto Shop:_____ Phone:_____

Their tow service:_____Phone:_____

Witnesses **(not actually involved in accident)**

	Tag #	State	Make	Model	Driver's Name	Phone:
1.						
2.						
3.						
4.						
5.						

Non-Driving Occupants **(of cars involved in accident - including your car)**

	Name	Address	Phone
1.			
2.			
3.			
4.			
5.			

Attending Officials:

EMTs: Unit#:_____ Hospital or Station:_____

Names: 1)_____ 2)_____ 3)_____

Fire: Unit#_____ Station:_____

Names: 1)_____ 2)_____ 3)_____

Law Enforcement: ❑ Sheriff ❑ City Police ❑ County Police ❑ State Patrol ❑ _____

Name:_____ Precinct:_____

Direct Phone:_____ Unit #:_____ Badge #: _____

Tow Truck
(The one that takes <u>your</u> car.)

Company:_____ Phone:_____Driver:_____

Note: To save time & effort, collect business cards where possible and attach along side near applicable section.

❑ **Detailed location of accident and description of events described on back.** ❑ **Photos or videos taken.**

The "Coming Home" Worksheet

Use this worksheet to gather important information about your home area after an evacuation to determine if it's safe to come home, and what to expect once you arrive.

On your "Important Contacts" and "Evac Atlas" forms, you should have your home-town contact information for:

❑ Sheriff – non emergency #   ❑ Police – non emergency #   ❑ Fire dept non emergency   ❑ County Health Dept.
❑ Local FEMA office   ❑ Local State _EMA office   ❑ Red Cross rep   ❑ City Hall
❑ County Commission office   ❑ Local hospitals   ❑ Other relief group   ❑ Local volunteer group

Call a minimum of three of the above offices to get answers to the following questions if other news sources or websites are not giving you the information you need. You'll want to call different offices to get a more accurate picture of what's actually going on, but you might not want to call all of them to conserve phone usage. Skip the questions you know the answer to.

Ask the following questions and record your answers here:

1. We live at (give your address). Was our home area damaged? _____

2. If so, how badly? What damage was done?_____

3. Is it safe to return? _____ If not, what is the estimated date of safe return? _____

4. On our return, are there any precautions we should take when repairing our property, such as debris _____ or hazardous materials?_____

5. We are currently at (give location). Which routes back to (your home town) are open?_____
_____

6. Are any local utilities inoperable? _____

7. If inoperable, what is the estimated date of repair? _____

8. Are any local services such as trash pickup inoperable? _____

9. If inoperable, what is the estimated date of return? _____

10. What is the current situation regarding grocery stores or food in general? _____

11. Do we need to check in with any office or service on our return? _____

12. Will we be required to undergo a medical examination? _____ If so, by whom? _____

13. To whom should we provide our current location or contact info?_____

14. Do you need a list of people we know to be safe ?_____ To whom should we give this list? _____

15. Do you need a list of those we know to be injured or worse? _____ To whom should we give this list? _____

16. Do you have information on (give name of missing loved one)? If not, who can we call? _____
_____

17. Are there any other offices or agencies we should contact to provide or receive information?_____
_____

Notes: _____
_____
_____
_____

Once you know it's safe to return, or have a definite date of safe return, call your: ❑ **Insurance agent** ❑ **Cleanup service.**

SUSPICIOUS ACTIVITY – ONE -- ACTIVITY AND LOCATION

Your Neighborhood Watch Group can use this to provide Police an organized description of suspicious activity.

Day of Week: Date: ___ / ___ / ___   Time of Activity: From: ____ : ___ AM PM   To: ____ : ___ AM PM

Total number of suspicious individuals: _____ (____ Male and ____ Female)   Weapons?: ❏ Y ❏ N (describe on back)

Location of Activity: Location is: ❏ Private residence ❏ Neighborhood or residential area ❏ Rural / unpopulated area

❏ Public gathering or establishment ❏ Private Commercial Facility ❏ Public Commercial Facility ❏ Government Facility

Name of location:

Address:

City:	County:	State: Zip:
Phone 1:	Phone 2:	Email:
Fax:	Website:	Other:

Contact person at this location: Was this person a witness to activity? ❏ Y ❏ N

Activity is primarily: ❏ Crime against person   ❏ Crime against property   ❏ Possible terrorist surveillance / activity

Activity was: ❏ Attempted   ❏ Actually perpetrated.

Description of Activity:

Direction of escape:

❏ **Other pages attached with more detailed description.**

Have each witness listed below to initial or sign the above description if possible:

Witnesses	Phone 1	Phone 2
1.		
2.		
3.		
4.		
5.		
6.		
7.		

❏ **Other pages attached to list additional witnesses, or additional contact info. ("Witness Information" pages.)**

This report submitted by: Name:

Address:

City:	County:	State: Zip:
Phone 1:	Phone 2:	Email:
Fax:	Website:	Other:

Attachments: ❏ Suspicious persons & vehicles page(s) ❏ "Witness Information" pages ❏ Photos ❏ Video footage

Note: <u>Do not touch physical evidence</u>. Point it out to Police.

Do <u>not</u> place yourself in any danger to gather this info. This is only meant to organize what you're able to see.

SUSPICIOUS ACTIVITY – ONE -- ACTIVITY AND LOCATION

Witness Information

Intvwng Agent:__________________________ Witness #:______Pg #:______

Full Name: SSN: Sex: DOB:___/___/___

Address: Apt/Ste Cmp/Sb:

City: St: Zip: -- Own/Rnt/Rsd Yrs:

Phones: (H): ( ) - "Subject Data File" started?

Phone: Pager: Cell Phone:

Fax: E-Mail: Website:

Place of Employment: Supervisor:

Address: Title:

City: St: Zip: -- Website:

Phone: - x Fax: E-Mail

Staple Business Card here

Vehicle: Type: Make: Model: Year: Colors: Body: Roof:

Marks/Features:

VIN: Tag: State: County:

Closest friend or relative: (if married, use for spouse) Name:

Relationship: SSN: Sex: DOB:___/___/___

Address: Apt/Ste: Cmp/Sb:

City: St: Zip: -- Own/Rnt/Rsd Yrs.:

Phones: (H): ( ) -

Pager: ( ) - Fax: ( ) -

Cell Phone: ( ) - E-Mail: Website:

Place of Employment: Supervisor:

Address: Website:

City: St: Zip: --

Phone: ( ) - Fax: E-Mail:

Other contact: Name:

Relationship: SSN: Sex: DOB:___/___/___

Address: Apt/Ste:

City: St: Zip: -- Own/Rnt/Rsd Yrs.:

Phone: Pager:

Cell Phone: E-Mail:

Fax: Website:

Place of Employment: Supervisor:

Address: Website:

City: St: Zip: --

Phone: ( ) - Fax: E-Mail:

Short synopsis of testimony:

Interviews: 1. ___/___/___ By:______ 2. ___/___/___ By: ______ 3. ___/___/___ By:______ 4. ___/___/___ By:_____

Full deposition made on: ___/___/___ Recorded by: Transcript filed in:

(This form courtesy of "The Case File," http://www.thecasefile.com)

SUSPICIOUS ACTIVITY -- TWO – SUBJECTS AND VEHICLES
Use this page to describe people and vehicles associated with a suspicious activity report.

SUBJECTS

Subject #___: Sex:　Approx. Age:　Height:　Weight:　Shoe size:　Eye color:　Race:

Color / complexion:　Nationality:　Language spoken:　Accent:

Hair color:　Hair style:　Hair length:　Wig or toupee? ❏ Y ❏ N　Shoe size:

Facial hair? ❏ Y ❏ N　If yes, describe:　❏ Left handed ❏ Right handed

Teeth: ❏ Normal ❏ Dentures ❏ Braces ❏ Retainer ❏ Other description:

Clothing:

Glasses? Describe::

Scars or Marks:

Tattoos:

Jewelry and/or Piercings:

Cologne, odor, or aroma:

Cigarettes or tobacco use:

Food consumed:

Subject #___: Sex:　Approx. Age:　Height:　Weight:　Shoe size:　Eye color:　Race:

Color / complexion:　Nationality:　Language spoken:　Accent:

Hair color:　Hair style:　Hair length:　Wig or toupee? ❏ Y ❏ N　Shoe size:

Facial hair? ❏ Y ❏ N　If yes, describe:　❏ Left handed ❏ Right handed

Teeth: ❏ Normal ❏ Dentures ❏ Braces ❏ Retainer ❏ Other description:

Clothing:

Glasses? Describe::

Scars or Marks:

Tattoos:

Jewelry and/or Piercings:

Cologne, odor, or aroma:

Cigarettes or tobacco use:

Food consumed:

VEHICLES

Vehicle 1: Make:　Model:　Year:

Tag #:　Year:　State:　County:　VIN #:

Body Color(s):　Roof Color(s):

Tires:　Gas or Diesel?:

Bumper stickers or decals:

Distinguishing marks, equipment, or damage:

Interior Decoration:

Vehicle 2: Make:　Model:　Year:

Tag #:　Year:　State:　County:　VIN #:

Body Color(s):　Roof Color(s):

Tires:　Gas or Diesel?:

Bumper stickers or decals:

Distinguishing marks, equipment, or damage:

Interior Decoration:

❏ **Other pages attached to describe additional subjects** ❏ **Other pages attached to describe additional vehicles.**

SUSPICIOUS ACTIVITY -- TWO – SUBJECTS AND VEHICLES

VOLUNTEER GROUP – MEETING WORKSHEET

Use this worksheet to plan your larger public meetings, education sessions, or recruiting drives.

PLANNING

Meeting purpose:_____

❏ Open meeting ❏ Members only

Speaker & Topic:_____, or

❏ Agenda and time schedule set.

Date set for meeting: ___ / ___ / ___ Time: From: _____ AM / PM To: _____ AM / PM

Meeting Budget: $_____

❏ Formal ❏ Informal

Presentation subject:_____

Agenda to be published in PSA announcement? ❏ Y ❏ N

❏ Outside vendors invited?

PREMISES

❏ Building / meeting room acquired or rented.

❏ Adequate, secure parking for number expected

❏ Adequate restroom facilities for number expected

❏ Adequate, functional HVAC system

❏ Kitchen or prep facilities for refreshments

❏ Security personnel scheduled

❏ Location accessible by public transportation

❏ Facility and grounds meet ADA compliance

❏ Adequate, accessible first aid and fire equipment

❏ Trash cans and sanitation

PROMOTION

Public Service Announcements placed:

❏ Radio Stations: _____, _____, _____

❏ Television Stations: _____, _____, _____

❏ Newspapers: _____, _____, _____

❏ Civic Organization Websites: _____

❏ Trade or professional publications and journals if associated.

❏ Invitations to local government officials and Law Enforcement representatives.

Announcements to group via:

❏ Newsletter on ___ / ___ / ___

❏ Website on ___ / ___ / ___

❏ Email groups on ___ / ___ / ___

❏ Phone tree on ___ / ___ / ___

❏ _____

❏ _____

PROCESSING

Expected number of attendees: _____ Current members _____ New visitors. Total expected: _____

❏ Registration table acquired ❏ _____ # of Registration packets printed ❏ _____ # of Information brochures ready

❏ Registration table signs or banners ❏ Fishbowl or basket if drawing for door prizes. ❏ Prize entry blanks or tickets.

❏ Nametags ❏ Sign-in sheet ❏ Pens and Pencils ❏ Display racks / holders for literature ❏ Notepads for attendees

PROVISIONS

❏ Restroom provisions if not supplied by facility.

❏ # of ____ Chairs and ____ tables for attendees.

❏ Door prize(s) if any.

❏ Refreshments and snacks.

❏ # of ____ Tables and ____ chairs for group officers.

❏ Tables for outside vendors if any.

PRESENTATION

❏ Lectern or podium

❏ Television

❏ Projector: Type: ___ Computer ___ Film ___ Overhead

❏ Computer with Power Point or similar.

❏ Audio cassette player

❏ Supplies and parts for wall displays

❏ Photocopy service on-site or nearby

❏ Public Address system

❏ VCR

❏ Projector screen and ❏ Laser pointer

❏ Audiovideo CD / DVD player

❏ Dry-Erase board ❏ Dry-Erase markers

❏ Easel ❏ Poster-sized paper pad ❏ Magic Markers

❏ Video and audio equipment to record meeting

PURCHASING

❏ Non-PSA ad costs: $_____

❏ Speaker / Presentation fees: $_____

❏ Facility supplies not provided: $_____

❏ Table and chair rental: $_____

❏ Printing / photocopy costs: $_____

❏ Newsletter and mailing costs: $_____

❏ Rental fee for facility: $_____

❏ Refreshments $_____

❏ Presentation equipment rental: $_____

❏ Facility cleanup fees: $_____

Attachments: ❏ Additional notes ❏ Receipts ❏ Copy of Agenda ❏ Copy of Promotional Literature ❏ Setup Diagrams

VOLUNTEER GROUP – MEETING WORKSHEET

Volunteer Application

The information you provide on this application will be kept in strictest confidence. It is for internal use only.		
Name: First:	Middle:	Last:
Address:	Apt/Ste	
City:	County:	St: Zip:
Phone 1:	Phone 2:	Email:
Cell phone 1:	Cell phone 2:	Email 2:
Pager:	Fax:	Website:

Staple Business Card here

Place of Employment:	Supervisor:
Nature of work:	Yrs there:
Address:	Title:
City: St: Zip: --	Website:
Phone 1: Phone 2:	Email:
Cell phone 1: Cell phone 2:	Email 2:
Pager: Fax:	Website:

Please list work skills you feel would be applicable. Preferably, please provide us with a full resume.

Fill out this information ONLY if your volunteer function involves driving others or involves using your vehicle.

Vehicle: Type: Make:	Model:	Year:	Colors: Body:	Roof:
VIN:	Tag:	State:	County:	

Drivers Lic. #:	Drivers Lic Type:	Any restrictions? (If yes, explain on back):

Are you a member of any other group that may be called upon in the wake of a disaster or attack? _____ If yes, explain.

Alternate Contact: Please give us the name of a friend or relative that can help find you if we need you:

Name: First:	Middle:	Last:
Address:	Apt/Ste	
City:	St: Zip: --	Own/Rnt/Rsd Yrs:
Phones: (1): ( ) -	Phone (2): ( ) -	
Phone 1:	Phone 2:	Email:
Cell phone 1:	Cell phone 2:	Email 2:
Pager:	Fax:	Website:

This person's relationship to you:

OFFICIAL USE ONLY: ❑ **"Skills & Assets"** form completed and attached ❑ A **"Find Me"** form completed and attached
❑ Check here if **full resume** provided by applicant. **This person's Volunteer #:**

Services, Skills, and Assets

All information will be kept in strictest confidence.

Name: First:	Middle:	Last:
Address:	Apt/Ste	
City:	County:	St: Zip:
Phone 1:	Phone 2:	Email 1:
Cell phone 1:	Cell phone 2:	Email 2:
Pager:	Fax:	Website:

Main type of service or goods you would like to volunteer:

Below, please check other types of assistance or assets you could provide as a volunteer.

Activities
- ❏ Search & Rescue (ie missing child) ❏ Neighborhood Watch ❏ Citizen Advisory group / Think Tank
- ❏ Organizer / Assistant at local shelter ❏ Food service at disaster scene or local shelter
- ❏ General labor (ie stacking sandbags, digging through rubble, clearing streets of debris)
- ❏ Secondary perimeter / Volunteer Command Post security ❏ Public information liaison
- ❏ Communications runner (ie phone systems are down)
- ❏ Future organization and operation of this volunteer group. ❏ Clerical operations of this volunteer group.

Skills
- ❏ Medical. Please explain:
- ❏ Former Law Enforcement. Please explain:
- ❏ Former Fire Department. Please explain:
- ❏ CDL ❏ CDL Passenger rating ❏ Heavy equipment. Types:
- ❏ Carpenter ❏ Mason ❏ Electrician ❏ Plumber ❏ Computer tech. ❏ _____
- ❏ Cook, food prep, mass menu organization.
- ❏ Pilot. Aircraft type: License type: License #:
- ❏ Interpreter. Language:_____

Services
- ❏ Transportation. I ❏ Own or ❏ Can operate a ❏ Passenger bus ❏ Recreation Vehicle ❏ _____
- ❏ Child care ❏ Shelter. I can house ____ people for up to ____ days. (Can guests bring pets? ❏ Yes ❏ No)
- ❏ Animals. I can care for and feed _____ (#) of _____ (type of animal) for up to _____ days.
- ❏ I am certified to teach: ❏ CPR ❏ General first aid ❏ Limited Firefighting ❏ _____
- ❏ Attorney (ie helping disaster victims deal with insurance). Explain:

Assets
- ❏ Electric generator ❏ Chainsaw ❏ Extendable ladder ❏ Welding equipment ❏ Blow torch equip.
- ❏ General power tools (saws, drills, etc.) ❏ General "yard" tools (chainsaws, shovels, picks, axes, etc.)
- ❏ Water well and pump ❏ Portable cooking equipment. Describe:
- ❏ Ham Radio operator. License: _____ Call sign:_____
- ❏ Heavy construction equipment: _____ ❏ Pickup truck ❏ Flatbed truck ❏ Tow truck ❏ Bus ❏ RV
- ❏ Aircraft. Type: ❏ Boat. Type:
- ❏ Trained dogs. Trained for:

Pledges
These are goods or services you pledge to provide in the event of an emergency or disaster.
- ❏ Food. Explain:
- ❏ Durable goods or equipment. Explain:
- ❏ Miscellaneous services or supplies. Explain:
- ❏ Monetary pledge. I pledge $_____ in event of disaster or attack. Initials: _____
- ❏ Meeting facilities for volunteer group meetings and/or training sessions.

OFFICIAL USE: ❏ Application attached ❏ "Find Me" info attached **Volunteer #:**_____

Neighborhood Watch – Needs and Intel			
This is all sensitive information and should only be given to group officers or first responders.			

The map below covers the following area(s):

Subdivision or Complex:

City Map Atlas page #: _____ , Grid #: _____ Location directions:

Contact Info:	Law Enforcement	Fire	EMS
Emergency Number			
Non-Emergency Number			
Contact Person			
Station / Precinct/Zone			

Attach a contact list for each utility company: ❑ Electric ❑ Gas ❑ Water/Sewage ❑ Telephone

Make a hand-drawn map of the residential area that makes up your Watch Group area.

Compass

Locate and show each of the following on your neighborhood map.

❑ Show houses and house #s	❑ List family name and phone #s	❑ Rendezvous area(s)
❑ Fire hydrants	❑ Community stockpile of gear	❑ Neighborhood escape routes
❑ Utility junctions / lines / pipes	❑ Special-needs residents	❑ Homes with pets
❑ HazMat potential homes	❑ Special-skills residents	❑ Special-assets residents
❑ Traffic obstacles (creeks, etc.)	❑ Neighborhood entrances / exits	❑ Group member houses

❑ A phone list of all members and copies of this map have been copied and distributed to all members.

(Also use this map to show location of non-ambulatory victims, utility damage, and structural damage after a disaster.)

Neighborhood Watch – Resident Information

Dear Resident, You are being asked to provide information that will help your Neighborhood Watch Group and First Responders serve you better in an emergency. Filling out this form is <u>completely</u> <u>voluntary</u>. You do not have to fill this out, and you are <u>not</u> required to fill in <u>all</u> blanks. All information provided will be kept in <u>strictest confidence</u>. In addition to this page, you will also be provided a "Volunteer Application" should you wish to join the Watch Group, a "Services, Skills, and Assets" form if there is aid you feel you can offer in an emergency, and "Find Me" sheets from Watch Group officers.

Owner / Resident

Name: First:	Last:	
Address:		Apt/Ste
City:	County:	St: Zip:
Phone 1:	Phone 2:	Email:
Cell phone 1:	Cell phone 2:	Email 2:
Pager:	Fax:	Website:

❑ **A "Find Me" sheet with more detailed contact info for our family has been filled out and attached.**

Occupants and Dependants (Important for fire and emergency reactions.)

We have _____ Adults and _____ children living at this residence. A list of names and ages ❑ is ❑ is not provided.

# of Pets: ____ Dogs ____ Cats ____ Birds, Other: _____

Special Needs People

❑ Mobility challenged. Details:

❑ Sight / hearing impaired. Details:

❑ Language barriers. Details:

❑ Medical condition: Details:

Do you wish us to help you register your special needs people with local officials? ❑ Yes ❑ No. **Initial:** _____

HazMats and Dangers

Please list anything in your home that may present a danger to Fire Fighters or other first responders.

❑ Medical oxygen in use	❑ Oxy – Acetylene torch in workshop
❑ Paints or solvents	❑ Dangerous pets or animals
❑ Propane tanks	❑ _____

Specific Assistance

Is there anyone specifically not allowed on your property? If so, describe on back:

❑ Resident has filled out and attached a "Suspicious Activity" form to describe the above.

Do you have household members who may need special assistance?

❑ Resident has been provided "First Responder Rescue Stickers" for doors and windows.

Please describe other ways we can enhance your security or assist you in an emergency:

❑ Resident has filled out select "Important Contacts" and "Family Member Data" pages to provide emergency info.

Dear Resident, Ask us about: ❑ **Children's "Safehouse"** ❑ **Rendezvous locations** ❑ **Safety training classes**

Attached Forms:	❑ **"Find Me"** ❑ **Volunteer Application** ❑ **Services, Skills, and Assets** ❑ **Medical Info**
	❑ **Suspicious Activity** ❑ **"Notify in Case of Emergency" list** ❑ _____

Thank you for participation in our mutual safety and security. Your Watch group members should provide you a "Find Me" sheet, , with their contact information so they'll always be reachable in the event of an emergency.

Neighborhood Watch – Damage Assessment

Watch Group Director: _____   **On Radio Frequency:** _____

Report prepared by: _____   **Cell Phone:** _____   **Time:** _____   AM / PM   Day: _____   Date: ___/___/___

Report checked by: _____   **Cell Phone:** _____   **Time:** _____   AM / PM   Day: _____   Date: ___/___/___

Triage area located at: _____

House # and Street	Family Name	Phone 1	Phone 2	OK	People I	People D	People M	Pets I	Pets D	Pets M	Structure S	Structure F	Structure U	Utilities E	Utilities G	Utilities W	Utilities T	C&C
1.																		
2.																		
3.																		
4.																		
5.																		
6.																		
7.																		
8.																		
9.																		
10.																		
11.																		
12.																		
13.																		
14.																		
15.																		
16.																		
17.																		
18.	**Totals This Page:**																	

Attachments:
☐ Family Member Data sheets, Identifiers and Medical, for the injured or missing ☐ Neighborhood Graphical Map
☐ Hardcopy photos of neighborhood damage ☐ Digital photos on disk ☐ Video of damage ☐ Utility company emergency contact list

Use this page to make your own assessment just after a disaster. This will be useful to hand off to officials in order to save time if they are somehow delayed.
OK = All family members accounted for, health is fine, no structural damage to home. **People & Pets: I, D, M** = Number of **I**njured, **D**ead, **M**issing.
Structure: S, F, U = Damage to **S**tructure, **F**ire Damage, or **U**tility Damage. **Utility: E, G, W, T** = Which utility is damaged? **E**lectric, **G**as, **W**ater / Sewage, **T**elephone.
For people and pets, put a number in the column. For structural and utility damage, put a ✓ to mean okay or an **X** to mean damaged or not working. You could also use
a damage score of 1 to 5 or 1 to 10 for minor to severe. Remember to maintain a listing of **emergency contact info for all utility companies.**
C&C = "Checked and Cleared." This is for officials to verify the information listed. They can use a ✓ or an **X**, or they can use their initials.
(Hint: Go ahead and fill out all street address, family name and phone info now so in an emergency filling in damage assessment can be done quickly.)

www.disasterprep101.com

Chart ideas, areas of focus, and discussion topics relating to Terrorist Operations Development steps.

Terrorist Operations Development	Involvement on a Federal, State & Local Level, Plus Input from Private Sector & Public				
	Prevention and/or Preemptive Mitigation				
1. Personnel	**Federal**	**State**	**County / Local**	**Private Sect.**	**Gen. Public**
A. Sleepers	1.	2.	3.	4.	5.
B. Illegal Entry	6.	7.	8.	9.	10.
C. Local Recruitment	11.	12.	13.	14.	15.
D. Legislative Gaps	16.	17.	18.	19.	20.
E. Domestic Terror	21.	22.	23.	24.	25.
2. Target Selection					
A. Primary Goal	26.	27.	28.	29.	30.
B. Secondary Goal	31.	32.	33.	34.	35.
C. Selection Criteria	36.	37.	38.	39.	40.
3. Tools / Weapons					
A. Hijacking	41.	42.	43.	44.	45.
B. Theft	46.	47.	48.	49.	50.
C. Smuggling	51.	52.	53.	54.	55.
D. Purchase	56.	57.	58.	59.	60.
E. Construction	61.	62.	63.	64.	65.
4. Gathering Intel					
A. Observation	66.	67.	68.	69.	70.
B. Infiltration	71.	72.	73.	74.	75.
C. Direct Instruction	76.	77.	78.	79.	80.
D. Internet	81.	82.	83.	84.	85.
E. Info Purchase	86.	87.	88.	89.	90.
F. Interrogation	91.	92.	93.	94.	95.
5. Planning					
A. Communication	96.	97.	98.	99.	100.
B. Meeting	101.	102.	103.	104.	105.
C. "Dry Run"	106.	107.	108.	109.	110.

(Execution of Attack)	Post-Execution Reaction			Post-Execution Mitigation		
6. Initial Execution	**Federal**	**State, Local**	**Pub/Private**	**Federal**	**State, Local**	**Pub/Private**
A. Conventional	111.	112.	113.	114.	115.	116.
B. Nuclear	117.	118.	119.	120.	121.	122.
C. Biological	123.	124.	125.	126.	127.	128.
D. Chemical	129.	130.	131.	132.	133.	134.
7. Secondary Effects						
A. Economic	135.	136.	137.	138.	139.	140.
B. Public Health	141.	142.	143.	144.	145.	146.
C. Panic / Public Order	147.	148.	149.	150.	151.	152.
D. Restoration	153.	154.	155.	156.	157.	158.
8. Non-Local Aftermath						
A. Refugees	159.	160.	161.	162.	163.	164.
B. Aid	165.	166.	167.	168.	169.	170.
C. Public Health	171.	172.	173.	174.	175.	176.
D. Panic / Public Order	177.	178.	179.	180.	181.	182.

www.disasterprep101.com

Our Two-CD Set Table of Contents

Welcome to our CDs, Table of Contents. We planned all along to offer a CD with our book in order to give you a softcopy of our various forms so you can more easily use them for your family's needs. It was decided to add, as a free bonus, many of the more pertinent files that are available for free download from all over the internet.

The files located in the "Appendix" folder (along with the "Studies" files and links collections) are ours and are copyrighted material. They're presented here for your personal use. All other information found on these CDs was specifically offered for free download from various unrestricted and non-subscription websites. We've included this material as a free bonus in order to save you the search and download time of the free material that you'll need to have in your library.

Also, we wish to point out that inclusion on our CD does not constitute our endorsement of any product or service. Let the reader beware. Our sole intention was to provide alternate viewpoints, additional educational material, and a free benefit. We also wish to thank our colleagues in these various fields for making this material available to the public.

This Table of Contents is arranged in the same order as the book's Table of Contents. We've done this so you could more easily locate files based on their subject or content. Each file is listed according to the section to which it best applies.

The folders on the CDs are arranged in the same fashion as the main book sections.

You'll find a table under each section header in this Table of Contents. The table will tell you the name of the file and a short description of what it's about.

Please note that not all sections will have corresponding files on the CDs.

*******SEE THE ELECTRONIC SOFTCOPY TABLE OF CONTENTS ON EACH CD.*******

I. INTRODUCTION

CD #1 hast the "FOUNDATION," "REACTION," AND "EVACUATION" sections.

CD #2 has the "ISOLATION," "COOPERATION," AND "APPENDIX" sections.

II. FOUNDATION.

All files in this section will be found under the main folder "1-FOUNDATION" on CD 1.

Health.

Health Improvement Considerations	
Filename	**Description**
09322.pdf	Health, nutrition, and aging.
09333.pdf	Health and dietary fiber
Aviator Nutrition Guide.pdf	Guide to diet and nutrition
DIETGD.pdf	USDA Dietary Guidelines for Americans booklet
Earlchildnutresguide.pdf	Guide to info sources for early childhood nutrition
Fast food guide.pdf	Guide to nutritional values of several fast food chains
Feb04.pdf	Heart-healthy nutrition article
FS787.pdf	Vegitarianism facts and tips
Hypertensiondietguide.pdf	Dietary guide for hypertension
Nutrition.pdf	Diet and alternative medicine vs western medicine
Nutrition.xls	Caloric intake guide
PM1082.pdf	Family nutrition guide

PM1404D.pdf	Food for the working family
Portions.pdf	Dietary portion flier
Usingtheguide.pdf	Pointers for using the USDA Dietary Guidelines for Americans booklet listed above

Health and Fitness Information	
BMI2.pdf	Body Mass Index chart
Fm21-20.exe	Physical Fitness Training
Fmfrp01b.pdf	Marine Physical Readiness Training
Kidstips.pdf	Healthy eating and exercise tips for kids
Mcrp302a.pdf	More Marine Corps Physical Readiness Training

Basic Training.

First Aid Background	
Filename	**Description**
!2001posteradult.pdf	CPR Poster – Adult
!2001posterchild.pdf	CPR Poster – Child
14295.pdf	Naval Hospital Corpsman Training Course
48000100.pdf	Air Force EMT procedural flowcharts – advanced info
CLS 08 STUDENT GUIDE.doc	Military short guide to shock, with quiz
CPR.pdf	CPR instruction chart
Diveslates.pdf	Diver first aid with CPR instruction
First Aid.pdf	Boy Scout First Aid test
First-aid.pdf	First aid quick reference booklet
Fm2111.pdf	First Aid for Soldiers (from 1988)
Fm4-02.6.pdf	The Medical Company
Fm4-25x11.pdf	Military First Aid Training Manual (supercedes FM21-11 above)
Fs315077.pdf	Scout's First Aid Training Pamphlet
Fwinfo.pdf	First aid info brochure
(Multiple files - subfolder)	FM 8-10-18 Veterinary Services
Woundcare.pdf	Wound closure manual

Fire Fighting Training	
(See "Fire" under "Basic Home Prep" and "Limited Fire Fighting" in "Life in Isolation" under "ISOLATION.")	

NBC Data	
2edmmrchandbook.pdf	Medical Management of Radiological Casualties
Anthraxbro.pdf	Red Cross pamphlet on Anthrax
Army_Chemical_Handbook.pdf	US Army's Field Management of Chemical Casualties
Batbooka.pdf	Medical NBC Battlebook
Biochem_threats.pdf	World Health Org's booklet on biochemical threats
Chemeffects.pdf	Chemical weapons info
Cia_cbrih.pdf	CBRN incident handbook with glossary
CTPR_Pocket_Card.pdf	Chemical weapon info card
Cwirp_guidelines_mass.pdf	Mass casualty decontamination
Cwmdc.pdf	WMD training course collection
Dod5100-52m.zip	Nuclear Accident Response Manual
Erg2000.pdf	Hazardous materials guidebook
Erg2004.pdf	Updated Hazardous Materials Guidebook
Ergo2000_e.exe	Hazardous materials guidebook
Ergo_e.exe	Hazardous materials guidebook
Fm105-6-2.pdf	Nuclear Play Calculator
Fm3_11.pdf	Procedures for NBC defense operations
Fm3_11x21.pdf	NBC Aspects of Consequence Mgt.
Fm3_11x22.pdf	WMD Civil Support Team

Fm3_11x4.pdf	NBC Protection
Fm3_3.pdf	Biochemical contamination avoidance
Fm3_6.pdf	Field Behavior of NBC Agents
Fm3_7.pdf	NBC Field Handbook
Fm3-11-4.pdf	NBC Protection
Fm3-11-9.pdf	Potential Military Chemical/Biological compounds
Fm3-3-1.pdf	Nuclear Contamination Avoidance
Fm3-4.zip	NBC Protection
Fm3-4-1.zip	Fixed Site NBC Protection
Fm4-02.7.pdf	Health Service Support in NBC environment
Fm4-02-283.pdf	Treatment of Nuclear and Radiological Casualties
Fm8_284.pdf	Treatment of Biological Warfare Agent Casualties
Fm8-10-7.zip	Health Service Support in and NBC Environment
Fm8-9.zip	NATO Handbook on Medical Aspects of NBC Operations
Microbesbook.pdf	Microbes and disease
Minnesotas-Nuclear-disaster-handbook.pdf	Radiological Emergency Preparedness Handbook
MMBCH4-0WordVer4-02withLinks.doc	Medical Management of Biological Casualties Handbook
Mmbch4AdobePDFVer4-02.pdf	Medical Management of Biological Casualties Handbook
(Multiple files - subfolder)	FM 8-285 Treatment of Chemical Agent Casualties
NBC-DEFENSE.pdf	Joint Doctrine for NBC Defense
NPG_only.zip	NIOSH Pocket Guide to Chemical Hazards
Nuclear.exe	E-book on surviving a nuclear attack.
Nuclear_Biological_Chemical_Defence.pdf	NBC Defense Pocket Handbook
Nuclearfaq.pdf	Details on radiation
Pcktcard.pdf	Pocket guide to terrorism with radiation
Pocket-card.pdf	Pocket info card for terrorism using ionizing radiation
RadEmergMgt.pdf	Radiological Emergency Management
RadiationProtectionManual.pdf	Technical publication on the nature and effects of radiation
Radiolog.pdf	Medical Management of Radiological Casualties
Radresourceguide.pdf	Officer's guide to radiation information sources
Rdo_course_manual.pdf	Radiological Defense Officer's course manual
Rdo_manual_tables.pdf	Effects of Nuclear Weapons and tables for the RDO officer's course
Survive.exe	E-book on surviving an NBC attack
TR-90.pdf	FEMA fallout exposure rate tables manual
WebNuke2.doc	Blast effects of various nuclear yields

Auto Repair Training

| Howcarswork.pdf | Basic info on how cars work |
| Mechanic_List.pdf | Maintenance and repair tips |

Outdoor Survival

Compmap.zip	How to use map and compass
Fire.zip	Lighting fires in the wilderness
Fishing.zip	Fishing in the wilderness
Fm21_31.pdf	Topographic Symbols
Fm21_76.pdf	Survival
Fm3_97x61c1.pdf	FM 3-97.61 Military Mountaineering (survival, rappelling, knots)
Fm31_70.pdf	Basic Cold Weather Manual
Fm3-25.26.exe	Map Reading and Land Navigation
Fm5-125.exe	Rigging and ropework
Knot_notes.pdf	Knot booklet with test
Knots A5.pdf	Knots for caving and rapelling
Knotss.pdf	Pamplet on knot tying
Map_compass.pdf	Brochure on using map and compass
Mcrp302h.pdf	Marine Corps Survival Manual
MonkeysFist.pdf	How to tie the Monkey's Fist knot

Nknots.pdf	Pamphlet on knot tying
Nucamp.zip	Tips on setting up a long-term camp in the wilderness
Pg51-56.pdf	Knots on the farm
Ropes&knots.pdf	Brochure on knots and ropes
Shltrall.zip	Making shelter in the wilderness
Sknots.pdf	Knots and lashing
Yarn08.pdf	Scouting brochure on knot tying and hut building
Information Research	
Alarm-en.pdf	Israeli Civil Defense booklet
Areyouready.pdf	FEMA's readiness guide
Areyouready_full.pdf	FEMA preparedness manual
ArticleIndex.pdf	Articles index for back issues of "Backwoods Home" magazine
Coloring_for_Kids.pdf	Disaster training book for kids in coloring book format
Dis-prep.pdf	General disaster preparedness info.
DP1 BookLinks.doc and .rtf	Our collection of links as they appear in the book.
DP1 General Links.doc and .rtf	A general collection of links searchable by category.
Ertss.pdf	FEMA's Emergency Response to Terrorism course
FEMA.pdf	FEMA manual on family disaster plans
G2_print.pdf	Web search tutorial
Is2-c_1.pdf	FEMA's emergency reaction explanation
Is2-tc.pdf	FEMA's emergency prep quiz
NESTHOMmcov2.pdf	NEST Individual and Family Disaster Planning Guide
Ostrich.pdf	LA FD Emergency Preparedness booklet
Search_engines101.pdf	A primer on types of search engines and how to use them
Survival.pdf	Carolina Preparedness brochure
Terrorism_handbook_2003.pdf	Personal Security Planning and Information Handbook
Miscellaneous Elective Training	
Tc21-24.exe	Rappelling

Frugality.

General Purchasing and Spending	
Filename	**Description**
2Handkbook.pdf	2003 Consumer Action Handbook
Budget01.xls	Excel spreadsheet for family budgeting
Budget3.xls	Household budget spreadsheet
Calculator.xls	Excel spreadsheet to calculate loan and mortgage payments
Fcs450.pdf	Planning for savings and financial wellbeing
HOUSEHOLDBUDGET.xls	Excel spreadsheet for managing a household budget
Personal_budget.xls	Excel spreadsheets for an individual's financial planning
Resource-guide.pdf	Comprehensive family financial planning resource guide
Consumables	
(Multiple files - subfolder)	FM 10-16 Fabric Repair (the US Army's sewing manual) © 2004 Paul Purcell
Miscellaneous Considerations	
Creditcar.pdf	Brochure on check and credit card fraud

Planning Ahead for Phase II.

Safeguards	
Filename	**Description**
A-NFIP-MythandFact.pdf	Information on the National Flood Insurance Program
Book.pdf	A primer on insurance

NaturalDisastersYourGuidetoInsurancePreparationandRecovery.pdf	A homeowner's guide to disaster and insurance

Services and Agencies	
E_owners.pdf	Forms of assistance for homeowners after a disaster

Supporting Documentation – The "Info Pack"	
209493.pdf	Family guide to identification through DNA
HomeOwnerInventoryChecklist.pdf	Brochure helping home owners catalog belongings
HouseholdandPersonalInventoryBook.pdf	Inventory guide for personal property

(See our forms under "APPENDIX.")

Emergencies and Disasters. An Overview.

Terrorism, Violent Crime, and Accidents	
Filename	**Description**
3 nuke scenarios.pdf	Three nuclear attack scenarios
Attack_on_america.pdf	Article about terrorism and 9-11
Doc_ruralcrime.pdf	Rural crime brochure
Mcrp302e.pdf	Marine Corps guide to understanding and surviving terrorism
Title3.pdf	EPA Consolidated List of Chemicals subject to "right to know"

Natural Disasters	
Apell_tsunamis.pdf	Brochure on tsunamis
EQ03 About NM Fault.pdf	About the New Madrid fault in central US
Eqhazards&risks.pdf	General data about earthquakes and their aftereffects
Hurrnoaa.pdf	Background info on hurricanes
Lightningawareness.pdf	Myth and fact about lightning risks
Lndmdfs.pdf	FEMA Landslide booklet with risk map
Noaaprep.pdf	Educational info on storms and bad weather
OFR97-513.pdf	Older Oregon booklet on volcanoes
OR-SHNMP_volcanic_chapter.pdf	Oregon EMA's booklet on volcanic activity and threats
SP23.pdf	Earthquake information pamphlet
Tsunami.pdf	Scientific guide to tsunami formation
Tsunami.pdf.pdf	A general guide to Tsunamis

Basic Home Prep.

Miscellaneous Considerations	
Filename	**Description**
Pest_Brochure.pdf	Everything You Want to Know About Household Pests

General Home Safety	
D000874.pdf	Homeowner chemical safety
GH6020.pdf	General home safety checklist
Homesafe2003.pdf	Home safety booklet
Homesafetychecklist.pdf	Home safety checklist
Home Safety Fact Sheet 2.pdf	Childproofing the home and poisons
Hzmthmfs.pdf	FEMA's home chemicals brochure
Ns-guide-10-01.pdf	A Homeowner's Guide to Nonstructural Earthquake Retrofit
PM1621.pdf	Home safety for parents
Poison_Safety_Tips_English_Brochure_01__04.pdf	Child poison safety in the home

Security	
19785.pdf	Personal security
19795.pdf	Personal security for the overseas traveler
Armed_robbery.pdf	Pamphlet on dealing with armed robbery

Brochure.pdf	Brochure on home security
Burglary_prevention.pdf	LAPD's Burglary Prevention brochure
Cp_environmental_design.pdf	Crime prevention through landscaping and decorative efforts
CPH7.pdf	Brochure on personal security
Doc_homesecurity.pdf	Home security checklist
Doc_homesecuritybook.pdf	Home security booklet
Hardware.pdf	Home security hardware brochure
Home_Security.pdf	Brochure on personal security at home
Homeland brochure 3.pdf	Homeland Security brochure on personal security
Homesec.pdf	Home security brochure
Homesecurity_4.pdf	Home security brochure
Homesecurityc.pdf	Booklet on home security
Homesecuritycl.pdf	Checklist for home security
Homesecurityl.pdf	Another checklist for home security
International_travel.pdf	Guide to personal health and safety while traveling internationally
Personal Security.pdf	Personal Security booklet
Personal_security.pdf	Personal security info flier
Personalsecurityathome.pdf	Staying safe from crime while at home or travelling
Robbery.pdf	Brochure on protection against robbery
Securityguide.pdf	Booklet on home security
Security-survey-home.pdf	Home security checklist
Vacasecu.pdf	Vacation security brochure
Structural Considerations	
126.pdf	Homeowner's Earthquake Retrofit Info
Agnote_810.pdf	Selecting plants for windbreaks and spray filters
Agstwnd.pdf	FEMA's pamphlet on strengthening homes against storm and wind
BAS-How-You-Can-Strengthen-Your-Home.pdf	LA's booklet on strengthening your home against earthquakes
Bluesky1.pdf	Strengthen your roof using structural adhesives
Building_safe_room.pdf	FEMA's brochure on building a safe room in your house
EQ Brochure.pdf	Securing fixtures in the home against earthquake damage
Ertdam.pdf	FEMA's homeowner's earthquake checklist
FEMA-ShelterFromTheStorm.pdf	FEMA's booklet on building a safe room
Fm3-34.471.exe	Plumbing, Pipefitting, and Sewerage
Fm5_428.pdf	Concrete and Masonry
Fm5-426.exe	Carpentry
Fortifiedforsaferlivingbuilder'sandhomebuyer'sguide.pdf	Guide to structural enhancement to protect homes from natural disaster
Guide.pdf	FEMA's guide to flood protection retrofit for the home
Hurdam.pdf	FEMA's hurricane checklist for homeowners
Increasing_Blast_and_Fire_Resistance_in_Buildings.pdf	Civil Defense booklet on increasing blast and fire resistance in buildings
In-Home Shelter Manual.pdf	Building a safe-room in your home
Military Basics Electronics Course.pdf	Course in electronics
Plumbing.pdf	Army plumbing manual
PLUMBINGtxt.pdf	Dos and Don'ts of home plumbing
Resshelter_bkgrdr.pdf	FEMA safe room background info
Retrofit.pdf	Earthquake retrofit tips for house foundations and support walls
Shutters.pdf	Window protection in hurricanes
Windbrochure.pdf	Protecting your home against hurricanes
Fire	
6940PortFireEx.pdf	Introduction to portable fire extinguishers
Apt. Fire Safety Information.pdf	Simple fire safety flier
AreyoufirewiseFlorida.pdf	Fire and wildfire prevention checklist

BrochureSprinklers.pdf	Brochure on home fire sprinkler systems
Canyonhills.pdf	Wildfire protection tips for canyon dwellers
Clistblank.pdf	Home fire protection checklist
Dcamanual.pdf	Florida's wildfire and community management book
D001102.pdf	Fire extinguisher selection, location, and use
Fire_extinguishers.pdf	Fire extinguisher descriptions and ratings
Firedotgov_fall_2003.pdf	Fire.gov newsletter copy
Firedotgov_spring_2004.pdf	Fire.gov newsletter copy
Firedotgov_summer_2002.pdf	Fire.gov newsletter copy
Fireextinguishersafety.pdf	Fire extinguisher descriptions and ratings
Firefs.pdf	Stats and emergency info about house fires
Firetips.pdf	Family home fire protection
Firewatch_cklist.pdf	Fire prevention checklist
FireXT.pdf	Fire extinguisher descriptions and ratings brochure
Freedownloads8-fireext.pdf	Tips and info on choosing a fire extinguisher
Homes.pdf	Arson prevention brochure

Basic Vehicle Prep.

General Considerations	
Filename	**Description**
Vehicle.pdf	Vehicle security brochure

The EAS Alerts.

Standard Alerts	
Filename	**Description**
CommunicationFinal.pdf	EAS and public warning info
Eas_draft.pdf	EAS info by the Partnership for Public Warning
PWHandbook.pdf	Introduction to Public Alerts and Warnings
Warning.pdf	Colorado EAS booklet
Getting the Word	
NOAA Brochure.pdf	An information pamphlet on NOAA emergency alert radio

Communication

Radio and Television	
Filename	**Description**
Commfaq.pdf	General info on communications with glossary.
FEMA_where_there_is_no_phone.pdf	FEMA guide to radio systems
Visual and Audible Signaling	
Alpha.bmp	Chart of nautical flags and semaphore alphabet
Captains_Code_-_Semaphore_Flags.pdf	Semaphore alphabet chart
Codes_rescue01.bmp	Ground to Air hand signal chart
Codes_rescue02.bmp	Ground to Air hand signal chart
Codes_rescue03.bmp	Ground to Air hand signal chart
Codes_tracking01.bmp	Chart showing on-the-ground tracking symbols
Flag_Semaphore.pdf	Semaphore alphabet
Fm21_60.pdf	Visual Signals
GTAsign.bmp	Ground to Air hand signal chart
Semaphore Alphabet Master.doc	Chart showing semaphore alphabet flag positions

III. REACTION.

All files in this section will be found under the main folder "2-REACTION" on CD 1.

Homeland Security Terror Threat Alert Levels

The Lower Alerts	
Filename	**Description**
HomelandSecurity.pdf	Red Cross brochure on terror threat alert levels

Basic Personal Safety

Gas Masks and "Moon Suits"	
Filename	**Description**
191520.pdf	Personal protection equipment selection for the first responder
191521.pdf	Personal protection equipment for the first responder volume 2
2003_rsg.pdf	Respirator selection guidelines
Book-PPE.pdf	Guide to Personal Protection Equipment
CF176.ch3.pdf	Performance and concerns of personal protective equipment
Cpc_lepo_ems_rtcw.pdf	Choosing chemical protective clothing for first responders.
Emergency_gas_mask.pdf	How to make an emergency gas mask
Faq.pdf	FAQs about respirators
Ffpe_scba.pdf	Firefighter's guide to self-contained breathing apparatus
Fumigating_mask.pdf	How to make a respirator for fumigating
lg25.pdf	PPE selection primer
Ornl_TM_2001_153_expedient.pdf	ORNL and FEMA on expedient respirators
Personal Protective Equipment.pdf	Guide to selecting personal protective equipment
Personalprotectiveequip.pdf	Guide to selecting personal protective equipment
Pk_3m_guide2respirators.pdf	3M's guide to respirator selection
Ppe.pdf	PPE for chemical emergencies
Ppe_manual.pdf	Manual on selecting personal protective equipment
PPESource.pdf	Sources of protective apparel and gear
Respirator.pdf	Respirator selection guide
Respirators.pdf	Respirator Basics
Respreference.pdf	Respirator selection and reference guide
Simpler_matrix_table_11.pdf	Matrix for emergency response to nerve agent event
Surviveair Selection Guide.pdf	Respirator selection guidebook
Basic Personal Decontamination	
Cm2506.exe	Military decontamination course
Fm3-5.pdf	NBC Decontamination
Fm3-5change1.pdf	Updates to FM 3-5 NBC Decontamination
Guide-F.pdf	Smallpox facility decontamination
Personalcleaningfacts.pdf	CDC on personal decontamination

(Also see sources listed under "NBC Data" under "Basic Training" in the "FOUNDATION" section.)

Manmade Misfortunes

Nuclear Detonation	
Filename	**Description**
11_steps.pdf	Canadian Nuclear Survival Pamphlet
Aussie_HouseholdHandbookNuclearWarfare.pdf	Australian Civil Defense pamphlet on nuclear war survival
Canada-cd-CivDefNotebook.pdf	Old Canadian Civil Defense booklet

Domestic_Nuclear_Shelters.pdf	A guide to nuclear detonation and public shelter creation and selection
Efx-nuke-war.pdf	The Effects of Nuclear War
FEMA-H20-ProtectionInNuclearAge.pdf	FEMA guide to nuclear war preparation and effects
H-6.pdf	Old civil defense booklet on nuclear aftermath
Kfm_inst.pdf	How to build the Kearny Fallout Meter
Minnesotas-Nuclear-Disaster-Handbook.pdf	Information on radiological attacks, accidents, and detection
Nuclearsurvival.pdf	Nuclear War Survival Skills by Kresson H. Kearny
Nukeblastinfo_a.pdf	Scientific details of a nuclear blast
Nukweaponseffect.pdf	More scientific details of a nuclear blast
Rso_handbook.pdf	Nuclear Weapons Effects Handbook
Survivedoomsday.pdf	You Will Survive Doomsday – nuclear war Q&A
Ywsd.pdf	You Will Survive Doomsday – early version

"Dirty" Bomb or Radiological Accident	
Dod5100-52m.zip	Nuclear Weapon Accident Response Procedures
Minnesotas-nuclear-disaster-handbook.pdf	Minnesota's guide to radiological disaster and emergencies
Pocket-card.pdf	Understanding terrorism with ionizing radiation
ZPotassium_Iodide.pdf	Frequently Asked Questions on the use of Potassium Iodide

Chemical Attack or HazMat Incident	
CPTR_Large_Format.pdf	Chemical Terrorism Preparedness and Response Card
MSDSInterpretationApp8.pdf	How to read Material Safety Data Sheets

Fire	
0424.pdf	Community water sources for fire protection
Cwpphandbook.pdf	Community Wildfire Protection Plan
Emergency-exit-routes-factsheet.pdf	OSHA pamphlet on emergency exits in the workplace
Evacuating-highrise-factsheet.pdf	OSHA pamphlet on evacuating highrise buildings
FWGlossary.pdf	Firewise glossary
Fwlistsz.pdf	Firewise landscaping checklist
GLOSSARY.pdf	Glossary of Wildland Fire Terminology
Patrolguide.pdf	Wildfire Prevention Patrol Guide
Resp.pdf	Fire Protection in the Wildland/Urban Interface
Wildfire Smoke.pdf	Guide to wildfire smoke for public health officials
WildfireSmokeGuide.pdf	Another public health guide to wildfire smoke

Bombing or Explosion	
Afttp(i)3-2.12.pdf	Military manual on unexploded ordnance
Bomb_broch.pdf	ATF's Letter & Package Bomb Detection brochure
Bombthreats.pdf	Bomb threat procedures
Explosive_devices.pdf	How to identify explosive devices

Weather Disasters

Hurricane	
Filename	**Description**
HurricaneTrackingChart(Atlantic).pdf	Tracking chart to map approaching hurricanes
Isabelfaq3n.pdf	Federal disaster assistance after hurricanes examples
Recovery.pdf	CDC guide to recovery after hurricanes
Tornado	
TR-83B.pdf	Tornado Protection – Selecting Refuge Areas in Buildings
Lightning	
Thunder.pdf	Tips on dealing with storms and lightning strikes

Other Natural Disasters

Tsunamis	
Filename	**Description**
OR-SNHMP_tsunami_chapter.pdf	Oregon brochure on tsunami warning and reaction
Tusnamiguidance.pdf	California brochure on tsunami warning and reaction
Volcano	
OHS-Volcanicash.pdf	Tips on dealing with volcanic ash

IV. EVACUATION.

All files in this section will be found under the main folder "3-EVACUATION" on CD 1.

Prepping the Gear.

Contents in Detail	
Filename	**Description**
Disaster 2 web.pdf	Emergency preparedness for large animals

Planning the Destinations.

Where Ya Gonna Go?	
Filename	**Description**
Fm3_19_40.pdf	Military Police Internment / Resettlement Operations

Getting There.

Out the Door	
Filename	**Description**
DANRGuide2.pdf	Guide to caring for animals in an evacuation
DISSEvac.pdf	Evacuation considerations for disabled employees
EvacuationguideforPWD.pdf	Evacuation guide for persons with disabilities
FA-235.pdf	FEMA manual for evacuation of people with disabilities
Mob-all.pdf	FEMA on the Mobility Challenged
Utilities.pdf	Shutting off your utilities in an emergency
Transportation	
((Multiple files - subfolder))	FM 55-501 - Four-Part Marine Crewman's Handbook

Life at the Destination.

Shelter Dynamics	
Filename	**Description**
Disruptive_people.pdf	Dealing with disruptive people and behavior
Managing_conflict.pdf	Brochure on managing conflict
Coming Home	
A-DryingOutYourHomeAfteraFlood.pdf	Tips on drying out a flood-damaged home

A-EnteringYourHomeAfteraFlood.pdf	Tips on entering your home for the first time after a flood
A-HosingtheHouseanditsContentsAfteraFlood.pdf	Hosing out your house after a flood.
AftertheFireReturningtoNormal.pdf	Cleanup and steps to take after suffering fire damage
D04-09.pdf	Meal prep and food handling after a flood
Is1692.pdf	More on food safety and cooking after a flood
PickingupthePiecesAteraDisaster.pdf	Hints, tips, and steps for recovery after a disaster
PickingupthePiecesAteraFire.pdf	Hints, tips, and steps for recovery after a fire
Ps8076.pdf	USPS Authorization to Hold Mail

Coming Home II – The Emotional Aftermath	
8pgBookletOctober5.pdf	Talking to kids about violence
Children.pdf	Red Cross on helping children cope with disaster
CLS 24 INSTRUCTOR GUIDE.doc	Military short guide to combat stress
Coping.pdf	Post Traumatic Stress Disorder pamphlet
Drjwtcspecial.pdf	Articles on lessons learned from the 9/11 aftermath
Exercise.pdf	The role of exercise in treating Post Traumatic Stress Disorder
Firstaid.pdf	Tips for trauma survivors
Fm22-51.pdf	Combat Stress Control
Fm6-22.5.pdf	Combat Stress
Mcrp611c.pdf	Marine Corps Combat Stress manual
(Multiple files - subfolder)	FM 8-51 Combat Stress Control
NCJ1902.pdf	Department of Justice brochure on Coping After Terrorism
Ptg1.pdf	Post-Traumatic Gazette sample issue
PTSDfirstaid.pdf	Emotional first aid for trauma survivors
Van_der_Kolk_2002_In_Terror's_Grip.pdf	PTSD and terrorism
When.pdf	Another sample copy of the Post-Traumatic Gazette

V. ISOLATION

THIS IS THE BEGINNING SECTION OF CD #2.

All files in this section will be found under the main folder "4-ISOLATION" on CD 2.

Advanced Home Prep.

Sealing the Home	
Filename	**Description**
TM_2001_154_duct_plastic.pdf	ORNL & FEMA on expedient sheltering in place
Fortified Shelters	
Basement_shelter.pdf	British fallout shelter booklet
Build_a_blast_shelter.pdf	Plans for building a home blast shelter
Cw_shelterh121.pdf	Underground fallout shelter plans
Cw_sheltrh122.pdf	Above ground fallout shelter plans
Cw_sheltrh12a.pdf	FEMA fallout shelter plan
Cw_sheltrh12c.pdf	Compact basement fallout shelter plans
Cw_sheltrh12e.pdf	Flip-up basement fallout shelter plans
Cw_sheltrh12f.pdf	Lean-to basement fallout shelter plans
Domestic_Nuclear_Shelters.pdf	British handbook on civilian fallout shelters and nuclear threats
Expfoutshelters.pdf	Construction of expedient fallout shelters
H-6.pdf	Civil Defense Fallout Protection Handbook
H-12-A-Basement-Fallout.pdf	Basement fallout shelter
H-12-C.pdf	Another basement fallout shelter
Home_Blast_Shelter.pdf	FEMA Home Blast Shelter booklet
HomeShelter.pdf	FEMA on home fallout shelters

Filename	Description
PLANS1.pdf	Fallout shelter plans
PLANS2.pdf	More fallout shelter plans
PSD_F-61-1.pdf	FEMA Basement fallout shelter plans
PSD_F-61-2.pdf	FEMA Basement fallout shelter plans
PSD_F-61-3.pdf	FEMA Basement fallout shelter plans
PSD_F-61-4.pdf	FEMA Basement fallout shelter plans
PSD_F-61-5.pdf	FEMA Basement fallout shelter plans
PSD_F-61-6.pdf	FEMA Basement fallout shelter plans
PSD_F-61-7.pdf	FEMA Basement fallout shelter plans
PSD_F-61-8.pdf	FEMA Basement fallout shelter plans

Alternative Energy	
Backup-and-emergency-power.pdf	Basics of backup electricity
Electrical.pdf	Army manual on electrical power supply
FEMA_wood_gas_generator.pdf	Manual on creation and use of wood-gas in a petroleum emergency
Generatorfaq.pdf	Generator selection info
Heat_wood_faq.pdf	Using wood as a heating fuel
Poweroutage.pdf	The CDC on power outages
SafelyInstallingYourStandbyGenerator.pdf	How to safely install a generator
Solar.pdf	A Manual on Home Solar Power Management

The Isolation Pantry

Water	
Filename	**Description**
82-93_Producing_more_water_Part_8.pdf	Australian publication on water production
Borderpact.pdf	Presentation on home water distillation
Cistern-water-saving.pdf	Making an outdoor cistern
Covingtonrainbarrels.pdf	How to make a rain barrel
Hand_pump_backup.pdf	How to make a hand-pump for a well
Pm1329.pdf	Coping with contaminated wells
Rain_barrel_guide.pdf	How to make a rain barrel
RainHarv.pdf	Texas Guide to Rainwater Harvesting
Rainwater-Harvesting.pdf	Article on rainwater collecting
Rws1c4.pdf	Making a roof catchment water collection system
Rws1d4.pdf	Designing roof catchment water systems
Rws3c1.pdf	Making a household sand filter for water filtration
Rws3c4.pdf	Contructing a home disinfection unit for water purification
Rws3d1.pdf	Basic household water treatment systems
Rws3d5.pdf	Water treatment in an emergency
Rws3m.pdf	Water treatment methods
Rws3o1.pdf	Operating and maintaining household water systems
Rws5c1.pdf	Constructing a household water cistern
Rws5d1.pdf	Designing a household water cistern system
Rws5m.pdf	Methods of storing water
Solar_distillation.pdf	How to make a solar still to produce drinking water
Update_on_SEA_Earthquake_Tsunami_emergencytreatment.pdf	Emergency treatment of drinking water
Water.zip	Pointers on water use and purification in the wild
WHO_Water_Supply_in_Emergencies.pdf	Wells, purification and water sources in emergencies

Food	
4078.pdf	Food canning basics
5perweek.pdf	Purchasing suggestions for stocking extra food

Cangui1.pdf	Guide to home canning
Canner_Cooker.pdf	Info tidbits and recipes for home canning
CanningFruit.pdf	Basics of how to can fruit items
Dry_fruit_faq.pdf	How to dry fruit
Dry_veggie_faq.pdf	How to dry vegetables
E-2297.pdf	Food storage brochure
Fcs3328.pdf	Canning vegetables and vegetable products
Fn500.pdf	Food storage info file
Fn502.pdf	Food storage info file
Fn503.pdf	Food storage info file
Fn579.pdf	Food storage info file
FN-SSB085.pdf	Food storage life chart
FN-SSB130.pdf	Food storage according to the food pyramid
Food2.pdf	Frozen and thawed food storage
Foodfacts_july_2004.pdf	Food handling and storage newsletter sample
Foodfacts_sep_2003.pdf	Food handling and storage newsletter sample
Foodprestrouble.pdf	Food preservation problems and troubleshooting
Foodshelf.pdf	Food shelf life and date code charts
Freezing animal products.pdf	Tips on freezing meats, dairy, seafood
Freezingfruit.pdf	Tips on freezing fruit items
Freezingpreparedfoods.pdf	Tips on freezing meals you've cooked at home
Fs273.pdf	Food storage and shelf-life info
Fs274.pdf	More on food storage and shelf-life
G1PrinciplesOfHomeCanning.pdf	Home canning basics
G4CanningVegetables&VegetableProducts.pdf	Canning veggies
G5Preparing&CanningPoultryRedMeats&Seafoods.pdf	Canning meats
G6Preparing&CanningFermentedFoods&PickledVegetables.pdf	Canning pickled veggies
G7Preparing&CanningJams&Jellies.pdf	Canning jams and jellies
GEEZERCB.pdf	Camping cookbook
Herbs and Spices.pdf	Herb pamphlet with recipes and spice substitution list
HGIC3000.pdf through HGIC3869.pdf (multiple files)	Clemson University series on canning and food preservation
Keeps.pdf	A short primer on the history of food preservation
IC407.pdf	Controling pests in food stores
L806.pdf	Kansas State University food shelf-life chart
L883.pdf	Reading and understanding food labels
Molds.pdf	A discussion of molds on food
NL0008.pdf	Introduction to LDS food storage newsletter
NL1 through NL12 *****(multiple files)*****	LDS food storage newsletter samples
OP_Rations.pdf	The history and FAQs on MREs
P1540.pdf	4H on canning vegetables
P1542.pdf	Canning fruits and tomatoes
PHLMNTCB.pdf	Philmont Country Cookbook
Pickles.pdf	How to make pickles
Poweroff.pdf	Keeping or discarding foods from fridge or freezer
SelectingPreparing&CanningFruit&FruitProducts.pdf	USDA on canning fruits and fruit products
SelectingPreparing&CanningTomatoes&TomatoProducts.pdf	USDA on canning tomatoes and tomato products
Smokingyourcatch.pdf	Tips on safely cooking fish by smoking
Texas_storage.pdf	Safe home food storage
Website.pdf	Websites for home food preservation info

Life Without Utilities

Climate Control	
Filename	**Description**
Cold weather Injuries Poster1.pdf	Cold weather injuries poster
Cold0102.pdf	Cold weather health and injury manual
Coldstress.pdf	Cold index and cold weather problem symptoms
Coldweatherinjuriesbookupdate.pdf	Cold weather injuries booklet
CWI RM Guide updated 2003.pdf	Cold weather hazards booklet
HeatEquationCard.pdf	Heat index and heat problem symptoms
HeatEquationPoster.pdf	Heat index and heat problem symptoms poster
HeatIndexTable.pdf	Heat problems with symptoms and treatment
Nu-cold.zip	Thoughts and tips on cold weather injuries and their prevention

The Manual Kitchen	
Boxoven.pdf	How to make a box solar oven.
Cook-it-Plans.pdf	How to make a solar oven
DSPC-CookerPlans.pdf	Solar oven plans
Fm8-34.pdf	Food sanitation
Food_microwave.pdf	Food safety while cooking with a microwave
Kitchen.pdf	Selection and use of kitchen thermometers
P-5010-1(2).pdf	Navy's Food Safety manual
PM1479.pdf	Serving safe meals in a chemically polluted environment
PM1523.pdf	Safe recipe checklist
Prepfood.pdf	Preparing food during a power failure
Solar Oven poster edited.pdf	How to make a solar oven
Solarbox.pdf	Building and using a solar box oven
Solaroven.pdf	Another how-to on building a solar box oven
Thermometer.pdf	Cooking thermometer selection and use

Privy Deprived	
Fm21-10.pdf	Field Hygiene and Sanitation
Fm4-25.12.pdf	Unit Field Sanitation Team
San1c5.pdf	Constructing bucket latrines
San1c7.pdf	Constructing and maintaining sumps, soakage pits, and soakage trenches
San1d5.pdf	Designing bucket latrines
San1d7.pdf	Designing sumps, soakage pits, and soakage trenches
San1m1.pdf	Simple methods of excreta disposal
San1o5.pdf	Operating and maintaining bucket latrines
San2m.pdf	Combined methods of wastewater and excreta disposal
WHO-Sanitation-in-Emergencies.pdf	Disposal of human wastes in an extended emergency

Life in Isolation.

Security and Defense	
Filename	**Description**
A Self-Help Guide to Combatting Terrorism.pdf	DOD guide to home, personal, and travel security for military personnel
Brochure.pdf	Brochure on home firearm safety
CPH7.pdf	Basic personal security pointers
Fm19_15.pdf	Civil Disturbances
Fm3-19.30.pdf	Physical Security
Fm3-22-40.pdf	Tactical Employment of Nonlethal Weapons
FM3-25.150.exe	Combatives (hand to hand combat)

FM_90-40_NLW.pdf	Field manual containing non-lethal weapons descriptions
GetTough.pdf	1943 military manual on hand-to-hand combat
MCRP 3-02B Close Combat.pdf	Marine Corps manual on hand to hand combat
(Multiple files - subfolder)	FM 22-6 Guard Duty
Text book of close combat.pdf	Unarmed self-defense
W3158.pdf	Tactical Deployment of Nonlethal Weaponry
Limited Firefighting	
4-17-02tools.pdf	Introduction to firefighting tools
6-8LessonPlan06.pdf	Red Cross lesson plan on behavior of fire
After the Fire.pdf	A guide to assist homeowners after a fire
Chapter01.pdf	Firefighting safety concerns
Essay3.pdf	Problems with flashover in fire situations
Fm5-415.exe	Fire-Fighting Operations
Forest_firefightin_manual.pdf	Tazmanian firefighting manual for forest fires
Cabin Fever	
Ornl_tm_2003_230.pdf	ORNL and FEMA on isolation Q&A

VI. COOPERATION.

All files in this section will be found under the main folder "5-COOPERATION" on CD 2.

Groups; to Form or Join

Base, Function, and Scope	
Filename	**Description**
Disaster.pdf	Disaster preparedness guidebook for community development professionals
GoodIdeasBook.pdf	FEMA book of volunteer group suggestions
NeighborhoodOrganization.pdf	Suggested functions of a neighborhood watch or volunteer group
NESTBook.pdf	Suggested functions of a Neighborhood Emergency Service Team
Nestcapt.pdf	Neighborhood Emergency Service Team Captain's Handbook
NESTHOMmcov2.pdf	Another version of the NEST handbook
Thandbook.pdf	Neighborhood volunteer group suggested projects and functions
Form and Organization	
092899-businesschecklist.pdf	Business Plan Checklist
700book.pdf	Local Emergency Committee Planning Handbook
All-Forms.pdf	Sample forms for volunteer watch groups **(Also see "APPENDIX.")**
B_chklst.pdf	Business Plan Checklist
Business_checklist.pdf	Business Plan checklist
BusPlanCheklist.pdf	Business Plan Checklist
Certnewsspring02.pdf	Sample of a CERT newsletter
Communicating-with-Volunteers.pdf	CERT paper on staying in communication with volunteers
File_2a.pdf	Meeting planner example
Mp-12.pdf	SBA's Checklist for Going into Business
(Multiple files - subfolder)	TC 25-30 A Leader's Guide to Company Training Meetings
R15_1.pdf	Army pamphlet on committee management
Rroo-reference-sheet.pdf	Robert's Rules of Order reference sheet
Rror.pdf	Full copy Robert's Rules of Order
Rules of Order2.pdf	Robert's Rules of Order quick sheet
Ssl_01rev.pdf	State Liability Laws for Charitable Organizations and Volunteers (*Downloaded from* http://www.citizencorps.gov/councils/liability.shtm.)
Volpro.pdf	Volunteer Protection Act of 1997

PR and Fundraising

Fm3-61.1.exe	Public relations manual
GroupFormCertification.pdf	Submission form for volunteer group awards
Mt-11.pdf	SBA Marketing brochure

The Group in Action

Neighborhood Watch

Filename	Description
DisasterResistantNeighborhoodHandbook.pdf	Red Cross brochure on setting up a Neighborhood Watch
Fs000301.pdf	Office of Victims of Crime Fact Sheet
Homesecurityn.pdf	Neighborhood Watch home security brochure
Nw0007.pdf	Example of a neighborhood watch newsletter
OperationCooperation.pdf	US DOJ's Guide for Law Enforcement & Private Security
Watchmanual.pdf	National Sheriff's Association Neighborhood Watch manual

The Service Functions

DisasterManual04-2002.pdf	St. John's Ambulance manual for volunteers
Emergency_signs_artwork.pdf	Schematics to make emergency signs
Fallout.pdf	Fallout shelter sign sample
H-16.pdf	Civil Defense Shelter Management Handbook
Openforbusiness.pdf	SBA guide to disaster planning for the small business
ProfitingThroughDisasterPreparedness.pdf	Example of a Florida business continuity project workbook
Shelter management manual.pdf	FEMA on shelter management (supercedes H-16.pdf)
Spontvol.pdf	Helping communities set up volunteer programs
TheyWillComePost-DisasterVolunteersandLocalGovernments.pdf	Booklet on handling walk-up volunteers
UnaffiliatedVolunteersInResponseandRecovery.pdf	Book on dealing with walk-up volunteers
Vulnerable Populations.pdf	How community organizations can help in disaster

The Think Tank

Council.pdf	FEMA on Citizens' Councils
Studies.doc and .rtf	A collection of brainstorm primer ideas and a threat planner.

Training and Equipment

Training

Filename	Description
78-Glossary.pdf	Glossary of official emergency reaction terminology
Basic_Rescue_Skills.pdf	Canadian civil defense rescue booklet
CERT_Manual.pdf	FEMA CERT training manual
CERTIG.zip	FEMA's CERT Instructor's Guide
CERTPM.zip	FEMA's CERT Participant's Manual
Civilian_sar.pdf	Civilian search and rescue manual
Compendium.pdf	Compendium of Federal Terrorism Training Courses contact info
COOP-Glossary.pdf	Government Emergency Management Acronym Glossary
CPR.pdf	Instructions for setting up a CPR and First Aid test and competition
E317smaCERTstudentmanualA.exe	CERT training manual part A
E317smaCERTstudentmanualB.exe	CERT training manual part B
General_Information_Helpinghelpers090105.pdf	PTSD and stress management for rescuers and volunteers.
General_Information_Managingstress090105.pdf	Pocket guide for rescuer PTSD management
H_courses.pdf	Disaster planning information and available training courses
Jp3_50.pdf	National Search and Rescue Manual Vol. I.

Jp3_50_1.pdf	National Search and Rescue Manual Vol. II
List_of_Guidelines_for_Health_Emergency_Caring-you__own.pdf	Helping rescue workers deal with stress
Mdgroundzero.pdf	An MD's Report from Ground Zero
MitigationIdeas.pdf	FEMA booklet on hazard and disaster mitigation
NationalSARManual_full_english.pdf	National Search and Rescue Manual
Newsletter22.pdf	Sample ODP newsletter
Newsletter5.pdf	Sample ODP newsletter
PM-CERT.zip	CERT Participant Manual
RescueHandbook(final).pdf	Australian Natural Disasters Organisation Rescue Handbook
Specialneeds.02.pdf	Handling special needs populations in disasters and emergencies
Terrorism.exe	FEMA's "Terrorism and Emergency Management" course
WMD-Courses.pdf	Listing of Federal training courses on terror threats.
Equipment	
Emergency_signs_artwork.pdf	Schematics for creating emergency signs
Fallout.pdf	Fallout shelter sign sample

External Liaison.

Federal Offices and National Groups	
Filename	**Description**
Fm100_19.pdf	Domestic Support Operations
Pd01018002c.pdf	Stormwatch and Tsunamiwatch affiliate and community programs
Policy_book.pdf	Overview of FreedomCorps

VIII. APPENDIX

All files in this section will be found under the main folder "6-APPENDIX" on CD 2.

Filename	Description	CD
These are the files for the forms in the "APPENDIX." These are copyrighted material and are not part of our collection of freely downloaded materials. However, you do have the right to reproduce these forms for your personal use.		
Calendars.doc and .rtf	Several planning calendars for setting your disaster prep goals.	
DP1Budget.doc and .rtf	Family budget, and household inventory / property forms.	
DP1Budget.xls	Excel spreadsheet containing several budget workbooks.	
EvacAtlas.doc and .rtf	Forms for planning your evacuation routes and recording locations.	
FamilyData.doc and .rtf	Important information records for family business and needs.	
ImpContact.doc and .rtf	"Address book" type forms for tracking people and places.	
Supply.doc and .rtf	Organizers to help you plan the locations of goods and gear.	
Volunteers.doc and .rtf	Info sheets to help organize Watch and Volunteer groups.	
WatchGroup.xls	Excel spreadsheet for organizing your Watch and Volunteer groups.	

Remember to check the electronic Tables of Contents on each CD. With those, you'll be able to use the "Document Map" function if you're using Microsoft Word to read them. Click on "View" on your toolbar and then click on "Document Map." This will let you search by section of the book so you can find pertinent files related to the section of the book you're reading. You can also use your wordprocessor's "find" command to search for specific files you might want to read by searching for specific words you think might be in the file's title or description.

This page intentionally left blank.

A GIFT COUPON FOR OUR READERS' FRIENDS.

Ready to help in one more little way? In your travels and conversations with people, you're bound to run across people that know about the dangers we all face, but don't know where to turn for help. This is why we're giving you a coupon. It's our way of helping you help others. The more people we help educate, the better off we'll all be. Thank you for your help!

This coupon good for 10% off of a single copy of <u>Disaster Prep 101</u>.

Disaster Prep 101 is the complete guide to civilian emergency preparedness. In it, we cover everything from tornadoes to terrorist attacks, and all that lies in between.

In this book you'll find:

- A **Foundation section** that will organize your basic training, provide you all the free sources of outside instruction you could ever use, and help you develop a "readiness mindset."

- **Reaction, Evacuation, and Isolation sections** that will teach you how to survive most any incident and then follow up with a well-planned secondary reaction such as an emergency evacuation, or an extended quarantine or "shelter-in-place" reaction.

- A **Cooperation section** that will detail the creation of an emergency volunteer group from the ground up, and help you develop its function and liaison with government entities.

- Over **70 pages of useful forms and worksheets** that will help your family record all your important personal information, create family member ID kits, budget and organize all your emergency goods and gear needs, and set yourself up for a speedy recovery from most any disaster.

- **Two CDs** containing not only these **70+** forms in useful wordprocessor format, but you'll find over **300 useful training publications**, military manuals, and other instructional files, too numerous to list here. However, we will tell you that this list includes such manuals as First Aid, and Outdoor Survival.

- You'll also find a **coupon collection** intended to not only give you yet another benefit of purchasing this book, but to help you along the way with your preparedness goals.

- -

Gift Coupon discount: This coupon good for **10%** off a copy of <u>Disaster Prep 101</u>.

Normally **$49.95** plus $4.95 S&H, your cost, including S&H, is only **$49**.

This copy ordered by: Name:_____

Billing Address:_____

City:_____ State:_____ Zip+4:_____-_____

Phone:_____ E-Mail:_____

Payment Method: ❑ Check or M.O. for **$49.00**   ❑ Visa ❑ Master Card ❑ Discover ❑ AmEx

Card #:_____ Expiration Date: ____/____/____

Signature:_____

Ship to a different address because: ❑ The above is a billing address ❑ This book is being sent as a gift

Please ship to:

Name:_____

Address:_____

City:_____ State:_____ Zip+4:_____-_____

Mail coupon to: InfoQuest, 6300 Powers Ferry Rd., Suite 600-294, Atlanta, GA 30339. Or contact us at DisasterPrep101@aol.com.

This page intentionally left blank.

Disaster Prep 101

VOLUME ORDERS AND RESELLER'S SCHEDULE

Our aim with this discount schedule is to help groups that would like to bulk-purchase copies of Disaster Prep 101 for their members, schools who would like to use this manual as a text in any sort of family education courses, or civic groups and the like who would like to use our book as a fund-raiser for their community-oriented disaster prep operations.

Ordered by: Name:_____

Billing Address:_____

City:_____ State:_____ Zip+4:_____ - _____

Phone:_____ E-Mail:_____

Payment Method: ❑ Check or M.O. for $_____ ❑ Visa ❑ Master Card ❑ Discover ❑ AmEx

Card #:_____ Expiration Date: ____/____/____

Signature:_____

Ship to a different address because: ❑ The above is a billing address

Please ship to:

Name:_____

Address:_____

City:_____ State:_____ Zip+4:_____ - _____

Pricing Schedule

	Volume	Pricing	Terms
❑	**10 to 20 copies**	$ 47.00 ea.	
❑	**21 to 40 copies**	$ 46.00 ea.	
❑	**41 to 60 copies**	$ 45.00 ea.	*Payment in advance, US funds, certified check or money order only. Shipping by FedEx Ground included, unless purchaser desires other shipping arrangements be negotiated.*
❑	**61 to 100 copies**	$ 44.00 ea.	
❑	**100 to 200 copies**	$ 43.00 ea.	
❑	**201 to 500 copies**	$ 40.00 ea.	
❑	**Over 501 copies**	Call for pricing	*Terms to be individually determined.*

Through bulk ordering and/or reselling, you can realize or share the following benefits:
- You can share a highly useful preparedness tool with people who may desperately need it.
- Your civic group can perform a positive public service while at the same time having a fund raiser.
- Any time you help prepare those around you, you invest in your own security.

For more info contact us at: InfoQuest, 6300 Powers Ferry Rd., Suite 600-294, Atlanta, GA 30339, or email us at DisasterPrep101@aol.com.

This page intentionally left blank.

The Complete Guide to Homemade Income

An older work, The Complete Guide to Homemade Income was written in 1984, but it's just as useful today as it was then. We're including this discount coupon to help you achieve the goals we outlined in the "Frugality" portion of the "Foundation" section, which is to help you improve your family budget by increasing your income.

The Complete Guide to Homemade Income focuses on simple income-producing home-based businesses or "cottage industries" that anyone can start on a shoestring budget with little to no outside assets.

You'll find over 100 businesses outlined in this publication, none of them requiring any major investment in time or money, and none requiring any outside memberships, franchises, instruction, or other publications.

Originally published at $12.95, this coupon offers you an over **50%** discount price of **$6.00** plus **$2.00** S&H.

The Complete Guide to Homemade Income

Ordered by: Name:_____

Billing Address:_____

City:_____ State:_____ Zip+4:_____ - _____

Phone:_____ E-Mail:_____

Payment Method: ❑ Check or M.O. for **$8.00**   ❑ Visa ❑ Master Card ❑ Discover ❑ AmEx

Card #:_____ Expiration Date: ____/____/____

Signature:_____

Ship to a different address because: ❑ The above is a billing address ❑ This book is being sent as a gift

Please ship to:

Name:_____

Address:_____

City:_____ State:_____ Zip+4:_____ - _____

Mail coupon to: InfoQuest, 6300 Powers Ferry Rd., Suite 600-294, Atlanta, GA 30339.

This page intentionally left blank.

About the Author

Paul Purcell is the co-owner, Vice President, and lead security analyst for InfoQuest Investigators in Atlanta, GA.

Coming from a background in the hospitality industry, Paul began his early investigative career in restaurant consulting which included a strong focus on internal and external security, and general risk management. He was approached in 1996 by an Atlanta area investigative agency to come on board as their security analyst. His early cases included an informal security analysis of the 1996 Summer Olympic Games in Atlanta conducted for a member of the Olympic Security Council.

In 1998, he, along with partners Frances Carter and Ben Willis, opened InfoQuest Investigators, Inc., and has been performing security analysis for corporate and industrial locations ever since. As an investigator, Paul created "The Case File" (see http://www.thecasefile.com), an in-the-field investigative case management system in use by Private Investigators, Law Enforcement, and Attorney firms across the country. He also served as the Legislative Committee chairman in 2002, and state Secretary in 2003, for the Georgia Association of Professional Private Investigators (see http://www.gappi.org).

As a post-9/11 contribution, Paul voluntarily submitted numerous vulnerability studies concerning the FAA, the CDC, the 2001 World Series Games, the 2002 Winter Olympics, and the 2002 Superbowl. Copies of these studies were shared with an agent of the Georgia Bureau of Investigation, and two weeks later, Paul was invited to join the DeKalb County Office of Homeland Security as a volunteer civilian security analyst. His duties included vulnerability studies of some of the county's sensitive infrastructure locations, the contribution of terror attack scenarios, and meeting attendance as a panel member at large.

During this same time period, he authored a regular civil-defense and homeland security issues column entitled "Paul's Corner" which appears on Col. David H. Hackworth's "Soldiers For The Truth" foundation website at http://www.sftt.us (the column's archives are at http://www.sftt.us/pc_archive.html) , and he serves as an SFTT board member. This column caught the attention of associates of Ross Perot's who invited Paul to attend a small homeland defense / civilian volunteer planning meeting in Baltimore, MD, in July, 2002. Paul's most recent project has been to assist the director of the Savannah area Public Health District in the formulation of emergency first responder protocols in preparation for the 2004 G8 Summit held on Sea Island, Georgia.

The culmination of these experiences led Paul to embark on his personal quest to create a detailed training guide aimed at improving the emergency preparedness and protection of the average civilian. While good information was to be found through FEMA and the Red Cross, Paul felt that a great deal of detail, desired by a very intelligent general population, was missing and needed to be added. Hence **Disaster Prep 101**.

After this project is able to stand on its own, Paul's next endeavor will be to complete **Disaster Prep 201,** which will be to business and government what "101" is to civilians.